Introduction to Biomedical Equipment Technology

Second Edition

Joseph J. Carr
Certified Clinical Engineer

John M. Brown
Formerly Associate Director
of Technical Instruction
Cooperative Education and Training
Program
The George Washington University
Engineering & Applied Science

Manager of Analog
Strategic Marketing
Burr-Brown Research Corporation

PRENTICE HALL CAREER & TECHNOLOGY
Englewood Cliffs, New Jersey 07632

Library of Congress Cataloging-in-Publication Data

Carr, Joseph J.
Introduction to biomedical equipment technology / Joseph J. Carr, John M. Brown.—2nd ed.
p. cm.
Includes bibliographical references and index.
1. Medical instruments and apparatus. I. Brown, John M. (John Michael) II. Title.
R856.C33 1993 92-30587
610'.28—dc20 CIP

Acquisitions: Holly Hodder
Editorial/production supervision and
interior design: Word Crafters Editorial Services, Inc.
Cover design: Marianne Frasco
Prepress buyer: Ilene Levy-Sanford
Manufacturing buyer: Ed O'Dougherty
Marketing manager: Ramona Baran

Printed in the United States of America
10 9 8 7 6 5 4 3

ISBN 0-13-014333-2

Prentice-Hall International (UK) Limited, *London*
Prentice-Hall Australia Pty. Limited, *Sydney*
Prentice-Hall Canada Inc., *Toronto*
Prentice-Hall Hispanoamericana, S.A. *Mexico*
Prentice-Hall of India Private Limited, *New Delhi*
Prentice-Hall of Japan, Inc., *Tokyo*
Simon & Schuster Asia Pte. Ltd, *Singapore*
Editora Prentice-Hall do Brasil, Ltda., *Rio de Janeiro*

Contents

Preface xi

Acknowledgments xii

Chapter 1
The Human Body: An Overview 1

1-1 Objectives, 1
1-2 Self-Evaluation Questions, 1
1-3 Introduction, 1
1-4 The Cell, 1
1-5 Body Fluids, 3
1-6 Musculoskeletal System, 3
1-7 Respiratory System, 3
1-8 Gastrointestinal System, 3
1-9 Nervous System, 6
1-10 Endocrine System, 6
1-11 The Circulatory System, 6
1-12 The Body as a Control System, 7
1-13 Summary, 8
1-14 Recapitulation, 8
1-15 Questions, 8
1-16 References, 9

Chapter 2
The Heart and Circulatory System 10

2-1 Objectives, 10
2-2 Self-Evaluation Questions, 10
2-3 The Circulatory System, 10
2-4 The Heart, 13
2-5 Bioelectricity, 16
2-6 Electroconduction System of the Heart, 17
2-7 Heart Problems, 22
2-8 Summary, 23
2-9 Recapitulation, 23
2-10 Questions, 23
2-11 Problems, 24
2-12 References, 24

Chapter 3
Electrodes and Transducers 25

3-1 Objectives, 25
3-2 Self-Evaluation Questions, 25
3-3 Signal Acquisition, 25
3-4 Electrodes for Biophysical Sensing, 25
3-5 Medical Surface Electrodes, 30
3-6 Microelectrodes, 34
3-7 Transducers and Other Sensors, 37
3-8 Strain Gages, 40
3-9 Inductive Transducers, 45
3-10 Quartz Pressure Sensors, 47
3-11 Capacitive Transducers, 48
3-12 Temperature Transducers, 50
3-13 Summary, 52
3-14 Recapitulation, 52
3-15 Questions, 52
3-16 Problems, 53
3-17 References, 53

Chapter 4
Bioelectric Amplifiers 54

4-1 Objectives, 54
4-2 Self-Evaluation Questions, 54

4-3 Bioelectric Amplifiers, 54
4-4 Operational Amplifiers, 55
4-5 Basic Amplifier Configurations, 58
4-6 Multiple Input Circuits, 68
4-7 Differential Amplifiers, 69
4-8 Signal Processing Circuits, 81
4-9 Practical Op-Amps: Some Problems Reviewed, 84
4-10 Bioelectric Amplifiers Reviewed, 87
4-11 Isolation Amplifiers, 88
4-12 Chopper Stabilized Amplifiers, 107
4-13 Input Guarding, 108
4-14 Summary, 119
4-15 Recapitulation, 119
4-16 Questions, 120
4-17 Problems, 120
4-18 References, 121

Chapter 5
Electrocardiographs 123

5-1 Objectives, 123
5-2 Self-Evaluation Questions, 123
5-3 The Heart as a Potential Source, 123
5-4 The ECG Waveform, 123
5-5 The Standard Lead System, 124
5-6 Other ECG Signals, 126
5-7 The ECG Preamplifier, 127
5-8 ECG Readout Devices, 144
5-9 ECG Machines, 146
5-10 ECG Machine Maintenance, 152
5-11 ECG Faults and Troubleshooting, 152
5-12 Summary, 156
5-13 Recapitulation, 156
5-14 Questions, 156
5-15 Problems, 157
5-16 References, 157

Chapter 6
Physiological Pressure Measurements 159

6-1 Objectives, 159
6-2 Self-Evaluation Questions, 159
6-3 Physiological Pressures, 159
6-4 What Is Pressure?, 159
6-5 Pressure Measurements, 161
6-6 Blood Pressure Measurements, 163
6-7 Oscillometric and Ultrasonic Noninvasive Pressure Measurements, 167
6-8 Direct Methods: H_2O Manometers, 168
6-9 Pressure Transducers, 169
6-10 Pressure Amplifiers, 170
6-11 Typical Calibration Methods, 173
6-12 Pressure Amplifier Designs, 176
6-13 ac Carrier Amplifiers, 177
6-14 Systolic, Diastolic, and Mean Detector Circuits, 180
6-15 Pressure Differentiation (dP/dt) Circuits, 184
6-16 Automatic Zero Circuits, 185
6-17 Practical Problems in Pressure Monitoring, 186
6-18 Step-Function Frequency Response Test, 192
6-19 Transducer Care, 194
6-20 Summary, 195
6-21 Recapitulation, 195
6-22 Questions, 195
6-23 Problems, 196
6-24 References, 197

Chapter 7
Other Cardiovascular Measurements 198

7-1 Objectives, 198
7-2 Self-Evaluation Questions, 198
7-3 Cardiac Output Measurement, 198
7-4 Dilution Methods, 199
7-5 Right-Side Heart Pressures, 205
7-6 Plethysmography, 207
7-7 Blood Flow Measurements, 209
7-8 Phonocardiography, 213
7-9 Vectorcardiography (VCG), 215
7-10 Catheterization Laboratories, 216
7-11 Summary, 217
7-12 Recapitulation, 217

7-13 Questions, 217
7-14 Problems, 218
7-15 References, 218

Chapter 8
Cardiac Stimulation and Life Support Equipment 219

8-1 Objectives, 219
8-2 Self-Evaluation Questions, 219
8-3 The Heart, 219
8-4 Defibrillators, 220
8-5 Defibrillator Circuits, 224
8-6 Cardioversion, 225
8-7 Testing Defibrillators, 227
8-8 Pacemakers, 227
8-9 Heart-Lung Machines, 229
8-10 Summary, 231
8-11 Recapitulation, 231
8-12 Questions, 231
8-13 Problems, 232

Chapter 9
The Respiratory System 233

9-1 Objectives, 233
9-2 Self-Evaluation Questions, 233
9-3 The Human Respiratory System, 234
9-4 Gas Laws, 234
9-5 Internal (Cellular) Respiration, 235
9-6 External (Lung) Respiration, 236
9-7 Organs of Respiration, 236
9-8 Mechanics of Breathing, 239
9-9 Parameters of Respiration, 241
9-10 Regulation of Respiration, 243
9-11 Unbalanced and Diseased States, 243
9-12 Environmental Threats to the Respiratory System, 245
9-13 Major Measurements of Pulmonary Function, 245
9-14 Summary, 245
9-15 Recapitulation, 246
9-16 Questions, 246
9-17 Problems, 247
9-18 References, 247

Chapter 10
Respiratory Instrumentation 249

10-1 Objectives, 249
10-2 Self-Evaluation Questions, 249
10-3 Respiratory System Measurements, 249
10-4 Respiratory Transducers and Instruments, 250
10-5 Spirometers, 255
10-6 Pulmonary Measurement Systems and Instruments, 257
10-7 Summary, 258
10-8 Recapitulation, 258
10-9 Questions, 258
10-10 Problems, 259

Chapter 11
Respiratory Therapy Equipment 260

11-1 Objectives, 260
11-2 Self-Evaluation Questions, 260
11-3 Diseased States Requiring Artificial Respiratory Therapy, 261
11-4 An Overview and Terms of Ventilation, 261
11-5 Historical Perspective of Artificial Respiratory Ventilation, 262
11-6 Medical Gases and Safety Systems, 262
11-7 Oxygen Therapy, 263
11-8 Intermittent Positive Pressure Breathing (IPPB) Therapy, 266
11-9 Artificial Mechanical Ventilation, 270
11-10 Accessory Devices Used in Respiratory Therapy Apparatus, 273
11-11 Sterilization and Isolation Procedures in Respiratory Therapy Units, 273
11-12 Typical Faults and Maintenance Procedures for Ventilators, 274
11-13 Summary, 275
11-14 Recapitulation, 276

11-15 Questions, 276
11-16 Problems, 277
11-17 References, 277

Chapter 12 The Human Nervous System 278

12-1 Objectives, 278
12-2 Self-Evaluation Questions, 278
12-3 Organization of the Nervous System, 278
12-4 The Neuron (Single Nerve Cell), 281
12-5 The Structure and Function of the Central Nervous System, 283
12-6 The Peripheral Nervous System, 290
12-7 The Autonomic Nervous System, 290
12-8 Behavior and the Nervous System, 293
12-9 Summary, 293
12-10 Recapitulation, 294
12-11 Questions, 294
12-12 Problems, 295
12-13 References, 296

Chapter 13 Instrumentation for Measuring Brain Parameters 297

13-1 Objectives, 297
13-2 Self-Evaluation Questions, 297
13-3 Instrumentation for Measuring Anatomical and Physiological Parameters of the Brain, 297
13-4 Cerebral Angiography, 298
13-5 Cranial X-Rays, 298
13-6 Brain Scans, 298
13-7 Ultrasonic Equipment, 299
13-8 Electroencephalography (EEG), 300
13-9 EEG Electrodes and the 10-20 System, 302
13-10 EEG Amplitude and Frequency Bands, 303
13-11 EEG Diagnostic Uses and Sleep Patterns, 307
13-12 Multichannel EEG Recording Systems and Typical External Controls, 308
13-13 The EEG System—A Simplified Block Diagram, 310
13-14 Preamplifiers and EEG System Specifications, 311
13-15 Visual and Auditory Evoked Potential Recordings, 314
13-16 EEG Telemetry System, 316
13-17 Typical EEG System Artifacts, Faults, Troubleshooting, and Maintenance, 316
13-18 Summary, 319
13-19 Recapitulation, 321
13-20 Questions, 321
13-21 Problems, 322
13-22 References, 322

Chapter 14 Intensive and Coronary Care Units 323

14-1 Objectives, 323
14-2 Self-Evaluation Questions, 323
14-3 Special Care Units, 323
14-4 ICU/CCU Equipment, 324
14-5 Bedside Monitors, 325
14-6 Bedside Monitor Circuits, 327
14-7 Central Monitoring Consoles, 333
14-8 ECG/Physiological Telemetry, 338
14-9 Summary, 347
14-10 Recapitulation, 347
14-11 Questions, 347

Chapter 15 Operating Rooms 348

15-1 Objectives, 348
15-2 Self-Evaluation Questions, 348
15-3 Surgery, 348
15-4 Types of Surgery, 349
15-5 OR Personnel, 349
15-6 Sterilization, 350

15-7 OR Equipment, 351
15-8 Summary, 352
15-9 Recapitulation, 352
15-10 Questions, 352

Chapter 16 Medical Laboratory Instrumentation 353

16-1 Objectives, 353
16-2 Self-Evaluation Questions, 353
16-3 Blood (Purpose and Components), 353
16-4 Blood Tests (Cells and Chemistry), 356
16-5 Medical Laboratory Department, 357
16-6 Overview of Clinical Instrumentation, 357
16-7 Colorimeter, 358
16-8 Flame Photometer, 360
16-9 Spectrophotometer, 361
16-10 Blood Cell Counter, 362
16-11 pH/Blood Gas Analyzers, 364
16-12 Chromatograph, 368
16-13 Autoanalyzer, 369
16-14 Summary, 372
16-15 Recapitulation, 373
16-16 Questions, 373
16-17 Problems, 374
16-18 References, 374

Chapter 17 Medical Ultrasound 375

17-1 Objectives, 375
17-2 Self-Evaluation Questions, 375
17-3 What Is Ultrasound?, 375
17-4 Physics of Sound Waves, 376
17-5 Absorption of Ultrasound Energy, 378
17-6 Biological Effects of Ultrasound, 379
17-7 Ultrasonic Transducers, 380
17-8 Transcutaneous Doppler Flow Detectors, 380
17-9 Flowmeters, 381
17-10 Ultrasonic Blood Pressure Measurement, 385
17-11 Fetal Monitors, 387
17-12 Echocardiography, 390
17-13 Echoencephalography, 391
17-14 Summary, 393
17-15 Recapitulation, 393
17-16 Questions, 393
17-17 Problems, 394
17-18 References, 394

Chapter 18 Electrosurgery Generators 396

18-1 Objectives, 396
18-2 Self-Evaluation Questions, 396
18-3 Electrosurgery Machines, 396
18-4 Electrosurgery Circuits, 397
18-5 Electrosurgery Safety, 402
18-6 Testing Electrosurgery Units, 404
18-7 Summary, 406
18-8 Recapitulation, 406
18-9 Questions, 406
18-10 Problems, 406

Chapter 19 Electrical Safety in the Medical Environment 408

19-1 Objectives, 408
19-2 Self-Evaluation Questions, 408
19-3 The Definition of Electrical Safety, 409
19-4 The Scope of Electrical Safety in Medical Institutions, 409
19-5 Major Organizations Producing Publications Pertinent to Electrical Safety, 410
19-6 Responsibilities of Hospital Personnel, 410
19-7 Preventive Maintenance Programs to Reduce Electrical Hazards, 411
19-8 Legal and Insurance Requirements of Electrical Safety, 412

19-9 Setting Up an Electrical Safety Program in the Hospital, 413
19-10 Physiological Effects of Electricity on the Human (Theory of Macroshock and Microshock), 415
19-11 Leakage Current, 417
19-12 Subtle Hazards and Cautions for Microshock in Hospitals, 420
19-13 Design Considerations for Reducing Electrical Hazard, 426
19-14 Line Isolation Systems, 428
19-15 Equipotential Grounding System in Reducing Electrical Shock Hazards, 429
19-16 Ground Fault Interrupters in Reducing Electrical Shock Hazards, 431
19-17 Proper Power Wiring, Distribution, and Ground System in Reducing Electrical Shock Hazards, 431
19-18 Specialized Electrical Safety Test Equipment, 432
19-19 Summary, 435
19-20 Recapitulation, 436
19-21 Questions, 437
19-22 Problems, 438
19-23 References, 439

Chapter 20
Waveform Display Devices 440

20-1 Objectives, 440
20-2 Self-Evaluation Questions, 440
20-3 Permanent Magnet Moving Coil Instruments, 440
20-4 PMMC Wiring Systems, 441
20-5 Servorecorders and Recording Potentiometers, 445
20-6 *X-Y* Recorders, 446
20-7 Problems in Recorder Design, 447
20-8 Maintenance of PMMC Writing Styluses and Pens, 449
20-9 Dot Matrix Analog Recorders, 452
20-10 Oscilloscopes, 454
20-11 Medical Oscilloscopes, 455
20-12 Multibeam Oscilloscopes, 456
20-13 Nonface Oscilloscopes, 459
20-14 Modern Oscilloscope Designs, 462
20-15 Summary, 463
20-16 Recapitulation, 463
20-17 Questions, 463
20-18 Problems, 464
20-19 References, 464

Chapter 21
Electro-Optics (Fiber Optics and Lasers) 465

21-1 Objectives, 465
21-2 Self-Evaluation Questions, 465
21-3 Fiber-Optic Technology, 465
21-4 Fiber-Optic Isolation, 466
21-5 History of Fiber Optics, 467
21-6 Review of Some Basics, 467
21-7 Fiber Optics, 469
21-8 Intermodal Dispersion, 470
21-9 Graded Index Fibers, 472
21-10 Losses in Fiber-Optic Systems, 473
21-11 Losses in Fiber-Optic Circuits, 474
21-12 Fiber-Optic Communications Systems, 476
21-13 Receiver Amplifier and Transmitter Driver Circuits, 477
21-14 Lasers, 479
21-15 Laser Classification, 480
21-16 Basic Concepts, 480
21-17 Types of Lasers, 482
21-18 Driver Circuits for Solid-State Laser Diodes, 486
21-19 Laser Diode Receiver Circuits, 490
21-20 Summary, 492
21-21 Recapitulation, 493
21-22 Questions, 493
21-23 Problems, 493

Chapter 22
Computers in Biomedical Equipment 495

22-1 Objectives, 495
22-2 Self-Evaluation Questions, 495

22-3 Introduction, 495
22-4 Computer Hardware and Software, 497
22-5 Computer Programming Languages, 500
22-6 Microprocessor/Microcomputer System, 502
22-7 Interface Between Analog Signals and Digital Computers, 504
22-8 Computers in Biomedical Equipment, 505
22-9 Summary, 508
22-10 Recapitulation, 509
22-11 Questions, 509
22-12 References, 510

Chapter 23
Radiology and Nuclear Medicine Equipment 511

23-1 Objectives, 511
23-2 Self-Evaluation Questions, 511
23-3 Types and Uses of X-Ray and Nuclear Medicine Equipment, 512
23-4 Origin and Nature of X-Rays, 514
23-5 Nature and Types of Nuclear Radiation, 516
23-6 Units for Measuring Radioactivity, 517
23-7 Health Dangers from X-Ray and Nuclear Radiation, 517
23-8 Generation of X-Rays in an X-Ray Tube, 517
23-9 Block Diagram and Operation of an X-Ray Machine, 519
23-10 Block Diagram and Operation of a Fluoroscopic Machine, 522
23-11 Block Diagram and Operation of a Nuclear Medicine System, 522
23-12 Computer Systems Used in X-Ray and Nuclear Medicine Equipment, 526
23-13 Calibration, Typical Faults, Troubleshooting, and Maintenance Procedures, 527
23-14 Summary, 527
23-15 Recapitulation, 528
23-16 Questions, 528
23-17 Problems, 529
23-18 References, 529

Chapter 24
Electromagnetic Interference to Medical Electronic Equipment 530

24-1 Objectives, 530
24-2 Self-Evaluation Questions, 530
24-3 Introduction, 530
24-4 Intermodulation Problems, 531
24-5 Dealing with TVI/BCI, 536
24-6 Dealing with Signal Overload Problems, 538
24-7 ECG Equipment and EMI, 544
24-8 EMI to Biomedical Sensors, 545
24-9 Summary, 549
24-10 Recapitulation, 549
24-11 Questions, 549
24-12 Problems, 550
24-13 References, 550

Chapter 25
Care and Feeding of Battery-Operated Medical Equipment 551

25-1 Objectives, 551
25-2 Self-Evaluation Questions, 551
25-3 Introduction, 551
25-4 Cells or Batteries?, 552
25-5 Nickel Cadmium (NiCd) Cells and Batteries, 552
25-6 Battery Capacity, 552
25-7 Battery-Charging Protocols, 553
25-8 NiCd Battery ''Memory,'' 554
25-9 Battery Maintenance, 555
25-10 Charging NiCd Batteries, 556
25-11 Multiple-Cell Batteries, 557
25-12 Other Batteries, 558
25-13 Summary, 560
25-14 Recapitulation, 561

25-15 Questions, 561
25-16 Problems, 561

Chapter 26
Medical Equipment Maintenance: Management, Facilities, and Equipment 562
26-1 Objectives, 562
26-2 Self-Evaluation Questions, 562
26-3 Introduction, 562
26-4 Types of MROs, 563
26-5 Levels of Capability, 563
26-6 Types of Organization, 565
26-7 Technical Personnel, 568
26-8 Management Approaches, 571
26-9 Summary, 574
26-10 Recapitulation, 574
26-11 Questions, 574
26-12 References, 574

Appendix A
Some Math Notes 576

Appendix B
Medical Terminology 579

Appendix C
Glossary 582

Index 585

Preface

When the first edition of *Introduction to Biomedical Equipment Technology* was published in 1981, it was greeted with an enthusiastic reception. In a very short time the book was used in a large number of courses around the country, and sales were stronger than either of us had dared hoped. We were most happy with that initial response, and have been even more so over the fact that the book continues to be the industry leader even today. A decade after the book was first published it is still being sold by national biomedical instrumentation professional organizations, and is still being used widely in college and technical institute courses.

But biomedical equipment technology has moved on, so it became necessary to revise the book to reflect advances in the field. This revision of the book reflects the latest in biomedical electronic and electromechanical device technology, while retaining some of those features that made *Introduction to Biomedical Equipment Technology* useful to instructors, students, and other readers.

The first step in undertaking this revision was a market survey made by Prentice Hall, who contacted a number of instructors who used the book. They were asked what they liked, what needed improvement, and what they would recommend be added or deleted from the text. The response was overwhelmingly favorable, and the authors deeply appreciate the insights provided by the review panel. The reviewers strongly recommended that the physiology and anatomy material be retained because it set a context and provided a single source for the type of material needed by the student. That recommendation was accepted.

The original concept was to reduce the amount of old material by the amount of new material added. While that's the ''standard wisdom'' for revised books, the opinions of the expert advisors—who actually teach the material—was that nearly all of the old material was needed, as well as some new material. So the old material was largely retained (only a small amount dropped), and much new material was added.

The new material includes lasers, electro-optics, and fiber optics. These subjects are of intense interest to the modern practitioner in the field of biomedical equipment. Also added is material on electromagnetic interference (EMI). This topic becomes more important every year both because of the proliferation of medical electronic devices and the huge increase in the number of licensed transmitters and radio frequency interference generating devices (e.g., computer products).

Material that is considerably strengthened in this second edition includes the discussions of biomedical electrodes, radio telemetry (including new digital telemetry), and bioelectric amplifiers. The telemetry discussion came about because it is new technology, and therefore the information is needed by technical personnel. The other material was strengthened both because of the opinions of our expert panel and because of personal contacts with the authors by readers and instructors.

As in the first edition, the theoretical discussions are balanced by practical information of immediate use in the laboratory or workshop. One of the strengths of the first edition was insight on practical matters such as troubleshooting, calibrating, and installing medical equipment. We have retained and strengthened that flavor. Both authors have tremendous theoretical and practical professional experience, and have endeavored to bring you the advantage of that experience. We are certain that you will find this strengthened second edition of *Introduction to Biomedical Equipment Technology* even more useful than the first edition.

Joseph J. Carr
John M. Brown

Acknowledgments

We gratefully acknowledge the contribution of numerous associates and professionals in the fields of technology, engineering, and science. We especially acknowledge the original and succeeding students of the Cooperative Education and Training Program, School of Engineering and Applied Science, George Washington University. Their success reflects a glow on our community.

We appreciate the cooperation of Burr-Brown in providing numerous diagrams of its products. Burr-Brown does not authorize or warrant any Burr-Brown product for use in life support devices and/or systems.

To our wives and children we owe sincere appreciation for their patience and support.

Chapter 1

The Human Body: An Overview

1-1 Objectives

1. Be able to list the major systems that make up the body.
2. Know how to describe the principal functions of the body systems.
3. Be able to describe how the body controls and regulates itself.
4. Be able to state the relationships between body systems.

1-2 Self-Evaluation Questions

These questions test your prior knowledge of the material in this chapter. Look for the answers as you read the text. After you have finished studying the chapter, try answering these questions and those at the end of the chapter (Section 1-15).

1. Two fluid transport systems in the body are the ___________ system and the ___________ system.
2. Define *homeostasis* in your own words.
3. List the principal organs in the gastrointestinal system.
4. Draw a rough sketch of the blood circulatory system.
5. List the principal systems in the body.

1-3 Introduction

The purpose of this chapter is to make you aware of the broad structure of the human body, but not in such detail as you would receive in anatomy and physiology courses. More detailed information on each system is given in later chapters of this book. Students wishing to attempt independent study should consult *Human Anatomy and Physiology*, 2nd edition, by James E. Crouch, Ph.D. and J. Robert McClintic, Ph.D. (John Wiley, 1976).

In this chapter we discuss the major systems of the body and how they work together to produce an essentially self-regulating machine. The body contains literally hundreds of feedback control systems that attempt to keep the body's internal environment *constant*. This process is called *homeostasis*, and it allows the body to respond to changes in the environment and to illness, as well as accomplish the regulation of levels of sugars, salts, water, acid-base balance, oxygen, carbon dioxide, and the other materials that make up the living organism.

1-4 The Cell

All mammals, including humans, are made up of basic building blocks called *cells*. Although many different types of cells are known, they differ according to *function*, but they are all similar in their basic constituents. The different types of cells perform different jobs and so have different gross structures. Figure 1-1 shows several different types of human and mammalian cells.

The size of cells also varies, ranging from 200 nanometers (1 nm $= 10^{-9}$ m) to several centimeters in length. Most cells, however, fall within the range of 0.5 to 20 micrometers (1 μm $= 10^{-6}$ m). An ostrich egg is a single cell that may reach 20 cm in length.

The cell contains material used in chemical reactions that keep the cell functioning. It is surrounded by a semipermeable *membrane*. This membrane not only contains the cell material but also allows selective passage of materials in and out of the cell. There may also be membrane structures inside of the cell that compartmentalize the various chemical reactions taking place.

The structure of most cells includes a *nucleus* inside of the cell, separated from the surrounding *cytoplasm* by its own membrane. The nucleus contains the genetic coding of reproducible cells.

Cells in the human body are quite numerous. It has been estimated that there are approximately *75 trillion* cells in the body, of which one-third (25 trillion) are red blood cells. The red blood cells are responsible for transporting oxygen to body tissues.

All cells in a many-celled animal retain certain powers or characteristics, such as organization, irritability (i.e., response to external stimuli), nutrition, metabolism, respi-

Figure 1-1 Diverse forms of mammalian cells (not to the same scale). From *Human Anatomy and Physiology*, 2nd edition, by James E. Crouch, Ph.D. and J. Robert McClintic, Ph.D. John Wiley & Sons, New York, 1976. Used by permission.)

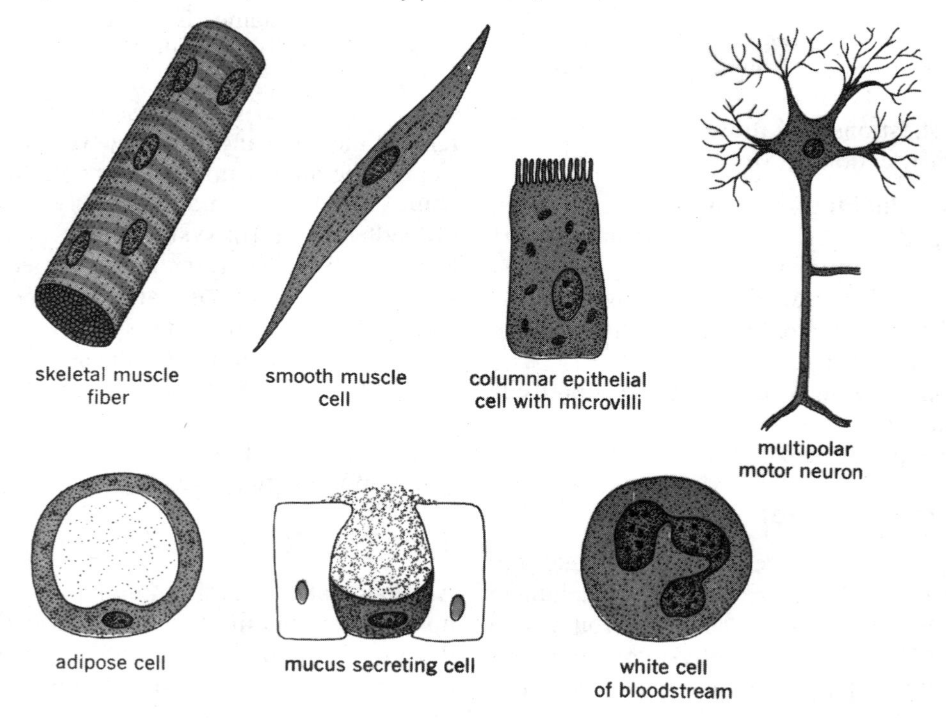

ration, and excretion. Some cells also possess the power of *reproduction*. All cells arise from preexisting cells through a process of cell division. In the process called *mitosis* there is a nonexact quantitative division of cell cytoplasm and an exact qualitative division of the nucleus material.

1-5 Body Fluids

The body is almost two-thirds fluid (actually, approximately 56 percent). Intracellular fluid contains large concentrations of potassium, magnesium, and phosphate ions, while the extracellular fluid contains significant concentrations of sodium, chloride, bicarbonate ions, oxygen, amino acids, fatty acids, glucose, and carbon dioxide.

1-6 Musculoskeletal System

The muscles and bones of the body provide *locomotion* (i.e., the ability to move around and manipulate our surroundings). If it were not for locomotion, humans would be more dependent on the local environment. The human would not be able to move to avoid danger, find food and water, or erect shelter from the elements.

Figure 1-2 shows the principal structures of the musculoskeletal system. The skeletal system (Figure 1-2*a*) consists mostly of *bones* and some *cartilage*. The bones are joined together to form *articulations* and *joints* and so are able to move with respect to each other. In general, muscles are connected between bones across a joint, so that the bones move with respect to each other when the muscle contracts.

1-7 Respiratory System

The respiratory system takes oxygen into the body and gives off carbon dioxide waste products from the cells. The respiratory system includes the mouth, nose, trachea or "windpipe," bronchii, and the lungs. Deoxygenated blood from the right side of the heart passes through the lungs such that only 0.4 to 2.0 microns of membrane separates the air-carrying *alveoli* from the pulmonary *capillaries* (i.e., tiny blood vessels). Gaseous oxygen diffuses across this membrane into the blood stream, while carbon dioxide comes out of the blood, into the alveoli, to be exhaled into the atmosphere.

1-8 Gastrointestinal System

The gastrointestinal (GI) system takes in raw materials in the form of food and liquids and then processes them so that they are absorbed into the body. Certain digestive organs are needed to chemically and physically process these raw materials: the liver, gall bladder, salivary glands, and pancreas, in addition to the stomach and intestinal tract. The system includes the mouth, esophagus, stomach, small intestines, and large intestines.

Digestion of food is the process of breaking down, liquefying, and chemically processing foodstuffs so that they can be used by the body. The process of digestion begins in the mouth, where the teeth and jaw *mechanically* break down the food material, and the saliva begins the chemical breakdown.

Both mechanical mixing and chemical breakdown occur in the stomach. Gastric juices mix with the food material to form a milky paste called *chyme*. Contractions called *peristaltic waves* existing in the stomach mix the foodstuffs together with the juices and occur approximately every 20 seconds. These waves can reach magnitudes of 50 to 70 cm H_2O. Every contraction will cause a few milliliters of stomach contents to enter the intestine.

Nutrients and fluid are absorbed by the body from the chyme as it moves through the intestine. The chyme is propelled at a

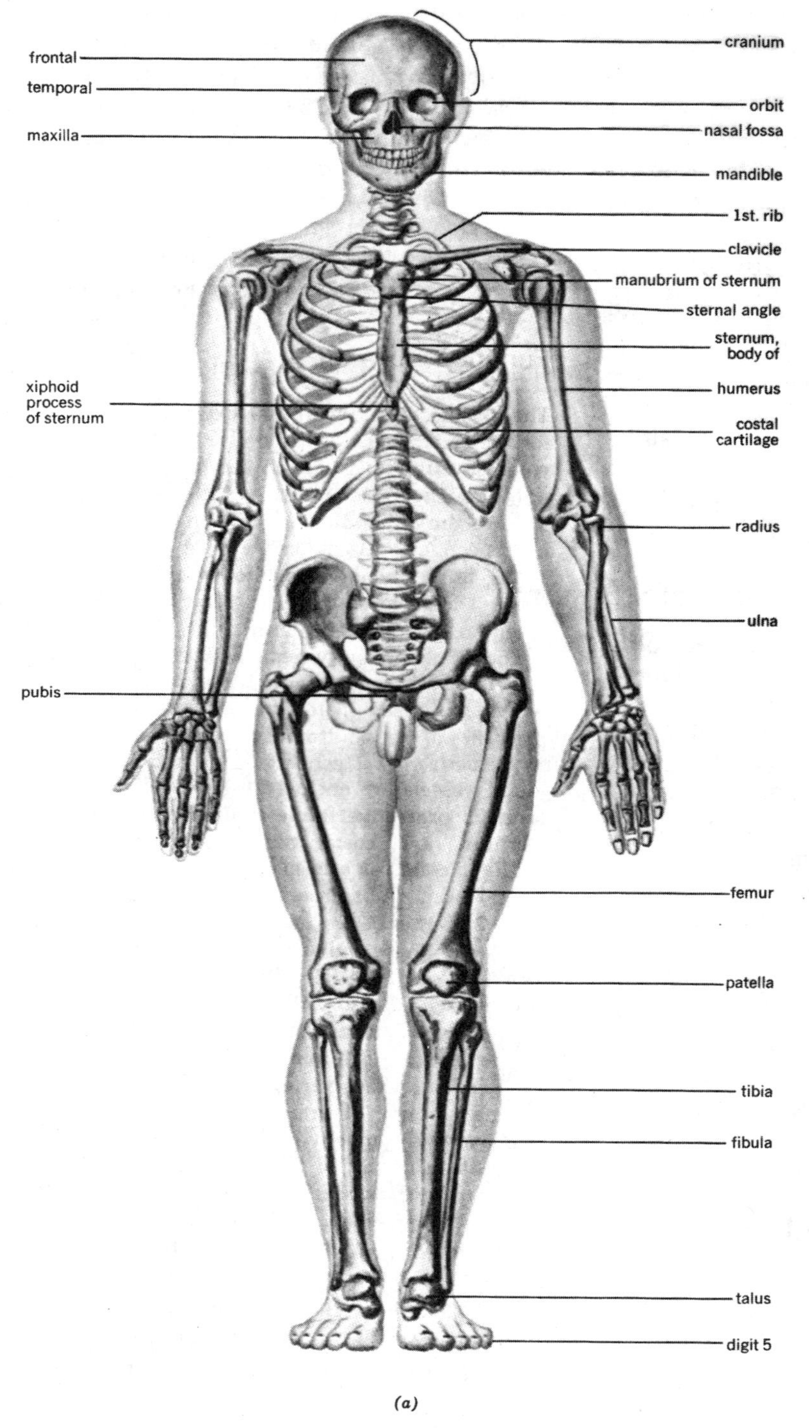

Figure 1-2*a* The human skeleton. (From *Human Anatomy and Physiology*, 2nd edition, by James E. Crouch, Ph.D. and J. Robert McClintic, Ph.D. John Wiley & Sons, New York, 1976. Used by permission.)

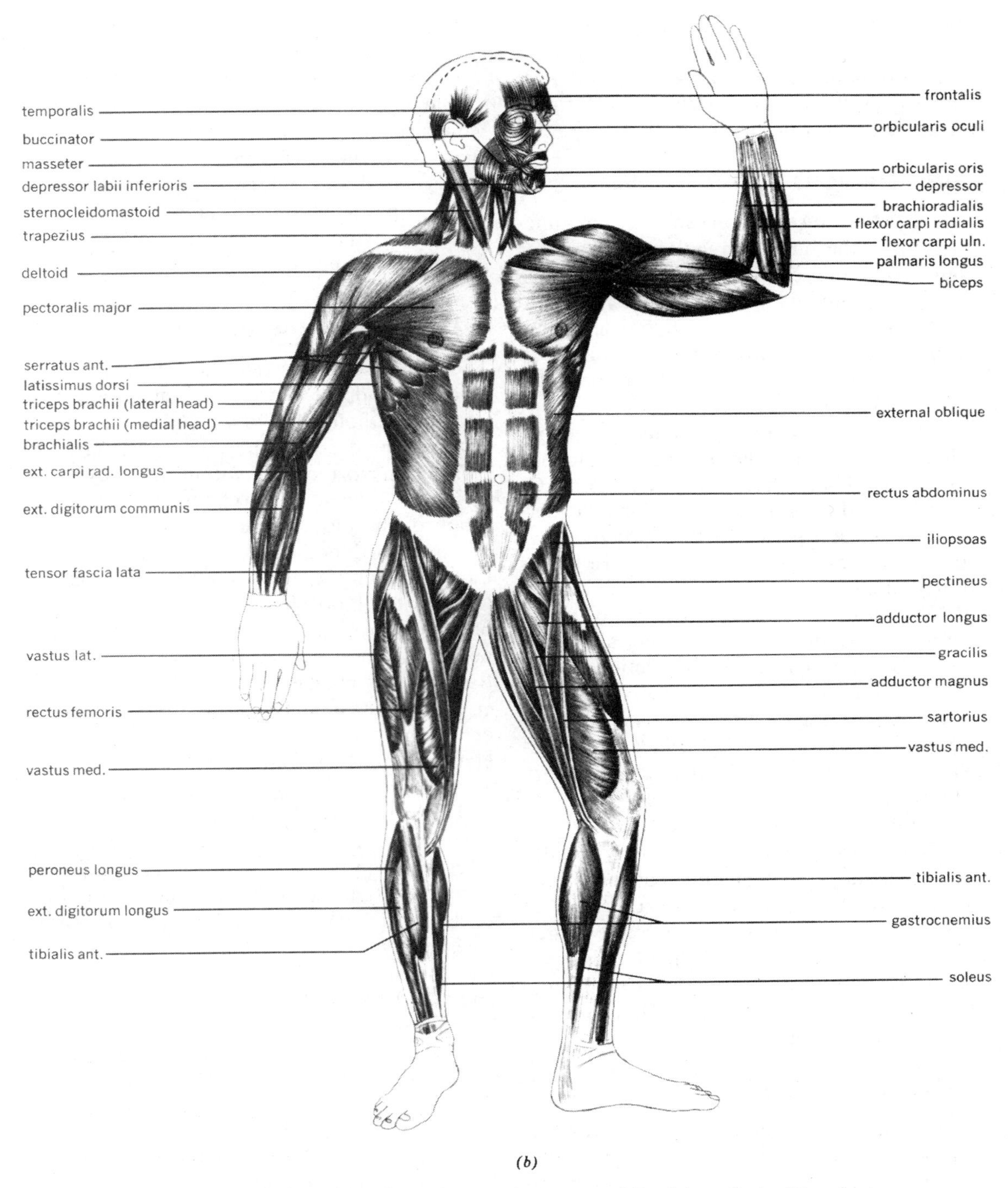

Figure 1-2*b* Muscular system. (From *Human Anatomy and Physiology*, 2nd edition, by James E. Crouch, Ph.D. and J. Robert McClintic, Ph.D. John Wiley & Sons, New York, 1976. Used by permission.)

rate of approximately 1 cm/min by weak peristaltic waves in the intestine.

Waste products and undigested foodstuffs are expelled from the body in the form of fecal material from the anus.

1-9 Nervous System

The nervous system is essential to the functioning of the human organism. It regulates our automatic control systems, integrates and assimilates data from the outside world and our internal organs, and regulates and controls the locomotor system. It has been compared to a computer with an electrical communications system.

The *autonomic nervous system* is responsible for regulating the automatic functions of the body—heartbeat, gland secretions, GI system, and so forth. The autonomic nervous system operates at a *subconscious level*—you are not generally aware of its functioning.

The *sensory nervous system* receives data from the outside world and certain internal organs through cells that function as *sensory receptors* (i.e., "transducers" in electrical terminology). The eyes and ears are sensory receptors for light and sound, respectively. But there are also other sensory structures that are sensitive to pain, heat, pressure, etc.

The *central nervous system* (CNS) gathers, assimilates, and integrates data from the outside world, information on the state of internal organs, etc. The brain is the principal organ of the CNS, and, like a computer, it can store, process, and generate information as well as react to stimuli. The CNS also includes the spinal cord.

1-10 Endocrine System

Where the CNS is an electrical communications and control system within the body, the endocrine system is a *chemical* communications/control system and aids in the regulation of internal body states.

Chemicals called *hormones* are secreted by the eight major endocrine glands into the bloodstream, and these are used as control agents to regulate various organic functions.

In general, the endocrine system controls slow-acting phenomena, mainly metabolic functions, while the CNS is responsible for fast-acting phenomenon.

1-11 The Circulatory System

The circulatory system transports body fluids around the body from one organ to another. Although the *blood* circulatory system is the most well known, there is also a *lymph* transport system within the body.

Figure 1-3 shows a schematic representation of the blood circulatory system. The transport of blood is caused by a pressure built up when the heart, a pump, contracts. Oxygenated blood from the left ventricle is pumped throughout the body, delivering needed oxygen to the various organs and tissues. It is claimed that the human blood circulatory system is so extensive that no cell in the body is more than one cell diameter (Section 1-4) from a small vessel called a capillary.

The oxygenated blood flows in *arteries* to the organs. The blood flowing into the GI tract picks up nutrients and water. The portion of the blood that flows into the kidneys is cleaned of impurities and waste products. The kidneys act as a blood *filter*. The blood will give up much of its oxygen to the tissues, and the deoxygenated blood returns to the heart in the *veins*.

Deoxygenated blood enters the right side of the heart at the right atrium. It is then pumped into the right ventricle and then out of the heart to the lungs. In the lungs the blood gives up its carbon dioxide and takes on a fresh supply of oxygen.

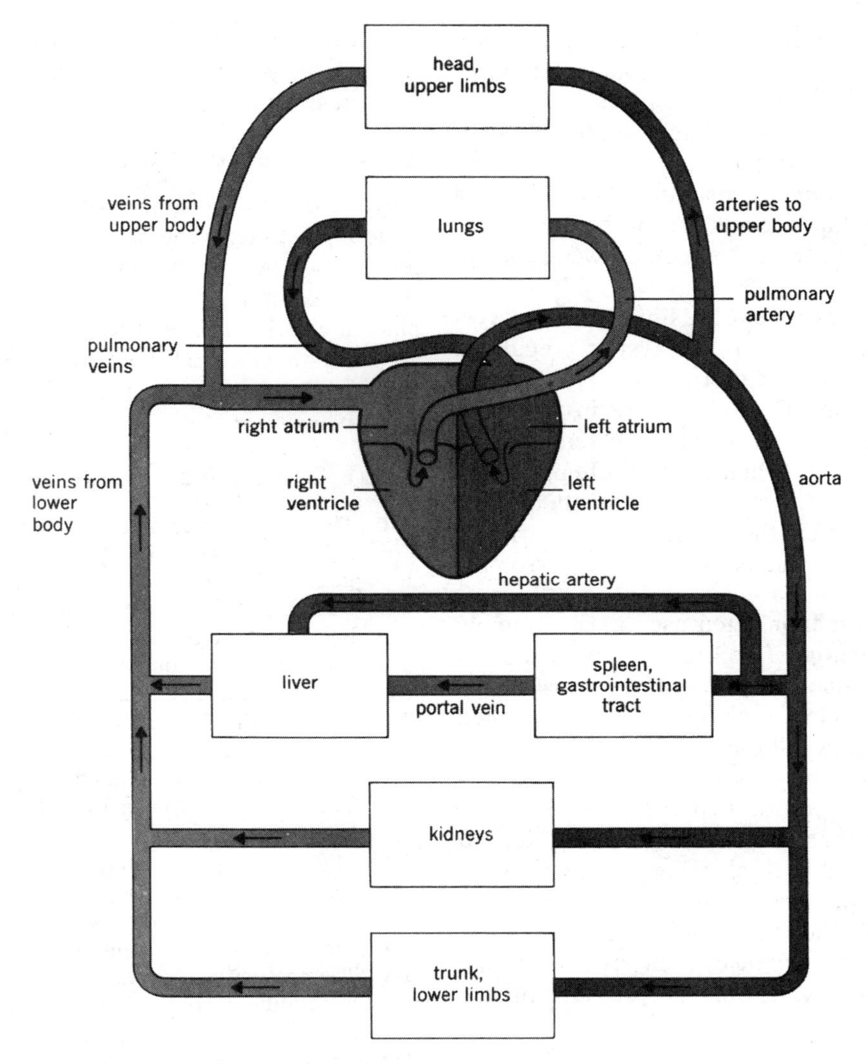

Figure 1-3 Circulatory system. (From *Human Anatomy and Physiology*, 2nd edition, by James E. Crouch, Ph.D. and J. Robert McClintic, Ph.D. John Wiley & Sons, New York, 1976. Used by permission.)

1-12 The Body as a Control System

Many functions of the body (estimates range from hundreds to thousands) are regulated in automatic *negative feedback loops*. Engineering and technology students should be at ease when studying these systems because they behave very much like the control systems we study in school. Conceptually, then, they are identical, if functionally different, and obey the same laws. Common electrical control systems include amplifiers, servomechanisms, and the old fashioned home furnace controller, the thermostat.

Any negative feedback control system compares *actual* conditions with the conditions that *should* exist and then causes a correction that cancels part of the difference, or *error* (i.e., the difference between the actual and the desired).

The usual model of a simple control system is the home furnace system, regulated by a simple on-off thermostat. The thermostat measures the temperature in the room and then compares it with the temperature set on the dial. When the room temperature drops below the set point, the thermostat closes a switch that turns on the furnace. The furnace remains on until the "error" is corrected.

A phenomenon often used as an example of a *physiological* control system is the automatic regulation of *blood pressure*. Pressure sensors in the circulatory system called *baroreceptors* tell the CNS of the conditions that exist. If the pressure drops below a certain normal point, then the brain issues a command that causes the blood vessels to *constrict*, and this will bring the pressure up. But if the pressure increases above a "normal" point, then the brain causes the vessels to *dilate* (i.e., increases their cross-sectional area, thereby reducing the pressure on the system).

1-13 Summary

1. The human body is a *homeostatic* mechanism; that is, it is self-regulating through a series of negative feedback control systems to maintain a constant internal environment.
2. The *cell* is the basic building block of the body. Different types of cells perform different functions.
3. Roughly two-thirds of the body is fluid.
4. Major body systems include the nervous, circulatory, musculoskeletal, respiratory, digestive, and endocrine systems.

1-14 Recapitulation

Now return to the objectives and self-evaluation questions at the beginning of the chapter and see how well you can answer them. If you cannot answer certain questions, place a check mark next to each, and reread appropriate parts of the text. Next, try to answer the following questions, using the same procedure.

1-15 Questions

1-1 The ability to maintain a constant internal environment in the body is called ______________.

1-2 ______________ are considered to be the basic building blocks of the body.

1-3 The sizes of various cells in humans range from ______________ to ______________.

1-4 One of the largest cells known is the ______________ ______________, and it may reach 8 in. in length.

1-5 The cell material is enclosed by a covering called a ______________.

1-6 There are approximately ______________ cells in the human body, of which ______________ are red blood cells.

1-7 List six properties common to all cells.

1-8 The process called ______________ is the qualitative reproduction of a cell by division.

1-9 The body is approximately ______________ percent fluid.

1-10 List the principal ingredients of intracellular fluid.

1-11 List the principal ingredients of extracellular fluid.

1-12 The musculoskeletal system provides ______________.

1-13 ______________ and ______________ are the principal types of tissue forming the skeleton.

1-14 Define *membrane* in your own words.

1-15 What is the function of the respiratory system?

1-16 List the principal components of the respiratory system.

1-17 What is the function of the gastrointestinal tract?

1-18 What is *chyme*?

1-19 The ______________ ______________ system integrates and assimilates data.

1-20 The ______________ nervous system provides information on the external and internal environments.

1-21 The ______________ nervous system operates mostly at the subconscious level.

1-22 The ______________ system is a chemical communications and control system within the body.

1-23 List the components of the circulatory system.

1-24 The two major fluid transport systems in the body are ______________ and ______________.

1-16 References

1. Crouch, James E. and J. Robert McClintic, *Human Anatomy and Physiology*, 2nd ed. Wiley (New York, 1976).
2. Guyton, Arthur C., *Textbook of Medical Physiology*, W. B. Saunders Company (Philadelphia, 1971).

Chapter 2

The Heart and Circulatory System

2-1 Objectives

1. Be able to state the biological principles behind the human cardiovascular system.
2. Be able to describe the anatomy of the heart.
3. Be able to describe the dynamics of blood flow.
4. Know how to explain the generation and propagation of bioelectric potentials in tissue.
5. Be able to describe the general details of the internal electroconduction system of the human heart.

2-2 Self-Evaluation Questions

These questions test your prior knowledge of the material in this chapter. Look for the answers as you read the text. After you have finished studying the chapter, try answering these questions and those at the end of the chapter (see Section 2-10).

1. Define *action potential*.
2. Name the four chambers of the heart.
3. Describe the location of the tricuspid valve.
4. What are the *sinoatrial* (SA) and *atrioventricular* (AV) nodes?
5. Describe the general path of the blood as it travels through the circulatory system.
6. Define the term *systole*.
7. What is the velocity of propagation of the action potential in the bundle branches following the AV node?

2-3 The Circulatory System

The circulatory system carries nourishment and oxygen to, and waste and carbon dioxide away from, the tissues and organs of the body. It may be considered as a closed loop hydraulic system, and indeed you will find that it possesses many of the properties of such a system.

2-3.1 Elementary Circulatory System

Figure 2-1 shows the human circulatory system in simplified form. The heart serves as a pump to move blood through vessels called *arteries* and *veins*. Blood is carried away from the heart in arteries and is brought back to the heart in veins.

In actuality the heart is a dual pump consisting of a two-chambered pump on both the left and right sides. The upper chambers are inputs to the pumps and are called *atria*

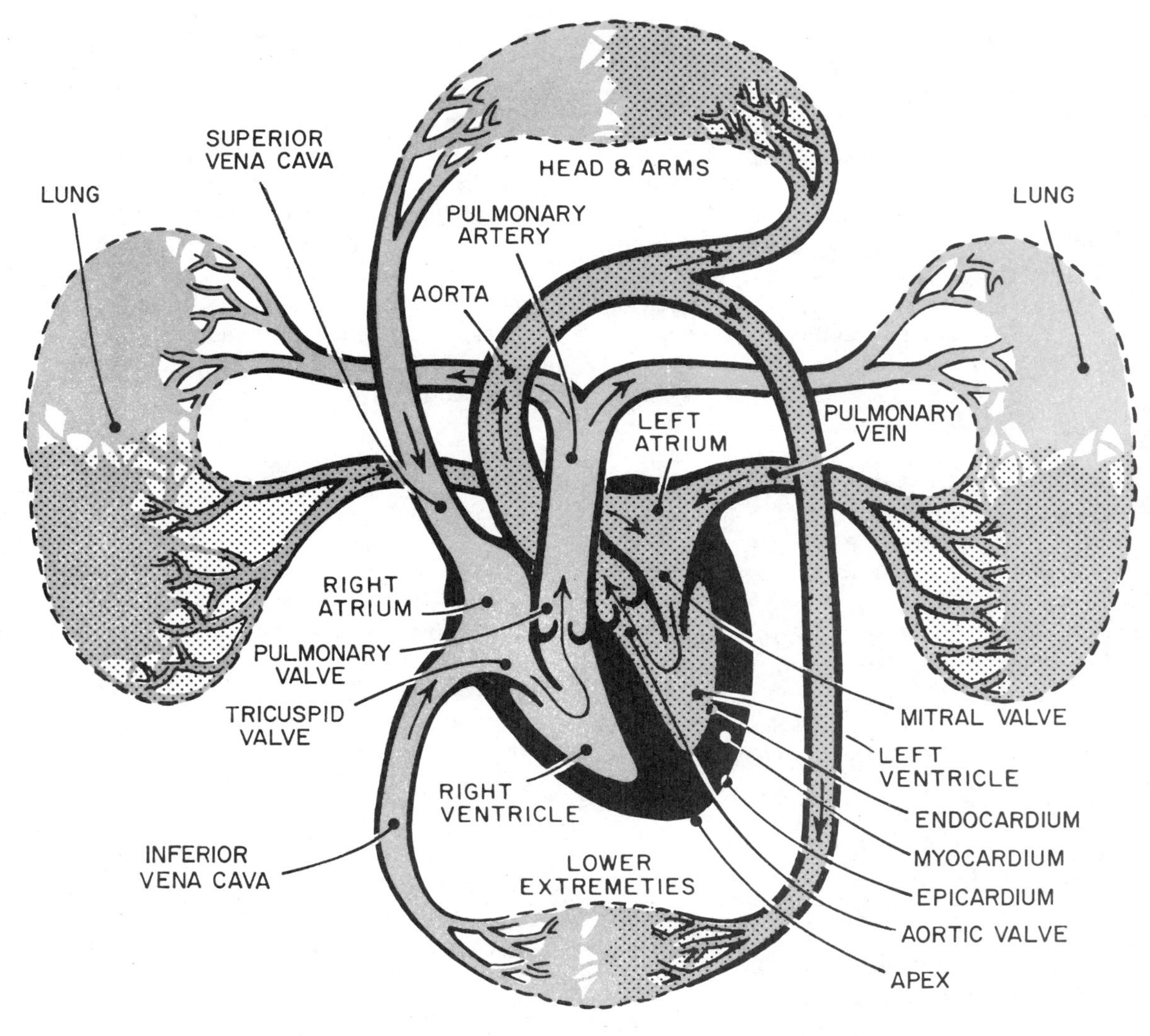

Figure 2-1 The human circulatory system. (Reprinted courtesy of Hewlett-Packard.)

(*atrium* singular). The lower chambers of the heart are called *ventricles* and are the pump outputs.

When blood is circulated through the body, it carries oxygen (O_2) and nutrients to the organs and tissues and returns carrying carbon dioxide (CO_2) and the various waste products that are to be excreted through the kidneys. The deoxygenated blood is returned to the right side of the heart via the venous system. Blood from the head and arms, as well as the rest of the upper portion of the body, returns to the heart through the *superior vena cava*, while blood from the lower portion of the body returns through the *inferior vena cava*. Note that the terms *inferior* and *superior* refer not to some qualitative assessment, but to the respective *positions* of the two vessels. The inferior is generally placed *lower* in the body than the superior. (Similarly, *great*, as used later in this section, means "large.")

Blood leaves the right atrium through the *tricuspid valve* to enter the *right ventricle*. From the right ventricle it passes through the *pulmonary semilunar valve* to the *pulmonary artery*. This vessel carries blood to the lungs, where CO_2 is given up and O_2 is taken on. This gas exchange is studied in greater detail in Chapter 6.

Blood returning from the lungs via the pulmonary vein reenters the heart through the *left atrium*. It then passes through the *mitral valve* to the *left ventricle*, and then back into the mainstream of the circulatory system via the *aortic valve*. The *great artery* attached to the *left* ventricle is called the *aorta*. Blood then circulates through the body to again return to the right side of the heart via the *superior* and *inferior vena cava*.

The blood flowing in the vessels of the body may be viewed in terms analogous to an electrical circuit. In your elementary electronics courses, you may have learned the old "plumbing analogy" of electricity, where an electric current is likened to water or fluid in a hydraulic system. Here we use the converse analogy, where blood flow and pressure are compared to an electrical circuit. We will even introduce an "Ohm's law" relationship for blood flow rate (Equation 2-1).

2-3.2 Blood

Blood is composed of the two main components: *cells* and *plasma*. Blood cells make up approximately 40 percent of the total blood volume, and the remaining 60 percent is plasma. Since approximately 99 percent of the cells are *red* cells, it may be said that 40 percent of the blood volume consists of red cells. White cells play only a small role in determining the physical properties and composition of blood.

Blood *flow rate* (measured in volume per unit time) in a blood vessel is described by two factors: the *pressure difference* along the vessel and the *resistance* offered by the vessel (a function of its cross-sectional area). Does this sound familiar? It should, because similar factors in an electrical circuit control the flow of current. In an electrical circuit we use Ohm's law to describe the relationship between potential difference (analogous to pressure) and current (analogous to blood flow rate). The same sort of relationship describes blood flow:

$$R = \frac{P}{F} \tag{2-1}$$

where

P is the pressure difference in millimeters of mercury (mm Hg)

F is the flow rate, in milliliters per second (ml/s) or cm^3/s

R is the resistance of the vessel in PRU or *peripheral resistance units* (1 PRU is the vessel resistance that allows a flow of 1 ml/s under a pressure of 1 mm Hg)

EXAMPLE 2-1

Find the resistance of a blood vessel whose flow rate is 1.7 ml/s at a blood pressure of 6.8 mm Hg.

SOLUTION

$$R = \frac{P}{F} \tag{2-1}$$

$$R = \frac{6.8 \text{ mm Hg}}{1.7 \text{ ml/s}} = \mathbf{4\ PRU}$$

Note that Equation 2-1 shows that a vessel that has a higher resistance (in PRU) would require a higher blood pressure to produce the *same* flow!

The actual situation in most cases is a little more complex than that given in Equation 2-1 because the resistance is *not* constant; that is, the vessel walls are *not* rigid (the radius of the vessel varies) and the blood itself is subject to changes in viscosity. The walls of the arteries and the veins are continuously distensible, so the pulsating blood flow will continuously vary some of the parameters. The flow quantity (F), for example, is more precisely described by Poiseuille's law, which gives the factors affecting flow rate or

$$F = \frac{P}{R} = P \times \frac{\pi r^4}{8\eta L} \quad (2\text{-}2)$$

where

η is the coefficient of blood viscosity in dyne-seconds per square centimeter (dyn-s/cm^2)
P is the pressure difference in dyn/cm^2
r is the vessel radius in cm
L is the vessel length in cm

EXAMPLE 2-2

A blood vessel has an average radius of 0.5 mm and a length of 20 mm. If the blood pressure is 7.2 mm Hg at an average viscosity of 0.01 dyn-s/cm^2, calculate the blood flow rate in (a) cm^3/s; (b) ml/s.
Hint: 1 mm Hg = 1330 dyn/cm^2.

SOLUTION

(a) $$F = \frac{P}{R} = \frac{P\pi r^4}{8\eta L} \quad (2\text{-}2)$$

$$= \frac{7.2 \text{ mm} \times 1330 \text{ dyn/cm}^2\text{/mm} \times \pi \times (0.5 \text{ mm} \times 0.1 \text{ cm/mm})^4}{8 \times 0.01 \text{ dyn-s/cm}^2 \times (20 \text{ mm} \times 0.1 \text{ cm/mm})}$$

$= \mathbf{1.175\ cm^3/s}$

(b) $1.175 \text{ cm}^3\text{/s} \times 1 \text{ ml/cm}^3 = \mathbf{1.175\ ml/s}$

The reader should perform a dimensional analysis to verify the units of flow rate (in cm^3/s) using the values substituted in Example 2-2a in Equation 2-2. When data are given in units other than those specified in Equation 2-2, it is first necessary to convert that data to appropriate units before substituting in Equation 2-2. This is done in Example 2-2a for units of P, r, and L, respectively, as shown. Note that Example 2-2 shows that a constricted blood vessel impedes the flow rate drastically. If the radius of the vessel is reduced by 50 percent (or one-half), the flow rate is reduced to 1/16 of its original value!

Blood is carried throughout the body in several *different* types of vessels. Those leading *from* the heart *to* tissue and organs are called *arteries*. Arteries tend to be *elastic*, allowing diameter changes to regulate the blood flow to various parts of the body. The diameter constrictions reflect ordinary changes in demand or emergency situations. Very small arteries are often called *arterioles*.

The *veins* carry blood back to the heart and lungs (where it is reoxygenated). Oxygen is transferred to cells of the tissue in *capillary beds* that permeate the entire body. *Capillaries* are small vessels connecting veins and arteries in a meshlike structure and have a diameter ($2r$) of only a few micrometers (μm). Capillaries are so numerous and widespread that it is claimed that no cell in the human body is more than its own diameter away from a capillary!

Note that blood flow rate is greatest in the aorta and least in the capillaries (Equation 2-2). The diameter of the capillaries is so small that blood cells must pass through them single file, one by one.

2-4 The Heart

The human heart is located in the upper middle portion of the chest (*thorax*). Although many people believe that the heart is clearly on the left side of the body, it is actually a little more centered, with the lower tip pointed toward the left hip. About one-third of the heart lies to the right of the midline of the body, while the rest lies to the left.

The size and weight of the heart vary from one individual to another. In most people the heart is approximately the size of their clenched fist, and the average weight of the heart is on the order of 300 grams.

The heart is a muscle encased in a sac called the *pericardium*. This double layer of tissue helps the heart stay in position and protects it from harm. The pericardium creates a lubricating fluid on its inside surface so that the friction between it and the heart wall is reduced, allowing the heart to beat freely within the walls of the sac.

A cutaway view of the human heart is shown in Figure 2-2. Besides two layers of pericardium (described before), there is an *epicardium* and a *myocardium*, the main muscle tissue of the heart. The thick myocardium accounts for approximately 75 percent of the heart wall thickness.

The heart contains four chambers, which are used to form two separate pumps. Each pump consists of an upper chamber (*atrium*) and a lower chamber (*ventricle*). The *high* pressure *output* side of each pump is the ventricle, so the myocardium *thickness* in the ventricular region is considerably *greater* than it is in the atrial region.

There are four *valves* in the human heart. The valve between the right atrium and the right ventricle is known as the *tricuspid valve*. It gets its name from the fact that it is formed of three cusp-shaped flaps of tissue arranged so that they will shut off and block passage of blood in the *reverse* direction (from ventricles back to the atrium).

These valves are attached at their bases to a fibrous strand of tissue ringing the opening between upper and lower chambers, and at their ends to objects called *chordae tendinae*. These structures are attached to the muscle tissue in the ventricle and keep the tricuspid valve closed as the right ventricular pressure builds up to force blood out of the heart into the pulmonary artery.

The valve between the right ventricle and the pulmonary artery is also given a name reminiscent of its shape: It is called a *semilunar* (half moon) valve. It also consists of three flaps, but it lacks the chordae tendinae of the tricuspid valve. It prevents reverse (regurgitation) flow of blood from the pulmonary artery to the right ventricle.

Blood returning to the heart from the lungs must pass through the left atrium and the mitral valve (also known as a *bicuspid valve* after its shape) to the left ventricle. This valve is formed of two flaps of cusp-shaped pieces of tissue.

The last valve is the *aortic valve*. It is shaped similar to the pulmonary valve and functions to prevent regurgitation of blood from the aorta back to the left ventricle.

The heart serves as a pump because of its ability to contract under an electrical stimulus. When an electrical triggering signal is received (Section 2-5) the heart will contract, starting in the atria, which undergo a shallow, ripplelike contracting motion. A fraction of a second later the ventricles also begin to contract, from the bottom up, in a motion that resembles wringing out a dishrag or sponge. The ventricular contraction is known as *systole*. The ventricular relaxation is known as *diastole*.

The heart in a resting adult pumps approximately three to five liters of blood per minute (3–5 l/min). This figure is called *cardiac output* (CO) and is defined as the product of heart rate in beats per minute (BPM) and the volume of blood ejected from the ventricles during systole.

$$CO = \textit{heart rate}\ (\text{BPM}) \times \textit{stroke volume}\ (\text{liters per beat}) \qquad (2\text{-}3)$$

EXAMPLE 2-3

Find the cardiac output for:

(a) A patient whose heart rate is 60 BPM if the stroke volume is 50 milliliters (ml) per beat.

(b) A heart rate of 90 BPM and a stroke volume of 80 milliliters (ml) per beat.

SOLUTION

$$\begin{aligned}\text{(a) } CO &= \textit{heart rate} \times \textit{stroke volume} \qquad (2\text{-}3)\\ &= 60 \text{ beats/min} \times 50 \text{ ml/beat} \times \frac{1\text{ l}}{1000\text{ ml}}\\ &= \mathbf{3\ l/min}\end{aligned}$$

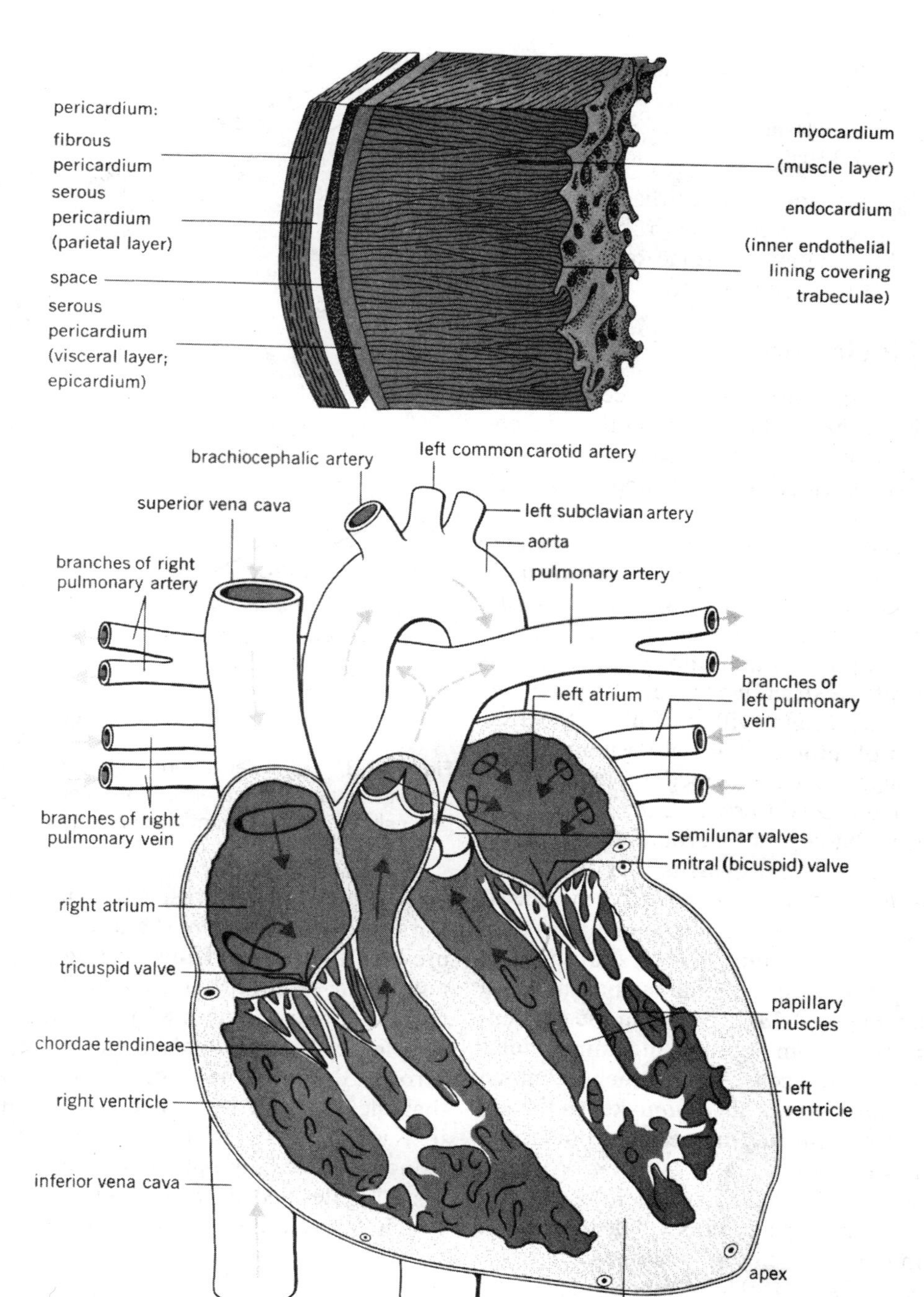

Figure 2-2 Cross-sectional view of the human heart. (From *Human Anatomy and Physiology*, 2nd edition, by James E. Crouch, Ph.D. and J. Robert McClintic, Ph.D. John Wiley & Sons, New York, 1976. Used by permission.)

(b) $CO = 90$ beats/min $\times$ 80 ml/beat

$$\times \frac{1\text{ l}}{1000\text{ ml}}$$

$$= \mathbf{7.2\ l/min}$$

Note that the values of cardiac output for the parameters given in the problem are extremes. Most human cardiac output values are in the 3 to 5 l/min range.

2-5 Bioelectricity

Ionic potentials are formed in certain cells of the body due to differences in the concentrations of certain chemical ions, notably sodium (Na^+), chloride (Cl^-), and potassium (K^+) ions.

The cell wall is a *semipermeable membrane*. Permeability is a measure of the ability of the membrane to pass certain ions. In the case of a semipermeable membrane, it is a selective process that allows some ions to pass while restricting or rejecting others. Such a membrane will not allow the free diffusion of all ions, but only a limited few. It is thought that this selective phenomenon is due to ion size differences, their respective electrical charges, and certain other factors. The end result, however, is that cell membranes at rest tend to be more permeable to some ions (i.e., potassium and chloride) than to others (i.e., sodium). As a result, the concentration of positive sodium ions inside a cell (see Figure 2-3*a*) is *less* than the concentration of sodium ions in the intracellular fluid *outside* the cell. A phenomenon known as the *sodium-potassium pump* keeps the sodium largely outside the cell and potassium ions inside.

Potassium is thus pumped into the cell while sodium is pumped out, but the *rate* of sodium pumping is roughly two to five times that of potassium. These rates result in a *difference* of ion *concentration* creating an *electrical potential;* and this causes the cell to be *polarized*. The *inside* of the cell is *less positive* than the *outside*, so the cell is said to be *negative* with respect to its outside. Various authorities give slightly different figures for the value of this *resting potential*, but all fall within the 70 to 90 millivolt (mV) range. Guyton (1) uses 85 mV; Crouch and McClintic (2) offer 70 mV as the figure; while Strong (3) uses 90 mV. All agree, however, that the cell polarity is negative, so in this text we will use −70 mV as the nominal value of the resting potential. The actual potential is derived from the Nernst equation, which is given in simplified form as

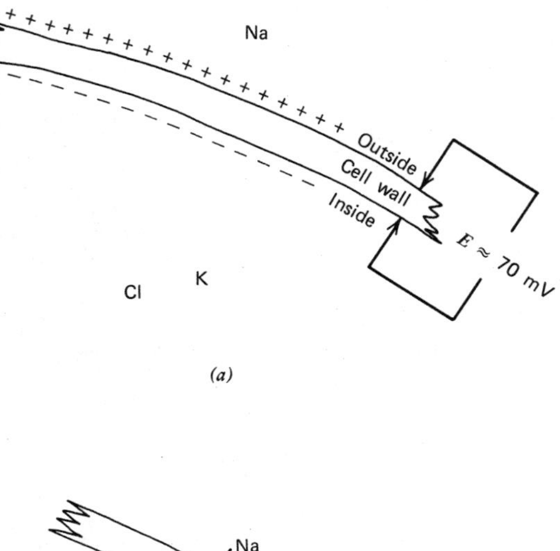

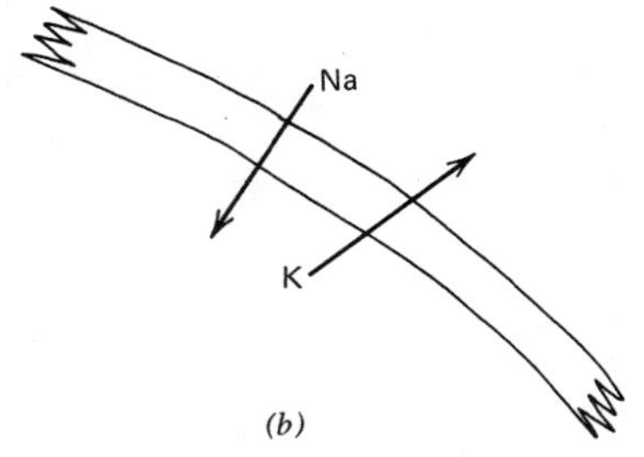

Figure 2-3 Cell polarization at rest and during stimulation. (*a*) Resting (diffusion) potential; polarized cell. (*b*) Action potential; depolarized cell.

$$E_{(\text{mV})} = \pm\, 60 \ln \frac{C_o}{C_i} \qquad (2\text{-}4)$$

where

E is the *resting* potential, in millivolts

C_i is the concentration inside of the cell in moles/cm^3

C_o is the concentration outside of the cell in moles/cm^3

ln indicates that the natural logarithm (base *e*) of the concentration ratio is used

EXAMPLE 2-4

The intracellular K^+ concentration of a group of cells average 20×10^{-6} moles/cm^3. The extracellular concentration of K^+ averages 5×10^{-6} moles/cm^3.

Calculate (a) The concentration ratio.

(b) Diffusion potential for K^+.

SOLUTION

$$\text{(a)} \frac{C_o}{C_i} = \frac{5 \times 10^{-6}\,\text{moles/cm}^3}{20 \times 10^{-6}\,\text{moles/cm}^3} = \frac{5}{20} = \mathbf{1/4}$$

$$\text{(b)}\ E^{k+} = 60 \ln C_o/C_i = 60 \ln 1/4 = \mathbf{-83.2\ mV}$$

When the cell is stimulated, the nature of the cell membrane wall changes abruptly and it becomes permeable to sodium ions. The sodium ions rush into the cell (Figure 2-3*b*) and potassium ions rush out. The result is an *action potential* (Figure 2-4) that sees the inside of the cell at a potential of 20 to 40 mV more positive than the outside (i.e., a polarity reversal lasting a few milliseconds).

The cell showing a resting potential is *polarized* (Figure 2-3*a*), but when it is generating an action potential it is said to be *depolarized.* There is a *refractory period* following depolarization during which the cell becomes repolarized (see Figure 2-4). In this period the cell is resistant to another depolarization. To the electronics student this action might resemble the monostable multivibrator: the action potential, once triggered, cannot be retriggered until the cell has again become repolarized.

Repolarization occurs when the cell membrane again changes its properties, forcing sodium out of the cell and drawing potassium ions inside of the cell wall.

Although ordinary ionic electrical conduction occurs, action potentials as such tend to be localized phenomena. But conduction does occur because depolarized cells trigger adjacent cells, causing them to produce an action potential. Returning to our multivibrator analogy from electronics, we could view this situation as a chain of monostable multivibrators in cascade so that the output of one will trigger the next.

2-6 Electroconduction System of the Heart

The conduction system of the heart (Figure 2-5) consists of structures called the *sinoatrial* (SA) *node, Bundle of His, atrioventricular* (AV) *node,* tissue called the *bundle branches*, and additional structures called *Purkinje fibers*.

The SA node serves as a *pacemaker* for the heart, and it provides the trigger signal mentioned earlier. It is a small bundle of cells (approximately 3 × 10 mm) located on the rear wall of the right atrium, just below the point where the superior vena cava is attached. The SA node fires electrical impulses through the bioelectric mechanism discussed in the previous section. It is capable of *self-excitation* (firing on its own) but is under control of the central nervous system so that the heart rate can be adjusted automatically to meet varying requirements.

When the SA node discharges a pulse, then electrical current spreads across the atria (auricles), causing them to contract. Blood in the atria is forced by the contraction through the valves to the ventricles. The velocity of propagation for the SA node action potential is about 30 centimeters per second (cm/s) in atrial tissue.

There is a band of specialized tissue between the SA node and the AV node, however, in which the velocity of propagation is faster than it is in atrial tissue, on the order of 45 cm/s (Figure 2-5). This internal conduction pathway carries the signal to the ventricles.

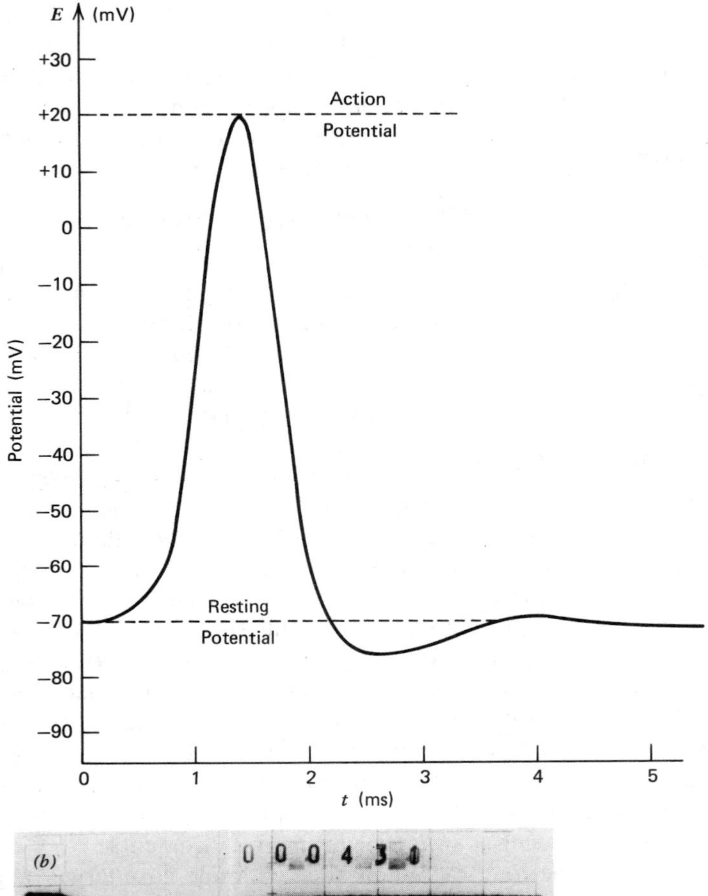

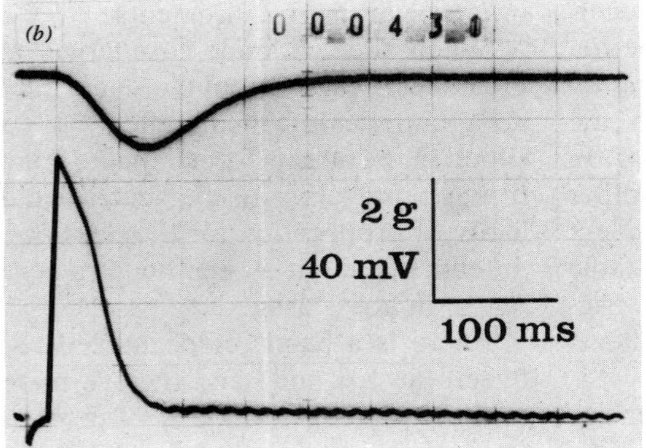

Figure 2-4 Action potential duration with time. (*a*) Typical cell action potential. (*b*) Contraction (upper trace) and action potential (lower trace) from a guinea pig myocardium. (Photo courtesy Dr. Martin Frank, Dept. of Physiology, The George Washington University Medical School.)

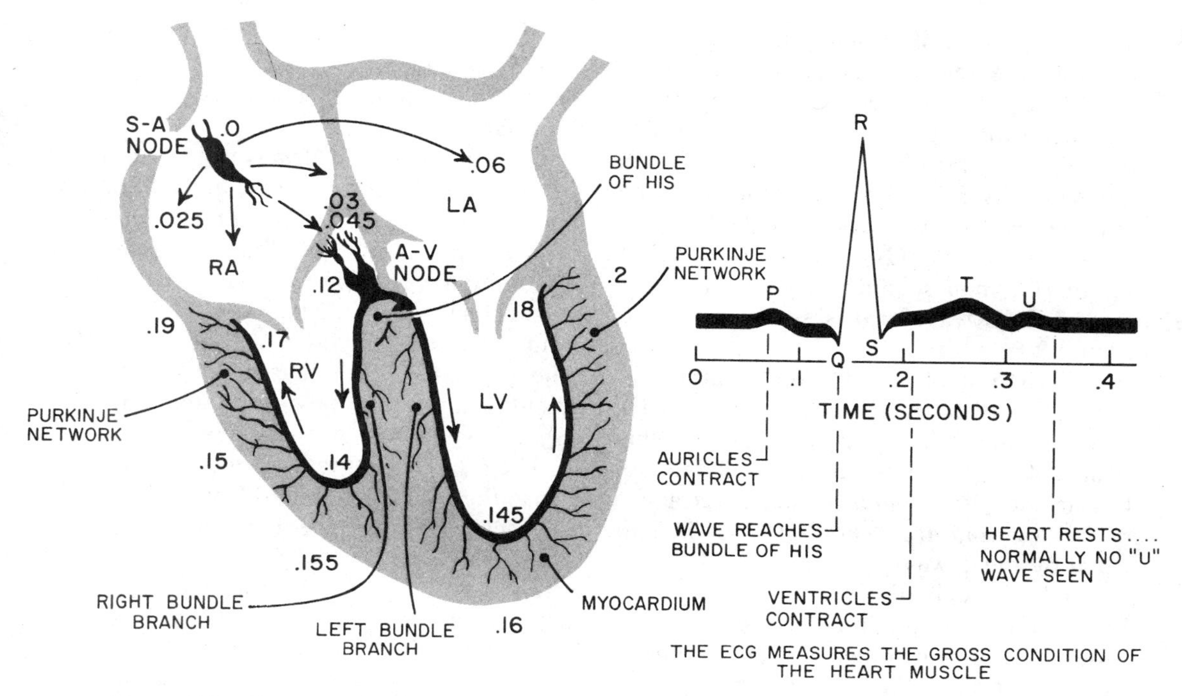

Figure 2-5 Electroconduction system of the heart and resulting ECG waveform. (Reprinted courtesy of Hewlett-Packard.)

It would *not* be desirable for the ventricles to contract in response to an action potential *before* the atria are empty of their contents. A *delay* is needed, therefore, to prevent such an occurrence; this is the function of the AV node. At 45 cm/s the action potential will reach the AV node in 30 to 50 milliseconds (ms) after the SA node discharges, but another 110 ms will pass before the pulse is transmitted from the AV node. The AV node, then, operates like a *delay line* to retard the advance of the action potential along the internal electroconduction system toward the ventricles.

The muscle cells of the ventricles are actually excited by the Purkinje fibers (Figure 2-5). The action potential travels along these fibers at a much faster rate, on the order of 2 to 4 m/s (i.e., 200 to 400 cm/s). The fibers are arranged in two bundles, one branch to the left and one to the right.

Conduction in the Purkinje fibers is very rapid. (Note that timing is given in Figure 2-5.) The action potential traverses the distance between the SA and AV nodes in about 40 ms and is delayed by the AV node for about 110 ms so that the contraction of the lower chambers can be synchronized with the emptying of the upper chambers. Conduction into the bundle branches is rapid, consuming only another 60 ms to reach the furthest Purkinje fibers.

The action potential generated in the SA node stimulates the muscle fibers making up the myocardium, causing them to contract. When the muscle is in contraction it is shorter, and the volume of the ventricular chamber is less, so blood is squeezed out.

The contraction of so many muscle cells at one time creates a mass electrical signal that can be detected by electrodes placed on the surface of the patient's chest or the patient's extremities. This electrical discharge can be mechanically plotted as a function of time, and the resultant waveform is called an *electrocardiogram* (ECG, or some use EKG after the German spelling). An example of a typical ECG waveform is shown as part of Figure 2-5.

Designation of the different features on the ECG waveform is by a letter system. The *P*-wave indicates *atrial* contraction; while ventricular systole occurs immediately following the *QRS complex,* and a refractory period (*resting* for *repolarization*) is indicated by the *T*-wave.

Oddly enough, the time duration of the ECG features is relatively constant over a wide range of heart rates. The *QRS* complex (see waveform in Fig. 2-5), for example, requires approximately 90 ms, the *P-R* wave roughly 150 to 200 ms, and the *S-T* segment about 50 to 150 ms. Right atrial pressure changes from a diastolic value of approximately 3 mm Hg to a systolic value of approximately 8 mm Hg.

By now you should be ready to correlate the ECG features, contraction of the heart, and the pulsatile flow of blood from the heart. Figure 2-6 shows the ECG and its relationship to the various pressures existing in the left and right sides of the heart. Notice that the ventricular pressure begins rising sharply as the heart begins to contract. This occurs in the period immediately following the *R*-wave of the ECG. It reaches a peak and then subsides, returning to its resting value. The peak pressure is known as the *systolic pressure* because it occurs during systole. The resting pressure is known as the *diastolic pressure*, and it occurs during the period called *diastole*. The period of systole lasts about 350 ms, while diastole is a little longer, on the order of 550 ms.

Also shown in Figure 2-6 are certain heart sounds labeled 1, 2, 3, and 4. These sounds are attributed to the mechanical action of the four valves.

Contraction of the atria is triggered by the SA node and begins immediately after the *P*-wave on the ECG. The pressure in the right and left atria will begin to rise as the contraction commences. Right atrial pressure changes from a 2 to 3 mm Hg diastolic value to a systolic value of 7 to 8 mm Hg. The pressure in the left atrium rises from about 3 mm Hg in diastole to about 10 mm Hg in systole.

The pressure in the atria does not actually cause the transfer of blood from the atria to ventricles. Valve openings, which allow blood transfer, are due mostly to changes in differential pressure. During diastole the pressure in the ventricles drops to less than the atrial pressure. This causes the valves (tricuspid on the right and mitral on the left) to open, allowing blood to be drawn into the ventricles under the influence of the pressure difference.

The ventricular contraction commences immediately after the *R*-wave on the ECG waveform. The ventricular pressure increases to a level greater than atrial pressure, forcing shut the tricuspid and mitral valves. These valve closures give rise to the first heart sound. The pressure in the right ventricle will rise from a pressure slightly less than atrial pressure during diastole to 28 to 30 mm Hg during systole. A pressure of 18 to 20 mm Hg is sufficient to overcome the reverse pressure in the pulmonary artery and open the pulmonary valve. The left ventricle, on the other hand, faces a higher pressure situation and so must attain a pressure of 75 to 80 mm Hg to open the aortic valve and reaches a peak pressure of 120 to 130 mm Hg.

The ventricles will begin to relax following the peak of systole, and the ventricular pressure will begin to drop. When the ven-

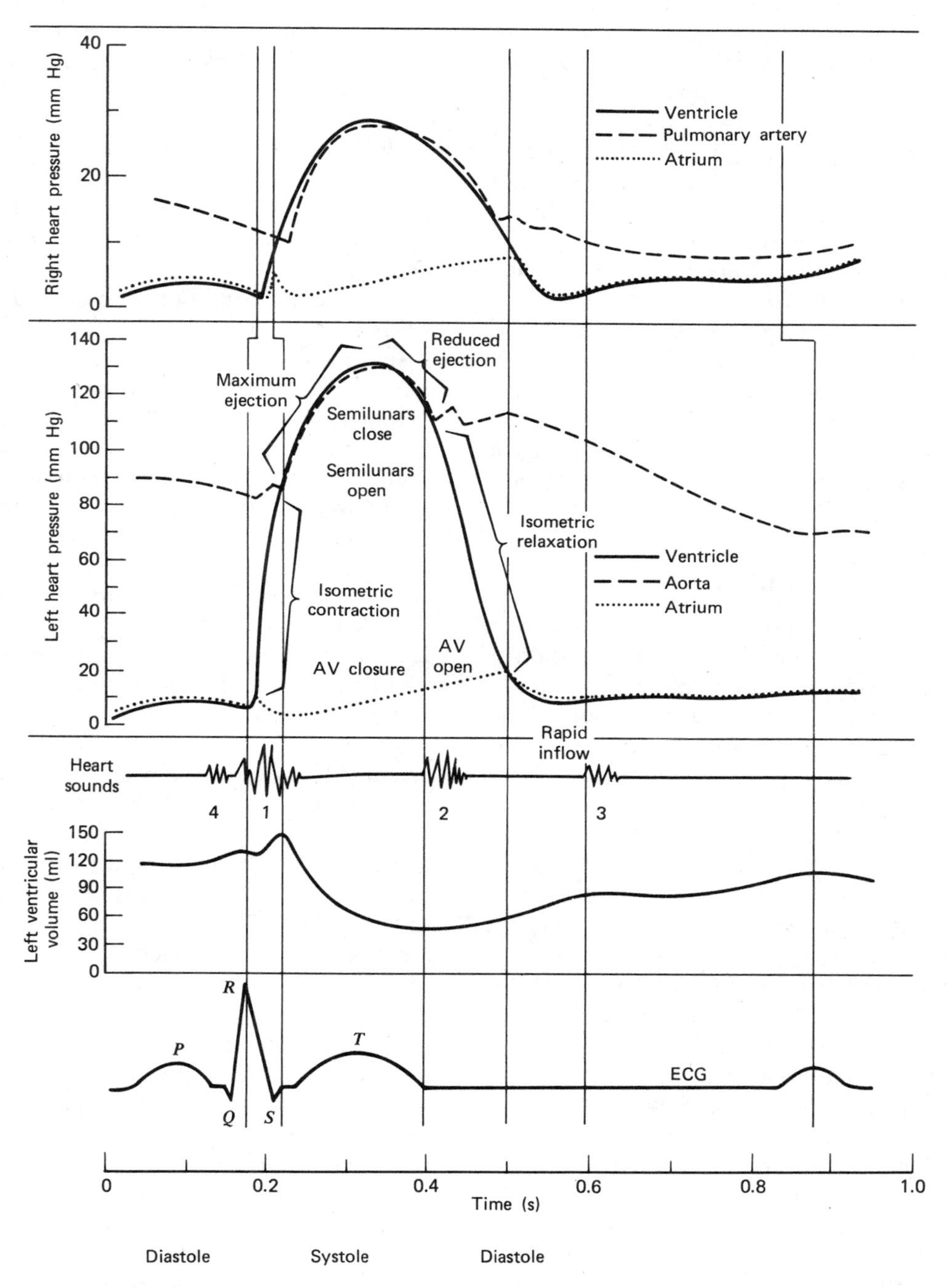

Figure 2-6 Heart sounds, ECG, and pressures at various locations.

tricular pressure is less than the pressure in the arteries, the respective valves will close. During this period of pressure reversal, note that blood attempts to flow back into the ventricles. The valve closure and blood flow dynamics at this time give rise to the second heart sound. Also noted is a *dicrotic notch*, occurring just past 0.4 s in Figure 2-6, in the aortic pressure waveform. Following valve closure is a period of relaxation during which the ventricles will again fill with blood.

Note that in all phases of the heart cycle, it is pressure changes that actuate the valve openings and closing. There is no fancy control system in operation here, just the results of ordinary pressure differences such as you may have studied in an introductory physics course.

2-7 Heart Problems

The physician will use the ECG and other factors to determine the gross condition of the heart. Although a complete discussion of heart problems is beyond the scope of this book, we can discuss some of the more common problems in generalized terms.

The heart is a muscle and, as such, must be *perfused* with blood in order to keep it healthy. Blood is supplied to the heart through the coronary arteries that branch off from the aorta close to where it joins the heart.

If an artery bringing blood to the heart becomes partially or totally occluded (i.e., blocked off), then the area of the heart served by that vessel will suffer damage due to the loss of the blood flow. That area of the heart is said to be *infarcted* and is dysfunctional. This type of damage is referred to as a *myocardial infarction*, or in medical jargon MI.

Another class of heart problem is the various cardiac *arrhythmias*. These are beats out of the normal and may be seen as ECG changes. Conditions under this classification include extremes in heart rate, premature contractions, heart block, and fibrillation.

The human heart rate varies normally over a range of 60 to 110 BPM. Rates faster than this are called *tachycardia*. Various authorities list slightly different figures as the threshold for official tachycardia, but most list 120 BPM, with the range being 110 to 130 BPM.

The opposite condition, too slow a heart rate, is called *bradycardia*, and again different sources list slightly different values, but all are within the 40 to 60 BPM range.

Premature contractions occur when an area of the heart becomes irritable enough to produce a *spurious* action potential at a time *between* normal beats. The action potential spreads across the myocardium in much the same manner as the regular discharge. Beats occurring at improper times are called *ectopic beats*. If they result in atrial contraction, then it is a *premature atrial contraction* (PAC), and if in the ventricle a *premature ventricular contraction* (PVC).

Heart block occurs when the internal electroconduction system of the heart is interrupted or significantly impeded. Among the more common forms is the atrioventricular block occurring at the junction of the atria and ventricles. More on this subject is given later when we discuss pacemakers.

Fibrillation is a condition where the muscle cells discharge asynchronously in a random manner. In *ventricular fibrillation* the major features of the ECG disasppear and the waveform takes on a low-amplitude, "jittery" appearance indicating that the ventricles are not contracting, but only quivering. Ventricular fibrillation is a fatal arrhythmia that will kill the victim in a few minutes if not corrected. Atrial fibrillation is less serious because the ventricles remain able to contract, as indicated by the existence of a

QRS complex in the ECG waveform, so blood supply to the body is not interrupted or seriously curtailed.

2-8 Summary

1. The heart is a dual two-chambered pump that supplies pressure to circulate the blood throughout the body.
2. Blood is carried *from* the heart in *arteries* and returns *to* the heart in *veins*.
3. Blood returns to the heart on the right side and is pumped from the heart on the left side.
4. An internal pacemaker (SA node) generates an electrical signal to initiate contraction of the heart. This signal is an action potential and is propagated by the electroconduction system to the ventricles.
5. Contracting heart muscle cells generate a mass action potential that may be picked up by electrodes on the surface of the body. The tracing of this potential is called an *electrocardiogram*.

2-9 Recapitulation

Now return to the objectives and self-evaluation questions at the beginning of the chapter and see how well you can answer them. If you cannot answer certain questions, place a check mark next to each and review appropriate parts of the text. Next, try to answer the following questions using the same procedure. When you have answered all of the questions, solve the problems in Section 2-11.

2-10 Questions

2-1 The upper chambers of the heart are called ______________.

2-2 The lower chambers of the heart are called ______________.

2-3 The superior and inferior ______________ return blood to the ______________ atrium.

2-4 The ______________ valve is located between the right atrium and the right ventricle.

2-5 Blood leaves the right ventricle via the ______________ artery.

2-6 The ______________ valve is located between the left atrium and the left atrium.

2-7 The great artery leaving the left side of the heart is called the ______________.

2-8 State Ohm's law for blood flow.

2-9 Blood volume is about ______________ percent red cells.

2-10 Very small vessels called ______________ have diameters in the micrometer range.

2-11 Blood velocity in the capillaries is (faster) (slower) than its velocity in the aorta.

2-12 The average human heart weighs about ______________ g.

2-13 The sac surrounding the heart is the ______________.

2-14 Heart muscle is called ______________.

2-15 The ______________ are the heart chambers that are subject to the highest pressures.

2-16 Name four valves in the heart.

2-17 The ______________ is the heart's pacemaker.

2-18 The time of contraction is called ______________.

2-19 An electrical potential is generated across a cell wall because of a difference in ______________ concentration.

2-20 The potential across a cell wall is called a ______________ potential.

2-21 The cell wall is a ______________ membrane.

2-22 A phenomenon known as the ______________ pump keeps ______________ ions predominantly outside of the cell wall.

2-23 Action potentials have an amplitude of about ______________ mV.

2-24 Write the simplified Nernst equation.

2-25 Resting potentials are about ______________ mV.

2-26 The "delay line" in the heart's electroconduction system is called the ______________.

2-27 The principal structures of the heart's electroconduction system are ______________, ______________, ______________, ______________, and ______________.

2-28 Action potentials travel down the conduction system at rates of ______________ cm/s between SA and AV nodes, and ______________ cm/s after the SA node.

2-29 The AV delay is about ______________ ms.

2-30 The ______________ is a recording of the heart's electrical activity.

2-31 A pressure of about ______________ mm Hg is required to open the aortic valve.

2-32 The ______________ notch in the aortic pressure waveform indicates the onset of diastole.

2-33 Too slow heart rate is called ______________.

2-34 Too fast heart rate is called ______________.

2-11 Problems

2-1 Find the resistance of a blood vessel whose flow rate is 2 ml/s at a pressure of 7.3 mm Hg.

2-2 Find the flow rate if the resistance is 6 PRU and the pressure difference is 8.5 mm Hg.

2-3 Find the resistance of a blood vessel that has a pressure of 4 mm Hg and a flow of 0.5 ml/s.

2-4 Calculate the flow rate in a blood vessel with an average diameter of 12 mm if the pressure drop over a 25-mm length is 8.4 mm Hg, and the blood viscosity is 0.015 dyn-s/cm^2.

2-5 Find the *resistance* in PRU of the vessel in Problem 2-4.

2-6 Find the flow rate in the vessel of Problem 2-4 if the diameter (a) increases to 18 mm and (b) decreases to 6 mm.

2-7 Calculate the cardiac output if the heart rate is 70 BPM and the stroke volume is 60 ml.

2-8 Find the cardiac output if the heart rate is 70 BPM and the stroke volume is 50 ml.

2-9 Find the diffusion potential across a membrane if the K^+ concentration on one side is 4×10^{-6} moles/cm^3 and 9×10^{-6} moles/cm^3 on the other side.

2-10 The K^+ concentration in the intracellular fluid is 6×10^{-6} moles/cm^3 and the K^+ concentration inside of a cell is 1.2×10^{-5} moles/cm^3. Find the diffusion potential.

2-12 References

1. Guyton, Arthur C., *Basic Human Physiology: Normal Function and the Mechanisms of Disease*, W. B. Saunders Company (Philadelphia, 1977).
2. Crouch, James E. and J. Robert McClintic, *Human Anatomy and Physiology*, 2nd ed. Wiley (New York, 1976).
3. Strong, Peter, *Biophysical Measurements*, Tektronix, Inc. Measurement Concept Series (Beaverton, Ore., 1976).

Chapter 3

Electrodes and Transducers

3-1 Objectives

1. Be able to list the problems associated with the acquisition of biopotentials.
2. Be able to list the different types of electrodes used to acquire biopotentials.
3. Be able to list and describe the types of transducers used to measure physiological parameters.

3-2 Self-Evaluation Questions

These questions test your prior knowledge of the material in this chapter. Look for the answers as you read the text. After you have finished studying the chapter, try answering these questions and those at the end of the chapter (see Section 3-15).

1. Define *halfcell* and *electrode offset potentials*.
2. Describe the *column electrode*. What is the principal advantage of the column electrode?
3. Describe the operation of a *Wheatstone bridge*. What factors determine the output voltage? What is the *null condition?*
4. What are the differences between *bonded* and *unbonded* strain gages?
5. List three types of temperature transducer.

3-3 Signal Acquisition

Most medical instruments are *electronic* devices and so *must* have an electrical signal for an input. In the cases where a biopotential must be acquired, some form of *electrode* will be used between the patient and the instrument. In other cases a *transducer* is used to convert some nonelectrical physical parameter or stimulus such as force, pressure, temperature (etc.) to an analogous electrical signal proportional to the value of the original stimulus parameter.

Def. A *transducer*, in this context, is a device that will *convert* some form of *energy* produced by a physical stimulus to an *electrical analog* of the stimulus.

3-4 Electrodes for Biophysical Sensing

Bioelectricity is a naturally occurring phenomenon that arises from the fact that living organisms are composed of ions in various different quantities. *Ionic conduction* is different from *electronic conduction*, which is perhaps more familiar in the ordinary experience of engineers and technologists. Ionic conduction involves the migration of ions—positively and negatively charged molecules—throughout a region, whereas electronic conduction involves the flow of electrons under the influence of an electrical field. In an *electrolytic solution*, ions are easily available. Potential differences occur when the concentration of ions is different between two points (Section 2-5).

When dealing with ionic conduction in depth, you will soon find that it is a very complex, nonlinear phenomena. But for small signal applications, where there is only a very small—indeed, minuscule—current flowing, modeling it as a flow of electrical current between points of potential difference is a fair first-order approximation. Chemists would find this model wanting, except in the most elementary classes, but their needs for understanding are greater than those of the instrumentation specialist. Keep in mind, however, that situations where a more substantial current flows change the situation entirely, and a higher-order model is needed.

Bioelectrodes are a class of sensors that transduce ionic conduction to electronic conduction, so that the signal can be processed in electronic circuits. The usual purpose of bioelectrodes is to acquire medically significant bioelectrical signals such as *electrocardiogram* (ECG), *electroencephalogram* (EEG), and *electromyogram* (EMG). Both clinical and research examples are easily found, although in many cases the two are the same. Most such bioelectrical signals are acquired from one of three forms of electrode: *surface macroelectrodes, indwelling macroelectrodes,* and *microelectrodes*. Of these, the first two are generally used in vivo, while the latter are used in vitro.

In this chapter we will discuss the acquisition of biopotentials by dealing with the types of electrodes commonly used in biomedical instrumentation. Again, it should be recognized that this discussion is generic and representative, not exhaustive, for the subject is quite complex (indeed, complete books have been written on bioelectrodes).

3-4.1 Electrode Potentials

The skin and other tissues of higher-order organisms, such as humans, are electrolytic and so can be modeled as electrolytic solutions. In some models, the solution is shown as saline, reflecting the fact that we humans are very similar to salt water in our composition. Imagine a metallic electrode immersed in an electrolytic solution (Figure 3-1). Almost immediately after immersion, the electrode will begin to discharge some metallic ions into the solution, while some of the ions in the solution start combining with the metallic electrodes. This is, incidentally, the chemical phenomenon on which the electroplating and anodizing processes are based.

After a short while, a *charge gradient* builds up, creating a potential difference, or *electrode potential* (V_c in Figure 3-1) or *half-cell potential*. Keep in mind that this potential difference can be due to differences in concentration of a single ion type. For example, if you have two positive ions (+ +) in one location (call it *A*), and three positive ions (+ + +) in another location (call it *B*),

Figure 3-1 Metallic electrode immersed in an electrolytic solution.

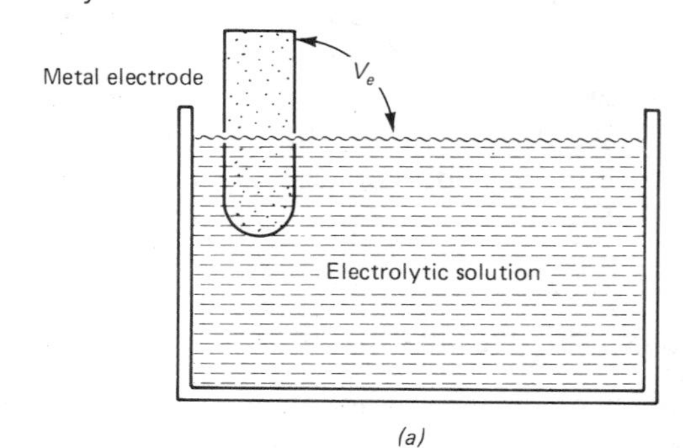

then there will be a net difference of 3 − 2, or 1, with point *B* being more positive than point *A*. Two basic reactions can take place at the electrode/electrolyte interface. An *oxidizing reaction* involves metal → electrons + metal ions; a *reduction reaction* involves electrons + metal ions → metal.

A complex phenomenon is seen at the interface between the metallic electrode and the electrolyte. Ions migrate toward one side of the region or another, forming two parallel layers of ions of opposite charge. This region is called the *electrode double layer,* and its ionic differences are the source of the electrode or halfcell potential (V_e). Different materials exhibit different halfcell potentials, as shown in Table 3-1.

By international scientific agreement, the zero reference point when making halfcell potential measurements is the *hydrogen-hydrogen* (H-H) *electrode,* which is assigned a halfcell potential of zero volts by convention. All other electrode halfcell potentials are measured against the H-H zero reference. The halfcell potentials cited for any given electrode are the differential potential between the actual electrode and the H-H reference electrode.

Now let's consider what happens when two electrodes (call them *A* and *B*), made of *dissimilar metals,* are immersed in the same electrolytic solution (Figure 3-2). Each electrode will exhibit its own halfcell potential (V_{ea} and V_{eb}), and if the two metals are truly dissimilar the two potentials will be different ($V_{ea} \neq V_{eb}$). Because the two halfcell potentials are different, there is a net potential difference (V_{ed}) between them, which causes an electronic current (I_e) to flow through an external circuit. The differential potential, sometimes called an *electrode offset potential,* is only a first-order approximation for the small signal case and is defined as

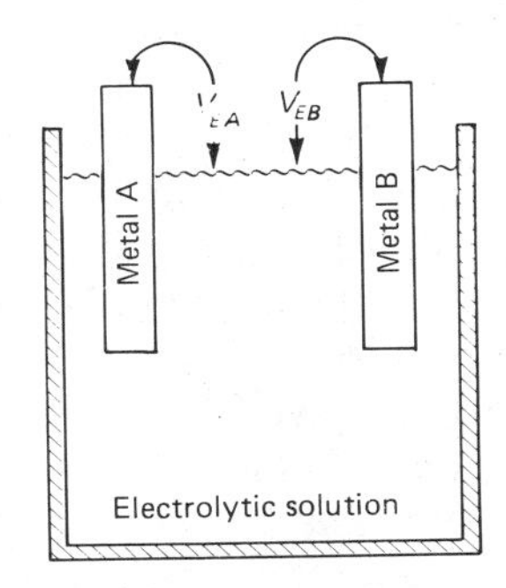

Figure 3-2 Dissimilar metals immersed in a common electrolytic solution produce differential potentials.

Table 3-1
Halfcell Potentials of Common Elements

Material	Halfcell Potential (Volts)
Aluminum (Al^{+++})	−1.66
Zinc (Zn^{++})	−0.76
Iron (Fe^{++})	−0.44
Lead (Pb^{++})	−0.12
Hydrogen (H^{+})	0
Copper	+0.34
Silver (Ag^{+})	+0.80
Platinum (Pt^{+})	+0.86
Gold (Au^{+})	+1.50

$$V_{ed} = V_{ea} - V_{eb} \tag{3-1}$$

For example, let's consider the case where a gold (Au^{+}) electrode is immersed in the same electrolyte as a silver (Ag^{+}) electrode. In that situation,

$$V_{ed} = V_{e(Au)} - V_{e(Ag)} \tag{3-2}$$

$$V_{ed} = (+1.50 \text{ volts}) - (+0.80 \text{ volts}) = +0.70 \text{ volts} \tag{3-3}$$

or, in the frequently seen case of copper (Cu^{++}) and silver (Ag^{+}), which can exist erroneously in electronic circuits that use copper connecting wires,

$$V_{ed} = V_{e(Ag)} - V_{e(Cu)} \tag{3-4}$$

$$V_{ed} = (+0.80 \text{ volts}) - (+0.34 \text{ volts}) = 0.46 \text{ volts} \quad (3\text{-}5)$$

The electrode offset potential will be zero when the two electrodes are made of identical materials, which is the usual case in bioelectric sensing.

Care must be given to the selection of materials when designing electrodes for bioelectric sensing. The choice of materials, as noted earlier, will affect the halfcell and offset potentials. Besides its initial materials dependency, the actual halfcell potential exhibited by any given electrode may change slowly with time. Some candidate materials look good initially but have such a large change with time and chemical environment that they are rendered almost useless in practical applications.

There are two general categories of material combinations. A *perfectly polarized* or *perfectly nonreversible* electrode is one in which there is no net transfer of charge across the metal/electrolyte interface; in these only one of the two types of chemical reactions can occur. A *perfectly nonpolarizable* or *perfectly reversible* electrode is one where there is an unhindered transfer of charge between the metal of the electrode and the electrode. Although these idealized situations are obtained in reality, care must be given to selecting the right electrode. In general, we need to select a reversible electrode such as silver-silver chloride (Ag-AgCl).

Body fluids are terribly corrosive to metals, so not all materials are acceptable for bioelectric sensing. In addition, some materials that form reversible electrodes (e.g., zinc-zinc sulphate) are toxic to living tissue and so are inherently unsuitable. For these reasons, materials such as the noble metals (e.g., gold and platinum), some tungsten alloys, silver-silver chloride (Ag-AgCl), and a material called platinum-platinum black are used to make practical biopotentials electrodes. In general medical use for simple surface recording of bipotentials, the Ag-AgCl electrode is used most often. Unless otherwise specified, you can generally assume that this material is used in clinical electrodes.

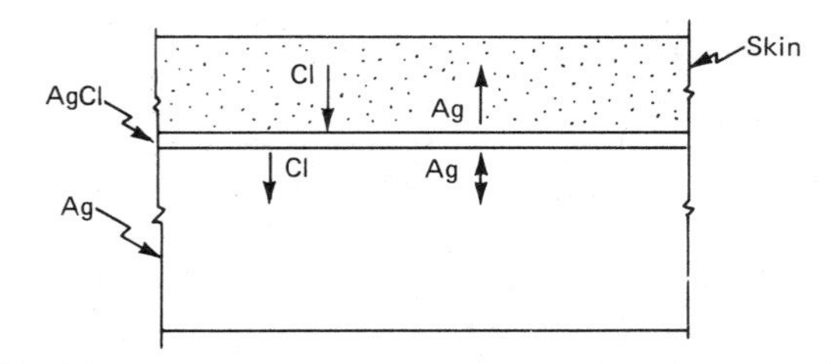

Figure 3-3 Ag-AgCl biomedical electrode.

Figure 3-3 shows why the Ag-AgCl electrode is so popular with medical instrument designers. These electrodes consist of a body of silver onto which a thin layer of silver chloride is deposited. The AgCl provides a free two-way exchange of Ag^+ and Cl^- ions so that no double layer forms. When manufacturing Ag-AgCl electrodes, it is necessary to use spectroscopically pure silver for the process. Such silver is 0.99999 fine (i.e., 99.999 percent pure), compared with ordinary jeweler's and silversmith's silver, which is 0.999 fine (i.e., 99.9 percent pure). *Note:* Sterling silver is 92.5 percent 0.999 fine silver, and 7.5 percent copper.

3-4.2 Electrode Model Circuit

Figure 3-4*a* shows a circuit model of a biomedical surface electrode. This model more or less matches the equivalent circuit of ECG and EEG electrodes. In this circuit a differential amplifier is used for signals processing and so will cancel the effects of electrode halfcell potentials V_1 and V_2. Resistance R_3 represents the internal resistances of the body, which are typically quite low. The biopotentials signal is represented as a differential voltage, V_3. The other resistances in the circuit represent the resistances

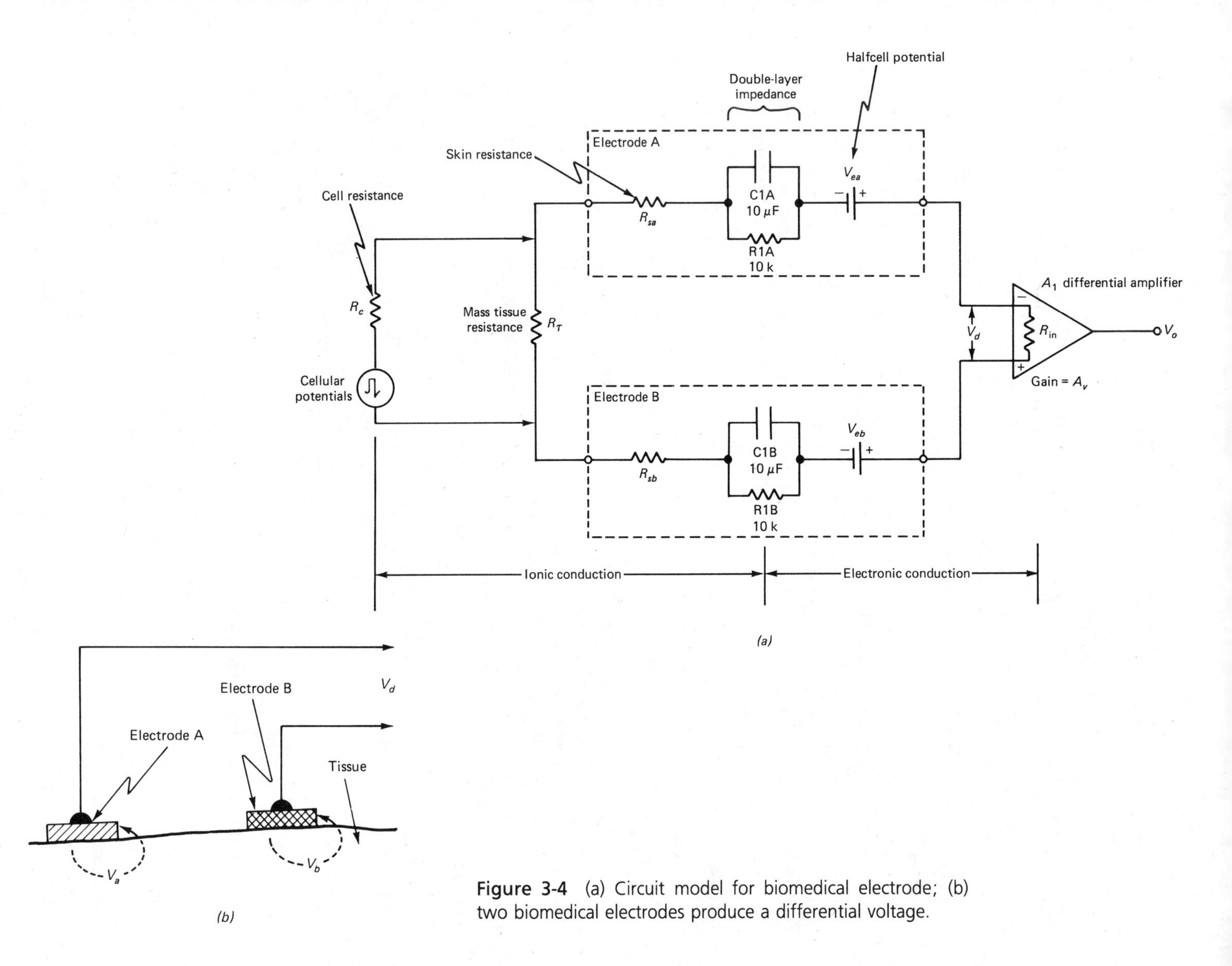

Figure 3-4 (a) Circuit model for biomedical electrode; (b) two biomedical electrodes produce a differential voltage.

at the electrode-skin contact interface. The surprising aspect of Figure 3-4*a* is the usual values associated with capacitors C_1 and C_2. While some capacitance is normally expected, it usually surprises people to learn that these contact capacitances can attain values of several microfarads (the value 10 μF is often cited).

When two or more electrodes are used together, as is almost always the case in physiological recording, then the differential voltage between them is the algebraic sum of the two. In Figure 3-4*b* there are two electrodes, A and B, producing voltages V_a and V_b. The differential voltage V_d is $V_a \pm V_b$.

3-4.3 Electrode Potentials Cause Recording Problems

The electrode halfcell potential becomes a serious problem in bioelectric signals acquisition because of the tremendous difference between these dc potentials and the biopotentials. A typical halfcell potential for a biomedical electrode is 1.5 volts, while biopotentials are more than 1000 times less than the halfcell potential. The surface manifestation of the ECG signal is one to two millivolts (1–2 mV), while EEG scalp potentials are on the order of 50 microvolts (50 μV). Thus, the halfcell electrode voltage is 1500 times greater than the peak ECG potential and 30,000 times greater than the EEG signal.

The instrument designer must provide a strategy for overcoming the effects of the massive halfcell potential offset. Because the halfcell potential forms a large dc component for the minute signal voltage, it is necessary to find an appropriate strategy that uses a combination of the following approaches:

1. We could use a differential dc amplifier to acquire the signal. If the electrodes are identical, then the halfcell potentials should be the same. Theoretically, at least, the equal potentials would be seen as a single common-mode potential and thus would cancel in the output. A limitation on this approach is that the gains required to process low-level signals also act on tiny differences between the two halfcell potentials. A difference of 1 mV between two halfcell potentials—that's only 0.1 percent of the total—looks like any other 1 mV dc signal the gain-of-1000 ECG amplifier.

2. The signals acquisition circuit must be designed to provide a counteroffset voltage to cancel the halfcell potential of the electrode. While this approach has certain appeal, it is limited by the fact that the halfcell potential changes with time and the relative motion between skin and electrode. Electrode motion can cause a wildly varying baseline.

3. We can ac-couple the input amplifier. This approach permits removal of the signal component from the dc offset. This option is, perhaps, the most appealing—especially where variations of the dc offset are of substantially lower frequency than the signal frequency components. In that case, the normal −3 dB frequency response limit can be of use to tailor the attenuation of variations in the dc offset.

In some biomedical applications, however, signal components are near-dc. For example, the frequency content of the ECG signal is 0.05 to 100 Hz. In medical ECG equipment, therefore, one can expect the baseline to shift every time the patient moves around in bed.

In most cases, the first and third options are selected for biopotentials amplifiers. The user will require an ac-coupled, differential input amplifier for signals acquisition.

3-5 Medical Surface Electrodes

Surface electrodes are those that are placed in contact with the skin of the subject. Also in this category are certain needle electrodes of such a size as to prevent their being inserted inside a single cell (which criteria defines a microelectrode). There is some basis for including needle electrodes under the rubric "indwelling electrodes," but that is not generally the practice in biomedical engineering.

Surface electrodes (other than needle electrodes) vary in diameter from 0.3 to 5 cm, with most being in the 1 cm range. Human skin tends to have a very high impedance compared with other voltage sources. Typically, normal skin impedance as seen by the electrode varies from 0.5 kΩ for sweaty skin surfaces to more than 20 KΩ for dry skin surfaces. Problem skin, especially dry, scaly, or diseased skin, may reach impedances in the 500 kΩ range. In any event, one must treat surface electrodes as a very high impedance voltage source—a fact that seriously influences the design of biopotentials amplifier input circuitry. In most cases, the rule of thumb for a voltage amplifier is to make the input impedance of the amplifier at least 10 times the source impedance. For biopotentials amplifiers this requirement means 5 MΩ or greater input impedance—a value easily achieved using either premium bipolar, BiFET, or BiMOS operational amplifiers.

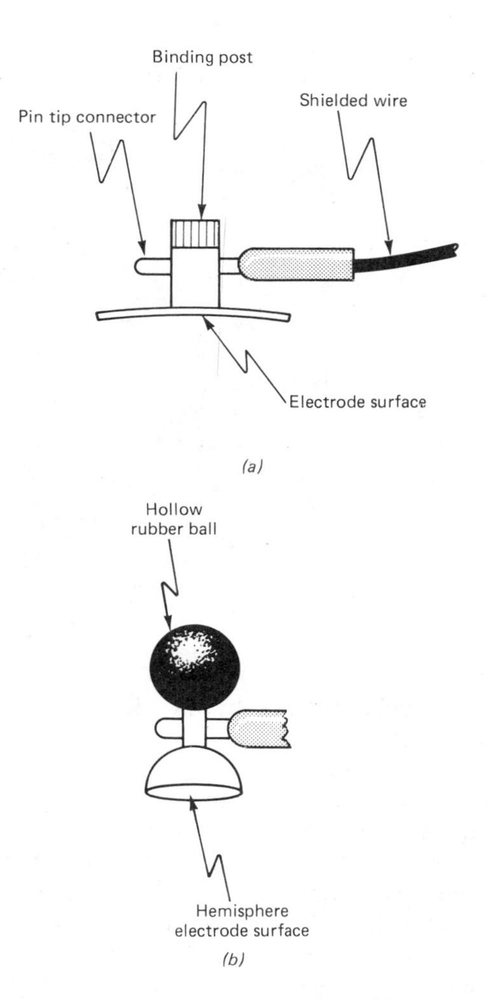

Figure 3-5 Typical ECG electrodes: (*a*) Strap-on electrode, (*b*) suction cup electrode.

3-5.1 Typical Medical Surface Electrodes

A variety of electrodes have been designed for surface acquisition of biomedical signals. Perhaps the oldest form of ECG electrodes in clinical use is the strap-on variety (Figure 3-5*a*). These electrodes are one- to two-square-inch (1–2 in.^{2}) brass plates that are held in place by rubber straps. A conductive gel or paste is used to reduce the impedance between the electrode and skin.

A related form of ECG electrode is the suction cup electrode shown in Figure 3–5*b*. This device is used as a chest electrode in short-term ECG recording. For longer term recording or monitoring, such as continuous monitoring of a hospitalized patient in a coronary or intensive care unit, the paste-on column electrode is used instead.

A typical column electrode is shown schematically in Figure 3-6*a*, while actual examples are shown in Figure 3-6*b*. The electrode consists of an Ag-AgCl metal contact button at the top of a hollow column that is filled with a conductive gel or paste. This assem-

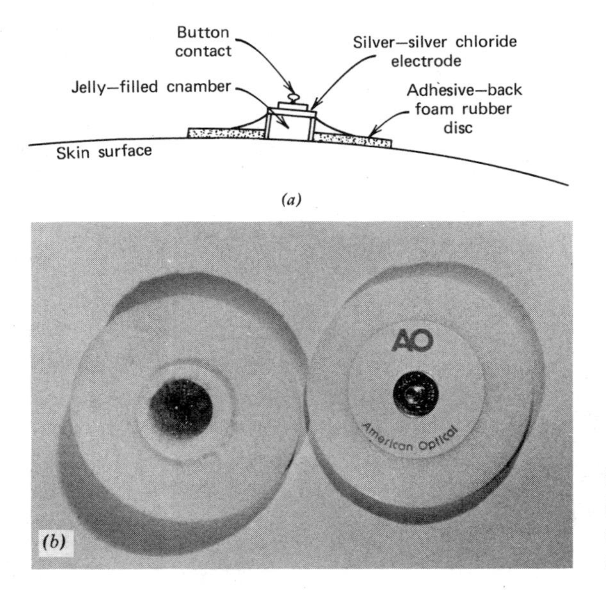

Figure 3-6 Column electrode. (*a*) Cutaway side view. (*b*) A popular type of foam-backed column electrode.

bly is held in place by an adhesive coated foam rubber disk.

The use of a gel- or paste-filled column that holds the actual metallic electrode off the surface reduces movement artifact. For this reason (among several others), the electrodes of Figure 3-6 are preferred for the monitoring of hospitalized patients.

A convenient form of column electrode that is often used in monitoring situations is the three-electrode pad. These adhesive pads have a surface area of 20 to 30 in.^{2} and contain three ECG electrodes (two differential signal pick-up electrodes and a reference electrode) in a single package. They provide a convenient electrode for monitoring, although for diagnostic use the more traditional electrodes are preferred. The three-electrode pad is a temporary disposable unit that is discarded after use.

3-5.2 Problems with Surface Electrodes

There are several problems associated with surface electrodes of all types. One of the problems with column electrodes is that the adhesive will not stick for long on sweaty or "clammy" skin surfaces. The user also must avoid placing the electrode over bony prominences. Usually, the fleshy portions of the chest and abdomen are selected as electrode sites. Various hospitals have different protocols for changing the electrodes, but in general the electrode is changed at least every 24 hours (it is often changed more frequently, as few last as long as 24 hours). In some hospitals, the electrode sites are moved—and electrodes changed—once every eight-hour nursing shift in order to avoid ischemia of the skin at the site.

Although nearly all forms of electrodes can be used in short-term recording situations, long-term monitoring is a little more difficult. One of the most significant problems is *movement artifact* (spurious signal component), which is generated by patient movements and consists of a small electrical component from the bioelectric signals in the patient's skeletal muscles and a large component from the change in interface between electrode and skin. Movement artifact becomes worse as times goes on and the paste or gel dries out.

For short-term recordings, movement artifact is of little practical importance because most patients can lay still long enough for the recording to be made. But in the intensive or coronary care units it is necessary to do long-term monitoring and recording of ECG signals, so the problem becomes more acute.

The most common mechanism that creates artifact signals is *electrode slippage*. If the electrode position slips, then the thickness of the layer of jelly or paste changes abruptly, and this change is reflected as changes in both the electrode impedance and the electrode offset potential. The outward effect produces an artifact in the recorded signal that could possibly obscure the real signal or be interpreted as a bioelectric event in its own right. In the former case, medical people will probably recognize the artifact; they are generally quite good at differentiating gross artifacts from similar-appearing genuine physiologically based anomalies. In the latter case, however, an artifact could lead to a misinterpretation of the waveform's informational content.

There have been several attempts at solving the movement artifact problem by securing the electrodes more tightly to the patient's skin. Sometimes adhesive tape is used to hold the electrode in place, and while this approach will work for a while, the electrodes inevitably work loose. In an hour or two the problem returns.

Another popular solution involves the use of a rough surface (i.e., "prickly") electrode that digs in below the scaly outer layer of skin. But these electrodes are often uncomfortable for the patient, and they fail to solve the problem completely.

Movement artifacts are particularly severe in ECG stress-testing laboratories. The patient walks on a treadmill while the monitoring system records the ECG waveform. The column electrode does fairly well in overcoming the motion artifact, but even so it is often necessary for the medical staff conducting the test to clean and gently abrade the skin at the site where the electrode is attached.

3-5.3 Needle Electrodes

The surface electrodes that we have discussed thus far are noninvasive types. That is, they adhere to the skin without puncturing it. Figure 3-7 depicts the needle electrode. This type of ECG electrode is inserted into the tissue immediately beneath the skin by puncturing the skin at a large oblique angle (i.e., close to horizontal with respect to the skin surface). The needle electrode is only used for exceptionally poor skin, especially on anesthetized patients, and in veterinary situations. Of course, infection is an issue in these cases, so needle electrodes are either disposable (one-time use) or are resterilized in ethylene oxide gas.

3-5.4 Indwelling Electrodes

Indwelling electrodes are those that are intended to be inserted into the body. These are not to be confused with needle electrodes (described earlier), which are intended for insertion into the layers beneath the skin. The indwelling electrode is typically a tiny, exposed metallic contact at the end of a long, insulated catheter (Figure 3-8). In one application, the electrode is threaded through the

Figure 3-7 Needle ECG electrode.

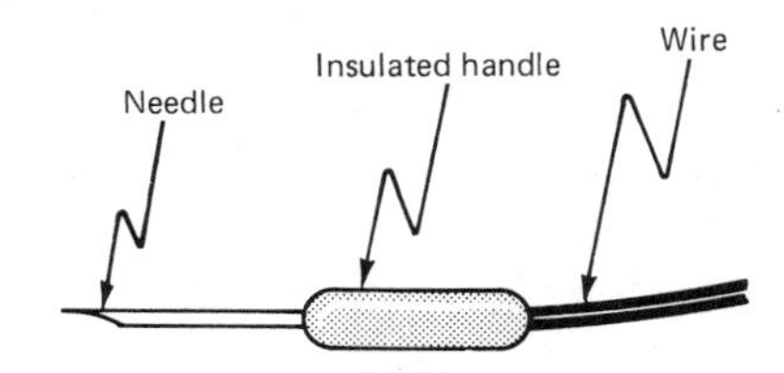

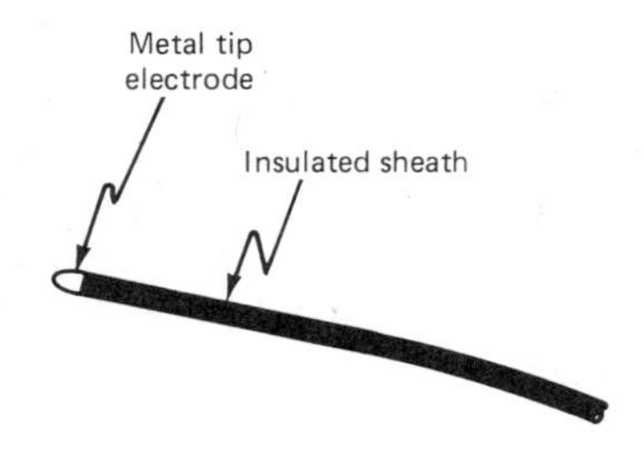

Figure 3-8 Indwelling electrodes.

victim's veins (usually in the right arm) to the right side of the heart in order to measure the intracardiac ECG waveform. Certain low-amplitude, high-frequency features (such as the Bundle of His element) become visible only when an indwelling electrode is used.

3-5.5 EEG Electrodes

The brain produces bioelectric signals that can be picked up through surface electrodes attached to the *scalp*. These electrodes will be connected to an electroencephalograph (EEG) amplifier that drives either an oscilloscope or strip chart recorder.

The typical EEG electrode may be a needle, as in Figure 3-2*c*, but in most cases it is a 1-cm diameter concave disc made either of *gold* or *silver*. The disc electrode is held in place by a thick paste that is highly conductive, or by a headband in certain monitoring applications.

3-6 Microelectrodes

The *microelectrode* is an ultrafine device that is used to measure biopotentials at the cellular level (Figure 3-9). In practice, the microelectrode penetrates a cell that is immersed in an "infinite" fluid (such as physiological saline), which is in turn connected to a reference electrode. Although several types of microelectrodes exist, most of them are of one of two basic forms: me-

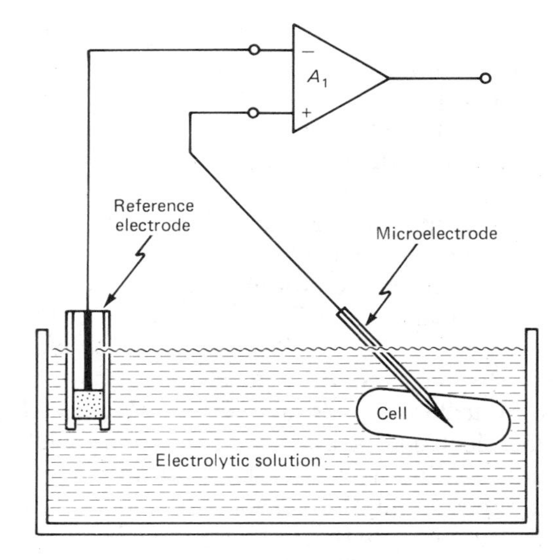

Figure 3-9 ECG microelectrode used to measure cellular potentials.

tallic contact or fluid filled. In both cases, an exposed contact surface of about 1 to 2 micrometers (1 μM = 10^{-6} M) is in contact with the cell. As might be expected, this fact makes microelectrodes very high impedance devices.

Figure 3-10 shows the construction of a typical glass-metal microelectrode. A very fine platinum or tungsten wire is slip-fit through a 1.5 to 2 millimeter (mm) glass pipette. The tip is etched and then fire-formed into the shallow angle taper shown. The electrode can then be connected to one input of the signals amplifier.

These are two subcategories of this type of electrode. In one type, the metallic tip is flush with the end of the pipette taper, while in the other there is a thin layer of glass covering the metal point. This glass layer is so thin that it requires measurement in Angstroms and drastically increases the impedance of the device.

The fluid-filled microelectrode is shown in

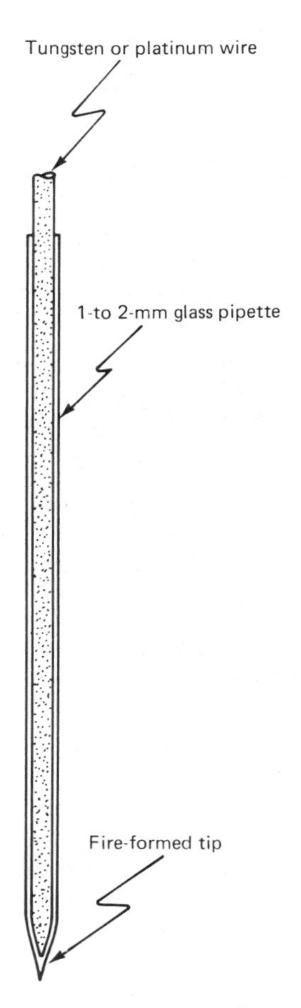

Figure 3-10 Glass-metal microelectrode.

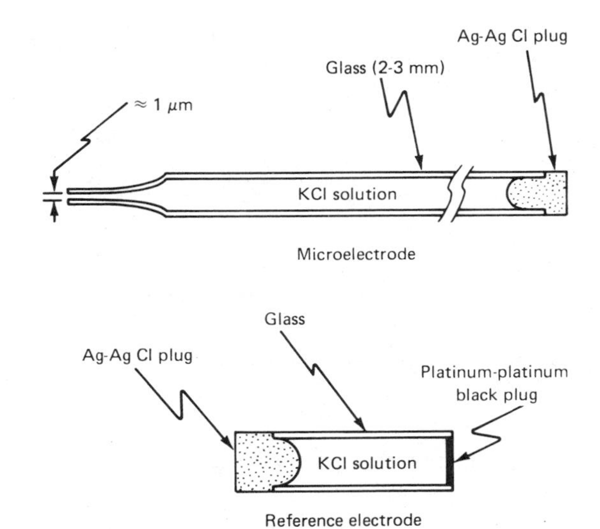

Figure 3-11 Fluid-filled microelectrode.

Figure 3-11. In this type, the glass pipette is filled with a 3-M solution of potassium chloride (KCl), and the large end is capped with an Ag-AgCl plug. The small end need not be capped because the 1-μM opening is small enough to contain the fluid.

The reference electrode is likewise filled with 3 M KCl but is much larger than the microelectrode. A platinum plug contains fluid on the interface end, while an Ag-AgCl plug caps the other end.

Figure 3-12 shows a simplified equivalent circuit for the microelectrode (disregarding the contribution of the reference electrode). Analysis of this circuit reveals the signals acquisition problem due to the RC components. Resistor R_1 and capacitor C_1 are due to the effects at the electrode/cell interface and are (surprisingly) frequency dependent. These values fall off to a negligible point at a rate of $1/(2\pi F)^2$ and are generally considerably lower than R_s and C_2.

Resistance R_s in Figure 3-12 is the spreading resistance of the electrode and is a function of the tip diameter. The value of R_s in metallic microelectrodes without the glass coating is approximated by

$$R_s = \frac{P}{4\pi r} \tag{3-6}$$

where

R_s is the resistance in ohms (Ω)
P is the resistivity of the infinite solution outside of the electrode (e.g., 70 Ω-cm for physiological saline)

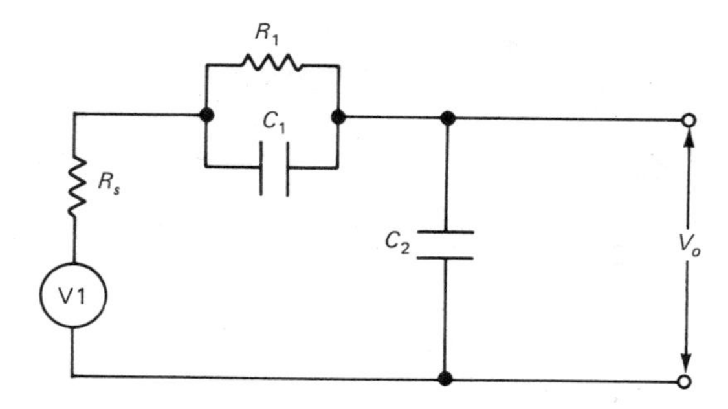

Figure 3-12 Microelectrode equivalent circuit.

r is the tip radius (typically 0.5 μM for a 1-μM electrode)

Assuming the aforementioned typical values, calculate the tip-spreading resistance of a 1-μM microelectrode.

$$R_s = \frac{P}{4\pi r} \tag{3-7}$$

$$R_s = \frac{70\ \Omega\text{-cm}}{(4\pi)\left(0.5\ \mu\text{m} \times \left(\frac{10^{-4}\ \text{cm}}{1\ \mu\text{m}}\right)\right)} \tag{3-8}$$

The impedance of glass-coated metallic microelectrodes is at least one or two orders of magnitude higher than this figure.

For fluid-filled KCl microelectrodes with small taper angles (π/180 radians), the series resistance is approximated by

$$R_s = \frac{2P}{\pi r \alpha} \tag{3-9}$$

where

R_s is the resistance in ohms (Ω)
P is the resistivity (typically 3.7 Ω-cm for 3 M KCl)
r is the tip radius (typically 0.1 μM)
α is the taper angle (typically π/180)

EXAMPLE 3-1
Find the series impedance of a KCl microelectrode using the aforementioned values.

SOLUTION

$$R_s = \frac{2P}{\pi r \alpha}$$

$$R_s = 13.5\ \text{meg}\Omega$$

$$R_s = \frac{(2)(3.7\ \Omega\text{-cm})}{(3.14)\left(0.1\mu\text{m} \times \left(\frac{10^{-4}\ \text{cm}}{1\ \mu\text{m}}\right)\right)\left(\frac{3.14}{180}\right)}$$

The capacitance of the microelectrode is given by

$$C_2 = \frac{0.55e}{\ln\left(\frac{R}{r}\right)}\ \frac{\text{pF}}{\text{cm}} \tag{3-10}$$

where

e is the dielectric constant of glass (typically 4)
R is the outside tip radius
r is the inside tip radius (r and R in the same units)

EXAMPLE 3-2
Find the capacitance of a microelectrode if the pipette radius is 0.2 μm and the inside tip radius is 0.15 μm.

SOLUTION

$$C_2 = \frac{0.55e}{\ln\left(\frac{R}{r}\right)}\ \frac{\text{pF}}{\text{cm}}$$

$$C_2 = \frac{(0.55)(4)}{\ln\left(\frac{0.2\ \mu\text{m}}{0.15\ \mu\text{m}}\right)}\ \frac{\text{pF}}{\text{cm}}$$

How do these values affect performance of the microelectrode? Resistance R_s and capacitor C_2 operate together as an RC low-pass filter. For example, a KCl microelectrode immersed in 3 cm of physiological saline has a capacitance of approximately 23 pF. Suppose it is connected to the amplifier input (15 pF) through 3 feet of small-diameter coaxial cable (27 pF/ft, or 81 pF). The total capacitance is (23 + 15 + 81) pF = 119 pF. Given a 13.5-megohm resistance, the frequency response (at the −3 dB point) is

$$F = \frac{1}{2\pi RC} \tag{3-11}$$

where

F is the −3 dB point in hertz (Hz)
R is the resistance in ohms (Ω)
C is the capacitance in farads (f)

EXAMPLE 3-3
For C = 119 pF (1.19×10^{-10} farads) and R = 1.35×10^{7} ohms, find the frequency response upper −3 dB point.

$$F = \frac{1}{(2)(3.14)(1.35) \times 10^{7}\,\Omega)(1.19 \times 10^{-10}\,\text{f})}$$

$$F = 99 \text{ Hz} \approx 100 \text{ Hz}$$

Clearly, a 100 Hz frequency response, with a −6 dB/octave characteristic above 100 Hz, results in severe rounding of the fast rise-time action potentials. A strategy must be devised in the instrument design to overcome the effects of capacitance in high-impedance electrodes.

3-6.1 Neutralizing Microelectrode Capacitance

Figure 3-13 shows the standard method for neutralizing the capacitance of the microelectrode and associated circuitry. A neutralization capacitance, C_n, is in the positive feedback path along with a potentiometer voltage divider. The value of this capacitance is

$$C_n = \frac{C}{A - 1} \tag{3-12}$$

where

C_n is the neutralization capacitance
C is the total input capacitance
A is the gain of the amplifier

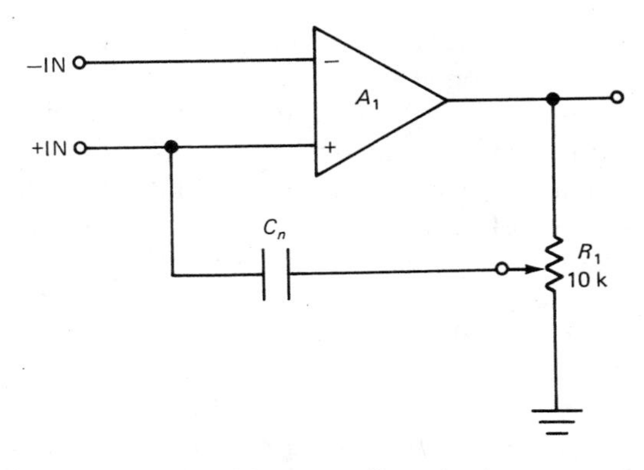

Figure 3-13 Capacitance nulling circuit.

EXAMPLE 3-4
A microelectrode and its cabling exhibit a total capacitance of 100 pF. Find the value of neutralization capacitance (Figure 3-12) required for a gain-of-10 amplifier.

SOLUTION

$$C_n = \frac{C}{A - 1}$$

$$C_n = \frac{100 \text{ pF}}{10 - 1}$$

$$C_n = \frac{100 \text{ pF}}{9} = 11 \text{ pF}$$

3-7 Transducers and Other Sensors

Transducers are part of an overall class of devices called *sensors*, which also includes biophysical electrodes. A general problem with the word *transducer* is that it also refers to devices such as loudspeakers and ultrasonic sender units. To understand this distinction, let us reiterate in different words the definition given in Section 3-3:

Def. A *transducer* is a device that converts energy from some other form into electrical energy for the purposes of measurement or control.

Transducers differ from electrodes in that they use some intervening transducible element to make the measurement, while electrodes directly acquire the signal. For example, a pressure transducer may use the change of resistance of a taut-wire element when it is flexed as a measure of pressure. Similarly, a thermistor relies on the change of electrical resistance of some materials undergoing temperature changes in order to measure temperature. Because many transducers use *piezoresistive* elements connected in a Wheatstone bridge, we will begin our study with these highly valuable circuits.

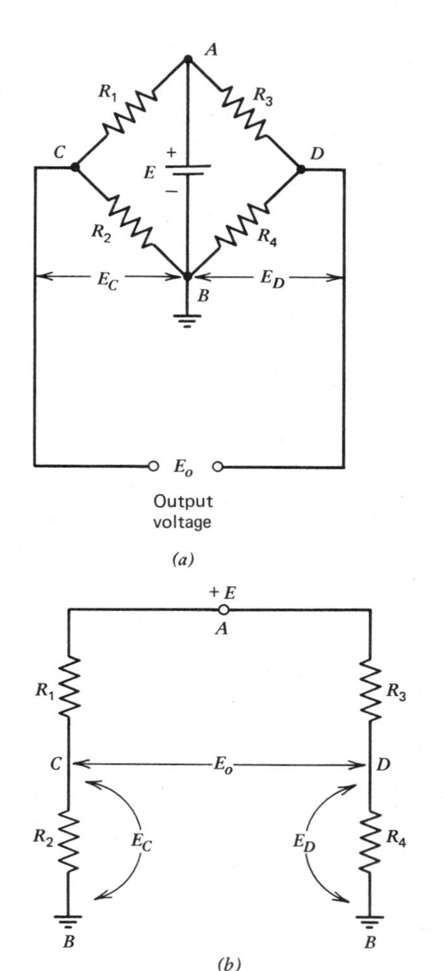

Figure 3-14 The Wheatstone bridge circuit. (*a*) Original circuit. (*b*) Original circuit redrawn.

3-7.1 The Wheatstone Bridge

Many biomedical transducers are used in a circuit configuration called a *Wheatstone bridge* (see Figure 3-14). Many transducers, not bridges in their own right, are often connected with other components to form a Wheatstone bridge. Any discussion of biomedical transducers, therefore, must begin with an introduction to the bridge circuit.

The basic Wheatstone bridge of Figure 3-14*a* uses one resistor in each of four arms. A battery (E) *excites* the bridge connected across two opposing resistor junctions (i.e., A and B). The bridge output voltage E_o appears across the remaining pair of resistor junctions (C and D).

The original circuit of Figure 3-14*a* is redrawn in Figure 3-14*b*, which simplifies analysis.

We may analyze the Wheatstone bridge circuit by initially breaking it into circuits across E: $R_1 - R_2$ and $R_3 - R_4$. Both of these networks are resistor voltage dividers. In fact, the Wheatstone bridge may be viewed as two resistor voltage dividers *in parallel* across the supply E. The output voltage (E_o) is the *difference* between the two ground-referenced potentials E_C and E_D produced by the divider networks. In equation form, this relation is

$$E_o = E_C - E_D \tag{3-13}$$

But E_C and E_D may be expressed in terms of the excitation potential E, using the simple voltage divider theorem

$$E_C = E \times \frac{R_2}{R_1 + R_2} \quad (3\text{-}14)$$

and

$$E_D = E \times \frac{R_4}{R_3 + R_4} \quad (3\text{-}15)$$

Substituting Equations 3-14 and 3-15 into Equation 3-13 we obtain the output voltage E_o as

$$E_o = \frac{ER_2}{R_1 + R_2} - \frac{ER_4}{R_3 + R_4} \quad (3\text{-}16)$$

$$E_o = E\left(\frac{R_2}{R_1 + R_2} - \frac{R_4}{R_3 + R_4}\right) \quad (3\text{-}17)$$

EXAMPLE 3-5

A Wheatstone bridge (Figure 3-14) is excited by a 12-V dc source and contains the following resistances: $R_1 = 1.2\ k\Omega$, $R_2 = 3\ k\Omega$, $R_3 = 2.2\ k\Omega$, and $R_4 = 5\ k\Omega$. Find the output voltage, E_o.

SOLUTION

$$E_o = E\left(\frac{R_2}{R_1 + R_2} - \frac{R_4}{R_3 + R_4}\right)$$

$$E_o = 12\left(\frac{3}{1.2 + 3} - \frac{5}{2.2 + 5}\right)$$

$$E_o = 12\left(\frac{3}{4.2} - \frac{5}{7.2}\right)$$

$$E_o = 12\,(0.714 - 0.694) = 0.02\ \mathbf{V}$$

Note from Example 3-5 and Equation (3-17) that the Wheatstone bridge voltage output is dependent on the resistances in the arms. Changing one or more of these resistances will change the output voltage. This phenomenon is the basis for many biomedical transducers.

The *null condition* in a Wheatstone bridge circuit exists when the output voltage E_o is zero. But from Equation (3-14), if E_o is zero, then either the excitation potential E must be zero (not true) or the expression inside of the brackets must be equal to zero (true). The null condition occurs when

$$E_C = E_D \quad (3\text{-}18)$$

$$E_{CB} = E_{DB} \quad (3\text{-}19)$$

and

$$E_{AC} = E_{AD} \quad (3\text{-}20)$$

So equals divided by equals are equal and

$$\frac{E_{CB}}{E_{AC}} = \frac{E_{DB}}{E_{AD}} \quad (3\text{-}21)$$

Since no current flows from C to D at the null and $E_C = E_D$, then from Figure 3-4*b*

$$\frac{I_{ACB}R_1}{I_{ACB}R_2} = \frac{I_{ADB}R_3}{I_{ADB}R_4} \quad (3\text{-}22)$$

So

$$\frac{R_1}{R_2} = \frac{R_3}{R_4} \quad (3\text{-}23)$$

Equation 3-23 gives us the sole necessary condition for the *null* condition in an excited Wheatstone bridge. Note that it is not necessary for the resistances to be equal, but only that the *ratios* (of the two *half-bridge* voltage dividers) be equal.

EXAMPLE 3-6

Show that the null condition exists in a Wheatstone bridge consisting of the following resistances: $R_1 = 2\ k\Omega$, $R_2 = 1\ k\Omega$, $R_3 = 10\ k\Omega$, and $R_4 = 5\ k\Omega$. (*Hint:* Equation 3-23 describes the sole condition necessary for null.)

SOLUTION

$$\frac{R_2}{R_1} = \frac{R_4}{R_3}$$

$$\frac{1}{2} = \frac{5}{10}$$

$$0.5 = 0.5$$

Since both sides of the equation evaluate to the same quantity, we may conclude that the bridge is in the null condition. A bridge in the null condition is said to be *balanced.*

In many biomedical transducers using the Wheatstone bridge, all four resistances are equal in the null condition. This is not a strict physical requirement, but it is the way many manufactuers choose to build their products. Values for R in the 150 to 800 Ω range are typical.

In most designs the null condition exists when the parameter stimulating the transducer is either at zero, or some predetermined value (i.e., atmospheric pressure) that is taken to be a *zero baseline*. The stimulus (i.e., parameter being measured) will cause any or all of the bridge resistance elements to change resistance by some small amount h (note that h is sometimes written "ΔR," meaning a *small* change in the parameter R). When the applied stimulus is zero, then all four resistors have a resistance of R and the output voltage is zero. The bridge is in the null condition.

When the stimulus is *not* zero, on the other hand, each arm takes on a resistance of $R \pm h$, and this unbalances the circuit to produce an output voltage that is proportional to the value of the applied stimulus.

3-8 Strain Gages

A *strain gage* is a resistive element that produces a *change* in its resistance proportional to an applied mechanical *strain*. A strain is a *force* applied in either *compression* (a push along the axis *toward* the center) or *tension* (a pull along the axis but *away* from the center).

Figure 3-15*a* shows a small metallic rod with no force applied. It will have a length L and a cross-sectional area A. Let changes in length be given by ΔL and changes in area by ΔA.

In Figure 3-15*b* we see the result of applying a *compression* force to the ends of the bar. The length *reduces* to $L - \Delta L$, and the cross-sectional area *increases* to $A + \Delta A$.

Similarly, when a *tension* force of the same magnitude is applied to the bar, the length *increases* to $L + \Delta L$ and the cross-sectional area *reduces* to $A - \Delta A$.

The resistance of a metallic rod is given in terms of the length and cross-sectional area in the expression

$$R = \rho\left(\frac{L}{A}\right) \tag{3-24}$$

where[1]

ρ is the resistivity constant unique to the type of material used in the rod in ohm-meters (Ω-m)

L is the length in meters (m)

A is the cross-sectional area in square meters (m^2)

EXAMPLE 3-7

Find the resistance of a copper bar that has a cross-sectional area of 0.5 mm^2 and a length of 250 mm. (*Hint:* The resistivity of copper is 1.7×10^{-8} Ω-m.)

SOLUTION

$$R = \rho\left(\frac{L}{A}\right)$$

$$R = (1.7 \times 10^{-8}\ \Omega\text{-M}) \times \frac{[250 \text{ mm} \times (1 \text{ m}/1000 \text{ mm})]}{0.5 \text{ mm}^2 \times (1 \text{ m}/1000 \text{ mm})^2}$$

$$R = \frac{1.7 \times 10^{-8}\ \Omega \times 0.25}{5 \times 10^{-7}} = \mathbf{0.0085\ \Omega}$$

Equation 3-24 tells us that the resistance varies directly as the length and inversely as the square of the cross-sectional area. Both of these phenomena are crucial to the operation of the resistive strain gage transducers.

[1] *Note:* In the British engineering system commonly used in the United States, the units of resistivity are ohms per circular mil foot. A *circular mil* is the area of a circle that has a diameter of 1 mil or $\frac{1}{1000}$ inch (in.).

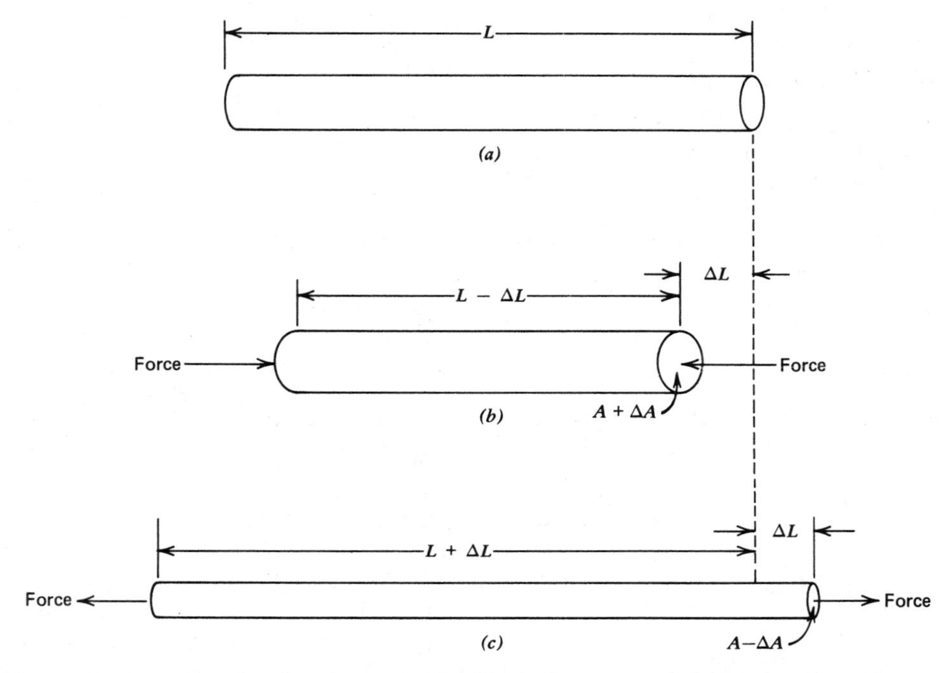

Figure 3-15 Mechanism for piezoresistivity. (*a*) The bar at rest (no force applied). (*b*) Under a *compression* force. (*c*) Under a *tension* force.

Example 3-7 illustrates the change of resistance with changes in size and shape. This phenomena is sometimes called *piezoresistivity*. The resistance of the rod will become $R + h$ in tension and $R - h$ in compression. If you examine Equation 3-24 carefully, you will note that changes in both length and cross-sectional area tend to *increase* the resistance in *tension* and *decrease* the resistance in *compression*. The resistances after force is applied are *in tension:*

$$(R + h) = \frac{L + \Delta L}{A - \Delta A} \tag{3-25}$$

in compression:

$$(R - h) = \frac{L - \Delta L}{A + \Delta A} \tag{3-26}$$

EXAMPLE 3-8

A thin constantan wire stretched taut has a length of 30 mm and a cross-sectional area of 0.01 mm². The resistance is 1.5 Ω. The force applied to the wire is increased such that the length increases by 10 mm and the cross-sectional area decreases by 0.0027 mm². Find the change in resistance h. (*Hint:* The resistivity of constantan is approximately 5×10^{-7} Ω-m.)

SOLUTION

$$(R + h) = \rho\frac{L + \Delta L}{A - \Delta A}$$

$$(R + h) = (5 \times 10^{-7}\ \Omega\text{-M}) \times \frac{[(30 + 10)\text{mm} \times (1\text{ m}/1000\text{ mm})]}{(0.01 - 0.0027)\text{ mm}^2 \times [(1\text{ m}/1000\text{ mm})]^2}$$

$$1.5 + h = \frac{(5 \times 10^{-7}\ \Omega)(40)(10^3)}{0.0073}$$

then $1.5 + h = 2.74\ \Omega$

so $h = 2.74 - 1.5 = \mathbf{1.24\ \Omega}$

The change in resistance will be approximately linear for small changes in dimensions provided that ΔL is much less than L. Of course, if too great a force is applied, modulus of elasticity is exceeded and the wire becomes permanently deformed. It is then useless as a transducer.

3-8.1 Gage Factor

The gage factor (GF) for a strain gage transducer is a means of comparing it with other similar transducers. The definition of gage factor is

$$GF = \frac{\Delta R/R}{\Delta L/L} \qquad (3\text{-}27)$$

where

GF is the gage factor (dimensionless)
ΔR is the change in resistance in ohms (Ω)
R is the unstrained resistance in ohms
ΔL is the change in length in meters (m)
L is the length in meters

EXAMPLE 3-9

A 20-mm length of wire used as a strain gage exhibits a resistance of 150 Ω. When a force is applied in tension, the resistance changes by 2 Ω and the length changes by 0.07 mm. Find the gage factor GF.

SOLUTION

$$GF = \frac{\Delta R/R}{\Delta L/L}$$

$$GF = \frac{2/150}{0.07/20}$$

$$GF = \frac{0.013}{0.0035} = \mathbf{3.71}$$

The gage factor gives us a means for evaluating the relative *sensitivity* of a strain gage element. The greater the change in resistance per unit change in length, the greater the *sensitivity* of the element and the greater the gage factor.

Equation 3-27 is sometimes given in the alternate form

$$GF = \frac{\Delta R/R}{\mathscr{E}} \qquad (3\text{-}28)$$

in which $\mathscr{E}$ (strain) is the factor $\Delta L/L$.

3-8.2 Types of Strain Gages

There are *two* basic forms of piezoresistive strain gage: *bonded* and *unbonded*. Figure 3-16*a* shows a crude example of the *unbonded* strain gage. The resistance element is a *thin* wire of a special alloy that is stretched *taut* between two flexible supports, which are in turn mounted on a thin metal diaphragm. When a force such as F_1 is applied, the diaphragm will flex in a manner that spreads the supports further apart, causing an increased tension in the resistance wire. This *tension* tends to *increase* the resistance of the wire an amount proportional to the applied force.

Similarly, if a force such as F_2 is applied to the diaphragm, the ends of the supports

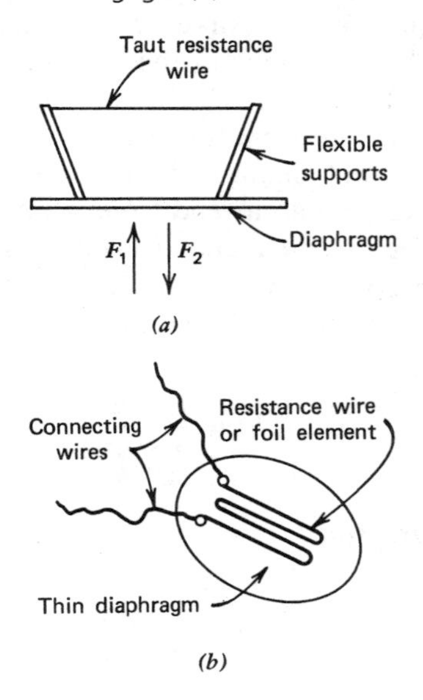

Figure 3-16 The Piezoresistive strain gage. (*a*) The unbonded strain gage. (*b*) The bonded strain gage.

move *closer* together, *reducing* the tension in the taut wire. This action is the same as applying a compression force to the wire. The electrical resistance in this case will *reduce* an amount proportional to the applied force.

A *bonded* strain gage is made by *cementing* a thin wire or foil element to a diaphragm, as shown in Figure 3-16*b*. Flexing the diaphragm *deforms* the element, causing a change in electrical resistance exactly as in the unbonded strain gage.

Unbonded strain gages can be constructed so that they are linear over a wide range of applied force but are very delicate. The bonded strain gage, on the other hand, is generally more rugged but is linear over a smaller range of forces. Note well, however, that *no* piezoresistive strain gage will take a large amount of abuse, and they should always be treated as *delicate instruments*.

Many biomedical strain gage transducers are of the bonded construction because the linear range is adequate and the extra ruggedness is a decidedly desirable feature in medical environments, where people cannot take the kind of precautions required if a more delicate type were used. Note, however, that the Statham P-23 series are of the unbonded type but are made in a very rugged housing. These are among the most common cardiovascular pressure transducers used in medicine.

Very few physiological strain gage transducers use a single element; most use four strain gage elements connected in a Wheatstone bridge circuit. In the unbonded types there will be four supports, one for each bridge junction. Two of the resistance elements will be connected to each support. In the bonded variety there will be foil or wire elements arranged in a bridge configuration.

Both types of transducers are found with an element geometry that places *two* elements in *tension* and *two* elements in *compression* for any applied force. Such a configuration increases the output of the bridge for any applied force and so increases the *sensitivity* of the transducer.

Figure 3-17*a* shows a Wheatstone bridge with strain gage elements for each of the four bridge arms. We may find that R_1 and R_4 are aligned parallel to each other along one axis of the diaphragm, while resistors R_2 and R_3 are parallel to each other and *perpendicular* to R_1/R_4.

Consider a force applied to the diaphragm

Figure 3-17 Wheatstone bridge strain gage. (*a*) Strain gage elements in a bridge circuit. (*b*) Mechanical configuration using a common diaphragm.

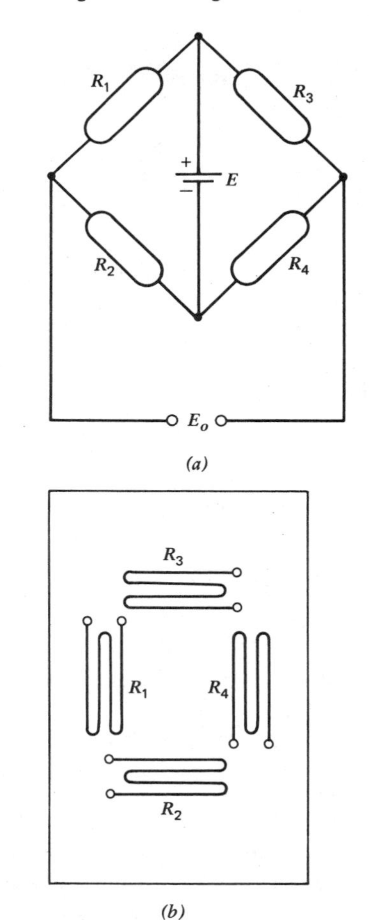

of a transducer such as Figure 3-17*b*. Resistors R_1 and R_4 will be in compression, while R_2 and R_3 are in tension.

Assume that all resistors are equal ($R_1 = R_2 = R_3 = R_4 = R$) when no force is applied to the diaphragm, and let $\Delta R = h$. When a force is applied, the resistance of R_1 and R_4 will be $R + h$, and the resistance of R_2 and R_3 will be $R - h$. From a rewritten version of Equation 3-17, we know that the output voltage is

$$E_o = E \times \left[\frac{(R-h)}{(R+h)+(R-h)} - \frac{(R+h)}{(R-h)+(R+h)}\right] \tag{3-29}$$

$$E_o = E \times \left[\frac{(R-h)}{2R} - \frac{(R+h)}{2R}\right] \tag{3-30}$$

$$E_o = E\left(\frac{h}{R}\right) = -E\left(\frac{\Delta R}{R}\right) \tag{3-31}$$

$$E_o = -E\left(\frac{\Delta R}{R}\right) \tag{3-32}$$

EXAMPLE 3-10

A strain gage transducer is constructed such as shown in Figure 3-17*b*. In the null condition each element has a resistance of 200 Ω. When a force is applied, each resistance changes by 10 Ω. Find that the output voltage if a 10-V excitation potential is applied to the bridge.

SOLUTION

$$E_o = -E \times \frac{\Delta R}{R}$$

$$E_o = -10 \text{ V} \times 10/200$$

$$E_o = -10 \text{ V} \times 0.05 = \mathbf{-0.50 \text{ V}}$$

3-8.3 Transducer Sensitivity, ϕ

The *sensitivity* (ϕ) of a transducer is the rating that allows us to *predict* the output voltage from a knowledge of the excitation voltage and the value of the applied stimulus. The units for ϕ are microvolts per volt of excitation per unit of applied stimulus (μV/V/U).

Let us consider a force transducer. A certain biomedical force transducer is usually calibrated in *grams*. Before you object that this must be wrong (*grams* are units of *mass*!), let us hasten to point out that the "force" in this case would be the earth's *gravitational attraction* on a mass of one gram. This convention allows calibration of force transducers using a simple metric weight set from a platform balance, as opposed to conventional force in dynes (1 gram-force = 980 dynes).

If the sensitivity factor ϕ is known for a transducer, then the output voltage may be calculated from

$$E_o = \phi \times E \times F \tag{3-33}$$

where

E_o is the output potential in volts (V)
E is the excitation potential in volts (V)
F is the applied force in grams (g)
ϕ is the sensitivity in μV/V/g

EXAMPLE 3-11

A transducer has a sensitivity of 10 μV/V/g. Predict the output voltage for an applied force of 15 g if the excitation potential is 5 V dc.

SOLUTION

$$E_o = \phi \times E \times F$$

$$E_o = \frac{10\ \mu\text{V}}{\text{V-g}} \times 5 \text{ V} \times 15 \text{ g}$$

$$E_o = 750 \text{ uV} = \mathbf{0.00075 \text{ V}}$$

Note that the sensitivity is important in both the design and repair of medical instruments because it allows us to predict the output voltage for a given stimulus level, and therefore the gain of the amplifier required for processing the signal (see Chapter 6).

3-8.4 Solid-State Piezoresistive Strain Gages

In the past, most strain gage transducers were made using wire elements or vacuum-deposited metallic elements. Today, however, many strain gage devices are based on solid-state silicon technology, in which all four elements of the Wheatstone bridge are formed of piezoresistive semiconductor material. Some are made similar to bonded strain gages (i.e., the material is deposited or diffused onto a diaphragm). Others use a cantilever design in which the semiconductor piezoresistive elements are supported between fixed supports.

3-9 Inductive Transducers

Almost any electrical property that can be made to vary in a predictable manner under the influence of a physical stimulus may be used for transduction of that stimulus. Inductance, for example, can be varied easily by physical movement of a permeable core within an inductor. Inductors, therefore, can be used to make transducers. There are, in fact, three basic forms of inductive transducer: *single coil, reactive Wheatstone bridges,* and *linear voltage differential transformers* (LVDT).

The first type, single coil devices, are rarely used in modern equipment. They are constructed much like the dynamic microphone, in which a diaphragm either affects the position of an iron or ferrite core inside of the coil or the field of a core formed from a permanent magnet. A force applied to the diaphragm creates a current in the winding of the latter and changes the inductance in the former.

An example of an inductive bridge transducer is shown in Figure 3-18*a*. The output function for this transducer is shown in Figure 3-18*b*. The Wheatstone bridge circuit consists of the inductive reactances of coils L_1 and L_2, plus the 200-Ω resistors.

Note that alternating current (ac) excitation is required because the reactance of a coil is zero when direct current (dc) is applied. Hewlett-Packard typically uses an excitation signal of 2400 Hz at 5 V(rms). Other manufacturers use as much as 10 V(rms) at frequencies between 400 Hz and 5000 Hz.

The Hewlett-Packard model 1280 transducer of Figure 3-18*a* is used for measurement of arterial and venous blood pressure in *millimeters of mercury* (*mm Hg*). (Note that the accepted units of pressure are Torr, 1 Torr = 1 mm Hg. In medicine, however, the obsolete term *mm Hg* is still used.)

The transduction occurs because of a change of *position* of the inductor's core. But this yields only position data *unless* the applied force operates against some other force such as a *spring*. The force required to compress or stretch a spring is given by Hooke's law: $F = -kX$, in which the term X is a displacement (i.e., change of position).

At zero gage pressure (transducer's diaphragm open to atmosphere) the diaphragm is not distended in either direction, so the armature core is displaced equally in both L_1 and L_2. Under this condition the inductive reactances of L_1 and L_2 are equal, so the bridge is *balanced*. There will be no output voltage.

When a pressure *above* or *below* atmospheric pressure is applied, the diaphragm becomes distended in one direction, and this forces the armature further into one coil than in the other. The respective inductive reactances of L_1 and L_2 are no longer equal, so the bridge is *unbalanced* and an output voltage develops. The amplitude of the ac output signal is proportional to the *magnitude* of the applied pressure, while its *phase* indicates whether the pressure is *positive* (a compression) or *negative* (a vacuum) (see Figure 3-18*b*). The sensitivity of the transducer in this case is about 40 uV/V/mm Hg.

Note in Figure 3-18*a* that the output voltage at pin A of the connector is taken from the wiper of a potentiometer. This sensitivity control is used to trim out *normal* differences between transducers so that pressure-moni-

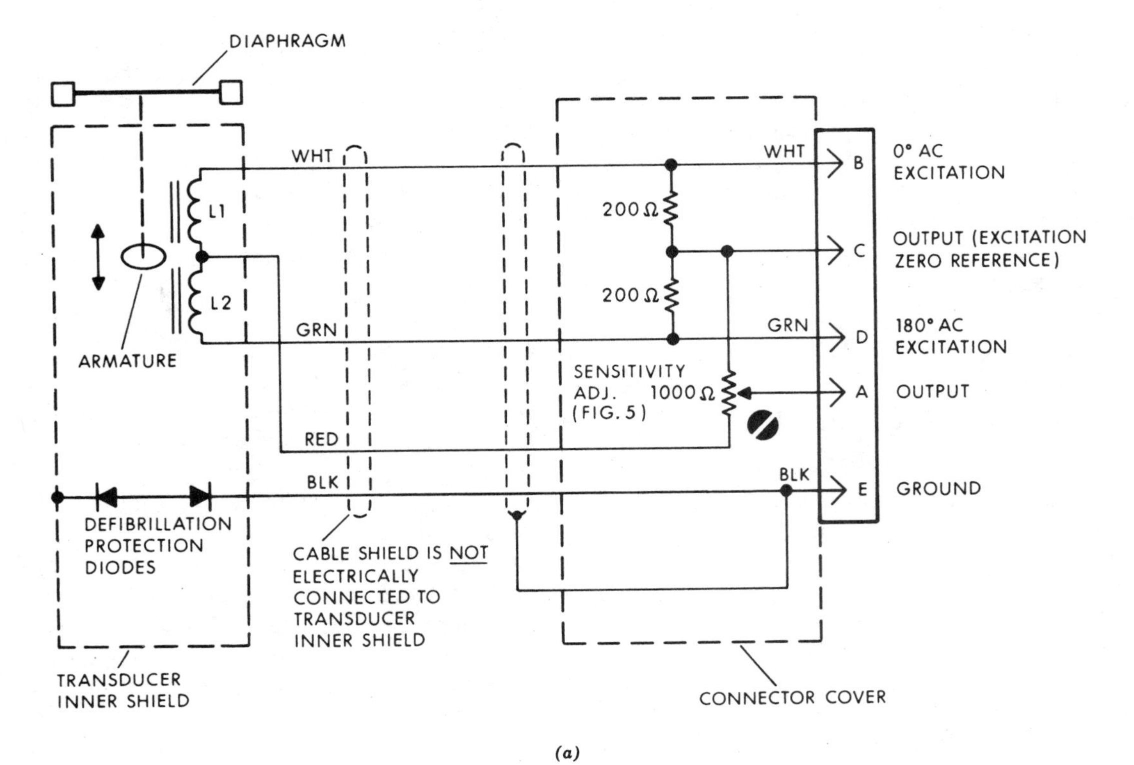

(a)

Sensitivity The sensitivity of the 1280 Series Transducers is expressed in terms of: Volts output, per volt of excitation, per unit of pressure applied to the transducer.

Transducer sensitivity is factory set to 40 microvolts per volt of excitation per millimeter of mercury, plus or minus one percent.

Since the transducer excitation voltage is commonly 5 Vac, the output voltage is:

200 μV per mmHg
20 mV per 100 mmHg
80 mV per 400 mmHg

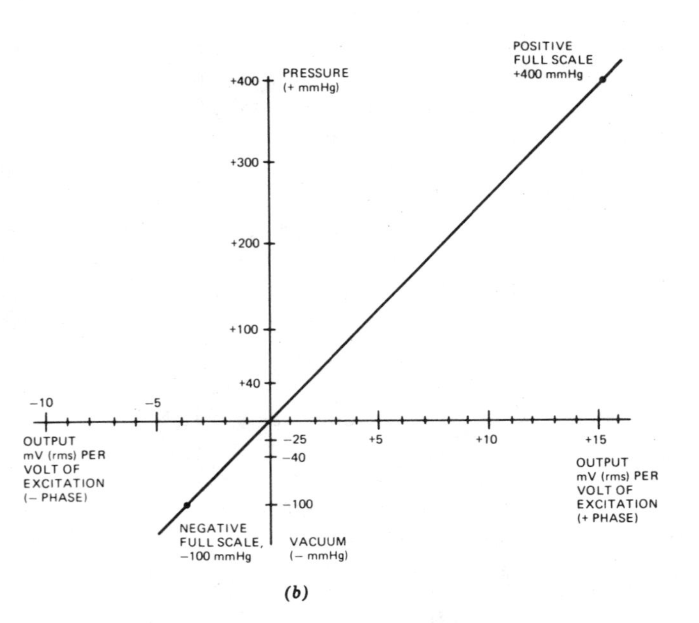

(b)

Figure 3-18 Inductive Wheatstone bridge strain gage. (*a*) Circuit for an H-P Model 1280. (Reprinted courtesy Hewlett-Packard.) (*b*) Output function. (Reprinted courtesy Hewlett-Packard.)

toring instruments can be easily calibrated by less skilled operators.

An example of an LVDT transducer is shown in Figure 3-19. It is a transformer with a primary (L_1) and *two* secondaries (L_2 and L_3). The secondaries are connected in *opposite sense* so that their respective currents tend to cancel each other. When the stimulus is zero, the core affects L_2 and L_3 *equally* so the current cancellation is *total* and the output voltage is therefore *zero*.

An ac excitation signal is applied to the primary. When the stimulus is applied to the diaphragm, it displaces the core. The inductive reactances of L_2 and L_3 are no longer equal, so their respective currents are no longer equal. Secondary current cancellation is less than total, so a current flows in the external load creating an output voltage signal. This output voltage has a magnitude proportional to the applied stimulus and a phase indicating the direction the core was moved. In the case of a pressure transducer, this tells us whether the pressure is positive or negative.

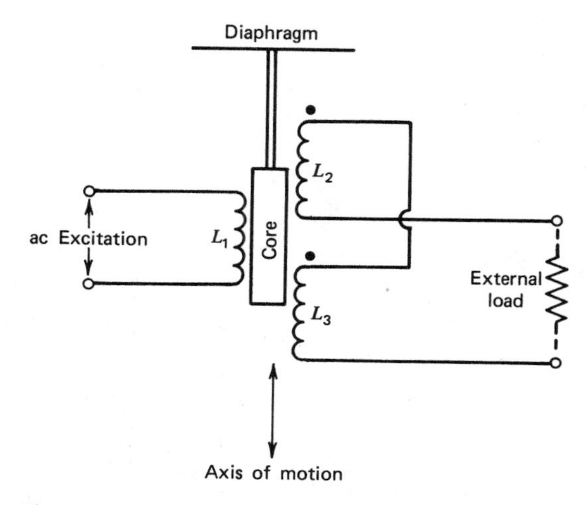

Figure 3-19 Linear differential voltage transformers.

3-10 Quartz Pressure Sensors

Another modern form of sensor, especially in medical pressure measurements, is the *quartz* transducer. These devices are basically capacitively based (see Section 3-11) but are made differently from other capacitive transducers. The pressure sensor capsule of these devices is made of homogeneous fused quartz (Figure 3-20). There are two capacitors in the capsule: a *pressure capacitor* (C_p) and a *reference capacitor* (C_{ref}). The capacitor plates are made of noble metals vacuum deposited onto their respective surfaces of the quartz capsule.

These capacitors are connected in a ratiometric series arrangement (see inset to Figure 3-20) so that differences in dielectric properties of the quartz material are compensated. The capacitors can be connected in a capacitive bridge circuit, a mixed RC bridge circuit (both similar to Wheatstone bridges), or in oscillator circuits (Fig. 3-21). Several of these circuits are discussed in Section 3-11.

Advantages of the quartz transducer include very low (some sources claim zero) hysteresis, very low slippage of the metals and alloys with respect to the crystal, very

Figure 3-20 Homogeneous fused quartz pressure transducer.

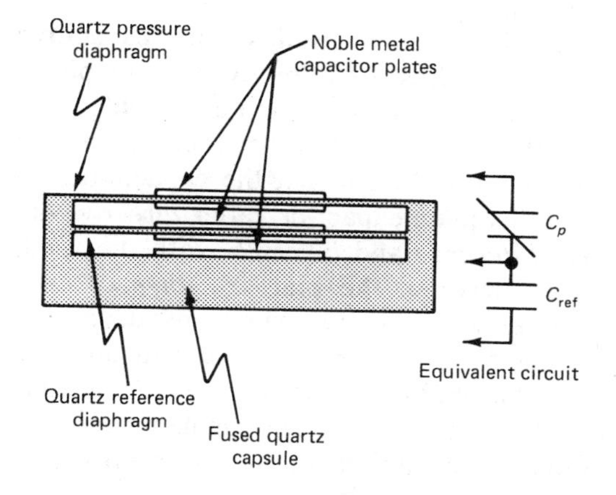

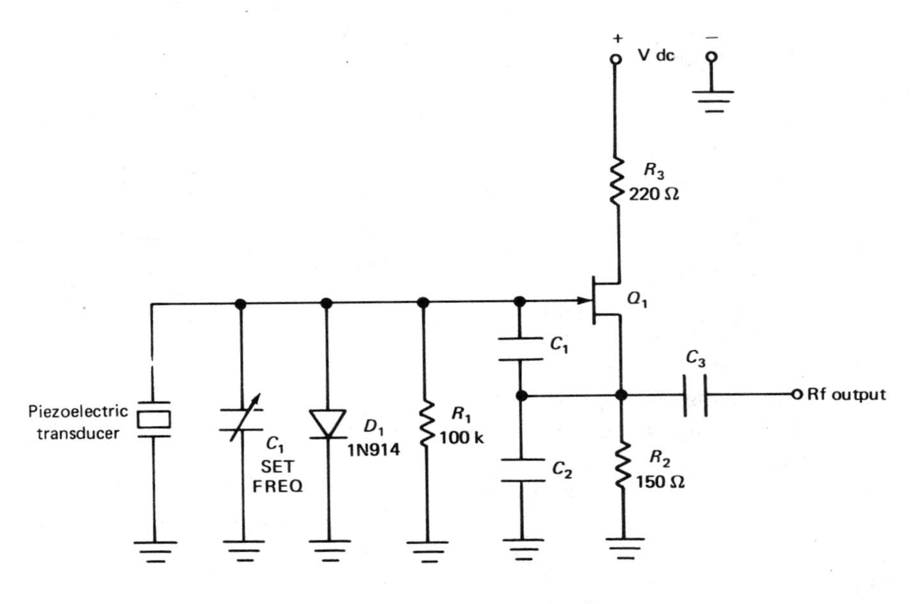

Figure 3-21 Oscillator circuit.

low temperature sensitivity, good elastic properties, and ruggedness.

3-11 Capacitive Transducers

Another capacitive transducer that is seen occasionally is the metallic plate *capacitive transducer*. These cause the capacitance of the transducer to vary with the stimulus. Since capacitance is used, ac excitation is necessary.

In almost all varieties, the capacitive transducer uses a stationary plate or plates attached to the housing and a movable plate that changes position under the influence of the stimulus. Recall that the capacitance of a parallel plate capacitor varies *directly* with the *plate area* and *inversely* with the *separation* between the plates. *Either* or *both* may be varied in any given transducer.

One form of capacitive transducer consists of a solid metal disc parallel to a flexible metal diaphragm; the two elements are separated by an air or vacuum dielectric (see Figure 3-22). This construction is very similar to that of the *capacitor microphone*, which is in fact a transducer for *sound waves!* When a force is applied to the diaphragm, it will either move closer to or further away from the stationary disc. This in-

Figure 3-22 Simple capacitance transducer.

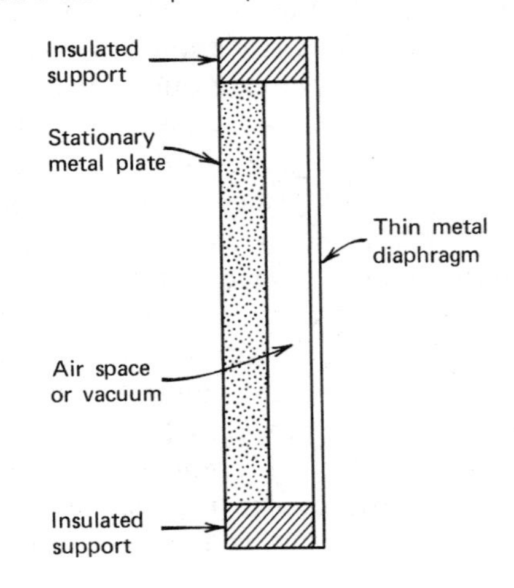

creases or decreases the capacitance, respectively.

Another popular form (see Figure 3-23) uses a stationary metal plate (i.e., *stator*) and a rotating movable plate. The movable plate is usually of a "butterfly" shape. The capacitance varies because the position of the rotor determines how much of the stator plate is shaded by the rotor. At only one position the shading will be greatest so the capacitance will also be greatest. At 90° rotation from that position the shading is least so the capacitance is also least.

Figure 3-24 shows still another form of capacitance transducer. In this type of transducer a movable metal plate (P_3) is placed between two stationary plates (P_1 and P_2). This forms a differential capacitor consisting of two sections (Figure 3-24*b*). Capacitor C_1 is the capacitance between plates P_1 and P_3, while capacitance C_2 is the capacitance between plates P_2 and P_3. When a force is applied to the diaphragm plate, P_3 will move closer to one end plate than the other. If the force is in the direction shown by the arrow, then P_3 is closer to P_2 than P_1, so capacitance C_2 is greater than C_1. In the opposite situation, the movable plate is closer to P_1 than P_2, so capacitance C_1 is greater than C_2.

There are several ways to use capacitance transducers in instrumentation circuits. One method, although rarely used in biomedical applications, is to have the transducer be part of an *L-C* resonant circuit controlling the frequency of an oscillator. Varying the capacitance under influence of a stimulus will vary the oscillator frequency (i.e., frequency modulate) an amount proportional to the stimulus.

Another way, shown in Figure 3-25, is called the electrometer technique. In this circuit the capacitance of the transducer is

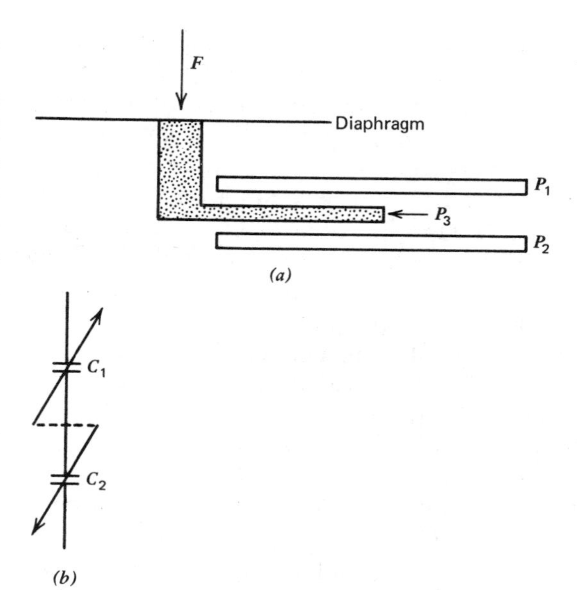

Figure 3-24 Differential capacitance transducer. (*a*) Mechanical structure. (*b*) Schematic symbol.

Figure 3-23 Butterfly plate transducer.

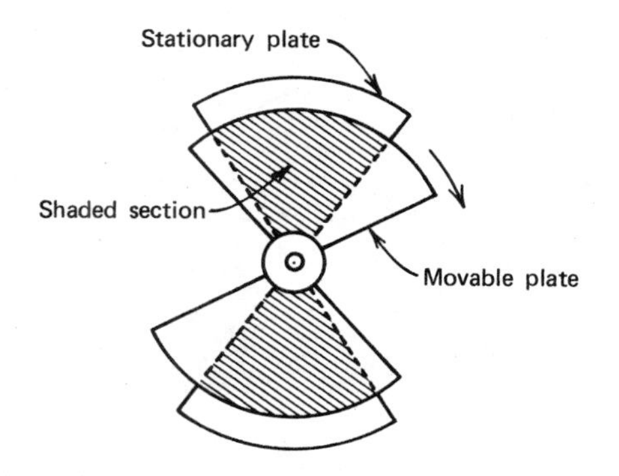

Figure 3-25 Electrometer transducer.

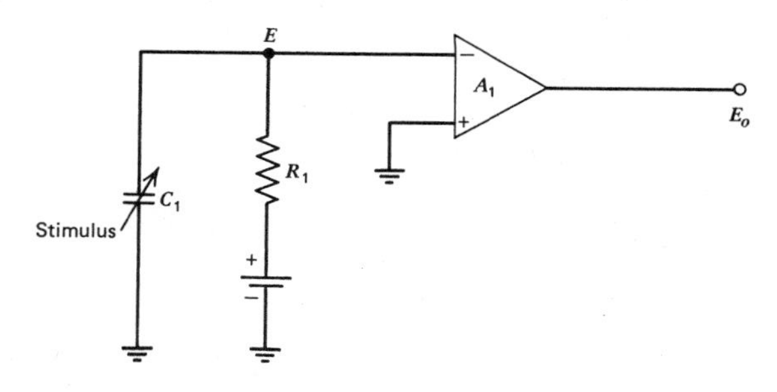

charged through a constant current source (R_1 and E). The voltage across the capacitance (i.e., the voltage applied to the input of the amplifier) depends on the capacitance, which is proportional to an applied stimulus.

One of the most common methods is to use the capacitor transducer in one arm of a Wheatstone bridge (see Figure 3-26). Two arms of the bridge are resistances, while two arms are capacitive reactances. Capacitor C_1 represents the capacitance of the transducer, while C_2 is the capacitance of a variable trimmer capacitor used to balance the bridge under zero stimulus conditions.

In some cases resistor R_2 will be the second section of a differential capacitor. Under zero stimulus conditions the capacitance of C_2 and the capacitance that replaces R_2 in Figure 3-26 will be equal, but under the influence of a stimulus the relation changes and balance is upset.

3-12 Temperature Transducers

There are three types of common temperature transducers: *thermocouples*, *thermistors*, and *solid-state pn junctions*. Of these, the latter two find the greatest use in clinical applications, while all three are used in biomedical and biophysical research applications.

A thermocouple (see Figure 3-27*a*) consists of two *dissimilar* conductors or semiconductors joined together at one end. Because the *work functions* of the two materials are different, there will be a potential generated when this junction is heated. The po-

Figure 3-26 Capacitive Wheatstone bridge.

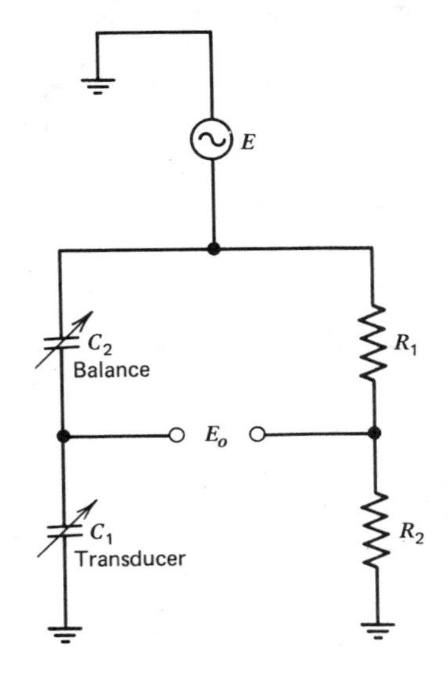

Figure 3-27 Three types of temperature transducer: thermocouple, thermistor, and *pn* junction. (*a*) Thermocouple. (*b*) Thermistor. (*c*) *pn* junction.

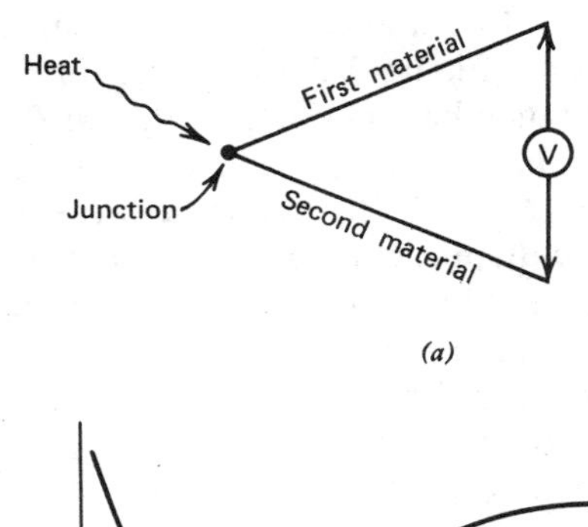

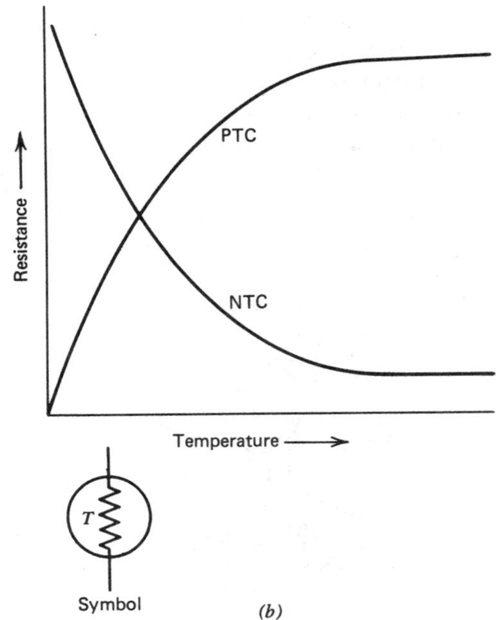

tential is roughly linear with changes of temperature over a relatively wide range, although at the extreme limits of temperature for any given pair of materials nonlinearity increases markedly.

Transistors (i.e., *thermal resistors*) are resistors that are designed to change value in a predictable manner with changes in temperature (Figure 3-27*b*). A positive temperature coefficient (PTC) device *increases* resistance with increases in temperature, while a negative temperature coefficient (NTC) device *decreases* resistance with increases in temperature.

Most thermistors have a nonlinear curve when the curve is plotted over a wide temperature range, but when limited to a narrow temperature range (such as human body temperatures) the linearity is better. When thermistors are used, it is necessary to insure that the temperature is not allowed to go into the ranges where the calibration is unknown or extremely nonlinear. Most medical temperature transducers are thermistors.

The last class of thermal transducer is the solid-state *pn* junction diode (Figure 3-27*c*). If you take an ordinary solid-state rectifier diode and connect it across an ohmmeter, then it is possible to observe this phenomenon. Note the forward biased resistance at room temperature, and then heat the diode temporarily with a lamp or soldering iron. The diode resistance drops as temperature increases.

Most temperature transducers, however, use the diode-connected bipolar transistor such as Figure 3-27*c*. We know that the base-emitter voltage of a transistor is proportional to temperature. For the differential pair in Figure 3-27*c* the transducer output voltage is

$$\Delta V_{be} = \frac{KT \ln(I_{c1}/I_{c2})}{q} \qquad (3\text{-}34)$$

where

K is Boltzmann's constant 1.38×10^{-23} joules/degrees Kelvin (J/°K)

T is the temperature in degrees Kelvin (°K) (*Note:* 0°C = 273 °K.)

q is the electronic charge 1.6×10^{-19} coulombs per electron

I_{c1} and I_{c2} are the collector currents of Q_1 and Q_2, respectively

The quantity K/q is a ratio of constants and is constant under *all* circumstances. The current ratio I_{c1}/I_{c2} is *held* constant by using constant current sources in the emitter circuits of Q_1 and Q_2. Of course, the *logarithm* of a *constant* is also a constant. So the only variable in Equation 3-34 is the temperature.

EXAMPLE 3-12

Find the output voltage of a temperature transducer constructed as Figure 3-15*c* if I_{c1} is 2 mA and I_{c2} is 1 mA and the temperature is 37°C. (*Hint:* 37°C is (37 + 273)°K, or 310°K.)

SOLUTION

$$\Delta V_{be} = (KT \ln (I_{c1}/I_{c2}))/q$$

$$\Delta V_{be} = \frac{(1.38 \times 10^{-23}\ \text{J/K})(310\ \text{K})\,[\ln (2\ \text{mA}/1\ \text{mA})]}{1.6 \times 10^{-19}\ \text{C}}$$

$$\Delta V_{be} = 1.85 \times 10^{-2}\ \text{J/C} = \mathbf{0.0185\ V}$$

Equation 3-34 yields a value of approximately 59.8 μV/°K. The student should verify that the result in this example is actually *volts*, since 1 V = 1 J/C.

The circuit in Figure 3-27*c* has one distinct advantage over others: It is widely linear over most temperatures (up to the point where the transistors are damaged), so the output voltage ΔV_{be} may be processed in a simple amplifier and requires no special circuitry to linearize the result. If an amplifier of suitable gain is chosen and provided, then the temperature and the amplifier output voltage can be made numerically equal, allowing a simple voltmeter readout device. Most such instrumental systems are scaled to produce an amplifier output voltage of 10 mV/°K.

3-13 Summary

1. When *metallic* electrodes are connected to skin, a *halfcell* potential is formed. The *combined* halfcell potentials of two or more electrodes form an *electrode offset potential. Polarization* occurs when dc flows through the electrode/skin interface.
2. *Plate* and *suction* cup electrodes are used for making short-term recordings of bioelectric potentials, but for longer recordings a *column* electrode is used. In all cases, however, an electrolytic paste or jelly is used between the electrode and the skin.
3. Many biomedical transducers are based on the Wheatstone bridge principle.
4. *Bonded* and *unbonded* strain gages use the principle of *piezoresistivity* to make pressure and force transducers.
5. Three basic types of temperature transducer are thermocouple, thermistor, and the solid-state *pn* junction.

3-14 Recapitulation

Now return to the objectives and self-evaluation questions at the beginning of the chapter and see how well you can answer them. If you cannot answer certain questions, place a check mark next to each and review appropriate parts of the text. Next, try to answer the following questions, using the same procedure. When you have answered all of the questions, solve the problems in Section 3-16.

3-15 Questions

3-1 Define *halfcell potential*. Describe the factors that give rise to the halfcell potential.

3-2 Describe the differences between *electronic conduction* and *ionic conduction*.

3-3 Define *oxydizing reactions* and *reduction reactions*.

3-4 What is the *electrode double layer*? Define in your own words.

3-5 Define *electrode offset potential*.

3-6 The two general categories of electrode material combinations are: ______________ ______________ and ______________ ______________. Which of these are preferred for biomedical electrodes?

3-7 List three techniques that might be used to overcome the effects of electrode offset potentials.

3-8 What is the approximate range of halfcell potentials in the materials normally considered for use in biomedical electrodes?

3-9 What material is normally employed in biomedical electrodes?

3-10 Draw the equivalent circuit for a pair of electrodes applied to the skin. Show the amplifier input resistance.

3-11 Sketch a diagram showing the electrode double layer phenomenon for (a) ordinary conductors and (b) silver-silver chloride materials.

3-12 What is the range of impedances normally found in medical electrodes?

3-13 Sketch a microelectrode based on platinum wire.

3-14 Sketch the circuit diagram of an amplifier that can compensate for microelectrode capacitance.

3-15 Name two forms of electrodes normally used for short-term ECG recording.

3-16 Describe or draw the diagram for a *column electrode*.

3-17 Long-term (more than a few minutes) monitoring of ECG is usually done with ______________ electrodes.

3-18 Why does the column electrode reduce movement artifacts?

3-19 State the most common cause of movement artifact in ECG recording.

3-20 Describe the typical EEG electrode.

3-21 EEG electrodes are used to record bioelectric potentials arising in the ______________.

3-22 Draw the circuit of a Wheatstone bridge.

3-23 Define *piezoresistivity*.

3-24 What are the differences between *bonded strain gages*, *unbonded strain gages*, and *solid-state semiconductor strain gages*?

3-25 What are the differences between quartz and moving plate capacitive transducers?

3-16 Problems

3-1 A copper wire is 40 cm long and has a cross-sectional area of 0.05 mm^2. Calculate its resistance. (*Hint:* the resistivity of copper is $1.7 \times 10^{-8}\ \Omega$.)

3-2 A copper bar is 4 in. long and has a cross-sectional area of 100 circular mils. What is its resistance? (*Hint:* the resistivity of any material can be converted to the British engineering units by dividing the mks units by 6×10^8.)

3-3 In the Wheatstone bridge of Figure 3-14, assume that $R_1 = 1\ k\Omega$, $R_2 = 980\ \Omega$, $R_3 = 990\ \Omega$, and $R_4 = 1.1\ k\Omega$. Find the output voltage if an excitation potential of +7.5 V dc is applied.

3-4 A Statham Model P-23Db pressure transducer is to be used to record arterial blood pressure. The manufacturer lists the sensitivity of this Wheatstone bridge strain gage as 50 uV/V/cm Hg. Find the output voltage when an excitation potential of 5 V dc is used and the applied pressure is 120 Torrs.

3-5 A 30-mm length of wire has a resistance of 90 Ω. When a tension force is applied, the resistance changes 0.75 Ω and the length becomes 30.02 mm. Calculate the *gage factor*.

3-6 A transducer with a sensitivity of 5 uV/V/g is used to measure a 120-g force. Find the output voltage if an 8-V dc excitation potential is used.

3-7 A dual transistor is connected as in Figure 3-15*c* to form a temperature transducer. Calculate the output voltage if $I_{c1} = 4$ mA, $I_{c2} = 1.5$ mA, and the temperature is 32°C.

3-8 Show that $\Delta V_{be} = KT\ [\ln(I_{c1}/I_{c2})]/q$. (*Hint:* for each transistor in Figure 3-15*c*: $V_{be} = KT[\ln(I_c/I_s)]/q$ where I_s is the theoretical reverse saturation current, a *constant* in matched transistors.)

3-9 Find the capacitance of a microelectrode if the pipette radius is 0.2 μm and the inside tip radius is 0.15 μm. Assume platinum wire is used inside a glass pipette.

3-10 For electrode capacitance of $C = 133$ pF (1.33×10^{-10} F) and an impedance of $R = 2.65 \times 10^7\ \Omega$, find the frequency response upper −3 dB point.

3-11 A microelectrode and its cabling exhibit a total capacitance of 140 pF. Find the value of neutralization capacitance required for a gain-of-20 amplifier.

3-17 References

1. Geddes, L. A. and L. E. Baker, *Principles of Applied Biomedical Instrumentation,* Wiley (New York, 1968).
2. Oliver, Frank J., *Practical Instrumentation Transducers*, Hayden Book Co. (New York, 1971).
3. Simmons, J. and Soderquist, D., "Temperature Measurement Method Based on Matched Transistor Pair Requires No Reference," Precision Monolithics, Inc., applications note AN-12 (1976).
4. Cobbold, Richard S. C., *Transducers for Biomedical Measurements*, Wiley (New York, 1974).
5. Fitzsimmons, J., "New Pressure Sensor Technology," *Sensors*, March 1986, pp. 18ff.
6. Perrino, Frank, "Solid-State Pressure Sensors," *Sensors*, July 1986, pp. 44ff.
7. "Disposable Low-cost Catheter Tip Sensor Measures Blood Pressure During Surgery," Product Feature column, *Sensors*, July 1989, pp. 25ff.
8. Bryzek, Janusz, "Silicon Low Pressure Sensors Address HVAC Applications," *Sensors*, March 1990, pp. 30ff.
9. Mulkins, Donald F., "The Quartz Capacitive Pressure Sensor," *Sensors*, July 1989, pp. 9ff.

Chapter 4

Bioelectric Amplifiers

4-1 Objectives

1. Be able to state the requirements for a *bioelectric amplifier.*
2. Be able to describe the basic principles of *operational amplifiers.*
3. Be able to draw several different bioelectric amplifier configurations.
4. Be able to state the principles of operation of isolation amplifiers.
5. Be able to describe the problems associated with the acquisition of bioelectric phenomena.

4-2 Self-Evaluation Questions

These questions test your prior knowledge of the material in this chapter. Look for the answers as you read the text. After you have finished studying the chapter, try answering these questions and those at the end of the chapter (see Section 4-16).

1. List the features required in a bioelectric amplifier.
2. List the basic properties of the operational amplifier.
3. Draw the circuit for an operational amplifier integrator, and describe its operation.
4. Find the voltage gain of an inverting follower that has a 100-kΩ feedback resistor and a 5-kΩ input resistor.
5. Describe an isolation amplifier. Why is it used as a bioelectric amplifier?
6. State the mathematical expression that gives the gain of an noninverting follower.

4-3 Bioelectric Amplifiers

Amplifiers used to process biopotentials are called *bioelectric amplifiers,* but this designation applies to a large number of different types of amplifier. The gain of a "bioelectric amplifier," for example, may be low, medium, or high (i.e., $\times 10$, $\times 100$, $\times 1000$–$\times 10{,}000$). Similarly, some bioelectric amplifiers are ac coupled, while others are dc coupled. The frequency response of "typical" bioelectric amplifiers may be from dc (or near dc, i.e., 0.05 Hz) up to 100 kHz.

Dc coupling is required where the input signals are clearly dc or change *very slowly* (some in vivo O_2 levels change in *mm Hg per minute* or *per hour*). But even at frequencies as low as 0.05 Hz, ac coupling may be used instead of dc. The reason for this is to overcome electrode offset potentials. In the ECG amplifier, for example, frequency components as low as 0.05 Hz might be processed. But the electrode-skin connection produces an electrode offset (dc) potential that will interfere with the ECG signal. The amplifier, therefore, must be ac coupled to block the dc offset in the input signal yet have a frequency response down to 0.05 Hz in order to faithfully reproduce the patient's ECG waveform.

The *high-frequency response* is the frequency where the gain drops 3 dB below its midfrequency value. In some cases the -3 dB high-frequency point will be a frequency as low as 30 Hz, but in most cases it is usually 10 kHz. Specialized models used to process specific waveforms may have a particular response. ECG amplifiers, for example, usually have a frequency response of 0.05 to 100 Hz.

A few general-purpose amplifiers have *adjustable* frequency response and are thus usable for a wide range of applications. In general, it is wise to use only the minimum frequency response needed to insure good reproduction of the input waveform. This practice permits rejection of high-frequency noise.

Low-gain amplifiers are those with gain factors between $\times 1$ and $\times 10$. The unity-gain ($\times 1$) amplifier is used mostly for isolation, buffering, and possibly impedance transformation between signal source and readout device. Low-gain amplifiers are often used for the measurement of action potentials and other relatively high-amplitude bioelectric events.

Medium-gain amplifiers are those that provide gain factors between $\times 10$ and $\times 1000$ and are used for the recording of ECG waveforms, muscle potentials, and so forth.

High-gain, or *low-level* signal amplifiers have gain factors over $\times 1000$, with some having factors as high as $\times 1{,}000{,}000$. This type of amplifier is used in very sensitive measurements such as the recording of brain potentials (EEG).

Two important parameters in bioelectric amplifiers, especially those in the high- and medium-gain classes, are *noise* and *drift*. *Drift* is the (spurious) change in output signal voltage due to changes in operating temperature (rather than input signal changes). *Noise,* in this case, normally is the thermal noise generated in resistances and semiconductor devices. Good design and prudent component selection reduce these problems to the negligible level in modern equipment.

All three classes of bioelectric amplifiers must have a very high *input impedance*. This requirement is the one commonality between *all* bioelectric amplifiers, because almost all bioelectric signal sources exhibit a high *source impedance*. Most bioelectric sources have an impedance between 10^3 and $10^7\ \Omega$, and ordinary engineering design practices dictate an amplifier input impedance that is at least an order of magnitude higher than the source impedance. Modern MOSFET and JFET amplifier (op-amp) devices have input impedances on the order of 1 *teraohm* ($10^{12}\ \Omega$).

The properties of the integrated circuit (IC) operational amplifier make it especially suited as a bioelectric amplifier. A discussion of operational amplifier theory follows. A greater understanding will be achieved as we discuss basic principles followed by practical circuits made from commercially available components. These parts help make biomedical instrumentation what it is today. In addition, circuit applications are shown to demonstrate the versatility of op-amps and to stress their importance in biomedical instrument circuits.

4-4 Operational Amplifiers

The *operational* amplifier is a device that behaves in a unique manner: The properties of the circuit containing an operational amplifier are determined by the properties of the *negative feedback loop*. For the elementary voltage amplifier circuit configurations, we require only the basic properties of the

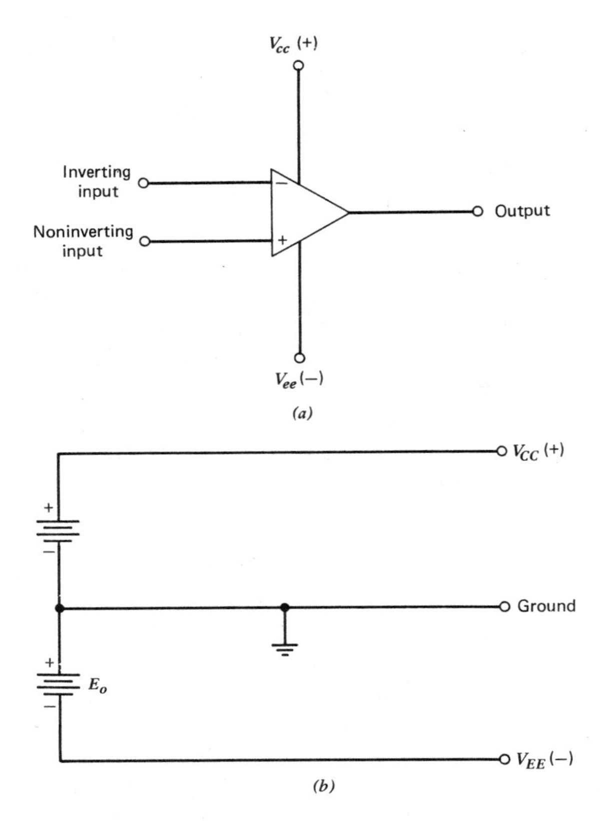

Figure 4-1 The operational amplifier (*a*) Circuit symbol. (*b*) Power supply configuration.

device, Ohm's law, and Kirchhoff's law to derive the transfer equations. The author of one book on operational amplifier circuit design[1] was reportedly tempted to title his book *Ohm's Law With Applications*. There are more elegant, more mathematical methods for analyzing the behavior of the voltage amplifier configurations, but none shows circuit action quite as vividly as the simple method presented as follows.

The *operational* amplifier gets its name from the fact that they were originally conceived to solve mathematical *operations* in analog computers. Although analog computers are no longer in widespread use, many electronic instruments use operational amplifiers and are, in effect, scaled-down, dedicated, or single-purpose analog computers.

Commercial operational amplifiers have been available since the early 1950s, and integrated circuit operational amplifiers since the mid-1960s. Prices of modern IC operational amplifier devices range from less than a dollar for low-quality units to dozens of dollars for high-grade, specialized units. Many premium grade op-amps are priced in the less-than-$20 range, and some are in the $5 range.

The circuit symbol for an operational amplifier is shown in Figure 4-1*a*. The amplifier must operate in all four quadrants, so the output terminal must be able to swing either positive or negative. The power supply (Figure 4-1*b*), therefore, must be *bipolar;* that is, it must consist of two supplies, one positive

[1] John I. Smith, *Modern Operational Circuit Design*, Wiley-Interscience (New York, 1971).

with respect to ground and a second negative with respect to ground.

The power supply shown in Figure 4-1*b* has two batteries, but an ac-mains operated bipolar supply will also work. Battery E_1 forms the V_{cc} supply and is *positive* with respect to ground. Battery E_2, on the other hand, forms the V_{ee} supply and is connected so that it is *negative* with respect to ground.

Note on the op-amp symbol shown in Figure 4-1*a* that there is no *ground* terminal on the operational amplifier. The only ground connection in this operational amplifier is formed at the junction of the two power supplies.

The operational amplifier has two inputs: the *inverting* input and the *noninverting* input. These are indicated by (−) and (+) signs, respectively.

The *inverting* input produces an output signal that is 180° out of phase with the input signal. This is called *inversion* of the signal.

The *noninverting* input produces an output signal that is *in phase* with the input signal. There is *no* phase inversion of the signal between input and output.

Both inverting and noninverting inputs offer the same gain, so we may conclude that these respective inputs have *equal* but *opposite* phase effect at the output.

The various signal voltages that can affect the operational amplifier output terminal are shown in Figure 4-2. E_1 is applied to the inverting input, while E_2 is applied to the noninverting input. As long as E_1 and E_2 are *not equal* and of the same polarity, the operational amplifier will see a *differential* input voltage consisting of $E_2 - E_1$. The output voltage will be proportional to the gain of the stage and the *difference* between E_1 and E_2.

Common-mode signal voltages are those that are *common* to *both* inputs, such as E_3, or where E_1 and E_2 are of equal magnitude and have the same polarity. In common-mode situations the differential voltage between the inputs is *zero,* so the output is zero.

The *common-mode rejection ratio* (CMRR) of an operational amplifier is an expression of how nearly any given device approximates the ideal situation where a common-mode signal has *no* effect on the output terminal voltage.

Figure 4-2 Signal voltage sources.

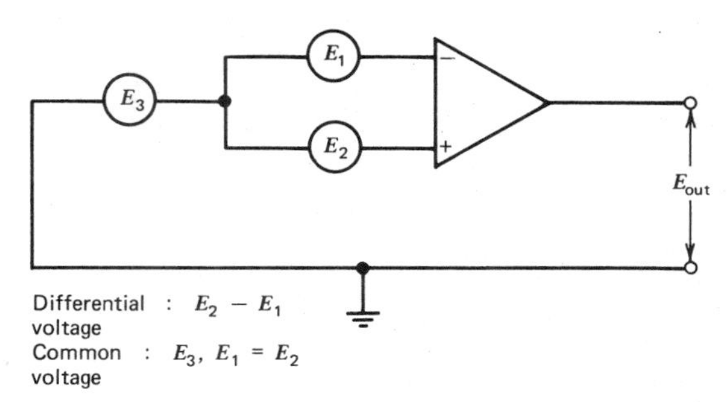

4-4.1 The Properties of Ideal Op-Amps

We can analyze the op-amp by considering the following ideal properties:

1. Infinite open-loop (i.e., no feedback) *voltage gain* ($A_{vol} = \infty$).

2. *Zero* output impedance ($Z_o = 0$).

3. *Infinite* input impedance ($Z_i = \infty$).

4. *Infinite* frequency response.

5. *Zero* noise contribution.

6. Both inputs *follow* each other in feedback circuits. That is, in a circuit with negative feedback, a voltage applied to one input allows us to treat the *other* input as if it were at the *same* potential.

These six properties will be cited frequently throughout this chapter.

Before continuing our analysis of the feedback voltage amplifier configurations, let us consider first some implications of these properties.

Infinite open-loop voltage gain. The open-loop voltage gain (A_{vol}) of any amplifier circuit is the gain *without* any feedback. In the *ideal* operational amplifier, this gain is defined as infinite. An implication of this property is that the closed-loop characteristics of the circuit are determined entirely by the properties of the feedback loop network and are independent of the amplifying device.

Zero output impedance, $Z_o = 0$, implies that the output is an ideal voltage source.

Infinite input impedance, $Z_i = \infty$, tells us that the input terminals neither *sink nor source any current,* nor load any circuit to which they are connected.

Property 6 is crucial to our circuit analysis: *The inputs tend to follow each other.* This means that we treat both inputs as if they were at the same potential. If a given voltage is applied to, say, the noninverting input, then we must treat the inverting input as if it were at the *same* potential. In fact, if a voltage is applied to one input, a voltmeter would *measure* the *same* voltage at the other input.

In this text we consider the operational amplifier as a special "black box" that has the previously cited six properties.[2]

4-5 Basic Amplifier Configurations

There are many circuit configurations using operational amplifiers as the active device, but only three basic classes of voltage amplifiers exist: *inverting follower, noninverting follower with gain,* and the *unity gain noninverting follower.* The following sections discuss simple circuits and applications of bioelectric amplifiers made from commercially available components.

4-5.1 Inverting Followers

Figure 4-3 shows the basic inverting circuit. It consists of an *operational amplifier,* an *input resistor* (R_1), and a *feedback resistor* (R_2).

The *noninverting* input is *grounded* in this circuit, so by property 6 we treat the *inverting* as if it were also grounded. This idea is sometimes called a "virtual ground," for want of a better phrase. Point A, the junction of the two resistors and the operational amplifier's inverting input, is properly called the *summing junction,* or *summation node.*

When an input voltage E_{in} is applied, it sees the point A end of R_1 as being at ground potential, so current I_1 flows and is equal to E_{in}/R_1.

By Kirchhoffs law we know that the sum of all currents entering and leaving the summing junction is zero. Property 3 tells us that no current flows into or out of the inverting input, so we may deduce that only input current I_1 and feedback current I_2 affect the junction. By Kirchhoff's current law, then, current I_2 must have a magnitude and polarity that exactly *cancels* I_1. We may view the operational amplifier as a servo system that generates an output voltage that permits I_2 to cancel I_1.

We can derive the *transfer function* of the inverting follower by the following procedure.

(a) By Kirchhoffs current law

$$-I_1 = I_2 \tag{4-1}$$

By Ohm's law

$$I_1 = \frac{E_{in}}{R_1} \tag{4-2}$$

$$I_2 = \frac{E_{out}}{R_2} \tag{4-3}$$

Substituting Equations 4-2 and 4-3 into Equation 4-1 gives us

$$\frac{-E_{in}}{R_1} = \frac{E_{out}}{R_2} \tag{4-4}$$

[2] For those desiring a deeper and more mathematical treatment, see the references at the end of the chapter.

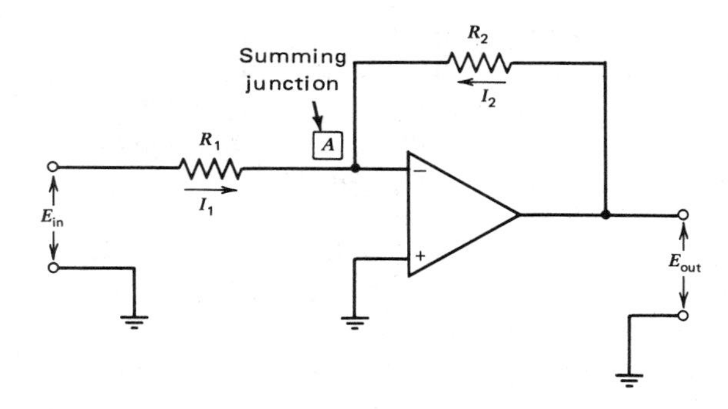

Figure 4-3 Inverting follower.

Solving Equation 4-4 for the transfer function E_{out} / E_{in} gives us the voltage gain of the inverting follower configuration.

$$\frac{E_{out}}{E_{in}} = \frac{-R_2}{R_1} = -A_v \tag{4-5}$$

The quantity R_2/R_1 gives us the magnitude of the voltage gain for this amplifier configuration, and the minus sign tells us that a 180° phase inversion takes place. The voltage amplification or gain expression is often represented by the symbol A_v.

Equation 4-5 is also frequently seen in two alternative but equivalent forms

$$E_{out} = -E_{in}\left(\frac{R_2}{R_1}\right) \tag{4-6}$$

and

$$E_{out} = -A_v E_{in} \tag{4-7}$$

EXAMPLE 4-1

Calculate the gain of an inverting follower if the feedback resistor (i.e., R_2) is 120 kΩ, and the input resistor (R_1) is 5.6 kΩ.

SOLUTION

$$A_v = \frac{-R_2}{R_1} \tag{4-5}$$

$$A_v = -(120\ \text{k}\Omega)/(5.6\ \text{k}\Omega) = \mathbf{21}$$

Figure 4-4 shows inverting operational amplifier circuit detail. Along with external circuit components is revealed some internal "guts" of the op-amp. All op-amps have plus

Figure 4-4 Inverting operational amplifier circuit detail.

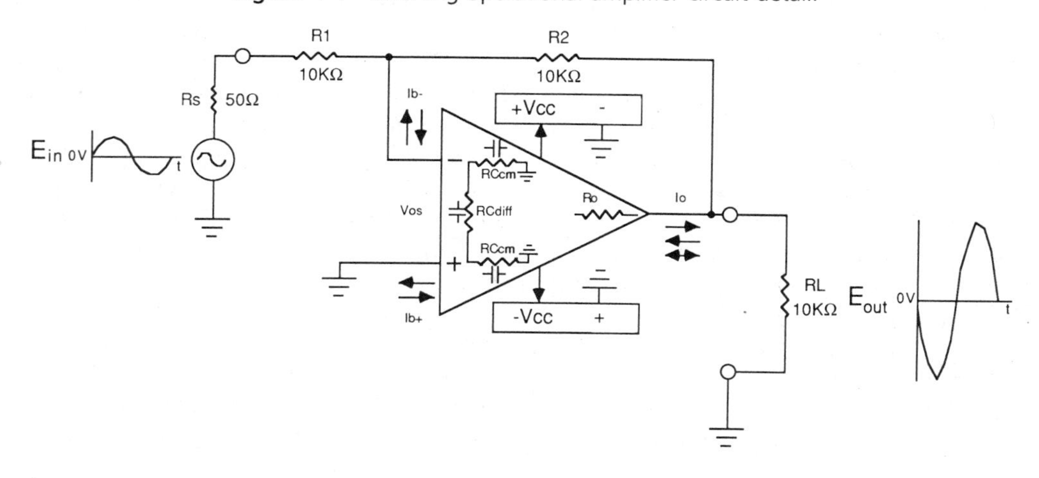

and minus bias current (I_{b+} and I_{b-}) and output load current (I_o). Also, all op-amps have three types of internal resistance and capacitance: (1) common-mode R_{cm} and C_{cm} (referenced to ground), (2) differential R_{diff} and C_{diff} (between op-amp terminals), and (3) output R_o. Notice there is no ground on the op-amp itself. The ground reference is actually the point between the power supplies where the minus of the $+V_{cc}$ power supply and the plus of the $-V_{cc}$ (sometimes called V_{ee}) come together.

Examining Figure 4-4 reveals that the external components interact with each other and with the op-amp to cause errors. One is a 0.5 percent gain error from the ideal of -1 V/V. It is caused by the external 50-Ω source resistance, R_s, and the input resistance, R_1. Essentially, R_s becomes part of a voltage divider with R_1 at the input, and some of E_{in} is undesirably dropped across R_s. This may not seem like much gain error, but over a number of op-amp stages the error could easily accumulate to several percent. Another error occurs at the output where R_o/A_{vol} acts as another voltage divider with the load resistance, R_L. Fortunately, R_o/A_{vol} equals about 1000 Ω/100,000, which is only 0.001 Ω or 1 mΩ. The quantity 0.001 Ω is so small compared to R_{load} or 10,000 Ω that the gain error is negligible.

How does internal op-amp resistance cause errors? R_{cm} on the minus op-amp terminal is in parallel with the input resistance, but R_{cm}, usually $>$ 1000 MΩ, causes only a very small error on R_1 = 10 kΩ. However, C_{cm}, usually $<$ 5 pF, can cause a gain error at higher frequencies. For example, at a frequency of 1 MHz, the reactance of C_{cm} is 32 kΩ, and this shunts the external resistance, increasing the gain error at 1 MHz. The differential resistance, R_{diff}, also enters the picture, but we will not calculate the error here. The intention of this discussion is to show "real-world" op-amp circuits.

One other error is bias current circulating through the feedback resistance, R_2. If I_{b-} = 10 nA, then 0.1 mV dc will be dropped across R_2, which shows up at the op-amp output as 0.1 mV. Assuming E_{out} = 10 V, a dc error of 0.001 percent (0.1 mV/10 V × 100) shows up, but it is actually rather small.

Examining op-amp circuit details shows that, although circuit gain is mostly dependent on external resistance such as R_1 and R_2, certain dc and ac errors are always present. Armed with this information, it is easier to understand and use op-amp circuits more fully.

Next is a special inverting op-amp circuit encountered frequently in biomedical instrumentation. It is the transimpedance amplifier or current-to-voltage converter shown in Figure 4-5. This circuit takes an input current, I_s, from a current source, and converts it into a voltage, E_o, at the op-amp's output. Notice that a positive input current pulse flowing into the op-amp's summing junction or negative input produces a negative output voltage pulse. The current flowing in the feedback loop is I_f, which is almost equal to I_{in}. The op-amp's bias current will either add to or subtract from I_{in} to produce I_f. If I_b is small, say 1/1000 of I_{in}, then the error is small. Ignoring op-amp error, a 10-nA input gives a 0.1 V output, which can easily be further amplified.

The most common bioelectric (bioelectronic used interchangeably) transimpedance amplifier circuit is the photodiode amplifier. Figure 4-6 shows a commercially available op-amp, in a real "light-to-voltage" circuit. Notice it is an inverting op-amp configuration. Light shining on the photodiode produces a current which flows into the summing junction of the op-amp and then through the feedback resistance (10,000 MΩ in this highly sensitive photodetector circuit). Remember, ignoring op-amp errors, the voltage across the feedback resistor equals the op-amp output voltage. An input light signal of 0.002 μW will give an output

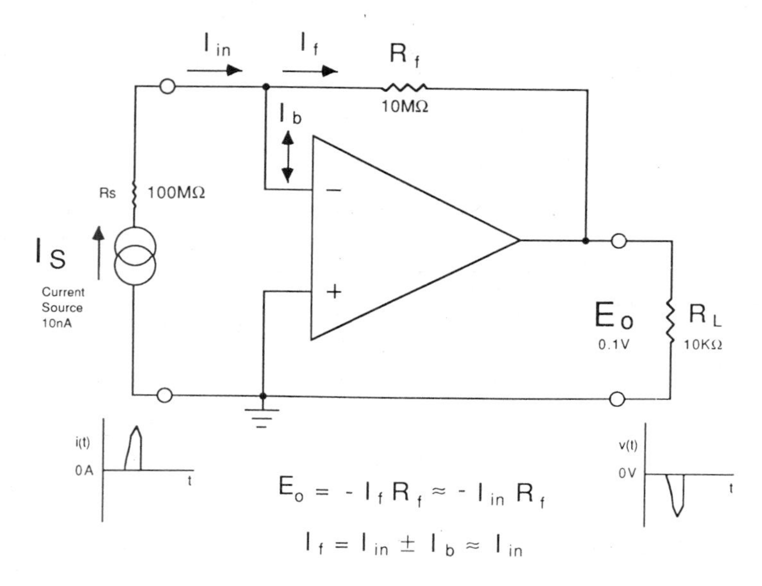

Figure 4-5 Basic trans-impedance amplifier (current-to-voltage converter).

of 10 V, because the overall gain is 5 × 1,000 MV/watt. The op-amp's analog output voltage can then be digitized through an analog-to-digital converter and used to record the exact light intensity. Notice that 10,000 MΩ is also placed from the positive input to ground. This tends to cancel the offset error caused by op-amp bias current, since + and − bias currents are nearly equal. Another major error is gain peaking caused by diode capacitance. To minimize this, a 1-pF capacitor is placed in the feedback loop.

Noise or random voltage fluctuations appearing in the output signal are the last error we will discuss here. This error is due to resistor noise, op-amp voltage and current noise, and noise pickup. To keep noise low, the circuit should use the smallest feedback resistor practical, a low-noise op-amp, and shielded op-amp input pins. After that, low-pass filtering must be provided.

Such photodetector circuits are used in optical pulse oximeters to measure human blood oxygen saturation, optical glucometers to measure human blood sugar levels, and medical lab spectrophotometers to measure blood plasma elements. This circuit shows up often.

Figure 4-6 Sensitive photodiode amplifier. (Courtesy of Burr-Brown under copyright 1989 Burr-Brown Corporation)

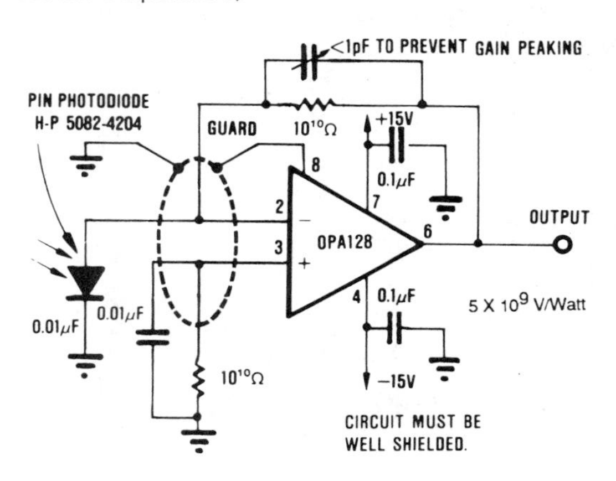

4-5.2 Noninverting Followers with Gain

An example of the *noninverting follower gain* amplifier is shown in Figure 4-7*a*. In this circuit the input voltage is applied directly to the *noninverting* input terminal of the operational amplifier. Feedback resistor

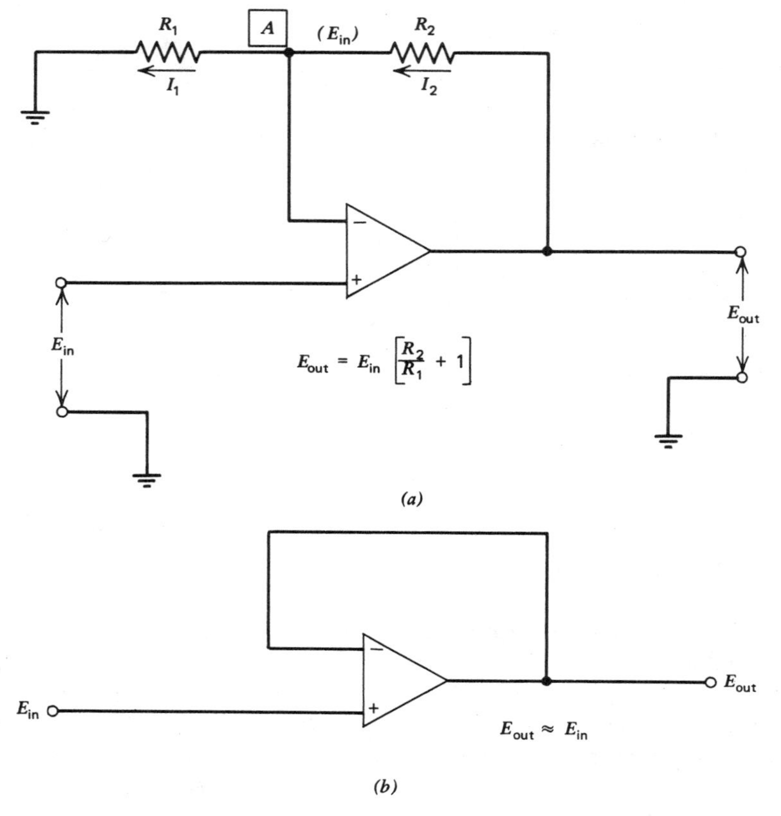

Figure 4-7 Noninverting followers. (*a*) With gain. (*b*) Unity gain.

R_2 and input resistor R_1 are the same as in the inverting follower, except that the other end of R_1 is grounded.

By property 6 we know that point A must be treated as if it were at a potential equal to E_{in}. We may derive the transfer equation for this circuit using the same technique that was used in the previous case.

By Kirchhoff's current law

$$I_1 = I_2 \tag{4-8}$$

By Ohm's law

$$I_1 = \frac{E_{in}}{R_1} \tag{4-9}$$

$$I_2 = \frac{E_{out} - E_{in}}{R_2} \tag{4-10}$$

Substituting Equations 4-9 and 4-10 into Equation 4-8 results in

$$\frac{E_{in}}{R_1} = \frac{E_{out} - E_{in}}{R_2}$$

$$\frac{E_{in}}{R_1} = \frac{E_{out}}{R_2} - \frac{E_{in}}{R_2}$$

$$\frac{E_{in}}{R_1} + \frac{E_{in}}{R_2} = \frac{E_{out}}{R_2}$$

$$\frac{E_{in}R_2}{R_1} + \frac{E_{in}R_2}{R_2} = E_{out}$$

$$\frac{R_2}{R_1} + 1 = \frac{E_{out}}{E_{in}} = A_v \qquad (4\text{-}11)$$

EXAMPLE 4-2

Calculate the voltage gain of a noninverting follower if $R_2 = 10\ \text{k}\Omega$ and $R_1 = 2.2\ \text{k}\Omega$.

SOLUTION

$$A_v = \frac{R_2}{R_1} + 1 \qquad (4\text{-}11)$$

$$A_v = 10\ \text{k}\Omega / 2.2\ \text{k}\Omega + 1$$

$$A_v = 4.6 + 1 = \mathbf{5.6}$$

At high gains (i.e., circuits with high R_2/R_1 ratios), the gains of the inverting and noninverting followers are very nearly equal, but at low gains a difference is noted.

4-5.3 Unity Gain Noninverting Followers

A special case of the noninverting follower is the *unity gain noninverting follower* of Figure 4-7*b*. The resistor network is not used in this circuit, and the output is connected directly to the inverting input, resulting in 100 percent negative feedback. By Equation 4-11 this yields a gain in Equation 4-11 of (0 + 1), or simply +1, *unity*.

The unity gain noninverting follower is used in applications such as output buffering and impedance matching between a high source impedance and a low-impedance input circuit.

A more complete understanding of the noninverting op-amp circuit is gained from the detail in Figure 4-8. Just as in the inverting case, the op-amp possesses internal resistance and capacitance. However, since the input signal is driving a very high impedance (op-amp common-mode and differential *R* and *C*), very little gain error occurs. The 50-Ω source resistance is much smaller than the noninverting op-amp input resistance of around 1000 MΩ. Again, as in the inverting case, the R_o/A_{vol} and R_{load} divider cause a small output gain error. Small op-amp bias currents also flow in the source and feedback resistances to cause small offset errors. Thus the circuit gain is 2, with small offset and gain errors created by the op-amp interacting with its surrounding circuit elements.

A main characteristic of op-amp circuits is that the op-amp's negative terminal is a reflection of its positive terminal. That is, when a dc and/or ac voltage is applied to the positive input, that same dc and/or ac voltage

Figure 4-8 Noninverting operational amplifier circuit detail.

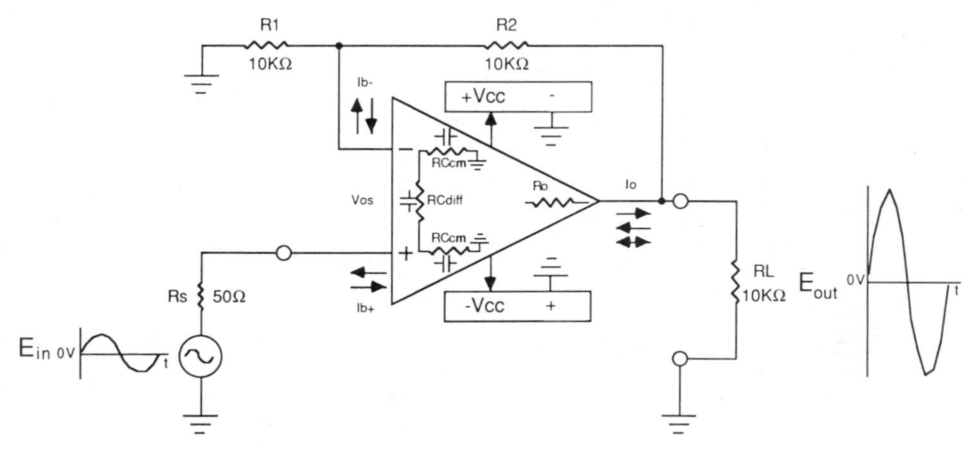

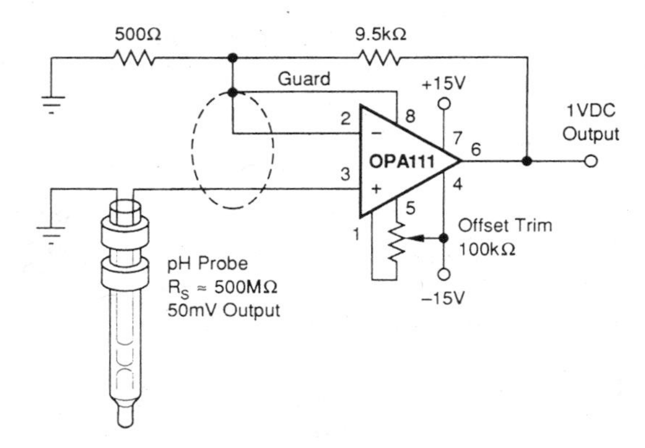

Figure 4-9 High-impedance (10^{14} Ω) pH probe amplifier. (Courtesy of Burr-Brown under copyright 1990 Burr-Brown Corporation)

will also appear on the negative input. For example, if a 1-V_{p-p} sinewave riding around 0 V dc drives the plus input, then a 1-V_{p-p} sinewave riding around 0 V dc will also be present on the negative input.

A common bioelectronic noninverting amplifier circuit is the pH probe amplifier. Figure 4-9 shows a commercially available op-amp connected in a gain of 20. The 50-mV probe output is amplified to produce 1 V at the op-amp's output.The 100-kΩ offset trim potentiometer will zero out the op-amp's offset voltage (say 250 μV) and the error caused by the noninverting terminal bias current times the large source resistance (say 0.5 pA times 500 MΩ = 250 μV). Gain errors are caused mostly by inexact feedback resistances.

In discussing a "real op-amp circuit," three important considerations show up again and again: bias current, voltage noise, and noise pickup. First, in Figure 4-10 we see from a typical product data sheet curve that bias current increases with temperature. At room temperature of +25° Centigrade the bias is 0.5 pA, but at +65° Centigrade it is 10 pA, which is 20 times bigger. This makes the upper temperature offset error (10 pA × 500 MΩ = 5 mV) much higher than the room temperature error (0.5 pA × 500 MΩ = 0.25 mV).

The second important consideration is high noise due to very high source resistance. In Figure 4-11, it is clear that at a source resistance of, say, 100 MΩ, the total circuit noise can be big. This noise has three parts: (1) op-amp voltage noise, (2) op-amp current noise times source resistance, and (3) resistor noise (resistors have noise all by themselves just sitting on the table). Sometimes noise is referred to as Johnson or ther-

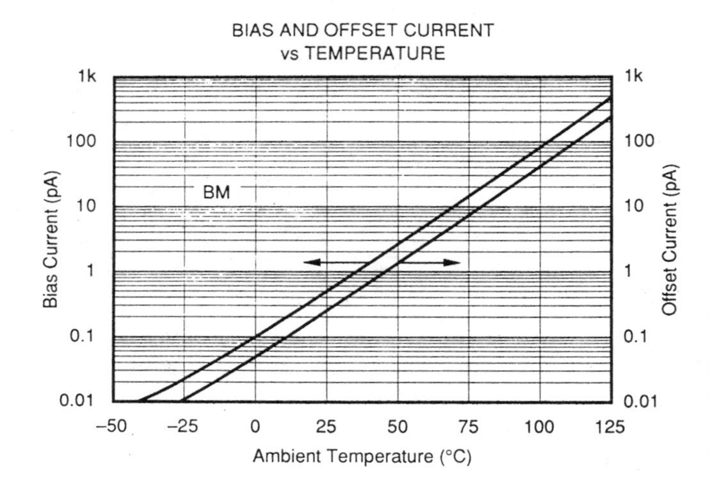

Figure 4-10 Important consideration: Temperature effect on bias and offset current from OPA111 FET input op amp. (Courtesy of Burr-Brown under copyright 1990 Burr-Brown Corporation)

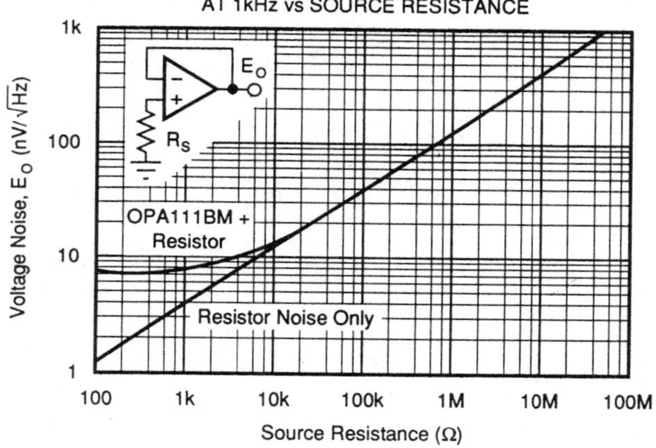

Figure 4-11 Important consideration: Source resistance effect on total noise from the OPA111 FET input op amp. (Courtesy of Burr-Brown under copyright 1990 Burr-Brown Corporation)

mal noise, because it is due to agitation of charges in materials. Noise on the typical product data sheet curve is 1000 nV per square root hertz (nV/√Hz). The purpose for showing this graph is to emphasize that unfiltered wideband white noise can be rather high in pH probe circuits. Seeing noise will come as no surprise when it is understood where it comes from.

The third important consideration is the noise pickup. The GUARD shown in Figure 4-9 is further explained in Figure 4-12. To keep noise low in this noninverting op-amp circuit, the input pins 2 and 3 must be guarded or shielded. The way to do this is to lay out the printed circuit board as shown. Notice that pins 2 and 3 are completely surrounded by metal traces and then connected to pin 8 (the op-amp metal case) of this commercially available op-amp. It works this way: When, say, that 60 Hz electromagnetic interference (EMI) noise cuts across the metal case, it tends to cause noise currents to flow in the input pins 2 and 3. However, since pins 2 and 3 are connected to the case through pin 8, they are at the same potential as the case (0V between case and pins). Hence, noise current cannot flow into the input, because the voltage is zero. We say that the input pins are guarded against leakage or stray resistance and capacitance.

Understanding dc and noise errors in pH probe circuits helps one appreciate and work more effectively with bioelectric amplifiers.

Figure 4-12 Important consideration: PC board layout and connection of input guard. (Courtesy of Burr-Brown under copyright 1990 Burr-Brown Corporation)

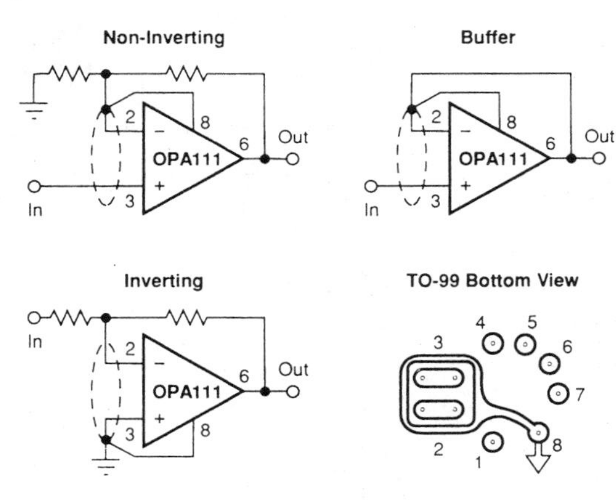

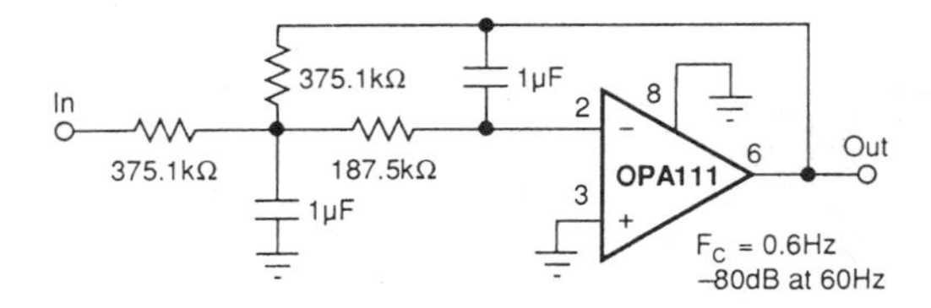

Figure 4-13 0.6 Hz two-pole (second-order) low-pass filter. (Courtesy of Burr-Brown under copyright 1990 Burr-Brown Corporation)

Inverting and noninverting op-amp circuits are also used to form filters which are commonly found in biomedical equipment. Figure 4-13 shows a 0.6-Hz inverting, low-pass filter using a commercially available op-amp. The circuit has two poles (also called second order) because there are two capacitors. One is in parallel with the signal path, and the other is in the feedback loop of the op-amp. This type of filter is often used to reduce wideband or white noise from op-amps and EMI noise from 60-Hz power lines. In fact, any 60-Hz noise present in the input signal will be attenuated by 80 decibels. This means that the 60-Hz noise at the output will be 10,000 times smaller than at the input. Getting rid of 60-Hz noise is important to many biomedical measurements, such as pH.

Multipole filters can be made from separate op-amps or can use one chip, as shown in Figure 4-14. This circuit is a noninverting, two-pole, 10-kHz low-pass filter. The quantity of 10 kHz is used in ECG circuits to filter out electrosurgery interference containing frequencies above 500 kHz. The reason 10 kHz is used is to get good high-frequency attenuation while avoiding phase disturbance of the actual 100-Hz bandwidth ECG signal. Keeping the cutoff frequency way above the highest ECG frequency minimizes phase shifts in the ECG signal.

No biomedical filter discussion would be complete without considering the 60-Hz notch filter. Figure 4-15 depicts a circuit which rejects just 60 Hz. It passes frequencies below and above the 60-Hz center. Al-

Figure 4-14 Two-pole, 10-KHz low-pass filter using a one-chip circuit. (Courtesy of Burr-Brown under copyright 1991 Burr-Brown Corporation)

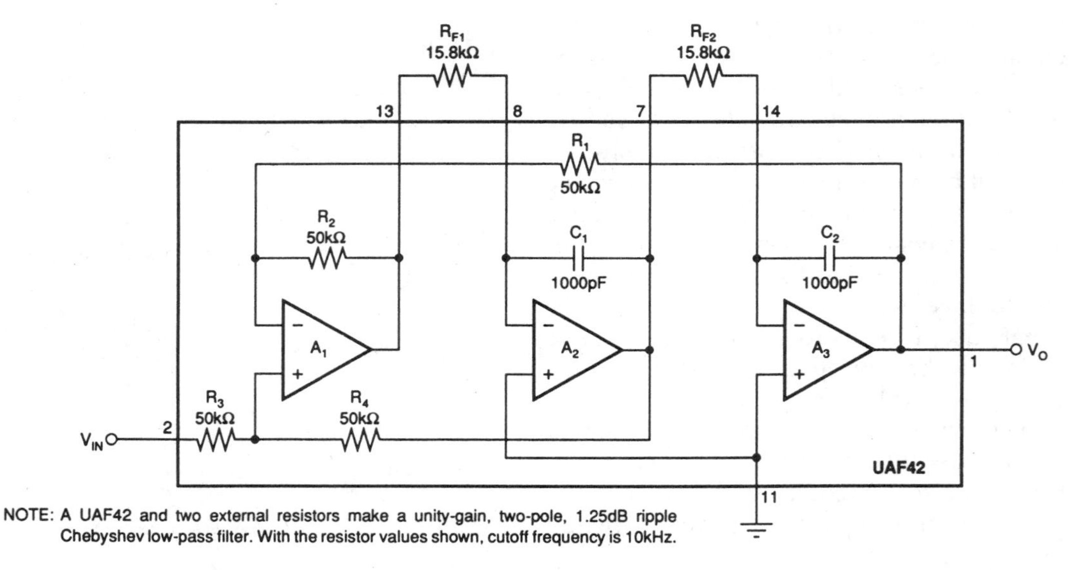

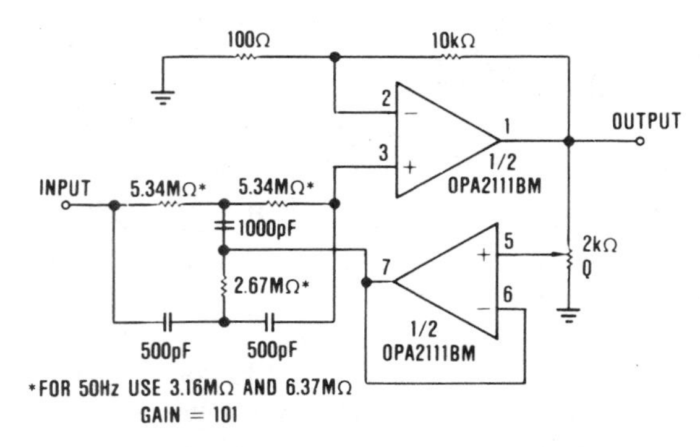

Figure 4-15 High-impedance 60-Hz reject filter with gain. (Courtesy of Burr-Brown under copyright 1988 Burr-Brown Corporation)

though the gain here is 101 V/V non-inverting, it can be set using different feedback resistances. This type of filter is used in ECG and other biomedical equipment for cutting down on residual interference that is within the bandwidth of interest. For example, 60-Hz noise is undesirable inside the 100-Hz ECG spectrum. However, since the notch does cause some amplitude and phase distortion, biomedical equipment is often designed to take measurements with the notch switched out and switched in.

So far, active filters have been discussed—those that use an amplifier connected to a power supply. Passive or unpowered filters can also be used where appropriate. An example is shown in Figure 4-16. Here a pyroelectric infrared heat detector circuit utilizes a simple R-C or single-pole filter. It cuts off at 0.16 Hz to reduce noise. After filtering, the signal is gained up by 101 V/V. Pyroelectric detectors are used in applications which require high sensitivity. Heat radiating from an object, such as a person, can be sensed at a considerable distance. A concentration of body heat can reveal tumors, especially if infection is present.

Figure 4-16 Balanced pyroelectric infrared (heat) detector. (Courtesy of Burr-Brown under copyright 1990 Burr-Brown Corporation)

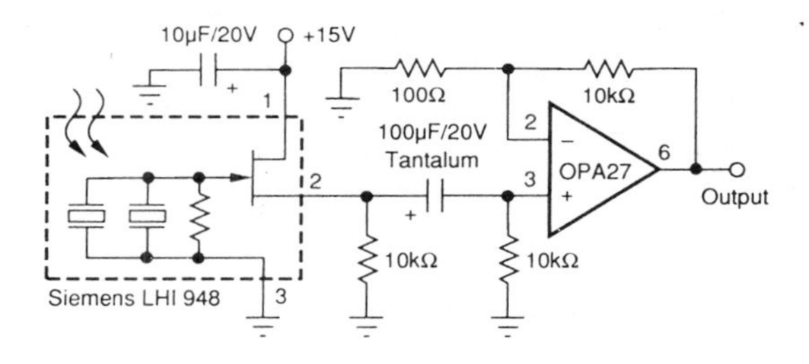

It is often desirable to put bioelectronics as close to the sensor as possible. However, when this is done the amplifier may have to drive a voltage signal over a long electrical cable. Cables have capacitance and, if too large, tend to make op-amps oscillate. Figure 4-17 shows how a resistor in series with the op-amp output cures this problem. A small value, say 20 Ω, buffers the op-amp from the load, and since it is in the feedback path it does not cause gain error. This circuit can drive a 5-nF or 5000-pF load. The 200-pF feedback capacitor also helps to maintain stability. Unstable bioelectric amplifier systems sometimes go undiagnosed. One of the only ways to troubleshoot such instability,

Figure 4-17 Driving large capacitive loads. (Courtesy of Burr-Brown under copyright 1990 Burr-Brown Corporation)

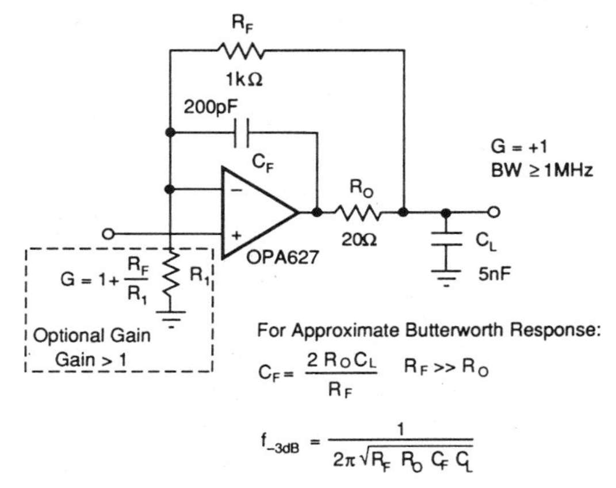

aside from observing erratic biomedical equipment operation, is to use an oscilloscope. Seeing a biophysical signal with high frequency riding on it indicates that the circuit is probably oscillating.

Op-amps, of course, can be used to create other bioelectronic circuits. The peak detector, shown in Figure 4-18, is frequently used to measure the peak of a biopotential such as an ECG signal. The first stage is a noninverting amplifier with a diode in the feedback loop and a diode in series with the output. When the input voltage goes negative (below ground), the series diode blocks signal transfer to the second stage. The feedback diode makes sure the input stage is in low gain. But when the input goes positive, the series diode conducts into the FET transistor diode, which applies the signal peak to the 0.01-μF holding capacitor. A low bias current FET op-amp is used for the second stage to make sure the peak captured on the capacitor does not change very much. The droop or amount of voltage decrease with time is just 0.1 mV/s. The output, therefore, is a constant voltage that represents the positive peak of the input voltage. A greater understanding of biomedical instrumentation hinges on recognizing circuits and knowing how they work.

4-6 Multiple Input Circuits

More than one input network may be used in an operational amplifier network, and the output voltage represents the summation of the respective input currents. An example of a multiple input inverting follower is shown in Figure 4-19. Three input networks are used in this circuit, and there are three input sources E_1 through E_3.

Again, the noninverting input is grounded, so we treat the inverting input as if it were also grounded. From a course of reasoning exactly as before, we know that the transfer equation of a multiple input circuit such as Figure 4-5 is

$$E_{out} = -R_4 \times \left(\frac{E_1}{R_1} + \frac{E_2}{R_2} + \frac{E_3}{R_3} + \cdots + \frac{E_n}{R_n}\right) \tag{4-12}$$

EXAMPLE 4-3

Find the output voltage in a circuit such as the one in Figure 4-19 if $R_1 = R_2 = R_3 = 10\ \text{k}\Omega$, $R_4 = 22\ \text{k}\Omega$, $E_1 = 100$ mV, $E_2 = 500$ mV, and $E_3 = 75$ mV.

Figure 4-18 Low-droop positive peak detector. (Courtesy of Burr-Brown under copyright 1991 Burr-Brown Corporation)

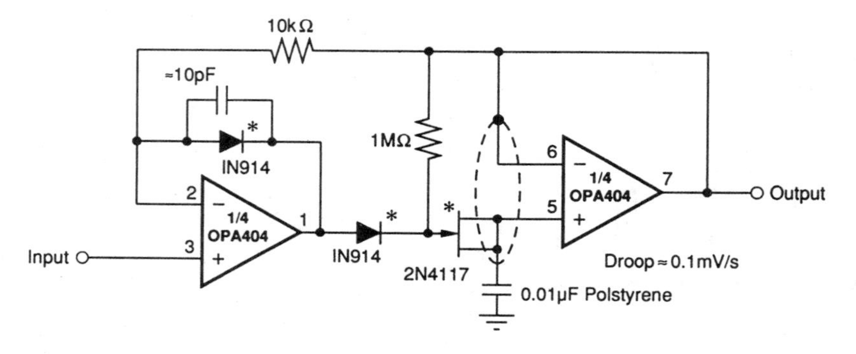

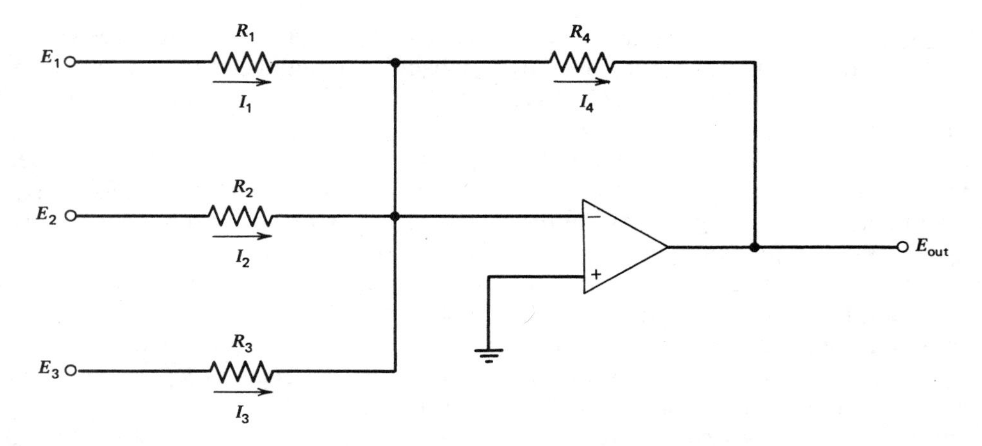

Figure 4-19 Multiple-input amplifier.

SOLUTION

$$E_{out} = -R_4 \times \left(\frac{E_1}{R_1} + \frac{E_2}{R_2} + \frac{E_3}{R_3}\right) \qquad (4\text{-}12)$$

$$E_{out} = -22\ \text{k}\Omega \times \left(\frac{100\ \text{mV}}{10\ \text{k}\Omega} + \frac{500\ \text{mV}}{10\ \text{k}\Omega} + \frac{75\ \text{mV}}{10\ \text{k}\Omega}\right)$$

$$E_{out} = -22\ \text{k}\Omega/10\ \text{k}\Omega \times 100\ \text{mV} + 500\ \text{mV} + 75\ \text{mV}$$

$$E_{out} = -2.2 \times 675\ \text{mV} = -1485\ \text{mV} = \mathbf{-1.485\ V}$$

Circuits such as Figure 4-19 are used in many medical instruments to *compute* a value (i.e., E_o) from several different input values (i.e., E_1 through E_3). It is also possible to use the noninverting input, in which case no polarity inversion takes place, and Equation 4-12 must be modified to reflect the contribution of signals applied to the noninverting input equation (4-11).

4-7 Differential Amplifiers

A *differential amplifier* produces an output voltage that is proportional to the *difference* between the voltage applied to the two input terminals. Since an operational amplifier has a pair of differential input terminals, it may be easily connected for use in a differential amplifier configuration.

In the most elementary form of dc differential amplifier (Figure 4-20) only a single IC operational amplifier is required. In this particular circuit the voltage gain for differential signals is the same as for inverting followers (i.e., $A_{vd} = R_2/R_1$), provided that the ratio

Figure 4-20 Differential input amplifier basics (single-ended output).

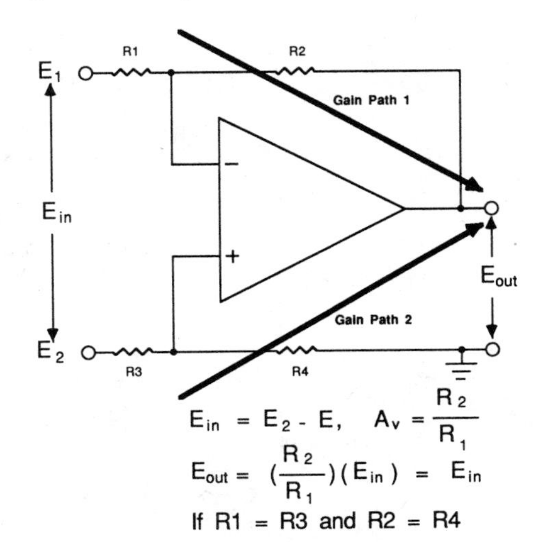

equality $R_2/R_1 = R_4/R_3$ is maintained. It is standard practice to insure this equality by stipulating that $R_1 = R_3$ and $R_2 = R_4$. Then gain path 1 equals gain path 2.

Figure 4-21 shows differential input amplifier circuit detail. Depicted is the external circuit and some internal "guts" of the diff-amp, which is really just an op-amp. All op-amps have plus and minus bias current (I_{b+} and I_{b-}), output load current (I_o), internal common-mode and differential resistance and capacitance, and output resistance, R_o. Of course, the ground reference is actually the point between the power supplies.

Examining Figure 4-21 reveals that the diff-amp is a combination of inverting and noninverting op-amp configurations. The external components interact with each other and with the op-amp to cause gain errors. For example, the external 50-Ω source resistance, R_s, in the inverting path causes a 0.5 percent gain error from the ideal of −1 V/V. Notice that the input resistance to the op-amp's inverting input is 10kΩ. The input resistance ratio on the noninverting input also has a 0.5 percent error. This is why low input resistance diff-amps require very low source resistance to achieve high gain accuracy. The nice thing about the op-amp's plus (+) input is that it is high impedance (resistance and capacitive reactance). It therefore does not load or cause an error on the R_3-R_4 resistor divider. We know that the internal op-amp resistance causes very little error, because R_{cm} is usually > 1000 MΩ. However, C_{cm}, usually < 5 pF, can cause a gain error at higher frequencies.

This discussion on op-amp circuit detail concludes that, although circuit gain is mostly dependent on external resistance such as R_1, R_2, R_3, and R_4, certain dc and ac errors are always present. This information helps us understand how real-world diff-amp circuits work.

Figure 4-22 shows a commercially available one-chip differential amplifier. It has internal thin-film resistors made of nichrome or nickel chromium. These resistors have been trimmed to high precision by cutting with a laser beam. The more the resistors are cut away by the automatic laser trimming machine, the higher the resistance (less conductive material). When precise resistor ratio accuracy is reached, cutting stops. In this component the gain error is 0.005 percent and the common-mode rejection (CMR) is 100 dB. This means that the negative and positive gain paths are very nearly equal.

Figure 4-21 Differential input amplifier detail.

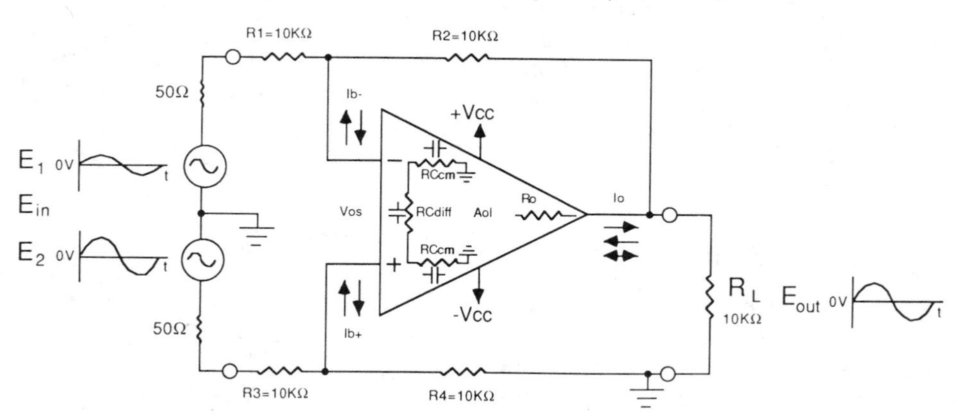

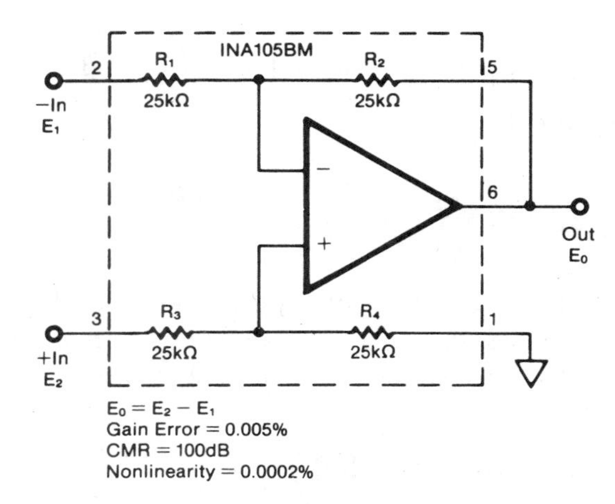

Figure 4-22 One-chip (monolithic) precision unity gain difference amplifier. (Courtesy of Burr-Brown under copyright 1988 Burr-Brown Corporation)

Also, the gain and CMR temperature drift is very low—approximately five parts per million per degree Centigrade (5 PPM/°C). It is really the resistor ratio that is changing slightly with temperature. This is much better than could be achieved by using separate external discrete resistors. Here the advantage of modern monolithic circuits becomes evident. All resistors on the same chip tend to behave exactly like one another. Their ratio stays constant.

The reason the diff-amp is so useful is that it rejects common-mode voltages while amplifying the differential signal of interest. For example, suppose equal 60-Hz noise is present on each input and one input is at 5 V dc and the other is at 2 V dc. The circuit in Figure 4-22 removes the noise and amplifies the 3 V dc differential signal. Remember, the common-mode rejection of this commercially available diff-amp is very high because on-chip resistors have been ratio matched to make the + and − gain paths nearly equal. The unity gain diff-amp output is a 3 V dc signal that has 60-Hz noise interference greatly reduced. The differential amplification removes noise, because equal common-mode noise is present on each input. The diff-amp just subtracts the equal noise voltage to give nearly zero while amplifying the difference in the unequal signals present on its inputs. How low noise becomes at the output depends on how high the diff-amp's common-mode rejection is.

The circuit shown in Figures 4-20, 4-21, and 4-22 suffers from the same restrictions on input impedance as the inverting follower because the input impedance is limited by R_1 and R_3. If a high gain is required, then a high R_2/R_1 ratio is required, yet practical circuit considerations limit the maximum and minimum values for these resistors.

The simple dc differential amplifier is used mostly in circuits where the source impedance is low. Strain gage transducers, for example, typically have element resistances below the 1-kΩ range. Wheatstone bridge strain gages with these resistor values have an equivalent resistance approximately equal to the resistance of each element arm (assuming all arms are equal). If the equivalent bridge resistance is as high as 1 kΩ, then the *minimum* value for R_1 and R_3 in the dc amplifier is 10 times as great, or 10 kΩ. Most low-cost IC operational amplifiers should not be used with feedback resistors greater than 1 MΩ. These limitations of real (as opposed to *ideal*) operational amplifiers limit the *practical* gain of a differential amplifier (Figure 4-20) to $10^6/10^4$ or 100.

The circuit of Figure 4-20, then, can be used with Wheatstone bridge strain gage transducers and other low-output impedance differential signal sources at gains up to ×100. If higher gains are required, then a premium cost amplifier (allowing feedback resistors greater than 10^6 Ω) must be used, or an additional stage of amplification is required. An alternate solution, one that provides a much higher input impedance, is to use the *instrumentation amplifier* circuit described in Section 4-7.1.

4-7.1 Instrumentation Amplifiers

The solution to both high-gain and high-input impedance problems is the *instrumentation amplifier* shown in Figure 4-23. This circuit uses *three* operational amplifiers, A_1 through A_3. The two input amplifiers (i.e., A_1 and A_2) are connected in the noninverting follower configuration, while the third amplifier is connected in the simple dc differential amplifier circuit of Figure 4-20. Let us initially simplify our circuit analysis by setting the gain of A_3 equal to unity (i.e., $R_4 = R_5 = R_6 = R_7$).

Let us also assume that E_1 is applied to the noninverting input of amplifier A_1, and that E_2 is applied to the noninverting input of amplifier A_2. Additionally, E_3 is the output of A_2 and E_4 is the output of A_1. Voltages E_1 and E_2 are also shown at the *inverting* inputs of A_1 and A_2, respectively, again reflecting property 6.

There are *two* contributing sources to E_3 and E_4. In the case of E_3:

$$E_3 = E_2 \times \left(\frac{R_3}{R_1} + 1\right) - \left(E_1 \times \frac{R_3}{R_1}\right) \tag{4-13}$$

and for E_4

$$E_4 = E_1 \times \left(\frac{R_2}{R_1} + 1\right) - E_2 \times \left(\frac{R_2}{R_1}\right) \tag{4-14}$$

If we set $R_2 = R_3$ (not essential, but it simplifies the analysis), and then combine Equations 4-13 and 4-14, we may write

$$(E_3 - E_4) = (E_2 - E_1)\left(\frac{R_2}{R_1} + 1\right) + (E_2 - E_1)\left(\frac{R_2}{R_1}\right)$$

$$(E_3 - E_4) = (E_2 - E_1)\left(\frac{R_2}{R_1} + 1 + \frac{R_2}{R_1}\right)$$

$$(E_3 - E_4) = (E_2 - E_1)\left(\frac{2R_2}{R_1} + 1\right) \tag{4-15}$$

Figure 4-23 Instrumentation amplifier.

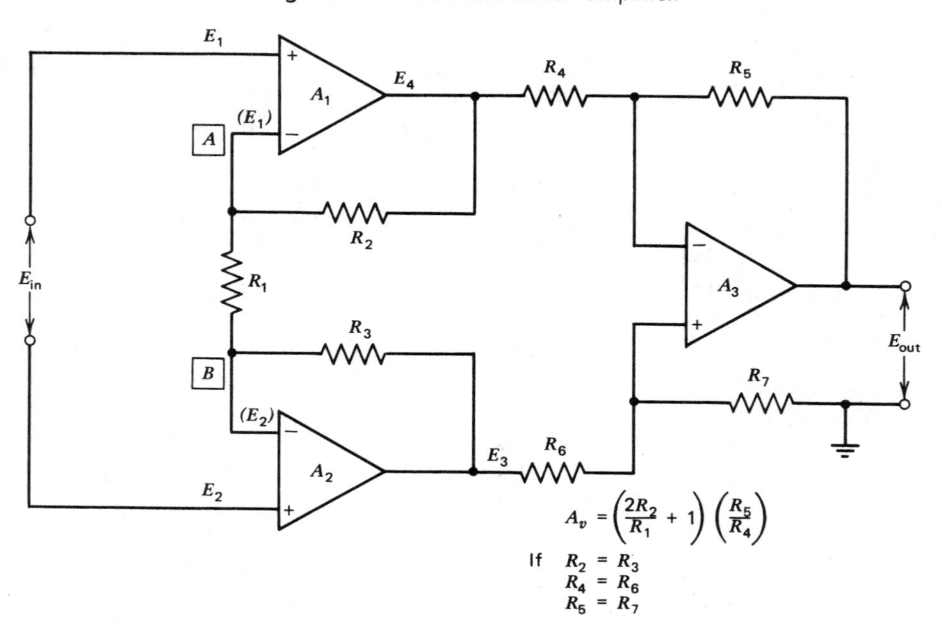

Therefore:

$$A_{vd} = \frac{2R_2}{R_1} + 1 \qquad (4\text{-}16)$$

The voltage gain of the A_1/A_2 section is given by Equation 4-16, but when the gain of A_3 is nonzero we must also include a term that accounts for this extra gain. The gain of an instrumentation amplifier such as Figure 4-23 is given by the transfer function

$$A_{vi} = \frac{E_{\text{out}}}{E_{\text{in}}} = \left(\frac{2R_2}{R_1} + 1\right)\left(\frac{R_5}{R_4}\right) \qquad (4\text{-}17)$$

EXAMPLE 4-4

Find the gain of an instrumentation amplifier (Figure 4-23) if the following resistor values are used: $R_2 = 10\ \text{k}\Omega$, $R_1 = 500\ \Omega$, $R_4 = 10\ \text{k}\Omega$, and $R_5 = 100\ \text{k}\Omega$.

SOLUTION

$$A_{vi} = \left(\frac{2R_2}{R_1} + 1\right) \times \left(\frac{R_5}{R_4}\right) \qquad (4\text{-}17)$$

$$A_{vi} = \left(\frac{2 \times 10\ \text{k}\Omega}{0.5\ \text{k}\Omega} + 1\right) \times \left(\frac{100\ \text{k}\Omega}{10\ \text{k}\Omega}\right)$$

$$A_{vi} = (40 + 1)(10) = \mathbf{410}$$

In ordinary practice the following equalities are observed: $R_2 = R_3$, $R_4 = R_6$, $R_5 = R_7$. Interestingly enough, a mismatch of R_2 and R_3 has little effect on the common-mode rejection ratio but does result in a differential gain error.

Instrumentation amplifiers are used in biomedical applications because of several factors: ability to obtain high gain with low resistor values, extremely high input impedance, and superior rejection of common-mode signals. Slight resistor mismatches in the circuit of A_3 can degrade common mode rejection; therefore, many designers use a potentiometer for R_7. The potentiometer is adjusted (while a high-level, common-mode signal is applied) for *minimum* output signal.

The physical form taken by an instrumentation amplifier might be a *discrete* operational amplifier circuit such as Figure 4-23, in which three IC operational amplifier devices are used. It may also take the form of a *hybrid-function* module in which chip-form IC operational amplifiers and the resistors are constructed on a thin ceramic substrate and then potted in a block of epoxy resin. The third form possible is a *monolithic* (IC) instrumentation amplifier. In both the hybrid and monolithic versions there may be a pair of external terminals for setting gain. These terminals are for an externally connected R_1 in Figure 4-23. The gain equation is usually a constant divided by the value of R_1 connected between the two terminals.

Modern instrumentation amplifier (IA) design can be tailored to meet specific biomedical applications. The circuit in Figure 4-24

Figure 4-24 Building three op-amp precision instrumentation amplifiers with NPN and FET inputs. (Courtesy of Burr-Brown under copyright 1988 Burr-Brown Corporation)

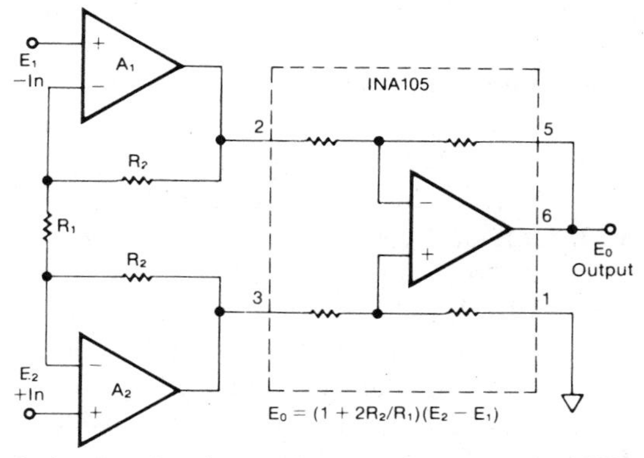

For low source impedance applications, an input stage using OPA37 op amps will give the best low noise, offset, and temperature drift performance. At source impedances above about 10kΩ, the bias current noise of the OPA37 reacting with the input impedance begins to dominate the noise performance. For these applications, using the OPA111 or Dual OPA2111 FET input op amp will provide lower noise performance. For lower cost use the OPA121 plastic. To construct an electrometer use the OPA128.

A_1, A_2	R_1 (Ω)	R_2 (Ω)	Gain (V/V)	CMRR (dB)	Max I_B	Noise at 1kHz (nV/√Hz)
OPA37A	50.5	2.5k	100	128	40nA	4
OPA111B	202	10k	100	110	1pA	10
OPA128LM	202	10k	100	118	75fA	38

shows how commercially available components are used to build IAs with specific front-end characteristics. Notice that the output stage is always a monolithic unity gain differential amplifier. Different IAs are designed by using different op-amps for the input stage. For example, a very low noise of 4 nV/$\sqrt{}$Hz can be achieved with low noise bipolar op-amps. The bias current is 40 nanoamps (40×10^{-9} amps) in this first example. If a low bias current of 1 picoamp (1×10^{-12} amps) is needed, then the input op-amps are changed to the FET type. Notice that the voltage noise goes up four times to 10 nV/$\sqrt{}$Hz, because FET op-amps have inherently more voltage noise than bipolar input op-amps. If extremely low bias current, say 75 femtoamps (75×10^{-15}) is required, the input op-amps need to be the electrometer type. Notice again that the voltage noise goes up almost four more times to 38 nV/$\sqrt{}$Hz. By making the right trade-offs, one can build an IA to suit specific biomedical needs. Just like differential amplifiers, IAs reject equal noise on their inputs while amplifying unequal signal voltages on their inputs.

To refine the discussion on noise, a brief description of how to use the nanovolt per square root hertz specification will be reviewed here. The reason for the strange nV/$\sqrt{}$Hz unit of measure is simple. It allows a designer to calculate noise in a specific bandwidth. If noise were given in, say, nV_{rms} over say a 1-kHz bandwidth, it would be difficult to find the noise in, say, a 100-Hz bandwidth. Here's an example. Consider a noise density of 38 nV_{rms}/square root hertz at 1 kHz. It is easy to calculate the total noise: Just multiply 38 $nV_{rms}/\sqrt{}$Hz by the square root of the bandwidth, namely 1000 Hz. The answer is 1.2 μV rms (38 nV/$\sqrt{}$Hz $\times$ 32 square root hertz). The total noise is, therefore, a little more than 1 μV rms from, say, 10 Hz to 1 kHz. Since noise is made up of many uncorrelated sinewave frequencies, the peak to peak is statistically about six times higher than the rms. That is, it is about 7.2 μV peak to peak. In biomedical equipment, like ECG machines, peak-to-peak noise is used to calculate dynamic range. In audio and video communications equipment rms noise is often used.

Figure 4-25 shows a commercially available monolithic instrumentation amplifier in which all three op-amps and resistors are one silicon chip. This particular component also has internal overvoltage protection diodes. It uses one external resistor to set the gain. You will see these types of single packaged parts used in modern compact biomedical equipment.

In fact, as discussed later, many biomedical systems are now being designed on one chip. All circuitry is contained on a single high-density integrated circuit. The name given to this newer approach is ASIC (Application Specific Integrated Circuit). ASIC manufacturers make available standard design tools for manipulating standard cells or

Figure 4-25 One-chip (monolithic) three op-amp precision instrumentation amplifier, INA101. (Courtesy of Burr-Brown under copyright 1988 Burr-Brown Corporation)

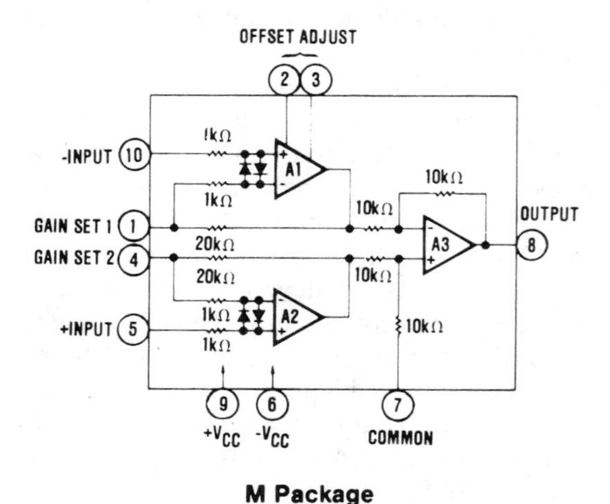

circuit functions. The advantage to the biomedical equipment designer is far reaching. A previously experienced printed circuit board engineer can now put together a complete system or subsystem on a chip even though he or she is not an expert in monolithic semiconductor technology. Just as this text describes available bioelectric amplifiers, ASICS lay out available design and manufacturing technology.

Not only is the circuitry small, the packaging is becoming smaller as well. The SOIC (small-outline integrated circuit) and the LCC (leadless chip carrier) packages are becoming dominant. They mount on the surface of a printed circuit board (PCB) instead of in holes drilled through the PCB. It is easy to see how this construction technique saves cost and space by picturing dozens of SOICs mounted on both sides of a PCB.

No matter how sophisticated instrumentation amplifiers become, they must still follow basic principles. Figure 4-26 shows an important consideration readers should know about. In this chapter we showed that real-world op-amps have bias current and that this current must flow somewhere in order for the circuit to work properly. Figure 4-26 indicates one should always provide a return path for bias current. That is, if a floating source such as a microphone, thermocouple, transformer, or the human body is used, resistors or some connection to ground or common must be present. If the return is disconnected, the output of the amplifier will slowly float up to the positive or down to the negative limit near the power supply voltage. The saturated output near the "rail" will no longer respond to the input signal. If this condition is seen in an instrument, the bias current return path should be checked.

The following shows some instrumentation amplifier applications to biomedical equipment. In older pH probes, one electrometer or femtoamp op-amp was used. Today, pH probes use electrometer instrumentation amplifiers, as shown in Figure 4-27. Here commercially available input stage op-amps connect to the high-impedance reference and sample electrodes. pH is automatically measured as solutions flow through specially designed electrodes. Such circuits are often isolated from earth ground to prevent noise pickup. This circuit will be encountered frequently. For example, pH is one of the measurements made in blood and body fluid analysis equipment.

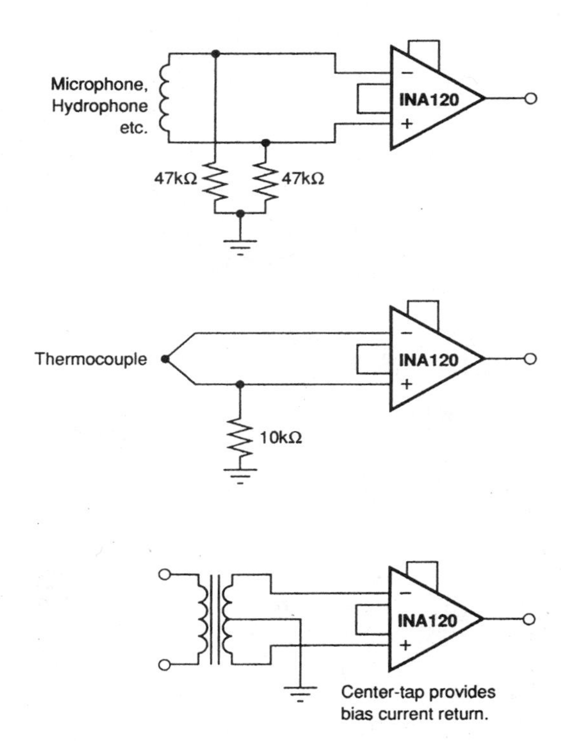

Figure 4-26 Important consideration: Providing an input bias current path. (Courtesy of Burr-Brown under copyright 1990 Burr-Brown Corporation)

Instrumentation amplifiers are also used to amplify the differential voltage from Wheatstone bridges, as shown in Figure 4-28. Since the source resistance is relatively low at 300 Ω, a designer would usually use a bipolar input IA with moderate bias current

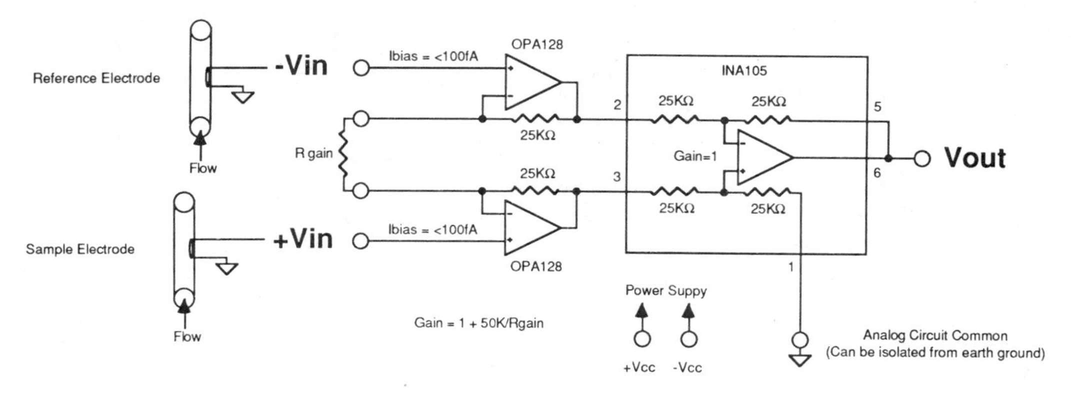

Figure 4-27 pH probe electrometer instrumentation amplifier.

of around 20 nanoamps (20×10^{-9} amps). But here a commercially available monolithic FET IA is used. What is the advantage? Large resistors, say 75 kΩ, and a small value capacitor, say 1 μF, for passive filtering can be used. Filtering is achieved without loading the source very much. Low offset errors ($I_{\text{bias}} \times R_{\text{source}}$) are still being maintained, because the bias current is very low, around 50 picoamps (50×10^{-12} amps). The dc error is only 3.75 μV (50 picoamps × 75 kΩ). Commonly these chips achieve gain by strapping two pins together where no external resistor is needed. Here the gain is 500

Figure 4-28 Bridge amplifier with 1-Hz low-pass filter. (Courtesy of Burr-Brown under copyright 1988 Burr-Brown Corporation)

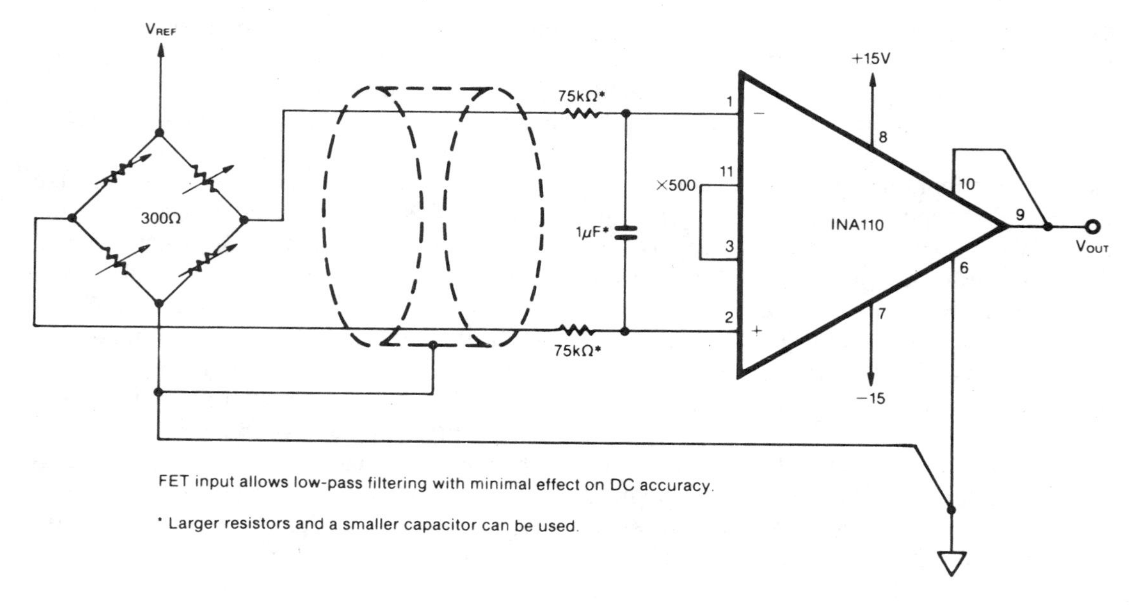

V/V. This example shows how modern components make biomedical instruments smaller and more reliable.

Instrumentation amplifiers usually amplify low-level signals in the tens of millivolt range. Unfortunately, noise can ruin the measurement. Figures 4-29 and 4-30 show two important considerations that relate to noise. First in Figure 4-29, common-mode rejection (CMR) comes into play. Most commercially available IAs do a good job of rejecting low-frequency noise, say 50 Hz or 60 Hz. However, when the CM noise contains high frequencies, rejection can be poor. The typical product data sheet performance curve in Figure 4-29 shows that 60-Hz rejection is around 120 dB when the IA is in a gain of 500. Even when the interference is at 100 kHz, the graph shows rejection is still at 80 dB. Therefore, this particular component is useful in rejecting higher frequencies from sources such as switching power supplies.

Why does high frequency interference disturb a low-frequency measurement? The answer is dc offset caused by rectification through semiconductor junctions inside the IA and filtering by parasitic capacitance around the junctions. Therefore, rejecting high frequency outside the biomedical bandwidth is important for maintaining baseline stability.

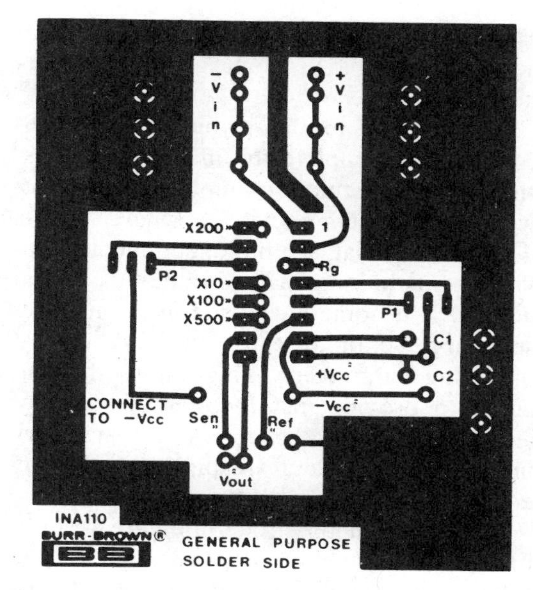

Figure 4-30 Important consideration: PC board layout for INA110. (Courtesy of Burr-Brown under copyright 1988 Burr-Brown Corporation)

Figure 4-29 Important consideration: Frequency effect on common-mode rejection (CMR), INA110. (Courtesy of Burr-Brown under copyright 1988 Burr-Brown Corporation)

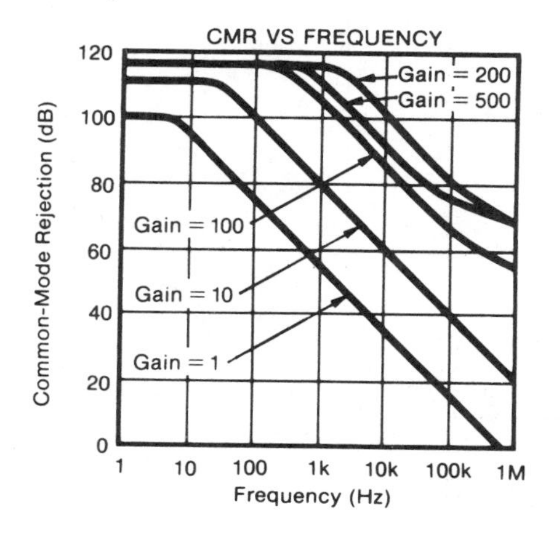

The second important consideration for instrumentation amplifiers is printed circuit board layout. The electrical environment external to the IA can disturb performance. Figure 4-30 shows that good grounding keeps noise low. Also, care must be taken to reduce stray capacitance, because external *R-C* mismatch on the two inputs degrades CMR. Furthermore, care must be taken to separate the output from the input when using a high-frequency IA. Notice that V_{out} is located at one end of the board and $+V_{in}$ is located at the other end. This prevents positive feedback that can cause oscillation. The reason for depicting good layout is to draw attention to proper construction practice.

This is another important ingredient in achieving high bioelectronic amplifier performance.

Examining a few more biomedical applications will complete the important discussion of instrumentation amplifiers. The hot-wire anemometer shown in Figure 4-31 is used in industrial, scientific, and medical environments to indicate flow rate of gases and liquids. It is sometimes used in artificial respiratory ventilators to measure patient breathing parameters. This allows the machine to provide just the right amount of assistance. (See the chapters on respiratory instrumentation and respiratory therapy equipment.)

The anemometer circuit in Figure 4-31 utilizes two IAs, one op-amp, and two voltage references. The first part of the circuit uses a thermistor temperature transducer. Its output is amplified by the noninverting op-amp to indicate air temperature of a patient's breath, usually +37° Centigrade. Accuracy is around plus and minus 1° Centigrade. The other part of the circuit produces an air flow voltage that is proportional to a patient's breathing volume and rate. It works this way. A current, driven through the bridge by the power transistor, heats the tungsten wire or filament to a few hundred degrees Centigrade. When the patient breathes, air flow increases and the wire cools off somewhat. This upsets the bridge balance and causes a differential voltage to appear, which is amplified by an instrument amp. The increasing IA output voltage causes more current to flow through the bridge, which reheats the filament. As the patient breathes, the power feedback loop tries to maintain the hot-wire at a constant temperature (also a constant resistance). As it does, the changing bridge voltage is amplified by the second IA. This

Figure 4-31 Hot wire anemometer-tnermistor circuit for measuring air flow.

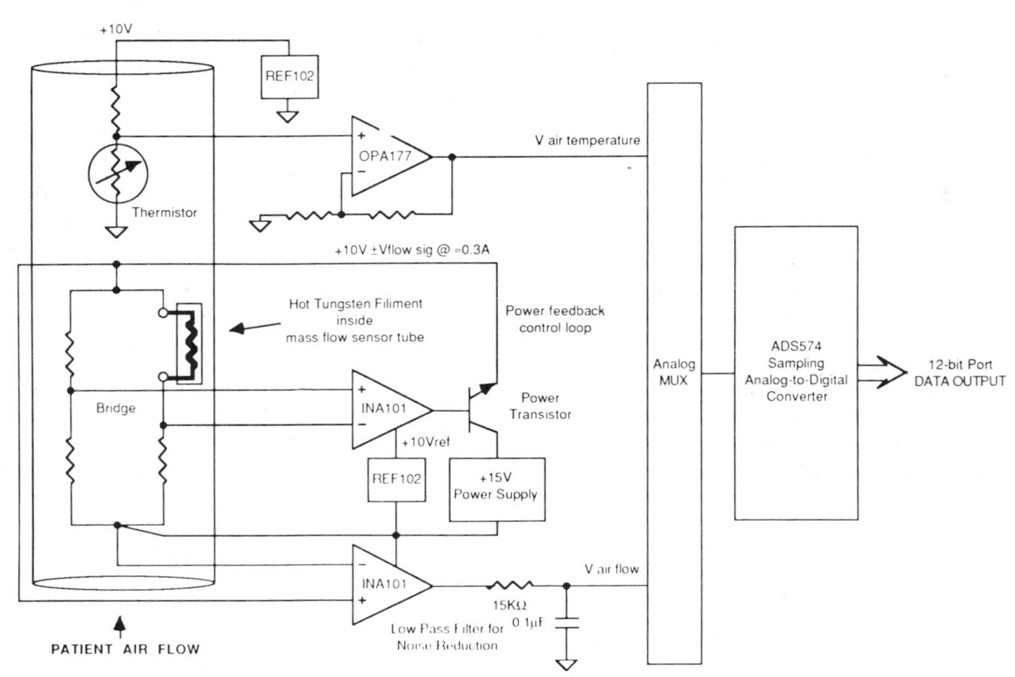

signal, which is actually measuring air speed, is low-pass filtered at around 100 Hz for noise reduction. Both the air temperature and flow voltages are multiplexed into a sampling analog-to-digital converter, where they are digitized and fed to a computer for analysis.

Feedback loops are commonly used in biomedical instrumentation. To further understand the hot-wire power feedback loop, think of the loop having two voltages. One is the fixed +10 V reference. This goes through the loop IA and power transistor to apply a constant +10 V to the bridge. The second voltage is the varying bridge signal. It is picked up from the bridge and fed back to the bridge through the loop IA and power transistor. It causes the bridge voltage to vary in accordance with air speed and attempts to keep the hot-wire at a constant temperature. In fact, it is this voltage that is amplified by the second IA at the bottom. Air speed is usually 0.1 to 200 liters per second. It can be calibrated to about 0.1 percent of full scale. It is clear that instrumentation amplifiers are the perfect companion to bridges, especially in biomedical equipment.

Another bridge amplifier is depicted in Figure 4-32. The clever technique shown here is how the amplified bridge signal is transmitted over a distance. If the IA output voltage were sent, say, 100 feet over twisted wire, it would pick up noise from the power lines before it reached the other end. An inexpensive technique for minimizing noise is to convert the voltage signal to a current and then transmit the current over the distance via a current loop. In Figure 4-32, an IA is used to drive a commercially available current transmitter. This monolithic component takes +2 V to +10 V and turns it into 4 mA to 20 mA, respectively. Why does the current signal in the loop pick up very little 60-Hz noise? The answer rests in how noise is picked up in the first place. Electromagnetic interference (EMI) is an electromagnetic wave, pulsating at 60 Hz, which cuts across wires and induces noise voltages. These voltages add up along the wires as they traverse power lines. The total noise voltage tries to drive a noise current. However it cannot drive much, because the loop is a high resistance. At one end of the loop is the current transmitter, which is actually a high-resistance current source, say 10 MΩ. At the other end is the load resistor, say 250

Figure 4-32 4 mA to 20 mA current loop bridge transmitter. (Courtesy of Burr-Brown under copyright 1988 Burr-Brown Corporation)

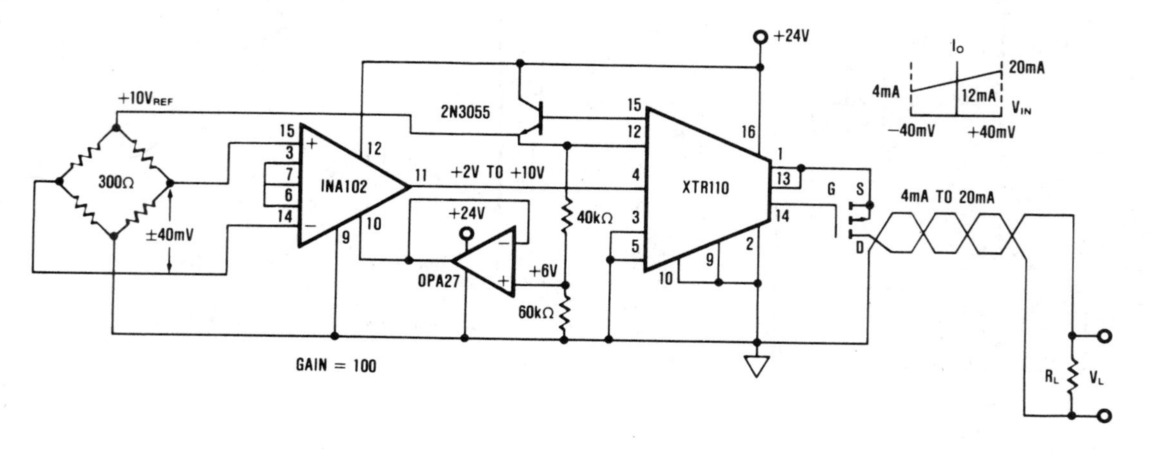

Ω, which turns the signal current back into a signal voltage. Even if the noise voltage is 10 V_{rms} at 60 Hz, the noise current is only 1 microamp rms (1.414 microamps peak) at 60 Hz. Compared to the minimum-scale signal of 4 mA peak, this amounts to only 0.04 percent peak error. This is a much lower error than you could achieve by sending a voltage over a 100-foot cable in an electrically noisy environment like a hospital.

Of course, nothing is perfect, not even precision instrumentation amplifiers. For example, a medical or scientific weighing scale might have to measure milligrams out of several kilograms. The dynamic range required can be tens of thousands to one. To do this you need to correct errors in the IA. Figure 4-33 shows how a precision +10 V voltage reference and ratio-matched resistors are used to automatically compensate for offset and gain errors. The zero (offset) calibration circuit derives 100 μV dc and differentially applies it to the IA through mechanical or electronic switches. Exactly 0 V dc could be used, but the 16-bit analog-to-digital converter shown has trouble digitizing 0 V ab-

Figure 4-33 Load cell weighing scale instrumentation amplifier. (Courtesy of Burr-Brown under copyright 1988 Burr-Brown Corporation)

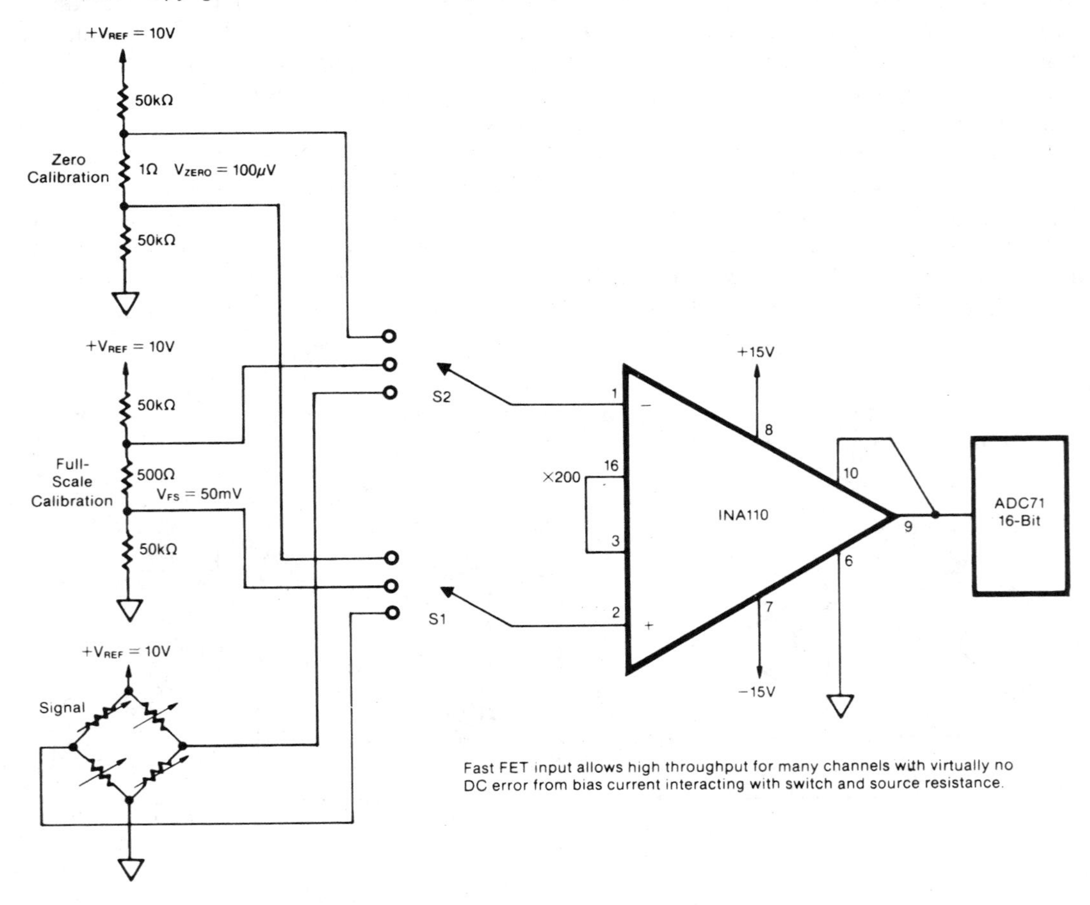

solutely. A voltage just above zero is acceptable. The full-scale (gain) calibration circuit derives 50 mV dc and applies it to the IA during a separate part of the calibration cycle. This technique allows IA and A/D converter errors to be stored in the computer. Then when the bridge signal is applied to the IA, the computer stores the signal and subtracts out offset and gain errors associated with the electronic circuit. Such calibration cycles are used frequently in biomedical instrumentation.

It is usually desirable to locate an instrumentation amplifer as close to a transducer as possible to minimize the chance of someone accidentally connecting overvoltages. However, when this is not practical, biomedical equipment designers include a front-end protection circuit like the one in Figure 4-34. Here two series resistors limit the input current that passes through clamp diodes. These diodes prevent the voltage on the IA input terminals from going more than one diode drop (about 0.7 V) above the positive or negative supply rail (V+ and V−). Therefore, should the common-mode input voltage (referenced to the common between power supplies) go above the IA's absolute maximum, no damage will occur. The external diodes clamp the overvoltage and take the current hit instead of the IA. How high is high? The answer depends on R_s. If R_s were 1 kΩ, for example, the overvoltage to the left of R_s could rise to plus or minus 100 V if the limit was to be bounded at 100 mA. This diode clamp protection circuit is the one most commonly used in the IAs that amplify ECG signals. It is the second line of defense against the 5000 V defibrillation voltage, which is first clamped by gas tubes or special high-voltage semiconductor diodes.

Figure 4-34 Important consideration: input protection circuit (Courtesy of Burr-Brown under copyright 1990 Burr-Brown Corporation)

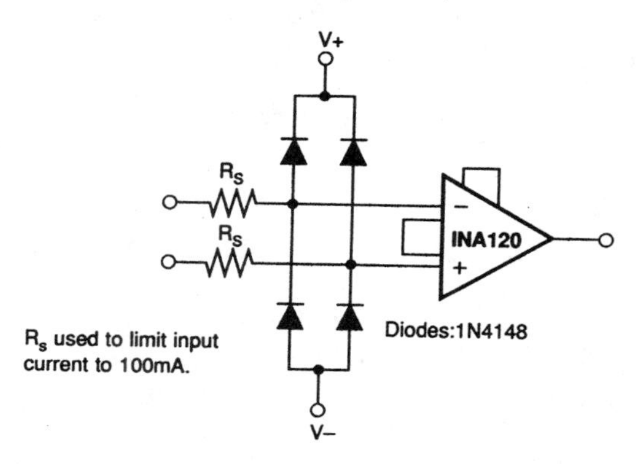

4-8 Signal Processing Circuits

Several special-purpose operational amplifier circuits are also used extensively in medical instrumentation. Among these are *integrators, differentiators, logarithmic amplifiers,* and *antilog amplifiers.*

Integrators and differentiators are analog circuits that perform the mathematical operations of integration and differentiation, respectively. If you are not familiar with these concepts, refer to Appendix A.

4-8.1 Integrators

Integration is a mathematical process that allows us to find the area under the curve defined by a function. If the function is a time-dependent voltage, such as might be found in an instrumentation circuit, then a circuit such as Figure 4-35 may be used to integrate the voltage function. The transfer equation of the analog integrator (Figure 4-35) is

$$E_{out} = -\frac{1}{R_1 C_1}\int_0^t E_{in}\,dt + E_{ic} \qquad (4\text{-}18)$$

where

E_{out} is the output potential in volts
E_{in} is the input signal potential in volts (V)
R_1 is the input resistance in ohms (Ω)

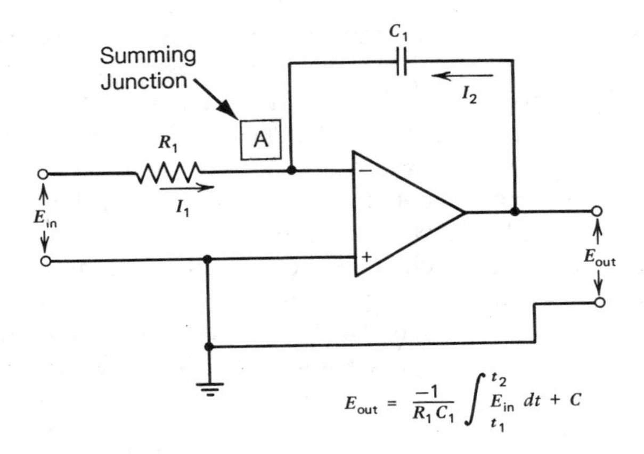

Figure 4-35 Integrator circuit.

C_1 is the feedback capacitance in farads (F)
t is the time in seconds (s)
E_{ic} is any initial condition output voltage already present at the integrator output when E_{in} begins (at $t = 0$)

EXAMPLE 4-5

An analog integrator (Figure 4-35) uses a 1-MΩ resistor and a 0.2-μF capacitor. Find the output voltage after 1 s if the input voltage is a *constant* 0.5 V.

SOLUTION

$$E_{out} = -\frac{1}{R_1 C_1}\int_0^1 E_{in}\, dt + E_{ic} \tag{4-18}$$

$$E_{out} = -\frac{1}{(10^6 \Omega)(2 \times 10^{-7}\text{ F})}\int_0^1 (0.5)\, dt + 0$$

$$E_{out} = -\frac{1}{2 \times 10^{-1}}(0.5)\Big|_0^1$$

$$= (-5)(0.5) = \mathbf{-2.5\ V}$$

The gain of the integrator is given by $-1/R_1C_1$, and if R_1 is less than 10^6 Ω, and C_1 less than μF, then the gain can become very large very quickly. For 10^5 Ω and 10^{-10} F, the gain is 10^5!

When a voltage is applied to the integrator input, current I_2 is generated and begins to charge C_1. The voltage at the output rises as C_1 charges (recall property 6, Section 4-4.1, the point A end of C_1 is effectively *grounded*). Charging continues in the manner dictated by E_{in}.

Integrators also function as low-pass filters. This may be deduced by considering the frequency-dependent behavior of the capacitive reactance of C_1. In some medical instruments, integrators are labeled as "low-pass filters," even though integration may be their real function.

4-8.2 Differentiators

The differentiator circuit produces a voltage output proportional to the *time rate of change* of the input signal voltage. Differentiation is the inverse process of integration (which is the *time average* of the input signal). The circuit is similar to the integrator, except that the resistor R_1 and capacitor C_1 have changed places (see Figure 4-9). The transfer function of a differentiator is

$$E_{out} = -R_1 C_1 \frac{d(E_{in})}{dt} \tag{4-19}$$

where

E_{out} is the differentiator output voltage in volts (V)
E_{in} is the input potential in volts (V)
R_1 is the feedback resistor in ohms (Ω)
C_1 is the input capacitance in farads (F)

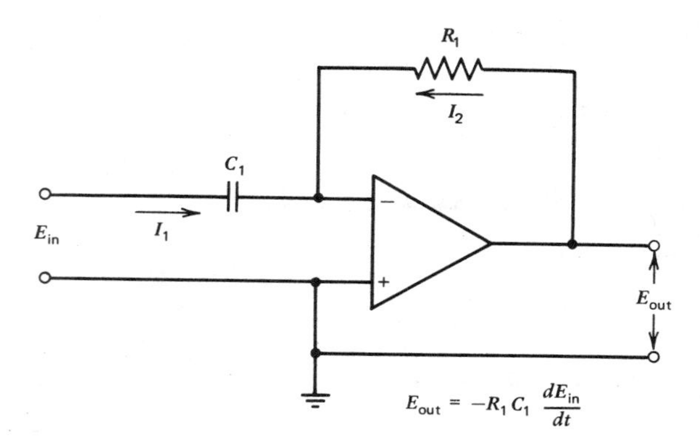

Figure 4-36 Differentiator circuit.

EXAMPLE 4-6

Find the output voltage produced by an operational amplifier differentiator (Figure 4-36) if $R_1 = 100\ k\Omega$, $C_1 = 0.5\ \mu F$, and E_{in} has a constant slope (i.e., is a ramp function) of 400 V/s.

SOLUTION

$$E_{out} = -R_1 C_1 \frac{dE_{in}}{dt} \quad (4\text{-}19)$$

$$E_{out} = -(10^5\ \Omega)(5 \times 10^{-7}\ \mu F)(400\ V/s)$$

$$E_{out} = -(5 \times 10^{-2}\ s)(400\ V/s) = 2 \times 10^1\ V = 20\ V$$

The R_1C_1 time constant, in seconds, must be *very short* compared to the time constant or period of the input signal.

The analog differentiator also functions as a *high-pass filter*. This feature follows from the same consideration of the reactance of C_1 that led us to the deduction that an integrator is a low-pass filter.

4-8.3 Log-Antilog Amplifiers

By now you should realize that many different transfer functions can be created using operational amplifiers. The designer need only know how to correctly manipulate the properites of the negative feedback network. Recall from our discussion of temperature transducers (Section 3-10) that the base-emitter voltage is proportional to the natural logarithm of the transistor collector current. The log dependence of I_c is used in circuits such as Figure 4-37*a* to form an amplifier that has a *logarithmic* transfer function.

Similarly, an *antilog* amplifier (Figure 4-37*b*) is formed by placing the transistor in the input circuit, between the input signal voltage and the inverting input of the operational amplifier.

The value of the logarithmic amplifier is best realized in applications where the dynamic range of an input signal is very large (ranging over several orders of magnitude). If a nonlogarithmic operational amplifier circuit is used in the "front end" of an instrument in such a case, then it is often found that it can handle the higher amplitude signals only at the expense of weaker signals. Alternatively, if the gain is set high enough to accommodate weak signals, then large signals are clipped by the amplifier, which is driven into saturation.

If a logaritmic amplifier is used in the front end, however, then the gain of the circuit is lower for larger signals than it is for smaller signals. This results in a *range compression.* At the output of the instrument an *antilog amplifier* returns the voltage signals to their proper relationship. Most simple logarithmic amplifiers can handle four decades of dynamic range in the input signal, while some exist that handle up to seven decades (i.e., a range from zero to 120 dB).

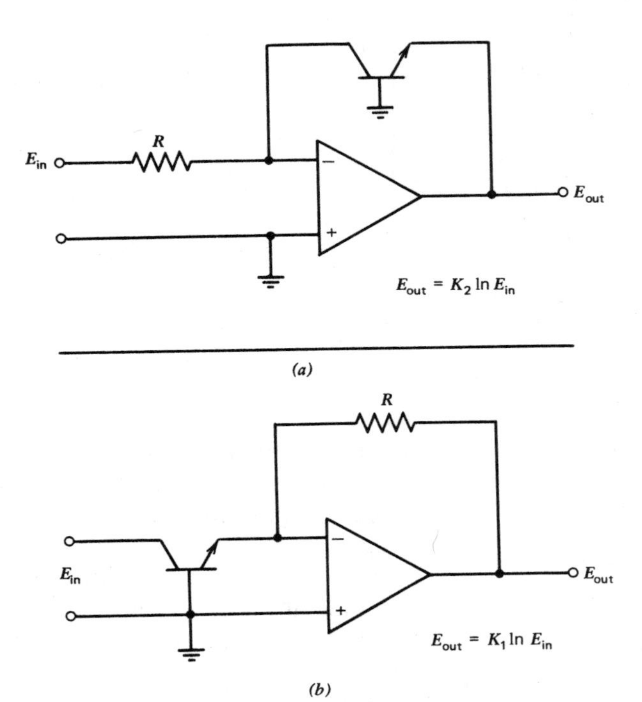

Figure 4-37 Log and antilog amplifiers. (*a*) Logarithmic amplifier. (*b*) Antilog amplifier.

From a practical point of view, how much accuracy can be obtained from commercially available logarithmic amplifiers? The answer depends on the range or number of decades of operation. For example, the log ratio component shown in Figure 4-38*a* is designed to handle six decades of input current (1 nA to 1 mA). Its output is $V_{out} = K \log_{10}(I_1/I_2)$. Laser trimming during manufacturing guarantees total error in V_{out} to be around plus or minus 25 mV, as shown in the typical performance curve of Figure 4-38*b*. This amounts to about 0.37 percent over five decades. However, if operated over six decades, the accuracy degrades to about 0.8 percent. Accuracy considerations for log ratio amplifiers are somewhat more complicated than for simple op-amp circuits. Errors include scale factor, bias current, log conformity, and offset voltage. The best a designer can hope for is something less than 0.5 percent at room temperature of $+25°$ Centigrade. At higher ambient temperatures, the log-amp becomes less accurate. This log-amp is very suitable for applications requiring, say, a few percent accuracy. For higher precision it is not. Most biomedical log functions are now done by computer software. But speed is sometimes an issue. It takes many machine cycles to get a log answer in hundreds of microseconds unless a more expensive high-speed computer is used. Clearly, the hardware log-amp is faster, giving an answer in 100 microseconds or so. However, the trend is unmistakable: software logging.

4-9 Practical Op-Amps: Some Problems Reviewed

The practical operational amplifier does *not* possess the ideal properties that we have shown thus far. The approximations of the ideal properties, however, apply to most

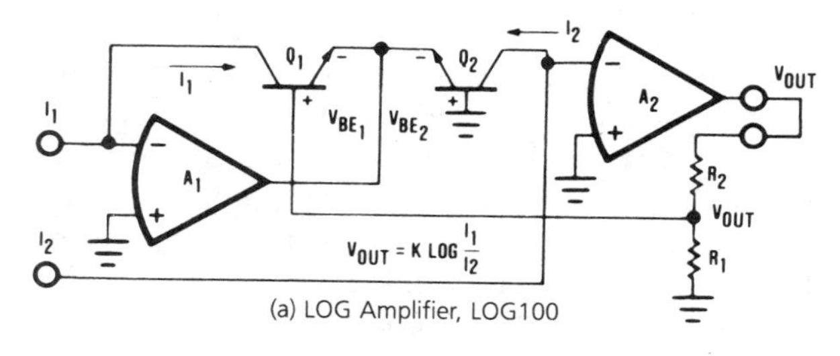

(a) LOG Amplifier, LOG100

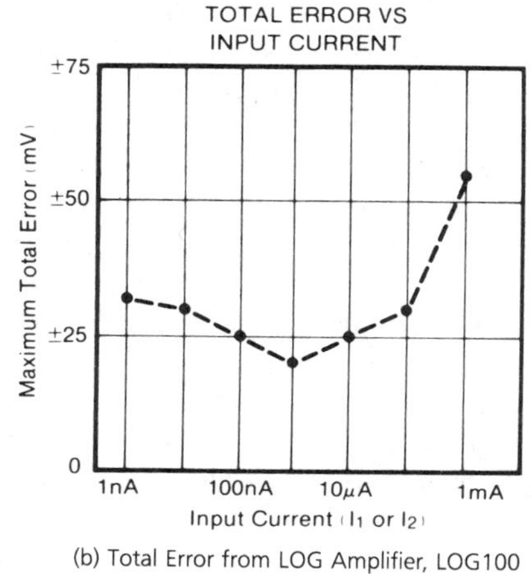

(b) Total Error from LOG Amplifier, LOG100

Figure 4-38 Accuracy obtainable from analog logarithmic and log ratio amplifier, LOG100. (Courtesy of Burr-Brown under copyright 1986 Burr-Brown Corporation)

devices. "Infinite" open-loop gain, for example, translates as "very high." Even the low-cost economy op-amp models in the 741-class offer gain figures in the 20,000 to 50,000 range. Premium grade operational amplifiers have gain figures from around 100,000 to 1,000,000!

Similarly, "zero output impedance" translates as "very low." All commercially available operational amplifiers have an output impedance under 200 Ω, and most are under 100 Ω!

The input impedance of practical operational amplifiers, while not infinite, is about 10^5 Ω in low-cost types to over 10^{12} Ω in models that use a MOSFET for the input stage. The RCA CA3130 through CA3160 series, for example, boasts an input impedance of 1.5 TΩ (1.5×10^{12} Ω)! Texas Instruments LM-662 (dual) and LM-660 (quad) are CMOS op-amps operating from a single dc power supply.

Of course, the closer the operational amplifier specifications reach ideal, the nearer its performance is to the ideal. In most cases, however, there may be several problems that must be recognized by circuit designers.

One problem is the existence of *input bias currents*. All operational amplifiers use some type of transistor in each side of the input stage, and these transistors will have bias currents. If bipolar NPN or PNP devices are used, then the bias current may be quite large. (For JFET and MOSFET stages it is much less.)

The input bias currents flow out of the respective input terminals to whatever resistances are connected to those inputs. In the inverting input circuit this will be the parallel combination of feedback and input resistors.

The bias current flowing in these resistors creates a *voltage drop* that is seen by the operational amplifier as a valid input signal.

The most obvious way to eliminate this voltage is placement of an equal *offset voltage* on the noninverting input. Since the same bias current flows in each input, we may cancel the voltage drop at the inverting input by placing a resistor to ground from the noninverting input that has a value equal to the parallel combination of the feedback and input resistors. The voltage drops applied to the two inputs have the same magnitude and phase, so by the basic properties of the operational amplifier, the resultant output voltage is zero.

There are other forms of problems that create output offset voltages. All may be solved by either of the offset null circuits shown in Figure 4-39. The circuit in Figure 4-39*a* uses a pair of *offset null* terminals found on many IC operational amplifiers. Each terminal is connected to one end of the null potentiometer. The wiper of the potentiometer is connected to the negative power supply (V_{ee}). The potentiometer is adjusted under zero input signal conditions so that the output voltage is also zero.

An alternative scheme is shown in Figure 4-39*b*. This circuit is used where there are no offset null terminals, or where a wider range than can be provided by the null terminals is required. The circuit of Figure 4-39*b* uses a potentiometer connected between V_{cc} and V_{ee}, with resistor R_4 connected between the potentiometer wiper and the inverting input of the operational amplifier. A current is set up in R_4 that cancels the output voltage.

Op-amp offset voltage error can continuously be removed by using the adaptive circuit shown in Figure 4-40. It works this way.

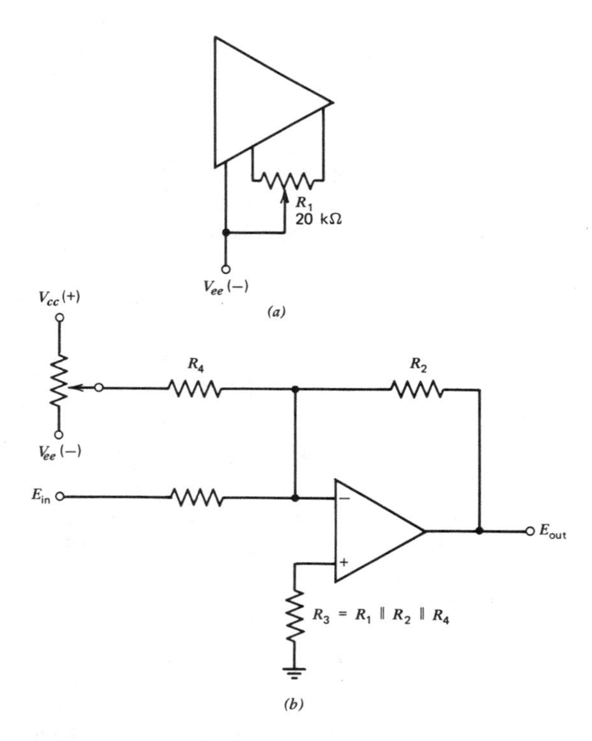

Figure 4-39 Offset null methods. (*a*) Using *offset adjust* terminals. (*b*) Using a null circuit.

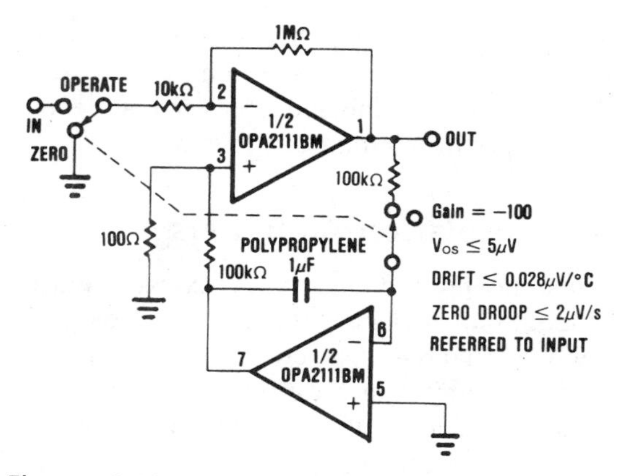

Figure 4-40 Auto-zero amplifier. (Courtesy of Burr-Brown under copyright 1988 Burr-Brown Corporation)

The op-amp configuration normally gives an inverting gain of -100 V/V. Automatic zeroing takes place when the input switch is placed in the zero mode and the feedback switch is closed on the op-amp's output. Any op-amp dc offset times the gain will be integrated by the feedback op-amp active integrator. It has a time constant of 0.1 seconds (100 kΩ × 1 μF). After 10 time constants or so, the integrator's output, which is the inversion of the offset error, is present on the main op-amp's noninverting terminal. This error corrector voltage is amplified by a factor of 101 V/V and essentially cancels the error caused by the main op-amp. When the input switch is placed in the operate mode, and the feedback switch is opened, normal dc-coupled signal amplification takes place.

The feedback error integrator will accurately hold the offset error for some time. Any droop or rise is caused by the integrator's bias current charging or discharging the integrator capacitor. But since bias is low, like 4 picoamps, it takes a long time to make any difference (Volt $= I_b/C$). The main op-amp's offset can be continuously corrected if the feedback switch remains closed and the integrator time constant is set properly. But then the main amplifier will essentially be ac coupled where it was dc coupled before. This type of adaptive correction is commonly used in ac-coupled ECG systems to zero out the offset potential of body electrodes. It is called *dc restoration*.

Of course, another way to get rid of offset voltage is to measure it, store it in computer software, and then subtract it out when the actual measurement is made. However, this requires the use of an analog-to-digital converter and a computer. It is obvious that the analog hardware approach can be less expensive.

4-10 Bioelectric Amplifiers Reviewed

The simple bioelectric amplifiers may be constructed of discrete components, IC operational amplifiers, and other IC devices. The unity of the operational amplifier is great enough that even those models that use discrete components (rather than ICs) are actually discrete versions of an operational amplifier circuit.

In most cases the bioelectric amplifier is a differential input circuit. In those few cases where a single-ended input is required, a differential amplifier may be used with one input grounded. However, many single-ended input noninverting buffers are used today ahead of the differential amplifier.

The properties desired in a bioelectric amplifier are:

1. Single-ended output, often differential input.

2. High common-mode rejection ratio.

3. Extremely high-input impedance.

4. Variable gain adequate to do the job intended. The following categories are generally recognized: low gain (×1 to ×10), medium gain ($\times 10$–$\times 1000$), and high gain (over $\times 1000$).

5. Frequency response suitable for the application. In the case of a "universal" bioelectric amplifier, the response should be variable through switch selection.

6. *Zero suppression.* This is an optional feature that allows shift about the zero baseline by nulling offsets inherent in the signal. This feature permits small varying signals superimposed on a larger dc signal (or dc offset) to be processed in the amplifier, using the *full gain* of the amplifier for the small varying signal only. For example, a 10-mV sinewave may be superimposed on a +1500-mV dc offset. This is seen as a 10-mV sinewave if a −1500-mV zero-suppression signal is summed with the actual signal at the input.

The designations of front panel controls in bioelectric amplifiers often reflect the vocabulary and jargon of the user, rather than standard electronic terms. For example, *gain* or *sensitivity* may be labeled *span* if the amplifier is designed for use by a life scientist or chemist. Similarly, the gain control may be divided into controls, *coarse* and *fine,* although the *fine* control might be labeled *span* and the *coarse* control (usually a rotary switch) is labeled *range*.

On a specialized amplifier, the coarse control may be labeled with physical units peculiar to the application, while on universal models the labeling will be more generalized. Since many amplifiers are designed for connection to a CRO (or strip-chart recorder), the units may be labeled as a vertical deflection factor in mV/mm or mV/cm.

4-11 Isolation Amplifiers

Some hospital patients are extraordinarily susceptible to electrical shock hazards. It is believed that 60-Hz ac currents small enough to be deemed harmless ordinarily may be lethal to a patient under certain circumstances (see Chapter 19).

To prevent accidental internal cardiac shock (see Chapter 19), the manufacturers of modern bioelectric amplifiers, especially those used in ECG recording, use *isolation amplifiers* for the direct patient connection. These amplifiers provide as much as 10^{12} Ω of insulation (isolation) between the patient connector and the ac power mains line cord.

The basic design of an isolation amplifier is shown in Figure 4-41. It is usually composed of an input amplifier, some type of modulator, an isolation barrier, a demodulator, and an output amplifier. Modulation schemes include amplitude, voltage-to-frequency, duty cycle, pulse width, flyback loading, etc. Barriers can be optical, magnetic transformer, capacitive, or even heat transfer. Notice that there is an input common and an output common that are electrically isolated from one another to the tune of millions of ohms. The iso-amp is really an

Figure 4-41 Basic design of an isolation amplifier.

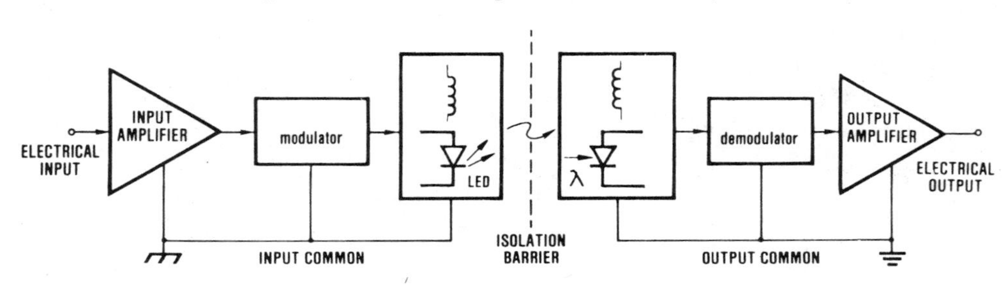

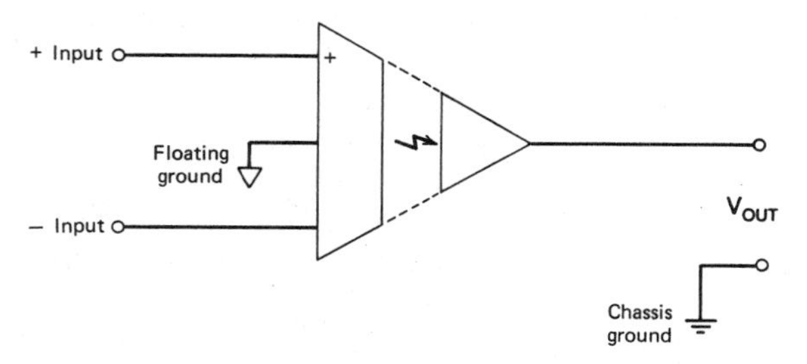

Figure 4-42 Symbol for an isolation amplifier.

energy converter. Electrical energy on the modulator side is converted to some "non-electrically conductive" energy in the barrier, and then converted back to electrical energy on the demodulator side. That's all there is to it. But what function does an iso-amp really perform?

A circuit diagram symbol that is often used to represent the isolation amplifier is shown in Figure 4-42. This symbol has not been standardized, so some manufacturers use variations of their own on their circuit diagrams.

Isolation amplifiers actually operate on the principle of attenuation. The high barrier impedance ($> 10^{12}\ \Omega$ in parallel with < 10 pF) acts in series between input and output or input and noise, as shown in Figure 4-43. Therefore, an interfering isolation-mode voltage (IMV) referenced to the iso-amp's output must go through the large barrier resistance before it can mix with the input signal. Hence, most of the interfering voltage or noise is dropped across the barrier; very little adds to the input.

Since the barrier is not an infinite impedance, some error is created. We say that an isolation-mode (IM) voltage (IMV), shown as V_{im} across the barrier, causes an error to appear in the output voltage, v_{out}. Figure 4-44 shows how this error is calculated. The measure of how well the iso-amp attenuates or rejects the IMV is called isolation-mode rejection (IMR). The isolation-mode rejection ratio in V/V is called IMRR. IMRR in V/V $= \log^{-1}(\text{IMR}_{dB}/20)$. This is the reverse of the equation shown in Figure 4-44, IMR in dB $= 20 \log_{10}(\text{IMRR}_{V/V})$. The error is gained up just like the input signal and results in some dc or ac voltage which adds to the normal signal. For example, if from a commercially available product data sheet IMR were 120 dB, then IMRR would be 1,000,000 V/V. If V_{im} were 1,000 V dc, the error at the

Figure 4-43 Reduction of interference in an IA is by cancellation. In an ISO amp, it can be thought of as attenuation.

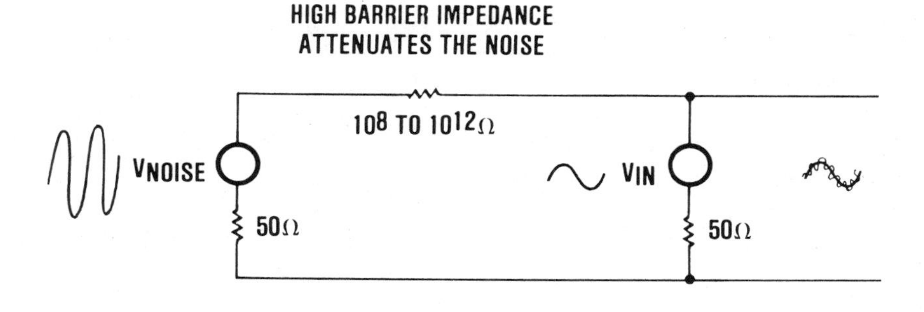

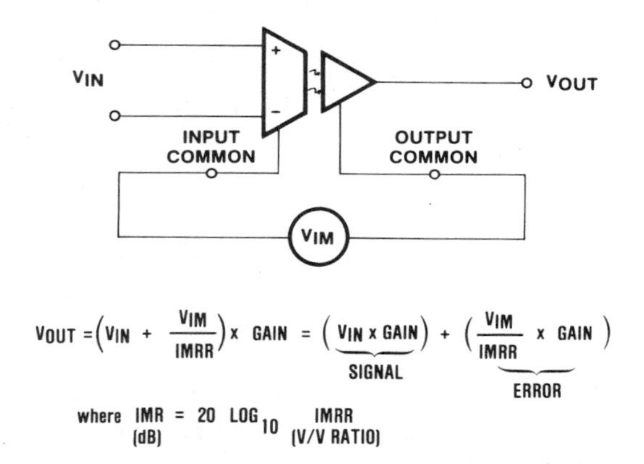

Figure 4-44 IMR error.

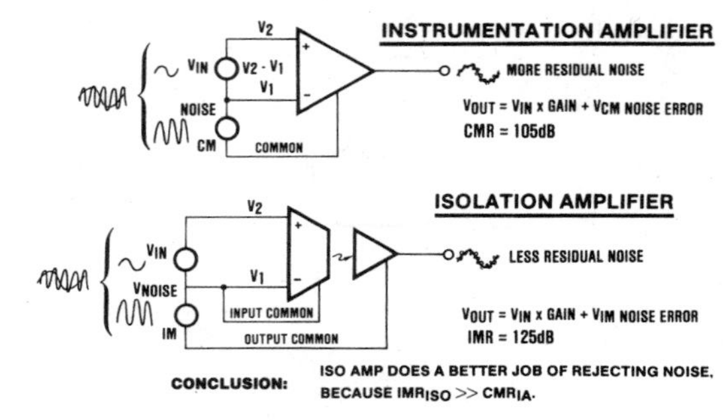

Figure 4-45 Comparison: CMR versus IMR.

iso-amp's input would be 0.001 V or 1 millivolt dc. This would be 0.1 percent error if the input signal were 1 V dc. If the iso-amp had a gain of 10, then the output signal would be 10 V dc and the error would be 10 millivolts dc. The percent error would still be 0.1 percent. It is the same provided that all voltages are taken with respect to the input (RTI) or with respect to the output (RTO). Note that the iso-amp rejects low-level dc or ac interfering voltages appearing across the barrier just like it rejects kilovolts. In fact, the iso-amp is just as good for attenuating noise as it is for attenuating high voltages.

In contrast, an instrumentation amplifier (IA) cancels or rejects common-mode voltage present on each of its two input terminals. A comparison of common-mode rejection (CMR) versus isolation-mode rejection (IMR) appears in Figure 4-45. CMR in an instrumentation amplifier is a measure of how well it rejects interference or noise. It does this by cancellation through its balanced gain paths. IMR in an iso-amp is also a measure of how well it rejects interference or noise. It does this by attenuation through its high-barrier impedance. If one were clever enough to reconfigure the IA circuit at the top to make the CM voltage in the IA appear like IM voltage in the iso-amp, then one would get better rejection of noise. Why is this so? Because IMR in an iso-amp is usually higher (125 dB) than CMR in an instrumentation amp (105 dB). It is just the way they are constructed that makes the difference.

Sometimes an iso-amp may have CM noise on its input as well as IM across its barrier. Figure 4-46 shows this. V_{cm} is noise on the two inputs with respect to the input common. V_{im} is noise on the input with respect to the output common. It is across the barrier. If an IA were inside the iso-amp

Figure 4-46 An ISO amp may have common-mode (CM) noise on its input as well as isolation-mode (IM) noise across its barrier.

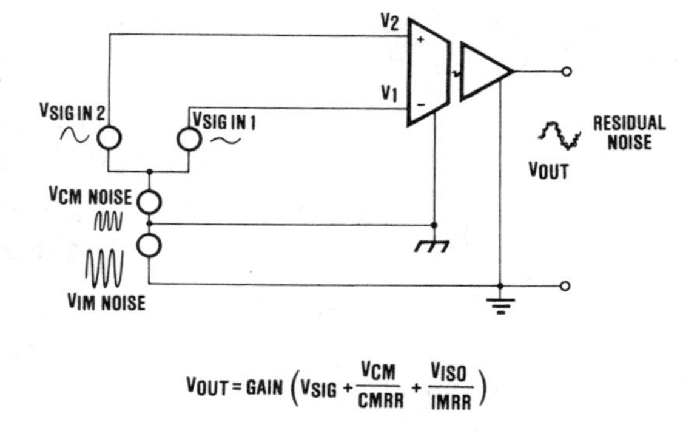

shown, then CM as well as IM interference would be rejected. Here,

$$V_{out} = \text{gain}(V_{sig} + V_{cm}/\text{CMRR} + V_{im}/\text{IMRR})$$

In summary, modern isolation amplifiers serve three purposes: (1) They break ground loops to permit incompatible circuits to be interfaced together while reducing noise, (2) they amplify signals while passing only low leakage current to prevent shock to people or damage to equipment, and (3) they withstand high voltage to protect people, circuits, and equipment.

Several approaches to the design of isolation amplifiers are used: *battery power, carrier, optically coupled,* and *current loading.*

4-11.1 Battery-Powered Designs

This approach is perhaps the simplest to implement, but it is not always the most suitable for the customer's convenience due to problems inherent in battery maintenance. A few products exist, however, that use a battery-powered, front-end amplifier, even though the remainder of the product is ac mains powered. Cardiac output computers (see Section 7-3) almost universally use this approach.

The bioelectric amplifier in this type of instrument is exactly like that in an ac-powered model. The sole difference is that it is powered from a battery pack.

This type of instrument must be totally self-contained. If any external instrument or device (such as an oscilloscope, strip-chart recorder, rate meter, battery charger, etc.) is used, then a model employing any one of the other isolation techniques must be used.

4-11.2 Carrier Amplifiers

Figure 4-47*a* shows an isolation amplifier using carrier technique. The circuitry inside of the dashed line is isolated from the ac power mains and the rest of the circuitry that is powered from the ac mains. In most cases, the voltage gain of the isolated section is in the medium-gain range (e.g., $\times 10$–$\times 500$).

The isolation is provided by separation of the ground, power, and signal paths in the two sections by transformers T_1 and T_2. These transformers have a core material that is very inefficient at 60 Hz but works well in the 20- to 250-kHz range. This feature allows the transformers to easily pass the carrier signal, but impedes any 60-Hz energy that might be present.

Although most models use a carrier frequency in the 50- to 60-kHz range, there are several types that use almost any frequency in the 20- to 250-kHz range.

The carrier oscillator signal is coupled through transformer T_1 to the isolated stages. Part of the energy from the secondary of T_1 goes to the modular stage; the remainder is rectified and filtered and then used as an isolated dc power supply. The dc output of this power supply is used to power the input amplifiers and modulator stages.

An analog signal applied to the input is amplified by A_1 and is then applied to one input of the *modulator* stage. This stage amplitude-modulates the signal onto the carrier.

Transformer T_2 couples the signal to the input of the *demodulator* stage on the nonisolated side of the circuit. Either envelope or synchronous demodulation may be used, although the latter is more popular. Ordinary dc amplifiers following the demodulator complete the signal processing.

An example of a *synchronous demodulator* circuit is shown in Figure 4-47*b*. These types of circuits are based on switching action. Although the example shown uses *PNP* bipolar transistors as the electronic switches, others use CMOS analog switches or FET transistors.

The signal from the modulator has a fre-

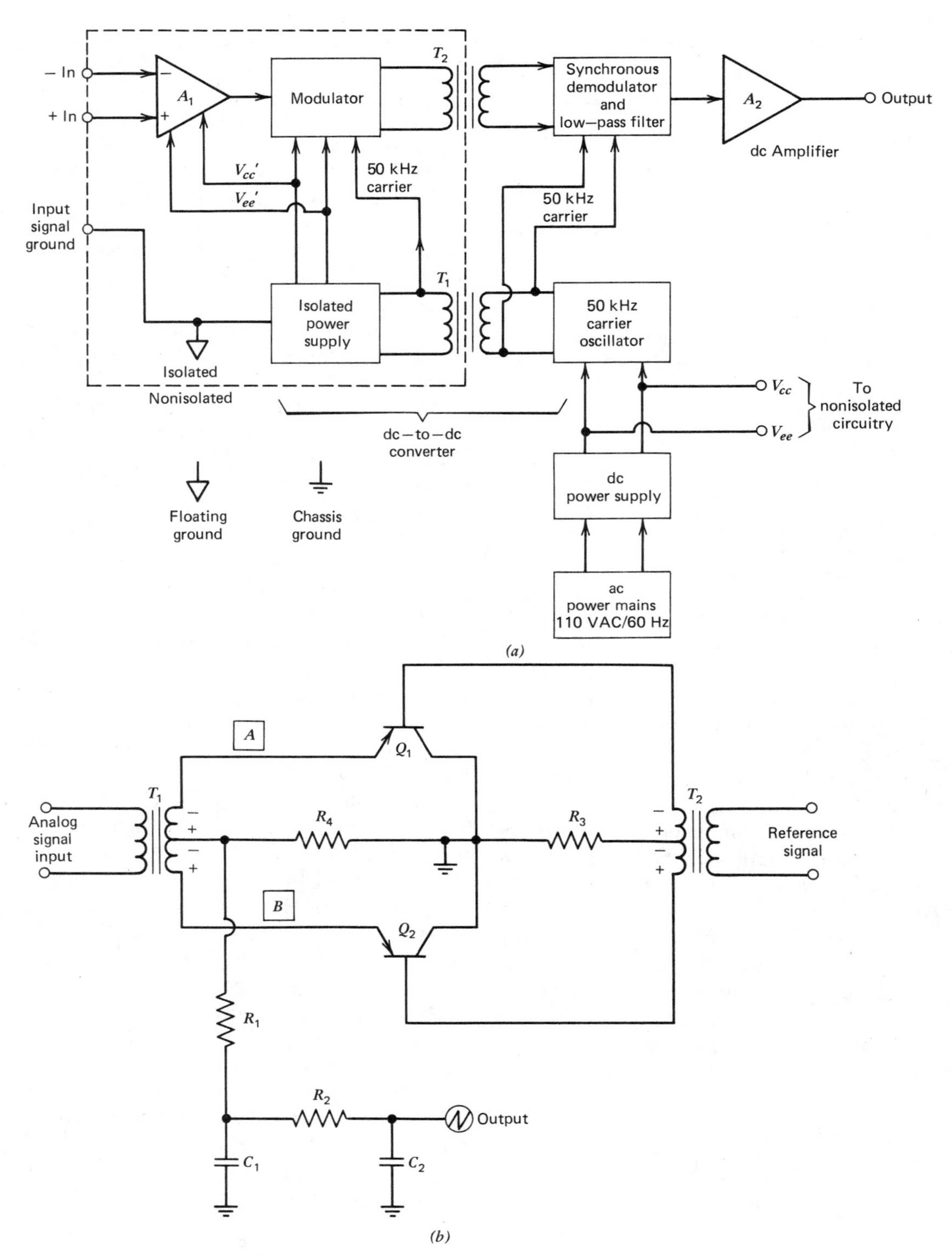

Figure 4-47 *(a)* Carrier-type isolation amplifier. *(b)* Synchronous demodulator.

quency of 50 kHz (up to 250 kHz, or even 500 kHz) and is amplitude modulated with the signal from the isolated amplifier. This signal is applied to the emitters of transistors Q_1 and Q_2 (via T_1) in *push-pull*. On one half of the cycle, therefore, the emitter of Q_1 will be positive with respect to the emitter of Q_2. On alternate half cycles, the opposite situation occurs: Q_2 is positive with respect to Q_1.

The bases of Q_1 and Q_2 are also driven in push-pull, but by the 50-kHz carrier signal. This action causes Q_1 and Q_2 to switch on and off, but out of phase with each other.

On one half of the cycle, we will have the polarities shown in Figure 4-47*b*. Transistor Q_1 is turned on. In this condition point *A* on T_1 is *grounded*. The voltage developed across load resistor R_4 is positive with respect to ground.

On the alternate half cycle, Q_2 is turned on, so point *B* is grounded. But the polarities have reversed, so the polarity of the voltage developed across R_4 is *still positive*. This causes a full-wave output waveform across R_4, which when filtered becomes a dc voltage level proportional to the amplitude of the input signal. This same description of synchronous demodulators also applies to the circuits used in some carrier amplifiers.

A variation on this circuit replaces the modulator with a *voltage-controlled oscillator* (VCO) that allows the analog signal to frequency modulate a carrier signal generated by the VCO. The power supply carrier signal is still required, however. A *phase detector, phase-locked loop* (PLL), or *pulse-counting detector* on the nonisolated side recovers the signal.

4-13.3 Optically Coupled Circuits

Electronic *optocouplers* (also called *optoisolators*) are sometimes used to provide the desired isolation. In early designs of this class a *light-emitting diode* (LED) was sandwiched with a photoresistor or phototransistor. Modern designs, however, use IC optoisolators that contain the LED and phototransistor inside of a DIP IC package.

There are actually several approaches to optical coupling. Two very popular methods are the *carrier* and *direct* methods. The carrier method is the same as discussed in Section 4-11.2, except that an optoisolator replaces transformer T_2.

The carrier method is not the most widespread in optically coupled amplifiers because of frequency response limitations of IC optoisolators. Only recently have these problems been resolved.

A more common approach is shown in Figure 4-48. This circuit uses the same dc-to-dc converter to power the isolated states as was used in other designs. This will keep A_1 isolated from the ac power mains but is not used in the signal coupling process.

The LED in the optoisolator is driven by the output of isolated amplifier A_1. Transistor Q_1 serves as a series switch to vary the light output of the LED proportional to the analog signal from A_1. Transistor Q_1 normally passes sufficient collector current to bias the LED into a linear portion of its operating curve. The output of the phototransistor is ac coupled to the remaining amplifiers on the nonisolated side of the circuit, so that the offset condition created by LED bias is eliminated.

4-11.4 Current Loading

A *current loading* isolation technique is used by Tektronix in their portable medical ECG monitors. A simplified schematic is shown in Figure 4-49. Notice that there is no *obvious* coupling path for the signal between the isolated and nonisolated sides of the circuit.

The gain-of-24 isolated input amplifier in Figure 4-49 consists of a dual JFET (Q_1) and

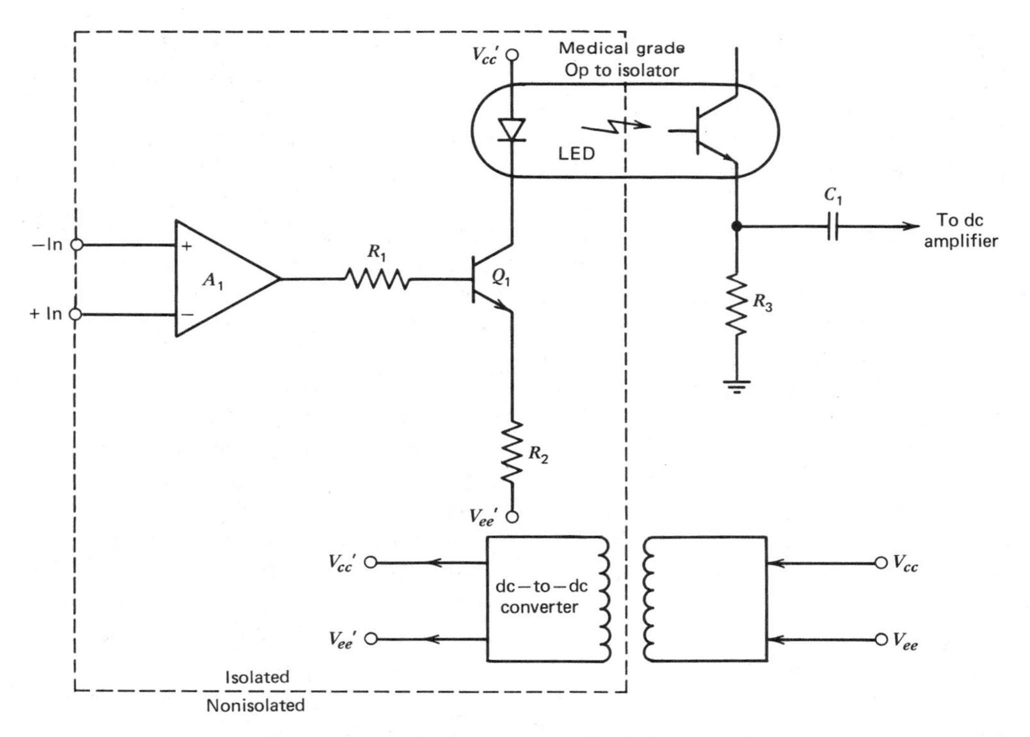

Figure 4-48 Optically coupled isolation amplifier.

an operational amplifier. This circuit illustrates use of JFETs to improve the input impedance of an operational amplifier.

The output of A_1 is connected to the isolated $V_{ee'}$ (i.e., -10 V dc power supply through resistor R_7). This power supply is a dc-to-dc converter operating at 250 kHz. Transformer T_1 provides isolation between the floating power supplies on the isolated side and the normal mains-powered supplies on the nonisolated side.

An input signal (i.e., an ECG waveform) causes the output of A_1 to vary the loading on the floating power supply ($B_1/C_1/C_2$), through resistor R_7.

Changing the loading on the floating power supply proportional to the analog signal causes variation of the T_1 primary current that is *also* proportional to the analog signal. This current variation is converted to a voltage waveform by amplifier A_2. An *offset null* control (R_{11}) is provided in the A_2 circuit to eliminate the offset at the output due to quiescent current flowing when the analog input signal is zero. In that case the loading on T_1 is constant.

Having discussed the basic principles of isolation, let's look at how it is implemented in some commercially available components. An optically coupled iso-amp in a single-wide, 18-pin dip package appears in Figure 4-50. This component accepts an input current, generated by a voltage source through R_{in}, and produces an output voltage across an isolation barrier. Notice that there are two optical paths. One is feedforward across the isolation barrier from the LED (light-emitting diode) to the output photodiode light detector. The other is negative feedback from the LED back to the input photodiode.

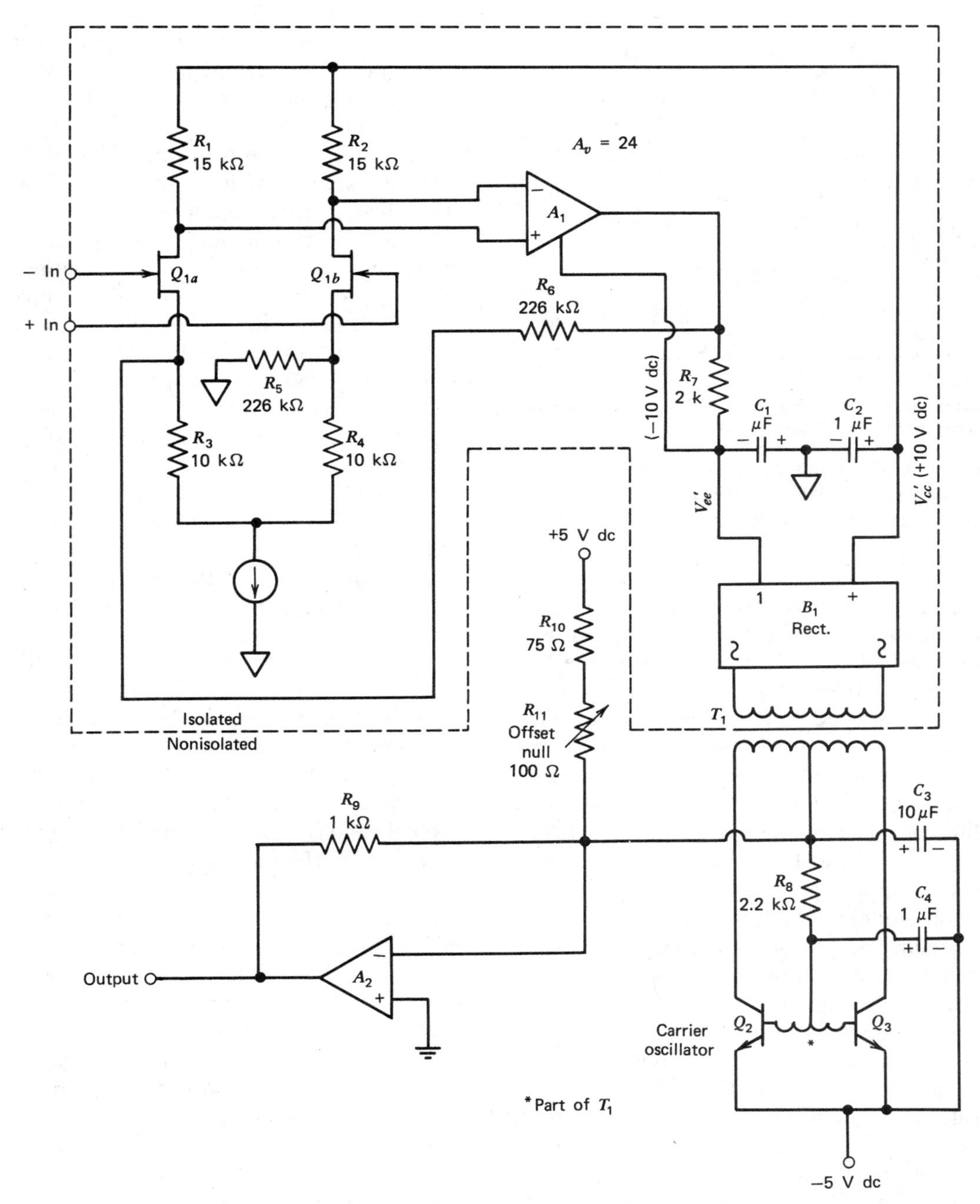

Figure 4-49 Current-loading type of isolation amplifier.

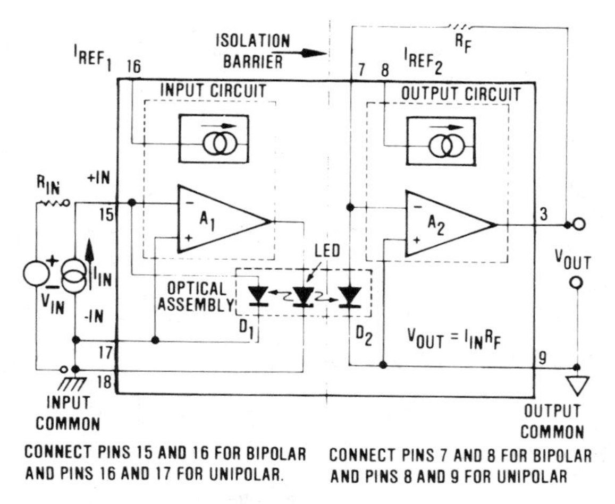

Figure 4-50 Single package optically coupled isolation amplifier, ISO100. (Courtesy of Burr-Brown under copyright 1987 Burr-Brown Corporation)

The latter linearizes the response of the LED in the input stage, which normally has quite a curve. That is, light output from the LED on its own does not exactly double when the voltage applied to it doubles. Interestingly, the two paths actually operate in a ratio fashion. This technique preserves accuracy as the LED slowly loses light intensity over its lifetime. The transfer function (V_{out}/I_{in}) depends on optical match rather than on absolute optical performance (LED to output photodiode). Laser-trimming improves matching and enhances accuracy. Discrete designs often struggle to achieve the good performance obtained from the single-packaged component.

Input and output current sources are used to provide offsetting for bipolar operation (plus and minus voltage swing around ground or common). This is necessary because LEDs and photodiodes are inherently unipolar. That is, there is no such thing as a "dark-emitting diode." The barrier of this device is tested at 2500 V dc, and it is rated at 750 V dc. Many standards such as AAMI, UL, and IEC recommend this type of testing at two times rated voltage plus 1,000 in volts.

Some biomedical applications, like defibrillator-protected ECG machines, require very high voltage ratings (e.g., 5000 V dc). However, some can live with much lower rating, like in EEG (brain waves) or industrial/home heart rate monitoring. The low leakage current of about 0.15 microamps at 120 V_{rms} provided by the optically coupled component shown in Figure 4-50 works well in such applications, even though the barrier rating is only 750 V dc. AAMI, UL544, and IEC601 Standards require that leakage be limited to less than 10 microamps through the patient at all times.

Another commercially available iso-amp is shown in Figure 4-51. This is a transformer coupled device containing a signal path and internal isolated power. It is housed in a low-profile SIL (single in-line) package. We say that this iso-amp is self-powered because it has its own 25-KHz oscillator, rectifier bridge, and filter capacitors. It also provides some external power on the isolated side. Signals get across the barrier as follows. The input op-amp drives an amplitude modulator (AM) which, on the other side of the barrier, is synchronously demodulated to minimize noise and interference. This iso-amp is rated at 750 V_{rms} continuous at 60 Hz and has 1-microamp leakage at 120 V_{rms}. It is suitable for low-barrier voltage medical and industrial applications, where low leakage current is necessary for patient or operator safety.

When medical circuits must withstand high defibrillation voltages, iso-amp barriers need to be specially designed. The barrier of the modern iso-amp shown in Figure 4-52 is constructed with high-voltage, differential ceramic capacitors. It is tested at 7900 V_{peak} and rated for 5000 V_{peak} continuous. It uses two 1-pF caps formed by firing tungsten on green ceramic. Hence, the barrier becomes part of the package. In fact, the barrier and

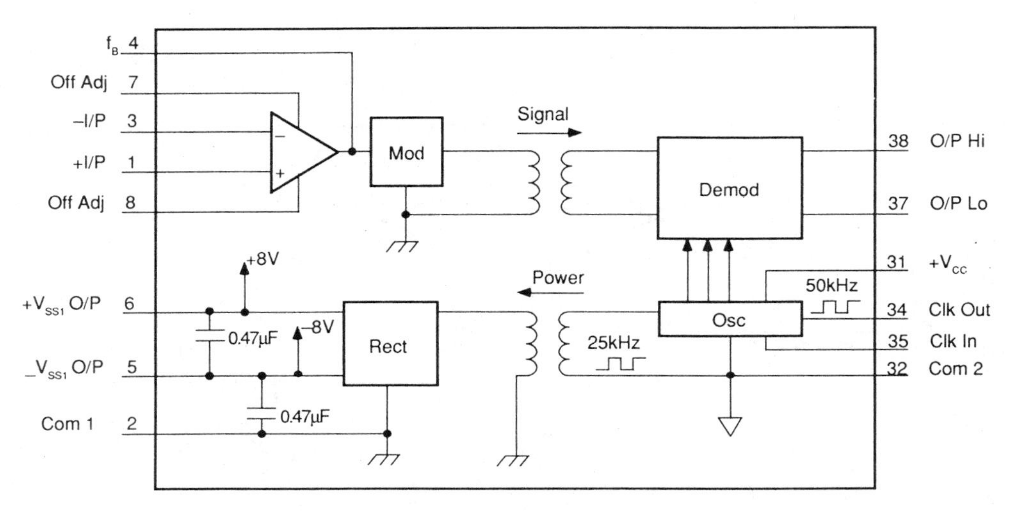

Figure 4-51 Single package transformer-coupled isolation amplifier, ISO212. (Courtesy of Burr-Brown under copyright 1989 Burr-Brown Corporation)

the electronics located on the ends of the 40-pin DIP are hermetically sealed. This makes the iso-amp internally immune to moisture, which improves reliability. The input is first duty-cycle modulated, then formed into opposite phase pulses, and transmitted digitally across the barrier. The small barrier caps essentially differentiate the edges of the digital signal to form spikes. The trick in recovering the digital signal is to differentially amplify the tiny spikes. This is done by the sense amp, which drives the output demodulator. The demodulator is just a low-pass filter with a sample-hold amplifier in the feedback to further remove ripple. Because the modulation is digital, the barrier characteristics do not affect signal integrity. This results in good high-frequency transient immunity across the barrier, which is important in rejecting high-frequency iso-mode noise.

An important consideration in isolation amplifiers is the frequency effect on isolation-mode rejection. Figure 4-53 shows that, for the iso-amp in Figure 4-52, the 1-Hz isolation-mode rejection (IMR) is 150 dB, and at 60 Hz it is 115 dB. This means that a 60-Hz sinewave appearing across the barrier will be attenuated at the output by 562,000 times ($1/\log^{-1}(115/20)$). Remember IMR = $20 \log_{10} (V_{\text{iso}(60\text{ Hz})}/V_{\text{out}(60\text{ Hz})})$ or IMR = $-20 \log_{10}(V_{\text{out}(60\text{ Hz})}/V_{\text{iso}(60\text{ Hz})})$. Only 0.0002 percent 60-Hz residual gets through. But what happens at higher frequencies? Most iso-amps leak heavily at high frequency due to barrier capacitance. This is why the component in Figure 4-52 is built with such low barrier capacitance (2 pF total). From the typical performance curve in Figure 4-53, it appears that the IMR at 10 kHz is 70 dB. At 300 kHz it is still 40 dB. While this is relatively good, it may not be high enough to reject electrosurgery unit (ESU) interference, which contains frequencies at several MHz up to 10 MHz. Other modern rejection techniques must used to get rid of ESU noise, including fancy input and barrier filtering.

Noise rejection is obviously important in isolation amplifiers. However, even more important is *high-voltage testing*. An isolation

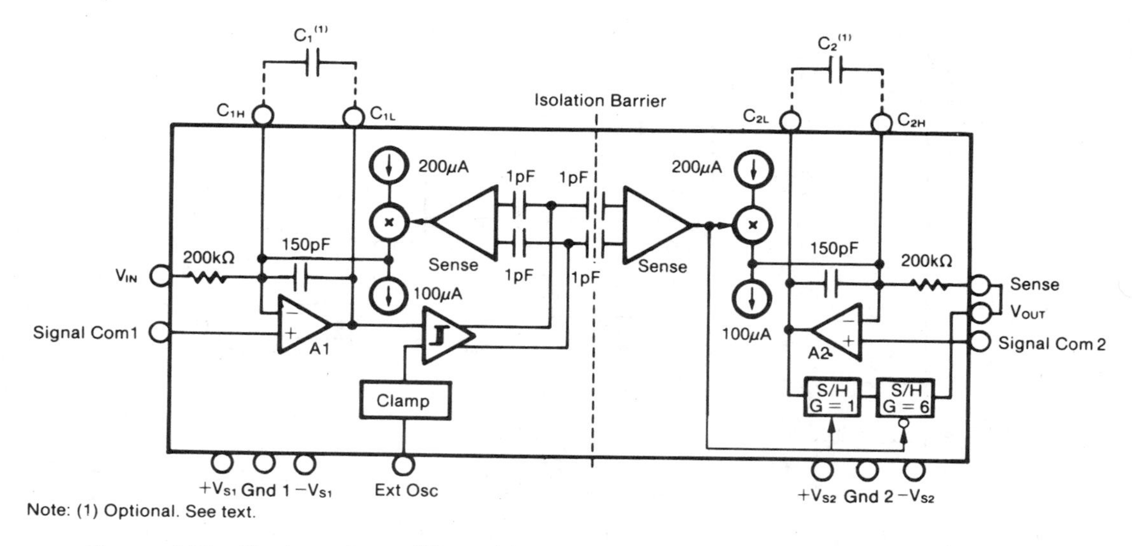

Figure 4-52 Single package differential-capacitative isolation amplifier, ISO121. (Courtesy of Burr-Brown under copyright 1990 Burr-Brown Corporation)

voltage rating describes the voltage a device can reliably withstand for long periods of time. Some iso-amps are stress tested first at a high voltage for short periods of time, and then functionally tested at its rated voltage for a longer fixed period of time. If the iso-amp survives, it is good. A new method has been established in the industry to assure higher barrier reliability. It is called *partial discharge testing*. It tests for internal microscopic voids in the barrier. Such defects display localized ionization when exposed to high voltage (HV). That is, intense electric fields build up across the voids, and at some point, called inception voltage, charges flow. At this point the void shorts itself out. As the barrier voltage decreases, charge flow stops. This is called the *extinction voltage*. This action redistributes charge within the barrier and is known as *partial discharge*. Today specialized production equipment can test for discharge under HV conditions. First, a one-second rated voltage test is conducted which looks for leakage current. Second, another one-second test checks for partial discharge at 1.6 times rated voltage per the VDE 0884 German standard. This takes into account the ratio of transient voltage to continuous rating. If partial discharge is less than 5 picocoulombs, the barrier has high integrity. In fact, this is the best test of all,

Figure 4-53 Important consideration: Frequency effect on isolation-mode rejection (IMR), ISO121. (Courtesy of Burr-Brown under copyright 1990 Burr-Brown Corporation)

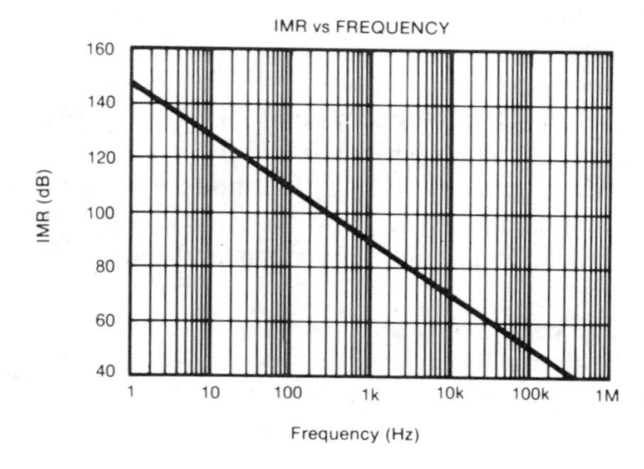

because it insures that the barrier is not being damaged by excessive partial discharge. The iso-amp shown in Figure 4-52 is 100 percent partial discharge tested.

Another important consideration in isolated systems is power supply feedthrough noise and carrier ripple reduction. Figure 4-54 shows how an *L-C* pi filter on the iso-amp's power supply pins attenuates the noise before it gets into the iso-amp. The noise is actually coming from the switching power supply oscillator, which is part of the isolated dc-to-dc converter. While the frequency is much higher than, say, a 100-Hz physiological signal, it can be rectified and filtered inside the iso-amp to cause a dc offset. What about carrier noise? Notice the active op-amp filter at the output. It removes noise associated with the carrier frequency, which is part of the modulation scheme inside the iso-amp. Keeping noise low in isolated bioelectric amplifiers is a never-ending battle.

So far we have looked at single-component optical, magnetic, and capacitive type isolation amplifiers. There is one more isolation technique which is actually the best of all. It is fiber-optic isolation, as shown in Figure 4-55. Here a transducer or perhaps a human body signal is first amplified by an instrumentation amplifer. It is then applied to a voltage-to-frequency converter (VFC). The signal information is now contained in frequency-modulated form and the actual voltage levels are digital. This is perfect for optical transmission. But this time the fiber-optic transmitter (FOT) energizes an LED and drives light down a fiber-optic cable. Since the glass or plastic cable responds only to light, it is not subject to electromagnetic interference (EMI). It can also withstand hundreds of kilovolts to provide the best

Figure 4-54 Important consideration: TT filter to minimize power supply feed-through noise and output two-pole filter to remove 500-kHz carrier ripple. (Courtesy of Burr-Brown under copyright 1991 Burr-Brown Corporation)

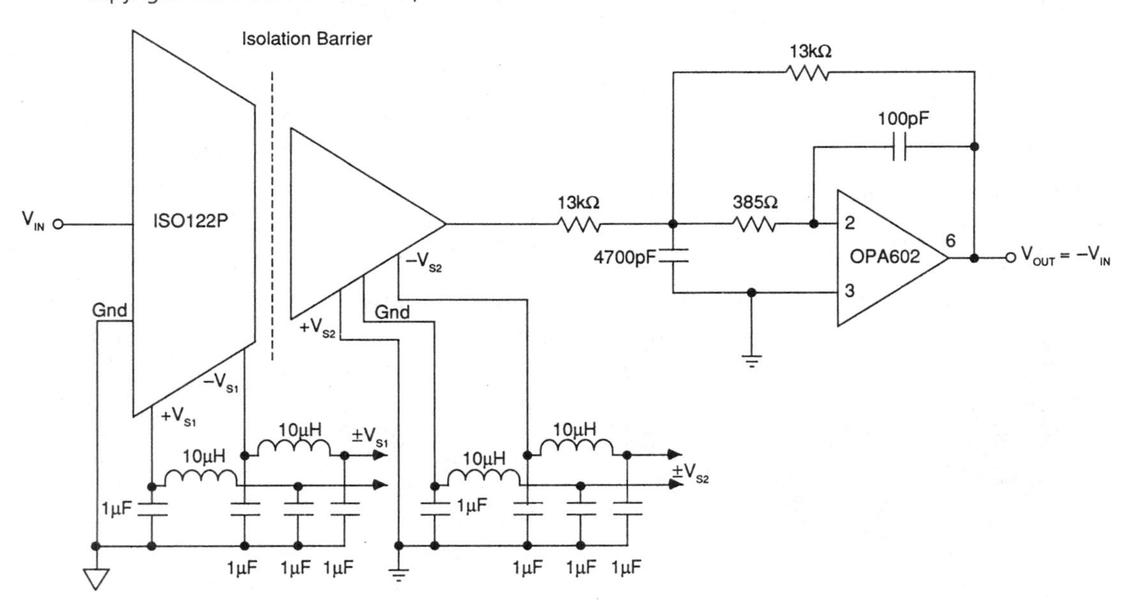

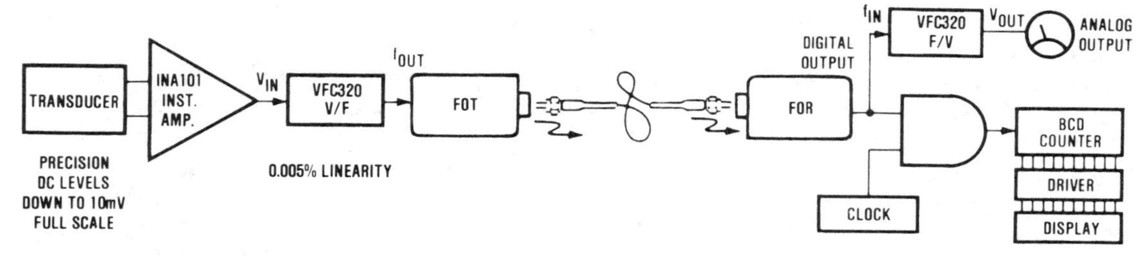

Figure 4-55 Fiber-optic isolation amplifier using voltage-to-frequency converter eliminates power supply feedthrough and noise coupling. (Courtesy of Burr-Brown under copyright 1988 Burr-Brown Corporation)

isolation imaginable. It also makes a wonderfully secure system, because fiber-optic cables cannot easily be tapped into. The fiber-optic receiver (FOR) recovers the electrical digital signal and feeds it to a counter, where it is turned into a clocked digital word of, say, 12 bits. This system is actually an analog-to-digital (A/D) converter that has been partitioned into a transmitter portion and digital recovery portion with isolation in between. Its drawback is that the VFC runs relatively slow (< 100 kHz), and hence it takes some milliseconds to complete an A/D conversion.

Another way to transmit an isolated signal after it has been digitized is to use a fiber-optic limited distance modem like the one shown in Figure 4-56. A modem (modulation-demodulation device) uses standard protocol to send and receive digital data. Transmit (T) and receive (R) connectors accept fiber-optic cables. This particular modem is powered by the host RS-232 port and provides complete EMI/RFI rejection, elimination of ground loops, reduced error rate, and data security against tampering.

Figure 4-56 Fiber-optic isolated limited-distance modem. (Courtesy of Burr-Brown under copyright 1986 Burr-Brown Corporation)

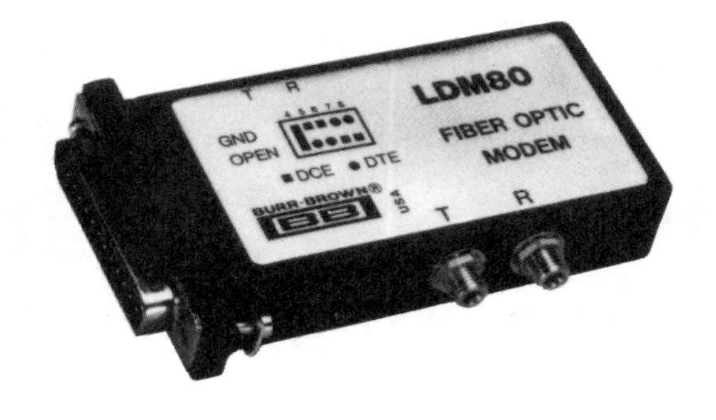

As we have shown, isolation applications can take many forms. Many bioelectric amplifiers use isolation to protect the patient. Other amplifiers use isolation around the hospital or medical environment to protect operators and expensive equipment.

Figure 4-57 shows an isolated bioelectric ECG amplifier. Notice that the NE-2H gas-filled neon tubes, series resistors, and clamp diodes protect the instrumentation amplifier (IA) from defibrillation voltages up to 5000 V_{peak}. This protection is not part of the isolation. It prevents high voltages from destroying the electronics only on the patient side. It clamps to the input common, not the output common. Before looking at how the isolation works, we will briefly discuss ECG front-end characteristics.

The commercially available IA shown derives the common-mode (CM) 60-Hz voltage internally and applies it to an inverting amplifier which drives the patient's right leg. This acts in a feedback loop (patient and electronics) to servo or drive the CM noise on the patient to a low level. While the 60-

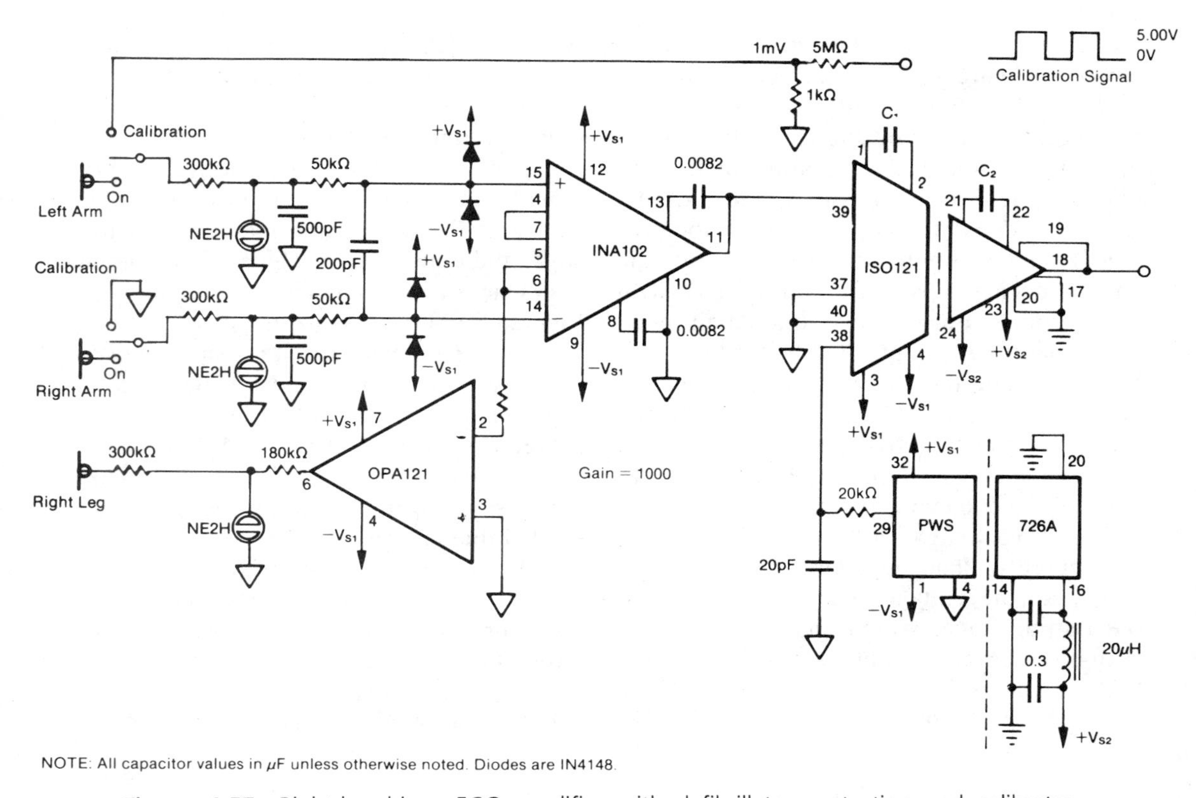

Figure 4-57 Right-leg-driven ECG amplifier with defibrillator protection and calibrator. (Courtesy of Burr-Brown under copyright 1990 Burr-Brown Corporation)

Hz noise on the left or right arm may become bigger with respect to the right leg, the 60-Hz noise on the IA input terminals (pins 15 and 14) with respect to IA common (pin 10) becomes smaller. Hence, the IA does not have to reject as much 60-Hz noise. The IA's common-mode rejection (CMR) then reduces noise even further. The isolation amplifier's isolation-mode rejection (IMR) also reduces noise by establishing 10^{12} Ω and 9 pF between the patient (pin 37) and earth ground (pin 17), as shown in Figure 4-57. Remember, barrier impedance is the parallel combination of the isolation amplifier and the isolated dc-to-dc converter. Isolation acts to attenuate noise (say, 1 V_{p-p} at 60 Hz) that is trying to mix with the low-level ECG input signal (say, 1 mV_{p-p}). Three actions reduce 60-Hz interference dramatically. They are right leg drive, IA CMR, and iso-amp IMR. A low-pass filter and/or 60-Hz notch filter following the iso-amp will further shrink noise. This circuit is dc coupled and is in high gain = 1000 V/V. It only works when the electrode offset potentials are low (< 10 mV) to avoid IA output saturation. DC restoration, covered in Chapter 5, allows much larger offsets to be tolerated.

Remember, this discussion is dedicated to explaining the uses of isolation in health care environments, so let's ask an interesting question. Is the only purpose of isolation to reduce 60-Hz noise pickup in the ECG example of Figure 4-57? The answer is no. The iso-amp also stands ready to protect semiconductor circuitry on the iso-amp's output when a high-voltage defibrillation pulse is applied to the patient. Actually, most of the time HV does not appear across the iso-amp's barrier. It only appears when, say, the patient accidentally comes in contact with earth ground. For example, if a patient lying on an electrically controlled bed is being defibrillated, and the patient's arm touches the metal bed rail, the "defib" pulse will be impressed directly across the iso-amp. In this situation the barrier must be able to withstand about 5000 volts peak. Fortunately, the barrier is very high impedance compared to the source or load impedance, and it drops nearly all the voltage. Very little is felt on the output stage, which shares a common with other circuitry hooked to it. Isolation in this case saves the electronics. It has nothing to do with protecting the patient from shock. After all, defibrillation is used to intentionally shock the patient to restart the heart.

What about other amplifiers that use isolation around the hospital or medical environment? They may not strictly be called bioelectric amplifiers, but they do amplify signals while protecting operators, medical staff, and expensive machines, like computers. One such isolated circuit is shown in Figure 4-58. It is an isolated power line monitor. An IA measures the small voltage across a 5-mΩ power resistor. This voltage represents the current in the *Y*-connected power transformer often used in hospitals. The low-bias current FET IA shown makes low-pass filtering with large resistors and a small value capacitor easy.

Why is isolation used for this application? The answer is noise reduction and protection under fault conditions. The iso-amp isolates (attenuates) the noisy transformer power ground from, say, an instrumentation or computer ground. If the iso-amp's output

Figure 4-58 Isolated power line monitor. (Courtesy of Burr-Brown under copyright 1990 Burr-Brown Corporation)

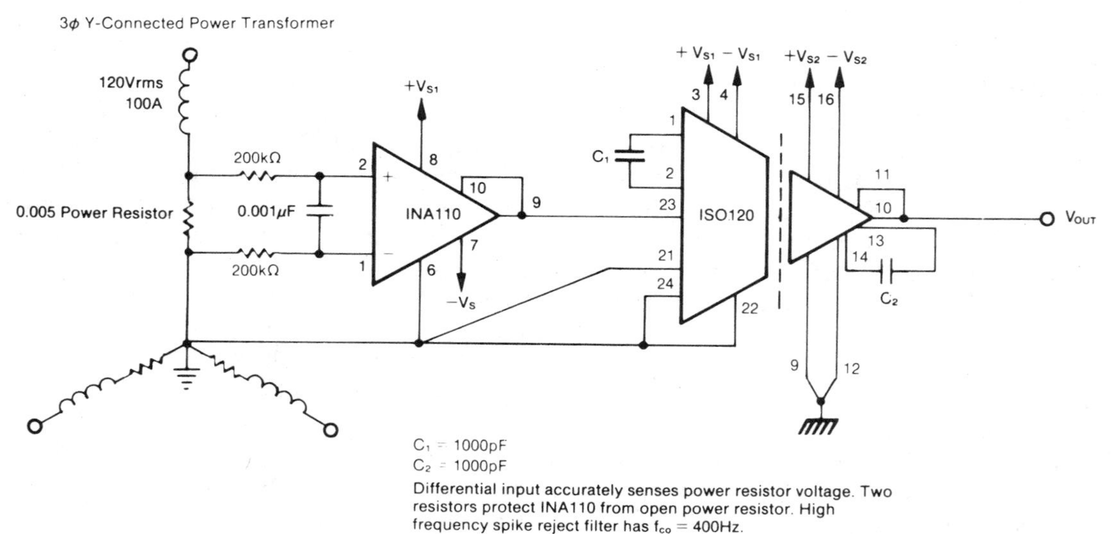

were digitized, a computer could accurately monitor transformer load current on a continuous basis. Should the power resistor burn open, the 120-V ac line voltage would not be impressed on the computer circuitry or a person touching the output. Isolation reduces noise to maintain clean measurements and protects humans from shock and equipment from damage should a fault occur.

A few more example applications will serve to solidify our understanding of isolation. In Figure 4-59, the thermocouple could be used to take a temperature measurement in a medical oven, autoclave, or boiler for instance. First let's discuss the front end and then examine why isolation is used. A thermocouple is a device made of two dissimilar metal wires spot welded together at one end. It produces a small voltage of, say, 5 mV in response to a temperature gradient across the wires. Table 4-1 shows materials used and how much voltage they produce for type E, J, K, and T thermocouples. Again the classical instrumentation amplifier is used to amplify the difference between the thermocouple voltage and a diode voltage which is mounted to an isothermal block. This technique is called *cold junction compensation.* The diode establishes a reference to the ice point or zero degrees Centigrade. The diode voltage changes with ambient air temperature and, hence, compensates for changes in ambient temperature of the electronic amplifier. No matter what the ambient goes to within reason, the thermocouple always takes the measurement relative to zero degrees Centigrade.

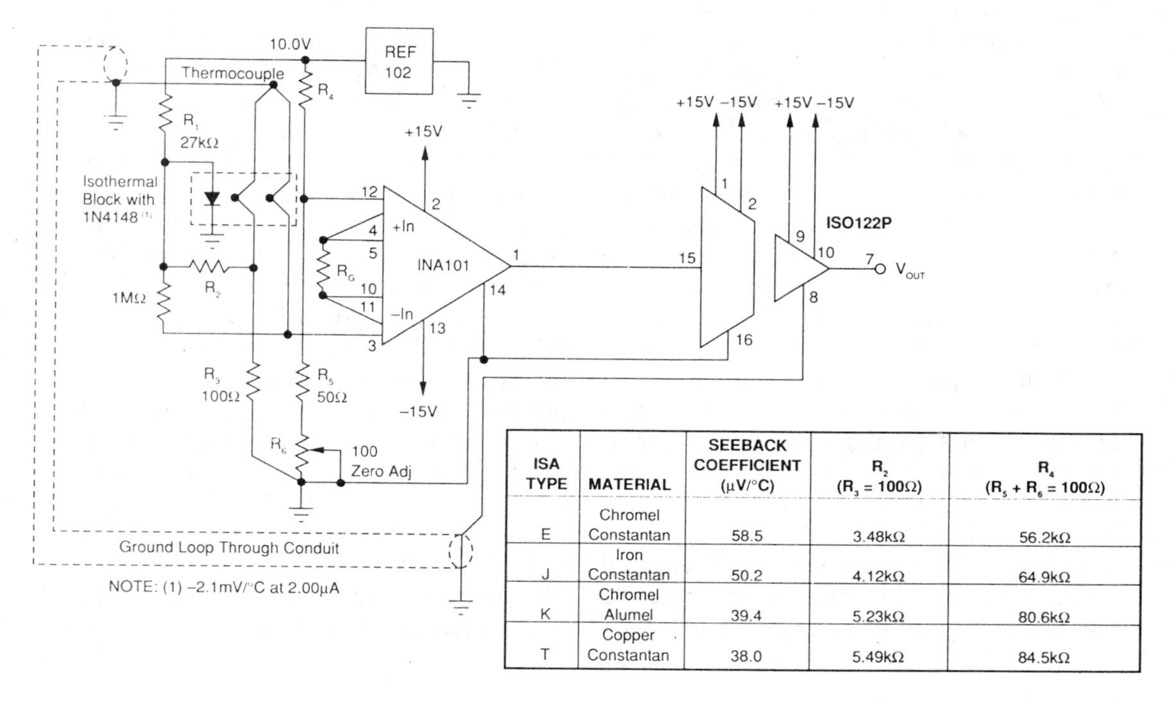

ISA TYPE	MATERIAL	SEEBACK COEFFICIENT (μV/°C)	R_2 (R_3 = 100Ω)	R_4 (R_5 + R_6 = 100Ω)
E	Chromel Constantan	58.5	3.48kΩ	56.2kΩ
J	Iron Constantan	50.2	4.12kΩ	64.9kΩ
K	Chromel Alumel	39.4	5.23kΩ	80.6kΩ
T	Copper Constantan	38.0	5.49kΩ	84.5kΩ

Figure 4-59 Thermocouple (temperature) amplifier with ground loop elimination, cold junction compensation, and upscale burnout indication. (Courtesy of Burr-Brown under copyright 1991 Burr-Brown Corporation)

Table 4-1

Thermocouple Type	Sensitivity (@ 20°C) μV/°C	Useful Range (°C)
Copper/ Constantan	45	−150 to +350
Iron/ Constantan	52	−150 to +1000
Chrome/ Alumel	40	−200 to +1200
Chromel/ Constantan	80	0 to 500

Now what purpose does isolation serve in Figure 4-59? It simply breaks the ground loop to reduce noise interference that can ruin the low-level sensitive temperature measurement. Although high voltage may not be present here, the thermocouple can be attached to a grounded metal heater or heated surface. This ground may be very "dirty" with respect to the instrumentation amplifier ground. That is, there may be large power or spike currents passing through the point where the thermocouple is attached. Therefore, a ground loop can be created through a conduit which eventually gets back to the input amplifier, as shown in Figure 4-59. Noise can then mix with the small temperature signal. So it should be obvious that the isolation amplifier with its high impedance prevents noise from interfering with the 1-mV differential input signal. Should the thermocouple become open, the IA's output will go to its positive P/S rail, around +12 V dc. This is called *up-scale burnout*. Isolation really has nothing to do with burnout, but the iso-amp must pass the overranged signal to allow, say, a computer to detect the fault condition.

Since we are on the subject of isolated temperature measurements, a final example will show a special way to transmit such measurements long distance. It is known as an isolated 4-mA to 20-mA current loop, as shown in Figure 4-60. These current loops pick up much less noise compared to sending voltages over long wires. Loops may be used without isolation, but when isolation is added the noise pickup drops even lower. In an electrically noisy hospital environment, this may be the only way to cost-effectively monitor a temperature remotely. Otherwise, special shielded wires and conduits have to be installed.

In Figure 4-60, the resistor temperature device (RTD) measures the temperature at its surface of, say, a medical incubator or perhaps a hospital room. The commercially available two-wire transmitter shown turns the temperature voltage into a current which circulates through a twisted-pair, two-wire loop on its way to the receiver. Here the current, 4 mA representing the minimum temperature and 20 mA the high temperature, is converted back into a voltage and applied to the iso-amp. The transmitter has two 1-mA current sources. One excites the transducer and creates a voltage drop across the RTD resistance. When temperature varies, RTD resistance varies, and the instrumentation amplifier inside the two-wire transmitter amplifies it. The other 1-mA current source provides a differential reference for the RTD. Notice that there is no power supply at the remote site of the RTD. Power is applied through the two-wire loop. In fact, the loop accepts power in one direction and delivers the signal in the other direction over the same wires. It is actually the power supply current in the loop that is changing in response to a changing temperature signal. Such isolated circuits allow modern biomedical systems to do a better job of measuring important temperatures.

The final example of isolation is one that does not really break a ground loop; it just adds a moderate resistance in series. We say

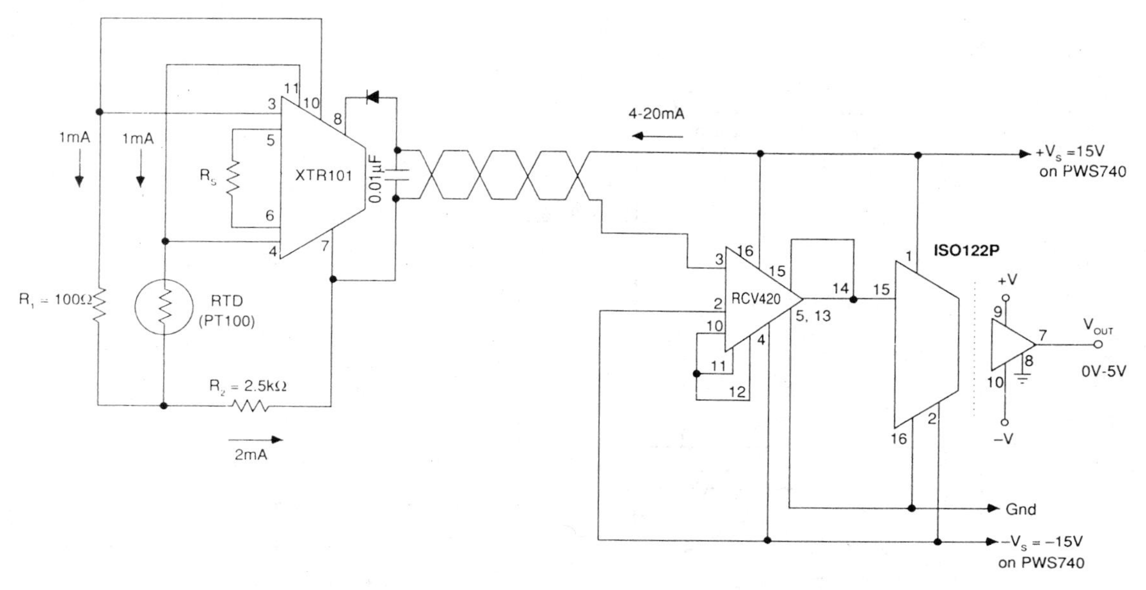

Figure 4-60 Isolated 4 mA to 20 mA RTD (temperature) instrumentation current loop. (Courtesy of Burr-Brown under copyright 1990 Burr-Brown Corporation)

that when a very large resistance is provided between two points in a circuit, they are isolated. But this is a matter of degree. The circuits shown in Figure 4-61 make the point. Here, a commercially available component takes a differential measurement, but its reference points (pins 1 and 5) are only about 400 kΩ away from either input (pins 2 or 3). Since normal semiconductors can handle around 20 V before breakdown, something special must be done to withstand 200 V. Here is what is done: The circuits in Figure 4-61 first attenuate the common-mode voltage on each input by a factor of 20. Two hundred volts are reduced to an acceptable 10 V. Then the reduced voltage is gained up by a factor of 20. The result is a unity gain differential amplifier with a unique characteristic. With 400 kΩ of input resistance, this circuit handles 10-V differential signals riding on as much as 200 V maximum.

Do the circuits in Figure 4-61 replace classical isolation amplifiers? The answer is sometimes. There is not enough isolation resistance to make such circuits suitable for ECG applications, but they are great for applications like monitoring battery cells. Here +200-V and −200-V power supplies are employed to charge, say, thirty 12-V lead-acid batteries. Uses range from supplies in portable X-ray equipment to emergency power backup in hospitals, for example. Battery cells near ground do not need a 200-V breakdown diff-amp, but cells near the maximum supplies require some type of amplifier like this. The multiplexer can select any channel, representing any battery cell, which can be sent to an A/D converter, digitized, and stored in a computer.

This completes the discussion on isolation amplifiers. From this presentation, the biomedical engineer and technician should now understand their varied use in medical environments.

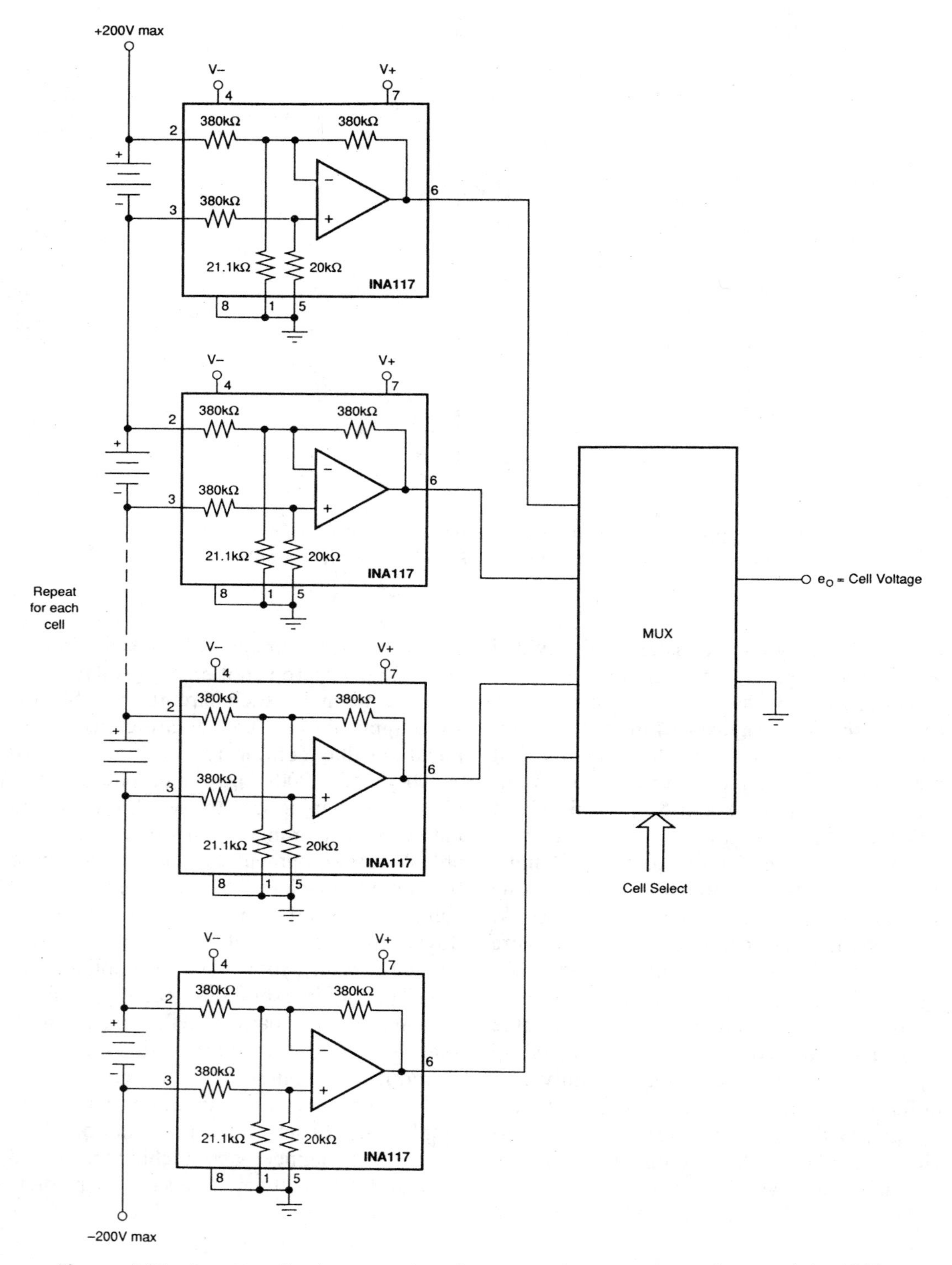

Figure 4-61 Battery cell voltage monitor. (Courtesy of Burr-Brown under copyright 1990 Burr-Brown Corporation)

4-12 Chopper Stabilized Amplifiers

Two problems arise when trying to record low-level biopotentials (i.e., EEG recording of brainwaves). These are *noise* and *dc drift*. Both these problems are made worse by the high-gain amplifiers needed to increase the very weak biopotentials to a readable level.

Noise is generated by almost every part of the recording apparatus including the patient's body; but the worse offender is the noise contribution of the amplifier itself!

Drift is the change in gain or dc offset (i.e., baseline) due to thermal effects on the amplifier components. Drift may be substantially reduced through the use of large amounts of negative feedback in an ac-coupled amplifier. The problem to be solved is to convert a dc (or near-dc, low-frequency analog) signal to an ac signal that will pass through the amplifier.

The solution is to sample, or *chop*, the analog signal at a frequency that *will* pass through the ac-coupled amplifier. Although most biomedical chopper amplifiers use a 400-Hz excitation signal for the chopper, there are also some models that use 60, 100, and 1000-Hz chopper frequencies.

An example of a simple chopper amplifier is shown in Figure 4-62*a*. The chopper is a vibrator-driven SPDT switch that grounds the amplifier input and output terminals on alternate swings of the switch.

The chopper vibrator coil is excited by a 400-Hz ac carrier signal. Figure 4-62*b* shows the analog waveforms for both the original and chopped versions. Only the chopped version of the signal will pass through the ac amplifier.

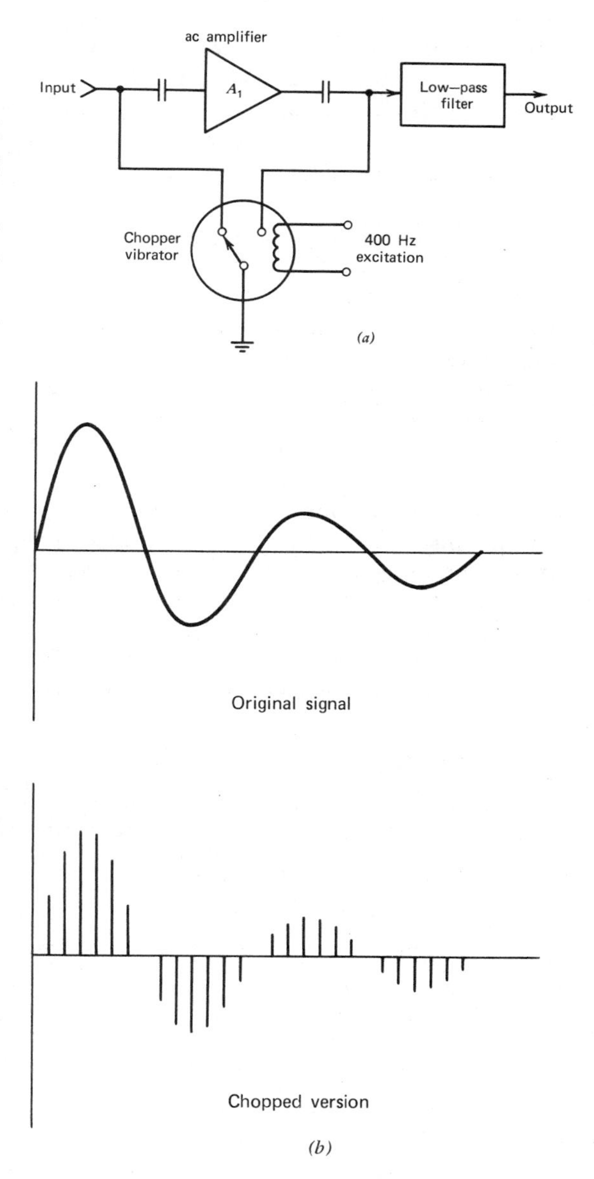

Figure 4-62 Chopper amplifier. (*a*) Simplified circuit. (*b*) Comparison of original and chopped waveforms.

The chopper technique not only gains stability from the ac-coupled amplifier but also provides low-noise operation. The sampling rate, by itself, tends to act as a *low-pass* filter for externally generated noise; although it puts some noise of its own on the system. Most manufacturers further limit the noise by making the ac-coupled amplifier a *band-pass amplifier* that passes only the narrow range of frequencies around 400 Hz. The *rms* value of the noise signal in any given system is proportional to (among other factors) the square root of the circuit bandwidth. By limiting the bandwidth, we also limit the noise amplitude.

A differential-input chopper amplifier is shown in Figure 4-63. In this circuit the chopper is on the input circuit only. Input transformer T_1 is connected so that its center tap becomes one terminal of the input connector, while the two winding extremities are connected to the chopper. The pole of the chopper switch becomes the other terminal of the input connector.

Most of the gain in this circuit is provided by ac-coupled amplifier A_1. The signal remains a chopped version of the input waveform until it is applied to the synchronous demodulator, where it is detected and filtered to recover the original waveshape.

The chopper amplifier technique of Figure 4-63 is typically used in EEG amplifiers and those universal bioelectric amplifiers that have gains in the $\times 1000$ and over range.

4-13 Input Guarding

Physiological signals tend to have low-level amplitudes. The -80-mV action potential is a powerhouse compared to the 1 mV of the ECG waveform and the 50 μV of the EEG waveforms.

In most cases the physiological signal of interest is accompanied by large common-mode signals. It is frequently the case that several hundred millivolts of 60-Hz signal will be coupled into the input cables of a bioelectric amplifier designed to sense and amplify low-level physiological signals.

This situation is shown in Figure 4-64. The circuit is shown in Figure 4-64*a*, where we have a differential amplifier connected to a differential signal source through a shielded

Figure 4-63 Differential chopper amplifier.

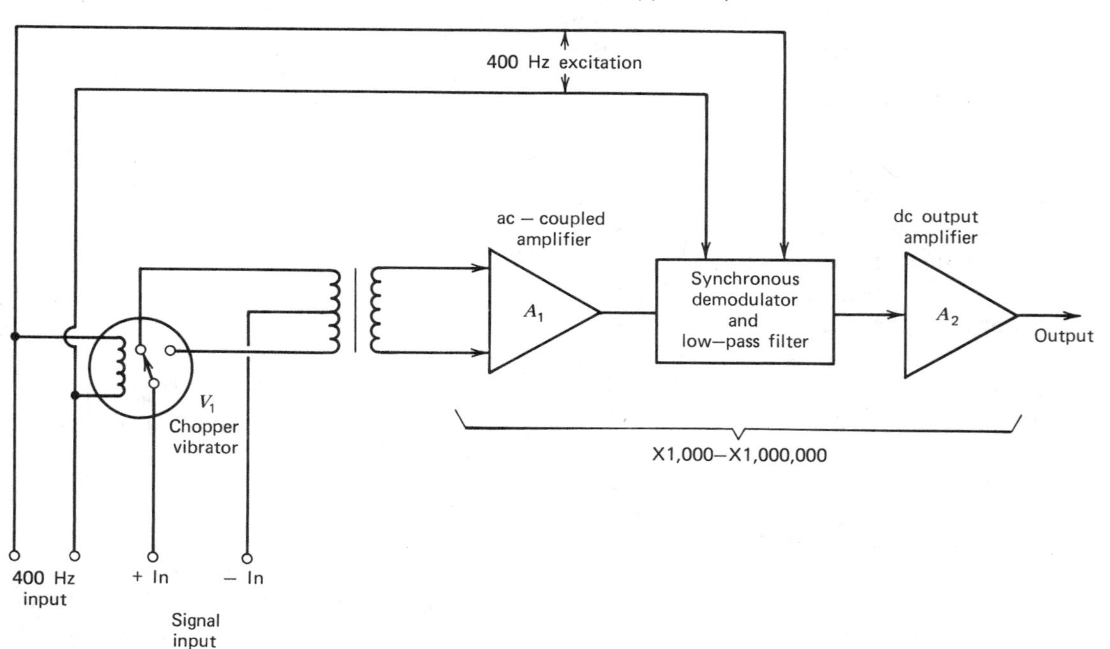

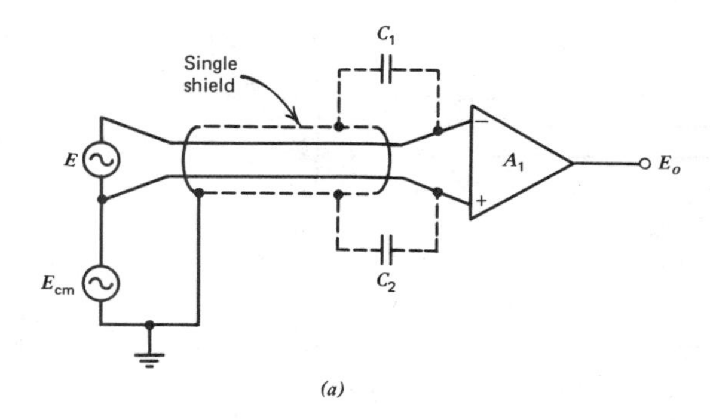

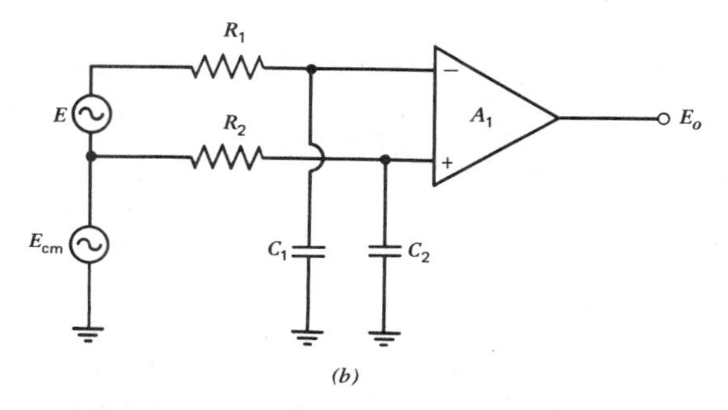

Figure 4-64 Bioelectric amplifier input circuitry. (*a*) Actual circuit. (*b* Equivalent circuit.

cable. Both differential (E) and common-mode (E_{cm}) signals are present. If the usual practice of grounding the shield is followed, then the equivalent circuit of Figure 4-64*b* is obtained. Resistances R_1 and R_2 are the sum of the cable resistances and the signal source output impedance. Capacitors C_1 and C_2 represent the capacitance of the shielded cable.

The networks R_1/C_1 and R_2/C_2 are of little consequence provided that $R_1 = R_2$ and $C_1 = C_2$, but if these equalities are not maintained the inputs become unbalanced to ground. Under that condition the circuit may *manufacture* a differential component from a common-mode signal. The amplifier will be unable to distinguish the artifact from the real signal because *both* are differential signals.

The solution to this problem is to use *input guarding,* as shown in Figure 4-65. The main concept here is to place the *shield* at the common-mode signal potential, which in effect places both sides of C_1 and C_2 at the same potential for common-mode signals.

Most bioelectric amplifiers use only single-shielded input cables, but Hewlett-Packard uses a double-shielded input as shown in the example. The H-P series 8800 bioelectric amplifiers are equipped with an input connector that separates guard and chassis grounds so that the circuit of Figure 4-65 can be used. The outer shield is especially useful where high-intensity common-mode interference is expected.

The drive source for the guard shield is derived by summing the +IN and −IN signals in the R_1/R_2 resistor network. In some cases the summation junction is connected directly to the guard shield, while in others it is passed through a unity gain buffer amplifier (A_2).

In ECG amplifiers the right leg of the patient is designated as the *common,* so the guard shield may be connected to the right leg.

Figure 4-66 depicts a commercially available instrumentation amplifier and single-ended op-amp used as the shield driver. Here the transformer-coupled signal is brought to the IA over two-wire shielded cable. The 60-Hz noise, equally cutting across each wire through the shield, tends to induce voltages. In fact, the distributed cable capacitance tries to charge and discharge in response to 60 Hz. This allows interference to take place. To minimize this effect, the shield is driven with the common-mode voltage. The capacitance is actually made up of metal wires as one plate, insulation as the dielectric, and metal shield as the other plate. By driving the shield with the same noise voltage that is induced on the wires, this parasitic capacitance cannot charge or discharge. It is effectively nulled out. This technique goes a long way toward reducing noise pickup in

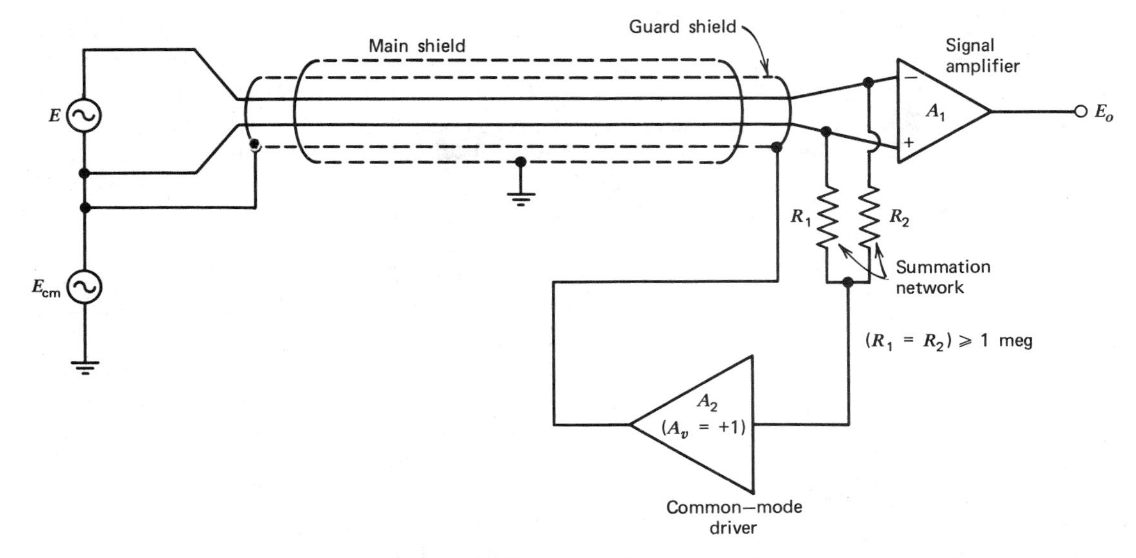

Figure 4-65 Guarded input circuit.

cables. It is better than just grounding the shield. Isolation would not help here, because the noise is on the input with respect to the input common. The shield cannot be isolated. The shield and wires it surrounds are intended to be part of the same circuit.

So far we have examined many different types of bioelectric amplifiers handling sig-

Figure 4-66 Amplification of transformer-coupled signal with shield driver. (Courtesy of Burr-Brown under copyright 1988 Burr-Brown Corporation)

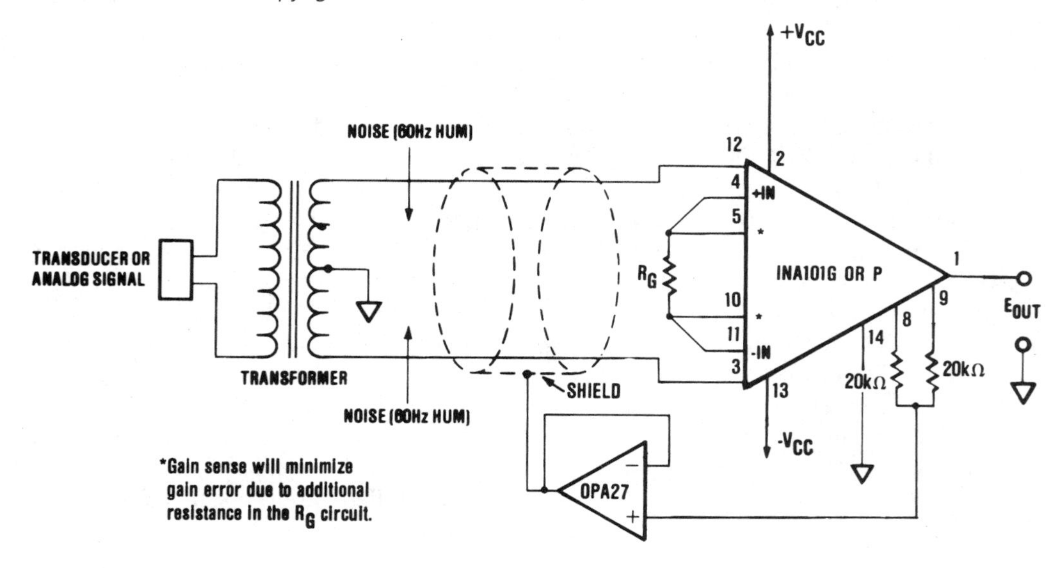

nals from a variety of transducers. Some signals were biophysical, arising directly from the human body. Others were from temperature and pressure Wheatstone bridges. Still others were from separate temperature, optical, and chemical devices. We have looked at filters and peak detectors, flow and weighing scale transducers. We have also discussed log amplifiers, instrumentation, and isolation amplifiers. With this working knowledge, it is easy to see how powerful these modern circuits have become in medical equipment.

The remaining circuits shown in this chapter are also commonly used in biomedical instrumentation. An understanding of them will expand one's view of how medical instrumentation works and how to troubleshoot problems.

First is the current source. Figure 4-67*a* shows the basic constant current source consisting of a low-resistance voltage source in series with a large resistor. The current is constant because, as the load varies from, say, 0 Ω to 10 kΩ, the current does not vary much. This is due to the fact that 10 kΩ is very small compared to 10 MΩ. The constant current source in Figure 4-67*b* is better, because the transistor makes the series resistance 100 MΩ instead of 10 MΩ.

Figure 4-68 shows how a current source is made from monolithic commercially available components. Notice the +10-V reference which, through R_1, produces a constant current into the load. The load, of course, is a much smaller resistance than R_1. The voltage follower forces the reference ground to be equal to the load voltage. This results in an accurate 10 V across R_1. Output current is easily calculated, I_{out} = 10 V/R_1.

Figure 4-69 shows a voltage-controlled current source with differential inputs. It can

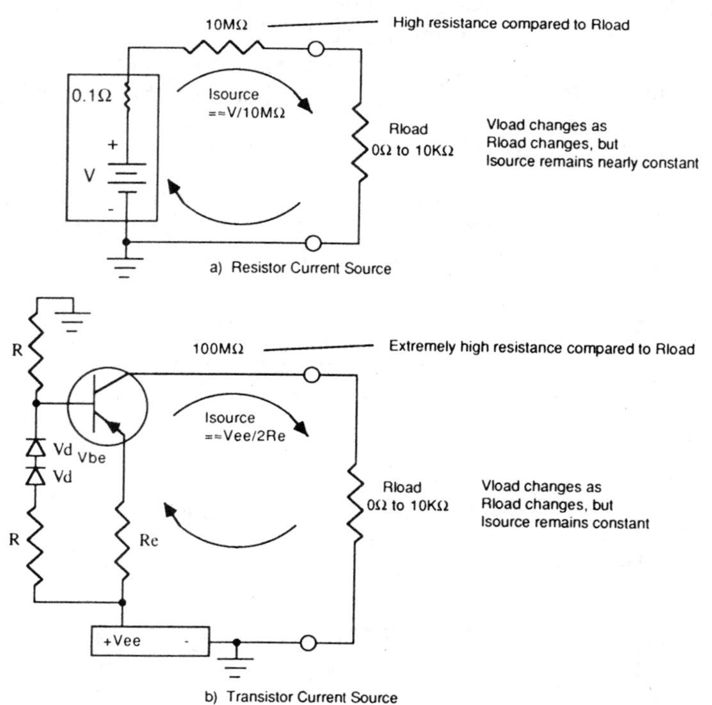

Figure 4-67 Basic current source.

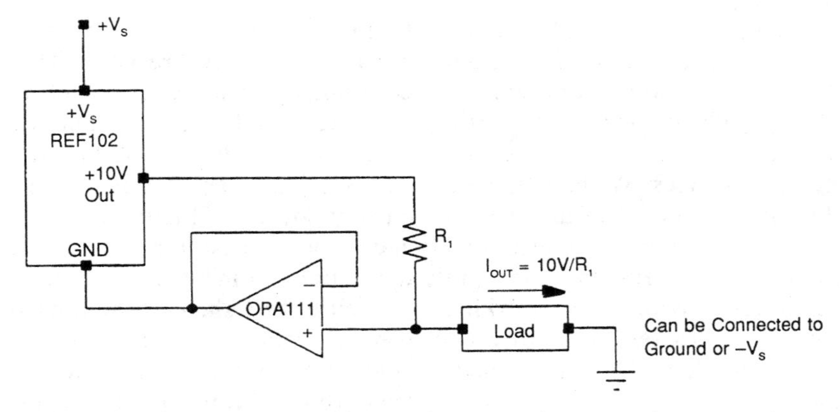

Figure 4-68 Current source using voltage reference and op amp. (Courtesy of Burr-Brown under copyright 1990 Burr-Brown Corporation)

accept a differential voltage and drive a proportional current. It is also very versatile, since grounding one input gives a noninverting or inverting transfer function. Notice that it is actually built from the basic diff-amp—that is, an op-amp plus four ratio-matched resistors. With the output connected as shown, this represents the modified Howland current pump.

Figure 4-70 shows a single chip (monolithic) dual 100-microamp current source component that is commercially available. It also has a current mirror, which is just a two-transistor circuit that takes a current, injected into one collector (pin 5), and reproduces the same current on the other collector (pin 4). This occurs because the base and

Figure 4-69 Voltage-controlled current source with differential inputs and bipolar output. (Courtesy of Burr-Brown under copyright 1990 Burr-Brown Corporation)

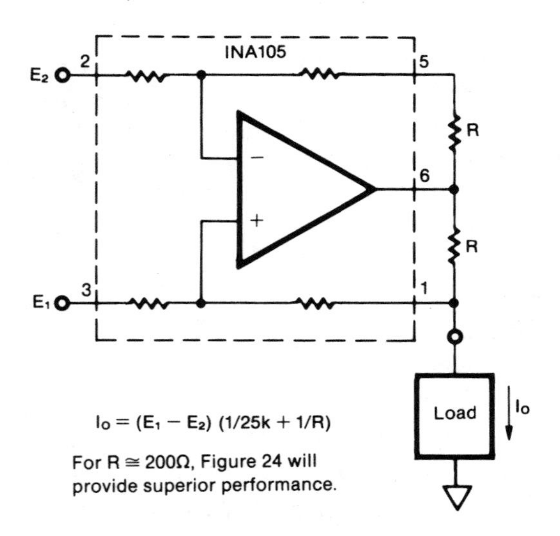

Figure 4-70 Single-chip (monolithic) dual current source with current mirror. (Courtesy of Burr-Brown under copyright 1990 Burr-Brown Corporation)

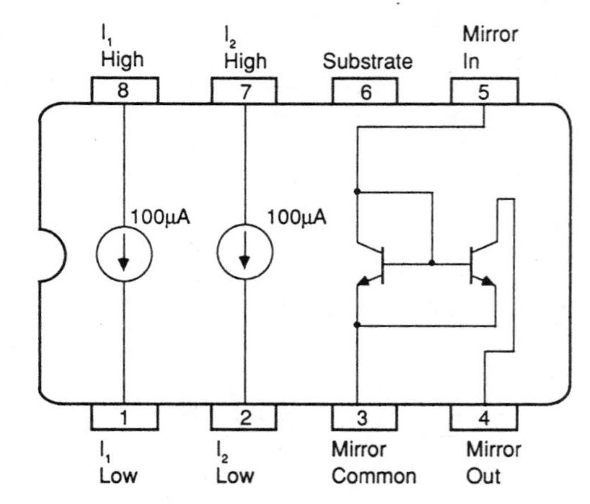

emitter currents are nearly equal (tied together), and the NPN transistors are matched. If I_b's and I_e's are equal, then I_c's must be equal. A current mirror is used extensively inside op-amps to reproduce biasing currents, which move together as the temperature changes. It can also be used in test equipment where the current from one circuit must be connected to several places, but loading effects must be eliminated. It delivers the same current while being impedance isolated from the original source.

How can current sources be used in biomedical equipment? One example is the slew rate limiter shown in Figure 4-71. Here a constant 100-microamp current drives a diode bridge. The 10-kΩ input and feedback resistors drive the left side of the bridge. The right side is connected to the inverting op-amp terminal. Virtual ground is maintained by the flow of current through the capacitor, *C*. But when that current exceeds 100 microamps, the bridge reverses, thus limiting the output slew rate (V/sec) to a value equal to 100 μA/C in farads. These types of circuits are very useful in ECG systems, where the slew rate of an artificial pacer pulse riding on the ECG signal must be limited and essentially removed. This is necessary to prevent successive amplifiers from saturating.

Figure 4-71 Slew rate limiter using current sources (REF200) and diode bridge. (Courtesy of Burr-Brown under copyright 1990 Burr-Brown Corporation)

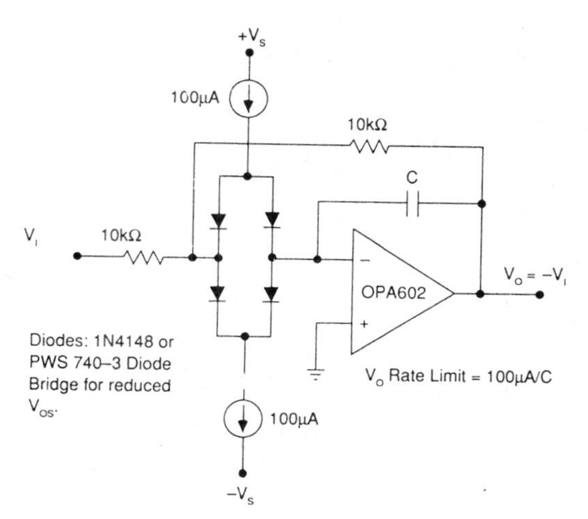

Another use of current sources is the measurement of switch on-resistance or perhaps patient cable resistance. In Figure 4-72, a floating current source is made of a commercially available current source (100 microamps), op-amp, and *N*-channel enhancement MOSFET (metal oxide semiconductor field effect transistor). This circuit can be used to test a switch. The on-resistance equals the voltage across the switch divided by I_o. I_o can be made as large as desirable [$I_o = (N + 1)$ times 100 microamps] to test the switch under actual current load conditions. Switches that are going bad change their on-resistance as current through them increases. The floating current source makes it easier to test switches while they are in the circuit. Sometimes, during troubleshooting, it is highly desirable to test switches in biomedical circuits.

Another circuit commonly used in biomedical systems is the sample-and-hold (S/H) amplifier, sometimes called track-and-hold. Figure 4-73 shows that it is composed of an input FET switch followed by a holding capacitor and noninverting amplifier. It works this way. When the sampling signal is at the ON-state (10 V), the input signal, V_{in}, charges the capacitor, *C*, through the on-resistance of the solid-state FET switch. When the sampling signal is taken to the OFF-state (−7.5 V), the input signal stops charging the capacitor. Whatever voltage was on the capacitor at that time will be held there. The only change will be due to the op-amp's bias current. This can cause droop or rise depending on whether the bias current is plus or minus. Biomedical instrumentation incorporates S/H amplifiers to make sure bioelectric signals remain constant during analog-to-digital conversion.

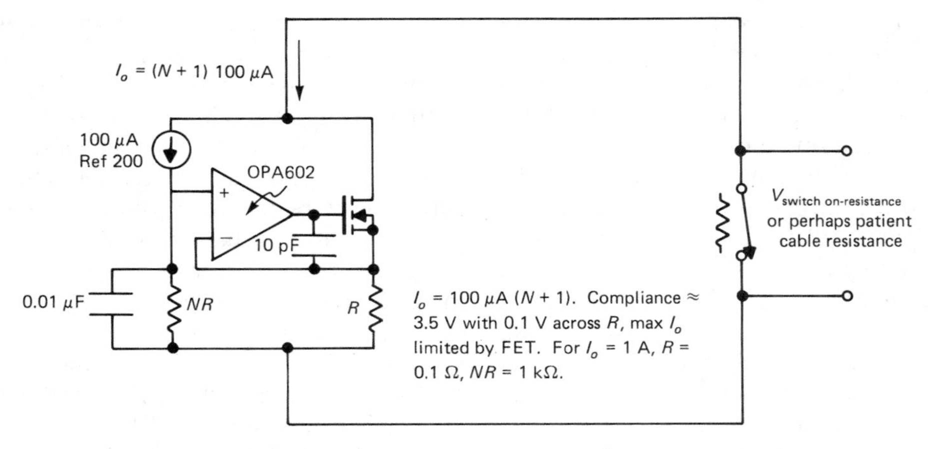

Figure 4-72 Switch-on resistance measurement using accurate current source.

Modern S/H amplifiers use monolithic circuits. Figure 4-74 depicts a component very common in the industry. It has the holding capacitor in the feedback loop of the output amplifier. An external holding capacitor can be added to reduce output voltage droop rate: Droop[V/second $=$ $I_{\text{discharge}}$(pA)/ C_{hold}(pF)]. This also reduces noise, because it provides greater filtering. Notice that this component has an input buffer amplifier to provide high-input impedance through a noninverting terminal. The S/H amplifier can also be connected in the inverting configuration by grounding the +Input and feeding a signal into the −Input through a resistor.

Sample-and-hold function and timing is shown in Figure 4-75. It works this way. The S/H has four phases: (1) acquisition time, (2) aperture delay time, (3) aperture uncertainty time, and (4) output amplifier settling time.

Figure 4-73 Basic sample-and-hold (S/H) amplifier.

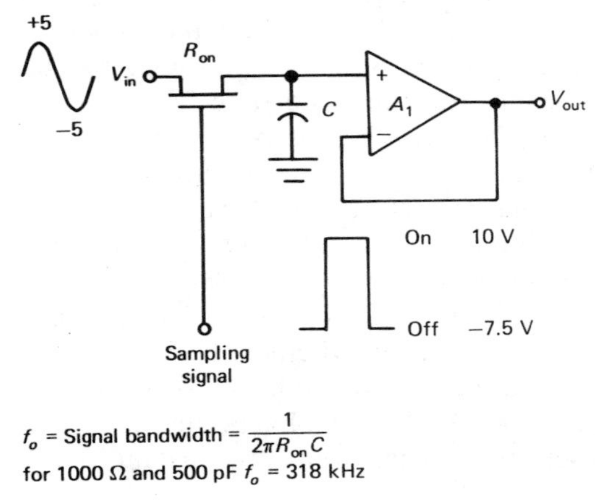

Figure 4-74 Single-chip (monolithic) noninverting unity-gain sample/hold, SHC5320. (Courtesy of Burr-Brown under copyright 1991 Burr-Brown Corporation)

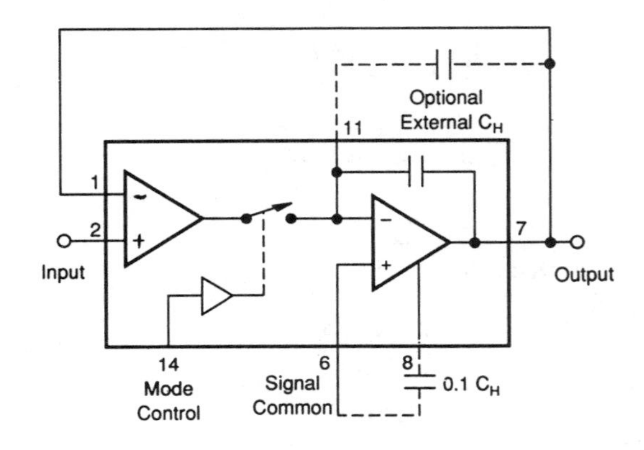

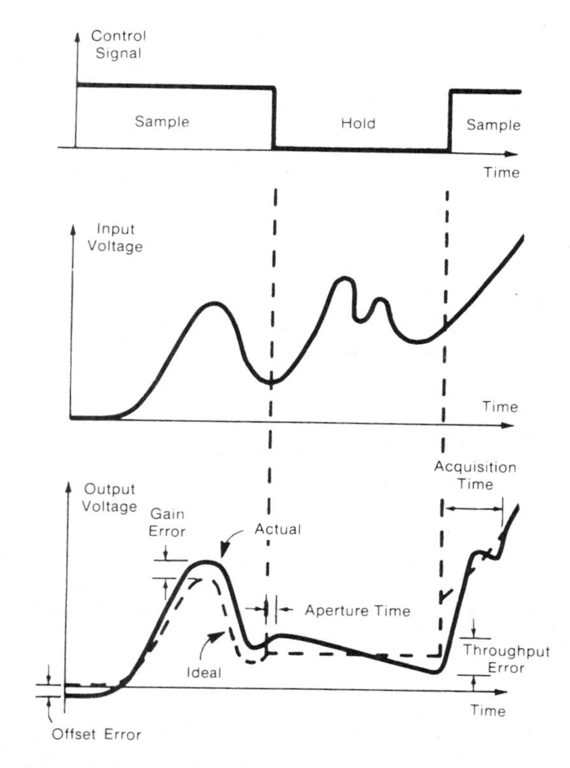

Figure 4-75 Sample/hold function and timing. (Courtesy of Burr-Brown under copyright 1991 Burr-Brown Corporation)

The control signal, at the top, with its two states, sample and hold, starts and stops the process. When in the sample or track state, the input voltage, shown in the second graph, is continuously charging the hold capacitor, or more correctly called the tracking capacitor at this point. This first phase is called the acquisition time, because the capacitor is acquiring the signal. Of course, there are imperfections in the S/H that cause offset and gain errors. This makes the acquired signal something different than ideal, but the difference is usually small. When the control signal goes to the hold state, the second phase begins. This is the aperture delay time. It occurs because the S/H switch does not respond instantly. It starts to open after some small delay, say 25 nanoseconds for the component shown in Figure 4-74. The third phase, aperture uncertainty, automatically follows. It occurs because of noise around the switching threshold. In other words, the exact time when the switch opens is statistically uncertain, with about a 300-picosecond variation. When many samples are taken, this single uncertain event turns into aperture jitter. When input signals are moving fast, this time jitter turns into amplitude noise at the S/H output. The fourth phase is the settling of the output amplifier, often called sample-to-hold transient settling time. S/H amplifiers are used extensively throughout medical instrumentation.

Not all amplifiers used in medicine are slow. The next example application is a video one. Figure 4-76 shows a commercially available high-speed amplifier in a noninverting gain of 2, acting as a video distribution amplifier. This component can drive 6 V peak to peak into a 50-Ω load. Three back terminated lines, so called because they are loaded, represent one-third of 150 Ω, or 50 Ω. These types of amplifiers are used to transfer bioelectric signals that have been modulated or translated into the video bandwidth.

As discussed earlier, many bioelectric signals are modulated on a carrier and then transmitted across an optical sensor, magnetic transformer, or capacitive barrier. Modern components, like the one shown in Figure 4-77, are actually high-speed analog multipliers. In this example, the multiplier is used as a single-chip balanced amplitude modulator. Notice the 120-kHz signal modulated on the 2-MHz carrier frequency. This circuit uses very few external components to accomplish its job.

The last application area to be discussed in this chapter is the Data Acquisition System (DAS). It is usually a multichannel system with some type of input signal amplification, followed by a sample-hold amplifier

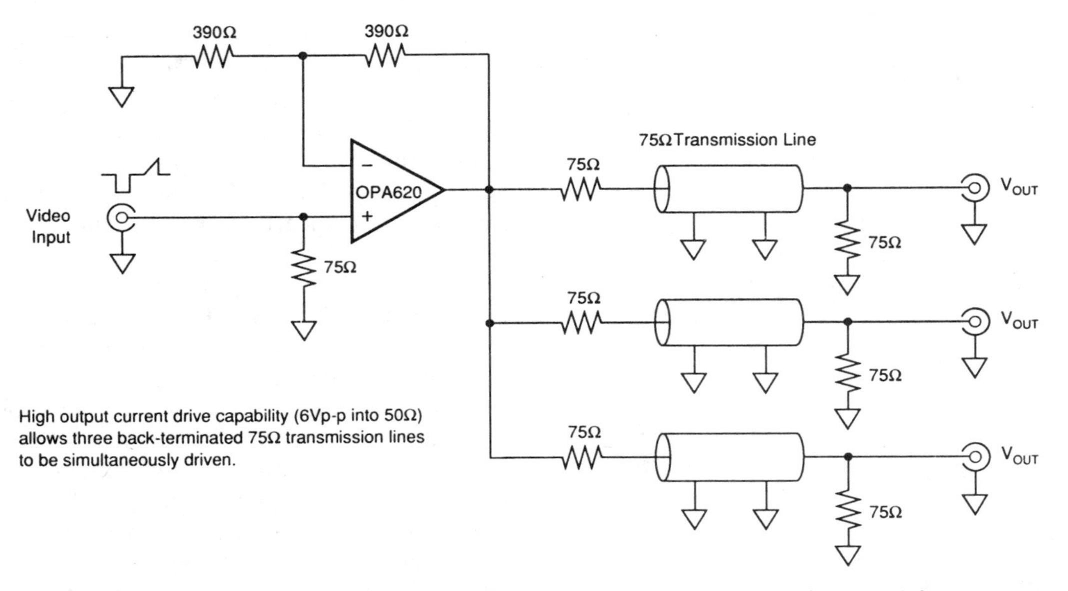

Figure 4-76 Video distribution amplifier. (Courtesy of Burr-Brown under copyright 1991 Burr-Brown Corporation)

which feeds into an analog-to-digital converter. Actually most biomedical systems are DASs. Figure 4-78 shows a rapid scanning-rate DAS with 5 microseconds settling time. The front-end differential input/differential output multiplexer steers a differential signal to the input of a fast settling time instrumentation amplifier (IA). Its output drives a familiar S/H. This system can quickly scan many channels. This is sometimes a require-

Figure 4-77 Single-chip (monolithic) balanced amplitude modulator. (Courtesy of Burr-Brown under copyright 1987 Burr-Brown Corporation)

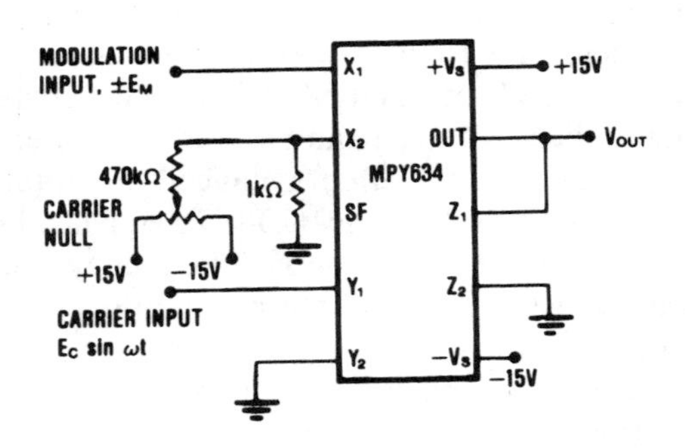

The basic multiplier connection performs balanced modulation. Carrier rejection can be improved by trimming the offset voltage of the modulation input. Better carrier rejection above 2MHz is typically achieved by interchanging the X and Y inputs (carrier applied to the X input).

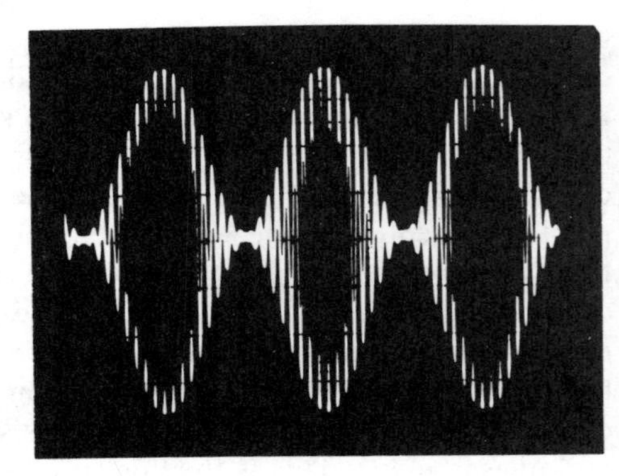

CARRIER: f_C = 2MHz, AMPLITUDE = 1Vrms
SIGNAL: f_S = 120kHz, AMPLITUDE = 10V peak

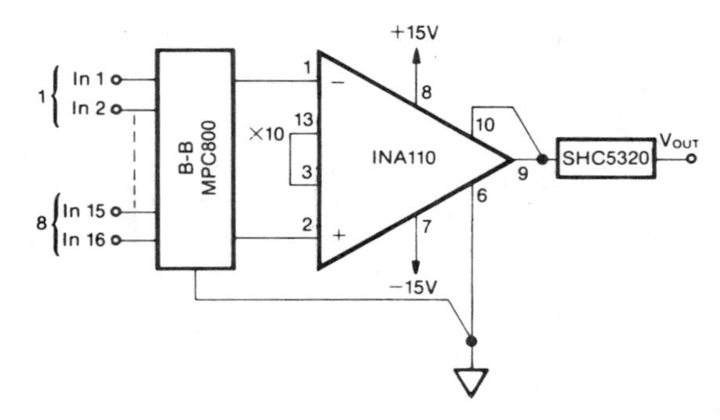

Figure 4-78 Rapid scanning rate data acquisition with 5-μs setting time to 0.01 percent, 1/2 LSB for 12-bit system. (Courtesy of Burr-Brown under copyright 1991 Burr-Brown Corporation)

ment in medical instrumentation if a large number of channels are included in a big system. Otherwise, a scanning time of 50 to 100 microseconds per channel is acceptable.

The input IA in a DAS often has to be put into different gains depending on the level of the input signal being amplified at the time. Figure 4-79 shows a commercially available one-chip, programmable-gain IA (PGIA). Although this PGIA is capable of higher frequencies, the input common-mode filter limits the response to 0.16 Hz. Under digital

Figure 4-79 One-chip (monolithic) AC-coupled programmable-gain instrumentation amplifier (PGIA) for frequencies above 0.16 Hz. (Courtesy of Burr-Brown under copyright 1991 Burr-Brown Corporation)

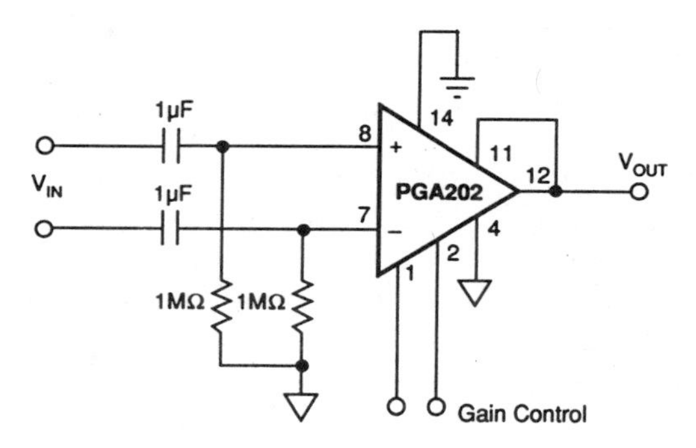

gain control, this component can be set to gains of 1, 10, 100, and 1000. When placed, for example, in front of an analog-to-digital converter, the PGIA essentially increases the system's dynamic range. That is, the system can digitize smaller input voltages with the PGIA than without. PGIAs are frequently used in biomedical equipment to automatically gain range under computer or software control.

The application shown in Figure 4-80 is a block diagram of an eight-channel differential input data acquisition system with programmable-gain front end. It contains an input multiplexer, PGIA, and sampling A/D converter. This particular A/D represents a trend in the industry. It has an internal S/H amplifier to capture and hold any channel's signal constant during the conversion time. Also, the A/D is the modern CMOS type, which requires very low quiescent power to operate. The parallel output data is composed of 12-bit digital words, each word representing each channel as it is selected through the multiplexer. Output 12-bit information on the data bus is analyzed by the microprocessor. Understanding medical instrumentation hinges on one's knowledge of modern data acquisition systems.

The last application diagram of this chapter is shown in Figure 4-81. It is a schematic of an eight-channel differential input data acquisition system (DAS) with channel-to-channel isolation. Notice that all the commercially available instrumentation amplifiers are on the input side. These are low cost, eight-pin mini-DIP precision IAs that use one external resistor to set their gain. They drive eight low-cost isolation amplifiers in 16-pin single-wide plastic DIPs, also on the input side. That is why this system has channel-to-channel isolation. Each channel can accept a low-level input signal with respect to its own common, which is isolated from all the other commons, including the output ground. Also, the eight isolated

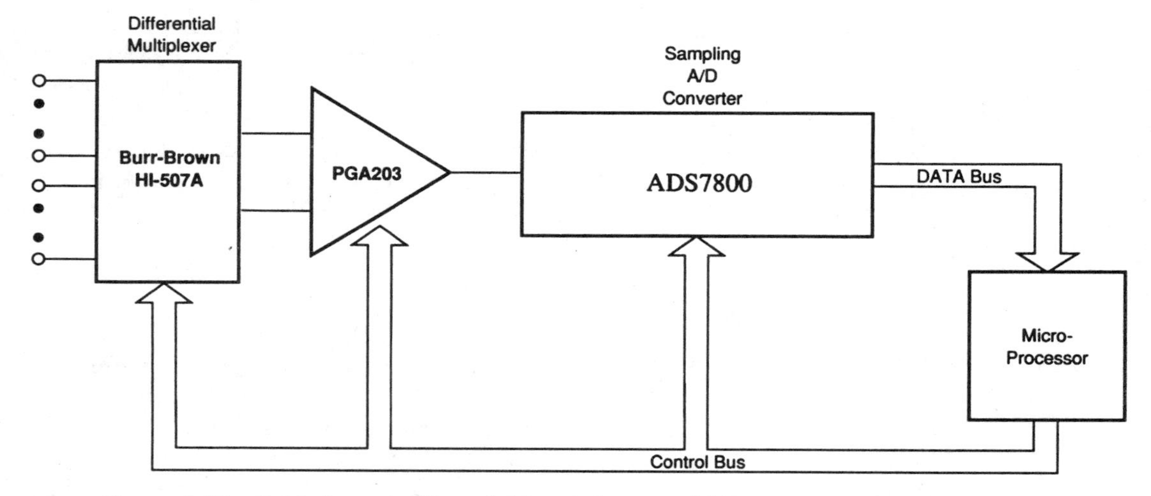

Figure 4-80 Eight-channel differential input data acquisition system with programmable-gain front end. (Courtesy of Burr-Brown under copyright 1991 Burr-Brown Corporation)

Figure 4-81 Eight-channel differential input data acquisition system with channel-to-channel isolation.

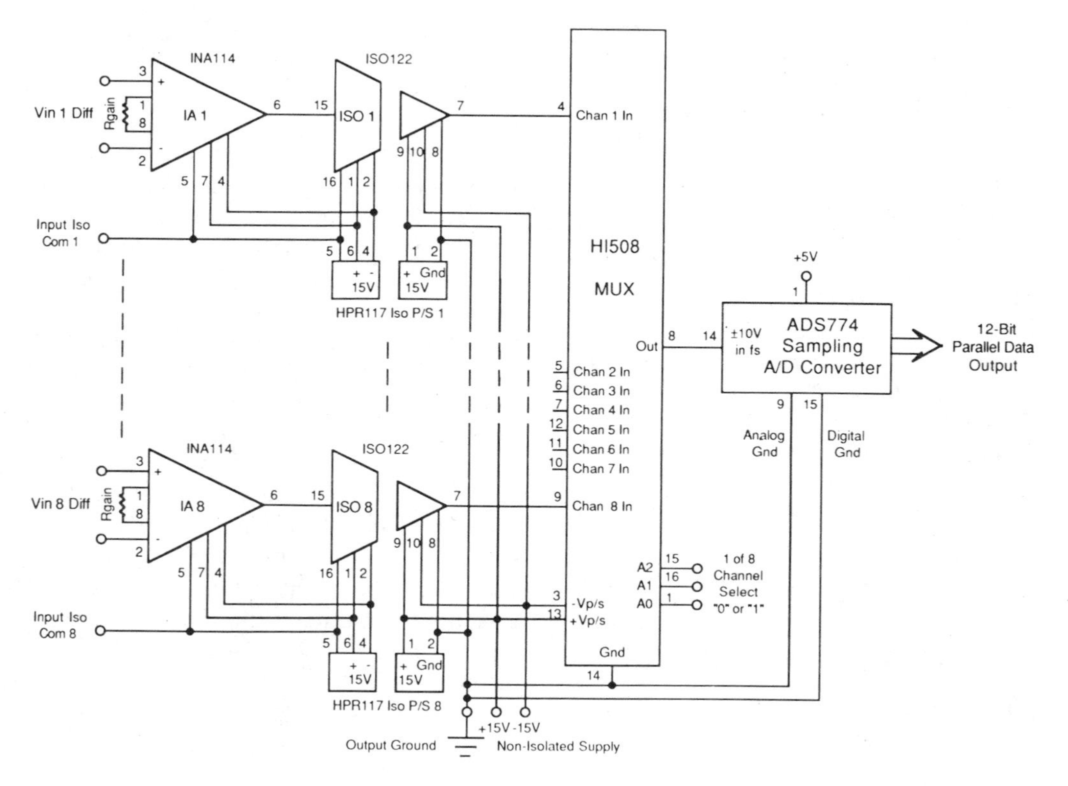

power supplies or dc-to-dc converters supply + and − 15 V separately to each channel. After crossing the eight isolation barriers, the power supply becomes the same. That is, + and − 15-V, non-isolated, is supplied to the output stages of each iso-amp and to the analog multiplexer. Remember that isolation resistance and capacitance is made up of eight iso-amp barriers and eight iso-P/S barriers, all in parallel. The isolation is specified for 750 V dc continuous rating. The sampling analog-to-digital converter in Figure 4-81 accepts each channel as it is selected, samples it in a sample-hold amplifier inside the A/D converter, and digitizes it into 12-bit digital words. The A/D converter shown is capable of sampling at a 117-kHz rate. This means that the system can switch to a new channel about every 8.5 microseconds. Also, the A/D shown is the modern low quiescent power CMOS type, operating from a single +5-V digital power supply. Notice that A/D analog and digital grounds are returned separately to the power supply point where ground enters the system. This insures that the analog signal at the output of the multiplexer is referenced to a "clean" ground. Mixing analog and digital ground currents too close to the A/D can cause digital noise to mix with the analog signal unless a ground plane is used.

The type of channel-to-channel isolated DAS depicted in Figure 4-81 is used in some medical and many industrial systems. Although isolation is necessary for most ECG monitoring, it does not require channel-to-channel isolation. That is, all channels on the input share the same isolated common (right-leg patient signal return), which is isolated from the output ground.

Many of the semiconductors described in this chapter have, to one degree or another, some static sensitivity, especially CMOS (complementary metal oxide semiconductor) chips. Special care in handling these types of components is necessary to avoid catastrophic failure (burnout) and latent failure (works but is erratic and eventually drifts out of specification and then fails). Body and machine models have been established which try to quantify how static charge buildup and discharge occurs through semiconductors. Keeping humidity high enough and using special insulation with antistatic sprays can help prevent static buildup, which can reach levels of thousands of volts. Using conductive wrist straps, standing on conductive floors, and grounding metal surfaces can drain off charge before it reaches dangerous levels. Static discharge damages semiconductors but rarely hurts humans, even though it can be felt as a sharp pin-prick sensation. Knowing how to deal with static is very helpful in troubleshooting and repairing medical electronic equipment. Otherwise one can cause more failures than one is trying to diagnose and fix.

The remaining chapters of this textbook discuss medical systems that use the bioelectric amplifier circuits and DASs described in this chapter.

4-14 Summary

1. The properties of operational amplifier circuits can be set by manipulating the negative feedback loop.
2. Operational amplifier voltage gain is set by the ratio of two external resistors.
3. Isolation amplifiers provided up to 10^{12} Ω of insulation between the patient connectors and the 120-V ac power mains.
4. Chopper amplifiers are used to process low-level signals because they possess superior noise and drift characteristics.
5. Input guarding is used to prevent formation of differential signals from common-mode signals.

4-15 Recapitulation

Now return to the objectives and self-evaluation questions at the beginning of the chapter and see how well you can answer them. If you cannot answer certain questions, place a check mark next to each and review appropriate parts of the text. Next, try to answer the following questions using the same procedure. When you have answered all of the questions, solve the problems in Section 4-17.

4-16 Questions

4-1 List the properties desired in a bioelectric amplifier.

4-2 List the terminals you would ordinarily expect to find on an operational amplifier.

4-3 Give the three types of signal voltage applied to operational amplifier input terminals.

4-4 List the six basic properties of an ideal operational amplifier, and give a brief definition or implication of each.

4-5 The ______________ input produces an output signal that is out-of-phase with the input signal.

4-6 What is a "virtual ground"? From which basic operational amplifier property does it follow (see Question 4-4)?

4-7 What is a "summing junction" and how does it differ from a virtual ground?

4-8 List the three basic classes of operational amplifier voltage follower configuration.

4-9 Write the transfer equations for: (a) inverting follower, (b) noninverting follower *with* gain, and (c) instrumentation amplifier.

4-10 Write the transfer equation for (a) an integrator, (b) a differentiator.

4-11 What property of transistors gives us the logarithmic and antilog amplifiers?

4-12 Draw two different types of *offset null* circuit.

4-13 Define *zero suppression* and describe how it is used.

4-14 A *span* control sets the ______________ of the amplifier.

4-15 What is an isolation amplifier? Why is it used?

4-16 List four different types of isolation amplifier circuit.

4-17 Describe *current loading* coupling.

4-18 Draw the block diagram for, and describe the operation of, a carrier type isolation amplifier.

4-19 What problems are encountered when low-amplitude signals are processed in high-gain dc amplifiers?

4-20 How does a chopper amplifier work, and how does it solve the problems discussed in Question 4-19?

4-21 Draw the simplified schematic for a chopper amplifier.

4-22 How can a *differential* component be formed from a common-mode signal in a shielded input cable?

4-23 Describe *input guarding*. How does it work and why is it sometimes needed?

4-24 What is a *common-mode amplifier?* How is it different from a *right-leg amplifier?*

4-17 Problems

4-1 Find the voltage gain of an inverting follower (Figure 4-3) if R_1 = 10 kΩ and R_2 = 220 kΩ.

4-2 Calculate the voltage gain of an inverting follower if R_1 = 500 Ω and R_2 = 5000 Ω.

4-3 Calculate the voltage gain of an inverting follower if (a) R_1 = 1 kΩ and R_2 = 1 Ω, and (b) R_1 = 1 kΩ and R_2 = 100 Ω.

4-4 What is the voltage gain of a noninverting follower (Figure 4-7*a*) if R_1 = 10 kΩ and R_2 = 100 kΩ?

4-5 Compute the gain of a noninverting follower if R_1 = 1.2 kΩ and R_2 = 100 kΩ.

4-6 Select resistors for an inverting follower with a gain of ×75 if the source impedance looking into the amplifier input is 200 Ω.

4-7 Compute the output voltage of a multiple-input circuit such as Figure 4-20 if R_4 = 10 kΩ, R_1 = 10 kΩ, R_2 = 5 kΩ, R_3 = 5 kΩ, and $E_1 = E_2 = E_3$ = 620 mV.

4-8 Find I_4 in Figure 4-20, using the parameters given in Problem 4-7.

4-9 Compute the *gain* of a simple dc differential amplifier such as Figure 4-21 if $R_1 = R_3$, $R_2 = R_4$, R_4 = 100 kΩ and R_1 = 9.1 kΩ.

4-10 Compute the *gain* of an instrumentation

amplifier (Figure 4-23) if $R_2 = 47\ k\Omega$, $R_1 = 1.8\ k\Omega$, $R_5 = 10\ k\Omega$, and $R_4 = 3.9\ k\Omega$. Assume that $R_2 = R_3$, $R_4 = R_6$, and $R_5 = R_7$.

4-18 References

1. Stout, David F. and Milton Kaufman (eds.), *Handbook of Operational Amplifier Circuit Design,* McGraw-Hill (New York, 1976).
2. Smith, John I., *Modern Operational Circuit Design,* Wiley-Interscience (New York, 1971).
3. Carr, Joseph J., *Op-Amp Circuit Design & Applications,* TAB Books (Blue Ridge Summit, Pa., 1976).
4. Faulkenberry, L. M., *An Introduction to Operational Amplifiers,* Wiley (New York, 1976.
5. Graeme, Jerald G., Gene E. Tobey, and Lawrence P. Huelsman, *Operational Amplifiers: Design and Applications,* McGraw-Hill (New York, 1971).
6. Strong, Peter, *Biophysical Measurements,* Measurements Concepts Series, Tektronix, Inc. (Beaverton, Ore., 1976).
7. *A Miniature Anemometer for Ultrafast Response,* H. Thurman Henderson and Walter Hsieh, University of Cincinnati, December, 1989, *Sensors* Magazine, 174 Concord St., Peterborough, N.H. 03458.
8. Product Data Sheets for components shown in this chapter. Burr-Brown Corp., 1991, P.O. Box 11400, Tucson, Ariz., 85734, 1-800-548-6132.
9. Application Bulletins, AB-001 through AB-035, Burr-Brown Corp., 1991.
10. Application Notes, AN-163 (Partial Discharge Testing, 1989), AN-165 (Current Sources and Receivers, 1990), Burr-Brown Corp.
11. *Design Update,* Volume 2, Number 3 (Filters and Temperature Sensor Circuit, 1991), Burr-Brown Corp.
12. "Test Methods for Static Control Products," chapter by James R. Huntsman and Donald M. Yenni, 1982; and *Basic Electrical Considerations in the Design of a Static-Safe Work Environment,* by Don Yenni, 1979, 3M, Static Control Systems, 223-2SW, 3M Center, St. Paul, Minn. 55101.
13. Standards and Organizations.

Consensus Standards (not law, referenced by legal organizations)

Specs can be obtained from the particular association or from the following company, which specializes in acquiring and selling standards: Global Engineering Documents, Division of Information Handling Services, 2805 McGraw Ave, P.O. Box 19539, Irvine, CA 92713, Tel: 714-261-1455.

1. Association for the Advancement of Medical Instrumentation (AAMI), 3330 Washington Blvd., Suite 400, Arlington, VA 22201, Tel: 703-525-4890.
a. *Standards and Recommended Practices,* Volume 1: *Biomedical Equipment.* Safe current limits for electromedical apparatus.
b. *Standards and Recommended Practices/American National Standard.* Safe current limits for electromedical apparatus.
c. *Biomedical Safety and Standards,* periodically issued.

2. Underwriters Laboratories (UL), 333 Pfingsten Road, Northbrook, IL 60062.
a. UL544, *Standard for Medical and Dental Equipment,* 1985.
b. UL508, *Standard for Industrial Control Equipment,* 1984.
c. UL1244, *Electrical and Electronic Measuring and Testing Equipment,* 1980.
d. UL1577, *Standard for Optical Isolators,* 1986.

3. International Electrotechnical Commission (IEC), France.
a. IEC601-1, *Medical Electrical Equipment, Part 1: General Requirements for Safety,* 1988.

4. Verband Deutscher Elektroteckniker (VDE), Germany.
a. DIN VDE 0884, UDC 621.372.2, 621.391.63, 620.1, 614.8, *Optocouplers for Safe Electrical Isolation—Requirements and Tests,* 1987.

5. National Electrical Code (NEC).
6. National Fire Protection Association (NFPA), Publication 70, Article 517, and Publication 99 from NEC.
7. Joint Commission on the Accreditation of Healthcare Organizations (JCAHO).
   a. Accreditation Manual for Hospitals, Requirements for 1992.
8. U.S. Public Health Service (PHS).
9. U.S. Occupational Safety and Health Administration (OSHA).
10. State and local government and public health service departments.

Chapter 5

Electrocardiographs

5-1 Objectives

1. Be able to describe the fundamentals of an ECG recording.
2. Be able to describe the basic ECG machine.
3. Be able to draw and describe the basic *lead system.*
4. Be able to list the causes and cures for most common ECG recording malfunctions.
5. Be able to list the basic *maintenance* procedures for the ECG machine.

5-2 Self-Evaluation Questions

These questions test your prior knowledge of the material in this chapter. Look for the answers as you read the text. After you have finished studying the chapter, try answering these questions and those at the end of the chapter (Section 5-14).

1. Which leads form the *Einthoven triangle?* Which limb electrodes are used?
2. How does a bioelectric amplifier *differ* from an ECG amplifier?
3. Which leads form the *unipolar limb leads?*
4. How do the unipolar limb leads differ from biopolar limb leads?
5. What is the standard speed in mm/sec of the basic ECG machine?
6. Which limb signals are summed to form the V_1 through V_6 leads?

5-3 The Heart as a Potential Source

The heart is a muscle formed in a way that allows it to act as a pump for blood. In Section 2-6 we learned that the heart contracts (i.e., pumps) under command from an electrical stimulus in the *electroconduction system.*

The heart pumps blood when the muscle cells making up the heart wall contract, generating their *action potential.* This potential creates electrical currents that spread from the heart throughout the body.

The spreading electrical currents create differences in electrical potential between various locations in the body, and these potentials can be detected and recorded through surface electrodes attached to the skin.

The waveform produced by these biopotentials is called the electrocardiogram (ECG or, after the German spelling, EKG), that is, a written record ("... gram") of the cardiac electrical potential waveform.

5-4 The ECG Waveform

An example of a typical ECG waveform is shown in Figure 5-1. This particular waveform is typical of a measurement from right

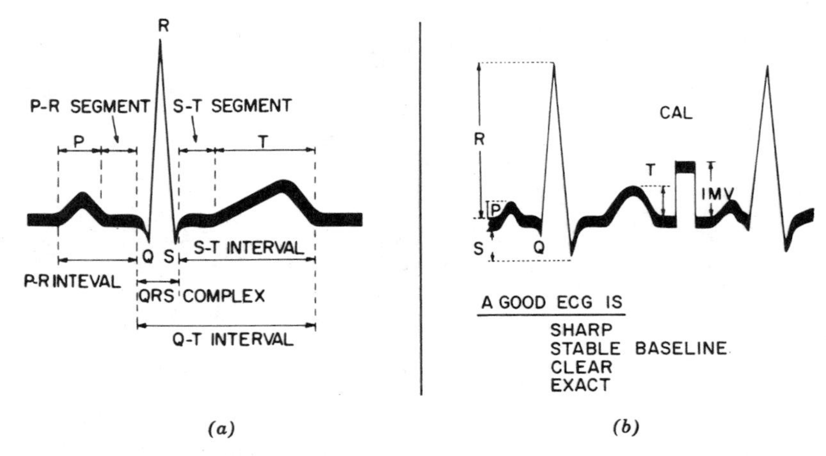

Figure 5-1 ECG time and amplitude measurements. (*a*) Time measurements. (*b*) Amplitude measurements. (Reprinted courtesy Hewlett-Packard.)

arm to left arm. In Figure 5-1*a* we see the various time intervals that are often measured by physicians examining the waveform, while Figure 5-1*b* shows the voltage amplitude relationships. A 1-mV *calibration pulse* is also shown in Figure 5-1*b*.

The low-level amplitudes normally encountered in ECG recording cause several problems that will be dealt with in Section 5-10.

5-5 The Standard Lead System

In standard ECG recording there are *five* electrodes connected to the patient: *right arm* (RA), *left arm* (LA), *left leg* (LL), *right leg* (RL), and *chest* (C). These electrodes are connected to the inputs of a differential buffer amplifier through a *lead selector switch*.

The recording obtained across different pairs of electrodes results in different waveform shapes and amplitudes; these different views are called *leads*. Each lead conveys a certain amount of unique information that is not available in the other leads. Figure 5-2 shows the electrical axis of the heart that is examined by six of the standard leads: I, II, III, AVR, AVF, and AVL. The physician is often able to diagnose the *type* and *site* of heart disease by examining these different views because waveform anomalies have been correlated with disease conditions in the past.

The electrical connections for the 12 standard leads are shown in Figure 5-3*a*. The ECG machine uses the patient's *right leg* as the *common* electrode, and the lead selector switch (deleted for sake of simplicity) connects the proper limb or chest electrodes to the differential amplifier input.

Figure 5-2 Cardiac axis viewed by different leads. (Reprinted courtesy Hewlett-Packard.)

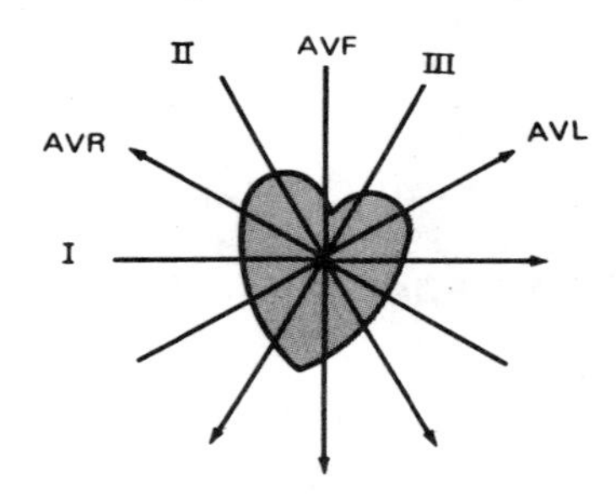

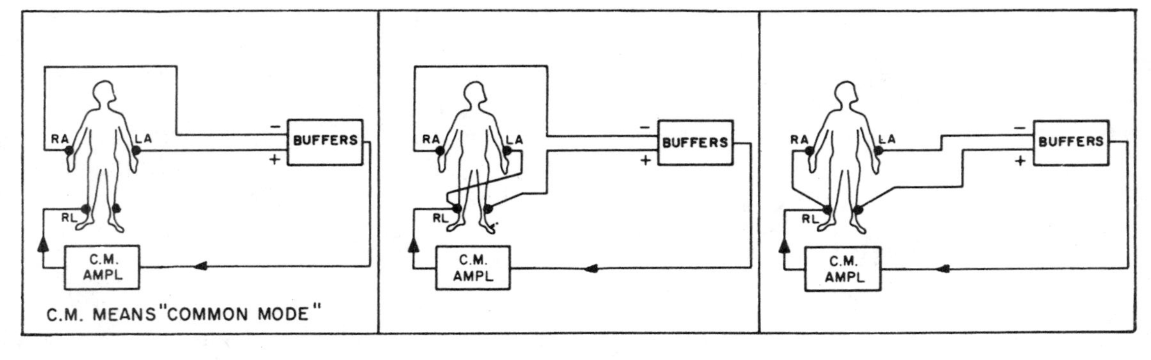

LEAD I　　LEAD II　　LEAD III

UNIPOLAR LIMB LEADS

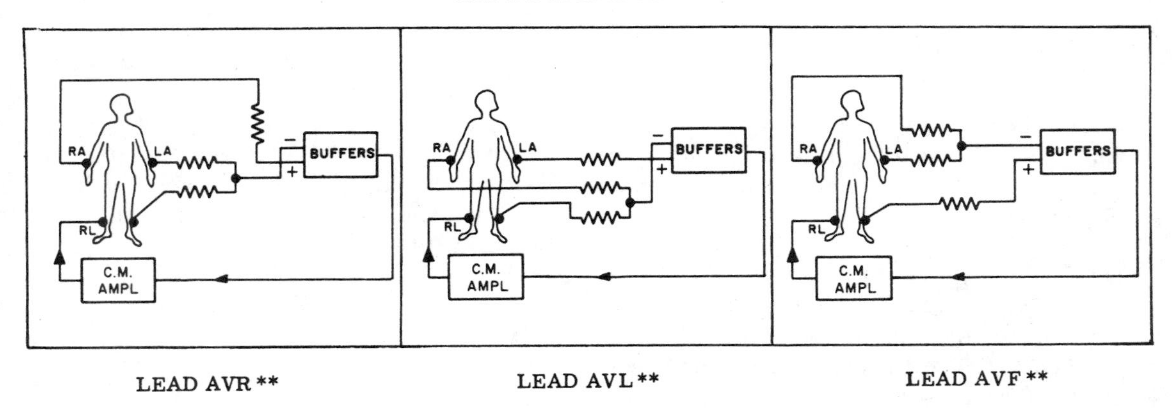

** Also known as "augmented" leads

UNIPOLAR CHEST LEADS

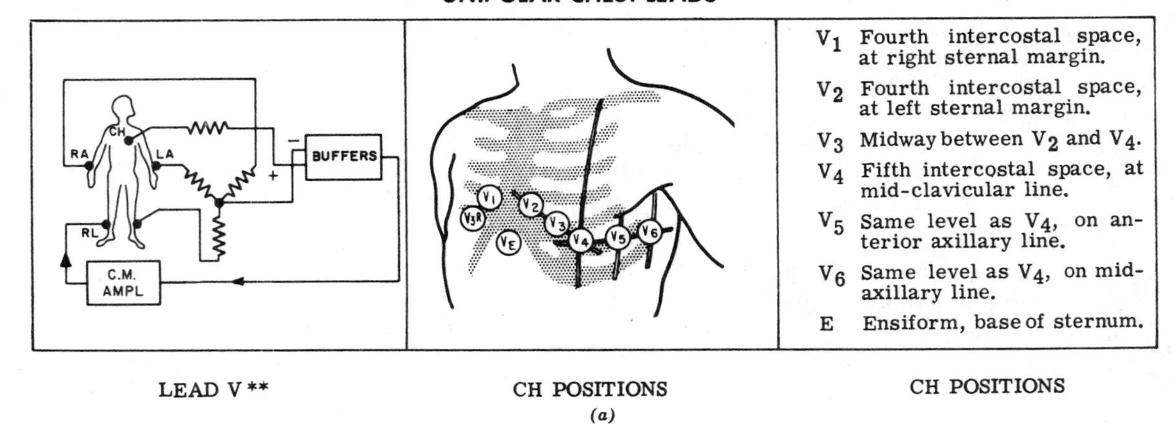

Figure 5-3 Standard leads and Einthoven triangle. (*a*) Limb and chest leads. (*b*) Einthoven triangle. (Reprinted courtesy Hewlett-Packard.)

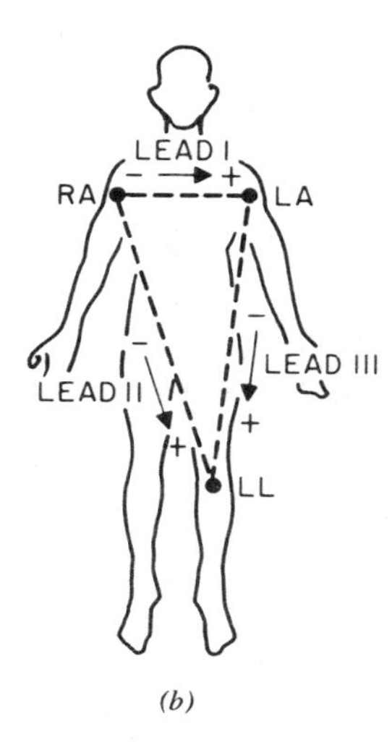

(b)

Figure 5-3 (continued)

The *biopolar limb leads* are those designated lead I, lead II, and lead III and form what is called the *Einthoven triangle* (see Figure 5-3).

1. *Lead I:* LA is connected to the amplifier's *noninverting* input, while RA is connected to the *inverting* input.
2. *Lead II:* The LL electrode is connected to the amplifier's *noninverting* input, while the RA is connected to the *inverting* input (LA is shorted to RL).
3. *Lead III:* The LL is connected to the *noninverting* input, while LA is connected to the *inverting* input (RA is shorted to RL).

The *unipolar limb leads*, also known as the *augmented limb leads*, examine the composite potential from all three limbs simultaneously. In all three augmented leads the signals from *two* limbs are summed in a resistor network and then applied to the amplifier's inverting input, while the signal from the remaining limb electrode is applied to the noninverting input.

1. *Lead AVR:* RA is connected to the *noninverting* input, while LA and LL are summed at the *inverting* input.
2. *Lead AVL:* LA is connected to the *noninverting* input, while RA and LL are summed at the *inverting* input.
3. *Lead AVF:* LL is connected to the *noninverting* input, while RA and LA are summed at the *inverting* input.

The *unipolar chest leads* (V_1 through V_6) are measured with the signals from certain specified locations on the chest applied to the amplifier's *noninverting* input, while the RA, LA, and LL signals are *summed* in a resistor "Wilson network" at the amplifier's *inverting* inputs (called the "indifferent electrode").

The chart in Figure 5-3*a* shows the locations for V_1 through V_6, plus some other locations that are also frequently used.

Figure 5-4 shows the waveforms from a single patient taken in the 12 different lead positions. The square pulse shown in some of the waveforms is a 1-mV calibration signal supplied by the ECG machine. Note the differences in shape and amplitude of the ECG signals.

5-6 Other ECG Signals

In addition to the conventional ECG signals discussed earlier, there are certain specific signals that are sometimes acquired. While the use of these is rare, the biomedical equipment technician or clinical engineer should be aware of them (see reference 3 listed in Section 5-16).

Interdigital ECG

This signal is taken between any two fingers. The interdigital ECG is used primarily in home monitoring of patients (especially those with implanted pacemakers—reduction of heart rate often precedes battery failure). One common ploy is to use the index finger of each hand as the signal source.

Esophageal ECG

In this type of ECG recording, an electrode set is placed close to the heart in the esophagus. Either a special ECG catheter containing a "pill electrode," an external bipolar pacing electrode, or a special electrode-equipped nasogastric tube is used to acquire the electrical signal. The electrode is placed in the esophagus very near the heart. A principal application of the esophageal ECG is to examine atrial activity of the heart. The relative amplitudes of the *P* and *R* waves are used to establish the position for atrial sensing.

Toilet Seat ECG

This type of recording uses two electrodes placed on either side of a toilet seat. The acquired signal is often connected to a computer in which arrhythmia detection software is running. The purpose is to detect cardiac arrhythmias that sometimes occur when the patient is straining during defecation.

5-7 The ECG Preamplifier

An ECG preamplifier is a differential bioelectric amplifier (Chapter 4). The input circuitry consists of the high-impedance input of the bioelectric amplifier, a lead selector switch, a 1-mV calibration source, and a means for protecting the amplifier against high-voltage discharges from defibrillators used on the patient.

The amplifier may be the bioelectric instrumentation amplifiers discussed in Chapter 4, although in all modern machines one of the *isolation amplifier* designs will be used for patient safety.

The simplest limited type of ECG amplifier is shown in Figure 5-5. Here a one-chip (monolithic) ECG amplifier with right-leg drive indicates that an instrumentation amplifier (IA) can be connected directly to a person through body electrodes. In this particular commercially available three op-amp IA, the output of the internal input amplifiers appears on pins 8 and 9. By connecting two 20-kΩ resistors to a single point, the common-mode voltage (CMV) can be derived. The CMV in the ECG case is composed of two components: (1) dc electrode offset potential and (2) 50- or 60-Hz ac-induced interference. Hum interference is caused by magnetic and electric fields from power lines and transformers cutting across ECG electrodes and patients. Hum currents flow in signal, common, and ground wires via capacitive coupling between the fields and the system. This type of noise seems to be ever present, and the battle to get rid of it seems to be never ending. Fortunately, modern noise reduction schemes are very successful at minimizing hum in the ECG recordings.

The technique in Figure 5-5 works as follows. First the common-mode rejection of the commercially available IA is very high and cancels some of the noise. The IA output is an ECG signal that has the 60-Hz noise greatly reduced. How does the IA do this? The differential amplification nature of the IA removes it, because equal common-mode noise is present on each IA input. The IA just subtracts equal noise voltages to give nearly zero while amplifying the difference in the unequal ECG signals present on its inputs. ECG signals on the left arm and right arm are different levels, because they come from different points on the body. How small the noise becomes at the output depends on how high the IA's common-mode rejection is.

The other noise reduction technique is right-leg drive. The CMV is inverted by the right-leg amplifier in Figure 5-5 and the resultant voltage applied to the patient's right leg. Just several microamps or less are actually driven into the patient. This is quite safe, since UL544 and VDE-0884 require a limit of 10-microamp maximum to prevent internal cardiac shock. Why is a noise-cancelling voltage applied to the patient in Figure 5-5? It is

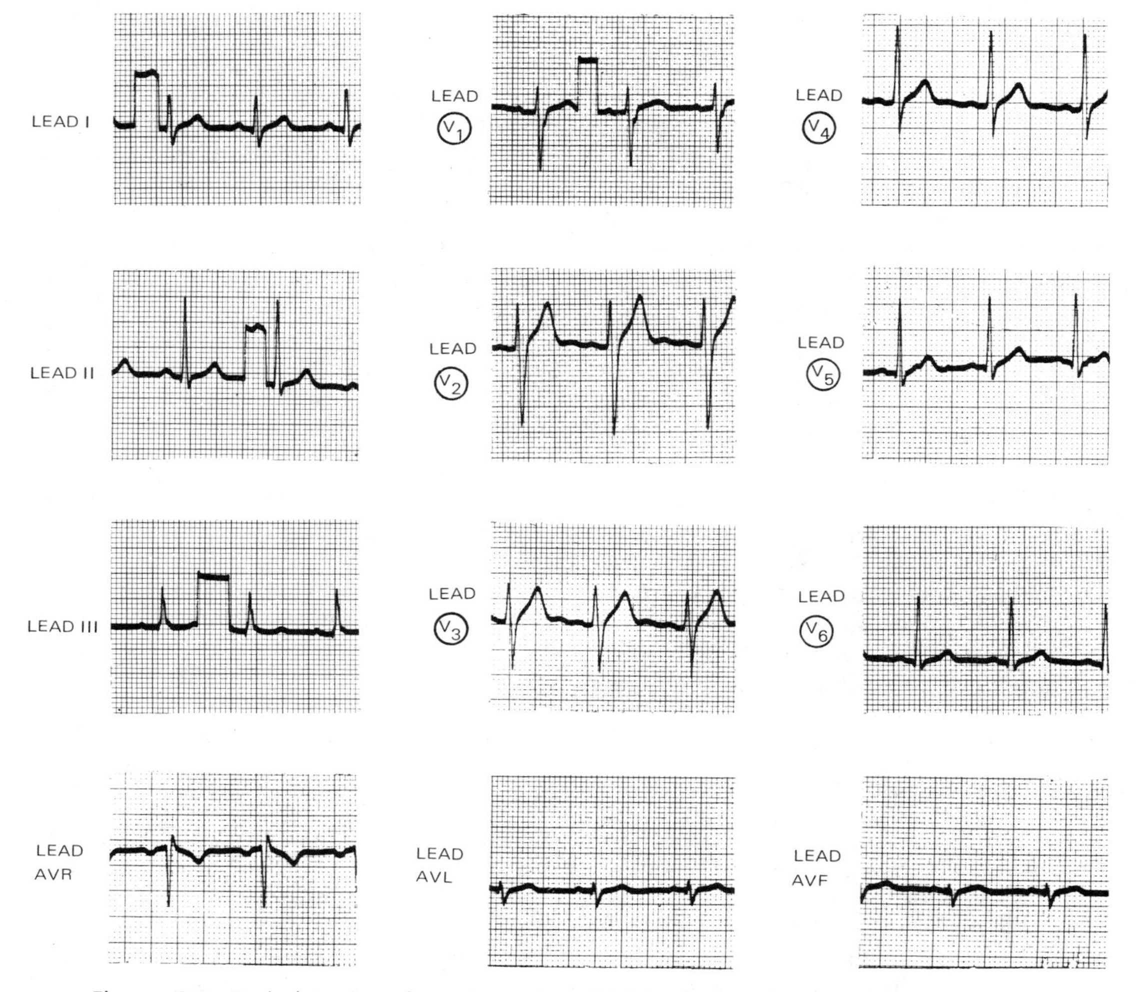

Figure 5-4 Typical tracings from the various ECG leads. (Reprinted courtesy of Hewlett-Packard.)

done to reduce 50- or 60-Hz noise. This circuit acts in a feedback loop (patient and electronics) to servo or drive the CM noise on the patient to a low level. Since the right-leg drive voltage is the inversion of the CM voltage (opposite phase), the right leg goes in the opposite direction compared to the CM voltage on the patient leads. While the 60-Hz noise on the left or right arm may become bigger with respect to the right leg, the 60-Hz noise on the IA input terminals (pins 12 and 3) with respect to IA common (pin 14) becomes smaller. Hence, the IA does not have to reject as much 60-Hz noise. The IA's common-mode rejection then reduces noise even further. Right-leg drive feedback circuits can oscillate if the phase shift through the patient flips the phase of the signal being

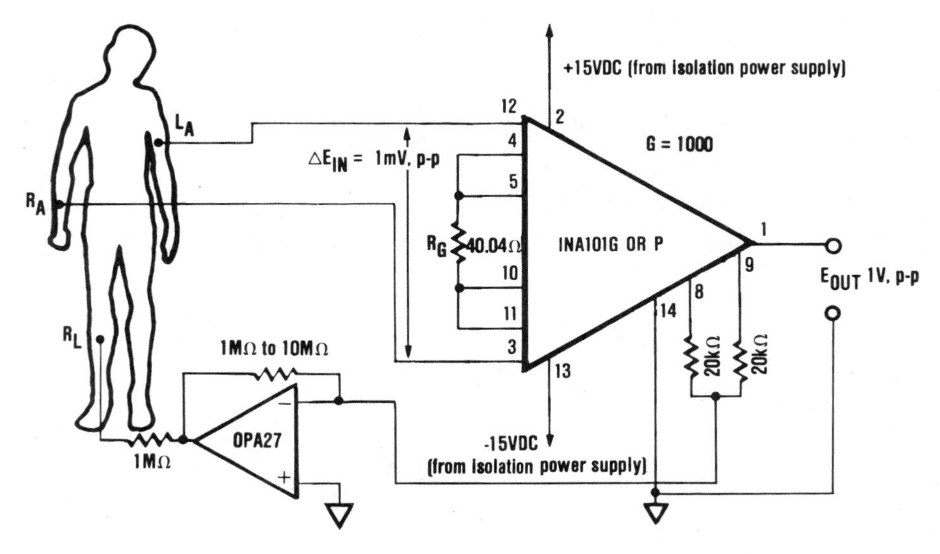

Figure 5-5 One-chip (monolithic) limited ECG amplifier with right-leg drive. (Courtesy Burr-Brown Corp.)

fed back to the electronics. If high-frequency oscillation is present on the ECG signal, positive feedback may be the problem.

Usually an isolation amplifier would follow the IA in Figure 5-5. The iso-amp's isolation-mode rejection (IMR) also reduces noise by establishing 10^{12} Ω and 9 pF between the patient and earth ground. Isolation acts to attenuate noise (say, 1 $V_{p\text{-}p}$ at 60 Hz) that is trying to mix with the low-level ECG input signal (say, 1 $mV_{p\text{-}p}$). Actually, three actions reduce 60-Hz interference dramatically: right-leg drive, IA CMR, and iso-amp IMR. A low-pass filter and/or 60-Hz notch filter following the iso-amp will further shrink noise. This circuit is dc coupled and is in high gain = 1000 V/V. It only works when the electrode offset potentials are low (< 10 mV) to avoid IA output saturation. DC restoration allows much larger offsets to be tolerated.

The standard −3 dB frequency response of the amplifier used in making *diagnostic* grade recordings is 0.05 to 100 Hz, while *monitoring* instruments have a response of 0.05 to about 45 Hz (varies from manufacturer to manufacturer).

ECG preamplifiers must be ac coupled so that artifacts from the electrode offset potential (Section 3-4) are eliminated. The low-frequency response of the amplifier, then, must not extend down to dc, but since certain features of the ECG waveform have a very low-frequency component, the response is very nearly dc (0.05 Hz).

Figure 5-6 shows how to overcome patient electrode offset potentials, which can rise as high as 300 millivolts according to AAMI standards. Sometimes they can go to plus or minus 500 millivolts. If high gain, say 500 or 1000 V/V, were provided immediately, the input amplifiers would saturate and be unable to amplify the ECG signal. To cure this problem, three things are done. First, the input stage buffer amplifiers (A_1 and A_2) are placed in low gain, like 10 V/V. Even if the

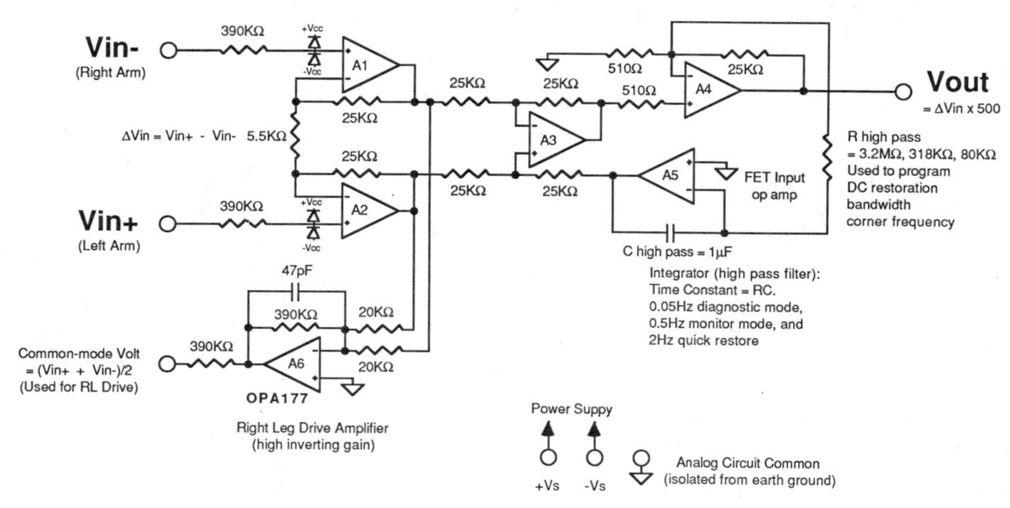

Figure 5-6 ECG amplifier schematic: one-channel front-end derives lead $I = LA - RA$.

electrode offset were 0.5 V, the output of the input amplifiers would only go to 5 V. This leaves plenty of ''head room'' or voltage to the plus or minus power supply ''rails'' for the amplified ECG signal.

Second, the next stage (A_3) in Figure 5-6 is a unity gain differential amplifier, so there are no worries here about ''head room.'' Third, a dc restorator amplifier is used in a feedback arrangement to null out the dc offset. It works this way. Assume the left arm electrode offset to be +300 mV dc, although it can be plus or minus. Also assume that the right arm electrode is 0 V dc. The dc differential input voltage is then 300 mV. This results in +3 V at the output of A_3 (remember the gain of 10). Now the output of A_4 would try to go to +150 V dc, because the A_4 is in a gain of 50. However, the circuit will not allow 150 V. Actually, it does not go very high, because as soon as it goes in the positive direction, the feedback integrator, A_5, applies a negative voltage to A_3 through the reference point (right side of 25-kΩ resistor). The gain from this point to A_3's output is unity (one). Therefore, we have what amounts to a linear summing effect. The positive 3-V offset goes through A_3 at the same time a negative correction voltage goes through A_3. This reduces the offset voltage that appears at A_3's output and hence A_4's output. In turn, this reduces the error fed to the integrator, which reduces the error correction voltage feedback. This negative feedback loop continues for about 10 RC time constants until it stops at the point where the offset at A_4's output is zero. Even the small offset contributed by A_4 as an op-amp error is removed. Since the feedback circuit in Figure 5-6 is an integrator or high-pass filter, it responds only to ac signals above its cutoff frequency. It can be set by R and C in the high pass filter for 0.05 Hz (diagnostic quality ECG), 0.5 Hz (monitoring), or 2 Hz (quick offset restore). The result of this dc restorator is to turn the original dc-coupled amplifier into an ac-coupled amplifier just as if a coupling capacitor were placed in series with the signal path. All frequencies below the cutoff are removed with either approach, but in the feedback dc restorator, the signal does not actually go through the capacitor.

The only advantage is that the feedback approach uses an active integrator, which is linear and more easily controlled compared to a passive *RC* coupling circuit.

Since the dc electrode offset voltage has been removed, the output amplifier (A_4) in Figure 5-6 can now gain up the signal by 50 V/V without becoming saturated. It is just amplifying the ac components of the ECG waveform. The dc component is gone. If the signal level from the left arm minus the right arm was 1 $mV_{p\text{-}p}$, then the output of A_3 would be 10 $mV_{p\text{-}p}$. The output of A_4, which is the output of the ECG amplifier, V_{out}, becomes 0.5 $V_{p\text{-}p}$.

ECG signal high-frequency components extend to 100 Hz, but interfering skeletal muscle signals also have significant components in this range and will generate *somatic artifact* in the ECG tracing.

It is usually easy to obtain the patient's cooperation for the few minutes required to make a diagnostic ECG recording; he or she will lie still, reducing the muscle artifact to a minimum. But during long-term *monitoring* of the ECG the patient will be unwilling or unable to cooperate, so there will be substantial amounts of somantic artifact in the ECG signal. Instruments used for monitoring, then, are designed to have a frequency response that extends only from 30 to 50 Hz. The lower frequency response limit used in monitoring instruments *distorts* the waveform (by elimination of harmonics) too much for diagnostic purposes but is sufficient to allow detection of the life-threatening arrhythmias that make monitoring a necessity.

The *lead selector* switch is a front-panel control that permits the operator to select the waveform to be displayed. It will be a rotary or multilevel pushbutton switch in manually selectable machines, and CMOS or JFET electronic switches in automatic machines. Some monitors do not have a lead switch, and in those instruments only the *common* and two *limb* electrodes are connected to the patient.

The *gain* of the ECG amplifier must be *standardized* when recording a waveform for diagnostic purposes. A 1-mV calibration pulse circuit is provided for this purpose.

In ECG machines, or other ECG systems using a strip-chart recorder as the display device, it is standard practice to adjust the gain to produce a tracing that is 10 mm high when the 1-mV button is pressed. The calibration pulse is also a powerful troubleshooting tool.

The *defibrillator* is a high-voltage electrical heart stimulator used to resuscitate heart attack victims. It is necessary to have an ECG monitor connected to the patient when using the defibrillator, so the ECG preamplifier input must be designed to withstand high voltages and high peak currents, even though the *normal* ECG waveforms are on the order of millivolts. The high-voltage (over a kilovolt) burst from the defibrillator will last 5 to 20 ms.

In some ECG preamplifiers the protection circuits are quite elaborate, while others—principally older machines—have very little protection. An example of various types of protection is given in Figure 5-7. This figure shows many forms of protection circuits, but most equipment uses only a few of them.

Most ECG preamplifiers use from two to nine neon glow lamps (i.e., type NE-2) across the input lines. Those shown in Figure 5-7 are representative of the most common configuration. Most preamplifiers also use series resistors R_1 through R_6, although in some models the input resistors are physically located inside of the patient cable or its connector.

The resistors serve to limit the current flow, while the glow lamps serve a voltage *bypass* function. These lamps consist of a pair of electrodes mounted in a glass envelope in an atmosphere of low pressure neon gas, or a mixture of inert gases. Normally,

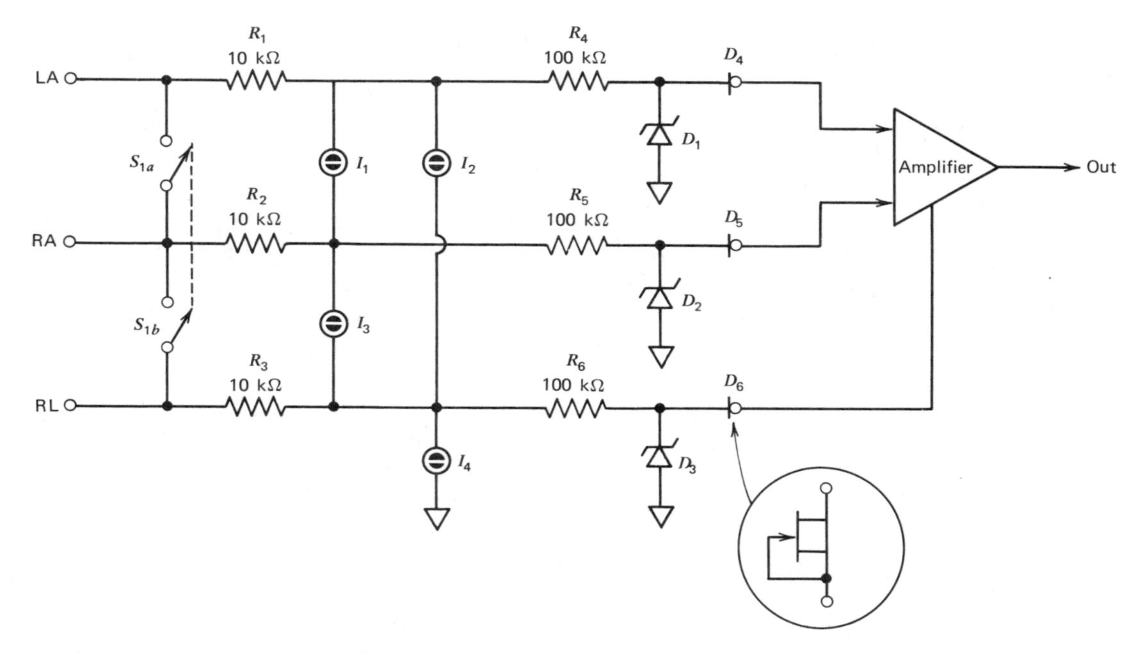

Figure 5-7 Defibrillator protection circuit.

the impedance across the electrodes is very *high*, but if the voltage across the electrodes exceeds the ionization potential of the gas, then the impedance drops suddenly to a very low value. For most lamps in the NE-2 class used in medical monitors, the firing potential is between 45 and 70 V.

Potentials normally encountered in ECG recording will not ionize the gas inside of the glow lamps, but the potentials from defibrillators *will* fire the lamps, bypassing most of the charge harmlessly to the ground.

Some monitors also use Zener diodes (D_1 through D_3) shunted across the amplifier inputs to ground. These diodes serve a function that is similar to the job performed by the neon lamps, but at a lower voltage.

Diodes D_4 through D_6 are called *current-limiting diodes*, even though they are not really *diodes* in the normal sense of the word, but are JFETs with the source and gate terminals tied together (see inset to Figure 5-7). Although the same current-limiting behavior is exhibited by an ordinary three-terminal JFET that is connected in the aforementioned manner, many manufacturers offer two-terminal JFET current-limiting diodes in which the connection is made internally. These diodes are usually rated as to the limiting current.

The current-limiting diode acts as a resistor (i.e., the JFET channel resistance) as long as the current level remains below the limiting point. If the current attempts to increase above that level, however, it is clamped and limited.

In some machines, *metal oxide variable resistors* ("varistors") are used in place of the shunt Zeners. These devices are similar to the surge protectors used on computers. They will maintain a high resistance until the voltage exceeds a critical threshold, but above that point the resistance drops. These devices thereby clip the high-voltage spike.

5-7.1 Types of Defibrillator Damage

The protection circuits available to an ECG preamplifier are never totally effective in all cases; after all, the voltage from the defibrillator is greater than six orders of magnitude greater than the normal working voltages. Some damage will occur.

Two forms of damage from defibrillator discharge are common, and they tend to produce different symptoms. If both amplifier inputs are blown out, then the readout device will show a *flat baseline* (i.e., no output signal and little noise). If, on the other hand, only *one* of the two inputs is affected, then the output waveform *may* be merely distorted. Typically, when this occurs, the output waveform will be unable to deflect above or below the baseline, depending on which input is damaged. This type of problem may very well be occult, that is, hidden, and not noticed by the medical staff unless they compare the display with an ECG taken from the same patient on *another* instrument.

The most common mechanisms for such failures are defective glow lamps and open Zener diodes. The glow lamps eventually lose their ability to protect the amplifier because the firing potential rises due to air leaks, recombination or absorption of the gases, and other defects. Some manufacturers recommend replacement of the glow lamps every one or two years, or more frequently if the machine is used in emergency rooms or ICU/CCU.

If a defibrillator discharge opens a Zener diode, as is sometimes the case, then a *subsequent* discharge may destroy the preamplifier input transistors.

5-7.2 Electrosurgery Unit Filtering

Preamplifier damage from defib high-voltage pulses is prevented by special clamping circuits. Likewise, preamplifiers must be protected from electrosurgery unit (ESU) high voltages. The ESU interference can range from hundreds of kHz to 100 MHz and up to several kilovolts. It can greatly disrupt the ECG signal. Why does this occur when the ECG bandwidth is only 100 Hz? The answer is (1) DC offsets and (2) obscuring the signal. ECG type instrumentation amplifiers may be relatively low bandwidth, but the junctions inside are capable of rectifying high frequency, like the ESU waveform. Then, parasitic capacitance around the junction filters the high frequency, resulting in dc offsets. The ECG baseline can actually move around when the ESU is triggered. Also, the high frequency goes right through amplifier stages, low-pass filters, and isolation stages to obscure the ECG as displayed on a CRT.

In the past, an ECG system just had to tolerate the ESU interference without being damaged. Then the requirement was to operate in some fashion. That is, the displayed ECG waveform had to be recognizable. Now medical professionals are requiring that ECG diagnostic quality remain in the presence of ESU noise.

Figure 5-8 shows a technique for reducing ESU noise in the front end of an ECG amplifier. It is composed of a three-stage *RC* filter arranged in a pi (π) configuration. An *LC* filter can also be used, but it is more difficult to maintain an *LC* match compared to *RC* match to common on one lead compared to the other lead. The common-mode effect results from series *R* or *L* and parasitic capacitance to patient common. Why does the common-mode time constant need to be matched? Because a mismatch in effect causes a 60-Hz differential error to appear. The effort in trying to reduce hundreds of kHz of ESU interferences unfortunately results in degraded total common-mode rejection (CMR) at 60 Hz. However, one can tolerate a little additional 60-Hz noise, because the right-leg drive still reduces 60 Hz, the IA still has its high CMR, and the isolation amplifier still has its high IMR to 60 Hz. It is

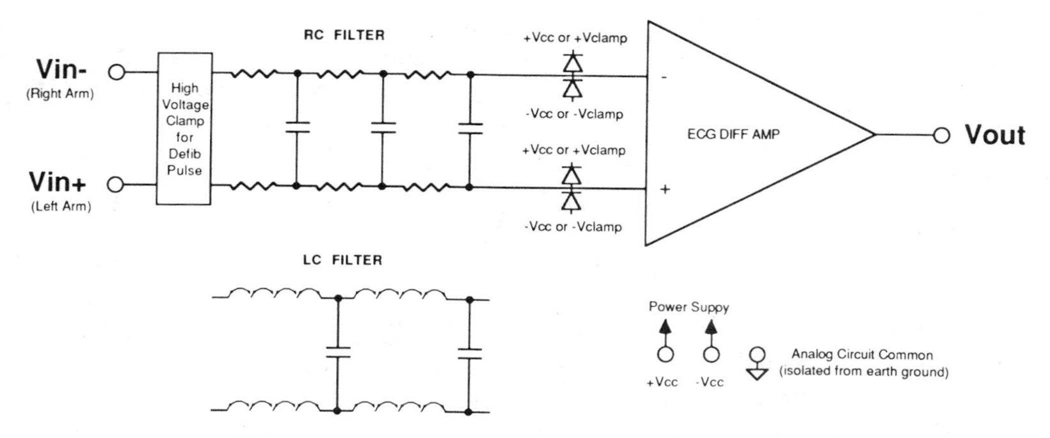

Figure 5-8 Electrosurgery unit (ESU) interference filter.

worth it to get rid of ESU noise. Also, some systems tune the isolation barrier to form a band reject filter around the ESU frequencies to further reduce ESU interference. The input filter shown has a cutoff frequency of 10 kHz. This is low enough to reduce ESU noise but high enough to prevent phase distortion of the ECG signal.

Wouldn't it be nice to get rid of ESU interference and retain as much 60-Hz rejection as possible? The circuit in Figure 5-9 can help to do this. It is a high-speed instrumentation amplifier designed with commercially available video op-amps. The gain is set to 5 V/V for the ECG signal, but the common-mode rejection is not only fairly

Figure 5-9 Wideband, fast-settling instrumentation amplifier with high common-mode rejection (CMR) at high frequency. (Courtesy of Burr-Brown under copyright 1991 Burr-Brown Corporation)

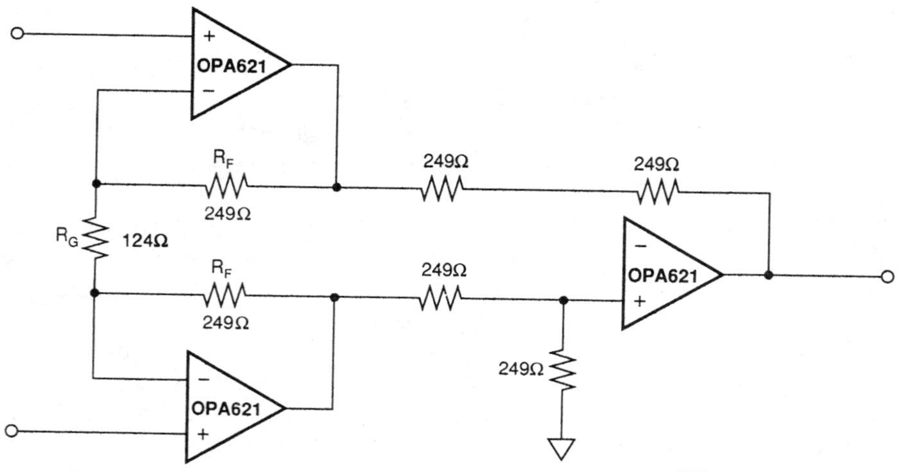

good at 60 Hz, it is also good at very high frequency.

The typical product data sheet shown in Figure 5-10 reveals that the CMR is about 75 dB at 60 Hz. This is not extremely high, but the op-amp CMR is still 40 dB at 100 MHz. Since two input op-amps are used to make the IA, their match might result in, say, 30-dB CMR for the complete instrumentation amplifier at 100 MHz and 60 dB at 1 MHz. That is high enough to reject ESU interference rather effectively. There is only one problem with the IA shown in Figure 5-9, aside from having to pay careful attention to printed circuit board (PCB) and system front-end layout: It has high bias current. The bias current of the high-speed op-amp shown is about 30 microamps. This exceeds the 10-microamp limit set by AAMI and UL544 standards. Also, the op-amp's input resistance is something less than 1 MΩ. This component was chosen to show that although it is possible, it is hard to find commercially available components that will have low bias current and high CMR at very high frequencies. To make the circuit shown in Figure 5-10 work for ECG systems, two high-speed, high-impedance, low-bias current input buffers would have to be added to the IA front end. Again, PCB layout and parasitic capacitance to common is critical in making the circuit function properly. This increases cost, but it may be worth it to get rid of ESU interference.

Figure 5-10 Important consideration: Frequency effect on common-mode rejection (CMR), OPA621. (Courtesy of Burr-Brown under copyright 1991 Burr-Brown Corporation)

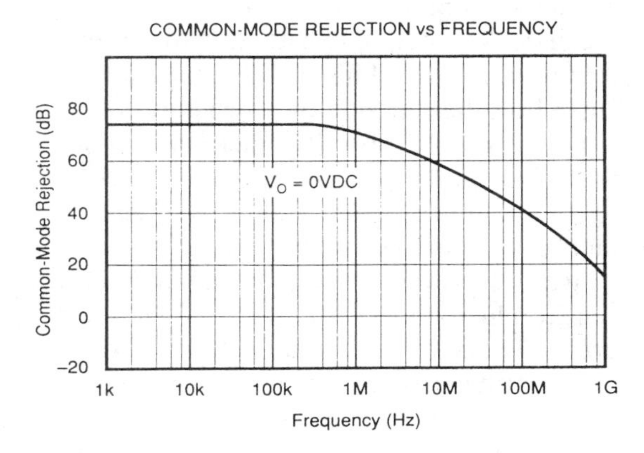

As we have seen, the ECG front-end amplifier is critical in achieving good performance. How does the ECG IA fit into the electronics of the entire monitoring channel? Figure 5-11 shows a block diagram of a multichannel physiological monitoring system. It accepts ECG, several channels of blood pressure (BP), body temperature, oxygen saturation, and perhaps other gases or body parameters. Notice the stray capacitance from the person to patient common and to earth ground. As mentioned earlier, this capacitance passes 60 Hz and other interference currents. These noise currents circulate in the patient and cables connected to the patient. Some efforts have been made to electronically model this phenomenon and predict how to minimize its effect. Stray capacitance should always be minimized to extract clean ECG signals from the human body. This involves connecting as little equipment to the patient as possible. The rest is magic.

The system in Figure 5-11 is composed of input buffers, an analog multiplexer (mux) plus Wilson network, programmable-gain instrumentation amplifier (PGIA), and sampling A/D converter with internal S/H amplifier. Notice that the digitized signals are in the form of serial data at the output. This allows isolation with just a few opto or isolated data couplers (one for data, one for clock, one for start conversion, one for data framing, etc.). If a 12-bit parallel output A/D converter were used, many more couplers would be needed (maybe 15).

The next few figures concentrate on how the ECG front end is implemented today. Although the complete ECG system uses 10

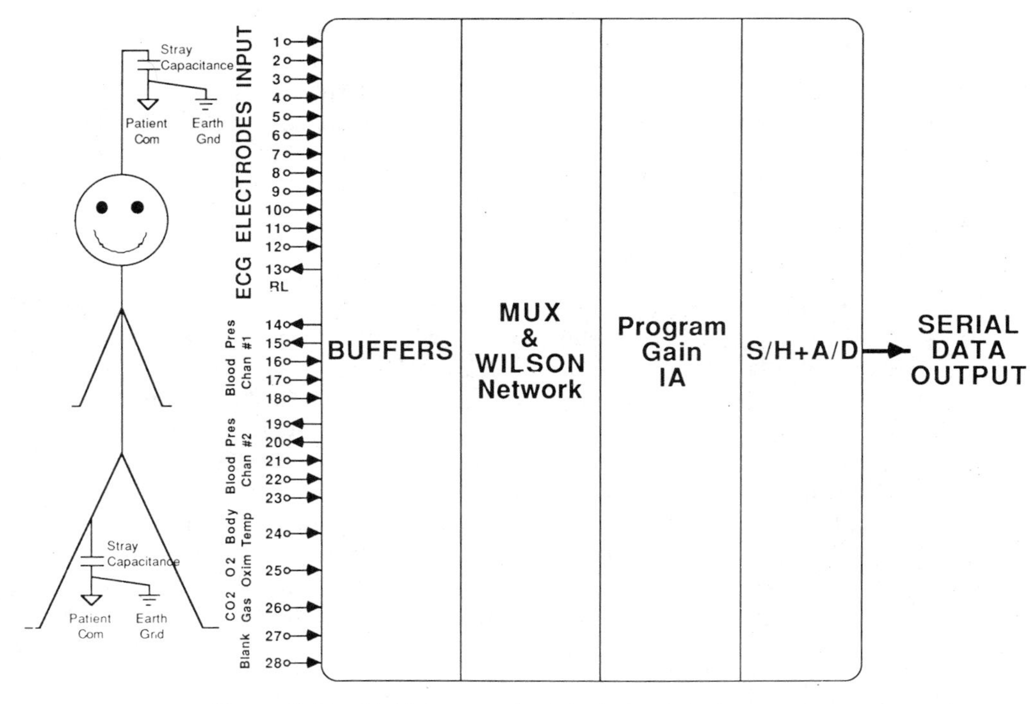

Figure 5-11 Multichannel physiological monitoring system.

to perhaps 14 patient electrodes, some more basic schemes use just three patient electrodes. These are right arm, left arm, and right leg. For this a simpler ECG amplifier composed of an IA, right-leg driver, and DC restorer can be used. Other ECG machines use five patient electrodes, as shown in Figure 5-12. Notice that all electrodes are buffered by a unity gain amplifier. These buffers have series protection resistors, on the order of tens to hundreds of kΩ, somewhere in the circuit and clamping diodes. This insures that amplifier inputs (from high defibrillation voltage) do not go more than one diode drop (about 0.7 V) above or below the plus and minus clamping voltage (sometimes just the power supply rails). Interestingly, modern AAMI and UL standards require that two or more faults occur before any current is driven through the patient. The fault current from the electronics is required not to exceed 50 microamps. Therefore, a monolithic circuit front end that has the series resistors on the chip might not qualify. One allowable fault on the chip could be the + input to the RA amplifier shorting to plus power supply of, say, 5 V. If patient resistance were 10 kΩ, 500 microamps would flow if it were not for the series protection resistors. If these resistors were on the chip, and the one fault occurred on the left side of a resistors to +5-V power supply, then 500 microamps would flow through the patient. This is not permitted. Off-chip resistors qualify as a separate fault condition. Therefore, series protection resistors should not be part of the monolithic semiconductor ECG front end shown in Figure 5-12. The only alternative is to make the clamp voltage so low that <50 microamps flow should one fault occur on the integrated

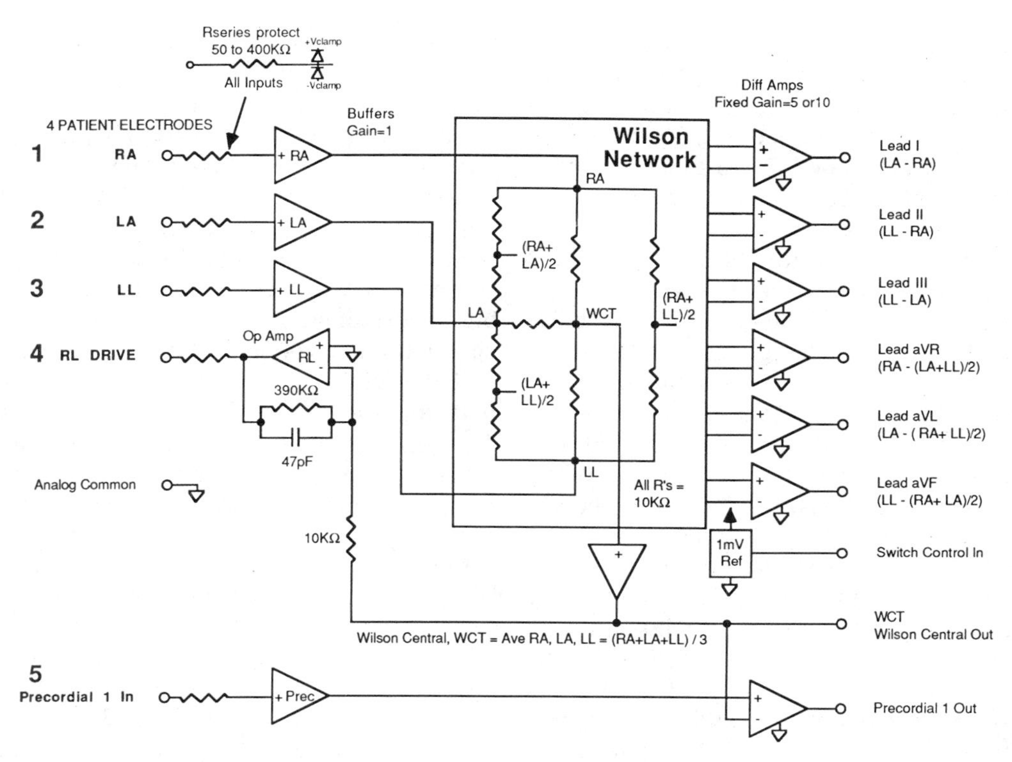

Figure 5-12 Five-patient electrode (6-Lead) ECG front end.

circuit. It would have to be 25 mV to limit current if patient electrode impedance were 500 Ω. This is less than one diode drop (700 mV), which means the clamp will never turn on. So it may not be practical to put resistors on the chip after all.

Another part of the circuit in Figure 5-12 is the Wilson network, which is just a passive resistor array that is used in deriving the six basic leads (I, II, III, AVR, AVL, and AVF). The tap points shown drive the bank of six differential input amplifiers each set to a gain of, say, 5 or 10 V/V. Gain must be low to prevent electrode offset voltage from saturating the amplifiers. Each diff-amp amplifies a tap or combinations of taps with respect to another tap as shown. For example, Lead I from the top diff-amp has left arm (LA) on its + input and right arm (RA) on its − input. Lead I, therefore, is (LA − RA). Other leads are derived similarly as shown. The output leads are ECG signals that have the 60-Hz noise greatly reduced. How does the circuit do this? The diff-amps remove it, because equal common-mode noise is present on each input of each diff-amp. The diff-amp just subtracts the equal noise voltage to give nearly zero while amplifying the difference in the unequal ECG signals (from different body locations) present on its inputs.

The Wilson central represents ECG zero and common-mode interference. It is derived from patient electrodes RA, LA, and LL and is actually the average of these (sum divided by 3). This is equal to the dc and 60-Hz common-mode voltage. The ECG signal from the Einthoven triangle is presumed to sum to zero. Of course, the right-leg drive is the inversion of the common-mode interference. The gain of the right-leg driver, RL, is usually set at around 30 to 50 V/V, but it could be higher. Actually, the higher the better, but care must be taken to assure the RL driver does not saturate. The 47-pF capacitor in the feedback loop limits the high-frequency gain and helps to prevent oscillation. How can oscillation occur? It breaks out if the patient's stray capacitance is such that the common-mode phase shift is enough to flip the total phase 180° around what was a negative RL-drive feedback loop. When this takes place, positive feedback occurs, and high frequency may appear on the ECG signal. The Wilson Central is shown as a buffered output in Figure 5-12, because it is used as one input to precordial diff-amps for removing 60-Hz interference. Phase shift through this buffer should be kept low.

No ECG front end would be complete without a 1-mV reference. The circuit shown in Figure 5-12 has a 1-mV reference connected to the input of the diff-amps. Upon manual switch or computer software command, this reference will be applied and appear on all output leads.

ECG systems often use QRS (left ventricular contraction) and pacer pulse (artificial pacemaker) detectors. Figure 5-13 shows a block diagram of such a circuit. Any of the six derived ECG leads or any of precordial signals can be used. Sometimes by examination the best one is chosen and used. The QRS detector is really just a differentiator. It measures how fast the ECG waveform is rising. If it is faster than, say, the *P*-wave but slower than, say, a pacer pulse, the circuit outputs a voltage transition to indicate that the QRS was present. The pacer pulse detector is also a differentiator set to detect faster-moving voltages. It outputs a voltage transition when the pacer pulse comes along during the ECG waveform. Actually, the pacer pulse is removed before further amplification is undertaken and reinstalled when the ECG signal is displayed. Amplitude accuracy is not necessary, because the pacer pulse is much bigger than the QRS complex. However, the time it occurred should be detected accurately. In Figure 5-13 both QRS and pacer pulse detect transitions are applied to comparators to make sure they really occurred. Then through output logic, digital levels indicate the presence of the respective signal. It is often very difficult to detect these signals correctly, because high noise may be present as well as ECG abnormalities like

Figure 5-13 QRS and pacer pulse detector block diagram.

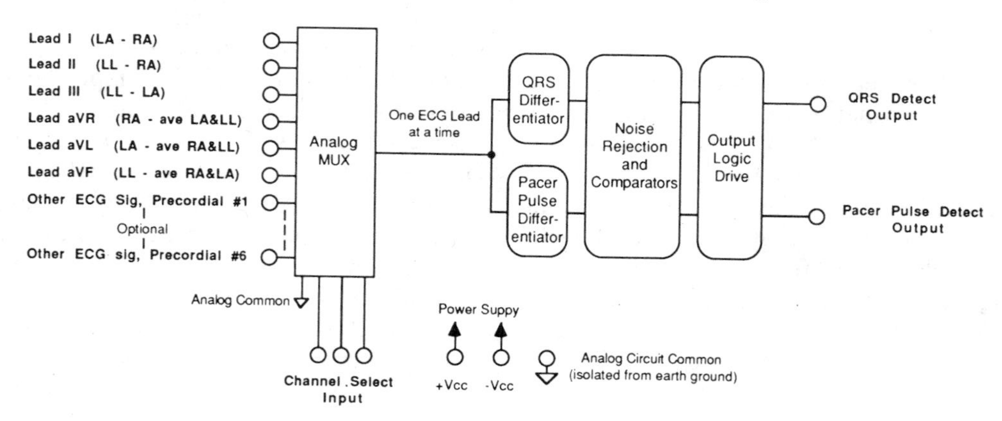

large *T*-waves or even muscle artifacts. AAMI Standard for Cardiac Monitors, Heart Rate Meters and Alarms, EC13, June, 1983, Revision—June, 1991, sets forth the amount of detection precision that is required.

How much circuitry is involved in a 10-patient electrode (12 Lead) ECG system? Figure 5-14 shows the block diagram. Notice that the four top patient electrodes involve buffers, Wilson network, and output diff-amps as described previously. The Wilson central, derived from the average of RA, LA, and LL signals, supplies a voltage to the negative inputs of the lower bank of six diff-amps set to a gain of 10. This acts to differentially cancel out the 60-Hz noise in the precordial signals. The upper bank of diff-amps remove 60-Hz noise and produce the six standard ECG leads, I, II, III, AVR, AVL, and AVF. These leads are applied to an analog multiplexer and selected one at a time for the QRS and pacer pulse detectors. They are also fed to a bank of 60-Hz notch filters, which can be switched in or out. Remember, notches remove residual 60-Hz noise but can cause phase shifts that disturb the normal ECG signal. That is why provisions for switching them out should be made.

The buffered precordial signals are also fed to the notch filters in Figure 5-14. Following the notches is a bank of 12 bandpass filters composed of high pass and low pass. The high-pass filters are selectable for 0.05 Hz (diagnostic quality ECG), 0.5 Hz (monitoring), and 2 Hz (quick dc restore). If just one high-pass filter were used, there would be a long time constant for it to charge to the new offset potential of a newly selected lead. Following the high-pass filters are programmable-gain amplifiers selectable for gains of 10, 20, 50, and 100 V/V. Then come low-pass filters, selectable to roll off at 40 Hz, 100 Hz, 150 Hz, and perhaps 3000 Hz. The bandpass filtered, amplified ECG signal is then applied to an sample-hold (S/H) amplifier that holds the ECG signal constant for the analog-to-digital (A/D) conversion time period. The serial data output from the A/D can be optically coupled for isolation to the computer. Also, the analog output from several of the 12 ECG signals can be simultaneously used for vector ECG. During transition between leads, the 1-mV reference can be switched in for calibration. The preceding discussion basically shows how a modern ECG system is constructed. New electronics available in the industry make the performance better and better with each passing year.

The diagrams in Figures 5-15 through 5-18 show trends in physiological monitoring, especially in ECG systems. The most significant change emerging is high-resolution digitization. Figure 5-15 shows the older approach, although it will still be used for some years to come. Basically, the ECG signal has to be digitized to the 10-bit to 12-bit level. That is, the resolution if 1 step out of 2^{10} or 1024 total steps. Each step is 0.1 percent of the total $[(1/2^{10})$ 100% in percent]. If the differential ECG input signal is 1 mV peak to peak, then one step must be 1 μV peak to peak, if it is digitized to 12 bits. This means that the input analog amplifiers must have noise that is somewhat lower, say 0.9 μV peak to peak. Since the rms noise is statistically about six times lower, the total rms noise of the input amplifiers in the 100-Hz ECG bandwidth must be about 0.3 μV rms. That is small but achievable with commercially available components.

The one difficulty in all this is the high gain required to get the ECG signal to a reasonable level, like 1 V peak to peak. As discussed earlier, electrode dc offset potentials must be removed or dc restored to zero before such high gain can be used. Figure 5-15 shows that 12-bit digitization can be done when analog hardware dc restoration is used. However, the trend is toward forgoing hardware dc restoration and using a high-resolution digitization as shown on the

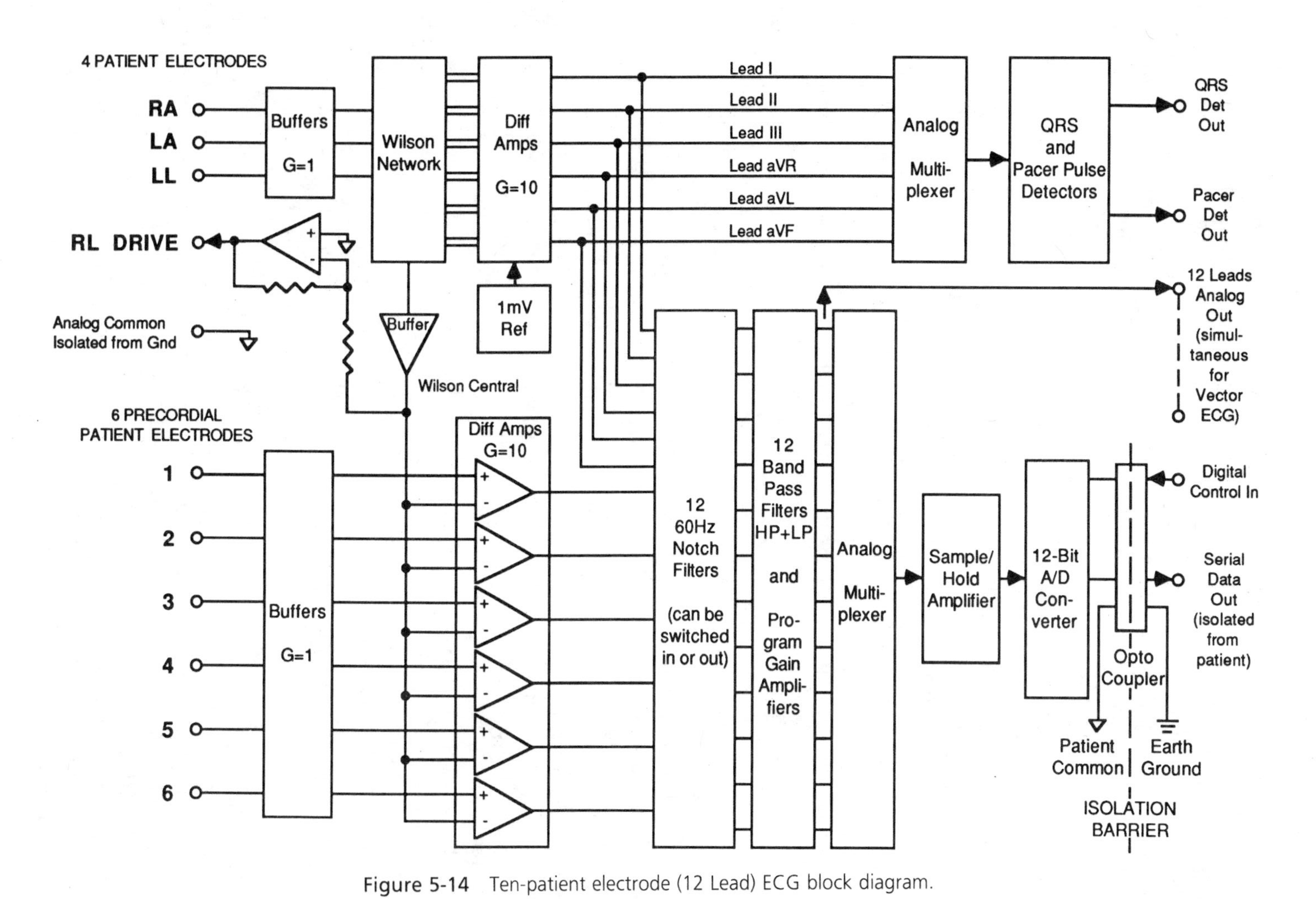

Figure 5-14 Ten-patient electrode (12 Lead) ECG block diagram.

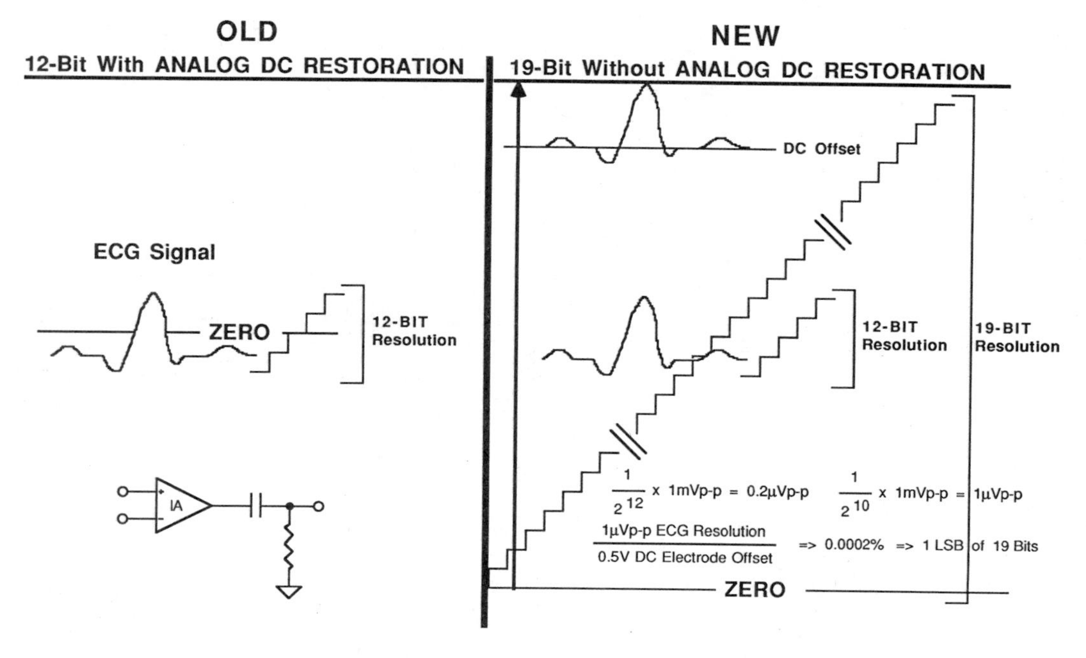

Figure 5-15 Digitization trend in physiological monitoring.

right. Here the 1-mV or even as small as 0.25-mV ECG signal rides on a 300-mV or even as big as 500-mV electrode offset. How many bits of resolution are necessary to do the job under the worse case condition? The answer is around 19 bits. Remember, the smallest ECG signal step for 10 bits on 1 mV is about 1 μV rms. Therefore, 1 μV ECG step/500 mV dc offset equals 0.0002 percent. This means that there must be 500,000 steps (1/0.000002). The nearest digital count is 19 bits, which is 2^{19} or 524,000 steps.

Some designers are even trying for 22 bits. The high-resolution digitization shown in Figure 5-15 allows, say a 1 mV ECG signal to be digitized to 10 to 12 bits while riding on a 500-mV electrode offset voltage. Once in the computer, the offset can be removed by the software and the ECG signal can be analyzed. This technique makes the analog front end simpler, although it places heavy demands on the A/D converter. This approach is used in industrial data acquisition systems also. One clear advantage is that no hardware dc restorer or high-pass filter bank is necessary. Digital offset subtraction and digital filtering take the place of hardware. There are no electronic high-pass filters with very low cutoff frequencies that have very long time constants. Therefore, one can switch from one channel to another without having to wait for the filter to charge to a new offset value.

How can this high-resolution digitization be implemented with commercially available components? Figure 5-16 shows an ECG system with 18-bit resolution. Notice how simple the analog front end is. It takes just three buffers, RA, LA, LL, a Wilson central resistor network, an amplifier with a gain of +1.25, and a bank of diff-amps to remove 60-Hz common-mode noise. The reason for this is clear when one examines how the RA,

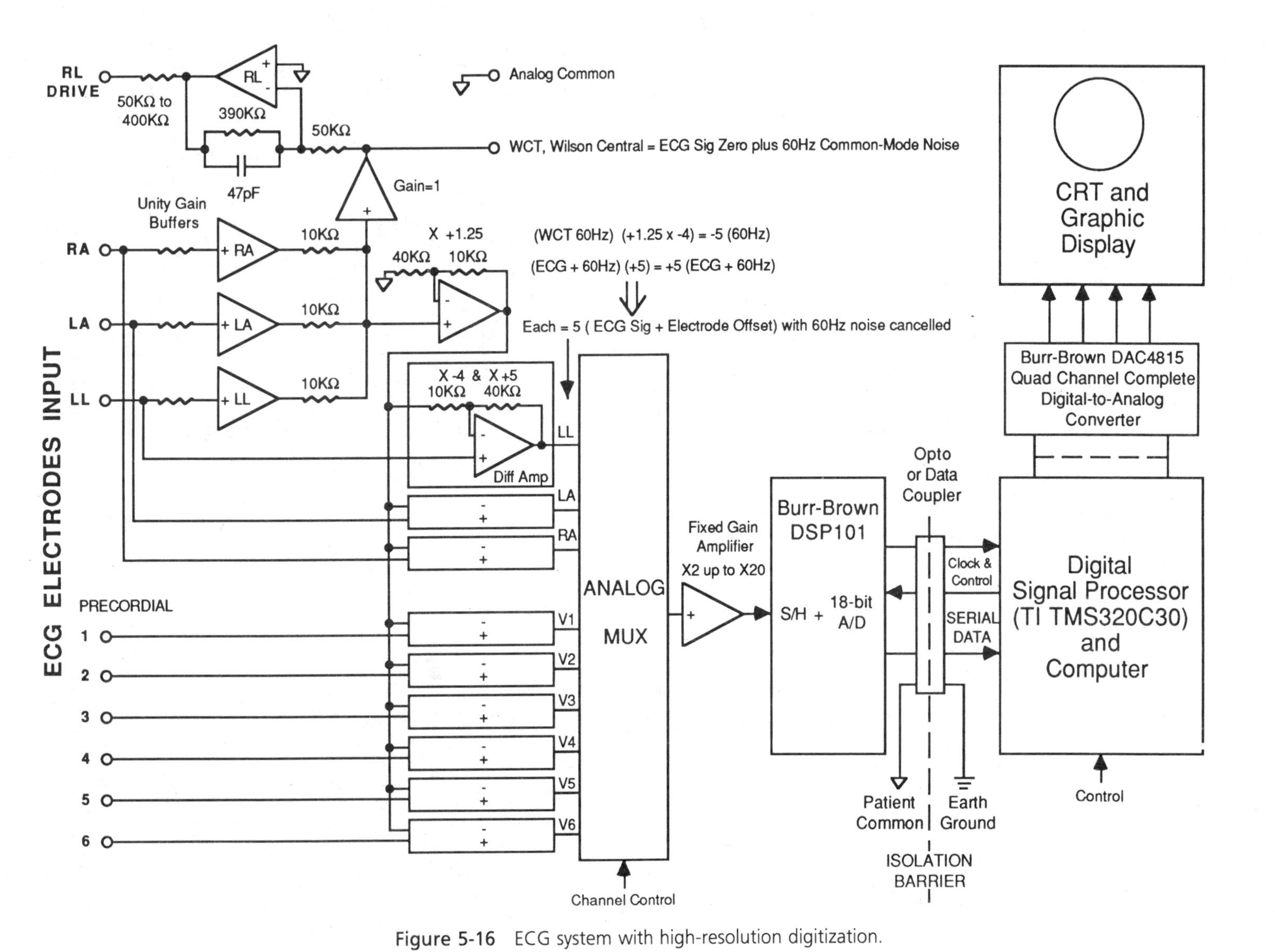

Figure 5-16 ECG system with high-resolution digitization.

LA, LL, and V_1 through V_6 ECG signals are amplified. The object is to amplify these signals of interest while rejecting the 60-Hz noise interference. The bank of nine very simple diff-amps does the job. The Wilson central 60-Hz common-mode noise times $+1.25$ V/V is gained up by -4 V/V to make it an inverted gain of -5 V/V at the output of the diff-amps. The signal with original phase 60 Hz riding on it is also gained up by $+5$ V/V ($1 + 40\text{k}\Omega/10\text{k}\Omega$) when it goes through the simple diff-amp. Hence, the original phase 60-Hz and inverted phase 60-Hz noise is cancelled at the diff-amp's output. Then all ECG signals are amplified by a factor of 5 V/V with common-mode noise greatly reduced at the diff-amp's output. Through a mux these signals can be further gained up by 2 V/V or perhaps as high as 20 V/V. The patient electrode offset potential, however, has not been removed. Hence the ECG signals plus electrode offset will be digitized by the high-resolution A/D.

The commercially available integrated circuit A/D converter shown in Figure 5-16 has an internal S/H amplifier, an 18-bit A/D converter, and glue logic to make it easily interfaceable to a digital signal processor (DSP). Its serial output, representing ECG signals riding on electrode offset voltages, is isolated through an opto or isolated data coupler and presented to the digital signal processor and computer. The ECG waveforms, after software has removed electrode offset, can be analyzed by the computer. ECG signals can be reconstructed into analog again through the commercially available quad channel digital-to-analog (D/A) converter shown. The analog ECG signal can be presented on a CRT or strip-chart recorder.

Since we are on the subject of high-resolution digitization, it would be interesting to discuss some future possibilities. Commercially available components are hard to come by to accomplish these system requirements. The block diagram shown in Figure 5-17 depicts the simplest analog hardware front end of all. It is just unity gain buffers. Through the analog mux, these voltages, representing ECG electrode signals, can be amplified in a

Figure 5-17 ECG system with high-resolution high-speed digitization: a possibility.

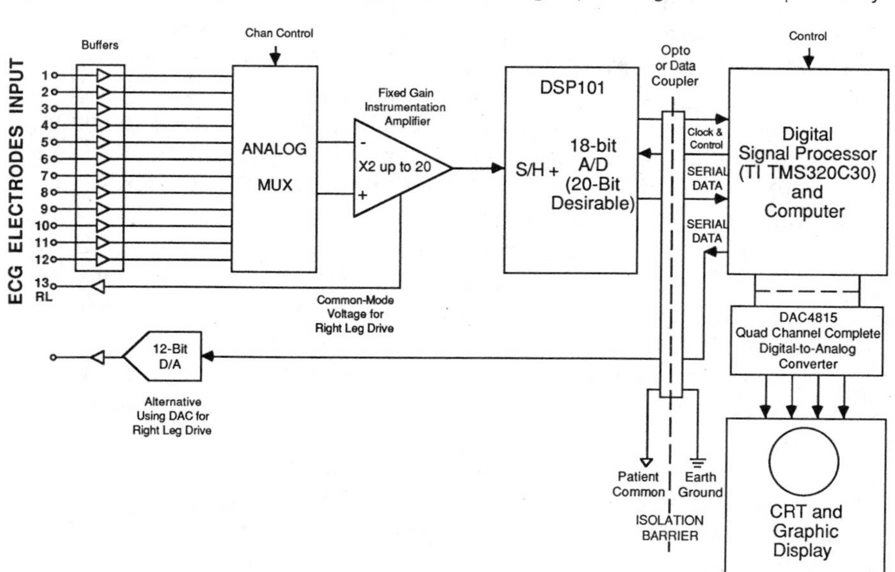

fixed-gain amplifier by a factor of 2 up to 20 V/V. The signals could be digitized by a high-resolution, high-speed sampling A/D converter and then fed to a microprocessor through opto isolation. If the A/D were fast enough, the 60-Hz noise interference could be digitized instantaneously along with the ECG signal. The software could then remove dc electrode offsets, 60-Hz noise, and derive all the leads. If the system were fast enough, it could even provide an analog signal through a digital-to-analog converter for the right-leg drive. Of course, right-leg drive could be done in hardware without looping through the computer.

This is all somewhat hypothetical, since low-cost, super-speed, super-high-resolution converters are not readily available. However, it is interesting to contemplate such a marvelous data acquisition system.

The last futuristic ECG system, in Figure 5-18, might be a little closer to realization. It uses simple analog circuity for the front end, followed by a bank of high-resolution delta-sigma A/D converters, one for each channel. A delta-sigma converter is really 1-bit digitization which essentially samples at a very high rate. The output is then decimated to reduce the data. *Delta* means that the input circuit is a difference amplifier, where one input is the analog signal and the other is a feedback signal from the sigma or integrator circuit. By going around many times, the converter accumulates counts in an up/down counter. The counter output is the digital output word, and through a D/A converter becomes the feedback signal for the input difference amp. This technique heavily oversamples the input analog signal, and therefore, the input antialiasing filter can be very simple. No brick wall or sharp cut-off filters are necessary. The digital output from each A/D can be digitally multiplexed into an opto coupler and then into the microprocessor.

Delta-sigma A/D and D/A converters are becoming very popular today, not just for ECG systems but for EEG as well. If the A/Ds in Figure 5-18 become less expensive and perhaps compact with several in one package, then this scheme could become reality.

5-8 ECG Readout Devices

Two forms of ECG readout device are commonly employed: oscilloscopes and strip-chart recorders.

Medical oscilloscopes are very much like any other CRO, except that the vertical amplifier bandwidth is severely limited, and the CRT *persistence* is very *long*. Most oscilloscopes made for ECG display use a horizontal sweep speed of 25 mm/s, while some *also* offer speeds of 50 mm/s and 100 mm/s.

Medical strip-chart recorders offer the same speeds; the 25-mm/s speed is standard, while the others are optional. Medical chart recorders *may* also offer 1 mm/s or 5 mm/s if they are to be used to make *trend* recordings of the arterial blood pressure wave-form (Chapter 6).

Standard ECG paper (Figure 5-19) has a grid pattern that is 50 mm wide. The *small* grid divisions are 1 mm apart, while the large grid divisions are 5 mm apart. These lines are used for making time and voltage measurements on the waveform. The vertical scale is calibrated at 0.1 mV/mm (i.e., 1 mV represented by two large divisions).

At 25 mm/s, the standard paper speed, the time interval between two small lines is 0.04 s (40 ms), and between two heavy lines it is 0.2 s (200 ms). The time intervals normally measured on the ECG waveform are shown in Figure 5-1*a*.

Heart rate in beats per minute can be measured on ECG paper because the rate (*frequency*) is the reciprocal of the *period*, which is measured on the recording using the aforementioned time intervals. Some ECG paper will have 3- or 6-s marks printed on the top margin, and on *any* paper 3-, 6-, or 10-s in-

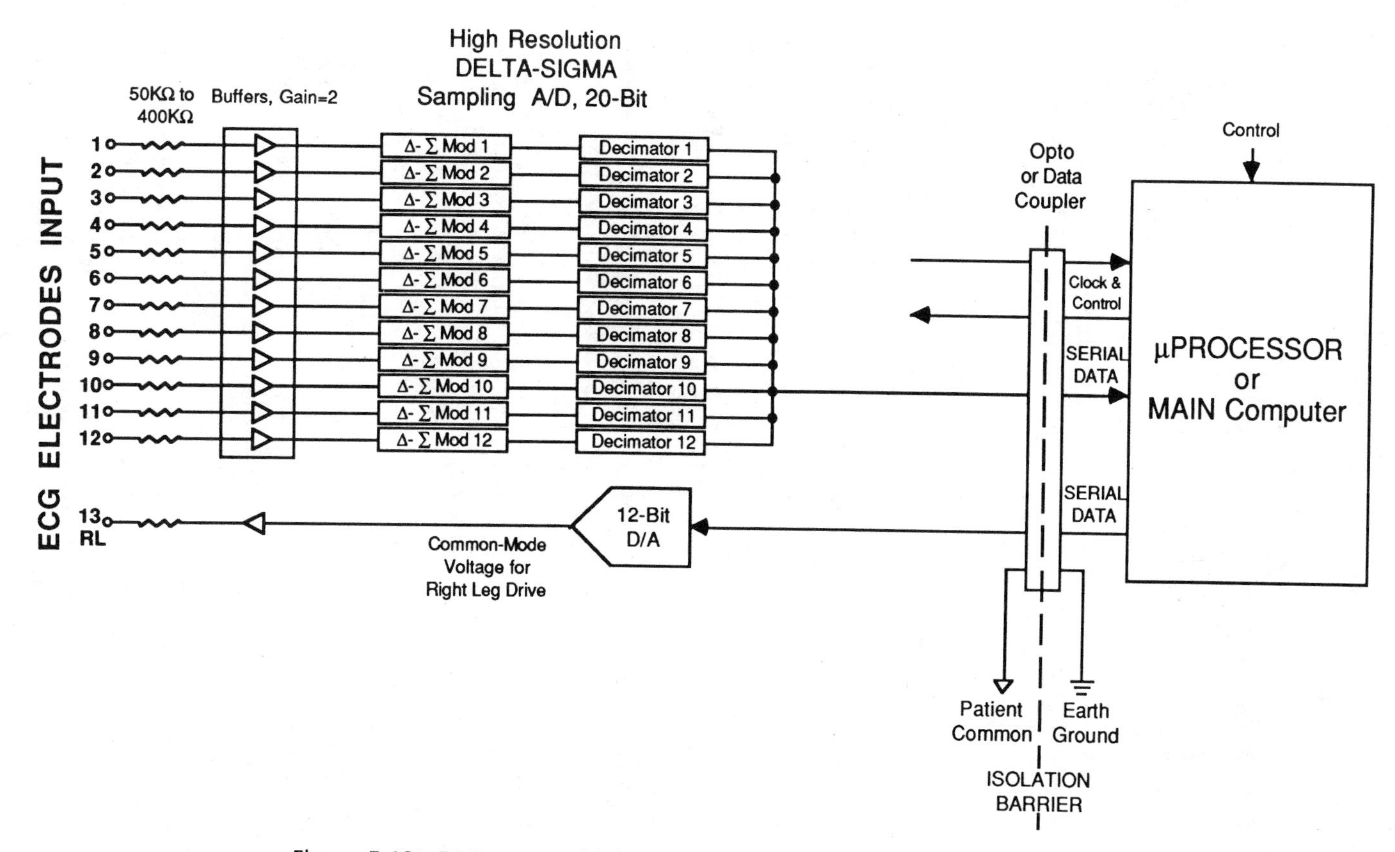

Figure 5-18 ECG system with high-resolution digitization per channel: a possibility.

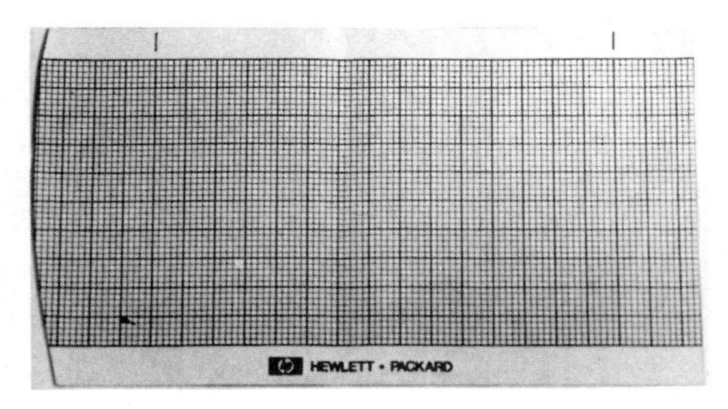

Figure 5-19 Standard ECG paper. (Photo reprinted courtesy Hewlett-Packard.)

tervals can be selected by counting heavy lines. Several methods for measuring the patient's heart rate are given next. Remember, however, that these are *sampling* methods that are only valid on patients with a *regular* heart rate (i.e., a constant *R-R* interval).

1. Count the number of *R*-waves in 3 s (15 large divisions) and multiply by 20.

2. Count the number of *R*-waves in 6 s (30 large divisions) and multiply by 10.

3. Count the number of *R*-waves in 10 s (50 large divisions) and multiply by 6.

4. Measure the time (t) between two *R*-waves, and divide it *into* 1500.

5. Use a millimeter ruler, or count the *small* divisions, to measure the *distance* between two adjacent *R*-waves. Find the heart rate on the standard chart (usually supplied free by medical suppliers, sales representatives and drug salespeople.)

6. Use a *heart rate ruler* (usually given away free by medical suppliers' sales representatives).

Some medical oscilloscopes use triggered sweep; the sweep cycle is initiated by the occurrence of an *R*-wave. In those scopes it is possible to read heart rate accurately by noting the distance between the left edge of the screen and the position of the first visible *R*-wave. Tektronix prints a heart rate scale on the upper margin of the screen to facilitate this measurement.

5-9 ECG Machines

5-9.1 A Typical ECG Machine

Figure 5-20 shows a typical ECG machine in portable (Figure 5-20*a*) and roll-around (Figure 5-20*b*) versions. These machines are identical, except for packaging. The portable is the Hewlett-Packard Model 1500B, while the roll-around model is the H-P Model 1511B.

The block diagram for a typical ECG machine is shown in Figure 5-21*a*. The electronics section will be an ECG preamplifier (see Section 5-6) and a power amplifier to drive the pen assembly or *galvanometer*.

A galvanometer is a *permanent magnet moving coil* (PMMC) assembly (see Figure 5-21*b*) that is similar in operation to the D'Arsonval meter movement.

A small bobbin wound with many turns of wire is mounted on a jewel bearing in the field of a permanent magnet. The writing pen is mounted, like a meter pointer, to the moving coil.

The PMMC assembly is constructed such that the pen will be at rest in the center of its travel when no current flows in the coil; but the pen will *deflect* in one direction when a current flows in the coil; the *direction* of the pen deflection is determined by the *polarity* of the current in the coil, while the *amount* of deflection is determined by the *amplitude* of the current. The deflection is caused because of the tiny magnetic field generated by the coil current interacting with the field of the permanent magnet.

Several different writing methods are used in strip-chart recorders, but most ECG recorders use *hot-tip*, or *thermal*, writing. The writing tip is not a *pen*, but is a *stylus* heated by a *resistance wire*. Early recorders fashioned the tip of the stylus from the heater

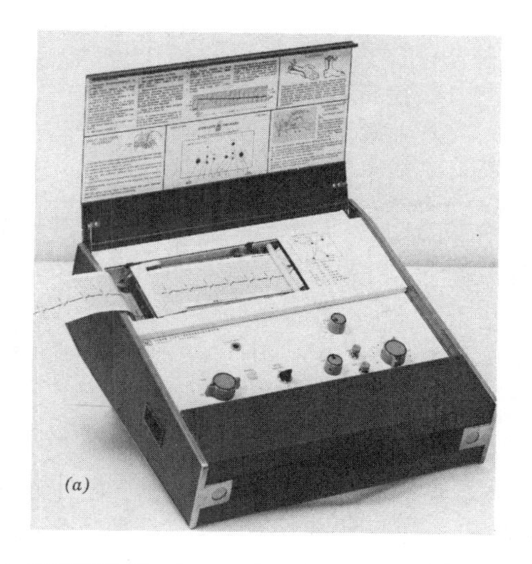

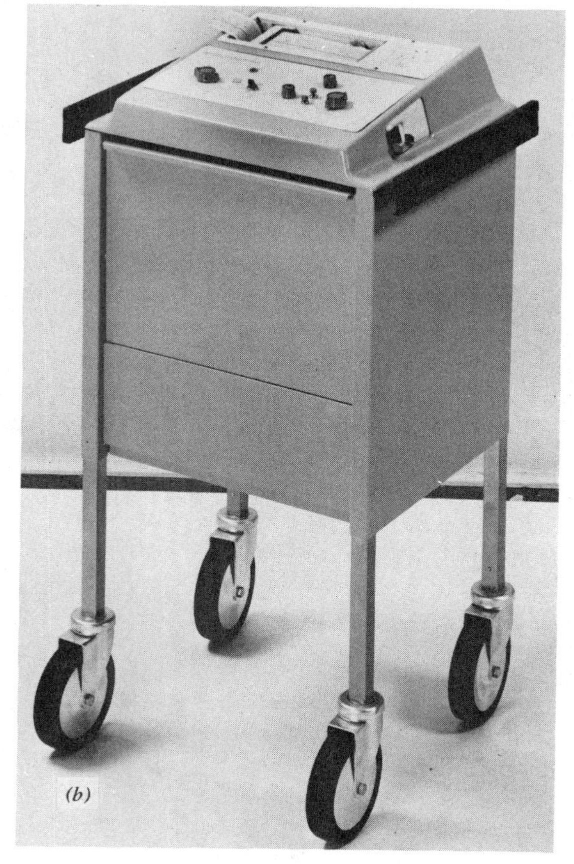

Figure 5-20 (*a*) Portable ECG machine. (*b*) Roll-around ECG machine. (Reprinted courtesy Hewlett-Packard.)

wire, while most modern types place the heater inside of a hollow metal cylindrical tube.

The paper used in thermal recorders is specially treated so that it turns black when it is heated. The hot tip of the stylus, then, will turn the paper black wherever it touches, thus recording the analog waveform applied to the galvanometer.

Very few ECG recorders actually use pens and pressurized ink to make the recording, and most of those that do use ink are in research applications.

The mechanical part of the ECG machine is also shown in Figure 5-21*a*. It should not be a surprise to note that the mechanics cause the largest share of faults in most models.

The ECG recording paper comes in either roll or "Z-fold" (folded flat) and is stored in a compartment just below the paper tray. A button release is usually found along one edge of the tray and allows the tray to swing up to give access to the paper storage compartment.

The paper is passed over the *writing edge* and the *paper tray* and between the drive roller and idler roller. Two forces act on the paper, and *both* are essential to the proper operation of the machine: a *forward* tension, provided by the drive/idler roller assembly, and a *reverse* tension. These forces act together to stretch the paper *taut* across the writing edge.

The reverse force is *weaker* than the forward force and may be provided by any of several ways. In some machines a *paper brake* is used, while in others *tension bars* or *tension rollers* are used. These structures will be located between the paper supply and the writing edge.

The writing (knife) edge in the thermal recorder is essential to producing an undistorted recording. The tip of a pen or stylus travels in an *arc*, so fast-rise-time waveforms such as a squarewave or ECG *R*-wave will

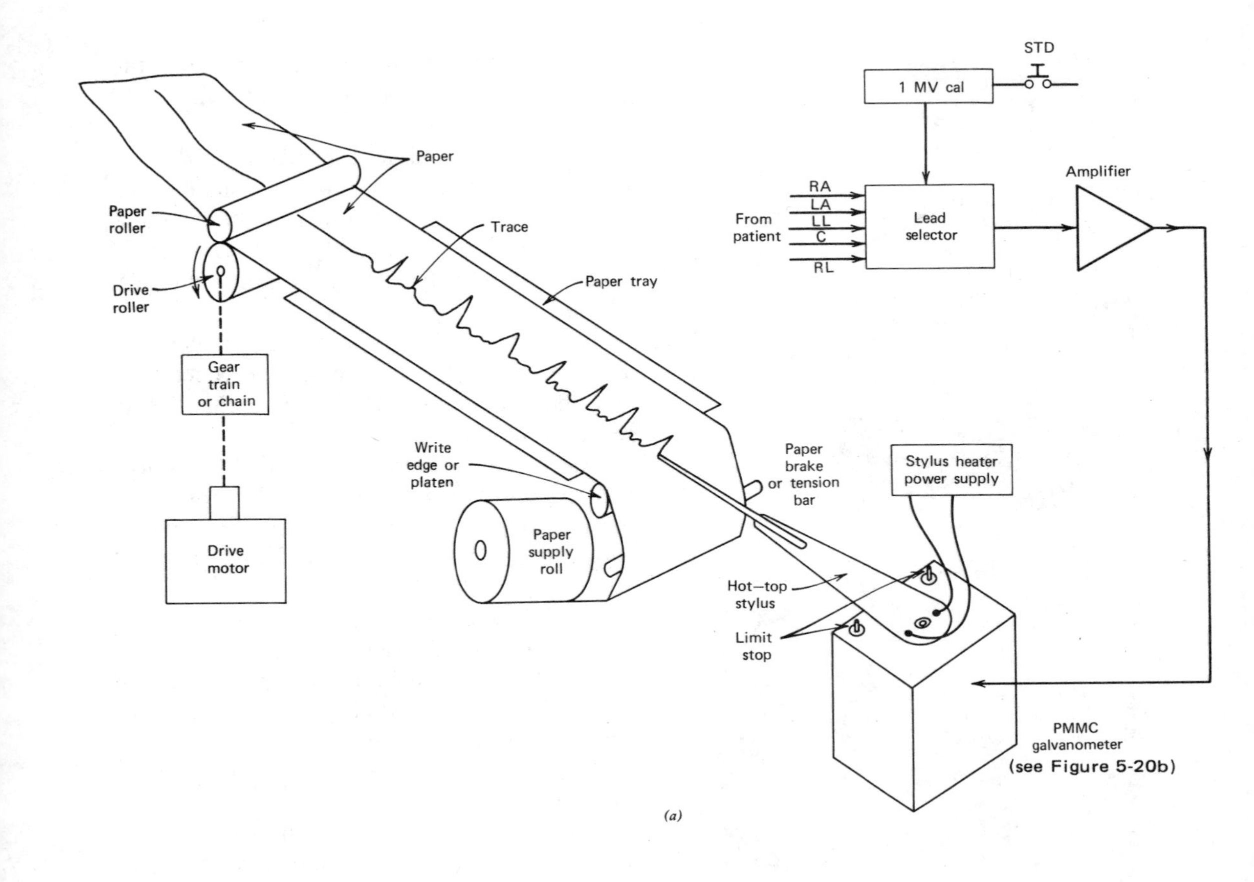

(a)

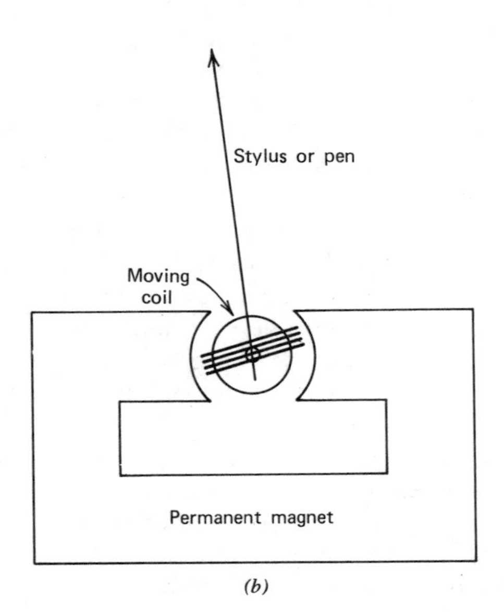

(b)

Figure 5-21 ECG machine mechanism. (*a*) Block diagram. (*b*) PMMC galvanometer. (Reprinted courtesy Hewlett-Packard.)

appear to be curved, or as an ECG machine user is likely to claim "it writes backwards." But in ECG recorders using a thermal stylus, the trace is made on the write edge and will appear *rectilinear*, as opposed to *curvilinear*, which is how noncorrected traces would appear. This type of recording is not truly rectilinear but is known as a pseudorectilinear recorder. The secret to the pseudorectilinear recorder is that different portions of the stylus touch the paper over the write edge in different portions of the paper.

The motor is connected to the drive roller through either a drive chain or a gear train. Most machines use ac motors, with tapped windings to select drive speed. Some ac motors are able to provide very accurate drive speeds that are synchronized to the ac power main's frequency—60 Hz in the United States.

Only a few models use dc motors, and those regulate speed by using a regulated dc power supply, and in some cases an alternator/tachometer on the motor shaft to provide negative feedback.

A large number of modern ECG recorders are based on the same principle as the dot-matrix printer used with computers. Even a 27-pin printer can make decent recordings of the waveform and print out numerical data such as heart rate, patient ID number, date, time, and the numerical data from other sensors (e.g., blood pressure, temperature, and so forth). These recorders will be discussed more fully when we discuss mechanical recorders.

5-9.2 Multichannel Machines

Some hospitals use three-channel ECG machines to make diagnostic recordings or in the exercise ECG laboratory. An example shown in Figure 5-22 is the Hewlett-Packard Model 1505A. This machine examines 3 of the 12 leads at a time and will automatically sequence through four groups of 3 leads each until the entire set of 12 leads are displayed. Each group is displayed on the paper for several seconds.

A few multichannel models are equipped with a transmitter device that converts the analog waveform, and digitally encoded numerical data such as the date, time, and patient ID number, to audio tones that can be sent to a central station in another part of the hospital, or even to another hospital. Tone-encoded systems allow a tentative diagnosis to be made by computer, followed by a manual reading by a trained physician.

The analog or digital-to-audio converter that allows communication of the analog ECG data to audio tones, for transmission over phone lines, is called a modulator-demodulator, although its acronym MODEM is now sufficiently widespread that it is considered a noun in its own right (*modem* instead of *MODEM*—just as *laser* and *radar* became nouns, the computer culture is making *modem* a noun also). The modem allows the physician to transmit the ECG signal to either a physician with superior knowledge of ECG reading or to a central computer where one of several validated diagnostic programs will evaluate it. Indeed, some of the computer diagnostics are superior to the abilities of most physicians—except those specially trained in electrocardiography. At one time, the modem-equipped ECG machines were rare and expensive, but currently it is possible to obtain portable ECG instruments with a modem option at only slightly higher cost than standalone machines.

5-9.3 Patient Cables

In one respect, the patient cable (Figure 5-23) is the most important part of the system; it is most frequently at fault when the machine fails to operate properly.

There are several different patient cable configurations used in ECG recording. Some are constructed of two pieces that plug to-

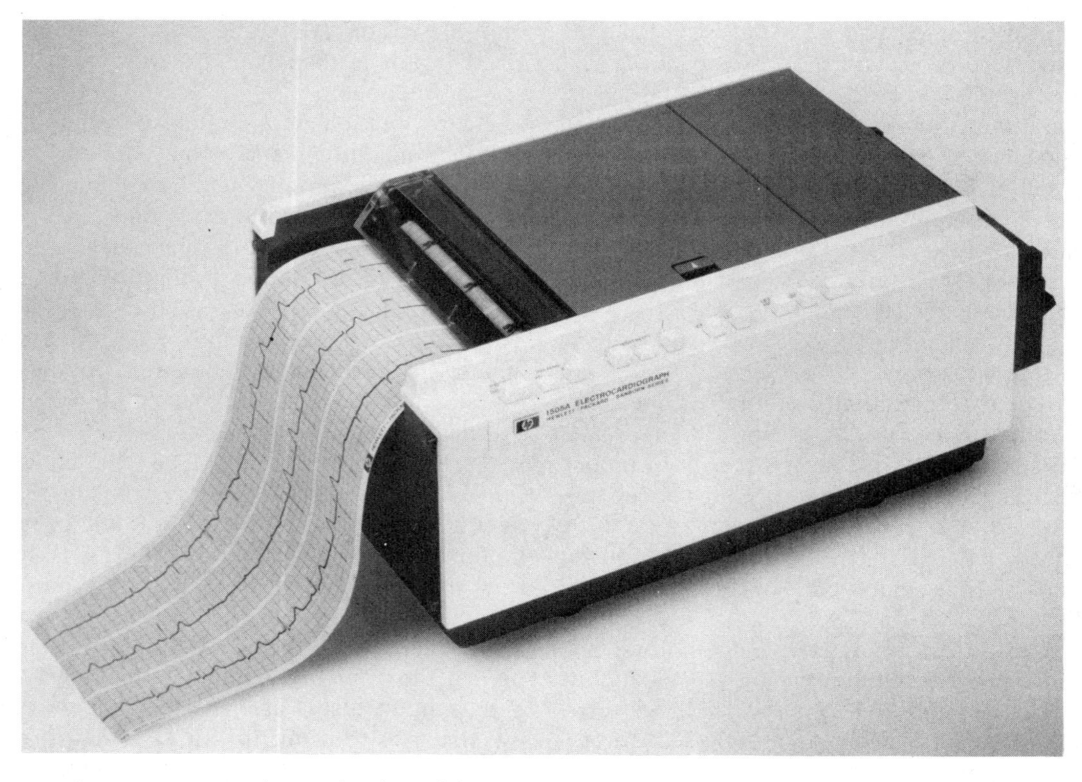

Figure 5-22 Hewlett-Packard Model 1505A automatic three-channel electrocardiograph. (Photo courtesy Hewlett-Packard.)

gether (Figure 5-23), while others are of one-piece construction. Two-piece units are generally more expensive initially but often prove more economical in the long run because breakage most frequently occurs at the electrode connector end of the cable. This end can be replaced at a cost that is somewhat lower than the cost of a one-piece cable. Additionally, the use of a two-piece cable allows the use of different types of electrodes. A single main cable (i.e., the machine end) and several adapters allow for the different electrode types.

There are three basic types of electrode connectors on the patient end, and these occur in various configurations; some are unique to one manufacturer's electrodes.

A *tip-end* cable is used to connect temporary, or short-term, electrodes such as the plate (Figure 3-2*a*) or suction cup (Figure 3-2*b*). These are available with standard *banana plug* ends or a special end that closely resembles a slightly oversized telephone tip plug (i.e., *pin tip*).

Another type of cable (the upper cable in Figure 5-23) has an Ag-Ag Cl cup electrode already attached. The cup is filled with electrode jelly every time that it is used and is secured to the patient's skin with adhesive tape, or adhesive patches intended specifically for this application.

The last type of cable uses a special *clip*, or *buttonsnap*, fastener that connects to the standard monitoring electrode (see Figure 3-

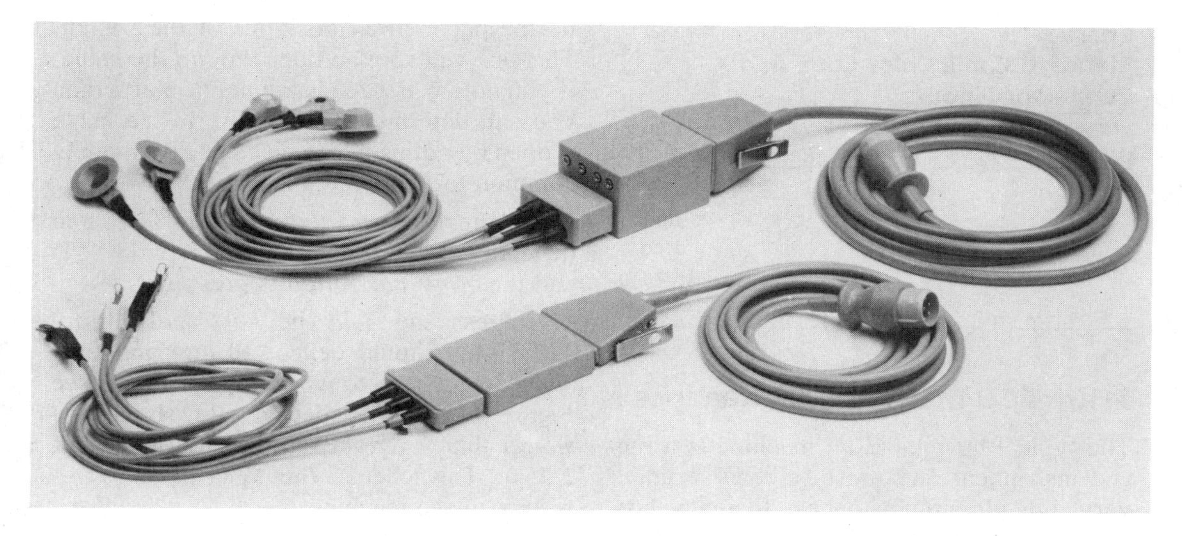

Figure 5-23 ECG patient cables. (Photo courtesy Hewlett-Packard.)

3). This the type of cable most often used in ICU/CCU monitoring.

Notice in Figure 5-23 that one of the cables has five patient electrode leaders, while the other has only three leads. A cable for a diagnostic machine would need all five electrodes in order to record all 12 leads, but in *monitoring* often only the *existence* of the ECG and certain gross *arrhythmias* are required, and these can be obtained through any lead. For those cases a three-electrode cable is sufficient. In most cases the three-electrode system is configured to produce lead I, although the nurse or physician can select *any* lead by proper electrode *placement.*

The upper cable in Figure 5-23 has a large block between the two halves of the cable. This device is a special *filter* to reduce the effects of *electrosurgery* equipment; that is, high-power rf generators (see Section 18-3). The filter is optional, and if not used, the two halves of the cable simply plug together as in the other cable shown.

Most ECG cables are made of shielded wire, so an ohmmeter connected between each electrode connector and the appropriate terminal on the main plug will read a short circuit. But some cables have a 1- to 10-kΩ resistor in series with each electrode to provide defibrillator protection. These resistors are usually located inside of the molded plastic connector that joins the two halves of the cable on the machine end.

There is no universal standard ECG input plug, and machines by the same manufacturer may have different input connectors. The plug shown on the cables in Figure 5-23, however, is the most widely encountered. Several companies claim credit for this plug, but most people seem to call it the *Sanborn* configuration after the company that popularized and may have invented it. Sanborn was bought by, and became the medical division of, Hewlett-Packard some years ago.

The semistandard Sanborn plug uses a five-pin, size-14, AN/MS military plug. Chart 5-1 shows the pinouts for the plug and the color code used to identify the electrode leaders.

Chart 5-1
Standard Cable Color Code and Connector Pinouts

AN/MS Plug Pins	Goes To	Electrode Color Code
A	RA	White
B	LA	Black
C	LL	Red
D	C	Brown
E	RL	Green

5-10 ECG Machine Maintenance

The typical hospital ECG machine is a rugged instrument and must be reliable under very difficult circumstances. In many hospitals the machines receive little care from operating personnel, but many hospitals now delegate an ECG technician or biomedical equipment technician to periodically inspect and perform minor repairs to the ECG machines.

A reasonable procedure for daily or weekly operational checks is as follows:

1. Turn machine on and allow it to warm up for a minute or so (longer on vacuum tube models).

2. Place the *function* switch in RUN, and the *lead selector* switch in STD. Observe whether a trace is present.

3. Press the *1-mV cal* button several times. Take note of (a) whether vertical edges of the pulse are visible and (b) whether the *sensitivity* control can be adjusted to provide at least 10 mm of deflection. (c) Is the pulse reasonably square?

4. Adjust the *position* control through its entire range and note whether the stylus is able to travel to the limits or stops at the top and bottom margins of the paper.

5. Short together all electrode connectors at the patient end of the cable, and then turn the *lead selector* switch through all 12 positions. You should see a quiet, stable baseline on the paper in all positions of the switch. This test will spot an open wire in the cable. If you note which leads are not "quiet," then you can determine which wire in the cable is open by noting which electrode is uniquely common to *all* affected leads.

6. Adjust the *sensitivity* for exactly 10-mm deflection when the *lead selector* is in STD and the *1-mV cal* button is pressed.

7. Press and *hold* the *1-mV cal* button. The stylus should deflect 10 mm and then *slowly* drop back to its original position (see Figure 5-24). The decay rate should be *slower* than 7 mm in 16 large divisions (i.e., 3.2 s). This checks the *low-frequency response* of the machine.

ECG machine manufacturers may specify a more elaborate procedure for half-yearly or yearly checks, but the foregoing procedure will locate and allow correction of the most commonly found faults, and it can be performed in only a few minutes per machine.

Additional safety inspections involving electrical leakage to the patient from the power mains are also specified (Chapter 19).

5-11 ECG Faults and Troubleshooting

Internal electrical or mechanical faults occur only occasionally in ECG machines, yet incidents of malfunction occur often enough that many hospital personnel become con-

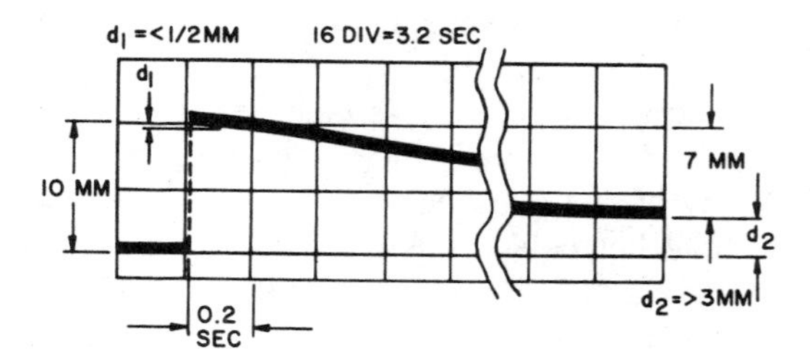

Figure 5-24 Self-check decay curve. (Reprinted courtesy Hewlett-Packard.)

vinced that ECG machines "are always on the blink." In *most* cases, however, the malfunction is an operator error or can be corrected by a simple adjustment or repair. The problems discussed in the following examples are common enough to occur *daily* in many large hospitals.

EXAMPLE 5-1

Symptom: Machine runs, but the thermal tip stylus does not write, or writes very lightly.

Possible Causes: (1) Too little heat on the stylus tip and (2) insufficient stylus pressure on the paper.

Troubleshooting (Machine running):

1. Using an *insulated* probe such as a screwdriver, *gently* press the stylus onto the paper.

2. If a dark line appears on the paper, then the problem is pressure, but if no dark line appears, then the problem is heat.

(*Note:* Some people use their finger to *quickly* touch the stylus to see if there is plenty of heat, but this is potentially painful. If the heat is sufficient to write on the paper, then it is also sufficient to cause second-degree burns.)

SOLUTIONS

1. For *no heat,* check the heater voltage at the stylus wires. If the voltage is *correct,* then change the stylus. If the voltage is *not correct,* then refer to the service manual for troubleshooting.

2. Adjust the stylus pressure (Figure 5-25). DO NOT GUESS at the proper pressure, as different models may require anything between 2 and 20 g. Use a stylus pressure gage and refer to the manufacturer's service manual for the correct value. On some models the pressure must be made at a specific heater voltage.

EXAMPLE 5-2

Symptom: Smeared trace (Figure 5-26*a*).

Possible Causes: Worn stylus or incorrectly loaded paper.

Troubleshooting: Check paper loading and, if proper, check stylus for wear, pitting, or other irregularities.

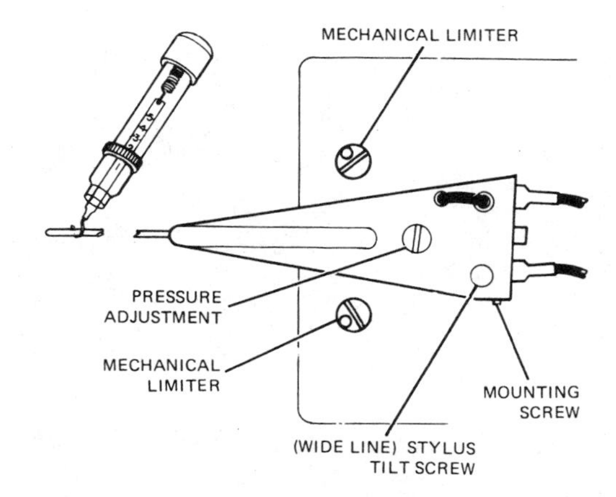

Figure 5-25 Measuring stylus pressure. (Reprinted courtesy Hewlett-Packard.)

Discussion: Incorrectly loaded paper is one of the most common faults, and in most cases the error resulted from *bypassing* the paper brake or tension bars (see Figures 5-26*b* and 5-26*c*).

EXAMPLE 5-3

Symptom: Poor recording.

Possible Causes: Electronic or mechanical problems, bad lead switch or input connector, bad patient cable, or improper connection to patient.

Troubleshooting

1. Place *lead selector* switch in STD, short all electrodes together, and press the *1-mV cal* button.

2. If normal calibration pulses appear, then the problem is the *connection to the patient*.

3. If problem persists, repeat step 1 using a known-good patient cable or a *deadhead plug* (i.e., ECG connector with all pins shorted together). If the problem clears up, then replace the bad patient cable, but if it persists, the fault is

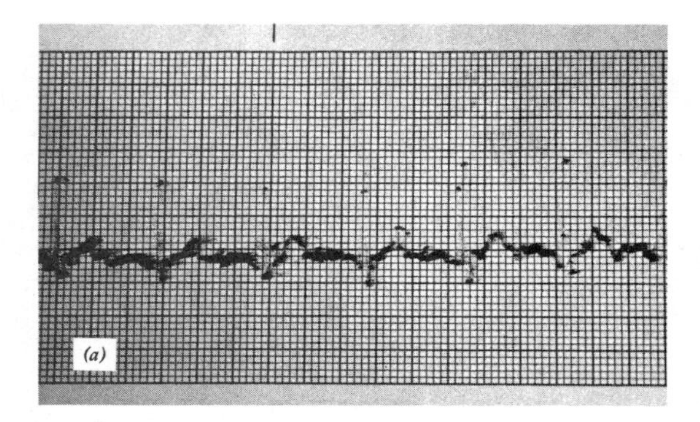

Figure 5-26 (*a*) Smeared trace. (*b*) Paper correctly loaded. (*c*) Paper incorrectly loaded.

inside of the machine. Refer to the service manual for troubleshooting.

The approximately 1-mV ECG biopotential must be recorded in a hostile environment; 60-Hz interference may exist, in addition to muscle potentials and other bioelectric artifacts. Figure 5-27 shows four common artifacts encountered in ECG recordings.

Figure 5-27*a* shows 60-Hz interference from nearby power mains. Ordinarily, the 60-Hz signals induced into the electrode wires form a *common-mode* signal because all electrodes and wires are affected equally; these signals usually do not interfere because the ECG preamplifier has differential inputs. But electrode defects, open patient cable wires, or poor contact to the patient result in an unbalanced input to the preamplifier and will thereby manufacture a differential signal from the common-mode 60-Hz signal.

This type of interference often results from a lack of electrode jelly or loose contact to the patient. This problem is especially prevalent on patients with moist or sweaty skin.

Another cause of 60-Hz interference is a loose or broken power mains ground on the ECG machine, or certain other instruments that may be connected to the patient. Additionally, certain dc power supply problems (i.e., an open filter or shorted voltage regulator) produce a similar artifact from 120 Hz ripple from the rectifiers.

The problem can be isolated by shorting together all electrodes of the patient cable and then checking each position of the *lead selector* switch.

1. If the interference ceases, then the problem is a bad electrode, no electrolytic jelly, poor skin preparation, and so forth.

2. If the interference exists on all positions of the lead selector switch, then the problem is probably internal to the machine.

3. If the problem exists only in certain lead positions of the selector switch, then suspect an open wire in the patient cable. Use an ohmmeter, conductance checker, or analyze the situation to determine which

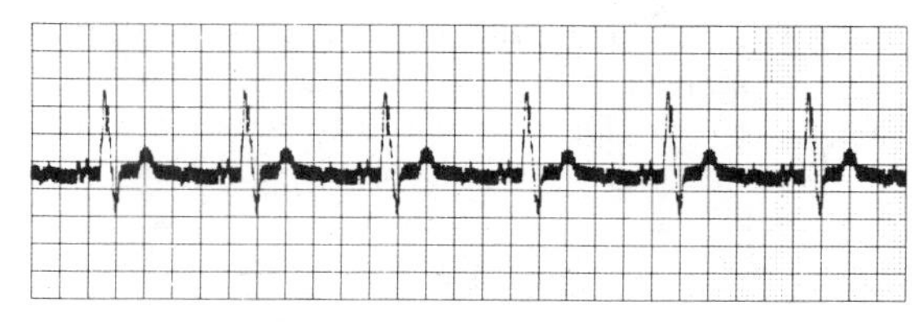

ALTERNATING CURRENT (AC) INTERFERENCE

(a)

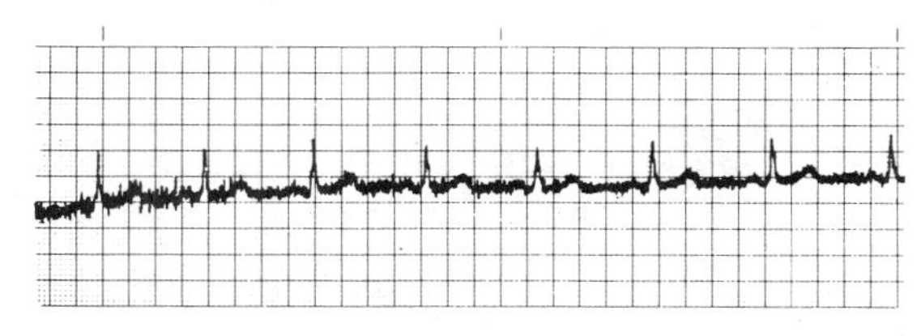

SOMATIC TREMOR

(b)

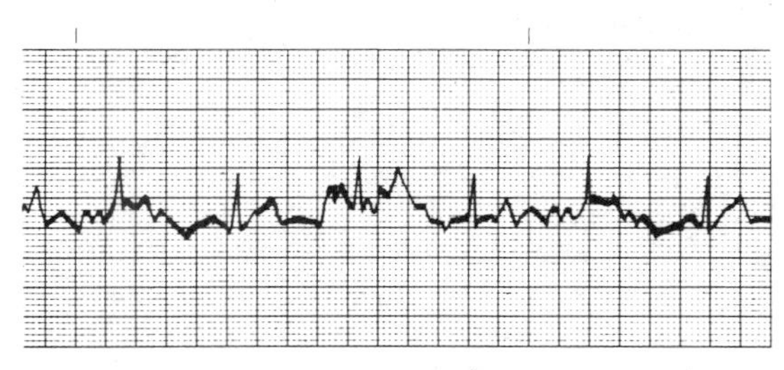

IRREGULAR BASELINE

(c)

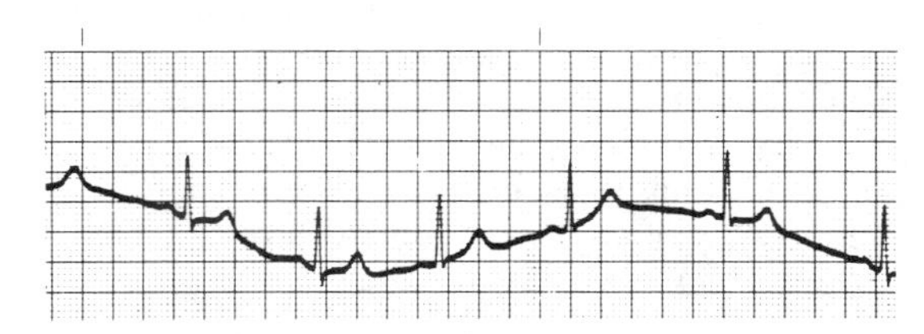

WANDERING BASELINE

(d)

Figure 5-27 ECG recording irregularities. (*a*) Alternating current (ac) interference. (*b*) Somatic tremor. (*c*) Irregular baseline. (*d*) Wandering baseline. (Reprinted courtesy Hewlett-Packard.)

electrode is uniquely common to the affected leads.

Muscle jitter, also called *somatic tremor,* is shown in Figure 5-27*b*. It is distinguishable from 60-Hz interference by its *lack of regularity* in both *amplitude* and *frequency* components and is caused by muscle bioelectric potential.

In some cases the symptoms can be eliminated by having the patient lie still, but if the patient is suffering from a tremor condition, or is being monitored on a long-term basis, then electronic *filtering* must be used. Most bedside monitors are designed with a frequency response of only 30 to 50 Hz in order to reduce the effects of somatic tremor signals.

Figure 5-27*c* shows an erratic baseline condition. This symptom results from dirty electrodes caused by dried jelly on the surface or metal particles on the patient's skin, for example. Both are cured by *cleaning,* in the former case of the electrodes and in the latter of the skin.

The wandering baseline of Figure 5-27*d* is almost always caused by *movement* of the electrode with respect to the surface of the skin. In theory, there are also electronic problems that could cause this problem, but these are very rarely encountered. In *most* cases, the wandering baseline is caused by *changes* in electrode contact impedance creating a *varying* electrode offset potential. Three mechanisms for wandering baseline are commonplace:

1. Loose or poorly fitted electrodes (may *also* create a 60-Hz artifact).

2. Patient cable dangling toward the floor, placing a *tension* force on the electrode.

3. Electrode or cable moving with the patient's respiration.

Securing the electrode will cure problem 1, and dressing the cable properly will cure problems 2 and 3.

5-12 Summary

1. The electrocardiograph machine takes 12 leads (I, II, III, AVF, AVR, AVL, and V_1 through V_6) to give the physician different views of the heart's activity.
2. There are three basic lead configurations: *bipolar limb leads* (I through III), *unipolar limb leads* (AVR, AVL, AVF), and the *unipolar chest leads* (V_1 through V_6).
3. *Diagnostic* ECG recordings require a bandwidth of 0.05 to 100 Hz, while *monitoring* requires a bandwidth of 0.05 to about 40 Hz.
4. An ECG preamplifier is a bioelectric amplifier with a *lead* selector switch, 1-mV calibration source, and defibrillator protection.
5. Strip-chart recorders and oscilloscopes are used as ECG readout devices. The standard speed is 25 mm/s, with some machines also offering 50 mm/s and 100 mm/s.
6. ECG strip-chart recorders must present *both* forward and reverse tension on the paper so that it is stretched taut across the *write edge*.

5-13 Recapitulation

Now return to the objectives and self-evaluation questions at the beginning of the chapter and see how well you can answer them. If you cannot answer certain questions, place a check mark next to each and review appropriate parts of the text. Next, try to answer the following questions using the same procedure. When you have answered all of the questions, solve the problems in Section 5-15.

5-14 Questions

5-1 The ECG waveform is created by ______, generated in the heart spreading throughout the body.

5-2 The peak amplitude on the ECG waveform is approximately ______________ mV.

5-3 The five electrodes used to make 12-lead ECG recordings are connected to the patient's ______________, ______________, ______________, ______________, and ______________.

5-4 The ______________ electrode is used as the common terminal in ECG recording.

5-5 Leads I, II, and III form the ______________ (*two words*).

5-6 Leads I, II, and III are called the ______ limb leads.

5-7 Leads AVF, AVR, and AVL are called the ______________ limb leads or ______ leads.

5-8 In leads AVF, AVR, and AVL the signals from two limbs are summed in a resistance network and applied to the ______________ input of the preamplifier.

5-9 Leads V_1 through V_6 are called the ______ leads.

5-10 State the principal features of an ECG amplifier. How does it *differ* from a bioelectric amplifier?

5-11 The frequency response of a diagnostic ECG machine is from 0.05 to ______________ Hz.

5-12 An ECG monitor usually has a frequency response of 0.05 to about ______________ Hz.

5-13 The chief artifact rejected by limiting the bandwith of an ECG monitor is ________.

5-14 Most ECG preamplifiers have an internal calibrator that injects a ______________ -mV pulse into the preamplifier input.

5-15 ECG preamplifiers require protection circuits against electrical potentials from a machine called a ______________.

5-16 A protection device (Question 5-15) used in most ECG preamplifiers is a ______________ lamp.

5-17 List two types of damage that can result from the phenomenon in Question 5-15.

5-18 List two types of ECG readout device.

5-19 What is (a) standard speed and (b) two optional speeds on an ECG machine?

5-20 Describe how heart rate is determined on a triggered sweep oscilloscope.

5-21 Most ECG machines write using a ______ stylus.

5-22 Both ______________ and ______________ tensions must be applied to the paper in an ECG recorder to pull it taut across the ______________.

5-23 List three types of electrode connector used in patient cables.

5-24 Why do some ECG patient cables use resistors in series with each electrode? What is the approximate range of values used for these resistors?

5-25 List the pinouts of the standard, or nearly standard, ECG connector.

5-26 Describe how the input circuitry and patient cable may be checked using *only* the machine itself.

5-27 What are the probable causes of *no trace* conditions on an ECG machine?

5-28 What is the principal cause of a smeared trace from an ECG machine? What are other possible causes?

5-29 What are the principal causes of 60 Hz interference?

5-30 What are the principal causes of erratic baseline recordings?

5-31 What are the principal causes of wandering baseline?

5-32 Give the quick method for isolating the cause of 60-Hz interference.

5-33 What can be done to reduce the effects of somatic tremor in ECG recordings?

5-34 Give a probable cure for (a) erratic baseline, (b) wandering baseline.

5-15 Problems

5-1 What is the *heart rate* in beats per minute of a patient with an R-R interval of 856 ms?

5-2 What is the heart rate of the patients whose ECGs are given in (a) Figure 5-27*a* and (b) Figure 5-27*b*. Assume that the recording speed was standard.

5-16 References

1. Strong, Peter, *Biophysical Measurements,* Measurement Concepts Series, Tektronix, Inc. (Beaverton, Ore., 1977).
2. *Technician's Guide to Electrocardiography,* 3rd ed. Hewlett-Packard applications note (Waltham, Mass., 1972).
3. Rawlings, Charles A., Ph.D, C.C.E., *Electrocardiography,* Spacelabs, Inc. (Redmond, Wash., 1991).
4. Service manual for the H-P model 1500B/1511B (Waltham, Mass., 1971).
5. *ECG Techniques,* Hewlett-Packard applications note AN721 (Waltham, Mass., 1972).
6. *ECG Measurement,* Hewlett-Packard applications note AN711 (Waltham, Mass., 1972).
7. *High Fidelity ECG Measurement,* unnumbered Hewlett-Packard applications note (Waltham, Mass., 1971).
8. Clinica, *World Medical Device and Diagnostic News,* Issue 464, July 31, 1991 (consistent ECGs possible with new standardizer device), 18/20 Hill Rise, Richmond, Surrey TW10 6UA, U.K., Tel: 081-948-3262; or 1775 Broadway, Suite 511, New York, NY 10019, Tel: 212-262-8230; also Dialog Information Services, Inc., 3640 Hillview Ave., Palo Alto, CA 94304, Tel: 1-800-334-2564 or 415-858-3810.
9. *Biomedical Instrumentation & Technology* magazine, January/February, 1991 (Article: "A BMET Study Program Should Ask Why"; Article: "Creating a Quality Measurement System for Clinical Engineering"; Article: "Biomedical Information Management for the Next Generation"; Article: "The Safe Medical Devices Act of 1989: A Burden on Medical Facilities, Device Manufacturers, and the FDA?"), publication of Association for the Advancement of Medical Instrumentation (AAMI), 3330 Washington Blvd., Suite 400, Arlington, VA 22201, Tel: 703-525-4890.
10. *AAMI News,* Volume 26, Number 5, September–October, 1991 (Information: "Patient Monitors in High Demand"; Standards, government, and European EC activities), publication of Association for the Advancement of Medical Instrumentation (AAMI), 3330 Washington Blvd., Suite 400, Arlington, VA 22201, Tel: 703-525-4890.
11. AAMI 1991 Resource Catalog, Publications & Services, publication of Association for the Advancement of Medical Instrumentation (AAMI), 3330 Washington Blvd., Suite 400, Arlington, VA 22201, Tel: 703-525-4890.
12. ANSI (American National Standards Insti-

tute)/AAMI (Association for the Advancement of Medical Instrumentation), *Standard for Cardiac Monitors, Heart Rate Meters, and Alarms,* EC13-June, 1983, Revision—June, 1991, 3330 Washington Blvd., Suite 400, Arlington, VA 22201, Tel: 703-525-4890.

13. *Biomedical Technology,* Issue 7 (Article: "Integrating Monitoring: Better Event Detection & Alarm Accuracy?"), 1991, Quest Publishing Company, 1351 Titan Way, Brea, CA 92621, Tel: 714-738-6400.
14. *Biomedical Safety & Standards,* Issue 7 (Article: "Safe Medical Devices Law: Hospitals Must Report Incidents to FDA" and Article: "Power Failure Leads to New Emergency Power Policies"), 1991, Quest Publishing Company, 1351 Titan Way, Brea, CA 92621, Tel: 714-738-6400.
15. *Occupational Biohazards Affecting Clinical Engineers & BMETs,* 1989, Quest Publishing Company, 1351 Titan Way, Brea, CA 92621, Tel: 714-738-6400.
16. *Hewlett-Packard Journal,* October, 1991, Hewlett-Packard Company, P.O. Box 51827, Palo Alto, CA 94303-0724; subscriptions to editor, *HP Journal,* 3200 Hillview Avenue, Palo Alto, CA 94304 (Articles on Physiological Monitoring, Introduction to the HP Component Monitoring System, Medical Expectations for Today's Patient Monitors, Component Monitoring System Hardware Architecture, Component Monitoring System Software, Component Monitoring System Parameter Module Interface, Measuring the ECG Signal with a Mixed Analog-Digital Application-Specific IC (ASIC), A Very Small Noninvasive Blood Pressure Measurement Device, Patient Monitor Two-Channel Stripchart Recorder, Patient Monitor Human Interface Design, Globalization Tools and Processes in the HP Component Monitoring System, The Physiological Calculation Application in the HP Component Monitoring System, Mechanical Implementation of the HP Component Monitoring System, An Automated Test Environment for a Medical Patient Monitoring System, Production and Final Test of the HP Component Monitoring System, Calculating the Real Cost of Software Defects.
17. *IEEE Transactions on Biomedical Engineering,* Vol. BME-20, No. 2, March, 1973 (Article: "60-Hz Interference in Electrocardiography," James Huhta and John G. Webster).
18. *IEEE Transactions on Biomedical Engineering,* Vol. BME-30, No. 1, January, 1983, (Article: "Reduction of Interference Due to Common Mode Voltage in Biopotential Amplifiers," Bruce B. Winter and John G. Webster).
19. *IEEE Transactions on Biomedical Engineering,* Vol. BME-30, No. 1, January, 1983, (Article: "Driven-Right-Leg Circuit Design," Bruce Winter and John G. Webster).
20. *International Medical Device & Diagnostic Industry,* September/October, 1990 (Article: "Surface Biomedical Electrode Technology," E. McAdams).
21. *Spectra,* February, 1991, Laurin Publishing Co., Inc., Berkshire Common, P.O. Box 4949, Pittsfield, MA 01202, Tel: 413-499-0514, Section in Medicine & Surgery, (Article: "The Evolution of the Endoscope").
22. Henderson Electronic Market Forecast for Medical Electronics, July, 1990, and August, 1991, Henderson Ventures, 101 First Street, Suite 444, Los Altos, CA 94022, Tel: 415-961-2900.
23. Product data sheets for components shown in this chapter, Burr-Brown Corp., 1991, P.O. Box 11400, Tucson, AZ 85734, 1-800-548-6132.

Chapter 6

Physiological Pressure Measurements

6-1 Objectives

1. Be able to define *pressure.*
2. Be able to describe how pressure is measured.
3. Be able to explain the various techniques used to measure human blood pressure.
4. Be able to list the principles of operation and calibration of medical pressures monitors.
5. Be able to list and describe some different types of pressure amplifier circuits.

6-2 Self-Evaluation Questions

These questions test your prior knowledge of the material in this chapter. Look for the answers as you read the text. After you have finished studying the chapter, try answering these questions and those at the end of the chapter (see Section 6-2).

1. Define *pressure* in your own words.
2. How does *hydrostatic pressure* affect the accuracy of transducer readings?
3. Describe the *auscultation* technique of measuring arterial blood pressure. Describe both the apparatus and the technique.
4. What is a *sphygmomanometer* and how is it used?
5. Describe the proper *placement* of an arterial blood pressure transducer when using an *intracardiac catheter.*

6-3 Physiological Pressures

The measurement of physiological *fluid* pressures is of interest to both biomedical researchers and medical clinicians. Given a rigorous definition of *fluid,* we would have to include both *liquids* and *gases,* but in this chapter our discussions center around liquids. The difference between liquids and gases is that the latter are *compressible* while the former are not; compressibility affects the measurement technique. Gas pressure measurement is discussed more fully in Section 10-6, on pulmonary instrumentation.

The most common pressure measurement is *arterial blood pressure,* which is almost routinely monitored by electronic instruments in ICU/CCU and other critical care areas. Also of interest, however, are the *central venous pressure* (CVP), *intracardiac blood pressure,* special pressures in the *pulmonary artery, spinal fluid pressures,* and *intraventricular* (brain) *pressures.*

6-4 What Is Pressure?

A group of students were asked to give definitions of *pressure.* Several gave the *correct* answer, but most wrote hazy, ambiguous statements that indicated that they did not really understand what is meant by that concept. Some students came close, indicating that *pressure* is a *force;* but this is still not correct. The *correct* definition is that *pressure is force per unit area.*

$$P = \frac{F}{A} \tag{6-1}$$

where

P is the pressure in newtons per square meter (N/m^2) or pascals (Pa); 1 N/m^2 = 1 Pa
F is the force in newtons (N)
A is the area in square meters (m^2)

EXAMPLE 6-1

A small coin has a diameter of 1 cm and a mass of 1.5 g. Find: (a) The gravitational force (weight of coin) in dynes and millinewtons. (b) The pressure this coin exerts lying horizontally on a flat table top in dynes per square centimeter and newtons per square meter (or pascals).

SOLUTION

$f = 1.5\text{g(a)}\ f = ma$ (Newton's second law) $\times\ 980\ \text{cm/s}^2 = 1470\ \text{g-cm/s}^2 = 1470$ dyn

$$f = 1470\ \text{dyn} \times \frac{1\ \text{N}}{10^5\ \text{dyn}}$$
$$= 1.47 \times 10^{-2}\ \text{N}$$
$$= 14.7\ \text{mN}$$

(b) $$P = \frac{F}{A} \tag{6-1}$$

$$P = \frac{1470}{\pi(1\ \text{cm}/2)^2} = \mathbf{1872\ dyn/cm^2}$$

$$P = 1872\ \text{dyn/cm}^2 \times 10^{-5}\ \text{N/dyn} \times (10^2\ \text{cm/m})^2$$

$$P = \mathbf{187.2\ N/m^2 = 187.2\ Pa}$$

Note that *pressure* can be *increased* by either *increasing* the applied *force* or by decreasing the area on which the force acts. Alternate units for pressure, as shown in Example 6-1, are dynes per square centimeter (dyn/cm^2) in the *cgs* system and pounds per square inch ($lb/in.^2$) in the British engineering system. The latter is often abbreviated *psi.*

When the force in a system under pressure is constant or static (i.e., unvarying), the pressure is said to be *hydrostatic.* If the force is *varying,* on the other hand, the pressure is said to be *hydrodynamic.* Most human physiological pressures are hydrodynamic, of which the most easily recognized is the pulsatile flow of arterial blood.

Pressure in a closed system obeys a physical law known as Pascal's principle (after French scientist Blaise Pascal, 1623–1662), which states that:

Def.: Pressure applied to an enclosed fluid is transmitted undiminished to every portion of the fluid and the walls of the containing vessel.[1]

If a pressure is applied to the stoppered syringe in Figure 6-1, then, the *same* pressure is felt throughout the interior of the syringe. *Changing* the pressure applied to the plunger causes the same change to be reflected at every point inside of the syringe.

Pascal's principle holds true in hydrostatic systems and also in *quasistatic* systems, where a *small* change is made and the turbulence is allowed to die down before subsequent measurements are made. Pascal's principle holds *approximately* true in those hydrodynamic systems where the flow is nonturbulent and the vessel lumen is small (except at the vessel wall boundaries). The study of pressure in turbulent or large lumen systems, or near the vessel wall, is a subject for engineering mechanics and physics courses. In this text we assume that Pascal's principle holds generally true but recognize that it results only in approximations when applied to the turbulent human circulatory system.

Pressure is exerted in the human circula-

[1] David Halliday and Robert Resnick, *Physics Parts I and II, 3rd ed.,* Wiley (New York, 1978), p. 430. Reprinted by permission of Wiley.

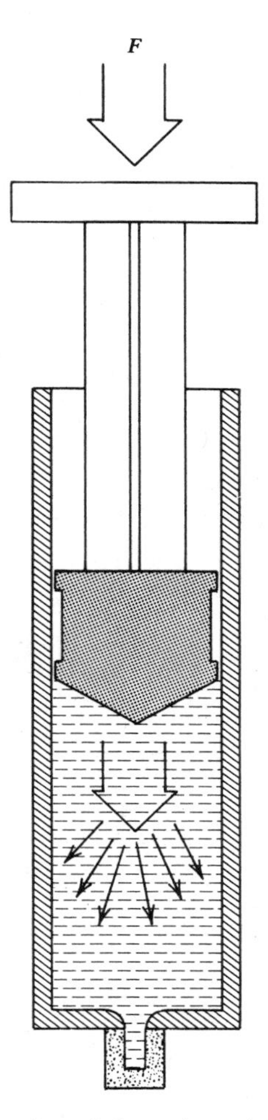

Figure 6-1 Sectioned view of a syringe. A pressure exerted by the plunger is transmitted through the fluid to all parts of the system.

tory system by the force created by the pumping heart transmitted through the fluid (i.e., blood) against the vessel walls. The circulatory system regulates blood pressure by constricting and dilating vessels, which causes changes in vessel surface area by changing vessel diameter. The pressure, as a result, is never constant, and our measurements always assume an *average* value.

6-5 Pressure Measurements

The air forming our atmosphere exerts a pressure on the surface of the earth. This pressure is usually expressed as being 1 atmosphere (1 atm) or approximately 14.7 lb/in.^{2}, or 1.013×10^6 dyn/cm^2, or 1.013×10^5 Pa at mean sea level.

EXAMPLE 6-2

A barometer measures a pressure of 750 mm Hg during a snowstorm. Convert this pressure to (a) atmospheres, (b) lb/in.^{2}, (c) pascals, (d) dyn/cm^2, and (e) N/m^2.

SOLUTION

(a) $750 \text{ mm Hg} \times \dfrac{1 \text{ atm}}{760 \text{ mm Hg}} = \mathbf{0.987 \text{ atm}}$

(b) $0.987 \text{ atm} \times \dfrac{14.7 \text{ lb/in.}^2}{1 \text{ atm}} = \mathbf{14.5 \text{ lb/in.}^2}$

(c) $0.987 \text{ atm} \times \dfrac{1.013 \times 10^5 \text{ Pa}}{1 \text{ atm}}$
$= \mathbf{1.00 \times 10^5 \text{ Pa}}$

(d) $0.987 \text{ atm} \times \dfrac{1.013 \times 10^6 \text{ dyn/cm}^2}{1 \text{ atm}}$
$= \mathbf{1.00 \times 10^6 \text{ dyn/cm}^2}$

(e) $1 \times 10^5 \text{ Pa} \times \dfrac{1 \text{ N/m}^2}{1 \text{ Pa}} = \mathbf{1 \times 10^5 \text{ N/m}^2}$

If a pressure is measured with respect to a vacuum (0 atm), then it is called an *absolute pressure;* and if measured against 1 atm it is called a *gage pressure*. Two gage pressures may be compared with each other in the form of a single number called *relative pressure* or *differential pressure* (i.e., the *difference* between two gage pressures). Pressures in the human circulatory system are measured against atmospheric and are gage pressures. In the pulmonary system, some pressures are gage pressures while others are relative pressures (see Section 10-5).

Figure 6-2*a* shows the Torricelli (after Evangelista Torricelli, Italian scientist, 1608–1647) *manometer* used to measure atmospheric pressure. An evacuated glass tube with a small lumen stands vertically, with the

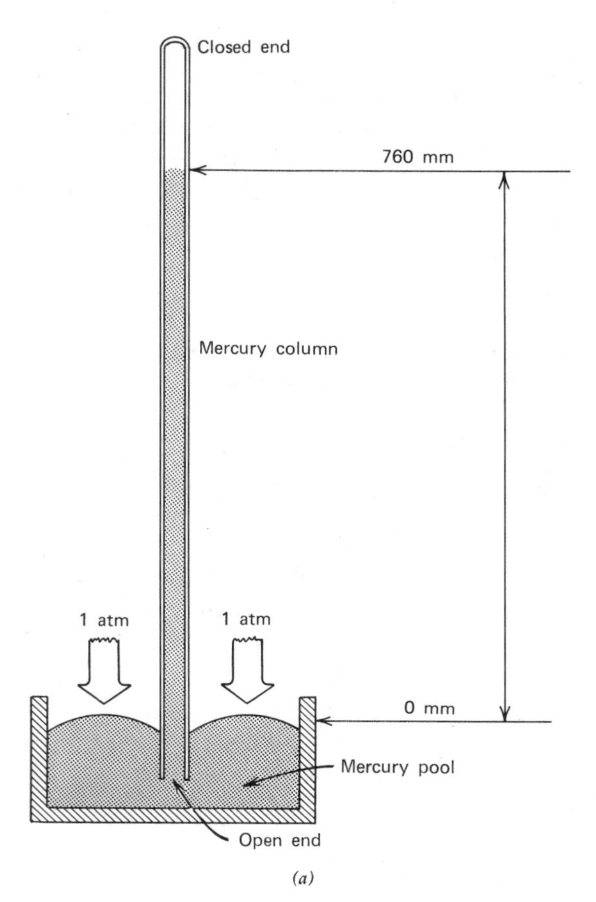

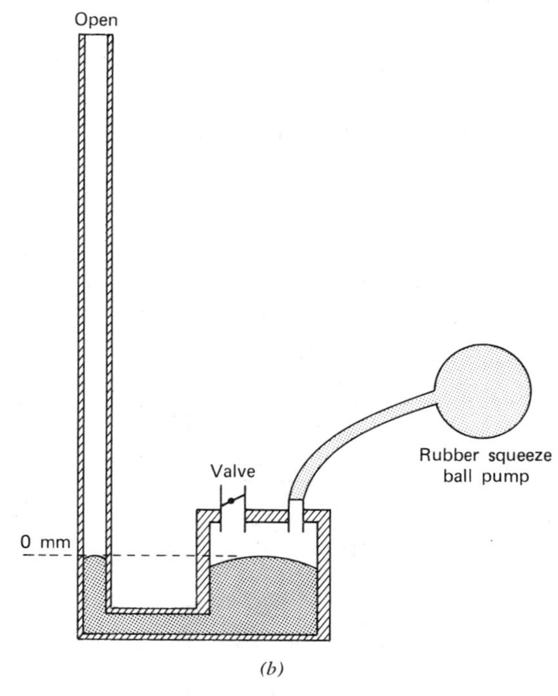

Figure 6-2 Mercury manometer. (*a*) Torricelli's manometer for measuring atmospheric pressure. (*b*) Gage pressure manometer.

open end immersed in a pool of mercury (Hg). The pressure exerted by the atmosphere on the surface of the mercury pool forces mercury into the tube, forming a column. The mercury column rises in the tube until its weight (i.e., a gravitational force) exactly balances the force of the atmospheric pressure. Torricelli found that the height of the mercury column that can be supported by atmospheric pressure is approximately 0.76 m or 760 mm. Atmospheric pressure, then, is frequently given in units of mm Hg, and 1 atm is 760 mm Hg.[2]

The *proper* unit of pressure, as established by scientists and adopted by *The National Bureau of Standards,* is the Torr (after Torricelli), where 1 Torr is equal to 1 mm of mercury (1 T = 1 mm Hg).[3]

Gage pressures are usually given in mm Hg above or below atmospheric pressure. A *manometer* is any device that measures *gage* pressure, although in commonly accepted jargon, gage pressures below 1 atm are said

[2] The *barometer* reading given in English units is usually inches of mercury (*in. Hg*), with 760 mm Hg being equivalent to 29.92 in. Hg.

[3] If convention were followed, then, we would quote physiological pressures in Torr, but it is common practice in medicine to use the archaic units mm Hg. To avoid confusion, we will follow this practice, but keep in mind that it is no longer accepted outside of the medical world.

to be measured on a *vacuum gage,* and pressures above 1 atm are said to be measured on a *pressure gage.* Both instruments are, however, examples of manometers. Some manometers, including most of the electronic instruments that we will study in this chapter, will measure both positive pressures (above 1 atm) and negative pressures (i.e., vacuums).

The *zero reference* in gage pressure measurements, therefore, is a pressure of 1 atm. Even though atmospheric pressure varies from one place to another, and in the same location over the course of a few hours, zero can be established at each measurement by setting the zero scale with the manometer open to atmospheric pressure. Figure 6-2*b* shows a mercury manometer similar to those used to make blood pressure measurements. The open tube is connected to a mercury reservoir that is fitted with a rubber squeeze-ball pump that can be used to increase pressure; a valve is used to either open the chamber to atmosphere or close it off.

If the valve is open to atmosphere, then the pressure on the mercury in the chamber is equal to the pressure on the column (i.e., 1 atm). The mercury in the column will have the same height as the mercury in the chamber, and this point is *designated* as having a pressure of 0 mm Hg. If the valve is *closed,* and the pressure inside of the chamber is *increased* by operating the pump, then the mercury in the column will *rise* an amount *proportional* to the increased pressure.

EXAMPLE 6-3

The mercury in a manometer such as Figure 6-2*b* rises to a height of 120 mm Hg. Find (a) the gage pressure and (b) the absolute pressure.

SOLUTION

(a) **120 mm Hg** (by definition)

(b) 120 mm Hg + 760 mm Hg = **880 mm Hg**

We use gage pressure because it is more easily referenced at zero and can be easily recalibrated at each use, and the absolute pressure confers no special advantage as to information content. The variation of atmospheric pressure from one location to another (dependent upon mean sea level) and over the course of a single day makes the use of gage pressures more advantageous.

6-6 Blood Pressure Measurements

The earliest recorded attempt at the measurement of arterial blood pressure was performed in 1773 by English scientist Stephen Hales, who used an open-ended tube inserted directly into an artery in the neck of an *un*anesthetized horse (presumedly tied down securely!). The tube was long enough that blood rose to a height where the weight of the blood exactly balanced the horse's arterial pressure.

According to Hales's observation, the blood pulsed its way up the tube, attaining a height of approximately 4 ft on the first pulse, but requiring an additional 40 or 50 pulses to attain a final height of just over 8 ft. After the blood in the manometer tube had stabilized to about the final height, it rose and fell approximately 2 or 3 in. on each pulse, due to the diastolic and systolic pressures.

Stephen Hales's technique is an example of *direct* measurement of blood pressure. *Routine* clinical measurements of blood pressure in humans, however, required the development of suitable *indirect* techniques without the painful and potentially hazardous surgical procedures performed by Hales.

Today, both *indirect* and *direct* methods are used to measure blood pressure in humans. The most popular *indirect* method, familiar to almost everybody whose blood pressure has been checked by a physician or nurse, involves *sphygmomanometry.* Currently used *direct* methods usually involve *electronic amplifiers* that process a signal from a *pressure transducer* that is coupled to the patient's artery or vein through a saline-filled *catheter* or *needle*. It is interesting to note that Stephen Hales's two-century-old method is still used in modern hospitals to measure *spinal fluid* pressure and *central venous pressure* (CVP). Almost every hospital stocks CVP and spinal-tap kits that contain an H_2O manometer not dissimilar to Stephen Hales's crude apparatus of the eighteenth century!

The indirect method routinely used by your doctor requires a device called a *sphygmomanometer* (see Figure 6-3), consisting of an inflatable rubber bladder called the *cuff,* a rubber squeeze-ball pump-and-valve assembly, and a manometer. The manometer might be an actual mercury column (as shown), or a dial gage. Professional-grade sphygmomanometers are based on an aneroid assembly or structure similar to the Bourdon tube, while many cheaper varieties offered as part of "blood pressure kits" to the general public use spring-loaded pressure gages. The spring-loaded types are usually just as accurate as more costly types initially but soon develop substantial errors as the spring wears out.

The procedure for using this apparatus is:

1. The *cuff* is wrapped around the patient's upper arm at a point about midway between the elbow and shoulder. The stethoscope is placed over an artery distal (i.e., downstream) to the cuff. This placement (see Figure 6-4) is preferred because the *brachial artery* comes close to the surface near the *antecubital space* (i.e., "inside" of the elbow) and so is easily accessible.

2. The cuff is inflated so that the pressure inside the inflated bladder is increased to a point greater than the anticipated systolic pressure. This pressure compresses the artery against the underlying bone, causing an *occlusion* that shuts off the flow of blood in the vessel.

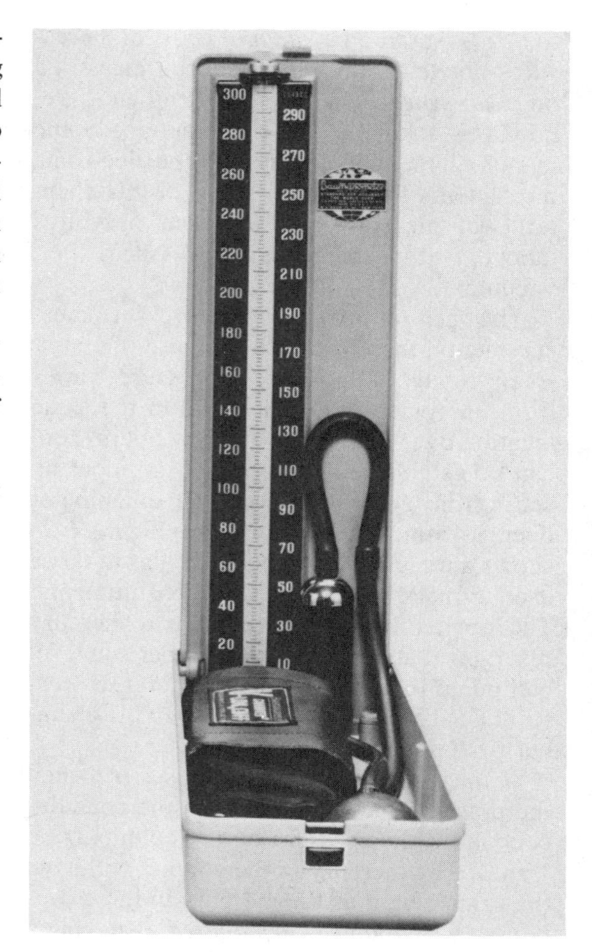

Figure 6-3 Mercury column sphygmomanometer. (Photo courtesy of E. Baum Co., Inc.)

3. The operator then *slowly* releases (i.e., reduces) the pressure in the cuff (about 3 mm Hg/s is usually deemed best) and watches the pressure gage or mercury column. When the systolic pressure first exceeds the cuff pressure, the operator begins to hear some crashing, snapping sounds in the stethoscope that are caused by the first jets of blood pushing through the occlusion. These sounds, called *Korotkoff sounds* (Figure 6-5), continue as the cuff pressure dimin-

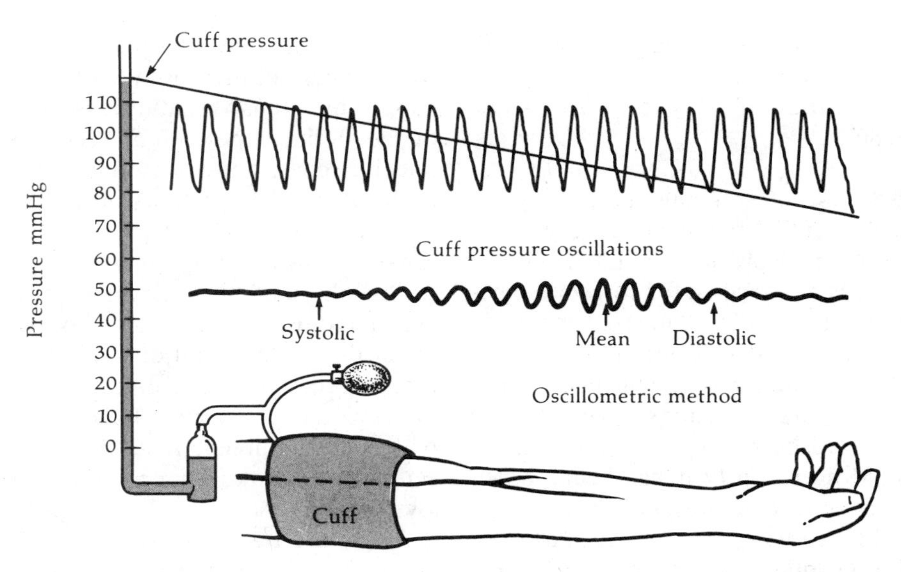

Figure 6-4 Cuff placement for the auscultatory method of blood pressure measurement.

ishes, becoming less loud as the blood flow through the occlusion becomes smoother. Korotkoff sounds disappear or become muffled when the cuff pressure drops *below* the patient's *diastolic* pressure. To read the blood pressure, the operator notes both the gage pressure at the *onset* of Korotkoff sounds (systolic) and also when the sounds

Figure 6-5 Diagram of auscultatory method of blood pressure measurement.

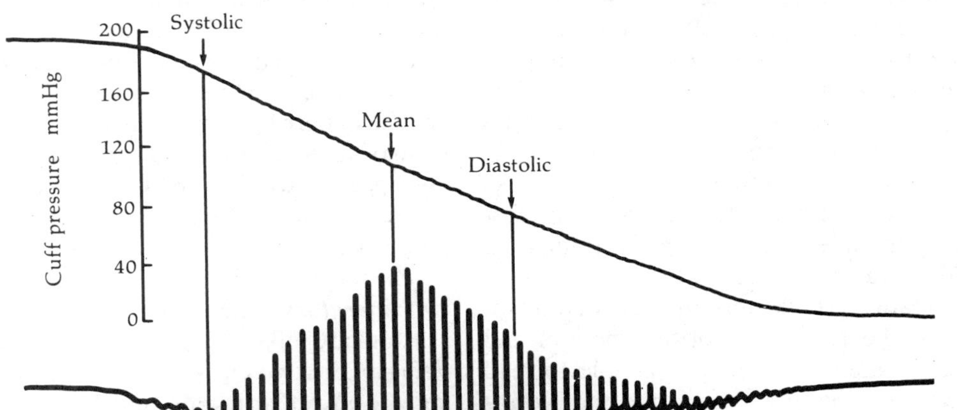

become muffled (or disappear) altogether (diastolic). These pressures are usually recorded in the ratio of systolic over diastolic (i.e., 120/80).

The first use of sphygmomanometry for the measurement of blood pressure was reported by N. S. Korotkoff in 1905, but the technique was not verified for correlation between indirect and direct measurements in animals until 1912. It was not until 1931 that a similar correlation was established for humans—that is, that variations of less than 10 mm Hg existed between direct and indirect methods. More recently it has been shown that indirect diastolic pressures are less in error if the reading is taken at the point where the Korotkoff sounds disappear. Yet most clinicians prefer to use the point where the sounds become muffled because this point can be recognized more consistently. The American Heart Association recommended in 1967 that muffling be used as the criterion for diastolic pressure but that *both* pressures (i.e., muffling and the cessation of Korotkoff sounds) be indicated if a significant difference between them exists. This measurement is recorded as a double-diastolic pressure (i.e., 120/80/77 mm Hg).

The use of Korotkoff sounds as the indirect indicator of blood pressure is also called *auscultation* (i.e., use of hearing) and is, by far, the most common indirect method used. It is accurate enough for ordinary clinical use and is simple enough that even nonprofessional personnel can be rapidly trained to "take blood pressures."

Limitations on the ausculatory method include the hearing acuity of the operator and how accurately the operator is able to read a *changing* pressure gage when the Korotkoff sound features are heard. In *hypotensive* (i.e., *low* blood pressure) patients, the event chosen to indicate the diastolic pressure may be either obscured or nonexistent.

Several modern instruments are available for the indirect measurement of blood pressure in hypotensive patients by replacement of the stethoscope with an electronic transducer. Some using ultrasound are discussed in Section 17-10. Others using *infrasound* (i.e., frequencies less than 50 Hz) are merely low-frequency microphones. The instrument will amplify and filter the microphone output signal and use it to turn on a beeper or lamp when the systolic and diastolic features are recognized. These instruments are also used in emergency rooms, ICUs, and CCUs where high ambient noise levels often obscure the Korotkoff sounds on nonhypotensive patients.

There are two other major indirect methods of blood pressure measurement: *palpation* and *flush*. Both use the cuff but differ in the respective methods used to detect the pressure points.

The palpation method uses the sense of *touch* to detect the patient's pulse in the radial artery (wrist). The cuff is inflated until the radial pulse *disappears*. The operator then slowly releases the pressure in the cuff until a pulse becomes palpable in the radial artery. The pressure at which this occurs is the *systolic* blood pressure.

Palpation can detect *only* the *systolic* pressure because there is no known palpable change occurring at the diastolic pressure. It is also noted that palpable changes tend to disappear below 75 or 80 mm Hg systolic, so the technique is often not useful on the hypotensive patient.

The flush technique requires *two cuffs* and *two operators*. The cuffs are placed on the arm and are inflated. The blood in the section *between* the two cuffs is massaged out, leaving the lower arm pale and blanched. The pressure in the upper cuff is then released slowly. The pressure at which a sudden red flush is noted in the blanched skin is recorded as the *mean arterial pressure*.

6-7 Oscillometric and Ultrasonic Noninvasive Pressure Measurements

The aforementioned ausculatory measurement method is the most widely used procedure for measuring blood pressures. It suffers, however, from at least two problems. First, the measurement is very intermittent because it takes time to accomplish. A medical person who constantly takes blood pressure readings has little time for anything else. Second, the Korotkoff sounds are normally in the range (< 200 Hz) where human hearing is not very acute. If long-term monitoring in intensive care is done, or if the ambient noise level is high, then either *oscillometric* or *ultrasonic* blood pressure measurement may be used.

6-7.1 Oscillometric Blood Pressure Measurement

The oscillometric method of blood pressure measurement is very similar to ordinary spygmomanometry, except that we measure small fluctuations (i.e., *oscillations*) in the cuff pressure rather than direct pressure (Figure 6-6*a*).

When blood breaks through the occlusion created by the inflated cuff, which occurs when the cuff pressure drops below the systolic blood pressure, the walls of the artery begin to vibrate slightly. These vibrations are related to the fact that the blood flow at this point is turbulent, rather than laminar, although the physiological basis is not well understood.

The flucuating walls of the blood vessel slightly alter the blood pressure, giving rise to oscillations in the cuff pressure (Figure 6-6*b*). The onset of the pressure oscillations correlates well with the systolic pressure, while the amplitude peak of the oscillations corresponds to the *mean arterial pressure* (MAP). The MAP is the *time-average* of blood pressure (see Section 6-7 for a discussion of MAP). The diastolic pressure event on the oscillation curve is somewhat less well defined compared to the systolic event but corresponds to the point where the rate of amplitude decrease suddenly changes slope.

Oscillometric blood pressure monitors are used extensively where monitoring is needed, but it is not desirable to do an invasive procedure required to perform *direct pressure measurements* (Section 6-8). A typical oscillometric blood pressure monitor is microprocessor controlled and is designed to periodically inflate and slowly deflate the cuff. The pressure sensor is designed to maximize *variations* in pressure rather than read the static pressure heads.

6-7.2 Ultrasonic Pressure Measurements

Ultrasonic waves are acoustical waves (like regular sound waves 30 Hz to 20 KHz) in the range above human hearing (> 20 KHz). Like all acoustical waves, ultrasonic waves are subject to *Doppler shift*—that is, a slight alteration of frequency (ΔF) when reflected from a moving object. If piezoelectric ultrasound sensors are placed over the artery, under the cuff, then they can perform Doppler detection of the blood flowing in the artery.

The principle of operation is a bit like radar. A transmit crystal sends a sinewave beam into the tissue. When it encounters a fluctuating vessel wall (see Section 6-7.1), some of its energy is reflected (''backscattered'') back to the receive crystal, which is located close to the transmit crystal. If ΔF is the Doppler shift, the $F \pm \Delta F$ describes the frequency content of the backscattered wave. The existence of the ΔF component alerts the circuit to the turbulent flow that corresponds to Korotkoff sounds, and it di-

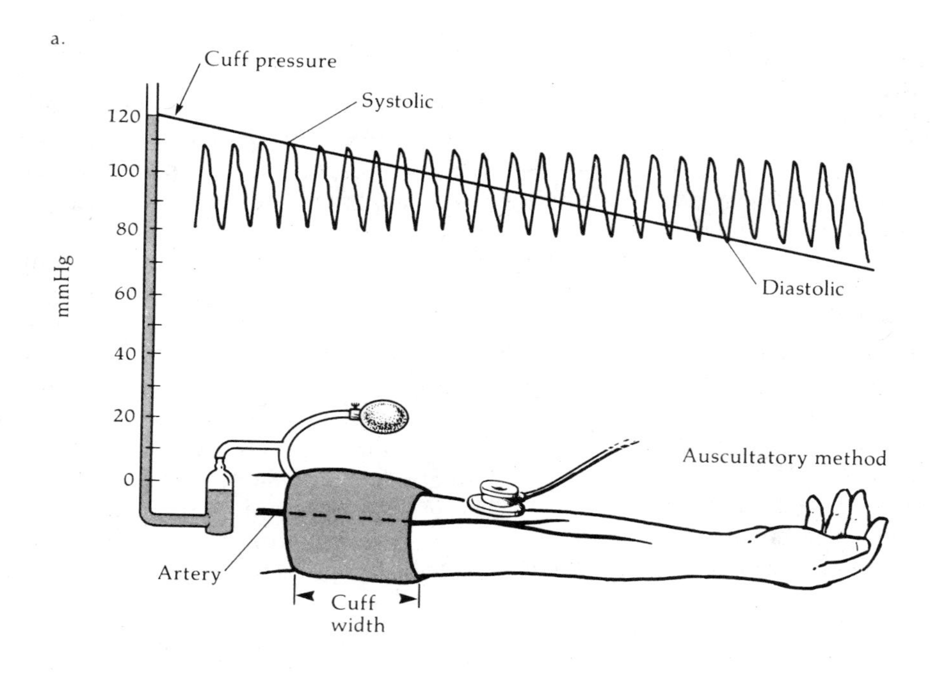

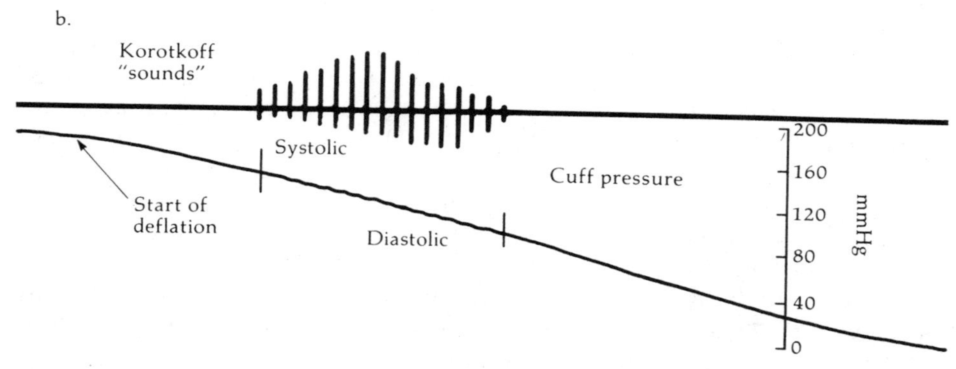

Figure 6-6 Illustration of the oscillometric method of blood pressure measurement.

minishes when near-laminar flow resumes ($P < P_{\text{diastolic}}$).

6-8 Direct Methods: H_2O Manometers

The method of Stephen Hales is still used in the measurement of spinal fluid pressures and central venous pressure (CVP). Most hospitals stock standard kits containing plastic tubes calibrated in centimeters of water (cm H_2O). Water rather than mercury is used for two reasons. One is that the thin manometer tube is introduced directly into the patient's body, so the use of a poisonous material like mercury as the pressure indicator is avoided. Water is physiologically compatible with the patient's body and so is a lot less hazardous. Second, the pressure in either CVP or spinal fluid measurements is low (only a few mm Hg). If a material, like water, that is less dense than mercury is used, then the column with a weight equal to the pressure force will be a lot higher. This will greatly improve resolution. Water has a specific gravity of one (by definition), and mercury has a specific gravity of approximately 13.5. The column of water produced by a given pressure will be 13.5 times higher than the column of mercury at the same pressure. CVP and spinal manometers are usually calibrated in cm H_2O, but a mm Hg value is easily found by dividing the H_2O value by 13.5 to yield a cm Hg value.

6-8.1 Direct Methods: Electronic Manometry

An electronic pressure transducer can be connected to the patient through a thin piece of tubing called a *catheter.* The catheter is introduced into the vessel through a thin, hollow tube called a *cannula.* The transducer's pressure diaphragm is coupled to the patient's bloodstream by a column of saline solution that fills the catheter.

The catheter to the patient must be placed inside a peripheral artery. There are two general methods for inserting the catheter: *percutaneous punction* and *arterial cutdown.* The percutaneous methods involve puncturing the skin over an artery and then using a needle and catheter assembly to insert the catheter into the artery. When the catheter is in place, the needle is withdrawn. The arterial cutdown method is a surgical procedure in which the tissue overlying the artery is cut and laid out of the way, revealing the artery. A puncture is then made, and a catheter is put into place.

Figure 6-7a shows typical apparatus used for measuring blood pressure with an electronic transducer. The transducer is mounted to a pole or support at bedside. There are two ports into the transducer's *pressure dome;* one is "deadheaded" and the other is connected to a hydraulic fitting called a *three-way stopcock.* One terminal of the stopcock goes to the catheter from the patient, while the other goes to a small syringe that is used to administer medications directly to the patient through the catheter, or to withdraw arterial blood samples for laboratory analysis.

6-8.2 Constant Flush Infusion System

When direct pressure measurements are made over a long time (e.g., several hours), as they are in intensive care units, there is a danger of coagulating blood stopping up the pressure catheter. In those cases, medical personnel may opt to use a *constant flush infusion system* (CFIS), such as in Figure 6-7b.

The CFIS consists of a special valve connected to a bag of intravenous (IV) solution; usually 0.9 percent saline with one or two units of aqueous heparin (a drug that prevents blood clotting) per cc of IV fluid. A constant low flow rate of about 3 cc/hr is set. Pressure is applied to the bag of IV solution by a standard IV pressure bag (a bladder) that is pumped to a pressure of about 300 mm Hg.

A *fast flush lever* is used to inject a sudden bolus of heparinized saline into the system in order to fill the transducer dome and clear blood clots. It can also be used to provide a squarewave stimulus to test the frequency response of the total system (see Section 6-18).

6-9 Pressure Transducers

Figures 6-8 and 6-9 show typical blood pressure transducers. A transducer is an electrical device that converts the pressure transmitted through the fluid-filled catheter to an electrical signal (recall Pascal's principle).

The basic components to a blood pressure transducer are shown in Figure 6-8. A thin, flexible metal diaphragm is stretched across

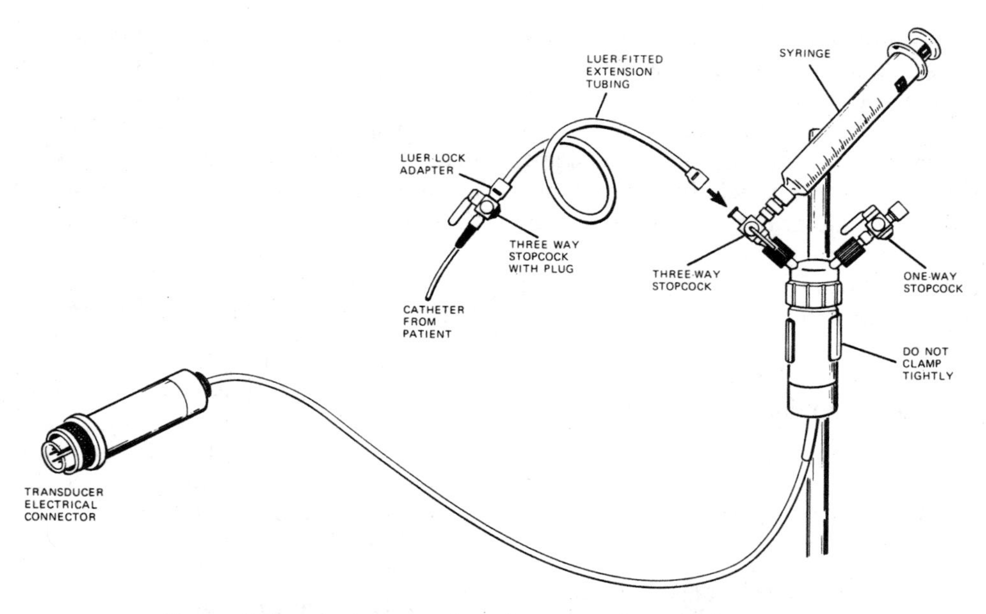

Figure 6-7a A representation of an arterial monitoring setup.

an opening at one end of the transducer body. This diaphragm is connected to an inductive bridge strain gage and will flex the strain gage an amount proportional to the applied pressure. Other models, incidentally, use a resistive Wheatstone bridge strain gage.

A clear plastic *pressure dome* fits over the diaphragm to contain the fluid and provide hydraulic connection to the catheter. The electrical connector is a little larger in this particular model because it houses part of the bridge circuit (i.e., two resistors) and a sensitivity control. The transducer shown in this example is the Hewlett-Packard model 1280 and, since it is inductive, requires an ac excitation potential.

The Statham model P23Id blood pressure transducer, shown in Figure 6-9, is then also very common in medical monitoring systems. This particular version of the P23 series is insulated to prevent damage to itself and injury to the patient during defibrillator discharge. The P23Id is an unbonded piezoresistive strain gage Wheatstone bridge of the type discussed in Section 3-8.2.

6-10 Pressure Amplifiers

There are four basic types of pressure amplifiers in common use: *dc, isolated dc, pulsed excitation,* and *ac carrier amplifiers*. The dc amplifiers work with resistance strain gages only, while the ac carrier amplifiers work with either resistive or inductive transducers. The pulsed excitation amplifiers may work with some inductive transducers but are generally used only with resistive transducers.

Regardless of the design, there are certain features common to all pressure amplifiers. Instruments intended for use in catheterization laboratories or research facilities should be capable of measuring a wide range of pressure values, and they must be very accurate and stable. Instruments in this class

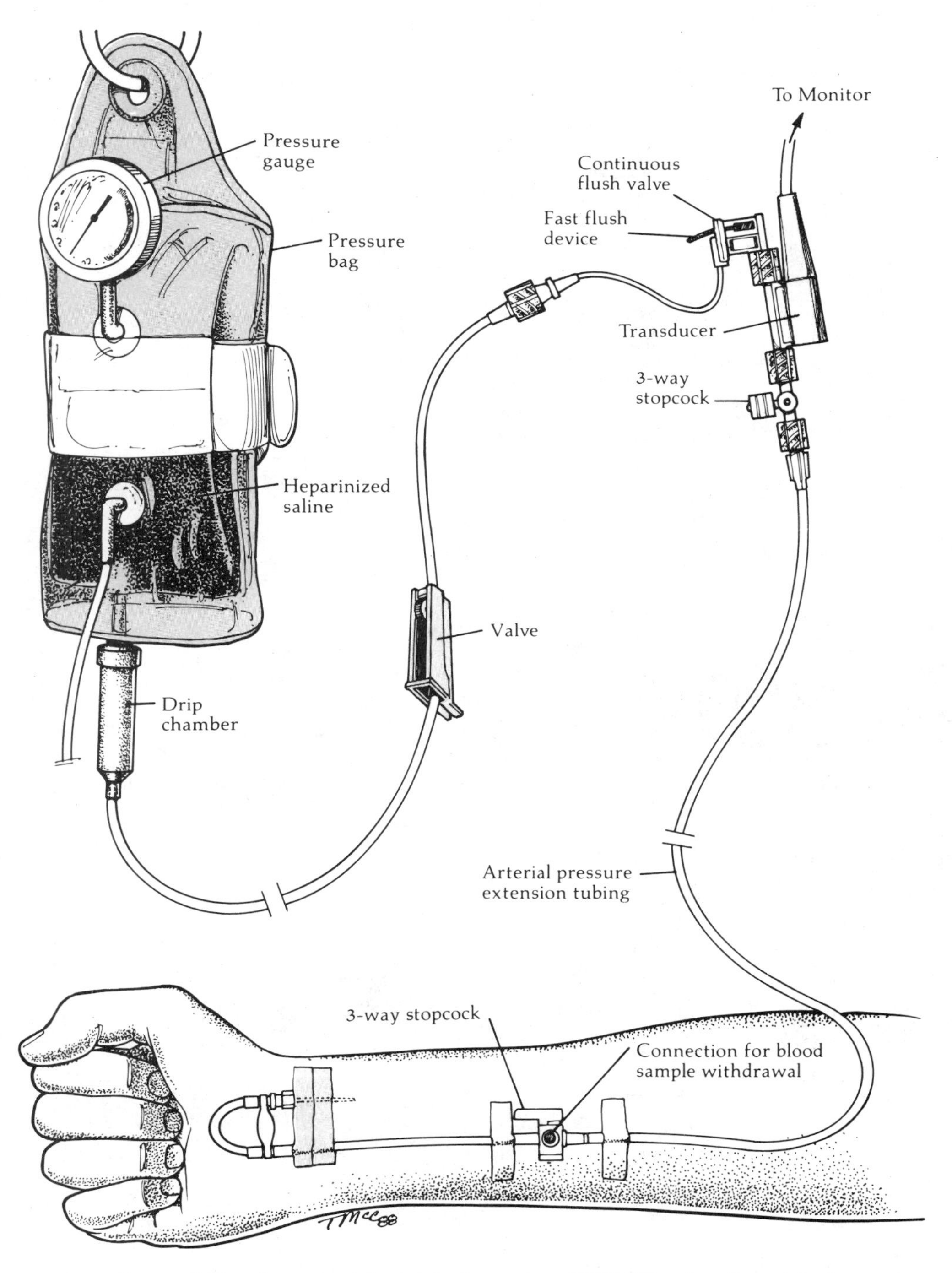

Figure 6-7b A constant flush infusion system (CFIS). (Courtesy Spacelabs.)

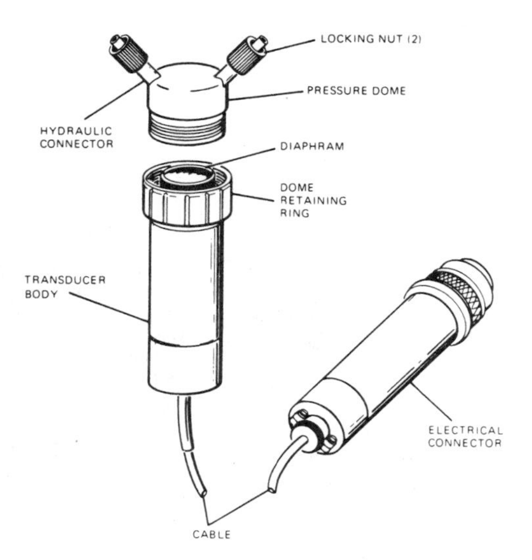

Figure 6-8 The components of blood pressure transducers. (Reprinted courtesy Hewlett-Packard.)

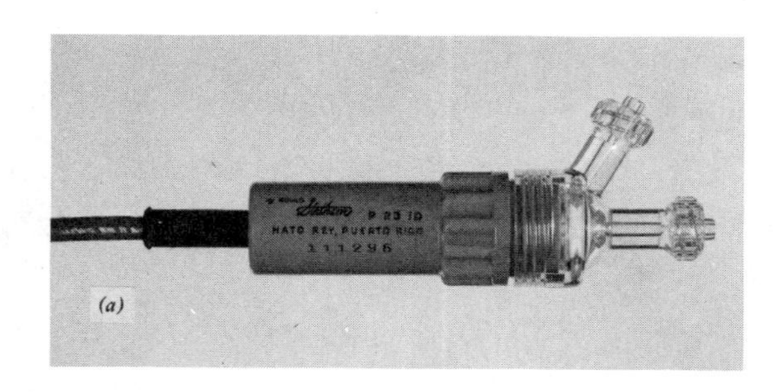

Figure 6-9 An arterial blood pressure transducer. (*a*) Photo. (*b*) Cross-sectional view. (Photo courtesy of Statham Instruments Division, Gould, Inc., Oxnard, Ca.)

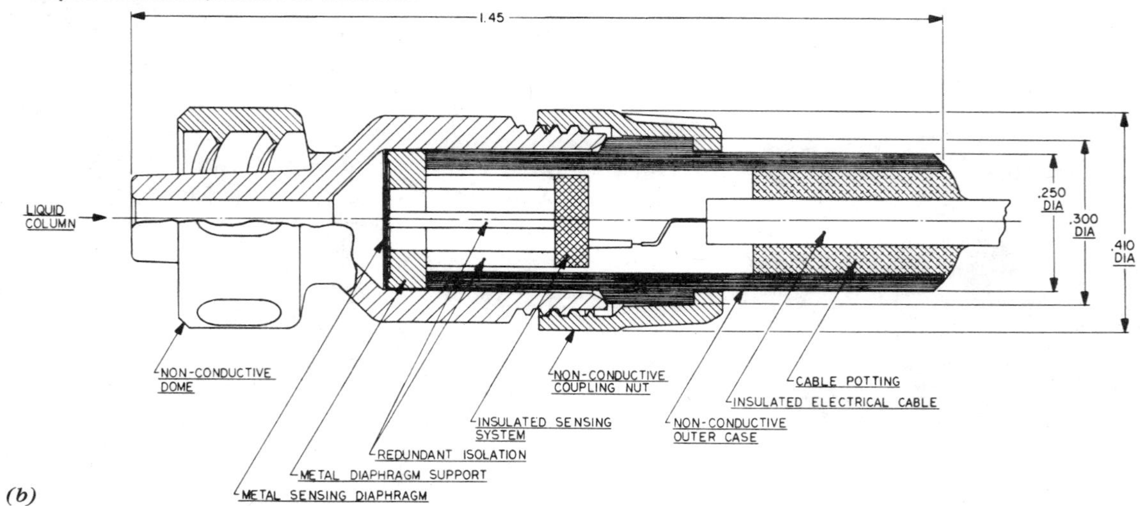

are generally more complex than clinical instruments and thus are not often found in bedside monitoring applications.

The pressure amplifier used for bedside applications is less complex and thus is less flexible and often somewhat less accurate. While its simplicity allows less well-trained personnel to successfully operate the instrument, it is more accurate than most indirect methods. In ICU/CCU/OR settings, the operator is usually a physician, nurse, or monitoring technician who has duties other than equipment operation and has limited time to calibrate pressure equipment. For these busy people, it is a valid trade-off to sacrifice a small amount of accuracy for simpler operation. As a result, clinical pressure amplifiers tend to have fewer front-panel controls and a simplified calibration procedure.

In applications where superior accuracy is needed, a mercury or aneroid manometer is used to calibrate pressure equipment every time it is used. Clinical pressure monitors have an internal calibration signal that is used for day-to-day operation, and it is checked periodically (usually monthly) against a manometer.

6-11 Typical Calibration Methods

The typical clinical pressure monitor will have the following controls: *zero* or *balance, sensitivity* or *gain,* and *calibrate* or a control yielding some specific pressure in mm Hg.

The zero control is used to adjust the amplifier output to zero volts under zero pressure conditions (i.e., 1 atm).

The sensitivity control adjusts the gain of the amplifier to produce the correct amplifier output voltage to represent a specific calibration pressure that is generated with a manometer/pump or supplied as a simulated pressure signal by an external transducer substitute or internal calibrate control.

The best accuracy is obtained when a manometer and squeeze-ball pump are used to calibrate the system. A mercury sphygmomanometer can be disconnected at the cuff and the loose end fitted with an appropriate Luer-Lock hydraulic connector so that it can couple to the transducer. It is essential that the transducer be disconnected from the patient when calibrating with a manometer, because we are going to introduce *air* into the system.

1. When the manometer is correctly connected to the transducer (see Figure 6-10), open the stopcock to atmosphere and adjust the amplifier zero balance control for an indication of zero on the meter (see Figure 6-7).

2. Next, close the stopcock and pump a standard pressure on the manometer, say, 100 mm Hg.

3. Adjust the sensitivity control for an indication of 100 mm Hg on the meter (Figure 6-7). It is prudent to check the agreement between the manometer and the meter at several points above and below the test pressure. If a transducer diaphragm has been strained beyond its safe limits, it may become *nonlinear.* Check for agreement at 50 percent and 200 percent of the test pressure, assuming that the transducer upper limit is not exceeded.

The same general procedure just described also works when an internal calibration signal is used to simulate a standard pressure. Some adjustment is necessary, however, to account for differences between several normal transducers. Even two transducers with the same model designation exhibit different sensitivity figures. One popular model, for example, has a *nominal* sensitivity specification of 5 μV/V/mm Hg. But calibration certificates supplied with each unit by the manufacturer attest that actual sensitivity figures range from 3.7 to 6.5 μV/V/mm Hg!

There are three approaches to standard-

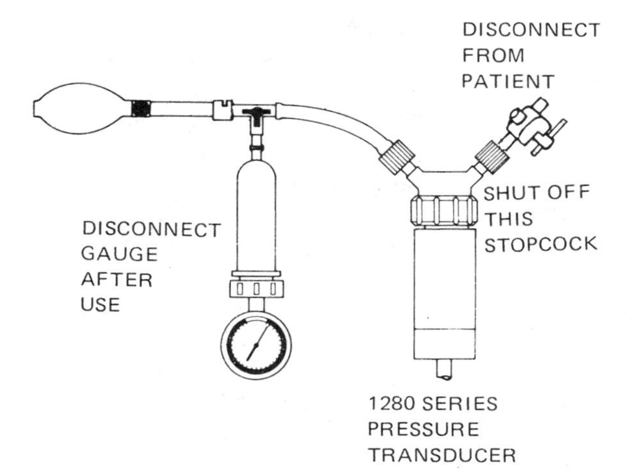

Figure 6-10 A manometer is used to calibrate the transducer. (Reprinted courtesy Hewlett-Packard.)

izing the adjustment procedure to accommodate different transducers: tight specification by the manufacturer, use of an internal transducer sensitivity adjustment, and the use of a *calibration factor* unique to each individual transducer used with a given pressure amplifier.

Some manufacturers of pressure-monitoring equipment solve the problem by requiring the transducer manufacturer to supply them with units that fall within very tight limits for sensitivity and *zero-stimulus offset error.* This approach works well in some situations, but it does not allow for changes in offset and sensitivity due to abuse and aging. It also forces the consumer to buy transducers at a premium price from the equipment manufacturer rather than directly from the transducer manufacturer at a lower price.

Other manufacturers, notably Hewlett-Packard, use a potentiometer inside the transducer electrical connector to *trim out* differences in sensitivity. The *actual* transducer sensitivity is always higher than that seen by the amplifier, but the effect of adjustment is that the amplifier always sees the *same* sensitivity regardless of which individual transducer is in use.

Some companies use a *calibration factor* for each transducer. The simplified circuit is shown in Figure 6-11. Each transducer must have its own calibration factor. Record the *cal factor* number or mark it on the transducer body. When that transducer is used again, the operator need only dial in the correct calibration factor.

Subsequent calibrations do not require a manometer test. The calibration factor for any given transducer remains valid for several months unless the transducer is physically abused. Of course, any transducer that is suspect should be checked out more frequently and immediately after an incident of abuse has been reported. Between periodical checks however, the following procedure is used:

1. Open the transducer stopcock to air; place switch S_1 in the 0 mm Hg position. Adjust balance for a 0-V output indication (i.e., 0 mm Hg on the meter).

2. Set switch selector to a position that is most convenient for the range being calibrated. In general, a setting that would put the meter pointer at mid-scale or above is best.

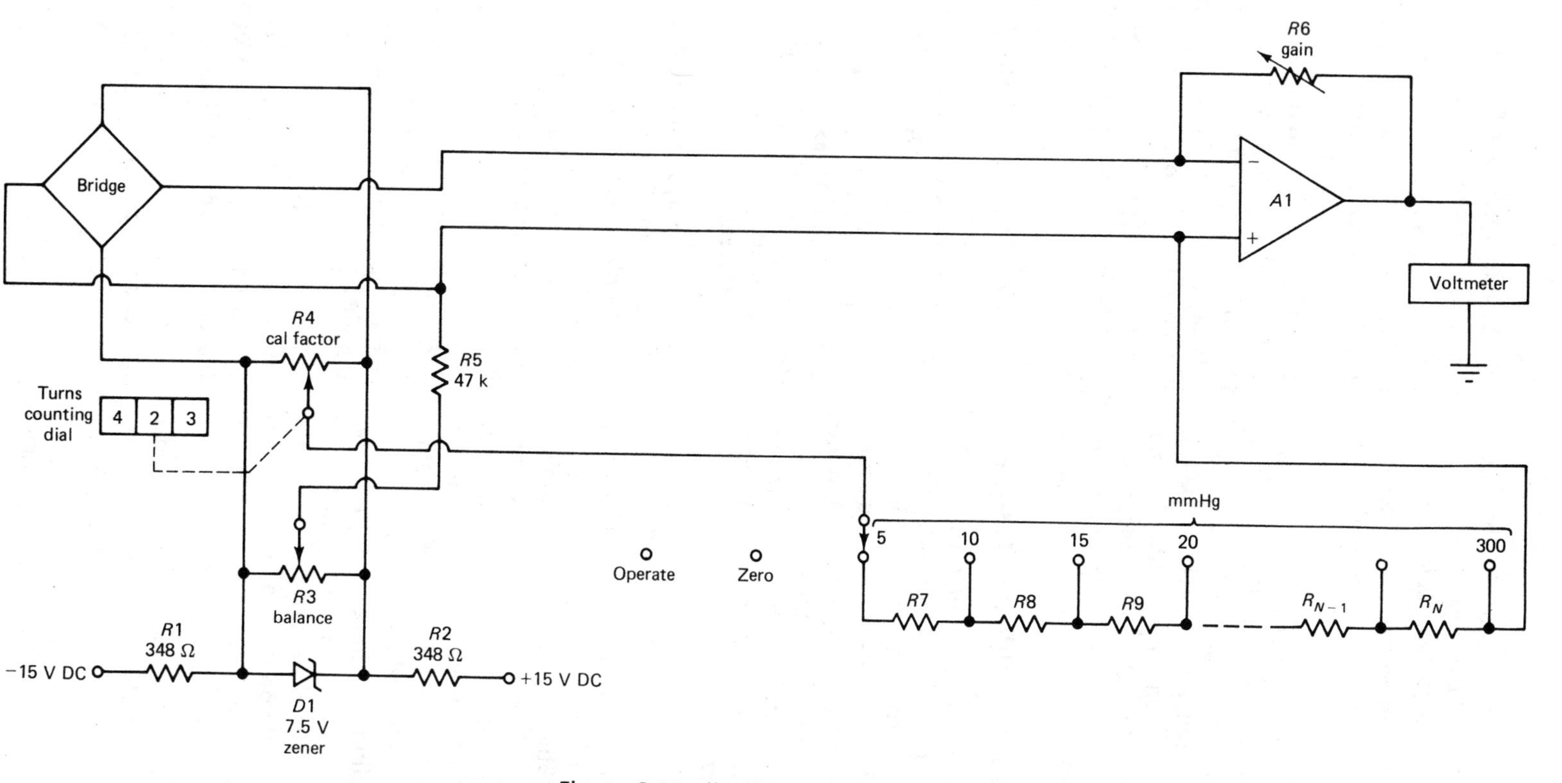

Figure 6-11 "Calibration factor" circuit.

3. Set the *cal factor* knob to the figure recorded previously (use a manometer if the calibration date is more than six months old).

4. Adjust the gain control until the meter reads the standard pressure used in the original calibration.

In general, it is preferable to settle on a single standard pressure for all calibrations. This avoids problems that could occur if one "standard" is used by one person making the initial calibration and another "standard" is selected by subsequent people performing the same job. For arterial monitors, 100 mm Hg or 200 mm Hg are good choices.

6-12 Pressure Amplifier Designs

There are four basic types of pressure amplifier design: dc, isolated dc, pulsed excitation, and ac carrier. Such amplifiers are implemented using transistors, operational amplifiers, special linear integrated circuits (LICs), or (in older equipment) vacuum tubes. The basic difference between the four classes is in the type of excitation applied to the transducer.

6-12.1 dc Pressure Amplifiers

Figure 6-11 shows the simplified circuit of a dc pressure amplifier that uses the calibration factor method. The pressure amplifier (A1) is a dc amplifier, so the pressure transducer is a resistive Wheatstone bridge strain gage. Diode D1 provides the 7.5-volt dc excitation to the transducer, and the potentials for the BALANCE and CAL FACTOR controls. The calibration factor for the transducer sometimes changes, so it is necessary to provide a procedure for measuring the new factor. The dc pressure amplifier is first calibrated with an accurate mercury manometer:

1. Set switch S1 to the OPERATE position, open the transducer stopcock to atmosphere, and adjust R3 for zero volts output (i.e., 0 Torr reading on the display).

2. Close the transducer stopcock and then pump a standard pressure (e.g., 100 Torr, or at least half-scale). Adjust gain control R6 until the meter reads the correct (standard) pressure. Check for agreement between the meter and the manometer at several standard pressures throughout the range (e.g., 50, 100, 150, 200, 250, and 300 Torr). This last step is needed to ensure the transducer is reasonably linear—that is, that the diaphragm was not strained by out-of-range pressures or vacuums.

3. Turn switch S1 to the position corresponding to the applied standard pressure, and then adjust CAL FACTOR control (R4) until the same standard pressure is obtained on the meter. The CAL FACTOR control is ganged to a turns-counting dial. The number appearing on the dial at the position of R4 that creates the same standard pressure signal is the Calibration Factor for that transducer. Record the turns-counter reading for future reference.

For a period of time (usually 6 months) the calibration factor need not be redetermined unless damage to the transducer occurs. The calibration factor is entered into the amplifier by turning R4 (or digitally in modern instruments). The following procedure is normally used:

1. Open the transducer stopcock to atmosphere, place switch S1 in the 0 Torr (0 mm Hg) position, and adjust R3 for a 0 volts output indication (0 Torr on the display).

2. Set switch S1 to a position that is most convenient for the range of pressures to be measured. In general, select a scale that places the reading mid-scale or higher.

3. Set the CAL FACTOR knob to the figure recorded previously. Adjust the gain control (R6) until the meter reads the standard pressure used in the original calibration.

The circuit in Figure 6-11 is an example of a dc pressure amplifier, while Figure 6-12 is a more detailed version of the actual amplifier block. Note that there are only two operational amplifiers in this circuit—little about the simple dc amplifier is complex. Amplifier A1 is the input amplifier, and it should be a low drift, premium model. Both gain and zero controls are provided, so the amplifier will work with a wide variety of transducers.

The excitation voltage of the transducer is determined (as a maximum) by the transducer manufacturer, with a value of 10 volts being common. In general, it is best to operate the transducer at a voltage lower than the maximum in order to prevent drift due to self-heating. Pressure amplifier manufacturers typically specify either 5 volts or 7.5 volts for a 10-volt (max) transducer.

The required amplifier gain can be calculated from the required output voltage that is used to represent any given pressure. Because digital voltmeters are used extensively for readout displays, the common practice is to use an output voltage scale factor that is numerically the same as the full-scale pressure. For example, one millivolt (1 mV) or ten millivolts (10 mV) per millimeter of mercury (mm Hg) is common. Let's assume a maximum pressure range of 400 mm Hg (common on arterial monitors), if we use a scale factor of 1 mV/mm Hg, when 400 mm Hg is represented by 400 mV, or 0.40 Volt. No further scaling of the meter output is needed.

6-12.2 Isolated dc Amplifiers

Patient safety considerations, identified in the past few years, have caused many manufacturers to redesign their pressure monitors to improve electrical isolation of the patient from the ac power mains supply. These amplifiers are built along the same lines as the isolated amplifiers in Section 4-11.

6-12.3 Pulsed-Excitation Amplifiers

Figure 6-13 shows the block diagram for a pulsed-excitation pressure amplifier. This amplifier uses a Wheatstone bridge strain gage transducer, although inductive types can be accommodated in some models. The excitation signal is a biphasic short-duration pulse. In one model the pulse (see Figure 6-13*b*) has a duration of 1 ms and has a 25 percent duty cycle.

Amplifier A_1 is a dc pressure amplifier, and amplifier A_2 is a unity gain summation stage. The output indicator is a digital voltmeter that will update the display only when the *strobe* line is high. Switches S_1 through S_5 are CMOS *electronic switches,* which *close* when the control line (C) is *high.* All circuit action is controlled by a four-phase clock. Phases ϕ_1 and ϕ_2 excite the transducer and operate the amplifier *drift cancellation* circuit. Phase ϕ_3 updates the display meter, and phase ϕ_4 resets the circuit following update.

All dc amplifiers tend to drive (i.e., create spurious offset voltages due to thermal changes). Switches S_2 and S_3, capacitor C_1, and amplifier A_2 serve as a drift cancellation circuit (Figure 6-13*a*).

The transducer is excited *only* when ϕ_1 is *high* and ϕ_2 is *low;* at all other times the transducer is *not* excited, which keeps transducer self-heating to a minimum. Amplifier A_1 will drift, however, because of its high gain and inherent offset voltages.

6-13 ac Carrier Amplifiers

The carrier amplifier requires ac excitation for the transducer, so it operates equally well with both inductive and resistive strain gages. Carrier frequencies are typically be-

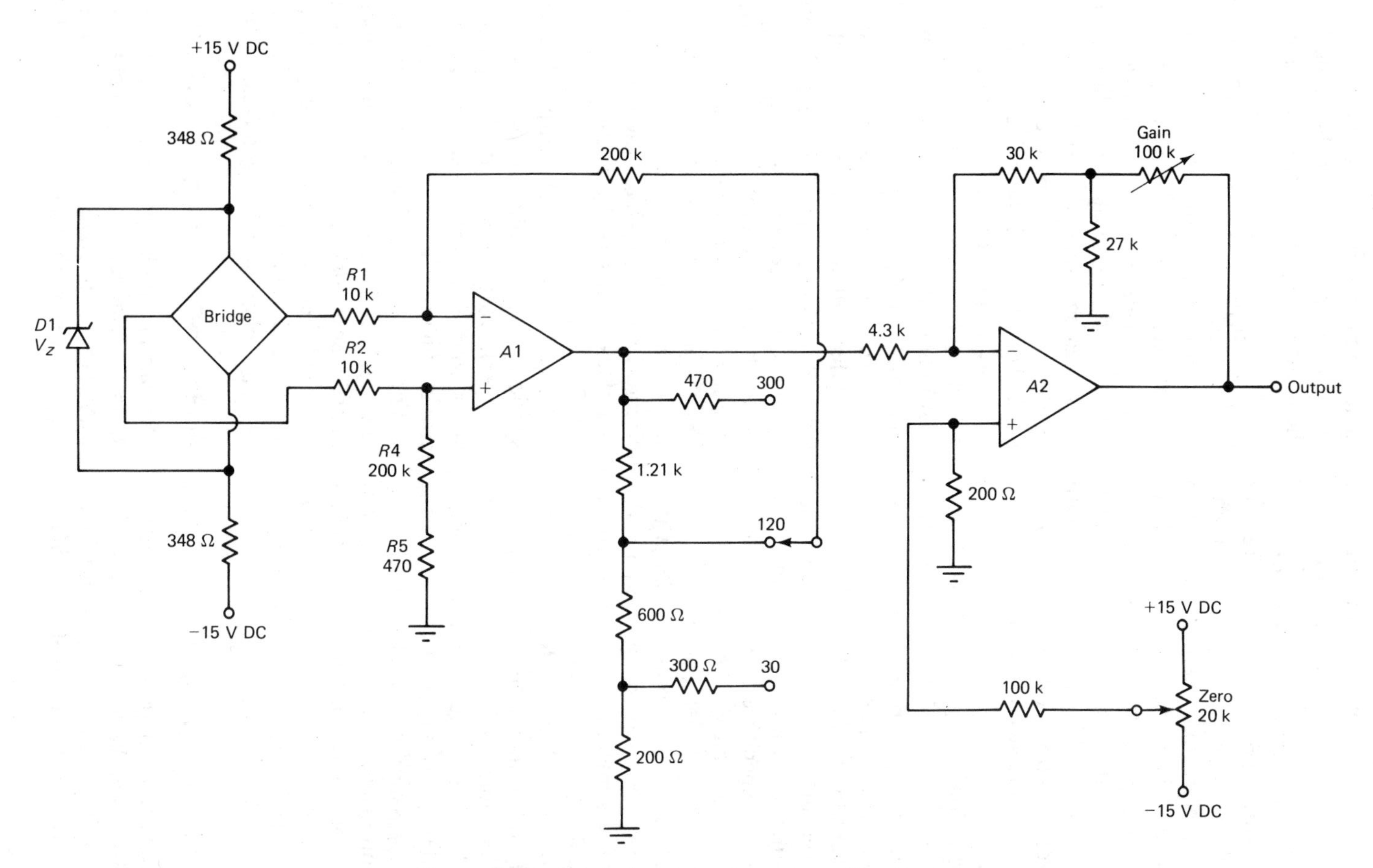

Figure 6-12 Detailed version of actual amplifier block.

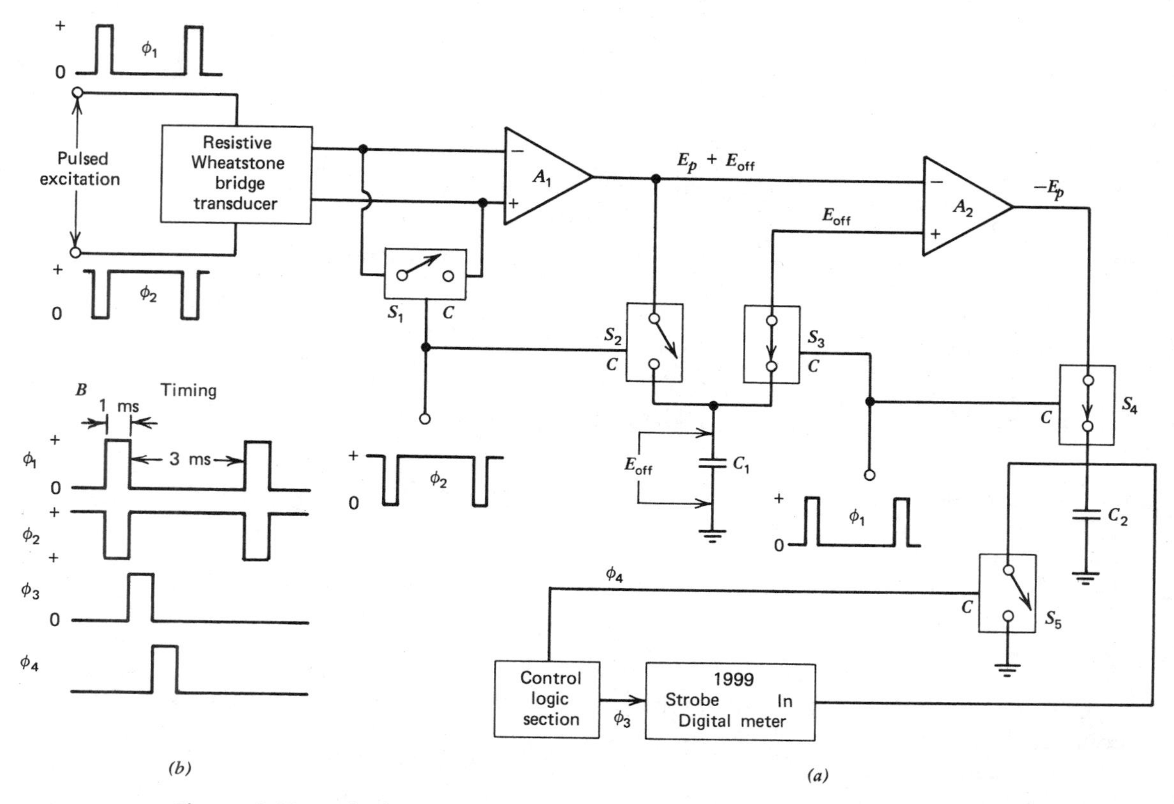

Figure 6-13 Pulsed excitation system. (*a*) Circuit diagram. (*b*) Timing diagram.

tween 400 and 5000 Hz, with amplitudes ranging between 5 and 12 V rms. Hewlett-Packard equipment, for example, is standardized on a frequency of 2400 Hz and an amplitude of 5 V rms.

Self-contained pressure monitors *will* contain their own carrier oscillator signal sources, while many *central* monitoring systems and rack-mounted catheterization laboratory instruments tend to use a *common* carrier source powerful enough to drive several carrier amplifiers.

Figure 6-14 shows the block diagram to a typical carrier amplifier. The carrier signal to the transducer is supplied through a push-pull transformer, so a 180° phase difference exists, allowing ground-referenced operation of the transducer.

1. Amplifier A_1 is a single-ended ac amplifier that is stabilized by heavy negative feedback. It is the stability of the ac amplifier that confers so much flexibility and quality on the carrier amplifier.

2. A calibration signal is supplied by switch S_1 and the voltage divider consisting of R_1 and R_2. This circuit introduces a small signal, equal to the output of the transducer at some standard pressure, into the ac amplifier. Calibration is performed with the transducer open to atmosphere, and the amplifier is prezeroed.

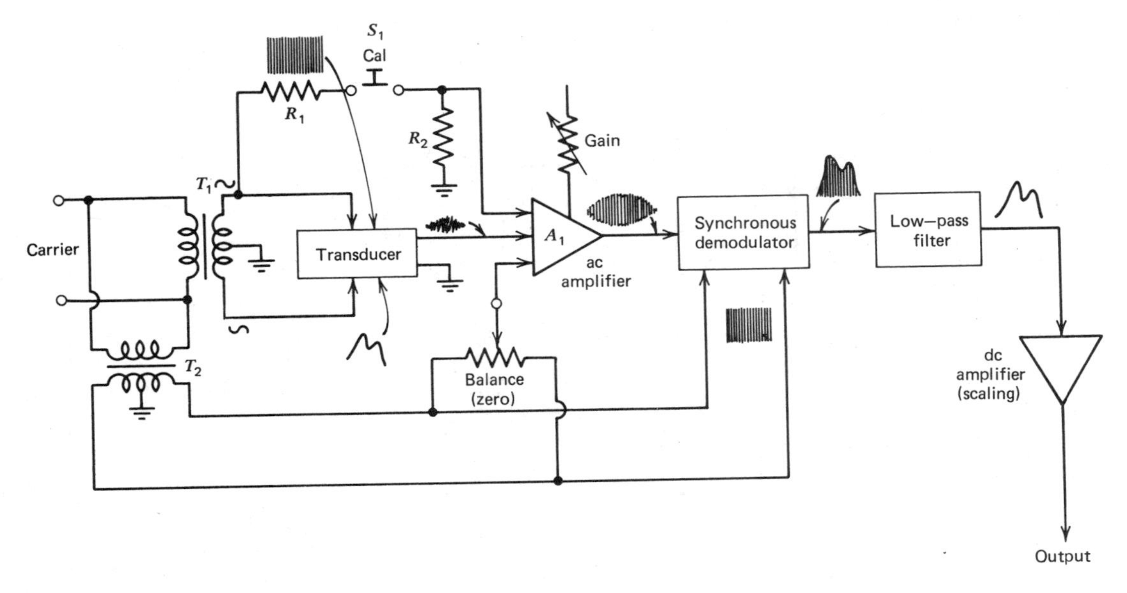

Figure 6-14 Carrier amplifier.

3. The balance control is a potentiometer connected to another pair of carrier signal lines that are also 180° out of phase. The wiper of the potentiometer is connected to an amplifier input, where it is *summed* with the transducer signal.

4. The *balance* control nulls the system by injecting a signal of *equal magnitude,* but *opposite phase,* into the system to algebraically sum with the offset signal to produce a net result of zero output.

5. The same carrier signal that is applied to the balance control is also sent to a synchronous demodulator (Section 4-11.2), where it converts the ac amplifier output signal to a varying dc signal. A low-pass filter following the demodulator removes any residual carrier signal, and a final dc amplifier buffers the output and provides any needed final scaling.

In many pressure amplifiers, the designers find it convenient to *scale* the output voltage so that it is *numerically* the same as the pressure it represents. Hewlett-Packard, for example, uses a scale of 0 to 3.0 V to represent 0 to 30 or 0 to 300 mm Hg. The scale factor is 10 mV/mm Hg. Using a numerically equal scale factor allows the use of simple 3 V dc voltmeters (scaled in mm Hg, or so labeled if digital) as the output indicator. In the case of a digital display instrument, it is *necessary* to use numerically equal scale factors because it is not possible to have an appropriate scale *printed,* as is often done in *analog* voltmeters. A digital voltmeter may be scaled by shifting the decimal point.

6-14 Systolic, Diastolic, and Mean Detector Circuits

The pressure amplifier produces an analog waveform with a peak amplitude representing the systolic pressure, and a minimum, or "valley," that represents the diastolic pressure. Additional circuitry is required that recognizes these points, and from them produce a steady dc voltage or current that can be used to drive a display meter.

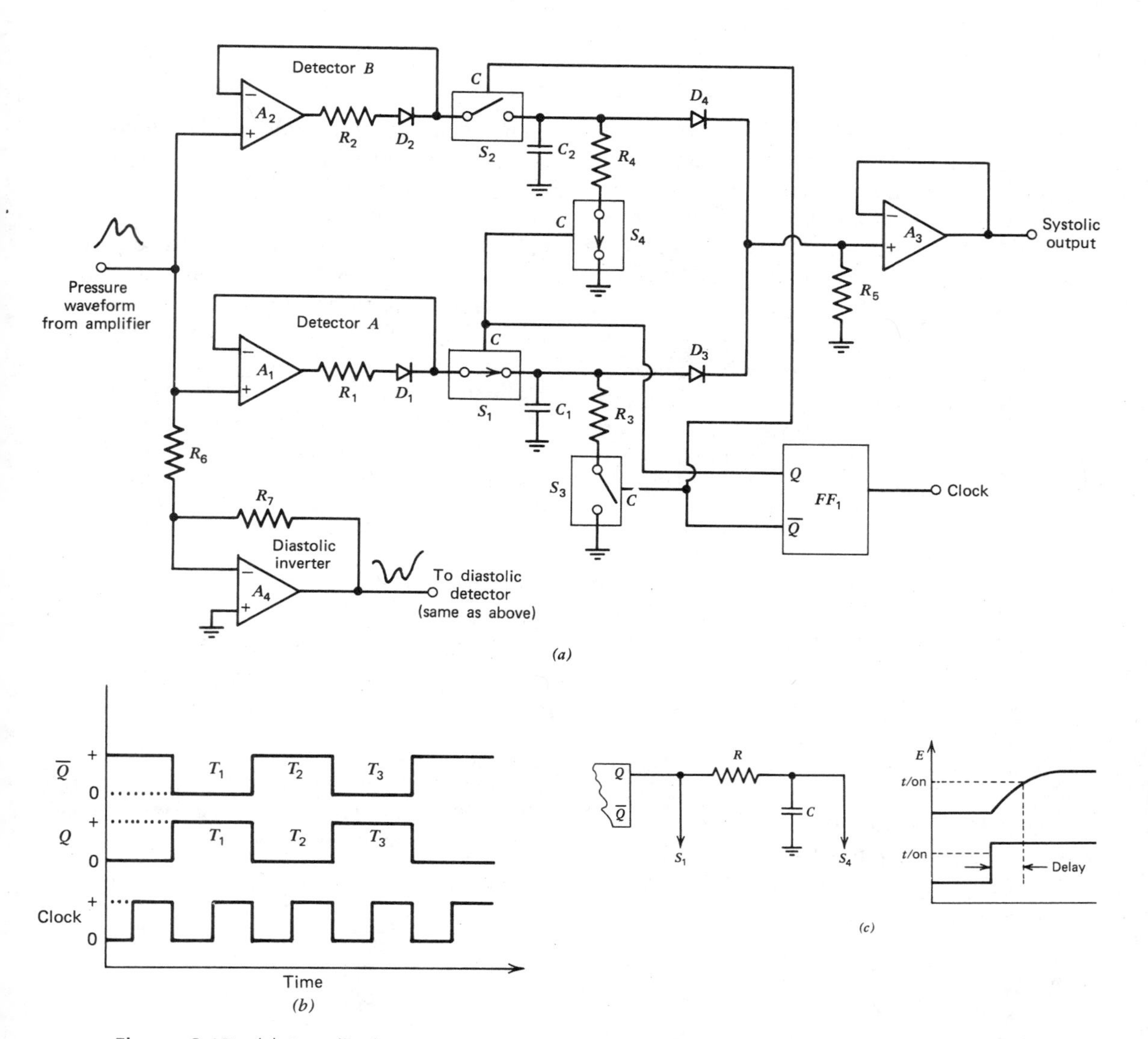

Figure 6-15 (*a*) Systolic detector circuit. (*b*) Timing diagram. (*c*) Turn-on delay circuit for S_4 and waveform.

The partial schematic of a pressure detector is shown in Figure 6-15*a*, while the timing diagram is shown in Figure 6-15*b*. Amplifiers A_1 through A_3 are operational amplifiers, while switches S_1 through S_4 are CMOS electronic switches. Note that the operational amplifiers may be integrated types, as shown, or may be constructed of discrete components in older models.

1. This circuit operates from a two-phase clock created by flip-flop FF_1. The Q and not-Q ($\overline{Q}$) outputs are complementary, so one will be *high* when the other is *low*.

2. Switches S_1 and S_4 are turned on when the Q is high, and switches S_2 and S_3 are turned on when the $\overline{Q}$ is high.

3. The analog waveform from the output of the pressure amplifier is applied simultaneously (i.e., in parallel) to the inputs of A_1 and A_2. The waveform, therefore, appears simultaneously at the outputs of A_1 and A_2.

4. During period T_1 (Figure 6-15*b*), the Q output of FF_1 is high, so switches S_1 and S_4 are closed. When S_4 is closed, capacitor C_2 is discharged, so it has no effect on the output. Closing switch S_1 allows the signal appearing at the output of A_1 to charge capacitor C_1 to the peak voltage, representing the *systolic pressure*.

5. The voltage across capacitor C_1 forward biases diode D_3, which then conducts and applies the voltage to the input of amplifier A_3. The output of amplifier A_3 goes to the *systolic* output meter.

6. The situation reverses during the time T_2: The not-Q of FF_1 becomes high and the Q is low. This turns on switches S_2 and S_3 and turns off S_1 and S_4. Closing S_2 causes capacitor C_2 to rapidly charge to the peak voltage of the input signal. This will occur within less than a second in most cases.

7. Switch S_3 closes to allow capacitor C_1 to discharge *slowly* through resistor R_3. The charge on C_2 will reach peak before any appreciable decay of the C_1 charge takes place. In some models each charge switch is operated *directly* from FF_1, while the discharge switch is operated by the same signal after it has passed through an *R-C* network delay line. (See Figure 6-15*c*.) This network insures that one capacitor is fully charged before the other begins to decay.

8. The voltage at the output of Figure 6-15*a* represents the *peak* of the waveform, which corresponds to the *systolic* pressure. The *same* circuit may be used to detect the diastolic voltage by *inverting* the waveform so that the diastolic feature becomes the peak.

The *mean arterial pressure* is found by taking the time average (i.e., *integrating*) the pressure waveform (Figure 6-16*a*). An example of a simple mean value integrator is shown in Figure 6-16*b*. Most pressure monitors use a simple *R-C* integrator with buffering amplifiers rather than regular operational amplifier integrators discussed in Section 4-8.2.

The mean reading may create a bit of confusion to some medical and nursing personnel, who were taught the *functional* definition of mean arterial pressure, which is

$$\overline{P} = P_d + \frac{P_s - P_d}{3} \tag{6-4}$$

where

$\overline{P}$ is the *mean* arterial pressure in mm Hg
P_d is the diastolic pressure in mm Hg
P_s is the systolic pressure in mm Hg

EXAMPLE 6-4

A patient's blood pressure is measured as systolic 120 mm Hg and a diastolic of 80 mm Hg. What is the mean arterial pressure?

SOLUTION

$$\overline{P} = P_d + \frac{P_s - P_d}{3} \tag{6-4}$$

$$\overline{P} = (80 \text{ mm Hg}) + \frac{(120 - 80)}{3} \text{ mm Hg}$$

$$\overline{P} = (80 \text{ mm Hg}) + \left(\frac{40}{3}\right) \text{ mm Hg} \cong$$

93 mm Hg

It must be recognized that Equation 6-4 is only an *approximation* of the correct integral. The use of Equation 6-4 finds the *functional mean pressure*, which is accurate only when the patient's arterial waveform has approximately the correct *shape*. Although a wide latitude exists, a significant error re-

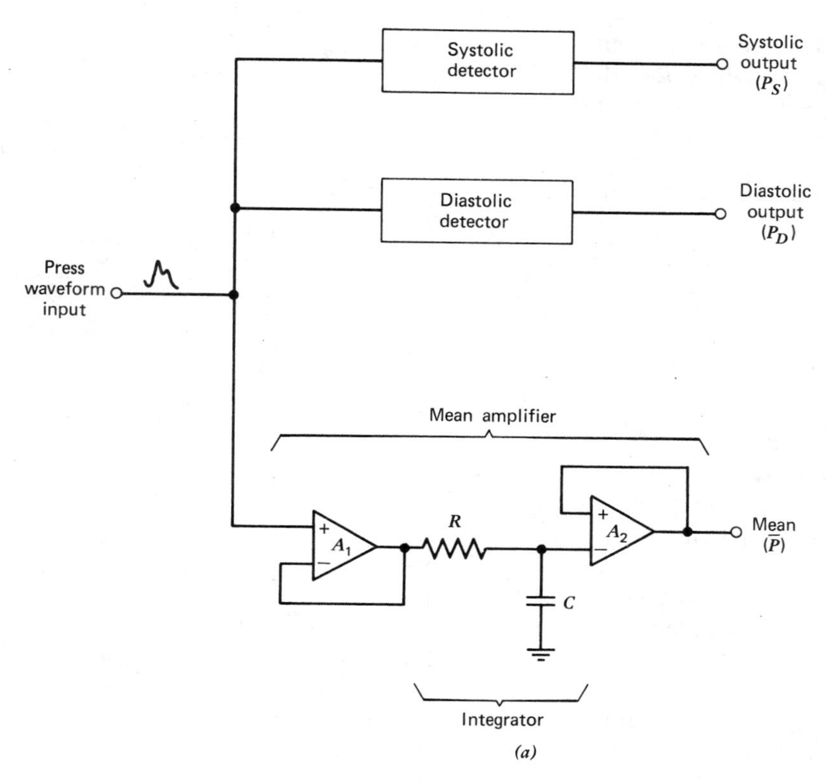

(a)

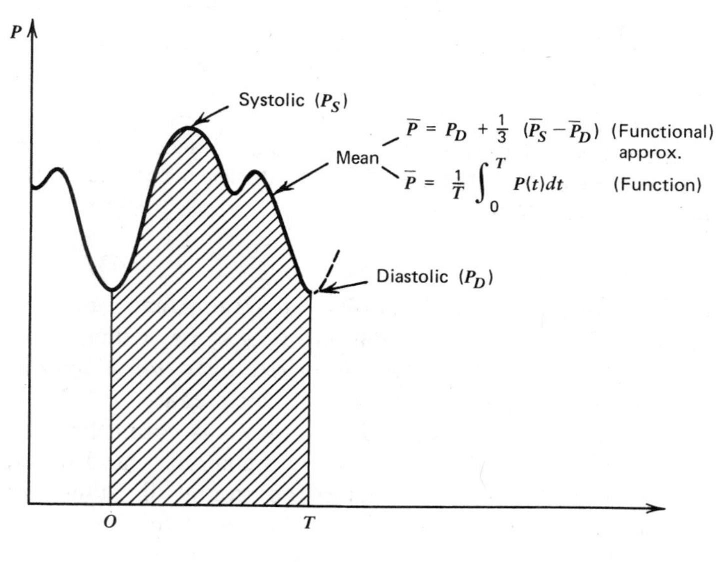

(b)

Figure 6-16 (*a*) Mean arterial pressure detector. (*b*) Graphical and mathematical meaning of "mean arterial pressure."

sults if the waveform is not normal. Some problems (see Section 6-17) will *distort* the arterial waveform, so the mean displayed by the *meter* is in *error*. In other cases, however, the patient's correct (actual) waveform is atypical, resulting in a discrepancy between the meter and functional values. In that case, however, the *meter* reading is *correct*.

6-15 Pressure Differentiation (dP/dT) Circuits

Most pressure amplifiers designed for research applications, and many clinical instruments, are equipped with a special output producing a signal representing the *derivative* (i.e., the *time rate of change*) of the pressure waveform. In calculus notation this is dP/dT. This symbol marks the derivative signal output jack on the pressure amplifier case or front panel.

Figure 6-17*a* shows a typical operational-amplifier *differentiator* used to differentiate physiological pressure signals. Resistor R_1 and capacitor C_1 are the actual differentiator components, while C_2 and R_2 are used to improve the stability of the circuit.

The *time constant* of a true differentiator must be very short compared with the *period* of the input signal. In the case of waveforms such as arterial pressure, it must be very short compared with the risetime of the waveform's *leading edge*. The *R-C* time constant of the circuit in Figure 6-17*a* is 10 ms, which is appropriate in dP/dT circuits. The same circuit principle is used in other physiological instruments with time constants as low as 25 μs.

Figure 6-17*b* shows the standard method for calibrating a differentiator. A *sawtooth,* or *ramp* signal, is applied to the input. The derivative of a signal quantifies its rate of change or *slope,* so the derivative of a ramp (i.e., *constant* rate of change) is a constant voltage.

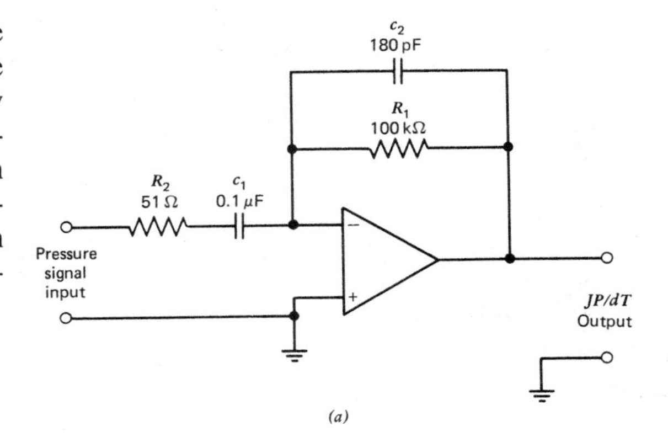

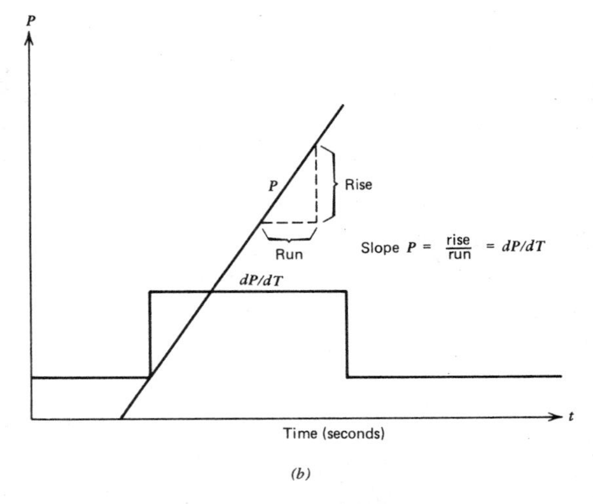

Figure 6-17 Differentiator to generate the *dP/dT* waveform. (*a*) Differentiator circuit. (*b*) Calibration with a linear ramp.

The derivative of a ramp can be found from computation of its slope, which is defined as the ratio of *rise* over *run* (*Y* over *X*, or simply *Y/X*). On a strip-chart recorder or oscilloscope display, we may count the number of divisions the signal rises and divide by the number of horizontal divisions that were required for the signal to rise to that height. The vertical divisions represent the pressure in mm Hg, while the horizontal divisions represent time in seconds. The units for dP/dT output, then, are mm Hg/s.

6-16 Automatic Zero Circuits

Many recent models of blood pressure monitors incorporate circuitry that automatically balances the amplifier to zero null. There is usually a front-panel pushbutton marked "zero." The operator opens the transducer stopcock to atmosphere, which is the zero pressure reference, and then presses the *zero* button. Any voltage existing at the output of the amplifier when the button is pressed is assumed to be an offset or error voltage, so it is nulled out.

Figure 6-18 shows the block diagram for an auto-zero circuit. The three major sections are *summation amplifier, ramp generator,* and *control logic*.

The summation amplifier receives the output of the pressure amplifier and the output of the ramp generator. Its output will be the *algebraic sum* of these two input signals.

The ramp generator produces a voltage that begins at zero and rises to a maximum voltage in a linear manner. The circuit shown in Figure 6-18 uses a digital-to-analog converter (DAC) to generate the ramp. A few obsolete instructions are still found using a glass dielectric (i.e., low leakage) capacitor charged from a constant current source to do substantially the same job.

The DAC produces an output voltage proportional to the binary word applied to its digital inputs. The binary word is created by a binary counter that is incremented by a 2.5-kHz clock. In this case the counter starts at 00000000_2 and increments once for each clock pulse until a count of 11111111_2 (i.e., 256_{10}) is reached, or the counter is halted by an external gate.

The control logic section consists of two monostable multivibrators (i.e., one-shots), a three-input NAND gate, and a ground-referenced voltage comparator. As long as any *one* input of the NAND gate is *low* (i.e. zero volts), its output will remain *high*. If *both* inputs 1 and 3 are high, then the clock pulses will pass through the gate to the input of the binary counter. The output of the voltage comparator will remain high as long as the summation amplifier output voltage E_o is greater than zero. Operation of the circuit is as follows:

1. The operator opens the transducer stopcock to atmosphere and presses the *zero* button. This action triggers the first one-shot (OS_1), which produces a 1-ms pulse.

2. The pulse from OS_1 resets the binary counter to zero and triggers the second one-shot (OS_2).

3. The output of OS_2 goes high, forcing the no. 3 input of the NAND gate to go high also, for a period of 500 ms.

4. If voltage E_o is greater than zero, then the output of the voltage comparator turns gate 1 of the NAND gate also high, allowing clock pulses to pass through to the binary counter.

5. The counter begins incrementing immediately, and this forces the DAC output voltage to rise. This voltage is applied to the input of the summation amplifier.

6. Output E_o will now be a summation of the error offset voltage and the ramp voltage. Since the ramp voltage has a polarity opposite that of the error voltage, it begins to cancel the offset component of E_o.

7. When E_o drops to zero, the voltage comparator output goes low, closing the NAND gate. The counter will stop, and its output digital word is that which existed at the instant E_o reached zero, plus or minus the usual one count.

8. The amplifier is now zeroed, so the transducer stopcock is closed. From now on, the voltage appearing at the output of the summation amplifier will represent *only* the pressure signal.

The capacitor-type circuits work in substantially the same manner, but suffer from

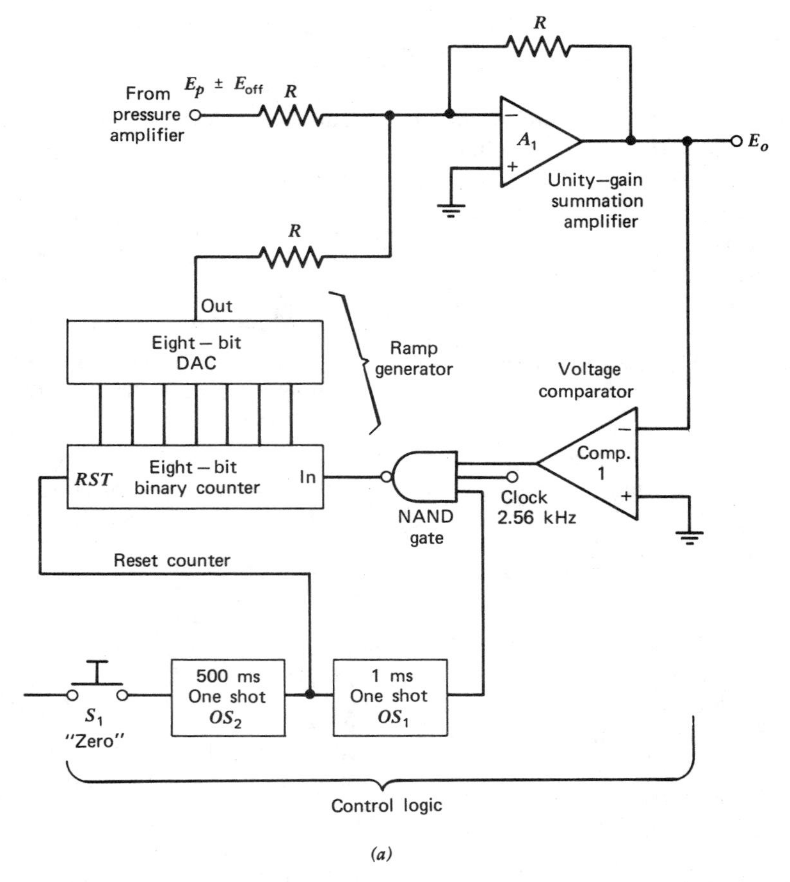

Figure 6-18 Automatic zero circuit. (*a*) Circuit diagram. (*b*) Timing and waveform diagram.

a type of zero drift caused by *droop* of the capacitor charge voltage (i.e., discharge of the capacitor through leakage and ordinary circuit impedances). The DAC method is preferred by most manufacturers because it holds the zero longer, and the cost of modern IC DACs is actually *lower* than the large glass capacitors formerly used in auto-zero circuits.

6-17 Practical Problems in Pressure Monitoring

In practical monitoring situations, there are a few problems that must be solved by either operating personnel or the people who deal exclusively with medical instrumentation (i.e., biomedical equipment technicians, clinical engineers, or monitoring/cardiovascular technicians).

6-17.1 Hydrostatic Pressure

The liquid in the transducer plumbing system has mass, so its weight can apply a force to the transducer diaphragm that is also interpreted as a pressure. This force creates an offset voltage at the transducer output.

Figure 6-19*a* shows a transducer with a *positive* hydrostatic pressure head caused by

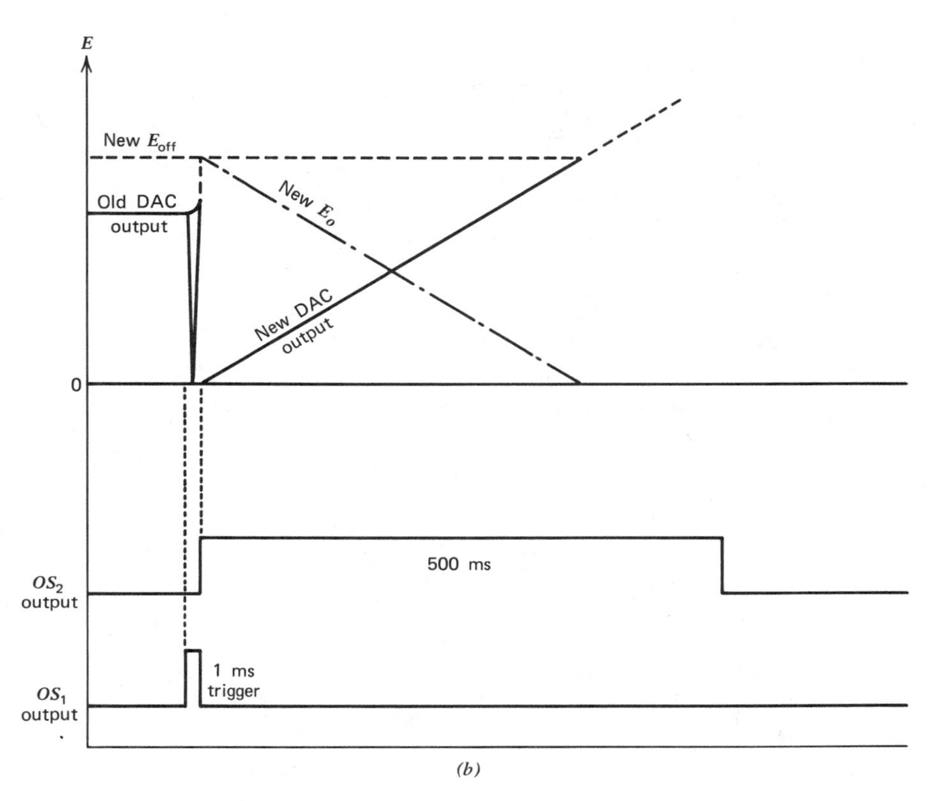

(b)

Figure 6-18 (continued)

the weight of the fluid in the tubing brought to the transducer from *above*.

Similarly, a *negative* pressure head is obtained when the tubing is predominantly *below* the level of the transducer diaphragm (see Figure 6-19*b*). The solution to this offset problem is shown in Figure 6-19*c*, where the tubing has approximately equal lengths above and below the level of the diaphragm. The transducer diaphragm serves as the guide point for positioning one end of the plumbing system, while the catheter tip inside of the patient serves as the reference for the other end.

The transducer is usually physically mounted on a special stand, or standard IV pole, next to the patient's bed. The transducer will be on a special mount that allows it to be moved up and down on the pole to adjust for different hydrostatic pressure situations. The diaphragm should be level and at a height equal to the height from the floor to the catheter *tip, not* the point on the patient's body where the catheter is inserted. In any given case, the catheter tip may be several centimeters above the insertion point.

Figure 6-20 shows the correct point for locating the transducer when intracardiac catheters are used: the midline of the chest as viewed from the side. Most authorities agree that a good approximation on *most* patients is 10 cm above the bed, when the patient is in the *supine* position. Older advice (referencing the midline to an anatomical point on the patient's chest) is used less often

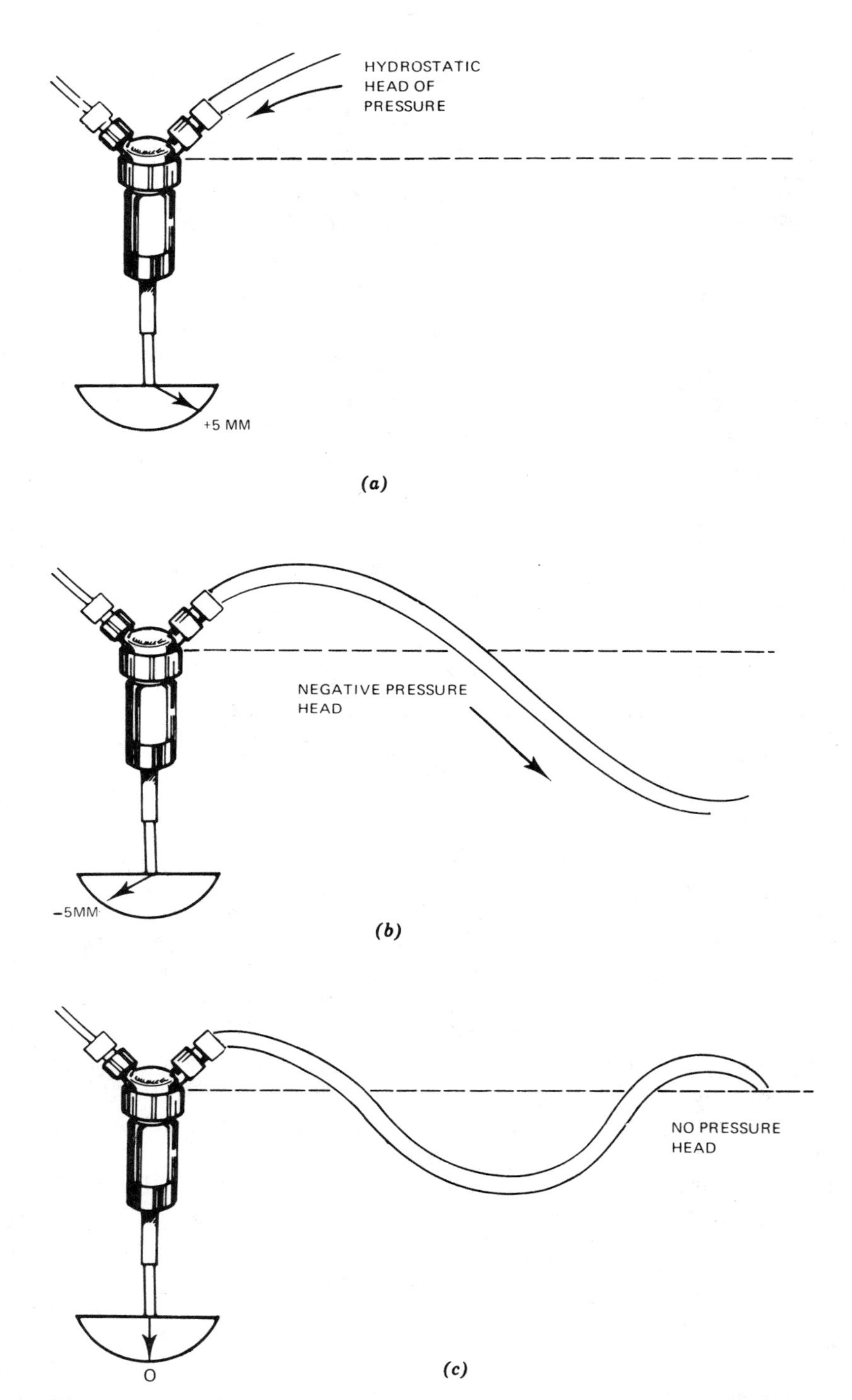

Figure 6-19 Hydrostatic pressure. (*a*) Positive pressure head. (*b*) Negative pressure head. (*c*) Balancing positive and negative pressure heads. (Reprinted courtesy Hewlett-Packard.)

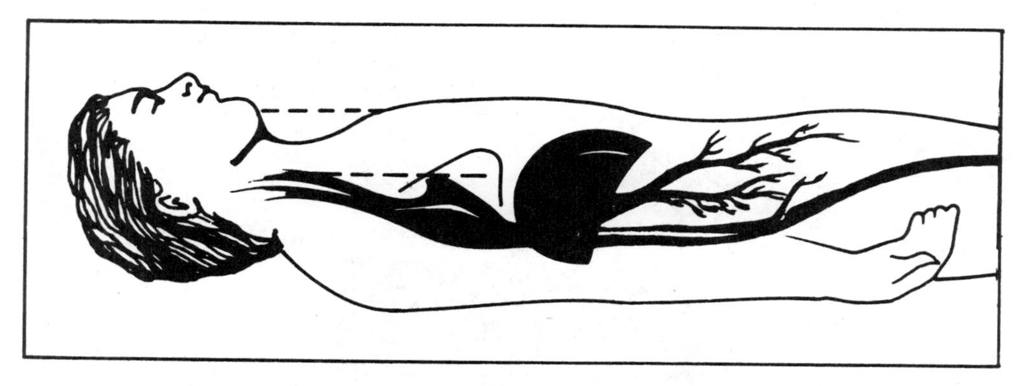

Figure 6-20 Proper transducer placement for measuring intracardiac pressures is along the midchest line. (Reprinted courtesy Hewlett-Packard.)

because it more frequently results in erroneous readings.

Hydrostatic pressure accounts for many *apparent* errors in pressure measurement systems. But these can be minimized by correct positioning of the transducer and need not affect a properly configured system.

6-17.2 Plumbing System Distortion

It is possible for mechanical problems in the transducer plumbing system to adversely affect the waveform and meter readings that are obtained. The properties of the system can create mechanical resonances and damping that distort the waveform. The arterial pressure waveform is *nonsinusoidal,* containing a fundamental frequency plus a number of *harmonics* (whole number multiples) of that fundamental frequency. If the plumbing system were perfect (which it never is), then the fundamental and all its harmonics would be transmitted through the system unattenuated in *amplitude* and *unchanged* in phase.

Figure 6-21*a* shows a good reproduction of the arterial waveform at the output of the transducer. This situation occurs when the plumbing system and the transducer have a frequency response high enough so that no significant harmonics of the arterial waveform are attenuated. The systolic and diastolic values determined by the peak holder circuits (Section 6-15) are then correct.

The lower waveform in Figure 6-21*b* shows a slightly *dampened* waveform. It is essentially a good representation of the original, although some loss of the high-frequency components occurs. This waveform is close to those ordinarily found in clinical pressure-monitoring situations. The most important thing to consider, however, is that the diastolic and systolic values are still very accurate, and essentially as good as those obtained from the waveform in Figure 6-21*a*.

An example of *ringing* in the waveform caused by *system resonances* is shown in Figure 6-21*c*. This problem inevitably causes systolic and diastolic readings that are *too high*.

The opposite problem, *excessive damping* of the waveform, is shown in Figure 6-21*d*. In the lower tracing all high-frequency components are obliterated, and the overall amplitude of the waveform is seriously *reduced*. The diastolic value obtained by the peak detectors will probably fall within reasonable bounds, but the systolic reading will probably be *very low.* Note that a similar waveform is obtained *normally* from hypotensive

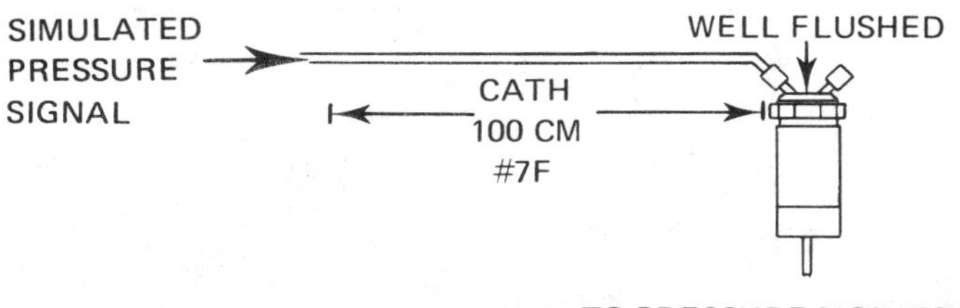

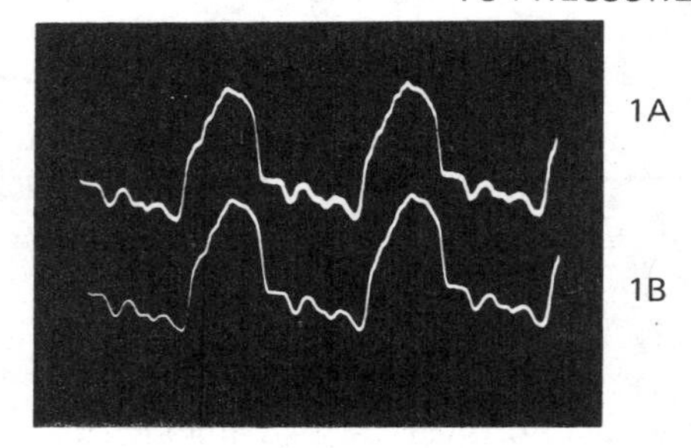

WAVEFORM A – SIMULATED ARTERIAL WAVEFORM

WAVEFORM B – A GOOD REPRODUCTION THROUGH PLUMBING

(a)

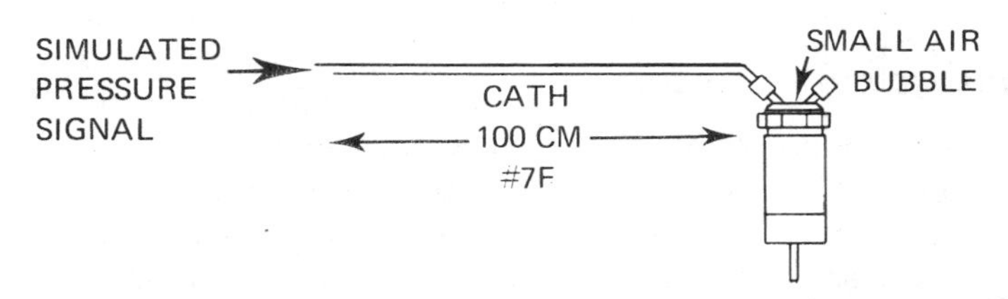

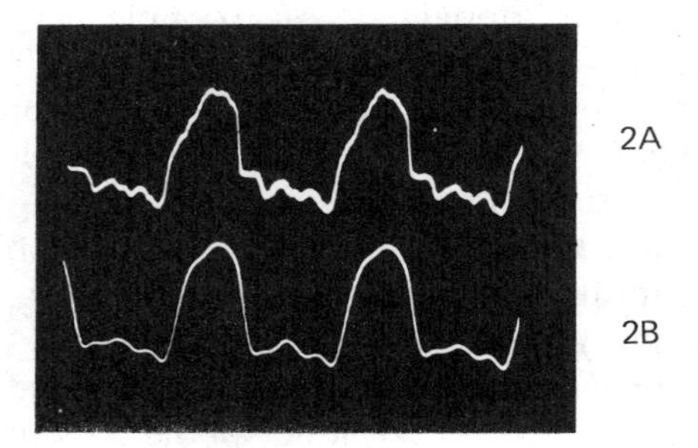

WAVEFORM A – SIMULATED ARTERIAL WAVEFORM

WAVEFORM B – SLIGHTLY DAMPED WAVEFORM
PRESSURE VALUES PROBABLY NOT LOST

(b)

Figure 6-21 Waveforms associated with several pressure monitoring system faults. (*a*) Proper reproduction of the waveform. (*b*) Mildly damped waveform. (*c*) Ringing in the waveform due to resonance in the system. (*d*) Badly damped waveform. (Reprinted courtesy Hewlett-Packard.)

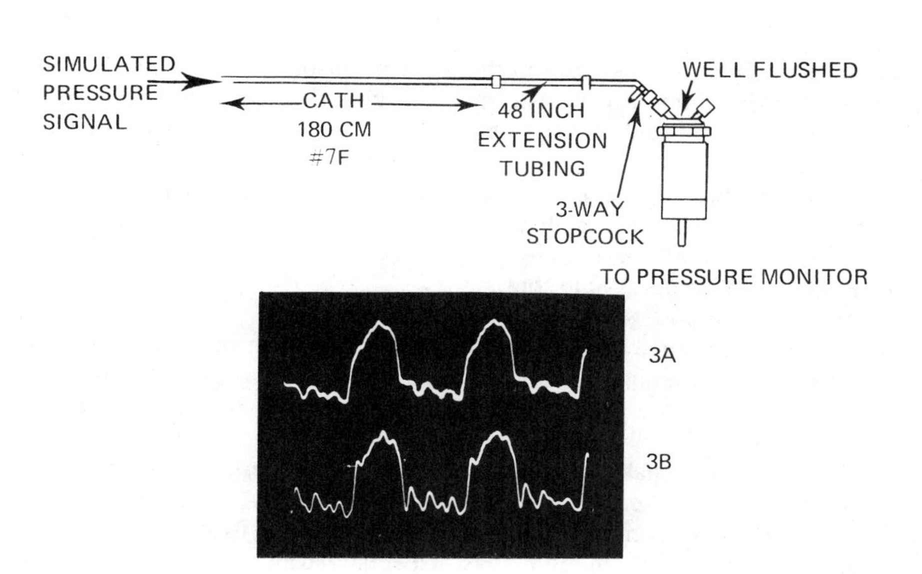

WAVEFORM A – SIMULATED ARTERIAL WAVEFORM

WAVEFORM 3B – ARTIFACT CAUSED BY RESONANCE
PRESSURE VALUES TOO HIGH

(c)

SIMULATED
PRESSURE
SIGNAL
CATH
100 CM
#7F
48 INCH
EXTENSION
TUBING
AIR BUBBLE
3-WAY
STOPCOCK
TO PRESSURE MONITOR
4A
4B

WAVEFORM A – SIMULATED ARTERIAL WAVEFORM

WAVEFORM B – BADLY DAMPED WAVEFORM
SYSTOLIC PRESSURE VALUES VERY LOW

(d)

Figure 6-21 (continued)

patients, even though there is no fault in the system. (In such cases, however, measuring the patient's pressure by auscultation reveals the low pressures.)

The example of Figure 6-21*b* showed slight damping of the waveform caused by the introduction of a *small* air bubble into the dome. Fluid is not compressible, but air (being a gas) is very compressible. As a result, an air bubble will reduce the frequency response of the system, attenuating the high-frequency components of the arterial waveform. The plumbing system, therefore, must be assembled in a way that reduces the chance of air bubble formation. A reasonable rule of thumb is that less complex systems are less prone to air bubble formation. It is also true that *metal* hydraulic fittings are *more* prone to such problems; metal fittings tend to leak because of wear and the effects of ethylene oxide gas used in sterilization on the rubber washers and lubricant. These washers must be replaced periodically, and the internal mechanism of the fittings must be relubricated using a product such as Dow-Corning *High-Vac* grease.

The waveform of Figure 6-21*c* resulted from extending the catheter tubing, which modifies the system resonance to a point closer to the frequency components of the arterial waveform. The higher harmonics are therefore transmitted to the dome *faster* than the lower harmonics, and this *enhances* the high-frequency features of the waveform.

The last waveform, Figure 6-21*d*, resulted from the formation of a *large* air bubble in the system. The cure is to purge the system of bubbles properly. This waveform can also result from the formation of the products of blood coagulation in the catheter line.

It is advisable to be careful as to the dimensions and properties of the tubing used in pressure-measuring systems. The catheter should have a large diameter, not less than the standard No. 7 French or 18-gage sizes, and should be stiff and noncompliant. Teflon tubing is preferred, while standard flexible IV tubing should *never* be used. The tubing should be less than 100 cm long, and the system should be as simple (i.e., fewest number of fittings) as possible.

6-18 Step-Function Frequency Response Test

The ability of a pressure system to respond to *changes* in pressure depends on its *dynamic response* characteristics, which are also its *frequency response*. All waveforms other than a perfect sinewave are composed of a *Fourier series* of sine and cosine harmonics of the fundamental frequency. The specific harmonics present, and their respective amplitude and phase shifts, determine the actual shape of the wave. If the harmonics are removed by any kind of filtering action, including damping (as in fluid pressure systems), then the waveform will be distorted. The electronic circuitry and fluid "plumbing" of the pressure measurement system, as well as the characteristics of the transducer, must be selected to pass frequencies from near-dc to about 20 Hz, or distortion of the blood pressure waveform will result.

Figure 6-22*a* shows a test apparatus that can be used to squarewave test the system. A water-filled catheter or needle to the system terminates in a syringe body, just below a port that accomodates a standard three-way stopcock. A small balloon is attached to the open end of the assembly with an O-ring. A sphygmomanometer squeeze-bulb pump is used to apply pressure to the system, inflating the balloon. Once a static pressure head is registering on the electronic pressure meter, a flame is used to pop the balloon, resulting in a sudden decrease in pressure that nearly simulates an actual negative-going squarewave. The result is a *free oscillation* of the pressure in the system with a frequency ω_o (Figure 6-22*b*) that dies out

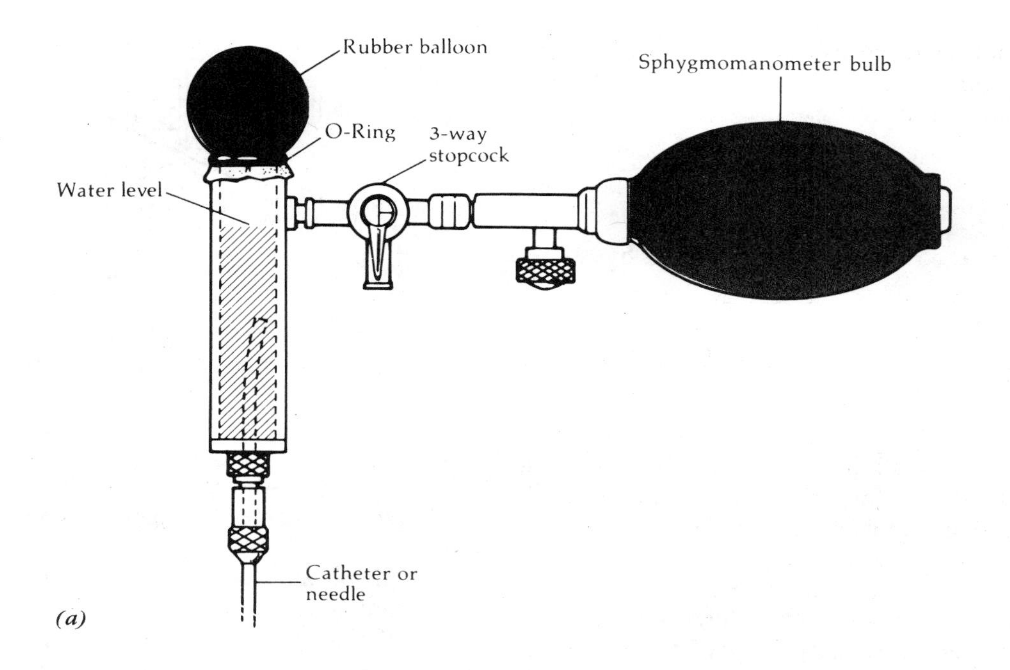

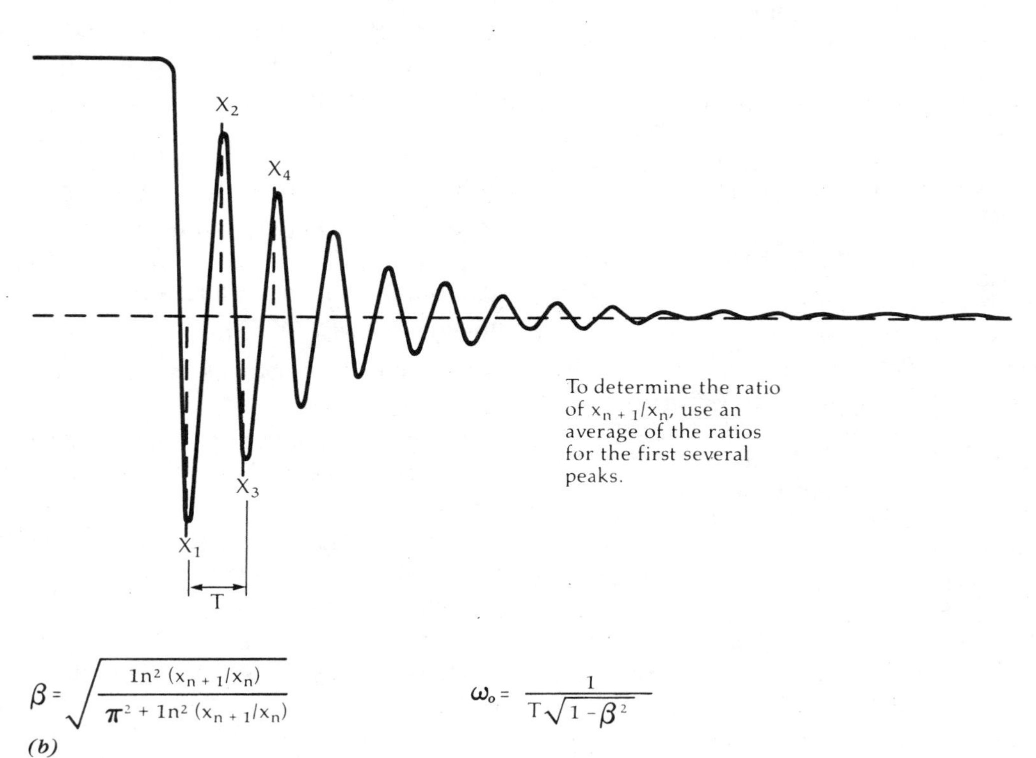

Figure 6-22 (*a*) An assembly used in testing the transient oscillatory response of the catheter-transducer system. (*b*) The transient oscillatory response of a catheter transducer system on application of a pressure square wave. (Courtesy Spacelabs.)

at an exponential rate determined by the *damping factor* (β). The oscillation frequency is given by:

$$\omega_o = \frac{1}{T\sqrt{1 - \beta^2}} \quad (6\text{-}5)$$

while the damping factor is given by:

$$\beta = \sqrt{\frac{\ln^2\left(\frac{X_{n+1}}{X_n}\right)}{\pi^2 + \ln^2\left(\frac{X_{n+1}}{X_n}\right)}} \quad (6\text{-}6)$$

It is usually desirable for β to be on the order of 0.7 for a frequency response of 20 Hz. This number provides *critical damping*. If the number is less by very much, then the system is *subcritically damped* and may *ring* (i.e., provide unnaturally high systolic readings due to a sharpening of the waveform caused by excess high-frequency components). Similarly, an *undercritically damped* system, with $\beta >> 0.7$, results in attenuation of high-frequency components, causing an unnaturally low systolic reading. The output of the system will correlate closely to actual pressure when β is about 0.7.

6-19 Transducer Care

Blood pressure transducers are delicate instruments. Physical abuse *will* damage a transducer. The diaphragm is especially sensitive and should *never be touched*. It may be cleaned by using a gentle solvent and a cotton ball or swab. Do not apply any pressure to the diaphragm while cleaning; if the dirt cannot be wiped off with the cotton, then leave it on!

Transducers are often sterilized between patients to prevent spread of disease. Steam sterilization in the autoclave must *never* be used because it destroys the transducer. Only *gas sterilization* may be used unless a disposable dome (see Figure 6-23) is used. The disposable dome is discarded after each use, so the transducer may be disinfected using a liquid disinfectant agent (such as Cydex).

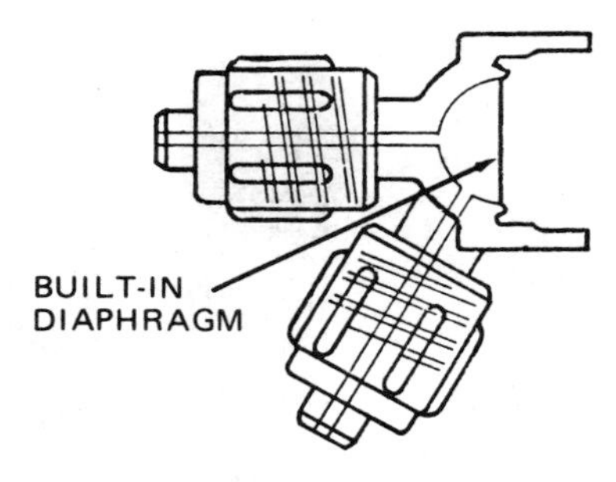

Figure 6-23 Disposable transducer dome. (Courtesy of Statham Instruments Division, Gould, Inc., Oxnard, Ca.)

The disposable dome uses a thin membrane to couple pressure inside of the dome to the diaphragm. Most designs require that a single drop of fluid be placed on the diaphragm before the dome is secured to the transducer body. The water or fluid drop insures good coupling.

6-19.1 Transducer Calibration/ Balance Procedure

Most arterial blood pressure transducers and amplifiers have two basic controls: *zero* and *calibration*. The procedure for adjusting the amplifiers is as follows:

1. Open the stopcocks on the transducer to atmosphere.

2. Adjust the *height* of the transducer to the level of the catheter tip inside of the patient's arm.

3. Adjust the *zero* control on the amplifier for a zero reading on the meter or for a zero baseline on the oscilloscope.

4. Press the *calibrate* button (this injects a simulated pressure signal into the amplifier, usually 100, 150, or 200 mm Hg. This button is often labeled with the pressure level simulated).

5. Adjust the *calibration* control (also called *span* or *sensitivity* control in many models) so that the meter readout, or oscilloscope, reads the indicated pressure level.

6. Close the transducer stopcock. The system should now be ready to use.

If there is any doubt as to the accuracy of the system, or if the transducer has been abused or is old, then it may be necessary to check the calibration. Also, there are many cases when the amplifier does not have a calibration signal built in. In those cases, modify the aforementioned procedure to allow the use of a mercury, or aneroid assembly, *manometer* to check the calibration. Most BMETs (biomedical equipment technicians) use a modified blood pressure gauge for this purpose (we prefer the use of mercury manometers instead of the aneroid types). Take an ordinary blood pressure set, and remove the cuff (there is sometimes a hose connector at the site where the hose enters the cuff; otherwise, *cut* the line). Install a Luer lock connector (usually available in most hospital central supply departments, or the anesthesiology department) in the end of the hose removed from the cuff. When this setup is used, we can check and adjust the amplifier. Once the zero is set, close the stopcocks and pump a pressure onto the system. The amplifier is then adjusted to indicate this pressure level. Check the linearity of the system by comparing manometer pressure against indicated pressure.

There are two ways to pressurize the transducer for this test. One is to place a fluid-filled blood transfusion bag on the system and pump it up to the required pressure level. The other is to use air. The former is preferred in clinical settings, while the latter may be used in bench testing pressure systems. One word of warning, however: Always disconnect the catheter from the patient. It is simply not acceptable to pump air into the patient!

6-20 Summary

1. *Pressure* is measured in *force per unit area*.
2. Human blood pressure is a gage pressure; that is, it uses atmospheric pressure as the zero reference.
3. An *indirect* method for measuring blood pressure is called *auscultation* and uses a device called a *sphygmomanometer* and a *stethoscope*. *Systolic* and *diastolic* pressures are indicated by the onset and disappearance of characteristic *Korotkoff* sounds, respectively.
4. A *direct* method for measuring blood pressure uses an indwelling *catheter, transducer, electronic amplifier,* and *display*.
5. Proper reproduction of the waveform in an electronic pressure monitor requires a plumbing system with a *high*-frequency response. Clotting of blood in the catheter, air bubbles, and improper or too long tubing alter the resonances sufficiently to distort the waveform.

6-21 Recapitulation

Now return to the objectives and self-evaluation questions at the beginning of the chapter and see how well you can answer them. If you cannot answer certain questions, place a check next to each and review appropriate parts of the text. Next, try to answer the following questions using the same procedure. When you have answered all of the questions, solve the problems in Section 6-23.

6-22 Questions

6-1 Define *pressure* (a) conceptually and (b) mathematically.

6-2 Define (a) *hydrodynamic* and (b) *hydrostatic* pressures.

6-3 State Pascal's principle in your own words, and give the limitations on it in *hydrody-*

namic systems. In which two situations does it apply absolutely?

6-4 What are the (a) SI and (b) English engineering units for atmospheric pressure?

6-5 Give the *standard value* for a pressure of 1 atm in (a) SI, (b) English engineering, and (c) cgs units.

6-6 A pressure of 1 *atm* will support the weight of a column of mercury approximately ______________ mm high.

6-7 One Torr equals ______________ mm Hg.

6-8 The *zero reference* pressure used for measuring a *gage pressure* is ______________.

6-9 Blood pressure measurements are referenced against ______________. So are ______________ pressures.

6-10 Stephen Hales's techniques for measuring arterial blood pressure is an example of a ______________ technique.

6-11 Indirect methods for blood pressure measurement use an instrument called a ______________.

6-12 The most common indirect method for measurement of arterial blood pressure is called ______________.

6-13 Describe the apparatus and procedure for using *palpation* to measure arterial blood pressure.

6-14 The palpation method measures only the ______________ value(s) of arterial blood pressure.

6-15 Describe the apparatus and procedure for using the *flush* method for determining arterial blood pressure.

6-16 The flush method obtains the ______________ value(s).

6-17 Describe the apparatus and procedure for using *auscultation* to measure arterial blood pressure.

6-18 Auscultation determines the ______________ value(s) of arterial blood pressure.

6-19 *Korotkoff sounds* occur in the ______________ method of measuring arterial blood pressure at the instant when ______________.

6-20 The ______________ pressure is indicated by the *onset* of Korotkoff sounds.

6-21 The ______________ pressure is indicated when the Korotkoff sounds ______________.

6-22 Direct electronic measurement of arterial blood pressure uses a fluid column in an indwelling ______________ between the patient and the transducer.

6-23 A ______________ manometer is used to measure central venous pressure (CVP).

6-24 Define in your own words (a) *mean arterial pressure* and (b) *functional mean pressure*. Use mathematical notation to describe and compare (a) and (b).

6-25 The two *basic* controls required on a clinical pressure monitor are ______________ and ______________.

6-26 Name four basic classes of pressure amplifier.

6-27 Describe in your own words the technique for calibrating a pressure amplifier using a mercury or aneroid manometer.

6-28 Describe in your own words how certain monitors use a *calibration factor* to calibrate the system. Is it permissible to use the *cal factor* for one transducer with another of the same model and similar serial number?

6-29 Draw the block diagram for an ac carrier amplifier.

6-30 Draw the block diagram for a pulsed-excitation pressure amplifier.

6-31 Draw the block diagram for an *auto-zero* circuit, and decribe how it operates.

6-32 Describe how a blood pressure monitor could indicate a mean arterial pressure that is considerably different from the functional mean. Discuss the cases where (a) there is a fault on the system, (b) there is *no* fault on the system, and (c) a method exists to tell you which is the case in any given situation.

6-33 The time constant of a pressure signal differentiator must be ______________ compared with the risetime of the pressure signal waveform.

6-34 Describe the method used for calibrating a dP/dT output on a pressure amplifier.

6-35 Describe hydrostatic pressure as it affects the pressure monitoring system.

6-23 Problems

6-1 A coin has a mass of 1.8 g and a diameter of 14 mm. How much (a) force in millinewtons and (b) pressure in pascals is exerted when the coin is lying horizontally on a perfectly flat table top?

6-2 A drinking glass has a bottom that is 2 in. in diameter. When empty the glass weighs 4 oz, and when full 16 oz. Calculate (a) the force in newtons and (b) the pressure in pascals exerted by the glass on a flat surface under *both* empty and full conditions.

6-3 A patient's arterial blood pressure is measured as 130/85. Calculate the *mean arterial pressure* ($\overline{P}$).

6-4 A patient's arterial pressure is 140/85. Calculate the *mean arterial pressure*.

6-5 A resistance Wheatstone bridge blood pressure transducer has a rated sensitivity of 50 μV/V/cm Hg. Calculate the *gain* required of the dc amplifier used with this transducer if the output scale factor is 10 mV/mm Hg, and the full-scale pressure is 300 mm Hg.

6-6 Solve Problem 6-5 for a full-scale pressure of 100 mm Hg and a sensitivity of 100 μV/V/cm Hg.

6-24 References

1. Halliday, David and Robert Resnick, *Physics Parts I and II,* 3rd ed. Wiley (New York, 1978).
2. *Guide to Physiological Pressure Monitoring,* Hewlett-Packard applications note AN-739 (Waltham, Mass., 1977).
3. *Operating Notes Series 1280 Transducers,* Hewlett-Packard operator's guide (Waltham, Mass., 1971).
4. Nara, Andrew R., et al., *Blood Pressure,* Space Labs, Inc. (Redmond, Wash., 1989).

Chapter 7

Other Cardiovascular Measurements

7-1 Objectives

1. Be able to list and describe the various techniques commonly used to make measurements of right-side heart pressures.
2. Be able to list and describe the most common cardiographics techniques.
3. Be able to describe how a cardiac output computer operates.
4. Be able to list the basic operations of specialized cardiovascular measurement systems.

7-2 Self-Evaluation Questions

These questions test your prior knowledge of the material in this chapter. Look for the answers as you read the text. After you have finished studying the chapter, try answering these questions and those at the end of the chapter (Section 7-13).

1. Draw a rough sketch showing the location of the first, second, and third heart sounds with respect to time.
2. What is the most common technique used to make cardiac output measurements?
3. A ______________ catheter may be used to directly measure pressures in the pulmonary artery.
4. The Frank lead system is used in ______________________ cardiography.
5. Describe the construction of a transducer used to make blood flow measurements.

7-3 Cardiac Output Measurement

Cardiac output (CO) is defined as the volume rate at which the heart pumps blood. Cardiac output is measured in units of *liters per minute* (l/min). In adults CO reaches a value between approximately 3 and 5 l/min.

One quantitative definition of cardiac output is that it is the product of *stroke volume* and *heart rate*. The stroke volume is the volume of blood ejected from the ventricle during a single contraction of the heart. Cardiac output, then, is

$$CO = V \times R \tag{7-1}$$

where

CO is the cardiac output in *liters per minute* (1/min)

V is the stroke volume in *liters per beat* (1/beat)

R is the heart rate in *beats per minute* (BPM) (i.e., beats/min)

EXAMPLE 7-1

Calculate the cardiac output for a patient with a heart rate of 86 BPM if the stroke volume is 42 ml/beat.

SOLUTION

$$\text{CO} = V \times R$$

$$\text{CO} = 42 \text{ ml/beat} \times 1\ 1/1000 \text{ ml} \times 86 \text{ beats/min} \quad (7\text{-}1)$$

$$= 3612/1000 \text{ L/min}$$

$$\text{CO} = \mathbf{3.61\ 1/min}$$

It is difficult, and usually impossible, to measure cardiac output using a technique based on Equation 7-1 in practical situations because of the difficulty in obtaining adequate stroke volume data.

7-3.1 Blood Flow Measurements

Blood flow measurements can yield cardiac output data. Such instruments measure the blood flow rate, and the cardiac output is determined by *integrating* the blood flow signal over a known period of time.

The problem with this method, however, is that most blood flow transducers capable of delivering meaningful quantitative data must be applied directly to the blood vessel being measured. It is not possible to obtain valid cardiac output data from blood vessels taken far downstream in the arterial system; thus none of the easily accessible peripheral arteries (i.e., an arm or leg) can be used. The measurement must be done either in the pulmonary artery or in the aorta immediately after these vessels leave the heart. This requirement limits cardiac output measurement via blood flow measurement to thoracic surgery procedures, whenever the vessels are normally exposed anyway. See Section 7-7 for a more detailed discussion of blood flow measurement.

7-3.2 Dilution Techniques

Most modern cardiac output measurement systems use a *dilution* technique in which a known concentration of some tracer material is injected into the bloodstream just before the heart. The diluted concentration is measured downstream, past the heart, thus yielding the cardiac output.

There are different types of tracer materials used in cardiac output measurement (the most common being indocyanine green, discussed in Section 7-4.2). These techniques are done almost routinely on certain patients in special hospital laboratories or even at bedside in intensive and/or coronary care units (ICU/CCU).

All dilution techniques used to obtain cardiac output information use an injectate that enters the heart at the atria in concentration. The diluted concentration is measured distal to the heart (on the output side), and the flow is given by

$$\text{CO} = \frac{\text{Injection rate (mg/min)}}{\text{Concentration (mg/1/}} \text{ (1/min)} \quad (7\text{-}2\text{a})$$

7-4 Dilution Methods

7-4.1 The Fick Method

An early—but still used—technique called the *Fick method* uses room oxygen inhaled by the patient during normal respiration as the indicator substance. The oxyen is injected into the system during respiration by the lungs.

The infusion rate is determined by the measurement of respiratory gases. It is necessary to subtract the oxygen concentration of the patient's exhaled air from the room air concentration, normally taken to be 21 percent.

Concentration is determined by measuring the oxygen concentration of arterial blood as it leaves the lungs. But this term contains an error term because returning venous blood contains some oxygen. This amount must be determined and subtracted from the arterial oxygen concentration. Repeated measurements made over a period of several minutes are averaged for the final result.

The Fick method has been automated, and is still used, but for the most part has been supplanted by other methods.

7-4.2 Dye Dilution

Dye dilution uses either an *optical* dye, such as indocyanine green, or a *radioactive* tracer dye. In the optical case the dilution curve is measured by a *light densitometer* and in the latter case by a *scintillation counter* or *gamma camera*.

Figure 7-1 shows the concentration at the point of measurement distal to the injection site. Shortly after injection, the concentration at the measurement site rises abruptly to a maximum value and then falls off exponentially as the injectate bolus passes the point. This exponential decay is due to the fact that the injectate bolus does not remain as a lump but is spread out (diluted) over a longer path as blood flows.

Figure 7-1 Dilution curve showing recirculation artifact.

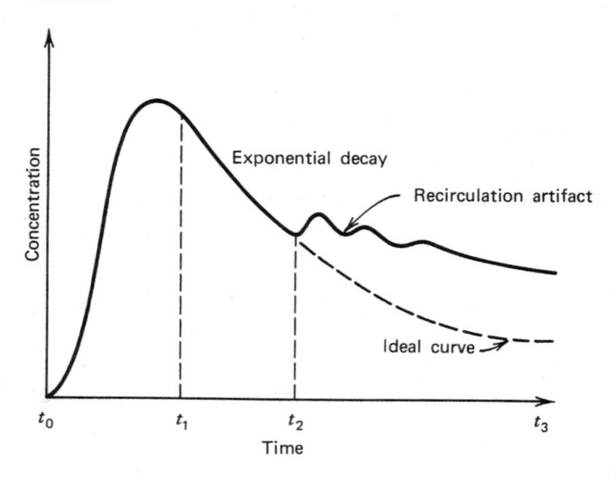

Cardiac output is then measured by integrating the concentration curve. This information is then used in an equation of the general form of Equation 7-2:

$$\text{Blood flow rate} = \frac{k \times M}{\int C\,dt} \text{ (ml/min)} \qquad (7\text{-}2\text{b})$$

where

k is a constant, usually 20 to 150 depending on injectate
M is the volume of injectate, in ml
C is the concentration mg/ml

A problem arises, however, in the form of a *recirculation artifact* (at time t_2 in Figure 7-1). The artifact changes the shape of the exponential decay curve between times t_2 and t_3 from the expected ideal shape.

In most early instruments the curve was traced out by a chart recorder. Graphical means were then used to extrapolate, and then integrate, the area under the curve in the exponential region. This procedure could be done with a mechanical graphics integrator device or by counting (i.e., summing) the number of squares under the idealized curve on the graph paper.

7-4.3 Thermodilution

Thermodilution has become one of the most common methods for measurement of cardiac output and forms the basis for most of the modern cardiac output computers on the market. The injectate used is ordinary intravenous solutions, such as saline or 5 percent dextrose in water (D_5W).

Most thermodilution cardiac output computers operate on a version of the following equation:

$$CO = \frac{KG_BG_IV_I(T_B - T_I)}{U_BU_I \int T'_B dt} \text{ } (l/min) \qquad (7\text{-}3)$$

where

CO is the cardiac output in *liters per minute* (l/min)

K is a constant, approximately 20 to 150

G_B is the density of human blood, in kg/m^3

G_I is the density of the injectate, in kg/m^3

V_I is the injectate volume, in liters (l)

U_B is the heat energy content of the blood in joules (J)

U_I is the heat energy content of the injectate in joules (J)

T_I is the preinjection temperature of the injectate (note: either *iced* or *room temperature* injectate may be used) in degrees Celsius (°C)

T_B is the preinjection temperature of the blood, in degrees Celsius (°C)

T'_B is the postinjection temperature of the blood at the measurement site.

Equation 7-3 looks rather formidable, but we find that it becomes a lot simpler once it is recognized that most of the terms are constants. Furthermore, many of the constants need not be measured, but assumptions can be made. Edwards Laboratories, for example, manufactures a cardiac output computer in which the following simplified version of 7-3 is used:

$$CO = \frac{(60)(1.08)(C_t)(V_I)(T_B - T_I)}{\int T'_B dt} \tag{7-4}$$

where

CO is the cardiac output in *liters per minute* (l/min)

60 is the conversion factor for seconds to minutes

1.08 is a composite of other constants, dimensionless

C_t is a dimensionless correction factor for the injectate temperature rise in the catheter and is published by the catheter manufacturer specific to each different type (the other terms are as defined in Equation 7-3.)

EXAMPLE 7-2

A special cardiac output test fixture enters a thermistor signal that represents a temperature change of 7°C. Find the expected cardiac output reading if the instrument is set to measure the following parameters: injectate volume of 10 ml, body temperature of 37°C, and an injectate temperature of 25°C. The catheter correction factor C_t is 52.5 and the computation time is 10 s.

SOLUTION

$$CO = \frac{(60)(1.08)C_t V_I(V_B - T_I)}{\int T'_B dt} \tag{7-4}$$

$$CO = \frac{60 \text{ s/min } (1.08)(52.5)[10 \text{ ml} \times 1 \text{ l}/1000 \text{ ml}](37 - 25)°\text{C}}{(7\,°\text{C})(10 \text{ s})}$$

$$CO = (60)(1.08)(52.5)(0.01)(12)$$

$$= (1/70 \text{ min}) \mathbf{5.81\ l/min}$$

The excitation of the Wheatstone bridge used to make the temperature measurement is done by a dc source and must be very stable, at least over the short term. Batteries usually meet this requirement because they maintain the same voltage level over the short term, even though deteriorating over the long term. In many cases, however, the isolated front-end dc power supply is dropped to a low level and then regulated by the drop across a germanium diode. This is an unusual procedure, but is justified because it limits the transducer excitation potential to approximately 200 mV, a value consistent with the electrical safety requirement of the design.

The recirculation artifact is small in thermodilution, so it is often the case that computers designed specifically for use in this technique may not have a means for compensating for the artifact.

Figure 7-2 shows the block diagram of one popular cardiac output computer. The bridge

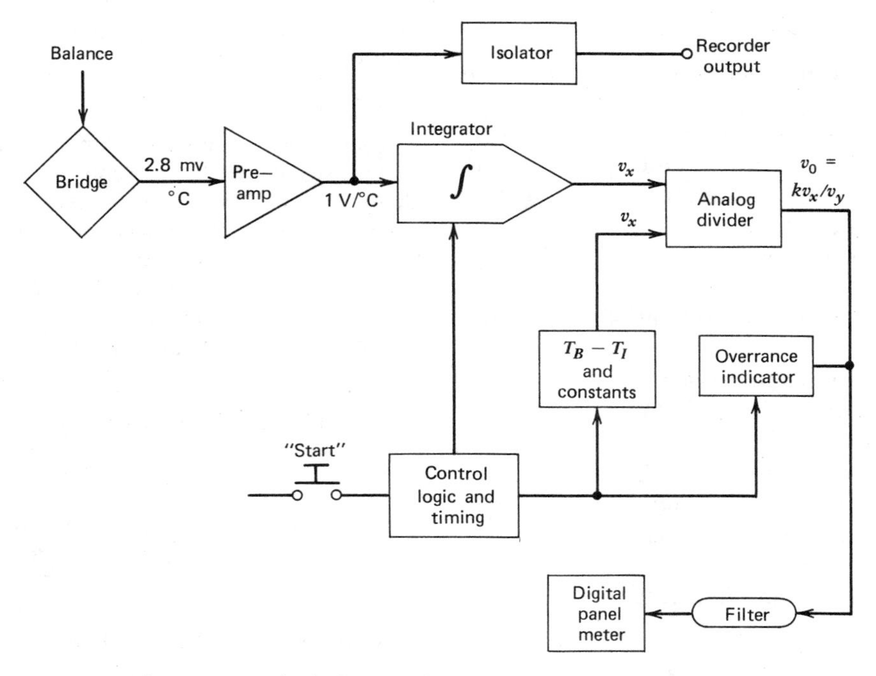

Figure 7-2 Block diagram for a cardiac output computer.

produces a zero output when the thermistor is at normal blood temperature, and the balance adjustment is made. After injection of the saline or D_5W, however, the bridge produces an output potential of 1.8 millivolts per degree Celsius (1.8 mV/°C). This signal is amplified in the preamplifier stage to a level of 1 V/°C. The high-level output signal is passed through an isolator to the remainder of the circuit. Part of the signal goes to an output jack so that it may be recorded on a strip-chart recorder, while it is simultaneously applied to the input of an operational amplifier electronic integrator stage. The integrator output, which supplies the denominator of Equation 7-4, changes by a factor of 1 V/°C-s. This stage provides the integral of temperature required by the instrument. The integrator output signal is applied to the denominator (i.e., V_y) input of an analog divider circuit.

7-4.4 Some Sample Computers

Most thermodilution cardiac output computers use a special catheter that has a *thermistor* in the tip to make the measurement. The catheter is introduced into the right side of the heart via a venous cutdown. It is threaded through the right atrium and right ventricle to the pulmonary artery. One lumen of the catheter carries the injectate, which is injected into the bloodstream just prior to the right atrium.

As the injectate bolus passes the thermistor tip of the catheter, the thermistor resistance changes an amount proportional to the temperature change. A Wheatstone bridge and an isolated preamplifier are used to actually acquire the temperature signal.

Figure 7-3 shows a typical "front-end" circuit for a thermodilution cardiac output computer. Resistors R_1 through R_4 form a

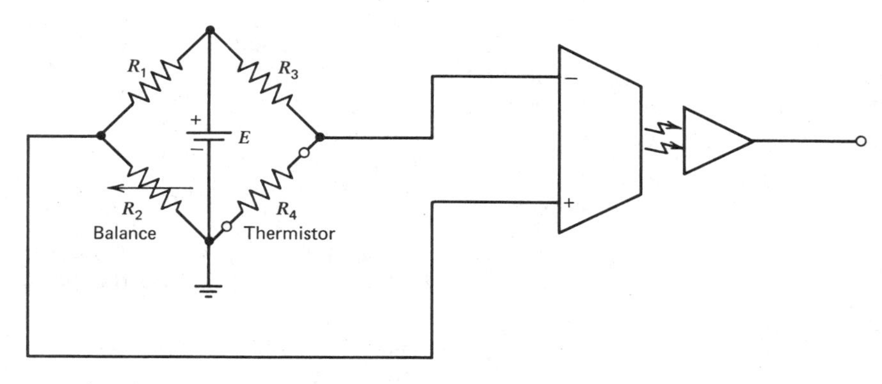

Figure 7-3 Input circuit for a thermodilution cardiac output computer.

Wheatstone bridge and may be either fixed or variable to allow the operator to compensate for differences between catheters. In the circuit shown, resistor R_4 is the thermistor in the catheter tip, and R_2 is the potentiometer used to adjust the circuit for null condition. This potentiometer should be adjusted only after the thermistor in the catheter tip has been in place for several minutes so that it has time to come into thermal equilibrium with the blood. At that time, R_2 is adjusted to produce a null or zero output from the preamplifier.

It is absolutely necessary to isolate the patient from any electrical hazards during any medical procedure, and in the case of a cardiac output computer special precautions are required because the catheter is introduced directly into the patient's heart (see Chapter 19). Any electrical defects could potentially be very serious. That is why it is standard procedure to operate the cardiac output computer from rechargeable batteries rather than the ac power mains. It is also why isolated input preamplifiers are universally used in these machines. A typical cardiac output computer operates as follows.

1. The numerator input of the divider is obtained from a stage that multiplies together the constants and a signal entered by the operator or, in more sophisticated models, taken from another electronic temperature measurement circuit that indicates the differences between blood and injectate temperatures. The output of the circuit is scaled at 10 mV/°C. The analog divided output signal is filtered and scaled before being applied to a digital panel meter that serves as the cardiac output indicator.

2. A control logic circuit is required to time the operation of the cardiac output measurement cycle. When the operator has the external controls (i.e., *balance* and *temperature difference*) adjusted and is ready to inject the solution, it is then necessary to press the *start* button. This action clears the integrator and resets the circuit to zero. Most computers have a beeper or audio tone that sounds as soon as the integrator is cleared, and this is the signal to immediately begin infusing the injectate. A short time later the cardiac output reading will appear on the digital panel meter. If the integration runs to overrange, then a panel light comes on to let the operator know that the data obtained is invalid. This problem is usually caused by too short a time selected for the measurement, a front panel adjustment easily made on the cardiac output computer. Skilled operators can spot this problem easily if the thermistor output curve is displayed on a strip-chart recorder. In any event, the data obtained during the measurement that caused the lamp to turn on is invalid and must be discarded; another measurement must be made to obtain valid data.

3. Cardiac output computers that are used in both dye and thermal dilution must have a circuit to compensate for the recirculation artifact during dye procedures. Most circuits that do this are based on the time constant of the exponential decay portion of the temperature curve. The time constant of any exponential decay curve is the time required for it to drop from its maximum value to 36.7 percent of the maximum value.

4. Two different techniques find common use in predicting the path of the ideal exponential decay curve in the presence of the recirculation artifact.

(a) The time period prior to the appearance of the artifact is used to predict the path of the ideal curve, and this is used to approximate the true integral.

(b) The technique of *geometric integration* (see Figure 7–4) is used to approximate the ideal curve.

Columbus Instruments has used the geometric integration method in their cardiac output computer. Figure 7-4 shows the dilution curve. The curve is *assumed* to enter the exponential decay portion when the dilution signal has passed the peak and then fallen to a value of 85 percent of the peak value. A voltage comparator is used to compare the concentration signal with the peak value stored in a peak holder circuit. The input of the electronic integrator stage will be connected to the thermistor or dye transducer from time t_1 when the computation begins until the curve enters the exponential portion at time t_2.

At time t_2 the input of the integrator is switched to another source and receives a signal equal to 85 percent of the peak value for a time period t_3 to t_2. A very good approximation of the area under the exponential decay portion of the dilution curve is

Figure 7-4 Geometric integration for determining the area under the exponential decay portion of the dilution curve.

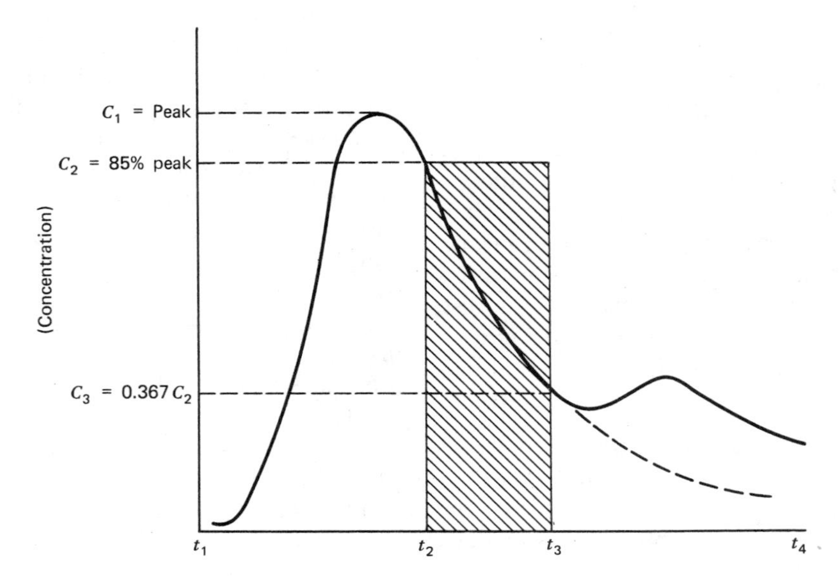

obtained by creating a rectangle with a height equal to the maximum value of the exponential curve, in this case 85 percent of the thermodilution curve, and a base equal to the time constant of the decay curve, defined as the time required for the curve to drop from its maximum value to 0.367 times its maximum. If the area of this rectangle is known, then the approximate area under the exponential portion of the curve is also known. In the case of the electronic cardiac output computer, the area under the rectangle is determined by connecting the integrator to a source equal to 85 percent of the peak for a period of time t_3 to t_2. The actual mathematical expression for the area under the curve, in the notation of calculus, is

$$\int_{t_1}^{t_4} C\,dt \approx \int_{t_1}^{t_2} C\,dt + \int_{t_2}^{t_3} C_2\,dt \qquad (7\text{-}5)$$

The Edwards Laboratories Model 9520 bedside cardiac output computer is shown in Figure 7-5. This model uses a liquid-crystal readout display to reduce battery drain to a bare minimum. Previous models required the operator to manually enter the temperature of the injectate and the patient's blood temperature, but this model measures those parameters automatically using thermistor probes. Additionally, this cardiac output computer is equipped with a self-test feature that allows the operator to check the instrument. A thumbwheel switch on the side of the unit allows the operator to enter the computation constant, which is a function of which particular catheter is selected.

Figure 7-5 Cardiac output computer. (Courtesy Edwards Laboratories, Inc.)

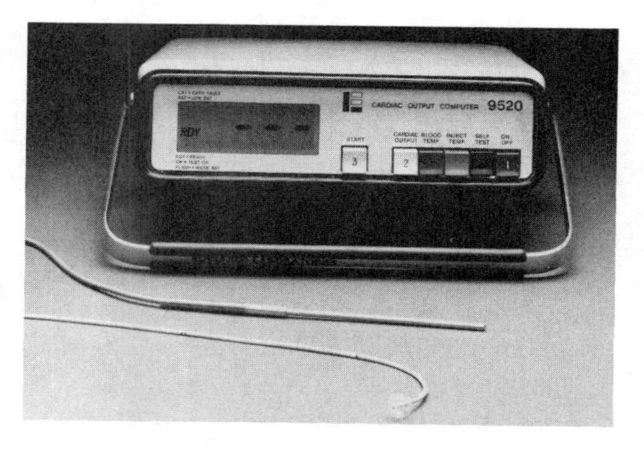

Almost any cardiac output computer can be checked by constructing a *dummy thermistor*. Some biomedical equipment technicians build a tester using the electrical fittings salvaged from a used thermistor, described in the following procedure.

1. Select a resistor with a value equal to the nominal value of the catheter thermistor at 37°C. Connect a second resistor of 100 to 500 Ω in series with the first resistor. The exact values depend upon the specific computer being tested, but 12 kΩ and 200 Ω work for many popular models.

2. A *normally closed* switch is connected across the low-value resistor, shorting it out. This assembly is connected to computer as if it were a thermistor.

3. Once the computer is adjusted for normal operation according to the manufacturer's instructions, the technician presses the computer's *start* button and opens the switch on the tester simultaneously. After about 5 to 8 s the tester switch is closed, and a few seconds later a cardiac output appears. This value will vary from one test to another because of differences in the technician's perception of "5 to 8 seconds" but indicates that the circuits are working. In actual practice, one of the authors found it consistently possible to produce readings between 4 and 5 l/min using a stopwatch as the timer. Undoubtedly, an electronic timer and switch to automate the tester would yield more reproducible results.

7-5 Right-Side Heart Pressures

The measurement of *central venous pressure* (CVP) and the other pressures on the right

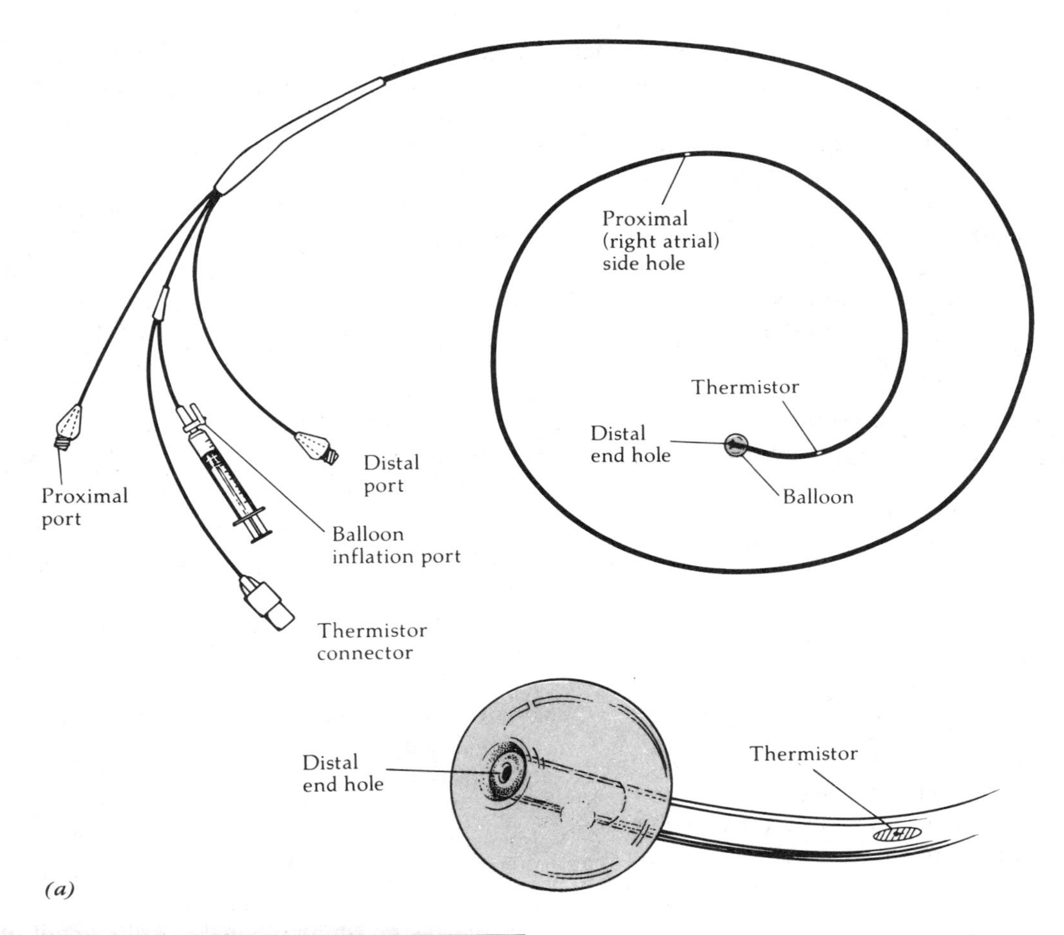

Figure 7-6 (*a*) The #7 French quadruple lumen, thermodilution pulmonary artery catheter illustrated in both drawings. (*b*) Balloon-tipped catheter. (Courtesy Edwards Laboratories, Inc.)

side of the heart are performed with a catheter such as the *Swan-Ganz* ® (registered trademark of Edwards Laboratories, Inc.), shown in Figures 7-6*a* and 7-6*b*. This catheter is a multilumen model with a thermistor tip, so it may also be used for cardiac output measurements by thermodilution.

To measure CVP the catheter tip is introduced into the right atrium of the heart (Figure 7-7), so this measurement is sometimes called the *right atrial pressure* (RAP). The catheter is inserted into the patient's body through one of the major peripheral veins (for example, the *jugular, brachial,* or *subclavian*).

The pressure-measuring instrument may

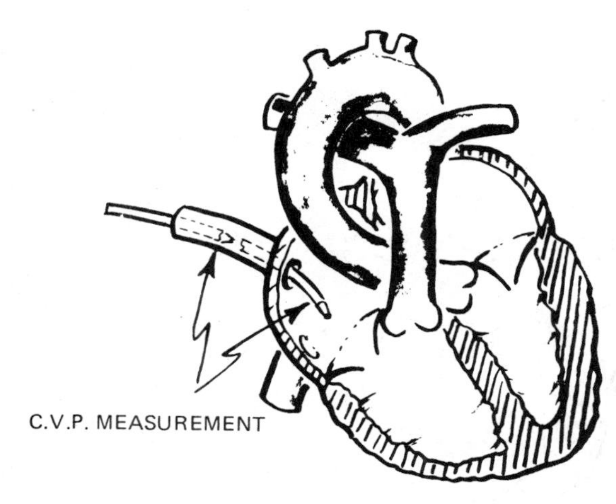

Figure 7-7 Catheter placement for measurement of CVP. (Reprinted courtesy Hewlett-Packard.)

be an electronic model (described in Chapter 6), or it may be a water manometer, calibrated in cm H_2O, not too dissimilar from Stephen Hales's eighteenth-century method. The modern CVP manometer, however, is made of plastic and is disposable.

The pulmonary artery wedge pressure can serve as an indicator of *left ventricular function*, the pumping efficiency of the heart. A flow-directed catheter with a balloon tip (shown in Figure 7-6*b*) is passed from a peripheral vein through the *vena cava, right atrium, right ventricle,* and the *pulmonary valve* to the *pulmonary artery*. If one of the lumens of the balloon-tipped catheter has an end, or tip, hole, and the balloon is *inflated* (Figure 7-8*a*), the catheter sees only the pressure ahead (i.e., distal) to the catheter tip. This pressure is called the *pulmonary artery wedge pressure* and is strongly correlated to *left* atrial pressure; this pressure is considered a good indicator of left ventricular function.

Inserting the catheter is a surgical procedure performed under sterile conditions but it may be done at bedside in the intensive or coronary care units. An X-ray fluoroscope can be used to follow the progress of the catheter through the body. Alternatively, an oscilloscope or strip-chart recording of the pressure waveform will also indicate the progress of the catheter through the heart. Figure 7-8*b* shows typical waveforms at various points in the heart. There are also coded distance markings on the body of the catheter. In the jargon this procedure is often called "inserting a Swan-Ganz" or, simply, "swan."

7-6 Plethysmography

Plethysmography is the determination of blood flow in a limb by measurement of *volume changes* of the limb. The plethysmograph produces a waveform that is similar to the arterial pressure waveform. To date, however, it has not proven possible to calibrate the plethysmograph waveform in terms of pressure units. The waveform is useful in measuring *pulse velocity* and to indicate arterial obstructions.

The photoplethysmograph (PPG), also called the pseudoplethysmograph, is constructed from a *photocell* and a *light source*. The circuit for the PPG is shown in Figure 7-9*a*, while the mechanical arrangement is shown in Figure 7-9*b*. In this example the light source is a light-emitting diode (LED), although in earlier models an incandescent lamp was used for the light source. The detector is a photoresistor (PC_1) excited by a constant current source. Changes in light intensity cause proportional changes in the resistance of PC_1. Since the current through the photoresistor is constant, the resistance changes produce voltage changes (E_o) at the output terminal.

The arterial pulse in the thumb causes the blood volume to change, changing the optical density of the blood. Therefore, the arterial pulse modulates the intensity of the light passing through the blood. Light from the LED is reflected into PC_1 by scattering and

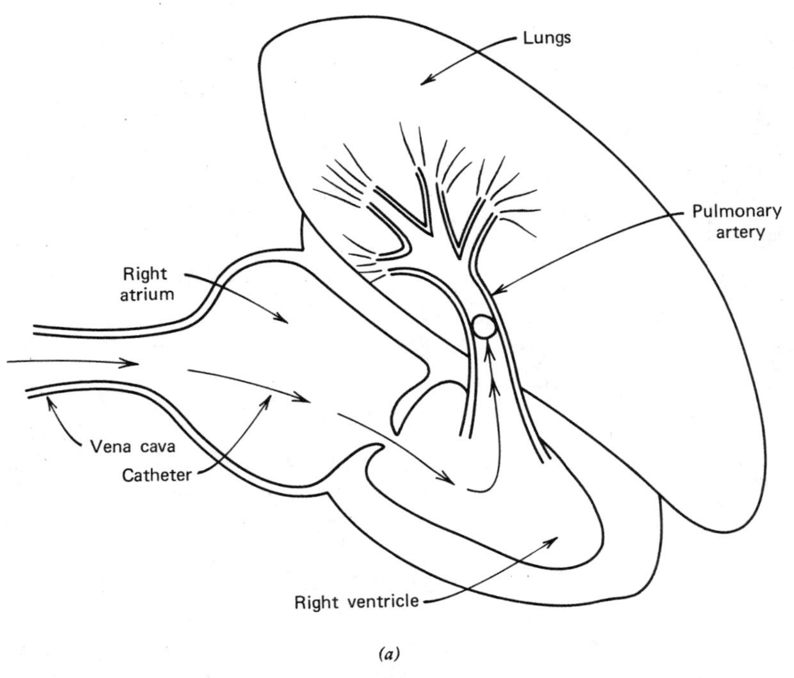

(a)

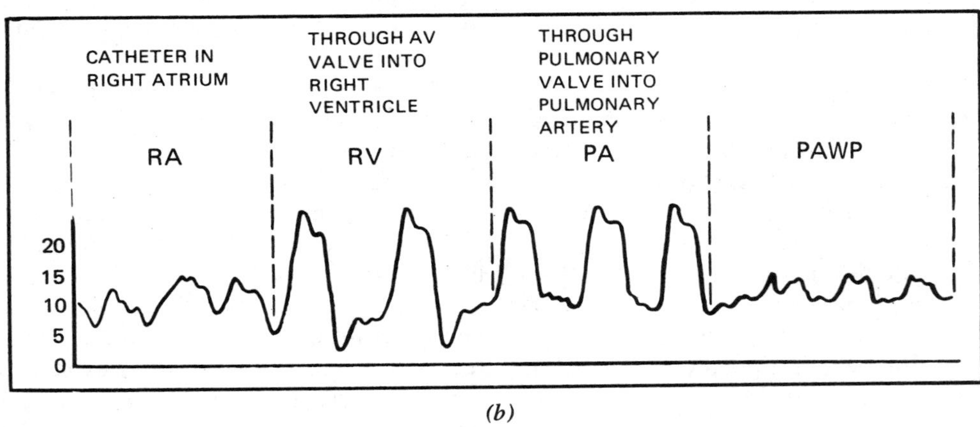

(b)

Figure 7-8 Placement of catheter for measurement of wedge pressure. (*a*) Placement in heart. (*b*) Pressure waveforms during insertion of catheter. (From R. S. C. Cobbold, *Transducers for Biomedical Measurements,* John Wiley & Sons, New York, 1974. Used by permission.)

by direct reflection from the underlying bone structures.

The PPG does *not* indicate *"calibratable"* volume changes. Its usefulness is limited to pulse-velocity measurements, determination of heart rate, and an indication of the existence of a pulse in the finger.

A "true" plethysmograph is a little more difficult to construct and operate but yields quantitative data. An example is shown in

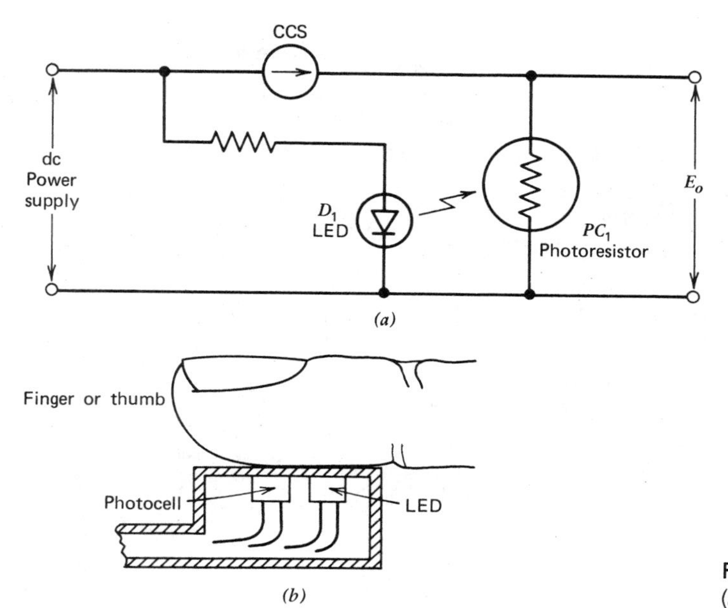

Figure 7-9 Photoplethysmograph. (*a*) Circuit. (*b*) Cutaway view.

Figure 7-10*a*. This device uses a sealed chamber surrounding the limb or digit being measured. The chamber is designed to have a constant volume, so volume changes in the limb are recorded as pressure variations inside of the chamber. The chamber pressure is measured electronically by a transducer and amplifier in the manner discussed in Chapter 6.

A small syringe is connected to the chamber and serves as a volume calibrator. Once the zero baseline has been established on the pressure amplifier readout device, the position of the syringe's plunger is changed to vary the volume of the system by a small, but precise, amount. The span control of an oscilloscope or strip-chart recorder used to display the output waveform may then be adjusted to properly represent the unit volume change. Note that most small syringes are already calibrated in cubic centimeters (cc or cm^3), which are units of volume.

The blood pressure cuff placed proximal to the chamber seal is inflated to a pressure slightly higher than the venous pressure. This condition permits the flow of arterial blood into the limb, while preventing the outflow of venous blood, allowing the limb volume to increase slightly following each systole. These changes (Figure 7-10*b*) are recorded on the strip-chart recorder. Note, however, that a saturation point is reached and the curve flattens out at some specific level. This phenomenon is caused by the trapped venous blood distal to the cuff; that is, the system "fills up."

Capacitance, mercury strain gages, and impedance plethysmographs are also used occasionally but mostly in research applications. In fact, most clinical plethysmography is done using the photoplethysmograph (PPG), despite its limitations.

7-7 Blood Flow Measurements

The measurement of *blood flow* is almost as important in many instances as the measurement of blood pressure. Although several

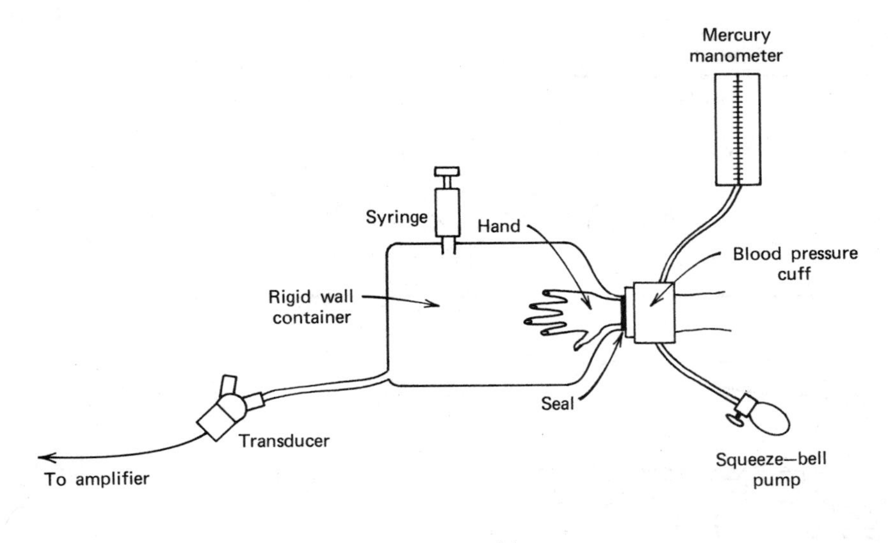

(a)

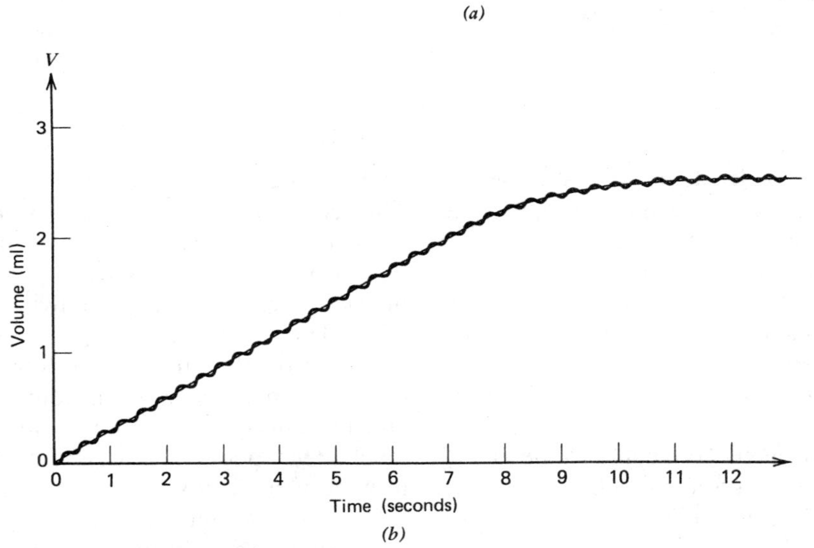

(b)

Figure 7-10 True plethysmograph. (*a*) Apparatus. (*b*) Output function.

techniques have been described,[1] the *electromagnetic* and *ultrasonic* methods have found the widest acceptance, although the latter is used less often clinically than the former. In this section we discuss the electromagnetic type, reserving the ultrasonic discussion for Chapter 17.

The popularity of the magnetic flowmeter is due to the following factors:

1. It measures volume flow rate independent of velocity.

[1] Richard S. C. Cobbold, *Transducers for Biomedical Measurements*, Wiley-Interscience (New York, 1974).

2. It produces accuracies up to ±5 percent.

3. The technique can accommodate blood vessels of diameters from 1 to approximately 20 mm diameter.

7-7.1 Theory

We know from basic electrical theory that a voltage is created when a moving conductor "cuts" the flux of a magnetic field. If that conductor is a blood-carrying vessel of diameter EE' (see Figure 7-11), the voltage generated will be

$$E = \frac{QB}{50\pi a}\ (\mu\text{V}) \tag{7-6}$$

where

E is the potential in *microvolts* (μV)
Q is the volumetric flow rate in *cubic centimeters per second* (cm^3/s) ($Q = \pi \bar{v} a^2$, where $\bar{v}$ is the average flow velocity over

Figure 7-11 Electromagnetic flow meter. (*a*) Cross-sectional view. (*b*) Block diagram. (From R. S. C. Cobbold, *Transducers for Biomedical Measurements,* John Wiley & Sons, New York, 1974. Used by permission.)

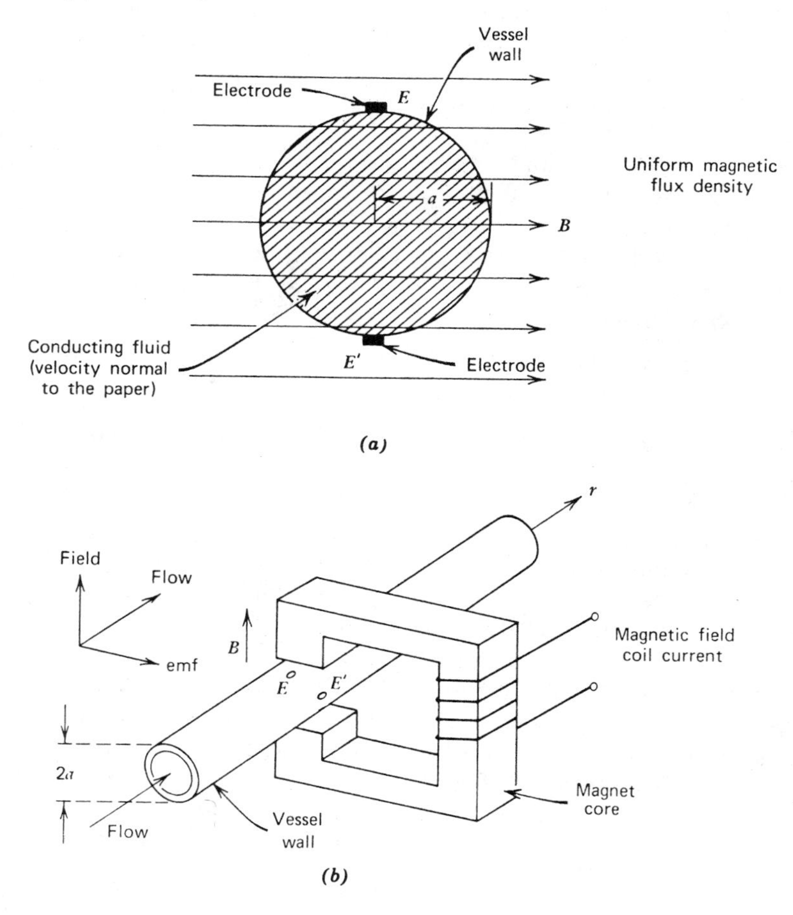

the region from the center of the vessel to the vessel wall)

B is the magnetic flux density in *gauss* (G)

a is the vessel radius in *centimeters* (cm)

EXAMPLE 7-3

Find the potential generated if blood flowing in a vessel with a radius of 0.9 cm cuts a magnetic field of 250 G. Assume a volume flow rate of 175 cm^3/s.

SOLUTION

$$E = \frac{QB}{50\pi a} \quad (7\text{-}6)$$

$$E = \frac{(175\ cm^3/s)(250\ G)}{(50)(3.14)(0.9\ cm)}$$

$$E = 309\ G\text{-}cm^2/s = \mathbf{309\ \mu V}$$

Figure 7-11 shows the typical transducer construction for magnetic flow measurements, while Figures 7-12*a* and 7-12*b* show commercially produced models for *in vivo* and *extracorporeal* use, respectively.

Most magnetic blood flow amplifiers use ac to excite the electromagnet coil in the probe. The electrode potential that exists in dc-excited systems introduces an offset artifact that proves difficult to discriminate from the actual voltage signal. The ac carrier usually has a frequency between 200 and 2000 Hz.

The block diagram to a magnetic flowmeter is shown in Figure 7-13*a*, with the associated waveforms shown in Figure 7-13*b*. This circuit uses a full-wave synchronous phase-sensitive demodulator circuit to extract the flow-rate information from the amplified electrode signal. This technique is used because the ac excitation produces an artifact in the electrodes by transformer coupling action. This signal has an amplitude that is several orders of magnitude *greater* than the desired flow signal and thus creates a tremendous level of interference if other types of detectors (i.e., simple envelope detectors) are used.

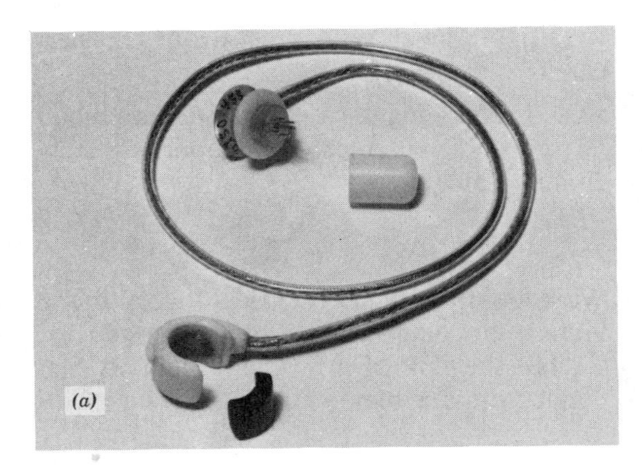

Figure 7-12 Electromagnetic flow meter transducers. (*a*) Clip-on type. (*b*) In-line type. (Photo courtesy Biotronix Laboratory.)

The detection system of Figure 7-13 is able to eliminate artifacts because of two factors: It is a *sampled* system, and the transformer emf is in *quadrature* (i.e., 90° out of phase with the excitation signal). The phase difference means that the artifact transformer emf signal is *zero* when the excitation signal is *maximum*. If a *short* sample of the waveform is taken when the transformer signal is near zero, then the magnitude of the artifact is reduced considerably. As with many pulse-sampling instruments, an *integrator* is used to extract the analog signal from the sampled ac signal.

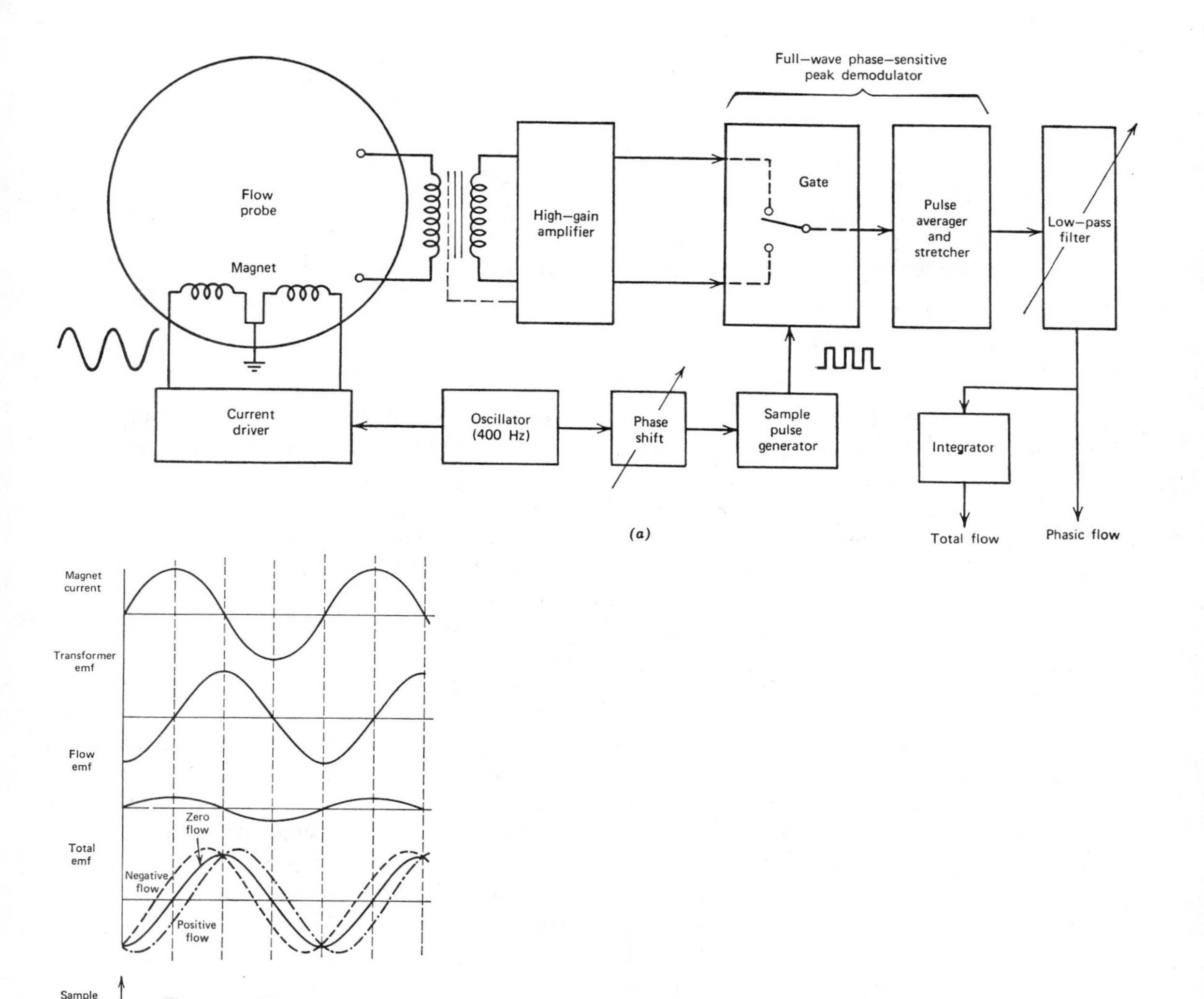

Figure 7-13 Electromagnetic flow meter circuit. (*a*) Block diagram. (*b*) Timing waveform. (Courtesy Biotronix Laboratory.)

7-8 Phonocardiography

Phonocardiography is the recording of *heart sounds*. The heart, like any mechanical pump, produces characteristic sounds as it beats. There are the sounds the physician hears with a stethoscope. Basic heart sounds occur mostly in the frequency range of 20 to 200 Hz. Certain heart murmurs produce

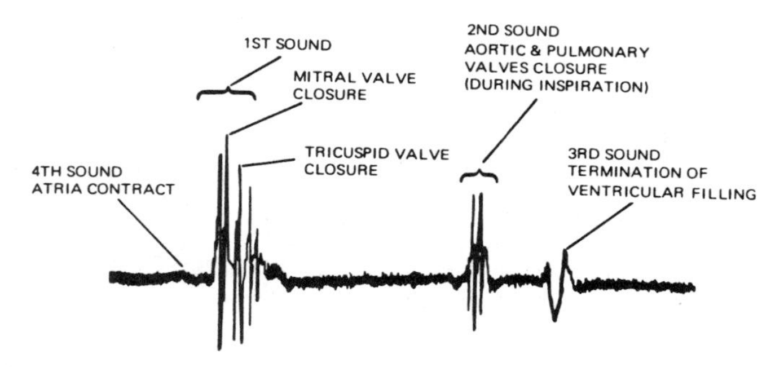

Figure 7-14 Basic heart sounds. (Reprinted courtesy Hewlett-Packard.)

sounds in the 1000-Hz region, and some frequency components exist down to 4 or 5 Hz.

The *first* heart sound (see Figure 7-14) is generated at the end of atrial contraction, just at the onset of ventricular contraction. This sound is generally attributed to movement of blood into the ventricles, the A-V valves closing, and the abrupt cessation of blood flow in the atria.

The *second* heart sound corresponds to the closing of the aortic and pulmonary valves, the *third* sound corresponds to the cessation of ventricular filling, and the *fourth* sound is correlated to the atrial contraction. This last sound has a very low amplitude and a low-frequency component (Figure 7-14).

The phonocardiograph transducer is a contact or air-coupled acoustical microphone held against the patient's chest (see Figure 7-15). Various types of microphones are used, but most are of the piezoelectric *crystal* or *dynamic* type of construction.

Figure 7-15 21050A sensor and accessories. (Reprinted courtesy Hewlett-Packard.)

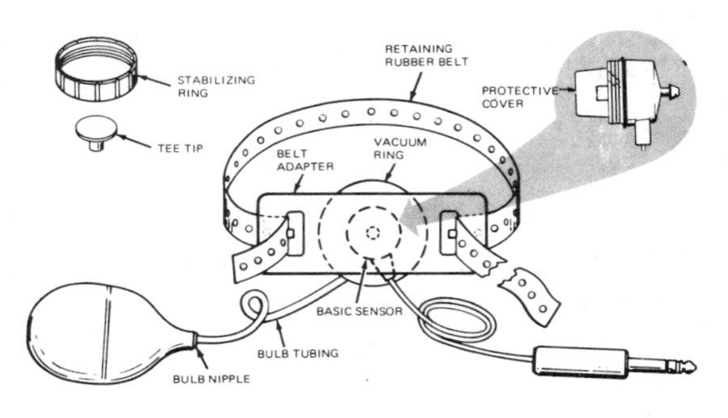

The crystal microphone generally costs less and is more rugged than the dynamic type. Also, the crystal microphone produces a larger output signal for a given level of stimulus.

The dynamic microphone uses a moving coil coupled to the acoustical diaphragm. The coil encircles a permanent magnet loudspeaker. The dynamic microphone is used when it is desired to have a signal frequency response similar to that of the medical stethoscope.

An *air-coupled* microphone with a 2-s *time constant* is often used in *apex phonocardiography* recordings. These microphones are generally crystal types and are coupled to the patient's chest through a column of air.

Although oscilloscopes are used to visualize heart sounds while they are being taken, hard-copy recordings are usually made on a strip-chart recorder. The frequency response of ink pen and thermal stylus recorders such as those used in ECG recording is only 100 to 200 Hz. Direct recording of the phonocardiogram requires a frequency response to 1000 Hz, so either an optical or high-velocity ink-jet model recorder is used.

Not all frequency components are important in phonocardiography, so some models use an *envelope recording* technique. In one

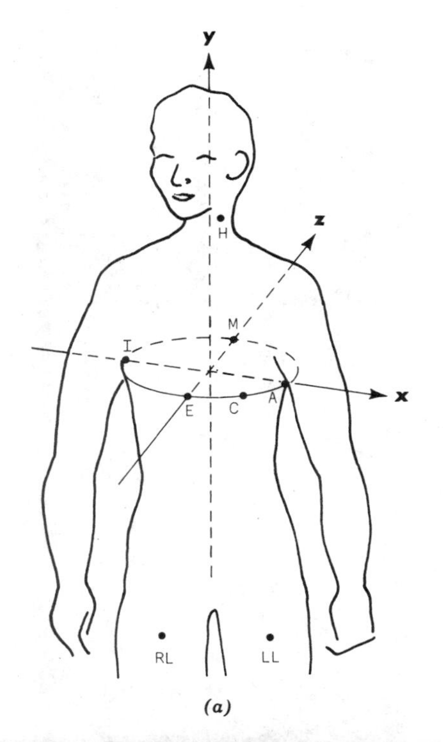

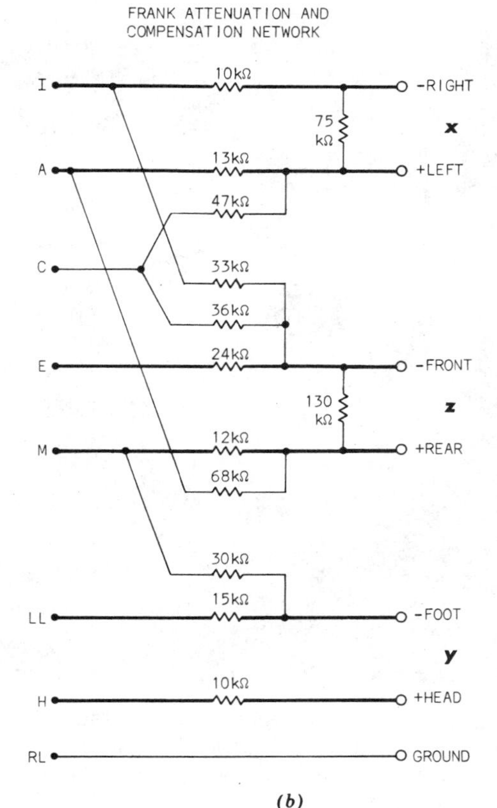

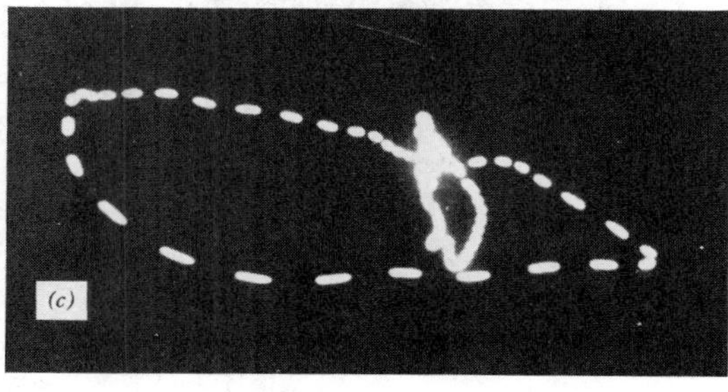

Figure 7-16 Vectorcardiography. (*a*) XYZ planes. (*b*) Frank lead network. (*c*) Vectorcardiogram. (*a, b* Reprinted courtesy of Hewlett-Packard.)

commercial product, the frequency components below 80 Hz are recorded directly, but the frequency components above 80 Hz are integrated (i.e., averaged) prior to recording. This technique allows the use of an ordinary low-frequency thermal or ink-pen type of strip-chart recorder.

A technique used in the Hewlett-Packard 1514 phonocardiograph is to *sample and hold* the positive and negative envelopes of the complex input signal from the microphone at an 85-Hz rate. This system is capable of detecting the shape, timing, duration, and intensity of the heart sounds plotted against time (50 mm/s or 100 mm/s). The sampled envelope signal can also be recorded on a low-frequency machine.

7-9 Vectorcardiography (VCG)

The vectorcardiograph (VCG) examines the ECG potentials generated along the three-dimensional axes of the body, that is, the *x*, *y*, and *z* planes. The *x*-vector is taken as the

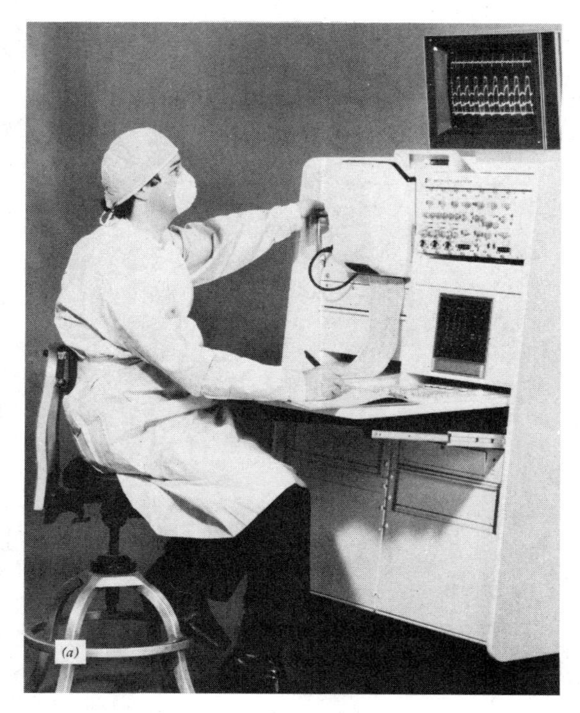

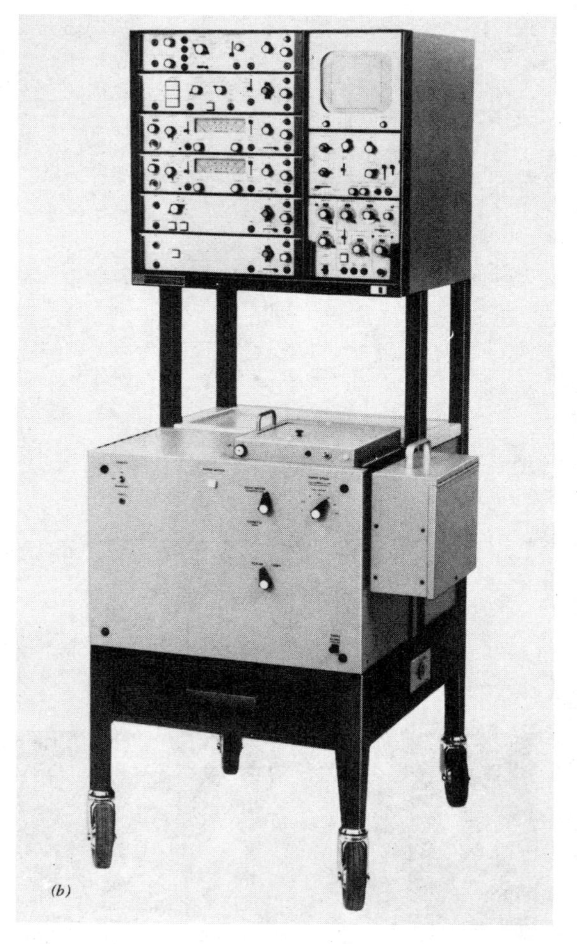

Figure 7-17 (*a*) H-P 8890B catheterization laboratory system. (Courtesy Hewlett-Packard.) (*b*) E-for-M VR-6 Simultrace Recorder. (Courtesy Electronics for Medicine.)

potential between two points under the arms (see Figure 7-16*a*), the *y*-vector is between the head and right leg, while the *z*-vector is from the front to the back of the body.

These vectors are not exactly orthogonal, and the amplitudes of the signals from the three planes are considerably different. The *Frank electrode system* of Figure 7-16*b* is used to normalize the input signals before applying them to the oscilloscope.

The oscilloscope must be an *x-y* model instead of the more common *y*-time type of display. A switching network at the output of the Frank lead system circuit will select the proper signal combinations for making a recording of *frontal, transverse,* or *sagittal* vectorcardiograms.

Figure 7-16*c* shows an oscilloscope VCG tracing. There are three individual loops corresponding to the *P*-wave, *QRS*-complex, and *T*-wave on the ECG waveform. These are joined at a single isoelectric point where all three vector components are zero. The dotted line effect is from *intensity* (*z*-axis) *modulation* of the CRT beam.

7-10 Catheterization Laboratories

Assessing the condition of a patient's heart may require data obtained from inside of the heart itself. The data are obtained by a process called *catheterization* and is used to make preoperative judgments on the need for surgery.

The catheter is inserted into the heart via the peripheral vascular system. It can record intracardiac pressures and allow withdrawal of blood from the heart chambers to measure oxygen and carbon dioxide levels.

In *right catheterization* the catheter is introduced through a peripheral *vein* through the vena cava into the right side of the heart.

A *retrograde catheterization* requires introduction of the catheter via an artery, usually the *brachial* or *femoral* artery. The catheter is threaded through the arterial system and enters the left ventricle via the aorta.

In *transseptal catheterization* a special large-diameter catheter is introduced through the femoral vein into the right atrium, where a special needle in the end of the catheter is used to puncture the septum wall dividing the right and left sides of the heart. A smaller catheter is then threaded through the large catheter and the needle to enter the left ventricle via the left atrium. This technique is used primarily when aortic stenosis prevents use of the retrograde technique.

The electronic equipment used in catheterization includes several channels of pressure amplifiers and ECG monitors; a chart recorder is used to make a permanent record of the pressure and ECG waveforms. Figures 7-17*a* and 7-17*b* show two types of catheterization laboratory equipment.

In addition to the measuring instruments, it is necessary to keep a *defibrillator* and other resuscitation equipment in the catheterization laboratory. In some cases, cardiac arrhythmias are generated when the catheter tip hits the ventricle wall. For this reason, the ECG waveform is monitored continuously during the catheterization procedure.

The equipment must be designed for very low ac (power mains) leakage operation. Because of the sensitivity of the heart to 25 to 60 Hz (in the United States it is 60 Hz) power mains current, leakage from that source must be kept to a minimum. This requirement mandates the use of isolated equipment.

7-11 Summary

1. Three methods for determining cardiac output are the Fick method (using respiratory oxygen), thermodilution, and dye dilution. Either optical or radio-opaque dyes are used.
2. A balloon-tipped catheter is used to measure right heart and pulmonary artery wedge pressures.
3. Plethysmography measures blood flow volume by changes in the physical volume of a limb.
4. Magnetic transducers are most often used in blood flow measurements.
5. The graphical recording of heart sounds is called phonocardiography.

7-12 Recapitulation

Now return to the objectives and self-evaluation questions at the beginning of the chapter and see how well you can answer them. If you cannot answer certain questions, place a check mark next to each and review appropriate parts of the text. Next try to answer the following questions, using the same procedure. When you have answered all of the questions, solve the problems in Section 7-14.

7-13 Questions

7-1 Human cardiac output is measured in units of ________ per ________.

7-2 The normal range of human cardiac output is ________ to ________.

7-3 Cardiac output can theoretically be measured by integrating ________ data.

7-4 Most clinical cardiac output computers use ________ techniques.

7-5 The technique using respiratory oxygen as a tracer is called the ______________ method.

7-6 Iced, or room temperature, IV solutions such as saline or D_5W are used in the ______________ method of cardiac output measurement.

7-7 In optical dye dilution techniques for cardiac output measurement a dye such as ______________ is used.

7-8 Most thermodilution cardiac output computers use a thermistor-tipped catheter placed in the ______________.

7-9 The input stage of most thermodilution cardiac output computers is a ______________ and an isolated differential amplifier.

7-10 The *recirculation artifact* can be ignored if the cardiac output computer uses ______________ integration to approximate the exponential decay of the temperature signal.

7-11 A multilumen balloon-tipped catheter can be used to measure ______________ and ______________ pressures.

7-12 The catheter discussed in Question 7-11 is introduced into the right side of the heart via the ______________ venous system.

7-13 A plethysmograph measures ______________ changes in a limb.

7-14 The ______________ is a simple type of plethysmograph but is of limited usefulness because it cannot be calibrated accurately.

7-15 A true plethysmograph may be calibrated using a ______________.

7-16 The blood pressure cuff used in plethysmography is inflated to a pressure greater than the ______________ pressure but less than the ______________ pressure.

7-17 Name two types of blood flow meter.

7-18 Which type of blood flow meter is most commonly used in clinical applications?

7-19 A ______________ is a recording of heart sounds.

7-20 Name two types of microphones that are often used in the recording of heart sounds.

7-21 The *first* heart sound is correlated to ______________.

7-22 The *second* heart sound is correlated to ______________.

7-23 The *third* heart sound is correlated to ______________.

7-24 The *fourth* heart sound is correlated to ______________.

7-25 An ______________-coupled microphone with a ______________-s time constant is used to record the apical heart sounds.

7-26 ______________ detection is sometimes used on phonocardiographs to allow recording on a low frequency strip-chart recorder.

7-27 A ______________ records the ECG in the form of three loops on a CRT.

7-28 The instrument in Question 7-27 uses the ______________ lead system.

7-29 List three different catheterization techniques and describe each.

7-14 Problems

7-1 Calculate the stroke volume in milliliters if the CO is 3.75 l/m and the heart rate is 76 BPM.

7-2 Calculate the blood flow if an injection rate of 3 mg/m results in a downstream concentration of 1 mg/l.

7-3 A cardiac output computer is tested by a thermistor simulator that uses a step-function change in resistance. Assuming that the correction factor for catheter temperature rise is 4.5, calculate the CO indicated if the integrator output rises 70°C-s. The computer controls are set as follows:
Injectate volume: 10 ml
Blood temperature: 37.6°C
Injectate temperature: 23°C

7-4 Calculate the input voltage in *microvolts* to a magnetic blood flow meter if the magnetic field is 325 G, the vessel diameter is 0.65 cm, and the flow rate is 200 cm^3/s.

7-15 References

1. Cobbold, Richard S. C., *Transducers for Biomedical Measurements*, Wiley-Interscience (New York, 1974).
2. Instruction/service manual for model 9520 cardiac output computer. Edwards Laboratories, Inc. (Santa Ana, Calif., 1974).

Chapter 8

Cardiac Stimulation and Life Support Equipment

8-1 Objectives

1. Be able to describe the principles behind, and the operation of, *defibrillators* and *cardioverters*.
2. Be able to describe how a pacemaker works and why it is necessary.
3. Know how to describe the principles behind the intraaortic balloon pump.
4. Be able to describe the operation of the extracorporeal blood pumps.

8-2 Self-Evaluation Questions

These questions test your prior knowledge of the material in this chapter. Look for the answers as you read the text. After you have finished studying the chapter, try answering these questions and those at the end of the chapter (Section 8-12).

1. A defibrillator ____________ the heart with an electrical shock so that all cells of the heart can enter their refractory periods at the same time.
2. What is *heart block?*
3. *Counterpulsation* is a technique employed in the ____________ pump.
4. An extracorporeal blood pump uses a ____________ pump head to transport blood.
5. Describe the operation and construction of the pump in question 4.

8-3 The Heart

In Chapter 2 we discussed the heart and its pumping operation. The heart is able to pump blood through the circulatory system only because the fibers that make up the heart muscle contract in a synchronous manner. A group of cells called the sinoatrial (SA) node is located on the rear wall of the right atrium. The SA node serves as a natural *pacemaker* for the heart by producing an electrical pulse output that stimulates contraction of the heart muscle fibers. The pacemaker pulse spreads across the atria, causing them to contract and force blood into the ventricles. The pulse spreading across the atrial tissue also flows to the electroconduction system of the heart to the ventricles. To be effective, ventricular contraction must follow atrial contraction by a fraction of a second so that atrial blood can fill the ventricular spaces. The heart's electroconduction system provides this delay. Part of the delay is due to the speed of propagation of the pulse down the electroconduction system, and part is due to the biological "delay line" called the *atrioventricular* (AV) node. The AV node is to the electroconduction system what a monostable multivibrator is to certain electronic circuits (i.e., pulse delay).

As long as the muscle fibers of the heart contract synchronously, the heart will function as an efficient blood pump. But certain problems can develop that disturb synchrony. One of these problems—*arrhythmia*—is called *fibrillation*, in which the muscle fibers of the heart quiver randomly and erratically instead of contracting together. If the atrial portion of the heart is in fibrillation, then it is called *atrial fibrillation*, and if the ventricles are involved, *ventricular fibrillation*.

When the heart is in atrial fibrillation it will still pump some blood because the ventricles are still able to contract—it is the ventricular contraction that maintains the system pressure. But when ventricular fibrillation occurs, the heart cannot pump, and death will occur within minutes unless the condition is corrected.

Figure 8-1 shows several ECG waveforms, including two arrhythmias and a normal wave-form for comparison. The normal waveform is Figure 8-1*a*. In this context, "normal" refers not necessarily to the total absence of any disease process but to the fact that all salient features of the waveform are present. Ventricular fibrillation is shown in Figure 8-1*b*. Notice the low amplitude and erratic aspects of the waveform, caused by the quivering heart muscle. The large sinusoidal-like waveform in Figure 8-1*c* is ventricular tachycardia.

Figure 8-1 (*a*) Normal waveform. (*b*) Ventricular fibrillation. (*c*) Ventricular tachycardia.

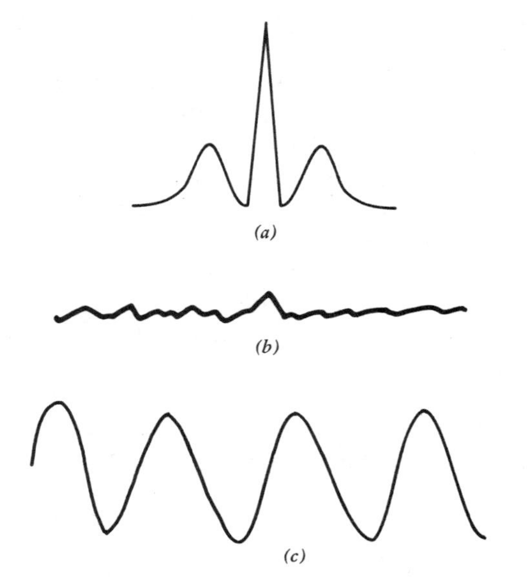

These waveforms can often be corrected by application of an *electric shock* to the heart. The electric shock will cause all of the heart muscle fibers to contract simultaneously, so they will all enter their refractory periods together, after which their normal rhythm will hopefully return.

8-4 Defibrillators

A *defibrillator* is a device that delivers electric shock to the heart muscle undergoing a fatal arrhythmia. Defibrillators prior to about 1960 were ac models. These machines applied 5 to 6 A of 60 Hz ac across the patient's chest for 250 to 1000 ms. The success rate for ac defibrillators was rather low, however, and the technique was useless for correcting atrial fibrillation. In fact, attempting to correct atrial fibrillation using ac often results in producing ventricular fibrillation, a much more serious arrhythmia.

Since 1960, several different dc defibrillators have been devised. These machines store a dc charge that can be delivered to the patient. The principal difference between dc defibrillators is in the waveshape of the charge delivered to the patient. The most common forms are the *Lown, monopulse, tapered (dc) delay,* and *trapezoidal* waveforms

In 1962, Dr. Bernard Lown of Harvard introduced the waveform that bears his name. The Lown waveform, shown in Figure 8-2, shows the voltage and current applied

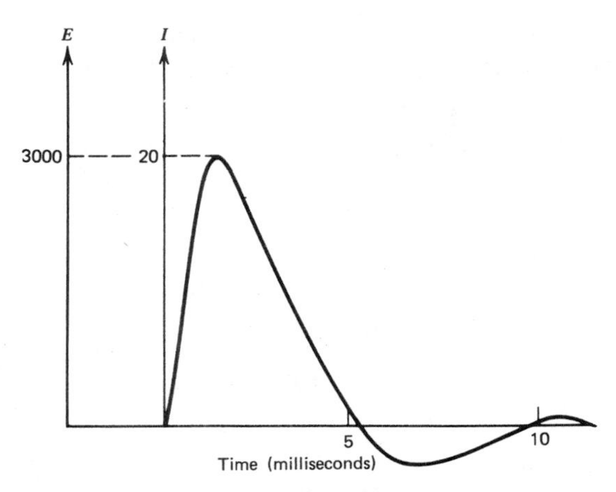

Figure 8-2 Lown defibrillator waveform.

to the patient's chest plotted against time. The current will rise very rapidly to about 20 A, under the influence of slightly less than 3 kV. The waveform then decays back to zero within 5 ms, and then produces a smaller negative pulse also of about 5 ms duration.

Figure 8-3 shows a simplified diagram of a Lown defibrillator. The charge delivered to the patient is stored in a capacitor and is produced by a high-voltage dc power supply. The operator can set the charge level using the *set energy* knob on the front panel. The knob controls the dc voltage produced by the high-voltage power supply and so can set the maximum charge on the capacitor. The energy stored in the capacitor is given by:

$$U = \tfrac{1}{2}CV^2 \qquad (8\text{-}1)$$

where

U is the energy in *joules* (J)
C is the capacitance of C_1 in *farads* (F)
V is the voltage across C_1 (V)

EXAMPLE 8-1

Calculate the energy stored in a 16-μF capacitor that is charged to a potential of 5000 V dc.

SOLUTION

$$U = \tfrac{1}{2}CV^2 \qquad (8\text{-}1)$$
$$U = \tfrac{1}{2}(1.6 \times 10^{-5}\ \text{F})(5 \times 10^3\ \text{V})^2$$
$$U = \tfrac{1}{2}(400) = \mathbf{200\ J}$$

Figure 8-3 Typical circuit for a Lown waveform defibrillator.

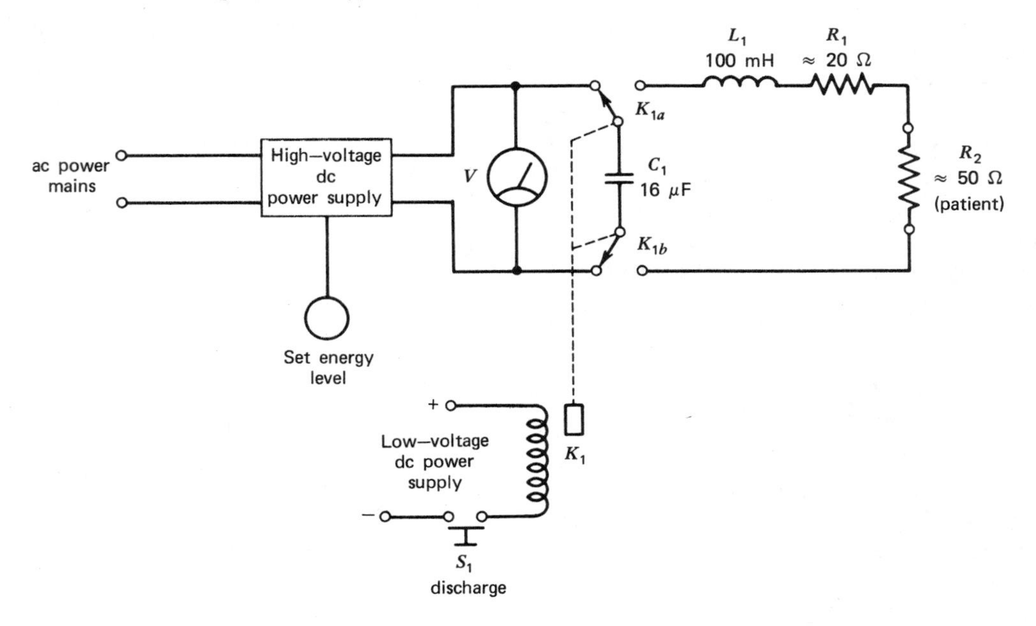

The stored energy is indicated by a voltmeter connected across the capacitor. The scale of the voltmeter is calibrated in energy units. It is common practice to use *watt-seconds* as the units of energy instead of joules, but this is really no problem because (by definition) one joule (1 J) is equal to one watt-second (1 W-s). Older dc defibrillators had energy meters calibrated in *stored* energy, but present regulations require that the *delivered* energy be indicated. Some energy is lost in the relay switching contacts and in the ohmic resistance of inductor L_1.

The capacitor charge is controlled by a relay switch, K_1. In early models SPDT relays were used, but in all recent models DPDT relays are used so that isolation of the patient circuit from ground is maintained. Although there are a few portable defibrillators that use open-air high-voltage relays, most use special sealed vacuum relays such as the Torr Laboratories type TMR-10. The vacuum relay is justified because of the high voltages used to charge capacitor C_1. If 16 μF is used (a common value), and 400 J are stored, then the potential across the capacitor will be greater than 7000 V dc.

The patient circuit of the Lown defibrillator consists of a 100-mH inductor (L_1), the ohmic resistance of L_1 (i.e., R_1), and the patient's ohmic resistance (i.e., R_2). It is the energy stored in the magnetic field of coil L_1 that produces the negative excursion of the Lown waveform during the last 5 ms. When the capacitor has discharged, the coil's field collapses, dumping energy back into the circuit. The sequence of events is described as follows.

1. The operator turns the *set energy* control to the desired level and presses the *charge* button (i.e., closes S_2).

2. Capacitor C_1 begins charging and will continue to charge until the voltage across the capacitor is equal to the supply voltage.

3. The operator positions paddle electrodes on the patient's chest and presses the *discharge* button (i.e., S_1).

4. Relay K_1 disconnects the capacitor from the power supply and then connects it to the output circuit.

5. Capacitor C_1 discharges its energy into the patient through L_1, R_1, and the paddle electrodes. This action occurs in the first 4 to 6 ms and gives rise to the high-voltage positive excursion of the waveform in Figure 8-2.

6. The magnetic field built up around L_1 collapses during the last 5 ms of the waveform, producing the negative excursion of the waveform in Figure 8-2.

The *monopulse* waveform shown in Figure 8-4 is a modified Lown waveform and is commonly found in certain portable defibrillators. It is created by a circuit such as Figure 8-3 but without inductor L_1 to create the negative second pulse. Consequently, the waveform decays to zero in the exponential manner expected of an *R-C* network.

Another form of dc defibrillator waveshape is the *tapered delay* shown in Figure 8-5. This waveform differs from the two pre-

Figure 8-4 Monopulse defibrillator waveform.

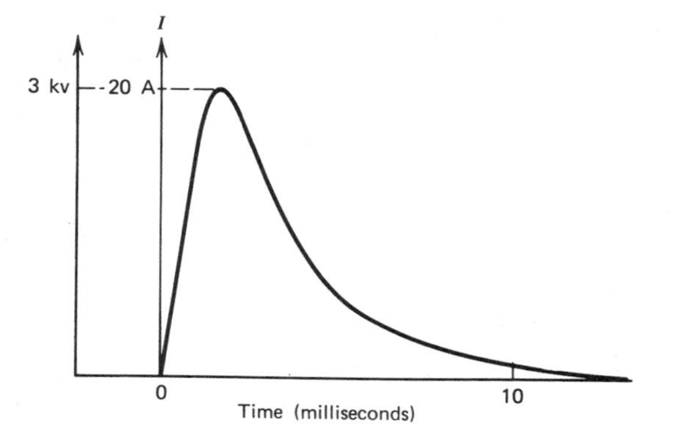

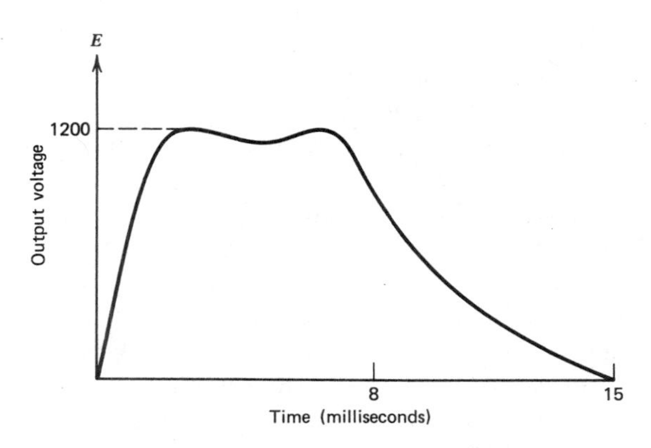

Figure 8-5 Tapered dc delay defibillator waveform.

vious pulses in that it uses a lower amplitude and longer duration to achieve the energy level. The energy transferred is proportional to the area under the square of the curve, so we may attain the same energy as in other waveforms. The double-humped waveform characteristic of tapered delay machines is achieved by placing two *L-C* sections, such as L_1/C_1 in Figure 8-3, in cascade with each other.

The trapezoidal waveform shown in Figure 8-6 is another low voltage/long duration shape. The initial output potential is about 800 V, which drops continuously for about 20 ms until it reaches 500 V, where it is terminated.

Figure 8-6 Trapazoidal defibrillator waveform.

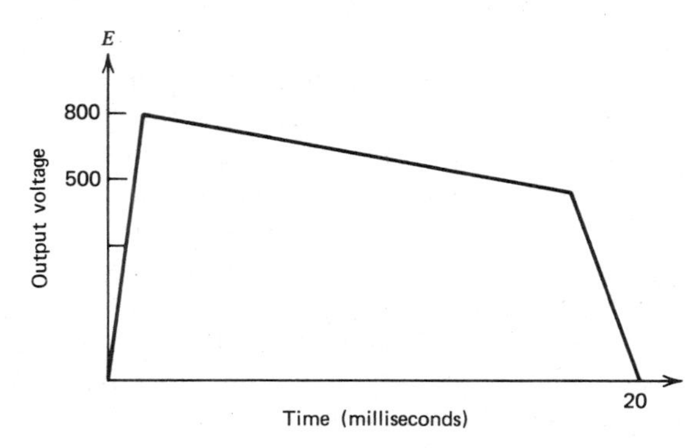

The energy from a defibrillator is delivered through a set of high-voltage paddle electrodes. Several popular styles are shown in Figure 8-7. The type shown in Figure 8-7*a* is called an *anterior* paddle. In this design the insulated handgrip is perpendicular to the metal electrode surface. The high-voltage cable enters from the side. A thumbswitch to control the discharge is mounted at the top of the grip. A defibrillator paddle and cable set using two of these electrodes is called an *anterior-anterior* set. To defibrillate, one electrode is placed on the chest directly over the heart, while the second electrode is placed on the left side of the patient's chest. A conductive paste is smeared on the electrodes to insure an inefficient transfer of charge and reduce any burning of the patient's skin.

A *posterior* paddle is shown in Figure 8-7*b*. This electrode is constructed flat and is designed so that the patient can lie on it. Posterior paddles are always paired with one anterior paddle to form an *anterior-posterior* pair.

A more modern anterior paddle is the *D-ring* type shown in Figure 8-7*d*. This type of paddle is used on most current model defibrillators and has been popular on portable models for some time.

One final form of paddle set is the *internal* type shown in Figure 8-7*c*. Internal paddle sets use two of these electrodes, but one may not have the thumbswitch. These paddles are used during open-heart surgical procedures to apply the electrical shock directly to the myocardium.

In Figure 8-3 discharge switch S_1 is used to fire the defibrillator by energizing the charge transfer relay. In some models, S_1 will be mounted on the front panel, but in most models it will be on one of the patient electrode paddles. Some manufacturers actually use two switches in series for the sake of safety, one switch mounted on each of the

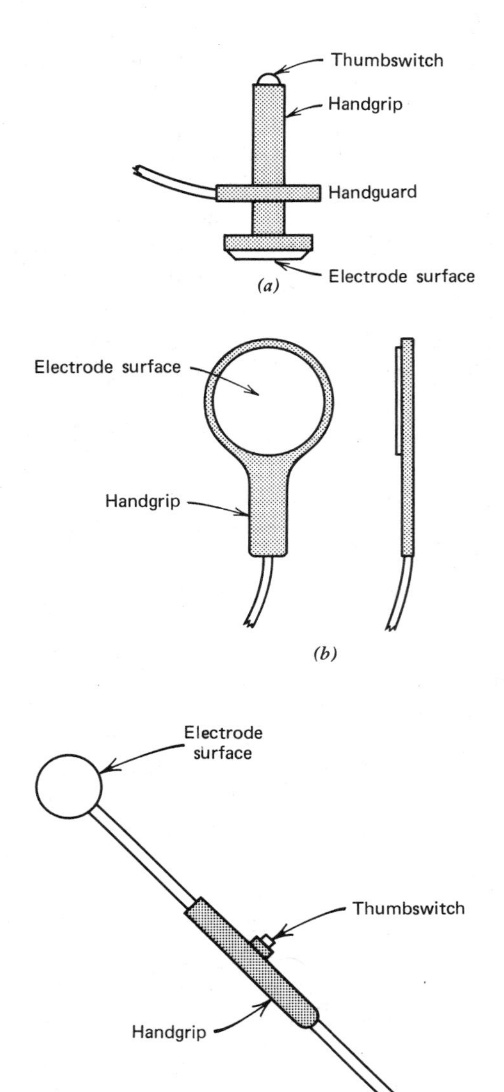

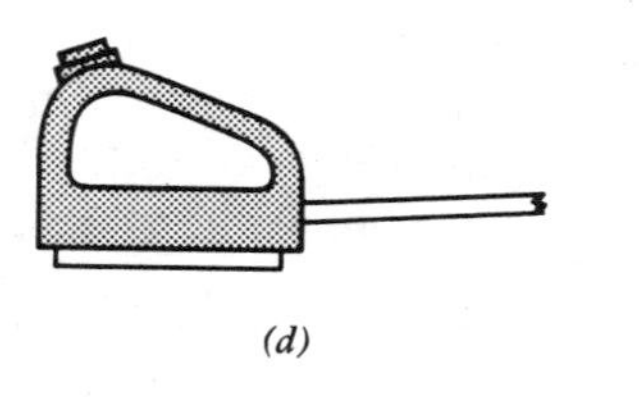

Figure 8-7 Defibrillator electrode handpieces. (*a*) Standard anterior. (*b*) Posterior. (*c*) Internal. (*d*) D-ring anterior.

paddles. Discharge of the defibrillator cannot occur unless *both* switches are closed. Some early models used a foot switch, but this proved to be too hazardous in the hectic and tense environment of a cardiac emergency.

It is absolutely essential that ventricular fibrillation is confirmed on an ECG machine, or monitor oscilloscope, before using a defibrillator. The nurse, physician, or other rescuer must first determine that fibrillation is in progress. Some defibrillator manufacturers have monitor oscilloscopes and ECG preamplifiers built into their products, but in others the user must provide an external ECG system. In some hospitals, an ECG or monitoring technician is responsible for bringing ECG equipment to the scene of a resuscitation attempt.

8-5 Defibrillator Circuits

The circuit shown in Figure 8-3 is inadequate to describe modern defibrillators, although for certain older models it is nearly complete. Two approaches to the design of the *set level* control circuit are evident. One technique uses a *variac* (variable autotransformer) in the primary of the high-voltage transformer in the dc supply. The operator, then, is actually adjusting the high-voltage dc supply when setting the energy level. The dc output voltage of the power supply is a function of the variac setting. When the *charge* button is pressed, ac power is applied to the transformer through the variac. The capacitor will continue to charge until its voltage is equal to the power supply voltage. When these two voltages are equal, the operator may discharge the defibrillator.

The other approach to control circuit design is shown in Figure 8-8. In this circuit, the dc output of the high-voltage power supply is fixed. A voltage comparator will turn the supply on and off depending on the voltages applied to its inputs. One input of the comparator is connected to voltage divider

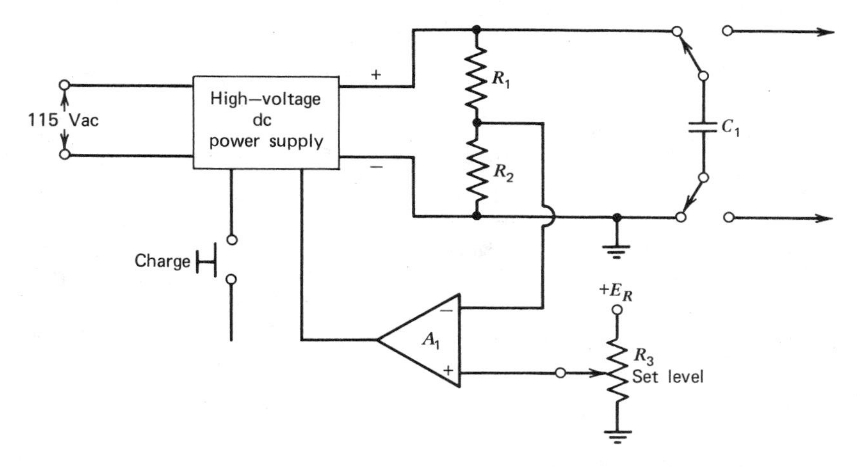

Figure 8-8 Electronic set change circuit.

R_1/R_2, which produces a low-voltage sample of the high voltage applied to capacitor C. The other input of the comparator is connected to a potentiometer designated as the *set level* control. When the operator selects an energy level, the voltage at the potentiometer wiper will represent the desired charge.

The operator initiates a *charge* cycle by pressing the *charge* button on the front panel. The voltage comparator sees the voltage from the *set level* control at one input, and zero voltage at the other, so its output will snap high. When the comparator output is high, the relay or digital IC logic controlling the high-voltage supply latches *on*.

As the capacitor charges, however, the voltage at the comparator's inverting input rises. When the voltage rises to the point where it is equal to the voltage from the *set level* control, the comparator shuts off, stopping the charge cycle.

The *discharge* circuitry is very much like the previous design. A switch on the patient paddle set energizes the coil of a high-voltage vacuum relay, causing discharge of the capacitor's stored energy into the patient's body.

8-6 Cardioversion

In certain types of arrhythmia (e.g., atrial fibrillation), the patient's ventricles maintain their ability to pump blood, as evidenced by the existence of an R-wave feature in the ECG waveform. These arrhythmias are also correctable by electrical shock to the heart, but it is necessary to avoid delivering the shock during the ventricles' refractory period (the T-wave feature of the ECG waveform), or the shock intended to correct the problem will actually create a much more serious arrhythmia such as ventricular fibrillation. The shock is usually timed to occur approximately 30 μs after the R-wave peak.

Human operators cannot be trusted to time the ECG waveform properly to avoid this problem, so an automatic electronic circuit is used. A machine equipped with the synchronizer circuit is called a *cardioverter*.

There will be a switch on the machine that allows the operator to select either *defibrillate* or *cardiovert* modes. In some machines, notably the Hewlett-Packard models, this control is labeled either *synchronized-instantaneous* or *sync-defib*.

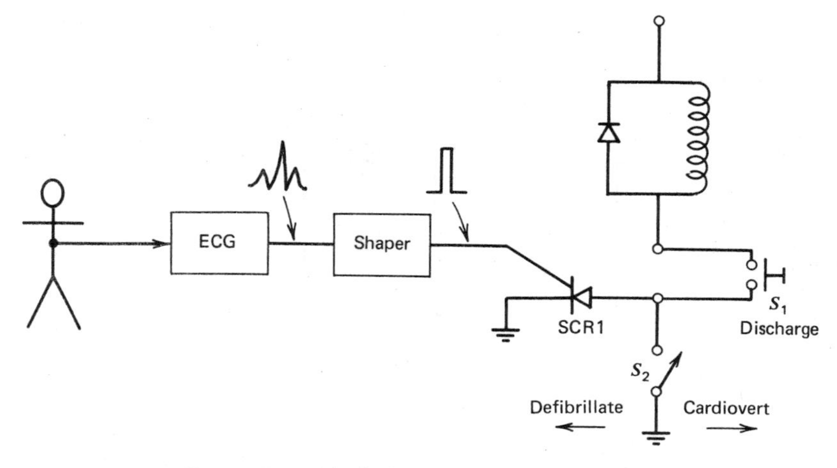

Figure 8-9 Block diagram for a cardioverter.

Figure 8-9 shows a partial schematic of a synchronizer circuit. Relay K_1 and switch S_1 are the same as in the previous circuits. When switch S_2 is in the *defibrillate* position, the circuit operates in the manner of other circuits; depressing S_1 will energize the relay, discharging the capacitor.

But when switch S_2 is in the *cardiovert* position, the relay is not energized unless both S_1 is closed and the silicon-controlled rectifier *SCR1* is turned on.

The SCR is turned on by an ECG *R*-wave. The ECG preamplifier acquires the signal from the patient. This amplifier may be internal to the defibrillator or an external patient monitor. Many bedside monitors have an output jack labeled *defibrillator* to provide such a signal.

Regardless of the ECG amplifier configuration, it is necessary to also provide circuits that discriminate against any feature other than the *R*-wave. In some early models, a simple threshold detector was used, depending on the fact that the *R*-wave is usually the highest amplitude feature of the ECG. But in some cases the *T*-wave amplitude would also exceed the threshold. To solve this problem, a *differentiator* circuit is used ahead of the threshold detector. The differentiator produces a much higher output on the *R*-wave than on the *T*-wave, because of the difference in their respective slopes (hence high-frequency content). The differentiator, therefore, insures a greater difference in amplitude between the *R*-wave and the other major features. It is also necessary that a pacemaker rejection circuit be included, because the pacer spike will appear to the discriminator much the same as the *R*-wave.

Some manufacturers apply the output of the threshold detector directly to the SCR gate. But other manufacturers use the detector output to trigger a monostable multivibrator, whose output pulse then triggers the SCR gate. In some models the monostable pulse is also applied to an output jack.

Figure 8-10 shows a commercial defibrillator, the Hewlett-Packard Model 78620A installed on a Model 78630 resuscitation cart. An internal battery provides up to 100 discharges at the maximum 400 W-s level, or 7 hours (h) of continuous monitoring. This instrument also includes a built-in patient monitor equipped with a nonfade oscilloscope.

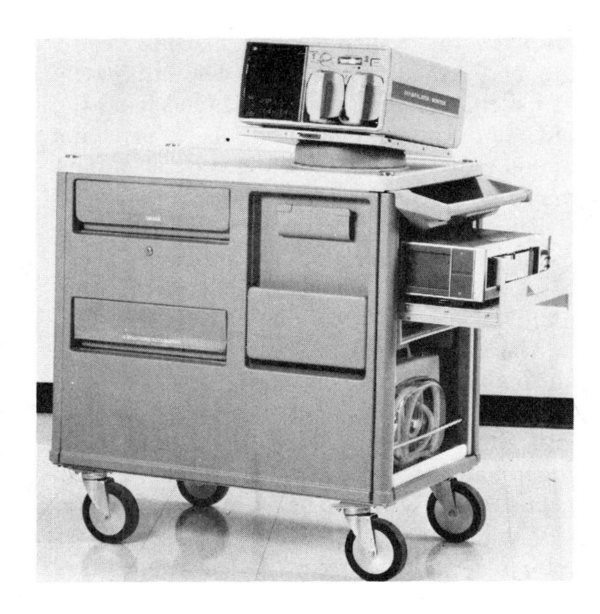

Figure 8-10 Typical defibrillator/crash cart. (Courtesy Hewlett-Packard.)

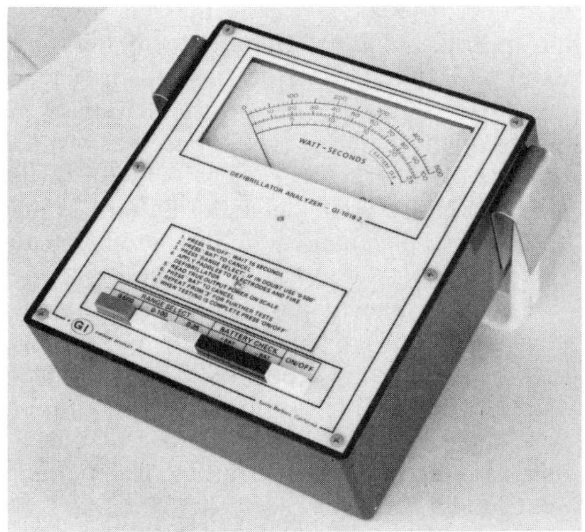

Figure 8-11 Defibrillator tester. (Photo courtesy of G.I. Medical Products, Inc., Santa Barbara, Calif.)

8-7 Testing Defibrillators

Defibrillators transfer large current charges at high electrical potentials. When trouble begins to develop, it develops rapidly. As a result, wisdom dictates frequent testing.

There are several defibrillator testers on the market. Most are basically integrating voltmeters that are calibrated in watt-seconds. An example of a defibrillator tester is shown in Figure 8-11. A 50-Ω dummy load is built into the tester and is connected between a pair of electrodes. The paddles are placed against the electrodes, and the capacitor is discharged into the load. The meter registers the delivered energy in watt-seconds.

It is preferable to specify a tester that has an oscilloscope output jack, so that the output waveform may be viewed on an oscilloscope. Most authorities agree that a proper evaluation of defibrillator performance requires *both* determination of the delivered energy and examination of the waveform. An oscilloscope camera is usually preferred for making a permanent record of the waveform, although at least one tester uses special digital circuitry that allows recording of the defibrillator output waveform on a 25 mm/s strip-chart recorder.

8-8 Pacemakers

In previous sections we have discussed the electroconduction system of the heart. An electrical impulse is generated by the SA node, located in the vicinity of the right atrium. The pulse propagates down the electroconduction system, incuding the AV node (which acts as a "delay line"), to the ventricles. The system splits into left and right bundle branches in the ventricles.

If an interruption occurs in the electroconduction system, causing a condition called *heart block,* then the heart's ability to pump blood is disrupted or stopped. Physicians can maintain electrical stimulation of the heart using a *pacemaker*—an electrical generator that delivers the needed pulse at an appropriate time. Some pacemakers are external and are worn on the belt or placed at bedside. Others are surgically implanted inside of the patient. It takes at least 10 μJ to pace the heart, while more than 400 μJ is likely to cause ventricular fibrillation. Many commercial pacemakers deliver about 100 μJ and are thus safely within both limits.

External pacemakers are a temporary measure that is used following open-heart surgery, for certain problems experienced by some myocardial infarction patients, and for patients who are to be evaluated as candidates for surgical implantation of a permanent model.

External models are usually adjustable over the range 50 to 150 BPM and produce fixed-duration, short-duty cycle pulses (i.e., 1.5 to 2.0 ms). The peak current amplitude is adjustable over the range of 100 μA to 20 mA.

Permanent pacemakers (Figure 8-12) are built into molded epoxy-silicone rubber packages, although some recent models include an outer titanium shield that guards against interference from radio frequency fields. The device is implanted subcutaneously in either the abdomen or in a region just below the collarbone.

Some implantable pacemakers have a single fixed rate usually about 70 BPM, while others are dual-rate models. The latter type of pacemaker can be programmed from the outside of the patient's body using a magnet or induction coil. Still others are programmable over the range 30 to 150 BPM.

Figure 8-12 Internal pacemaker. (Courtesy Medtronics, Inc., Minneapolis, Minn.)

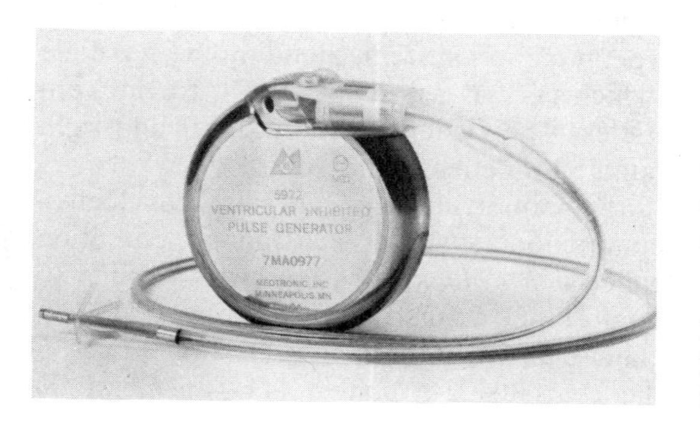

Two types of pacemaker lead wire are used: *endocardial* and *myocardial*. These categories are also broken down further into *unipolar* and *bipolar* varieties.

The endocardial lead is inserted through an opening in a vein and then threaded through the venous system, right atrium, and into the right ventricle of the heart. On the other hand, the myocardial leads are connected directly to the heart muscle.

The pacemaker shown in Figure 8-12 uses a bipolar lead; both electrodes are inside of a single catheter. The distal tip is one electrode, while the second is located a short distance behind the tip. These electrodes are made of plantinum-iridium alloy to prevent interaction with body fluids.

The principal power source for *implantable* pacemakers is the *mercury cell,* although many new models are using the *lithium iodine cell.* Mercury pacemaker batteries are theoretically able to operate for as long as 4 to 5 years (yr), but it is more usual to find service periods of 1.5 to 3 yr. Pacemaker manufacturers go to great lengths to screen pacemaker batteries, even to the point of X-ray examination of the cell to spot assembly defects known to shorten life expectancy. The battery remains the weak point in the pacemaker, however. Some hospitals operate pacemaker surveillance clinics to spot premature battery failure. It is known that the pulse rate drops with decreased battery voltage. The heart rate of the patient, therefore, can sometimes serve as an early warning indicator of battery failure.

Some work has been done on a nuclear power source for the pacemaker. Early reports were optimistic of a 10-yr life expectancy for a nuclear power pack. This type of power pack uses the heat generated by the natural decay of certain radionuclides to generate electricity in a thermocouple. At the present time, however, the mercury and lithium battery cells seem the most viable energy source for implantable pacemakers.

8-8.1 Pacemaker Classifications

There are several ways to classify pacemakers besides the external versus internal scheme of Section 8-8. Pacemakers may also be classified by the type of output pulse that they produce. Some models produce a monopolar output pulse, while others produce a *biphasic* pulse (i.e., a high-amplitude pulse of one polarity, followed by lower-amplitude pulse of the opposite polarity).

There are four general categories of pacemaker: *asynchronous, demand, R-wave inhibited,* and *atrial-ventricular synchronized.*

The asynchronous pacemaker produces pulses at a fixed rate in the 60- to 80-BPM range. The standard rate is 70 BPM, but rates within the specified range are obtainable on special order.

A demand pacemaker adjusts its firing rate to the patient's heart rate. It contains circuitry that senses the ECG *R*-wave and measures the *R-R interval.* During the first quarter of this period, the pacemaker is dormant in order to prevent response to the *T*-wave feature of the ECG. But during the last three quarters of the *R-R* interval it is in a *sense* mode. If an *R*-wave is not sensed within this period, then the pacemaker emits a pulse.

An *R*-wave inhibited pacemaker is similar to the demand type, except that it does not emit pulses during normal heart activity. The triggering circuits are inhibited for a period of time following each *R*-wave. The pulse is enabled, however, if no *R*-wave occurs for a preset period.

The atrial-ventricular synchronized pacemaker responds to the ECG *P*-wave—the ECG feature created by contraction of the heart's atria. The atrial pacer circuitry contains a *P-Q* delay circuit that simulates the propagation time in the heart's electroconduction system. Once the pacer senses a *P*-wave, and the *P-Q* delay of about 120 ms has elapsed, then the pacer will fire a pulse to stimulate the ventricles into conduction. The AV pacer has the advantage that it will follow the changing heart rate demands of the body. Pacers synchronized to the ventricles only can result in ventricular contractions independent of the atrial heart rate set by the SA node. The atrial-ventricular pacer will usually track heart rates in the 50- to 150-BPM range but will revert to fixed rate if the atrial rate exceeds the 150 BPM upper limit. This feature prevents atrial triggering should atrial flutter or atrial tachycardia occur.

8-9 Heart-Lung Machines

The heart is unable to maintain circulation during surgery to either itself or the great vessels from the heart. During these types of surgical procedures, perfusion of the body tissues with blood is maintained by an *extracorporeal* (i.e., outside of the body) pump called the *heart-lung machine.*

The heart-lung machine also oxygenates the blood. The lungs normally operate due to a partial vacuum inside of the thoracic cavity created when the diaphragm drops, thereby increasing the volume of the cavity. The pressure inside of the lung is essentially atmospheric pressure, and the pressure in the interpleural space outside of the lung is slightly less than atmospheric. This differential pressure is sufficient to cause the lungs to expand and fill with air. But during certain thoracic procedures, the chest is open, so *both* sides of the lung are at a pressure equal to atmospheric pressure.

Figure 8-13 is a diagram of a heart-lung machine *pump head.* This type of mecha-

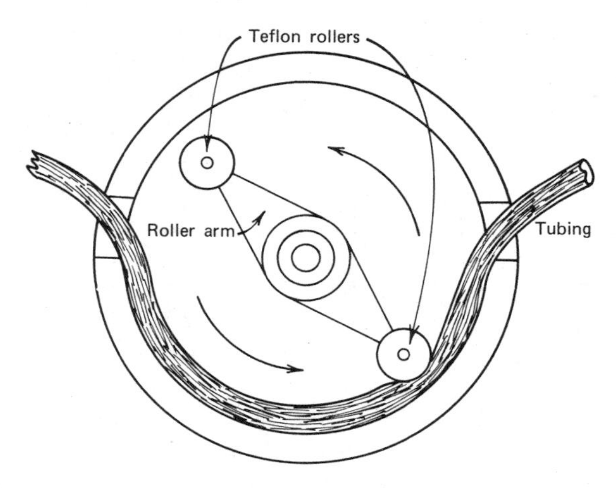

Figure 8-13 Peristaltic pump head.

nism is called a *peristaltic pump*. The blood to and from the patient's body is carried in a length of sterile, clear plastic tubing called a cannula. Cannulas with appropriate fittings to accommodate the tubing are inserted into the vessels to take blood from, and deliver blood to, the patient. Pumping action occurs because the rollers on the rotating arm compress the tubing carrying the blood, forcing the blood ahead of the compressed section. This *peristaltic action* produces a wavelike, pulsatile flow of blood through the tubing.

The heart-lung machine diagrammed in Figure 8-14 uses five pump heads: one for perfusion of the body, two "suckers," and two for perfusion of the coronary arteries.

The main perfusion system includes pump head 1 and a combination *heat exchanger/oxygenator.* Some models separate these units, having separate oxygenator and heat exchanger devices. The heat exchanger consists of water coils isolated from, but thermally coupled to, the blood. A temperature controller permits the pump operator to keep the blood at a proper temperature and compensate for heat loss through radiation from the lines.

Figure 8-14 Schematic representation of the heart-lung machine.

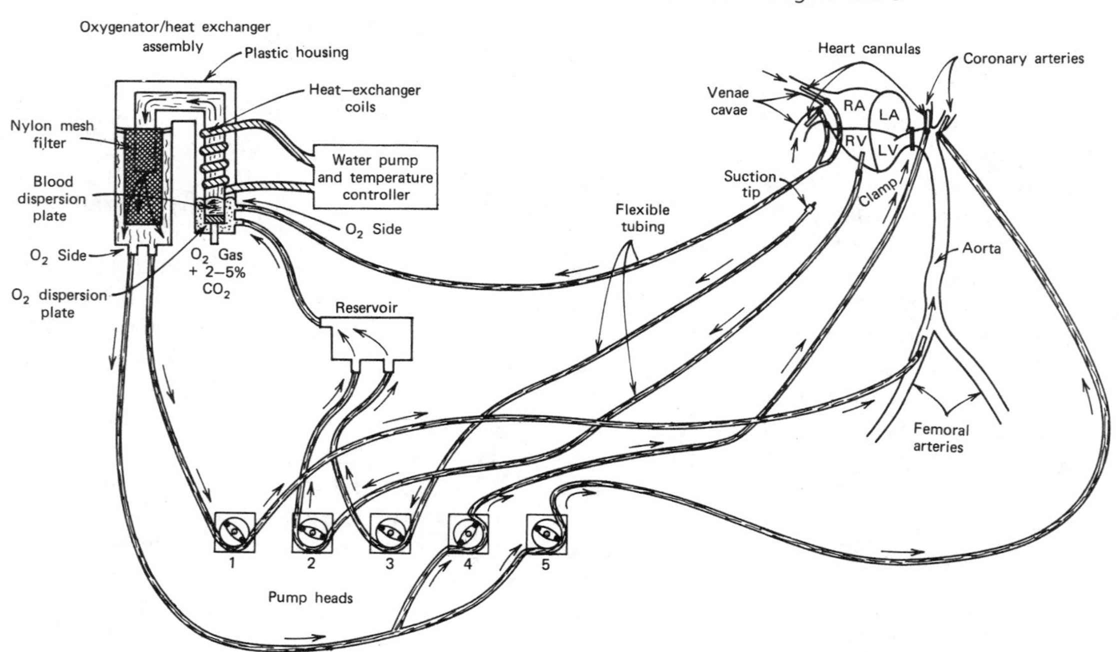

The input port of the oxygenator/exchanger assembly is called the *O_2-minus* side, while the output port is called the *O_2-positive* side.

Blood is taken from the patient's venae cavae. A cannula is placed into the superior vena cava and another into the inferior vena cava. These lines are joined into a single piece of tubing through a Y-adapter.

The blood flows from the venae cavae through the tubing to the O_2-minus side of the oxygenator/exchanger assembly. Another length of tubing carries blood from the O_2-positive side of the assembly, through the pump head, and back to the patient through a cannula inserted into the femoral artery.

Pump heads 2 and 3 are used as "suckers." Pump head 2 has a *vent* function to perform in the heart, while head 3 is shown as a suction device. The surgeon can use the suction tip to collect blood that pools during surgery; this blood is ordinarily lost unless it is autotransfused back to the patient. Blood from the suckers is delivered to a reservoir tank, and then transferred into the oxygen/exchanger assembly on the *O_2-minus* side.

The purpose of pump heads 4 and 5 is to perfuse the coronary arteries. A piece of tubing is routed from a port on the *O_2-positive* side of the oxygenator through the pump heads to the cannulas placed in the coronary arteries.

8-10 Summary

1. A *defibrillator* corrects certain cardiac arrhythmias by applying an electrical shock from a charge stored in a capacitor.
2. Four common defibrillator waveshapes are the *Lown, monopulse, tapered dc delay,* and *trapezoidal*.
3. Most defibrillators include a *cardioversion* synchronizer circuit that allows the machine to discharge only on the patient's *R*-wave.
4. A *pacemaker* is a device that supplies an electrical impulse to the heart to stimulate contraction.

8-11 Recapitulation

Now return to the objectives and self-evaluation questions at the beginning of the chapter and see how well you can answer them. If you cannot answer certain questions, then place a check mark next to each and review appropriate parts of the text. Next try to answer the following questions, using the same procedure. When you have answered all of the questions, solve the problems in Section 8-13.

8-12 Questions

8-1 The heart's natural pacemaker is the ________________.

8-2 The ________________ system of the heart controls synchronization of the heart's pumping by controlling the distribution of the pacemaker impulse.

8-3 The ________________ of the heart acts analogous to an electronic delay line.

8-4 List three cardiac arrhythmias that can be corrected by an electrical shock.

8-5 Which of the arrythmias of Question 8-4 cannot be corrected by an ac defibrillator?

8-6 Which of the arrhythmias in Question 8-4 is corrected by using the *cardioversion* mode?

8-7 The purpose of using electrical shock to correct arrhythmias is to ________________ the heart, so that all cells enter their refractory period together.

8-8 List four dc defibrillator waveforms.

8-9 A cardioverter is a dc defibrillator that is ________________to the ________________ feature of the patient's ECG waveform.

8-10 A cardioverter contains three stages: ________________, ________________, and ________________.

8-11 Testing of a dc defibrillator requires both measurement of ________________ and examination of the ________________ ________________.

8-12 What condition is the artificial pacemaker designed to correct?

8-13 What is the *minimum* energy required from a pacemaker?

8-14 Pacemaker output energy levels of ________________ or more may cause ventricular fibrillation.

8-15 Most implantable pacemakers are packaged in molded ________________, while some have an outer shield of ________________ to guard against radio frequency interference.

8-16 Two types of pacemaker lead wire are ________________ and ________________.

8-17 Two batteries often used in pacemakers are ________________ and ________________.

8-18 Describe *biphasic* pacemaker pulses in your own words.

8-19 External pacemakers usually have a heart rate range of ____________ to ____________ BPM.

8-20 The output current of external pacemakers may be varied from ________________ to ________________ mA.

8-21 List four general pacemaker types.

8-22 What is a principal advantage of the atrial-ventricular synchronized pacemaker? How is this advantage achieved?

8-23 The ________________ ________________ machine uses ________________ circulation of the blood to bypass the heart and pulmonary arteries while maintaining perfusion of the body.

8-24 A ________________ pump is used at each pump head on a heart-lung machine.

8-25 List the function of the five pump heads that might be used in a heart-lung machine.

8-26 List the major parts of the perfusion system in a heart-lung machine.

8-13 Problems

8-1 Find the energy stored in a 16-μF capacitor that is charged to 6000 V.

8-2 Calculate the energy stored in a 25-μF capacitor that is charged to 2800 V.

8-3 Find the potential across a 16-μF defibrillator capacitor that is used to store 400 W-s.

8-4 Find the open-circuit (i.e., no load) voltage across the patient paddle electrodes of a defibrillator using a 16-μF capacitor charged to 200 W-s (a) normally, (b) when the *discharge* button is pressed.

8-5 Calculate the energy delivered to a 50-Ω, noninductive load resistor by a Lown defibrillator charged to 300 W-s. Assume 25 Ω internal resistance.

Chapter 9

The Respiratory System

9-1 Objectives

1. Be able to introduce the biological principles underlying the respiratory system.
2. Be able to list and describe the gas laws.
3. Be able to describe internal (cellular) respiration.
4. Be able to describe external (pulmonary) respiration and pulmonary function (physical, chemical, and exchange of gases).
5. Be able to list the organs of respiration.
6. Be able to list and discuss mechanics of breathing and typical parameters of respiration (lung compliance, lung volumes/capacities, intraalveolar pressure, airway resistance, and intrathoracic pressure).
7. Be able to describe the regulation of respiration.
8. Be able to list and describe unbalanced and diseased states (hypoventilation, hyperventilation, dyspnea, hypercapnia, and hypoxia and apnea).
9. Be able to list the main threats of environmental pollution on the respiratory system.
10. Be able to list major measurements of pulmonary function.

9-2 Self-Evaluation Questions

These questions test your prior knowledge of the material in this chapter. Look for the answers as you read the text. After you have finished studying the chapter, try answering these questions and those at the end of the chapter (see Section 9-16).

1. List two purposes of the respiratory system.
2. State Boyle's, Charles's, Dalton's, and Henry's laws.
3. Describe internal respiration.
4. Which gas law describes gas exchange across cellular membranes?
5. Describe external respiration (inspiration and expiration) in terms of physical and chemical phenomena.
6. List and construct a block diagram of the organs of respiration and describe the function of each.
7. Describe the mechanics of breathing.
8. List and describe the major parameters of respiration.
9. Name and describe two control systems effecting the regulation of respiration.
10. State five unbalanced and diseased states of the respiratory system.
11. List and describe the major environmental threats to the respiratory system.

12. List the major measurements of pulmonary function.

9-3 The Human Respiratory System

The *human respiratory system* is critical to immediate survival. The main organs effecting pulmonary function are the lungs, and these are surprisingly delicate given that they interact with the external environment.

The purpose of the respiratory system includes a means of acquiring oxygen eventually distributed to bodily cells for metabolism and eliminating carbon dioxide accumulated from cellular metabolic activities.

Respiratory organs provide maximum *surface area* (alveolar spaces) for diffusion of O_2 and CO_2; the means of constantly renewing gases in contact with this surface (known as ventilation); the means of protecting surface membranes from harsh environmental factors, such as airborne toxic particles, microorganisms, drying, and extreme temperature; and a method of counteracting sudden shifts in pH of the body and body fluids.

9-4 Gas Laws

The understanding of respiratory function lies in the gas laws. These include *Boyle's law, Charles's law,* and *Henry's law.*

Boyle's law states that the volume of a gas varies inversely with the pressure if the temperature is held constant. The mathematical expression is:

$$\frac{V_2}{V_1} = \frac{P_1}{P_2} \tag{9-1}$$

where

V_1 = original volume
V_2 = final (new) volume
P_1 = original pressure
P_2 = final (new) pressure

EXAMPLE 9-1

Oxygen gas occupies 6 l under a pressure of 720 mm Hg. Calculate the volume of the same mass of gas at a pressure of 760 mm Hg (standard). The temperature remains constant.

SOLUTION

$\frac{V_2}{V_1} = \frac{P_1}{P_2}$ $\quad V_1 = 5\text{ l}$

$P_1 = 720\text{ mm Hg}$

$P_2 = 760\text{ mm Hg}$

$V_2 = \text{unknown}$

$$V_2 = \frac{V_1P_1}{P_2} = \frac{6\text{ l} \times 720\text{ mm Hg}}{760\text{ mm Hg}}$$

$V_2 = \mathbf{5.68\ l}$

Charles's law states that the volume of a gas is directly proportional to its absolute temperature if the pressure is held constant. The mathematical expression is:

$$\frac{V_2}{V1} = \frac{T_2}{T1} \tag{9-2}$$

where

V_1 = original volume
V_2 = final (new) volume
T_1 = original temperature
T_2 = final (new) temperature

EXAMPLE 9-2

Oxygen gas occupies 220 cc at 110°C. Calculate its volume at 0°C. The pressure remains constant.

SOLUTION

$\frac{V_2}{V_1} = \frac{T_2}{T_1}$ $\quad V_1 = 220\text{ cc}$

$V_2 = \text{unknown}$

$T_1 = 100°\text{C} + 273\text{ K} = 383\text{ K}$

$T_2 = 0°\text{C} + 273\text{ K} = 273\text{ K}$

$$V_2 = \frac{V_1T_2}{T_1} = \frac{220\text{ cc} \times 273\text{ K}}{383\text{ K}}$$

$V_2 = \mathbf{156.81\ ml}$

Dalton's law states that the total pressure exerted by a mixture of gases is equal to the sum of the partial pressures of the various gases. The partial pressure of a gas in a mixture is equal to the pressure of that gas if it were alone in the container. The mathematical expression is:

$$P_{total} = P_1 + P_2 + P_3 + \cdots + P_n \quad (9\text{-}3)$$

where

P_{total} = combined pressure
P_1 = partial pressure of first gas
P_n = partial pressure of nth gas

EXAMPLE 9-3

Atmosheric pressure at sea level is 760 mm Hg. Calculate the partial pressure of oxygen, nitrogen, and carbon dioxide.

SOLUTION

$P_{total} = P_{O_2} + P_{N_2} + P_{CO_2}$
P_{O_2} by volume = 20.96% × 760 mm Hg
= **159.30 mm Hg**
P_{N_2} = 79% × 760 mm Hg = **600.40 mm Hg**
P_{CO_2} = 0.04% × 760 mm Hg = **0.03 mm Hg**
P_{total} = 159.30 + 600.40 + 0.03
P_{total} = 760 mm Hg

Henry's law states that if the temperature is held constant, the quantity of a gas will go into solution proportional to the partial pressure of that gas. The gas with the greater partial pressure will have more mass in solution.

EXAMPLE 9-4

Oxygen has a partial pressure of 120 mm Hg and carbon dioxide has 40 mm Hg above a liquid with no initial dissolved gases. Which gas will have more mass in solution?

Answer: Oxygen will have more gas in solution due to its higher partial pressure.

9-5 Internal (Cellular) Respiration

Respiration is the interchange of gases between an organism and the medium in which it lives. *Internal respiration* is the exchange of gases between the bloodstream and nearby cells. Figure 9-1 shows a body cell exchanging gases with its environment (an adjacent capillary). During its passage

Figure 9-1 Internal respiration—exchange of O_2 and CO_2 between the capillary and the body cell.

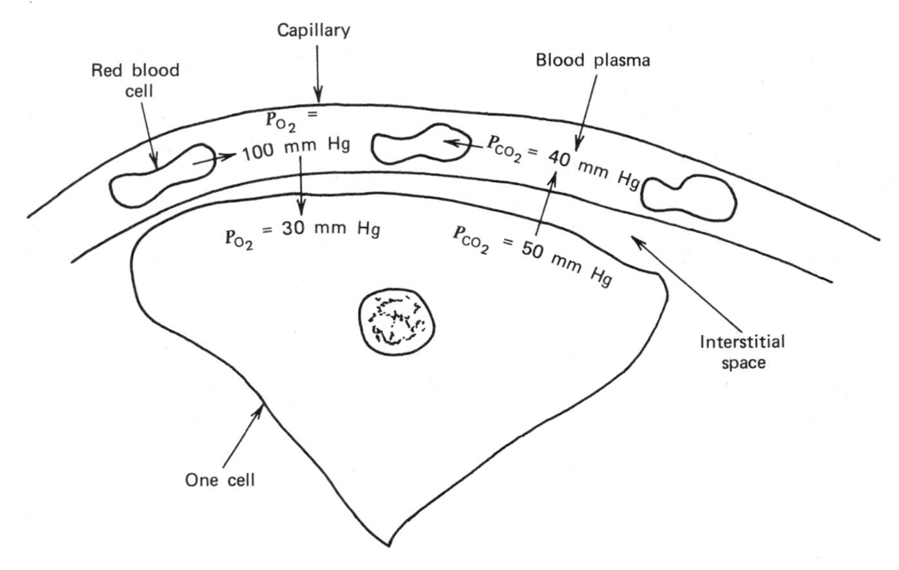

through the body tissues, blood gives up approximately 5 to 7 *volumes percent* (number of cc of a gas contained in 100 cc of blood) oxygen and takes up 4 to 6 volumes percent carbon dioxide. When the temperature or acidity increases, more O_2 is released to the tissues. Most of the 0_2 (95 percent) is carried by the red blood cell (RBC) hemoglobin and upon release to tissues, the RBCs still remain 75 percent saturated. However, the RBCs carry only about 30 percent CO_2, and the remainder is carried in the plasma. The exchange of O_2 and CO_2 is dependent upon *Dalton's law of partial pressures* (see Section 9-4). For example, when the partial pressure of O_2 (P_{O2}) is 100 mm Hg in a capillary surrounding a cell, and the internal cellular partial pressure is 30 mm Hg, O_2 moves into the cell. Similarly, CO_2 moves out of the cell into the capillary. It is the concentration of O_2 and CO_2 blood gases that is critical for sufficient gas exchange. The *oxygen dissociation curve* shows this condition in Figure 9-2. Notice that the arterial blood is 100 percent oxygenated containing oxyhemoglobin (oxygen chemically combined with hemoglobin within RBCs) and venous blood is normally 75 percent oxygenated.

Figure 9-2 Oxygen dissociation curve.

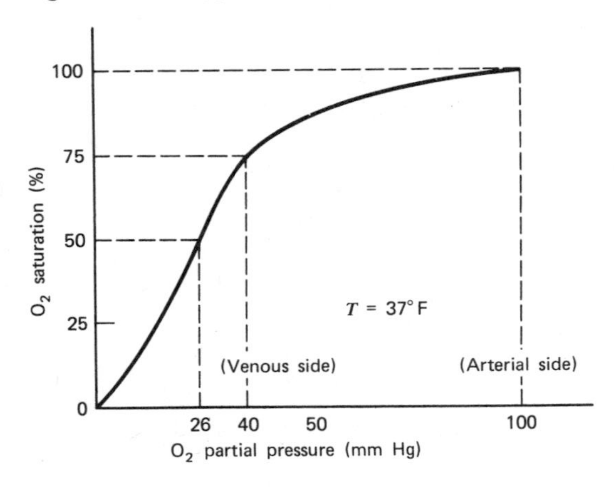

9-6 External (Lung) Respiration

External respiration is the exchange of gases between the lungs and bloodstream. Most biomedical respiratory apparatus is concerned with measuring or treating conditions of external respiration. Hence, the remainder of this chapter describes respiratory organs, physiology, and parameters.

Basically, external respiration includes *inspiration* (intake of air—79 percent N, 20.96 percent O, 0.04 percent CO_2) and *expiration* (exhaust of waste gases—79 percent N, 17 percent O, 4 percent CO_2). Pulmonary function involves *physical* processes (mechanics of breathing) and *chemical* processes (reaction of gases with liquids/exchange of gases). All of these can be explained through the laws of physics and chemistry (gas laws).

9-7 Organs of Respiration

The organs of respiration shown in Figure 9-3 are typically divided into the following:

1. Conducting division—containing thick walls (no gas exchange to capillaries) and including the nasal cavities, pharynx, larynx, trachea, bronchi, and bronchioles.

2. Respiratory division—containing thin walls (permitting gas exchange to blood capillaries) and including respiratory bronchioles, alveolar ducts, atria (space from which the alveoli of the sacs arise), and alveolar sacs.

Both divisions function through the muscles of respiration (diaphragm and intercostal/chest muscles), ribs, and sternum.

A block diagram of the passage of air from the nose to the capillaries is shown in Figure 9-4.

Specifically, the organs of respiration include the following:

1. *Nose and nasal cavities*—facial organ serves for sense of olfaction (smell) and to warm, moisten, and filter air for respiratory tract.

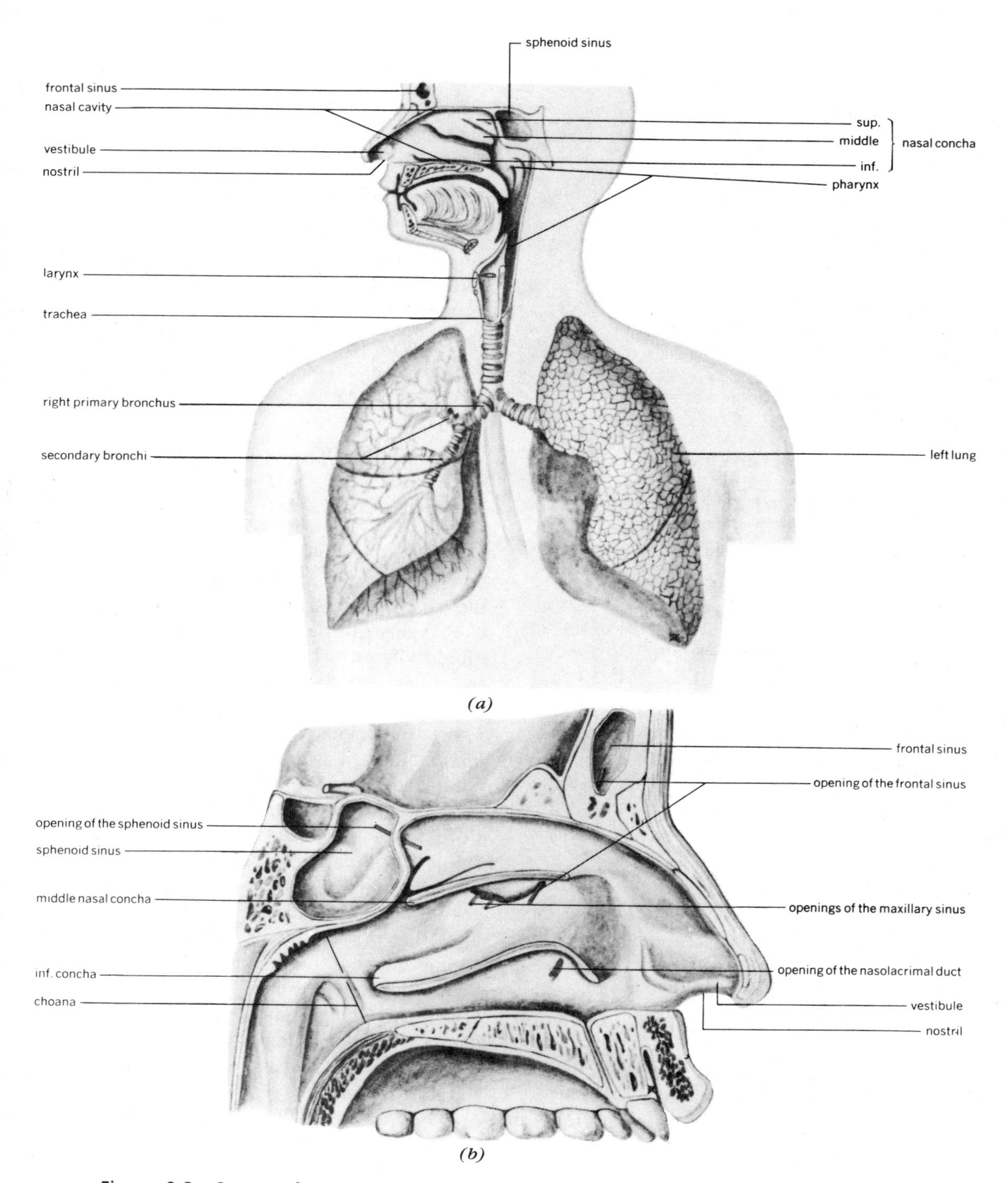

Figure 9-3 Organs of respiration. (*a*) Organs. (*b*) Left nasal cavity. (From *Human Anatomy and Physiology,* 2nd edition, by James E. Crouch, Ph.D. and J. Robert McClintic, Ph.D. John Wiley & Sons, New York, 1976. Used by permission.)

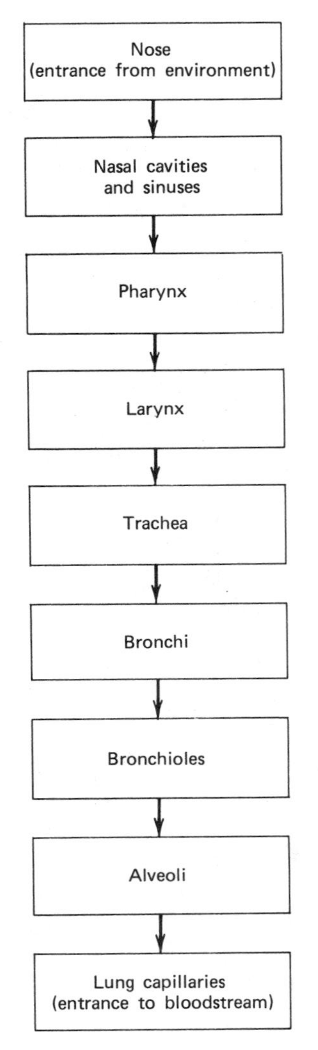

Figure 9-4 Block diagram of air pathway from nose to lung capillaries.

2. *Pharynx (throat)*—there are three divisions:

(a) *Nasopharynx* (near nose) including adenoids (mass of lymphatic tissue).

(b) *Oropharynx* including tonsils (mass of lymphatic tissue).

(c) *Hypopharynx or laryngopharynx.*

3. *Larynx (voice box)*—houses the vocal cords that vibrate when air is forced upward.

4. *Trachea (windpipe)*—vertical tube kept open by rings of cartilage to allow passage of air to and from the lungs.

5. *Bronchi*—two branches of trachea that descend into each *lung.*

6. *Bronchioles*—smallest of the bronchial branches that form a network of tubes throughout the lungs.

7. *Alveoli (air sacs)*—air cavities (one cell thick) at the end of the bronchioles that trap air and allow exchange of gases to the blood capillaries.

8. *Lung capillaries*—thin tubes carrying blood that surround the alveoli and allow exchange of gases.

The *lungs* (shown in Figure 9-5) consist of two cone-shaped spongy organs that contain the alveoli (air sacs) that trap air for gas exchange with the blood. Three lobes separated by fissures make up the right lung, and two comprise the left. An indentation appears in the left lung (slightly larger than the right) to provide room for the heart. The *hilum,* through which respiratory structures enter and leave the lung (lymphatics, blood vessels, and bronchi), is located posteriorly. Both lungs lie within two lateral (side by side) *pleural cavities* of the thorax. A serous (moist) membrane, the *visceral pleura,* covers the lung surface and the *parietal pleura* lines the thoracic cavity. These membranes contact one another and attach at the root of the lung. This fluid-lined potential space between two membranes accounts for easy slippage between lung and chest wall during breathing. The cohesive effect helps to keep the lung expanded.

Blood in need of oxygenation enters both lungs via the *pulmonary arteries* (from the heart's right ventricle). Oxygenated blood

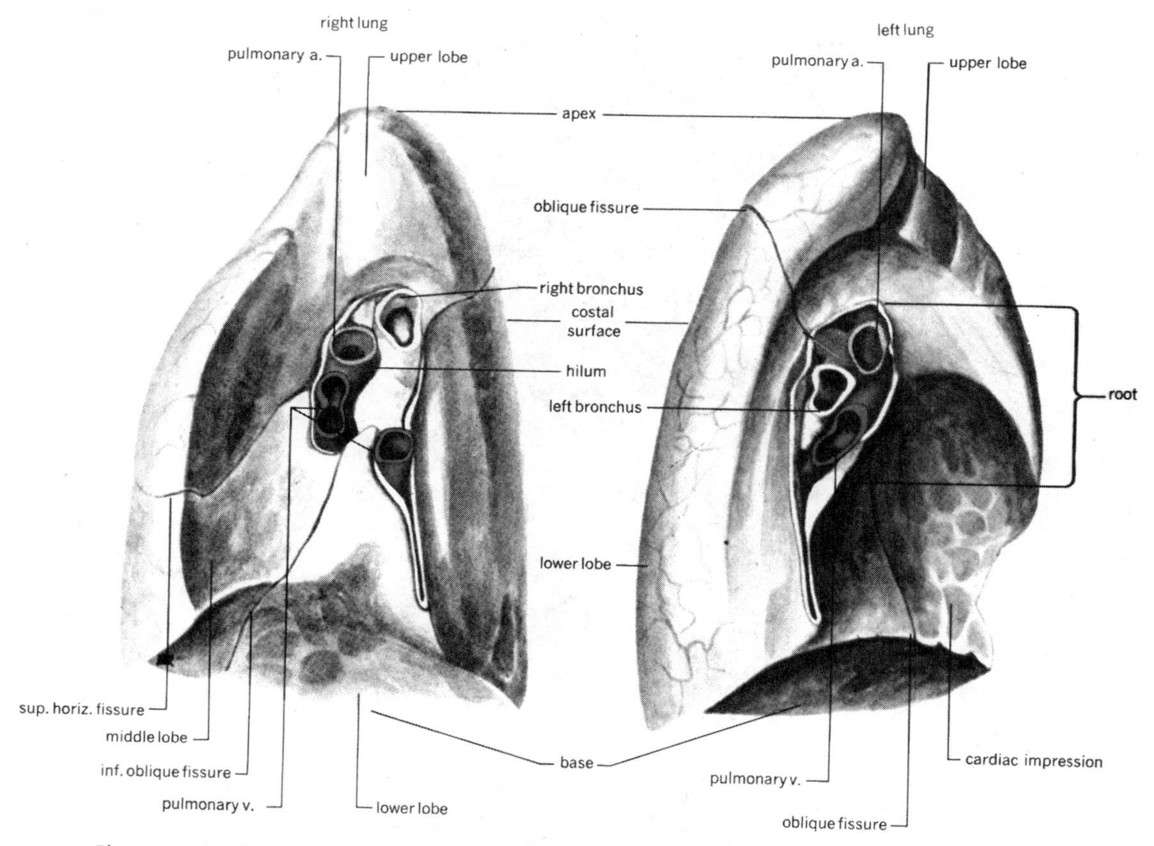

Figure 9-5 The lungs. (From *Human Anatomy and Physiology,* 2nd edition, by James E. Crouch, Ph.D. and J. Robert McClintic, Ph.D. John Wiley & Sons, New York, 1976. Used by permission.)

leaves the lungs through the *pulmonary veins* (to the heart's left atrium).

Air inspired through the nose, passed through the trachea, and branched to the bronchi eventually enters terminal bronchioles that separate into *respiratory bronchioles*. There, *alveoli or air sacs,* each about 0.2 mm in diameter, are attached as shown in Figure 9-6. An estimated 300 million alveoli are contained in the lung generating about 70 m^3 of surface area (average tennis court). This gives rise to a total lung capacity of 3.6 to 9.4 l in the adult male and 2.5 to 6.9 in the normal female.

9-8 Mechanics of Breathing

The *mechanics of breathing (respiration)* involve muscles that change the volume of the thoracic cavity to generate *inspiration* (intake) and *expiration* (exhaust).

The two sets of muscles involved are the *diaphragm*—the wall separating the abdomen from the thoracic (chest) cavity that

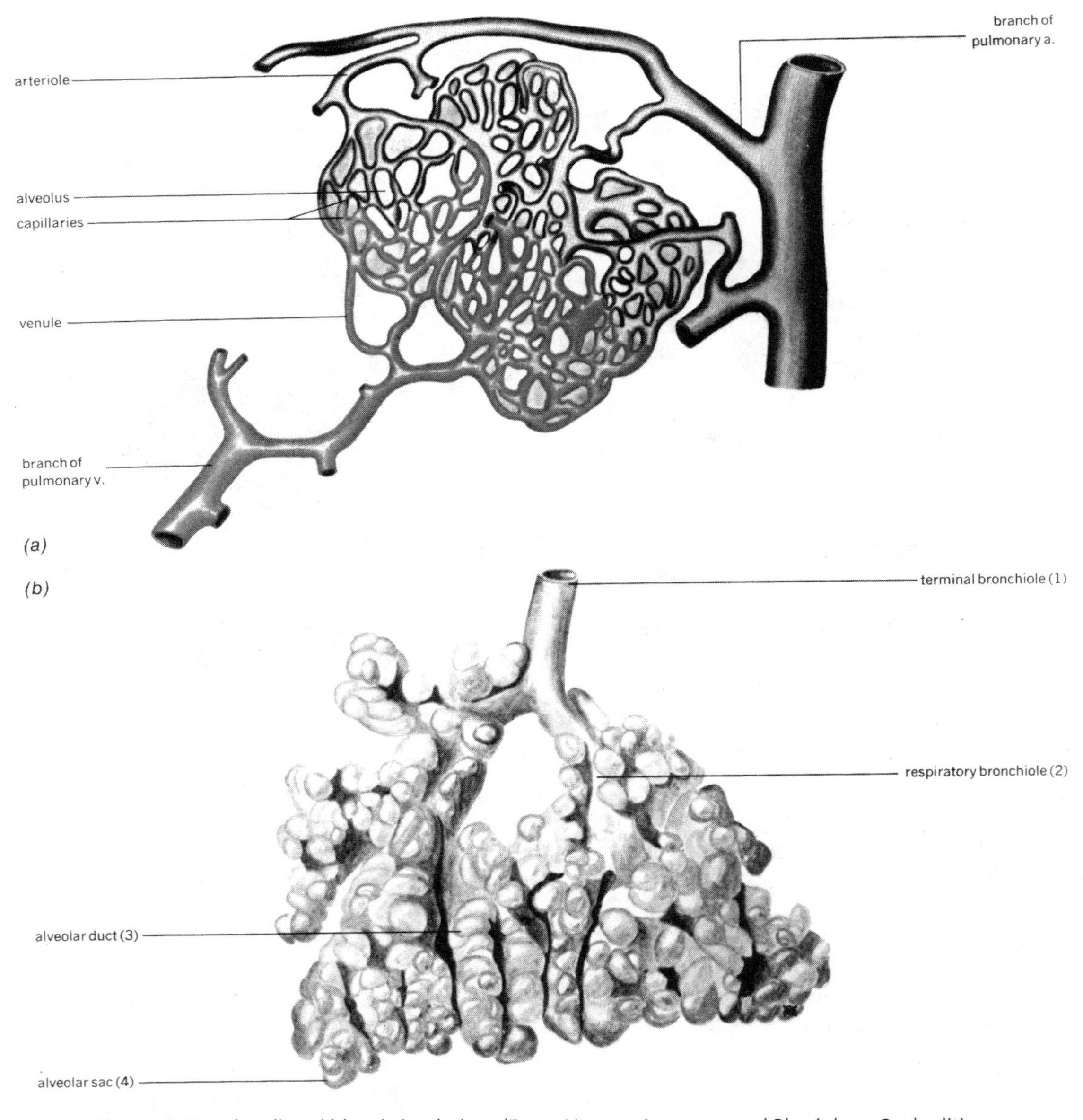

Figure 9-6 Alveoli and blood circulation. (From *Human Anatomy and Physiology,* 2nd edition, by James E. Crouch, Ph.D. and J. Robert McClintic, Ph.D. John Wiley & Sons, New York, 1976. Used by permission.)

moves up and down—and the *intercostal muscles*—muscles surrounding the thoracic cavity that move the rib cage in and out.

As shown in Figure 9-7, *inspiration* results from contraction of the diaphragm (downward movement) and intercostal muscles (rib cage swings up and outward). The enlarged cavity housing the lungs undergoes a pressure reduction (−3 mm Hg) with respect to that outside the body. Since the lungs are passive (no muscle tissue), they expand due to the positive external pressure. If external environmental pressure is 760 mm Hg at sea level, the lung pressure is 757 mm Hg upon inspiration. It is the closed nature of the thoracic chamber that allows air to enter the lungs from one external opening.

Figure 9-7 Expiration and inspiration (thoracic dimension changes.) (From *Human Anatomy and Physiology,* 2nd edition, by James E. Crouch, Ph.D. and J. Robert McClintic, Ph.D. John Wiley & Sons, New York, 1976. Used by permission.)

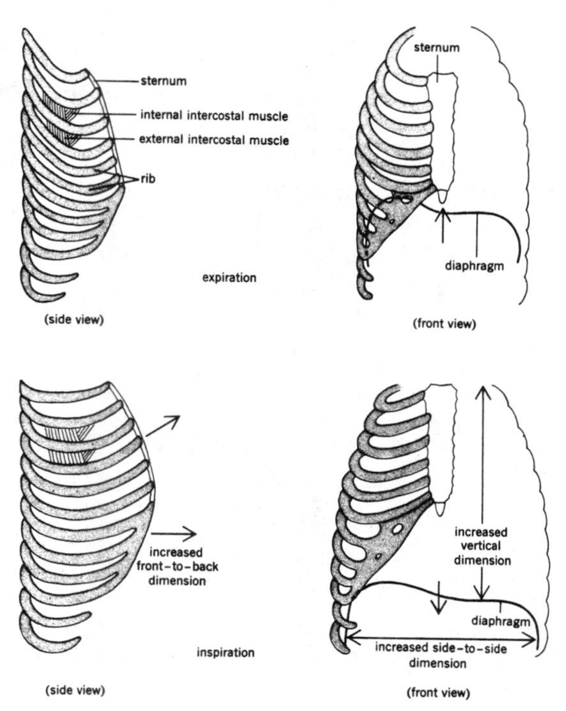

Expiration (Figure 9-7) results from the relaxation of the diaphragm (upward movement) and intercostal muscles (inward and downward). The elastic recoil of the lungs creates a higher-than-atmospheric intrapulmonic pressure (+3 mm Hg) that forces air out of the lungs.

Mechanical and electrical analogies have been constructed to demonstrate and study respiratory function. One uses masses sliding along surfaces restrained by springs. The elasticity of the spring indicates lung compliance (ability to stretch). Another uses a volume-pressure piston pumping system. A bellows reflects compliance and bronchial airway resistance. A third is constructed of electrical components. A set of capacitors represents bronchial resistance and gas compressibility, while sets of resistors and capacitors give lung tissue elasticity. The muscles of respiration are simulated by ac generators.

9-9 Parameters of Respiration

The parameters of respiration are measurements that indicate the state of respiratory function, including: lung volumes and capacities, airway resistance, lung compliance and elasticity, and intrathoracic pressure.

Only a portion of the air entering the respiratory system actually reaches the alveoli. The volume of air that is not available for gas exchange with the blood resides in the conducting spaces. This is known as *dead air* and fills *dead space* consisting of 150 ml. Due to uneven distribution of ventilation (exhaust) and perfusion (spread), *"wasted ventilation"* results in *"wasted blood flow."* The total dead space results in less than 30 percent of the total volume.

Important volumes to consider are shown in Figure 9-8. They are for a standard 70 kg-

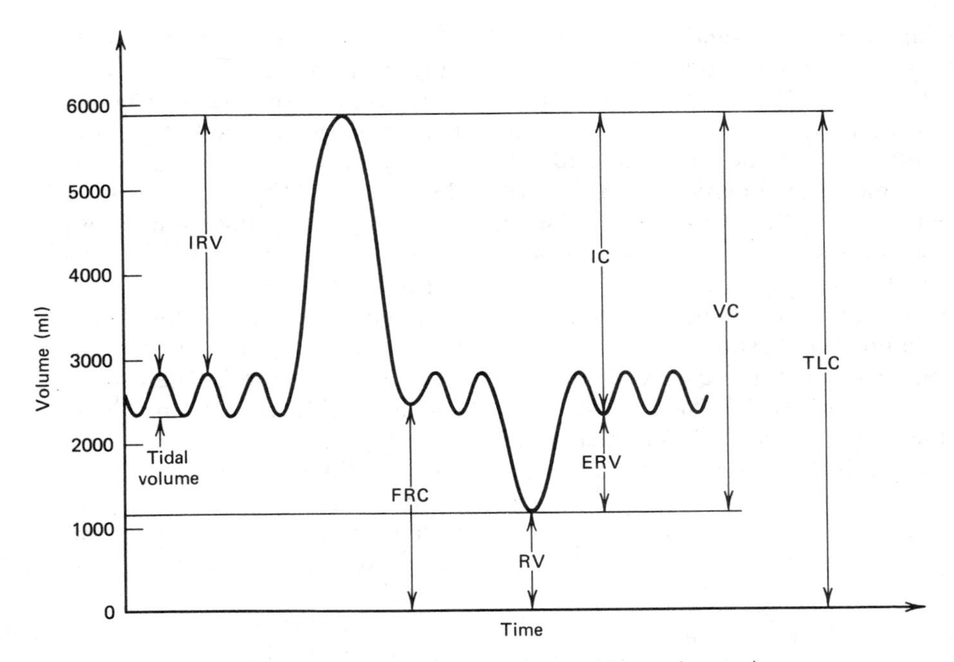

Figure 9-8 Lung volumes and capacities—reiterated.

male breathing at rest. The *tidal volume (TV)*—500 ml—is the depth of breathing or the volume of gas inspired or expired during each respiratory cycle. *Inspiratory reserve volume (IRV)*—3600 ml—is the maximal amount of gas that can be inspired from the end-inspiratory position (extra inspiration from the high peak tidal volume). *Expiratory reserve volume (ERV)*—1200 ml—is the maximal amount of gas that can be expired from end-expiratory level (extra expiration from the low peak tidal volume). *Residual volume (RV)*—1200 ml—is the amount of gas remaining in the lungs at the end of maximal expiration (amount that cannot be squeezed out of the lung).

Important capacities (addition of various volumes) to consider are shown in Figure 9-7. *Total lung capacity (TLC)*—6000 ml—is the amount of gas contained in the lungs at the end of maximal inspiration. *Vital capacity (VC)*—4800 ml—is the maximal amount of gas that can be expelled from the lungs by forceful effort from maximal inspiration. *Inspiratory capacity (IC)*—3600 ml—is the maximal amount of gas that can be inspired from the resting expiratory level. *Functional residual capacity (FRC)*—2400 ml—is the amount of gas remaining in the lungs at the resting expiratory level (end-expiratory position is used as a base because it varies less than the end-inspiratory state).

It is interesting to note the amount of volume reserve available with respect to total lung volume. This equals:

$$\frac{RV}{TLC} \times 100 = 20 \text{ percent in 25-year-old (70 kg) male} \tag{9-4}$$

$$= 40 \text{ percent in 55-year-old (70 kg) male}$$

The *work of breathing* involves airway resistance, lung compliance, and lung elasticity.

Airway resistance relates to the ease with

which air flows through tubular respiratory structures. Higher resistances occur in smaller tubes, such as the bronchioles and alveoli, that have not emptied properly.

Lung compliance is the ability of the alveoli and lung tissue to expand on inspiration. The lungs are passive, but they should stretch easily to insure sufficient intake of air.

Lung elasticity is the ability of the lung's elastic tissues to recoil during expiration. The lungs should return to their rest (unstretched) state easily to insure sufficient exhaust (ventilation) of gas.

Intrathoracic pressure is the positive and negative pressure occurring within the thoracic cavity. These are critical to proper inspiration (negative internal pressure) and expiration (positive internal pressure).

Intraalveolar pressure is of importance in maintaining proper respiration and gas exchange to and from the blood.

9-10 Regulation of Respiration

Respiration rate and depth are controlled by two things: the nervous system and chemical concentration of CO_2 in the blood.

Respiration results from *involuntary neuronal activity* (see Chapter 12, the human nervous system) modified by *chemical influences*. Voluntary control is also possible but is limited by internal body homeostasis. For example, voluntary deep breathing for a prolonged time may result in temporary unconsciousness to permit blood chemistry (pH level) to return to normal.

Respiratory centers in the brain are located within the *medulla and pons* of the brain stem. Nerve cells from the brain send out streams of impulses which stimulate the *diaphragm* and *intercostal muscles* to contract and effect *inspiration. Pneumotaxic centers* in the pons receive impulses that rise to a maximum at inspiration peak from *inspiratory centers* in the medulla. Messages relayed to the *expiratory centers* (located dorsally to inspiratory centers in medulla) initiate expiration by sending out streams of pulses that inhibit inspiration. Muscles of inspiration relax, and *expiration* follows passively. This *feedback system* maintains *rhythmic breathing rate and depth (tidal volume).*

Respiratory activity is also affected by the chemistry and temperature of the blood passing through the brain. Changes in concentration of carbon dioxide in the blood changes respiratory rate. The *acid base balance of the blood* (normally pH of 7.4) arises from the following chemical reaction of carbon dioxide waste from cells with water in blood plasma.

$$CO_2 + H_2O \rightarrow H_2CO_3 \rightarrow H^+ + HCO_3^- \quad (9\text{-}5)$$

A *reflex mechanism* regulates breathing as shown in Figure 9-9. For example, if the accumulation of blood CO_2 from body cells should increase, more H^+ and HCO_3^- ions will be produced, leading to stimulation of brain respiratory center. This increases the depth and, eventually, rate of breathing. The CO_2 level in the blood drops due to ventilation, and brain respiratory centers decrease breathing rate. The constitutes a *negative feedback loop*, and blood pH is maintained within normal limits (7.36 to 7.44). Stimulation and inhibition act through *baroreceptors* (stretch sensors in the lung), O_2 *chemoreceptors* (cells in the aorta), and *respiratory centers* (cells in the brain) to insure the balance of the partial pressure of O_2 and CO_2 in the blood.

9-11 Unbalanced and Diseased States

Unbalanced states of the respiratory system include the following: *Hyperventilation* is alveolar ventilation in excess of metabolic needs for CO_2 removal. Partial pressure of CO_2 in blood falls below 40 mm Hg. This

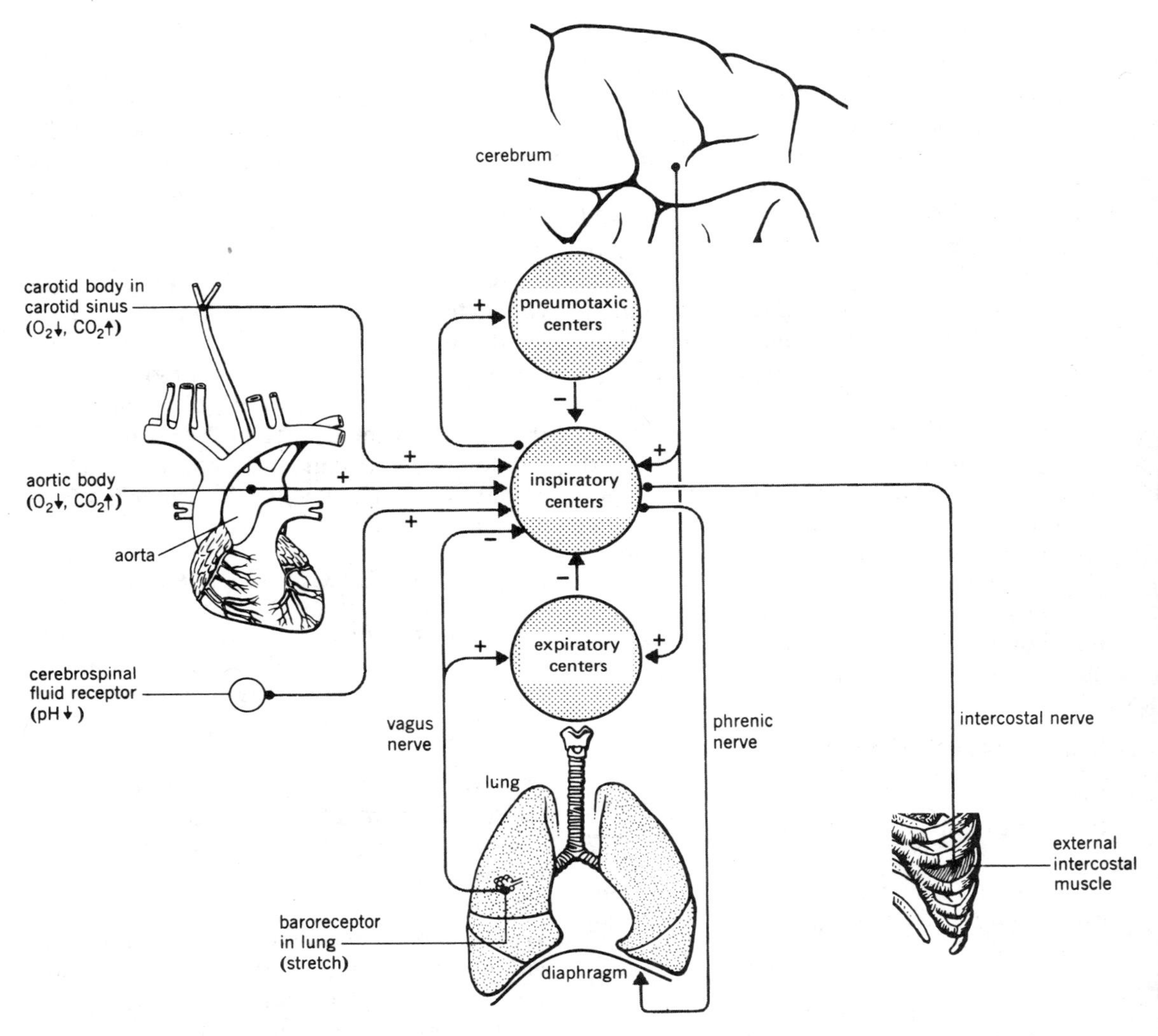

Figure 9-9 Breathing reflex mechanisms. (From *Human Anatomy and Physiology,* 2nd edition, by James E. Crouch, Ph.D. and J. Robert McClintic, Ph.D. John Wiley & Sons, New York, 1976. Used by permission.)

results from voluntary or involuntary rapid or deep breathing ridding the blood of excessive CO_2. *Hypoventilation* is alveolar ventilation inadequate for CO_2 removal. Partial pressure of CO_2 rises above 40 mm Hg. This results from voluntary or involuntary shallow breathing causing excessive buildup of CO_2 in the blood.

Diseased states of the respiratory system include the following: *Hypoxia* is low oxygen content in the blood resulting in excessively reduced partial pressure of O_2 in the blood to the point of death. This results from damage to respiratory neurons, alveolar damage, respiratory tissue damage, or inadequate O_2 transport. *Apnea* is the cessation of breathing, usually temporary. This results from reduced stimulus to respiratory centers or brain center damage. *Hyperpnea* is the increased tidal volume with or without increased breathing rate, which reduces alveolar and blood partial pressure of O_2. *Dyspnea* is labored breathing resulting from acidosis (low blood pH), pneumonia, cardiac failure, hemorrhage, or fever. *Polypnea* (tachypnea) is an accelerated breathing rate without increase in breathing depth resulting from fever or hypoxia. *Hypercapnia* is decreased ventilation (excessively low partial pressure of CO_2 in the blood) resulting from central nervous system disorders, disease of nerves or respiratory muscles, metabolic disorders, or respiratory obstruction.

9-12 Environmental Threats to the Respiratory System

The respiratory system, particularly the lungs, must withstand the insults of the external *environment*. Smoking cigarettes and inhaling gases, fibers, and liquids from occupational environments cause lung tissue damage. The National Institutes of Health has issued programs (since 1970) to manage the detection and prevention of respiratory disease resulting from environmental factors.

Mortality rates among cigarette smokers average about 70 percent greater than among nonsmokers. Smog formed through chemical reactions of sulfur oxides and hydrocarbons with ozone in the atmosphere causes respiratory damage. Acids that form deep within the lung alveoli destroy respiratory membranes and cause swelling. Particulate matter such as smoke and dust embeds within lung surfaces and causes insufficient oxygenation. Carbon monoxide (250 million metric tons annually) and other machine exhaust gases cause hypoxia and, ultimately, death.

Environmental impact assessment has become an essential part of industrial advance. Air quality took on a legal aspect with the introduction of the *Clean Air Act* in 1970. This act establishes quality air guidelines that must be met by all industry in 1982. Part of assessing air quality deterioration includes *identifying* pollutants, describing existing air quality levels, summarizing meteorological data, gathering current air quality standards, and determining air pollution caused by new industrial construction.

Disease resulting from organic or environmental factors often requires *artificial respiratory ventilation*.

9-13 Major Measurements of Pulmonary Function

Given all of the respiratory problems that can exist (organic and environmentally induced), *measurement of pulmonary function* is essential. This includes the following items: *Maximum voluntary ventilation* (*MVV*) is the deep rapid breathing measurement on a spirometer (respiratory volume measurement device). *Forced expiratory volume*$_1$ *timed* (*FEV*$_1$) is the rapid inhalation-exhalation as measured on a spirometer. *Maximum expiratory flow rate* (*MEFR*) is the forcible inhalation-exhalation measured on a pneumotachometer (flowmeter). *Intraalveolar pressure* is pressure in the alveolar sac as measured by a body plethysmograph (pressure recording device). *Blood gas measurement* is partial pressure of O_2 and CO_2 in the blood as measured by a blood gas analyzer. *Acid-base balance measurement* is the quantity of CO_2 in the blood as measured by a pH meter.

9-14 Summary

1. The human respiratory system is critical to immediate survival. This system provides a means of *acquisition of oxygen* and *elimination of carbon dioxide*.
2. The development of the respiratory system begins with the nose and nasal

passages and ends with the laryngotracheal tube and alveoli within the lungs.

3. *Major gas laws* of concern in the study of respiration are Boyle's law, Charles's law, Dalton's law, and Henry's law. Dalton's law of partial pressures explains gas exchange between lungs and blood and blood and body cells.
4. *Internal respiration* is the exchange of gases between the bloodstream and nearby cells.
5. *External respiration* is the exchange of gases between the lungs and bloodstream. This involves the process of *inspiration* (intake of air) and *expiration* (exhaust of gases). Physical (mechanics of breathing) and chemical (gas exchange) phenomena underlie respiration.
6. The organs of respiration are separated into a *conducting* and *respiratory* division and include the nose, nasal cavities, pharynx (throat), larynx (voice box), trachea (windpipe), lungs, bronchi, bronchioles, alveoli, and lung capillaries.
7. The *mechanics of breathing* involve the diaphragm and intercostal (rib cage) muscles. Through these, the thoracic cavity (chest) enlarges (negative internal pressure generated on inspiration) and relaxes (positive internal pressure generated on expiration).
8. *Parameters of respiration* include lung volumes/capacities, compliance/elasticity, airway resistance, and intrathoracic pressure.
9. *Regulation of respiration* rate and depth depends upon *nervous system* and *chemical* controls. Carbon dioxide concentration in the bloodstream indicates ventilation level to brain respiratory centers. Negative feedback loops exist between lungs, heart, and brain centers to maintain constant blood pH (acid-base balance).
10. *Unbalanced and diseased states* of the respiratory system are hyperventilation, hypoventilation, hypoxia, apnea, hyperpnea, dyspnea, polypnea, and hypercapnia.
11. *Environmental threats* to the respiratory system include smoking cigarettes and inhaling liquids, vapors, fumes, smoke, and dust from occupational environments.
12. *Major measurements* of pulmonary function are maximum voluntary ventilation (MVV), forced expiratory volume_1 timed (FEV_1), maximum expiratory flow rate (MEFR), intraalveolar pressure, blood and gas measurement, and acid-base balance (pH) measurement.

9-15 Recapitulation

Now return to the objectives and self-evaluation questions at the beginning of the chapter and see how well you can answer them. If you cannot answer certain questions, place a check mark next to each and review appropriate parts of the text. Next, try to answer the questions in Section 9-16 using the same procedure. When you have answered all of the questions, solve the problems in Section 9-17.

9-16 Questions

9-1 Two purposes of the respiratory system are ______________ of ______________ and ______________ of ______________ ______________.

9-2 The respiratory system counteracts sudden changes in blood ______________.

9-3 Boyle's law states that the volume of a gas varies inversely with the ______________.

9-4 Charles's law states that the volume of a gas is directly proportional to its absolute ______________.

9-5 Dalton's law states that the total pressure exerted by a gas mixture equals the sum of the ______________ ______________.

9-6 Henry's law states that if the ____________ remains constant, a quantity of gas that goes into solution is proportional to its ______________.

9-7 Internal respiration is the ______________ of ______________ between the ________ and nearby ______________.

9-8 Blood is ______________ percent oxygenated in the ______________ vein and ______________ percent oxygenated in the ______________ artery.

9-9 The term oxyhemoglobin refers to oxygenated ______________.

9-10 What is the percentage oxygen, nitrogen, and carbon dioxide in inspired air and expired gases?

9-11 List eight major organs of respiration.

9-12 What is the approximate surface area of the internal lung alveoli?

9-13 Mechanical motion of breathing occurs through which two sets of muscles?

9-14 State the following approximate volumes of these lung parameters: tidal volume (TV), residual volume (RV), total lung capacity (TLC), and vital capacity (VC).

9-15 Define *lung compliance*.

9-16 What two major factors control the rate and depth of respiration?

9-17 Do positive or negative feedback systems control respiration?

9-18 State the normal regulated pH range of the blood.

9-19 Define *unbalanced states of hyperventilation* and *hypoventilation*.

9-20 Describe the respiratory diseases of hypoxia, apnea, and hypercapnia.

9-21 How does environmental pollution affect respiratory function?

9-22 State four major measurements of pulmonary function.

9-17 Problems

9-1 Given a 4-l container of oxygen gas under a pressure of 780 mm Hg, calculate the volume of the same mass of gas at a pressure of 760 mm Hg.

9-2 Given 200 cc of oxygen gas at 100°C, calculate the volume at 0°C if the pressure remains constant.

9-3 Given expiratory gas (79 percent N, 17 percent O, and 4 percent CO_2) at 40 mm Hg, calculate the partial pressure of oxygen, nitrogen, and carbon dioxide.

9-18 References

1. Crouch, James E. and Robert J. McClintic, *Human Anatomy and Physiology*, Wiley (New York, 1976).
2. Guyton, Arthur C., *Textbook of Medical Physiology*, W. B. Saunders Co. (Philadelphia, 1976).
3. McNaught, Ann B. and Robin Challander, *Illustrated Physiology,* Churchill Livingstone (New York, 1975).
4. Koa, Frederick F., *An Introduction to Respiratory Physiology*, American Elsevier Publishing Co., Inc. (New York, 1972).
5. Peters, Richard M., *The Mechanical Basis of Respiration*, Little, Brown and Co. (Boston, Mass., 1969).
6. *A Multicompartment Analysis of the Lung, 12* (4): 405–414 (July 1974).
7. Moore, Francis D., John H. Lyons, Jr., Ellison C. Pierce, Jr., et al., *Post-Traumatic Pulmonary Insufficiency*, W. B. Saunders Co. (Philadelphia, 1969).
8. Cole, R. B., *Essentials of Respiratory Disease*, J. B. Lippincott Co. (Philadelphia, 1972).
9. Lee, Douglas H. K., *Environmental Factors in Respiratory Disease*, Academic Press (New York, 1972).
10. Hodges, Laurent, *Envrionmental Pollution—A Survey Emphasizing Physical and Chemical Principles*, Holt, Rinehart and Winston (New York, 1973).
11. Canter, Larry W., *Environmental Impact Assessment*, McGraw-Hill (New York, 1977).
12. *Environmental Quality*, Annual Reports, Council on Environmental Quality, U.S. Government Printing Office (1975–1978).

13. Comroe, Julius J., Robert E. Forster, Arthur B. Doubis, et al., *The Lung—Clinical Physiology and Pulmonary Tests*, Year Book Medical Publishing, Inc. (Chicago, Ill., 1962).

14. Hunsinger, Dorris, Karl J. Lisnerski, Jerome J. Maurize, et al., *Respiratory Technology, A Procedural Manual*, Reston Publishing Co., Inc. (Reston, Va., 1976).

Chapter 10

Respiratory Instrumentation

10-1 Objectives

1. Be able to list the principal pulmonary parameters measured.
2. Be able to describe the operation of various respiratory transducers.
3. Be able to list and describe the major instruments used in respiratory system measurements.

10-2 Self-Evaluation Questions

These questions test your prior knowledge of the material in this chapter. Look for the answers as you read the text. After you have finished studying the chapter, try answering these questions and those at the end of the chapter (Section 10-9).

1. A *pneumotachometer* is used to measure ____________.
2. Define *inspiratory reserve volume*.
3. Flow volume may be measured by using a ____________.
4. Relationships between the various respiratory volumes are called ____________ ____________.
5. What is an *apnea* alarm?

10-3 Respiratory System Measurements

The respiratory system is responsible for bringing oxygen into the body and for discharging waste carbon dioxide from the body. The basic principles of this system were covered in Chapter 9 and will not be repeated here except where added emphasis is desirable.

There are actually several different transducers used in respiratory measurements, although only a few different types of measurement are made. One class of instruments known as *pneumographs* is used to *detect* respiration, but these instruments do not deliver quantitative data about the system. These devices are, however, often paired with a *pneumotachometer* (i.e., respiration rate meter) to perform *monitoring* jobs in intensive care units.

Instruments devised to quantitatively measure lung volumes are known as *spirometers;* both mechanical and electronic models are available. The measurements are as follows:

1. *Tidal volume* refers to the volume of air flowing while breathing *normally*, at rest. In the 70-kg "standard" male subject this volume reaches approximately 500 ml.

2. *Inspiratory reserve volume* (IRV) is the volume of air that may be inspired during exercise over and above the tidal volume.

This value is between 2000 and 3000 ml in the standard male subject.

3. *Expiratory reserve volume* (ERV) is the volume of gas that may be forcibly expired after a normal expiration. In the standard male subject this volume is approximately 1100 ml.

4. *Residual volume* is the volume of gas remaining in the lungs after a maximal forced expiration. In the standard male subject this is about 1200 ml. Even in a collapsed lung there is a *minimal volume* of 500 to 600 ml.

5. *Minute volume* is the volume of air breathed normally for a period of 1 min.

Relationships exist between the various volumes, and these are expressed as various *pulmonary capacities*. The more commonly used capacities are:

1. *Vital capacity* (VC). The *sum* of the inspiratory reserve volume. The VC is also defined as the volume of gas expired during a maximal forced expiration following a maximal forced inspiration.

2. *Functional residual capacity* (FRC). The sum of expiratory reserve volume and residual volume.

3. *Inspiratory capacity* (IC). The sum of the tidal volume and the inspiratory reserve volume.

4. *Total lung capacity* (TLC). The sum of all four volumes as defined above.

Blood gases and CO_2 expired at the end of a normal tidal volume are also considered to be respiratory measurements, but only the latter are considered in this chapter because the others are also considered in Chapter 16.

10-4 Respiratory Transducers and Instruments

Respiratory instruments, like most other measurement instruments, tend to be little more than extensions of the transducers used to acquire the data from the subject. In some cases no more than a simple dc amplifier is needed. It is, therefore, practically impossible to distinguish the transducers from the instruments. In this section we discuss the transducers of respiratory instrumentation along with the instruments used to process the signals.

An *impedance pneumograph* uses the fact that the ac impedance across the chest of a subject *changes* as respiration occurs. This technique is used in many neonatal *respiration monitors* and *apnea alarms*. Figure 10-1*a* shows the block diagram for an impedance pneumograph. A low-voltage, 50- to 500-kHz ac signal is applied to the chest of the patient through surface electrodes of the same type as used in ECG monitoring. In fact, many of these monitors are also ECG monitors, using a common set of electrodes and a single pair of lead wires. High-value fixed resistors connected in series with each electrode create a constant ac current source. The signal voltage applied to the differential ac amplifier is the voltage drop across the resistance representing the patient's thoracic impedance (see Figure 10-1*b*):

$$E_o = I\,(R \pm \Delta R) \qquad (10\text{-}1)$$

where

E_o is the output potential in *volts* (V)
I is the current through the chest in *amperes* (A)[1]
R is the chest impedance *without* respiration, in ohms (Ω)
ΔR is the change of chest impedance due to respiration, in ohms (Δ)

The current passed through the patient's chest is very small and is nearly constant without respiration because the source volt-

[1] Although the ampere unit is used in Equation 10-1, the actual current would be in the microampere range.

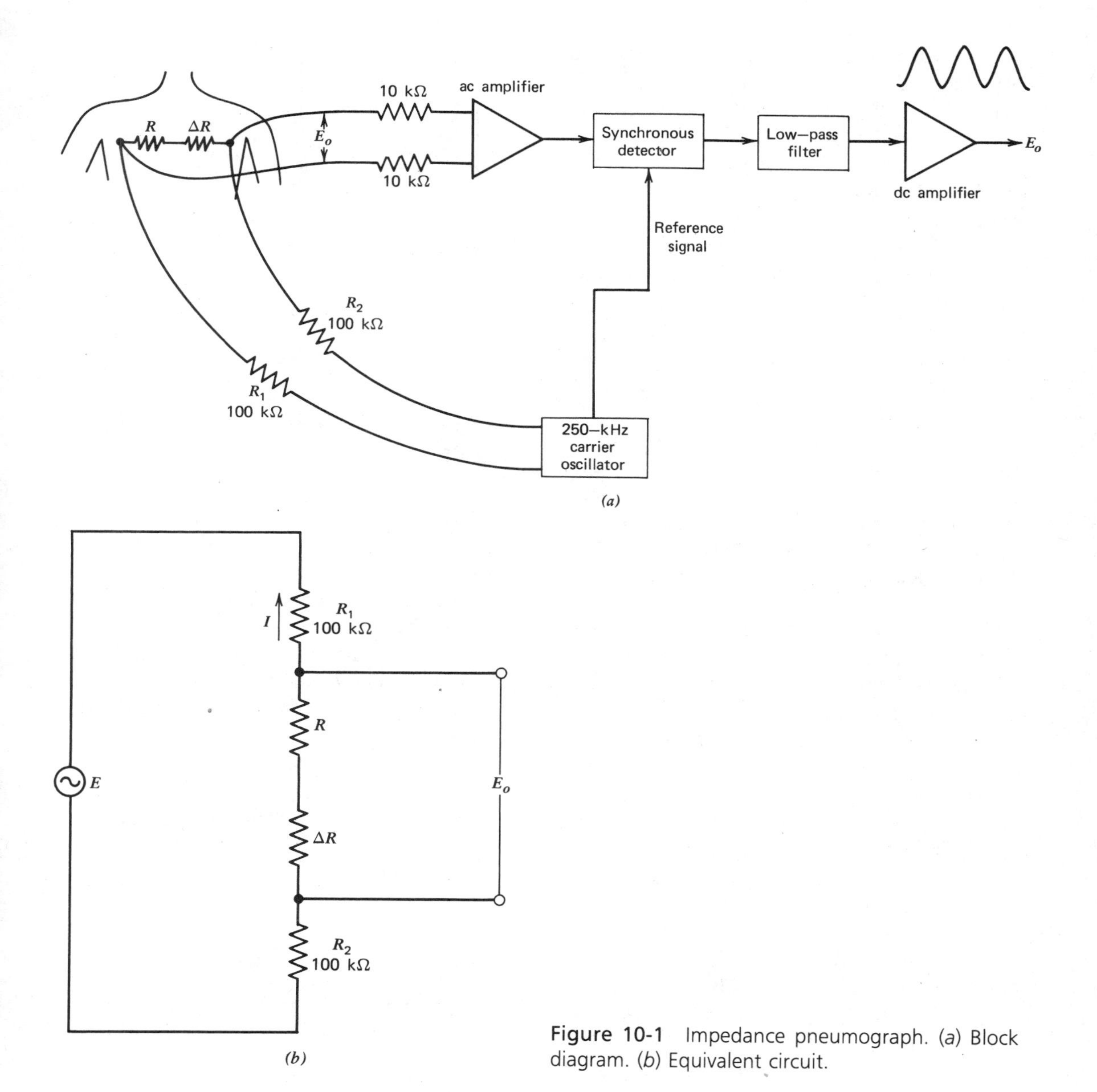

Figure 10-1 Impedance pneumograph. (*a*) Block diagram. (*b*) Equivalent circuit.

age E is constant and the term ΔR is very small with respect to the sum $R_1 + R_2 + R_3$.

The signal E_o is amplified and then applied to a synchronous AM detector; the respiration waveform is contained within amplitude variations in E_o due to ΔR. A low-pass filter following the detector removes residual carrier signal, and a dc amplifier scales the output waveform to the level required by the display device or pneumotachometer following this circuit.

The output of the impedance pneumograph contains only *rate* data and the *existence* of respiration; hence its use in monitoring and apnea alarm devices.

These are two types of pneumograph that use piezoresistive strain gage transducers to detect respiration. One type, now largely obsolete, is the *mercury strain gage*. In this type of instrument a very thin elastic tube filled with mercury (Hg) is stretched across the patient's chest. The tube is typically 0.5 mm inside diameter, and perhaps 2 mm outside diameter. Most are about 3 cm long. The ends of the tube are typically plugged with amalgamated copper, silver, or platinum. Modern versions of this elastic strain gage use copper sulphate, or an electrolytic paste, instead of mercury. These materials have a higher resting resistance, therefore reducing the electrical current needed to create a usable output voltage.

The other type of piezoresistive strain gage transducer uses the same type of wire, foil, or semiconductor piezoresistive devices as discussed in Chapter 3. In pneumograph applications, however, the strain gage element is attached between two elastic bands. When this assembly is stretched across the patient's chest, the strain gage element will change resistance with movement of the patient's chest. As the chest rises and falls with respiration, therefore, a ΔR component is created, and this component is translatable into a changing voltage signal. Again, no flow volume data is obtained; only rate and existence information is contained within the output signal.

Both of these strain gages, as well as the thermistor transducers to follow, may be used in either a half, or full, Wheatstone bridge (shown in Figures 10-2 and 10-3).

In Figure 10-2 the resistive strain gage element is in series with constant current I. Output voltage E_o represents the respiration signal and is also given by Equation 10-1 (used previously in the discussion on the impedance pneumograph). But this circuit has one serious drawback: Voltage E_o is *not* zero when there is no respiration but is equal to $I \times R$. Pneumographs that use this circuit, therefore, are either ac coupled (the lower -3 dB frequency is 0.05 Hz!) or have an *offset control* to compensate for the static value of E_o.

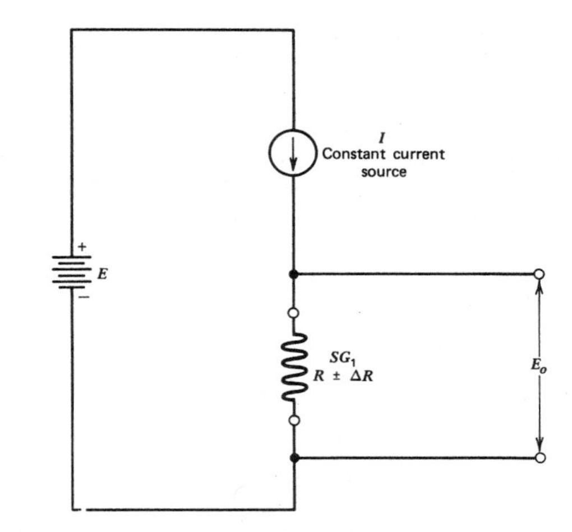

Figure 10-2 Half-bridge circuit.

The full Wheatstone bridge circuit of Figure 10-3 eliminates the offset problem; the static value of E_o in this circuit is zero volts. In some cases, R_2 will also be a transducer element connected in opposition to SG_1; that

Figure 10-3 Full-bridge circuit.

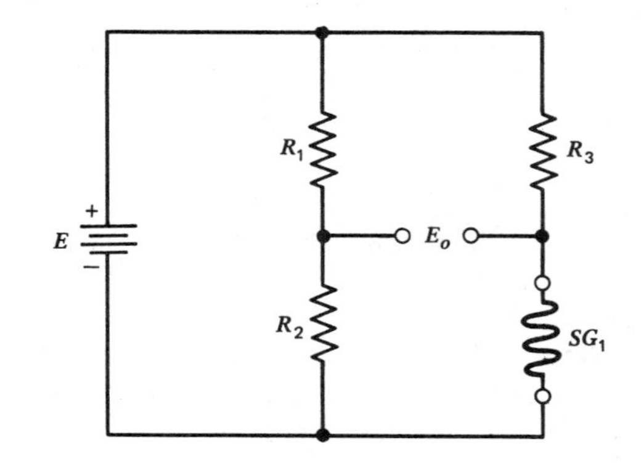

is, it changes $+\Delta R$ when SG_1 becomes $-\Delta R$, and vice versa.

Thermistors are also used as flow detectors in some pneumographs. One type of transducer consists of a bead thermistor placed just inside of the patient's nostril. A constant current is passed through the thermistor (i.e., Figure 10-2), but its value is limited to the current required to barely allow self-heating of the thermistor. This level will be 5 to 10 mA in most thermistors. The power dissipation is usually limited to less than 40 mW in order to avoid injury or discomfort to the patient. The thermistor changes resistance due to the temperature difference between inspired and expired air.

A thermistor transducer that can be used on a patient that is fitted with an endotracheal tube or is on a respirator or ventilator is shown in Figure 10-4. Two thermistors are mounted inside of a *tee piece* (a standard piece of apparatus used in respiratory therapy). Thermistor R_1 is in the flow of inhaled and exhaled gases, while thermistor R_2 forms a reference point by being placed in nonturbulent gas dead space. External resistors R_3 and R_4 form the other half of a Wheatstone bridge. Again, the current flowing through the thermistors is limited to the point of self-heating.

Output voltage E_o is normally zero when there is no gas flow but takes on a nonzero value when gas flows through the tube. A voltage waveform representing respiration is created because thermistor R_1 responds to the difference in temperature between inhaled and exhaled air.

In some transducers the thermistor is replaced with a thin platinum wire (Figure 10-5) stretched taut across a short section of

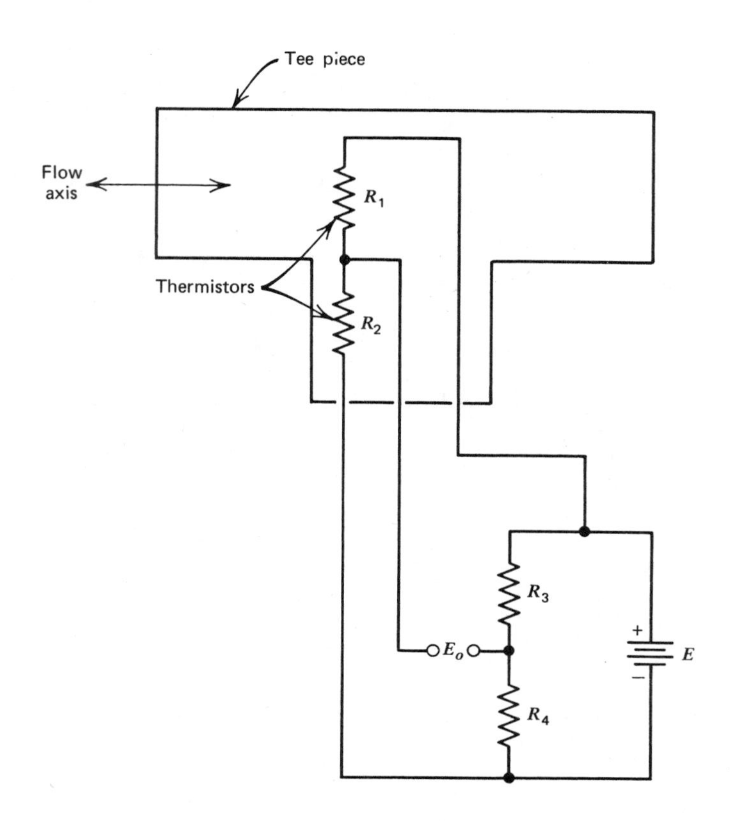

Figure 10-4 Thermistor airway flow detector.

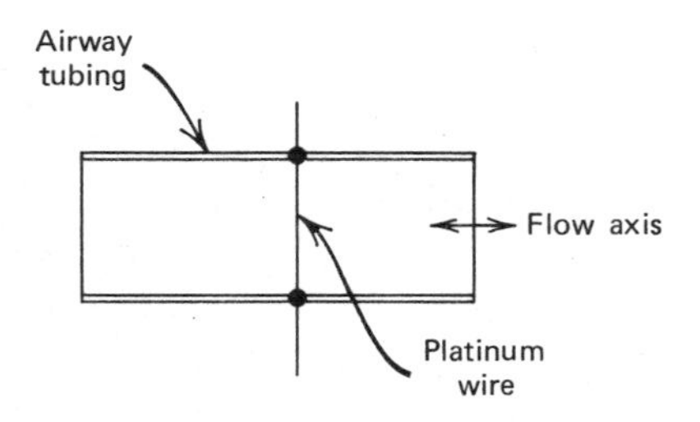

Figure 10-5 Platinum wire flow detector.

tubing. As in the previous cases, the wire is treated as a resistor at the point of self-heating. Also, as in thermistor types, the platinum wire does not yield numerical flow data, only respiration rate and existence.

Figure 10-6 shows a flow volume transducer that is capable of quantitative measurement. The flow volume is measured in *liters per minute* (l/m).

The transducer assembly consists of a differential pressure transducer (see Chapter 3) and an *airway* containing a *wire mesh* obstruction.

When a mesh is placed in an airway it causes a *pressure drop* that is measured as a differential pressure across the mesh. The pressure transducer is connected such that this difference can be measured.

It is necessary to keep the pressure drop to less than 1 cm H_2O, or normal breathing will be affected. The standard transducer, therefore, offers a 50-mm diameter mesh that has a grid density of 158 wires per centimeter (i.e., 400 per inch). This mesh offers a pressure difference of approximately 0.09 cm H_2O per 10 l/min gas flow (i.e., 9×10^{-3} cm H_2O/l).

The manufacturers of respiratory instruments using the mesh transducer calibrate the instrument to the transducer, but the calibration may be checked in the field by using a precision pneumatic flow meter to create a known flow volume.

Air flow rate is flow volume per unit time and may be obtained by integrating the volume-time signal for a known time period. One version of this type of measurement frequently seen in medical equipment is the *minute volume* measurement, in which the inspiratory volume only is measured for a period of one minute. This is usually obtained by integrating the volume signal for a one-minute period.

Figure 10-6 A common form of flow rate transducer.

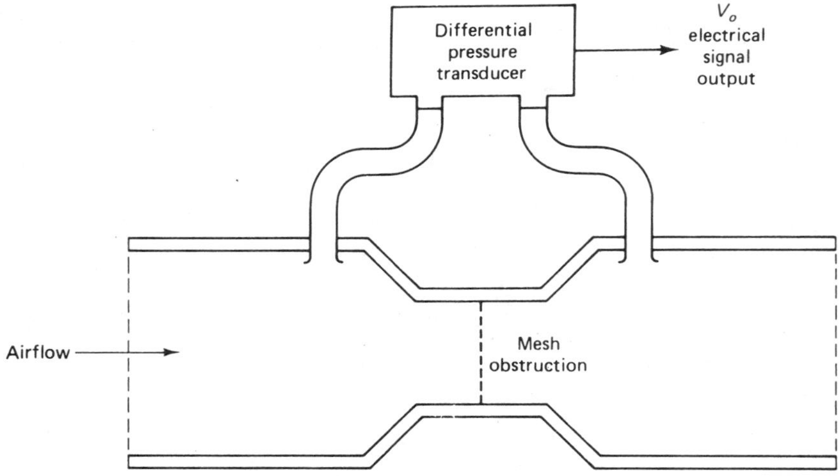

10-5 Spirometers

A conventional spirometer is shown in Figure 10-7. This instrument uses a bell jar, suspended from above, in a tank of water. An air hose leads from a mouthpiece to the space inside of the bell, but above the water level. A weight is suspended from the string that holds the bell in such a way that it places a tension force on the string that exactly balances the weight of the bell at atmospheric pressure. When no one is breathing into the mouthpiece, therefore, the bell will be at rest with a fixed volume above the water level. But when the subject *exhales*, the pressure inside of the bell *increases* above atmospheric pressure, causing the bell to rise. Similarly, when the patient *inhales*, the pressure inside of the bell *decreases*. The bell will *rise* when the pressure increases and *drop* when the pressure decreases.

Changing bell pressure changes the volume inside of the bell, which also causes the position of the counterweight to change. We may record the volume changes on a piece of graph paper by attaching a pen to the counterweight or tension string.

The chart recorder is a rotary drum model called a *kymograph*. It rotates slowly at speeds between 30 and 2000 mm/min.

Some spirometers also offer an electrical output that is the electrical analog of the respiration waveform. Most frequently the electrical output in generated by connecting the pen and weight assembly to a linear potentiometer. If precise positive and negative potentials are connected to the ends of the potentiometer, then the electrical signal will represent the same data as the pen. When there is no one breathing into the mouthpiece, E_o will be zero. But when a patient is breathing into the tube, E_o will take a value proportional to the volume and a polarity that indicates inspiration or expiration.

There have also been some ultrasonic flow meters on the market. These instruments

Figure 10-7 Bell-jar mechanical spirometer.

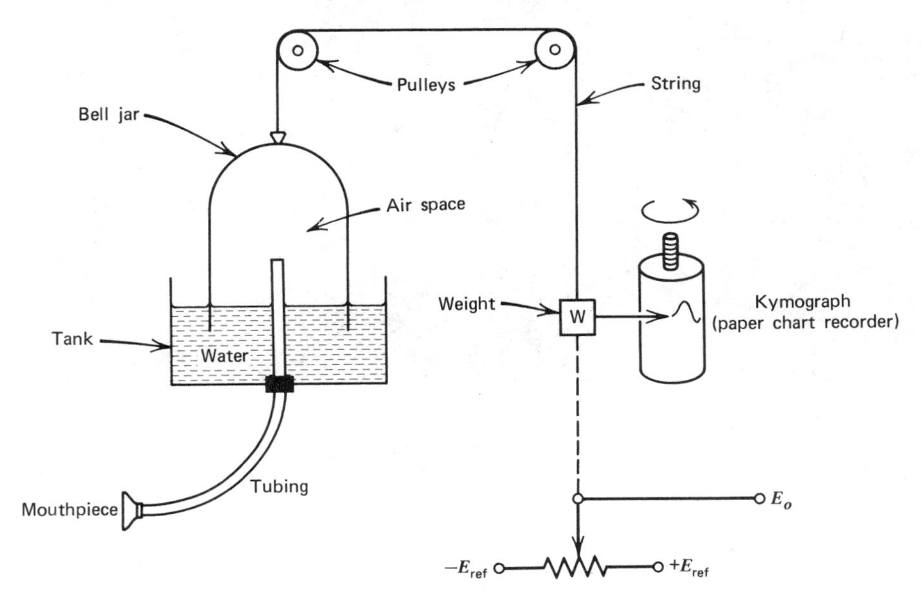

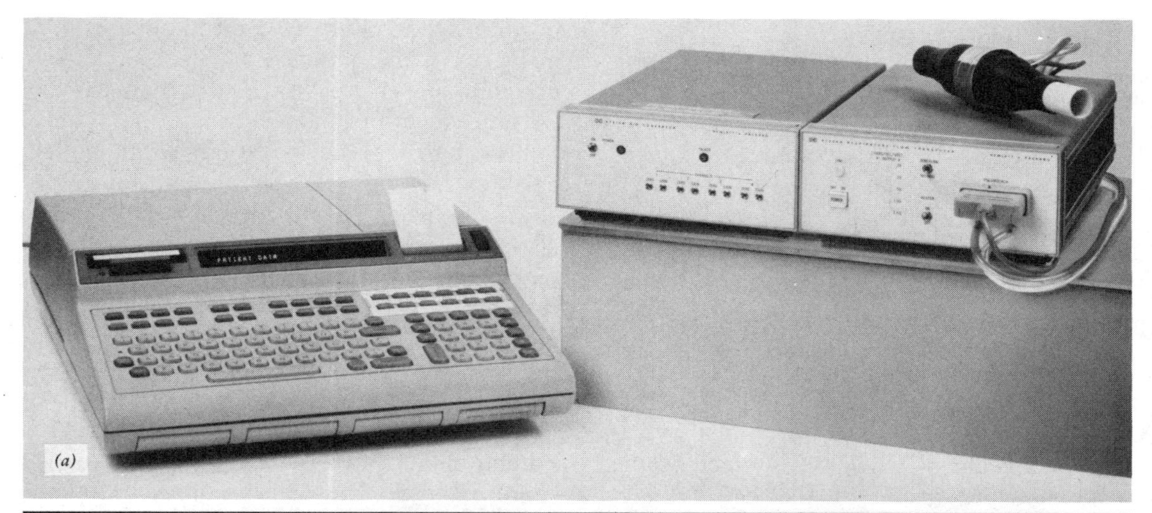

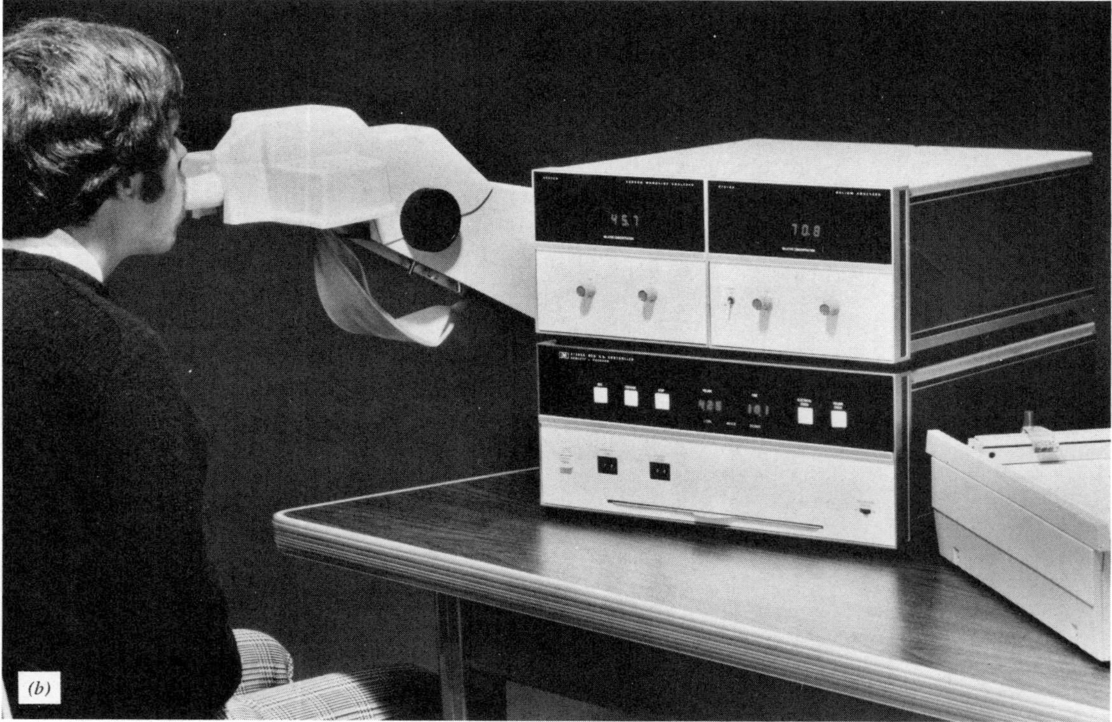

Figure 10-8 Pulmonary function analyzers. (*a*) H-P Model 47804A pulmonary data acquisition system. (*b*) H-P Model 47404A single-breath diffusion system. (Photo courtesy Hewlett-Packard.)

work on the Doppler shift difference noted on ultrasonic waves traveling with and against the direction of flow. The principles of operation for this instrument are covered in Chapter 17.

A few instruments use a pinwheel transducer. A pinwheel is either connected to an alternator or interrupts a photocell light path to produce an output frequency that is proportional to the gas flow rate.

10-6 Pulmonary Measurement Systems and Instruments

For many years pulmonary instrumentation was limited to the bell spirometer and a few other pieces of equipment. This was not due to a lack of importance, only a certain difficulty in making the measurements. Cardiovascular measurements (i.e., the ECG and pressures) are easier to make and so were developed first. Today, however, the range of pulmonary instrumentation available to hospital pulmonary function laboratories, ICU/CCU/OR, researchers, and even individual physicians in private practice has increased dramatically.

Figure 10-8 shows two pulmonary measurement systems that are used in pulmonary function laboratories in hospitals and medical centers. The instrument in Figure 10-8*a* is the Hewlett-Packard Model 47804A Pulmonary Data Acquisition System, while that in Figure 10-8*b* is the H-P Model 47404A Single-breath Diffusion System. The 47804A uses a micro-computer terminal (at left) to process the data.

The various volumes and capacities are often regarded as the only pulmonary measurements. But also considered in this class

Figure 10-9 CO_2 analyzer. (Courtesy Electronics for Medicine, Pleasantville, N.Y.)

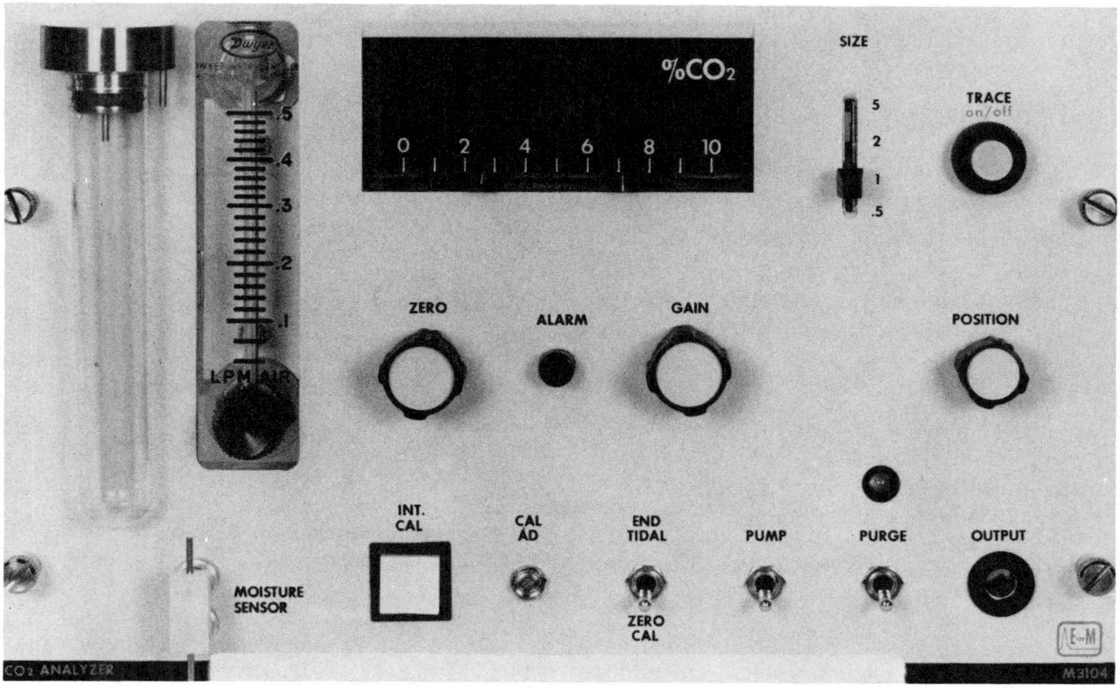

are P_{CO_2} and P_{CO_2} measurements on arterial and venous blood and on the exhaled gases. Blood gas analyzers are discussed in Chapter 16 and are not covered here. We will, however, discuss a gaseous CO_2 analyzer.

Figure 10-9 shows an instrument that measures the percentage of carbon dioxide in a sample of gas. When the sensor is placed in the exhalation line of a respirator or anesthesia machine, it will measure the *end tidal volume CO_2* content.

The device on the left of the CO_2 analyzer is a cold trap that reduces the moisture content of the incoming gas. The device immediately to the right of the cold trap is a flowmeter. An internal pump creates a reduced pressure so that a sample of the exhaled gases is drawn into the instrument. The flowmeter allows the operator to determine whether the flow rate is sufficient to permit an accurate measurement.

The typical clinical CO_2 analyzer uses the fact that CO_2 will absorb infrared energy, while O_2 and N will not. The gas sample is made to pass through a transparent cell that is in the light path between an infrared source and a photocell or phototransistor that is a filter for, or sensitive only to, infrared energy.

Calibration is performed by adjusting the *zero* control for a zero reading on the $\%CO_2$ meter when ordinary room air, or some other gas free of CO_2, is being drawn into the instrument. Once the zero is obtained, the gain can be adjusted by drawing a sample from a calibrated gas cylinder into the machine. A relatively common standard gas available from most compressed gas distributors is 95 percent O_2 and 5 percent CO_2. If this gas is used, then the *gain* control is adjusted until the $\%CO_2$ meter reads 5 percent.

10-7 Summary

1. Nonquantitative volume transducers may use transthoracic impedance changes, an elastic strain gage, or a thermistor to acquire a respiration signal. Only the existence of respiration and respiration rate, however, can be determined with these transducers.
2. Two common transducers that are capable of yielding quantitative data are the *wire mesh* and *pinwheel.*
3. Most end tidal CO_2 analyzers used in clinical work use infrared absorption, while a few research instruments use mass spectometry.
4. A mechanical device called a bell-jar spirometer is used to record respiration volumes. This instrument will record the volumes on a chart recorder called a kymograph.

10-8 Recapitulation

Now return to the objectives and self-evaluation questions at the beginning of the chapter and see how well you can answer them. If you cannot answer certain questions, place a check mark next to each and review appropriate parts of the text. Next, try to answer the following questions using the same procedure. When you have answered all of the questions, solve the problems in Section 10-10.

10-9 Questions

10-1 What are the functions of the human respiratory system?

10-2 A ______________ can be used to detect the presence of respiration but is not able to make quantitative volume measurements.

10-3 Respiration rate can be determined from the instrument in Question 10-2 if the output signal is applied to a ______________.

10-4 What general class of respiration instruments is used in ICU/CCU monitoring applications?

10-5 A ______________ is capable of making quantitative lung volume measurements.

10-6 Define *tidal volume*.

10-7 Define *inspiratory reserve volume* (IRV).

10-8 Define *expiratory reserve volume* (ERV).

10-9 Define *residual volume*.

10-10 Define *minimal volume*.

10-11 Define *minute volume*.

10-12 Define *vital capacity*.

10-13 Define *functional residual capacity* (FRC).

10-14 Define *total lung capacity* (TLC).

10-15 List three different types of transducer that might be used with a simple pneumograph.

10-16 List materials other than mercury that might be used in an elastic strain gage.

10-17 Elastic strain gages operate on the principle of ____________.

10-18 An impedance pneumograph uses the change in ____________ __________ as the patient breathes.

10-19 When thermistors and platinum wires are used to detect respiration, they are energized with just enough current to cause ____________ ____________.

10-20 The wire-mesh transducer produces a signal proportional to flow ____________ in liters per minute.

10-21 The wire-mesh transducer depends on the ____________________ across the mesh when gas flows through the mesh.

10-22 The approximate density of a typical wire mesh used in the transducer of Question 10-20 is ____________ per centimeter.

10-23 The ____________ volume can be measured by integrating a flow signal for a period of 60 s.

10-24 Describe in your own words (a) a bell jar spirometer, (b) the operation of the bell jar spirometer.

10-25 The chart recorder that is used with a mechanical bell jar spirometer is called a ____________.

10-26 The bell jar spirometer will produce an analog electrical output representing flow volume if a ____________ is mechanically connected to the weight or pen of the spirometer.

10-27 Most clinical end tidal CO_2 analyzers, especially those intended for bedside use, operate on the principle of ____________ absorption.

10-10 Problems

10-1 Calculate the IRV of a patient in which the VC is 4130 ml, tidal volume is 480 ml, and the ERV is 1156 ml.

10-2 Find the tidal volume of a patient in which the IRV is 2100 ml and the IC is 2580 ml.

10-3 Find the *total lung capacity* for a patient in which the TV is 500 ml, IRV is 2300 ml, ERV is 1050 ml, and RV is 1200 ml.

10-4 Calculate the current required in an impedance pneumograph if the output potential is to be 1 mV when the patient's chest resistance varies $\pm 20\ \Omega$ with respiration from a resting resistance of 15 kΩ.

10-5 Find the output potential of a half-bridge circuit energized by a 500-nA constant current source if the resting resistance is 30 kΩ.

10-6 Calculate the amplitude of the respiration signal in Problem 10-5 if the resistance changes $\pm 90\ \Omega$ as the patient breathes.

Chapter 11

Respiratory Therapy Equipment

11-1 Objectives

1. Be able to describe the physiological basis (diseased states) for requiring artificial respiratory therapy.
2. Be able to describe medical gases and safety systems (cylinders, high-pressure control).
3. Be able to describe the procedure and equipment used in oxygen therapy (regulators, flowmeters, humidifiers/nebulizers, cannulas, oxygen masks, and oxygen tents).
4. Be able to describe the procedure and equipment used in intermittent positive pressure breathing (IPPB) therapy.
5. Be able to describe the procedure and equipment used in artificial mechanical ventilation.
6. Know how to describe accessory devices used in respiratory therapy apparatus (respiratory monitor/alarms, oxygen analyzer).
7. Know how to describe sterilization and isolation procedures used in respiratory therapy units.
8. Be able to list typical faults and maintenance procedures for artificial respiratory ventilators.

11-2 Self-Evaluation Questions

These questions test your prior knowledge of the material in this chapter. Look for the answers as you read the text. After you have finished studying the chapter, try answering these questions and those at the end of the chapter (Section 11-15).

1. List two general respiratory diseases that may require artificial respiratory therapy.
2. What is the goal of artificial ventilatory control mechanisms?
3. What is the difference between controlled artificial ventilation and mechanically assisted ventilation?
4. List six regulating agencies involved in compressed gas cylinder safety (manufacture, transportation, storage, and usage).
5. What safety systems are used throughout the hospital for oxygen, carbon dioxide, helium, and nitrous oxide?
6. Define *oxygen therapy*.
7. Name four specific diseases requiring oxygen therapy.
8. How should respiratory therapists and physicians orient patients to respiratory therapy?
9. Describe gas regulators, flowmeters, humidifiers, nebulizers, and oxygen masks and tents.
10. Define IPPB therapy and list its objectives.

11. Describe the hardware and operation of an IPPB unit (include block diagram of Bennett valve and accessories).
12. Describe the operation of two major types of artificial respiratory ventilators (include block diagram and external controls).
13. Describe respiratory therapy equipment sterilization and isolation procedures, including special sterilization equipment.
14. List typical faults and maintenance procedures for artificial respiratory ventilators.

11-3 Diseased States Requiring Artificial Respiratory Therapy

Four basic *pulmonary abnormalities* (see Chapter 9) that require *artificial respiratory ventilation* are abnormalities of ventilation (hypoventilation), ventilation/perfusion problems, membrane permeability defects (diffusion impairment), and arteriovenous shunts (abnormal bypassing of blood).

Essentially, two respiratory diseases cause respiratory (ventilation) failure. These are:

1. *Hypoxia*—low oxygen content in the blood due to improper ventilation. This is caused by pulmonary emphysema (stretch alveoli—loss of lung elasticity), chronic bronchitis, pulmonary tumors, aspiration pneumonia, interstitial fibrosis, or pulmonary infarction (tissue death from lack of blood supply).

2. *Hypercapnia*—poor alveolar ventilation causing an accumulation of carbon dioxide in the blood. This results from central nervous system disorders, diseases of nerves and muscle weakness, metabolic diseases, and pulmonary emphysema, chronic bronchitis, or lung obstruction.

The goal of an *artificial ventilatory control mechanism* (correction of pulmonary abnormalities) is to adjust alveolar ventilation to changing body needs of oxygen intake and carbon dioxide removal. Alveolar ventilation depends on the relationship between respiratory rate, tidal volume (TV), and dead space.

In addition, *tracheostomy* (surgical practice to overcome obstructions of the upper respiratory tract) may be performed to permit prolonged artificial ventilation.

11-4 An Overview and Terms of Ventilation

Various types of mechanical devices can assist the patient in the ventilation process. However, when *lung damage* or *blood perfusion* (pouring of blood over the lung tissue) occurs, cures to ventilation problems cannot always be complete. Essentially, the following ventilating processes are common.

1. *Mechanically assisted ventilation*—the sum of all ventilation by mechanical means.

2. *Controlled artificial ventilation*—process in which the patient plays no role in initiating the respiratory cycle.

3. *Patient-cycled artificial ventilation*—process in which the patient initiates the inspiratory cycle.

4. *Volume-cycled ventilation*—process by which inspiration is terminated after a preset volume is delivered.

5. *Pressure-cycled ventilation*—process by which inspiration is terminated when a preset pressure is reached.

Terms of ventilation relate to the parameters of pulmonary function (see Chapter 9), including the following:

1. *Peak pressure*—maximum airway pressure reached in distending the lungs during artificial ventilation.

2. *Respiratory failure*—failure of the respiratory system to meet body needs for O_2 uptake and CO_2 removal.

3. *Ventilatory failure*—failure of the respiratory system to meet the body needs for CO_2 removal.

4. *Venous return*—minute volume of blood flow into right atrium. Normal minute volume is the tidal volume (500 ml) times the breaths per minute (about 16). This equals 8000 ml. Respiratory minute volume is known as pulmonary ventilation.

5. *Lung volumes* (see Chapter 9)—TLC, VC, RV, IC, IRV, ERV, FRC.

11-5 Historical Perspective of Artificial Respiratory Ventilation

The understanding of pulmonary physiology followed the *discovery of oxygen* by *Joseph Priestley* (1733–1804) and *Joseph Black* (1728–1799).

In 1790, *Thomas Beddoes* (1760–1808) founded the *Pneumatic Institute* for the treatment of diseases by inhalation of gases. The era of inhalation therapy took form when *Sir Arbuthnot Lane* invented the nasal catheter in 1907, and *John Haldane* (1860–1936) introduced the oxygen mask (for World War I gas-poisoned victims with pulmonary edema). The oxygen tent constructed by *Sir Leonard Hill* in 1920 also showed promise. Later, the foot bellows *Fell-O'Dwyer respiration apparatus* and *compressed air techniques* paved the way for invention of the modern artificial respiratory ventilator.

11-6 Medical Gases and Safety Systems

A variety of *medical gases, cylinders, and regulating equipment* is used by the respiratory therapist. The following *regulating agencies* are important to the therapist:

1. National Fire Protection Agency (NFPA)—voluntary agency (National Consensus Standards) that promotes prevention and detection of fire hazards.

2. U.S. Department of Transportation (DOT)—regulates gas cylinder construction, testing, and maintenance.

3. Compressed Gas Association (CGA)—develops safest methods for handling compressed gases.

4. Federal Food, Drug, and Cosmetic Act—regulates shipment of medical gases in interstate commerce and provides standards through the Pharmacopeia of the United States (USP).

5. Federal Occupational Safety and Health Act of 1970 (OSHA)—provides for federal inspections of facilities used for storage of flammable gases.

6. Local and state agencies—providing safeguards for their communities from hazards of storage, transportation, and use of compressed gases.

Gas cylinders are made from seamless tubing (brazed or welded) or from flat sheets of drawn steel shaped into a cylinder. Some common types are ICC-3A (seamless high pressure, 150 to 15,000 psig), ICC-3AA (seamless high pressure above 3A rating), ICC-3B (seamless, 150 to 500 psig), ICC-3E (seamless 2 in. maximum diameter, 24 in. maximum length, 1800 psig), and ICC-8 (seamless low pressure, 250 psig for acetylene service). *Gas cylinders are dangerous* and if mishandled may result in fire and explosion.

Markings on gas cylinders are required. For example, first line: ICC spec 3AA, 2015 psig; second line: H 396042; third line: BAP (manufacture marking); and fourth line: 8-70 (month and year qualification test).

Piping systems, as shown in Figure 11-1, are used throughout the hospital. Central oxygen cylinders supply oxygen to specific hospital areas through a pressure regulating valve, which maintains a pressure between 50 and 100 psi at the service outlet. The main supply must have a shutoff valve and an

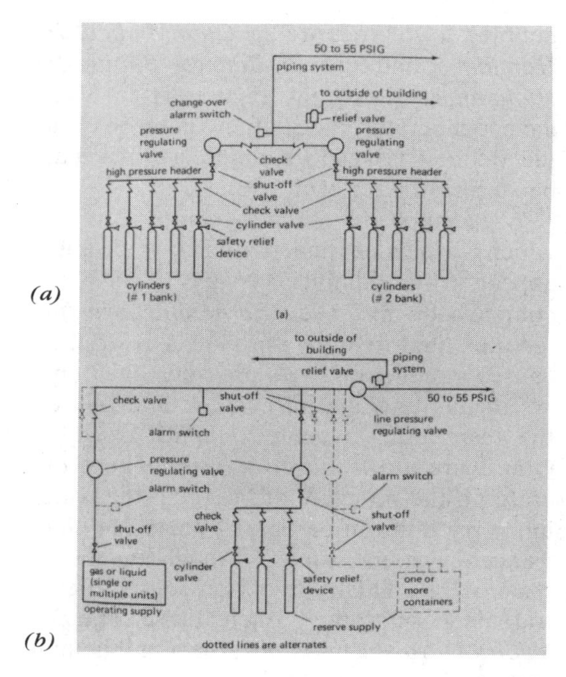

Figure 11-1 Oxygen supply systems. (*a*) Oxygen cylinder system. (*b*) Oxygen bulk system. (Reprinted with permission of Prentice Hall.)

alarm system. Separate station manual or automatic shutoff valves are also required. All personnel must be educated as to the placement and use of these valves.

Therapeutic gases fall into these areas: oxygen—essential to life (normally 21 percent); carbon dioxide—important to the control of respiration and circulation (normally 0.025 percent); helium—lightweight inert gas used in cases of respiratory obstruction; nitrous oxide—inorganic gas used as an anesthetic agent; helium-oxygen mixtures—low-density gas used on asthmatic patients; and oxygen-carbon dioxide mixtures—used to stimulate deep breathing and to relieve cerebral vascular spasm.

Safety systems for compressed gases in hospitals are essential. The *Diameter-Index Safety System* (DISS) was developed by the *Compressed Gas Association, Inc.* to provide interchangeable threaded connections. Each connection of DISS (make-and-break threaded connections) consists of a body, nipple, and nut. The *Pin-Index Safety System* (two-pin approach) prevents incorrect interchange of medical gas cylinders with flush-type valves. Ten combinations are possible, of which eight are in current use. With the two-pin approach, no two incompatible cylinders can be attached together.

11-7 Oxygen Therapy

Oxygen therapy is the administration of oxygen for the treatment of conditions resulting from oxygen deficiency. These conditions result from pneumonia, pulmonary edema (swelling), obstruction to breathing, congestive heart failure, coronary thrombosis (blood blockage), and complications following surgery. Oxygen is administered through nasal catheters, masks (nasal oronasal), funnels or cones, oxygen tents, or special oxygen chambers. A typical oxygen mixture is 70 to 100 percent by volume. Oxygen is usually introduced together with medicines, water vapor, other gases (carbon dioxide or helium), and anesthetics.

Oxygen therapy is accomplished through the use of special procedures and equipment such as gas regulator/flowmeter control devices, humidifier/nebulizer conditioning units, and oxygen mask/tent administering systems.

Gas regulators are used to reduce cylinder pressure to a safer level such as 50 psig. A flowmeter adjusts the gas flow in liters per minute. Cylinder regulators are typed as *preset* (reduction to working pressure of 50 psig through one chamber and one safety valve), *adjustable* (reduction to 50–100 psig through one chamber and one safety valve), and *multiple stage* (reduction to 50 psig through two or three chambers and the same number of safety valves). Figure 11-2 shows a cylinder

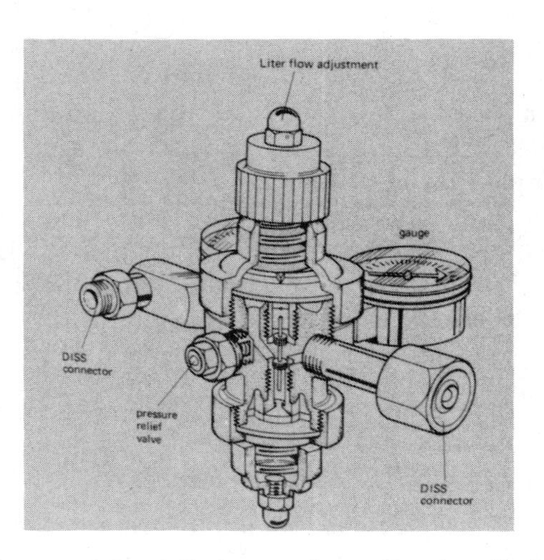

Figure 11-2 Cylinder regulator. (Reprinted with permission of Prentice Hall.)

regulator, with a flowmeter attached, and a flow-adjusting valve.

A *flowmeter* is a device that contains a calibrated tube to indicate oxygen flow in liters per minute and a valve to control the rate of flow. It must be in the upright position to obtain an accurate reading. Due to the back pressure generated in the flow line by these devices, back pressure compensated flowmeters must be used to give correct indications. Figure 11-3 shows the uncompensated Thorpe Flowmeter (needle valve control), Bourdon Flow Gauge (needle-valve, fixed-size orifice control), and pressure-compensated flowmeter (rising ball that partially drops from back pressure in the line-variable area flowmeter).

Wall outlets and adapters are extensive in number. Some standard ones appear in Figure 11-4.

Humidifiers add water vapor to medical gases administered to a patient. This is necessary since alveolar-capillary gas transfer membranes require high humidity to be effective. This is accomplished by passing the gas through sterile water, which creates tiny bubbles. Two types are the *Ohio Medical Jet Humidifier* and *Puritan-Bennett Bubble-Jet Humidifier*, shown in Figure 11-5. Some have temperature regulation units, such as the Bennett Cascade Humidifiers. All must be sterilized after each use.

Nebulizers are aerosol therapy units in which suspended fine particles or droplets appear in the administered gas. This is accomplished by the *Bernoulli principle* (greater fluid velocity through a restriction causes a reduction in its pressure). Nebulizers of this type are all purpose and rely on the restriction to convert liquid to vapor. Others are *ultrasonic nebulizers*. These operate on the principle of breaking the water into very uniform particles by passing high-frequency sound waves through the liquid (1.35 MHz). This type of aerosol can penetrate deeper into the bronchial tree. The *DeVilbiss Ultrasonic Nebulizer* is a mobile unit that can be cleaned easily.

Oxygen therapy is primarily administered through oxygen masks or tents. *Oxygen masks* are devices that fit over the patient's face and allow oxygen to pass to the nose and mouth. They are dangerous since suffocation will result if oxygen is cut off and the patient cannot remove the mask. Nevertheless, they are more effective than the *nasal cannulas* (simple inexpensive flexible tubes at each nostril opening for 22 to 30 percent oxygen) or *nasal catheters* (plastic tube placed into a nostril for 30 to 35 percent oxygen). *Oxygen tents,* such as the *Bunn Tent,* provide the patient with a temperature-humidity-oxygen concentration regulated environment. A canopy surrounds the patient, gas is pumped in, and waste gases are removed. *Oxygen hoods,* such as the *Olympic Oxyhood,* deliver high concentrations of oxygen to infants. Special *infant incubators,* such as the *Ohio-Armstrong Isolation Incubator,* supply heated, filtered, humidified air for newborns.

It is important to remember that technol-

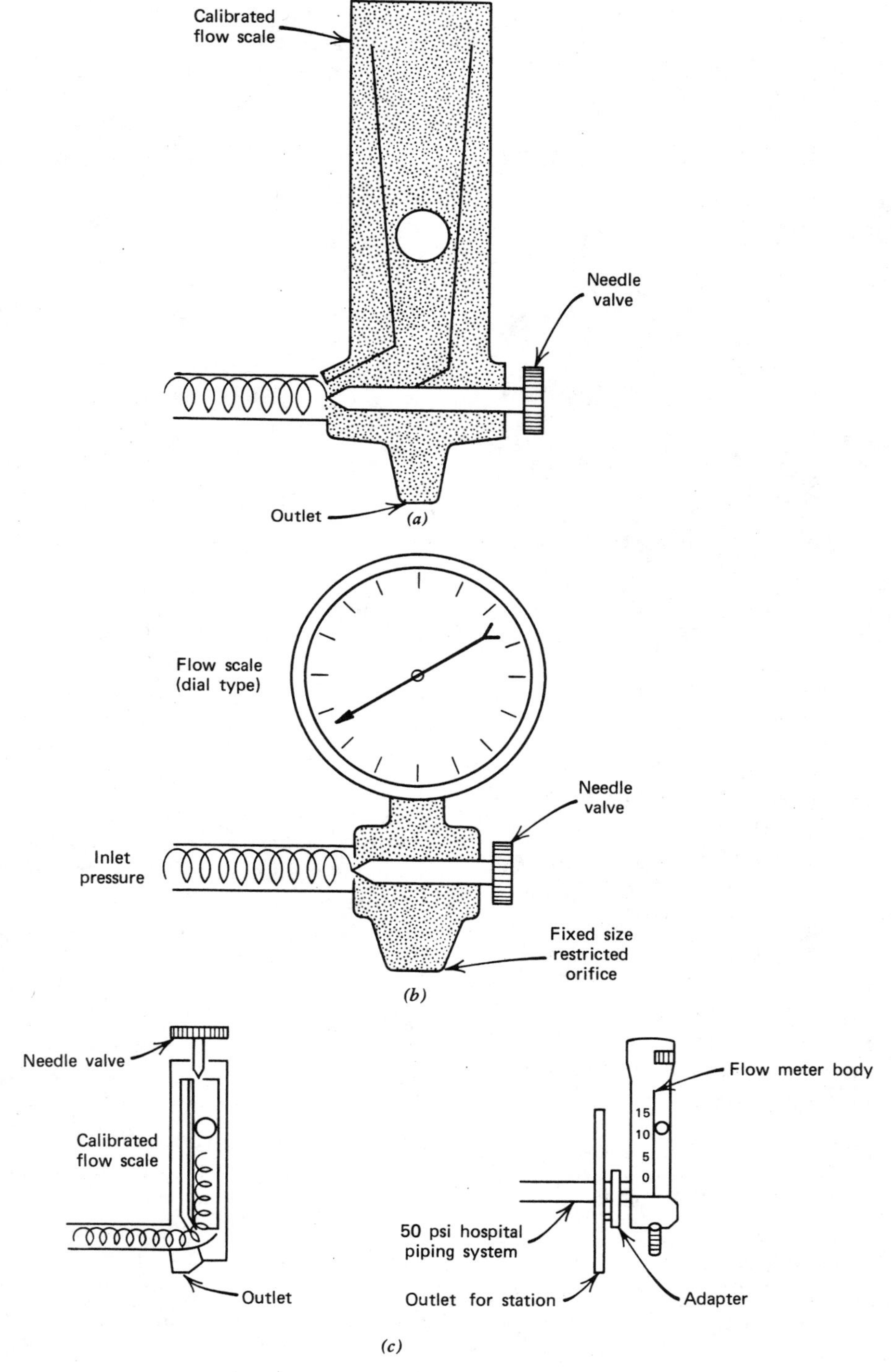

Figure 11-3 Flow gages. (*a*) Thorpe uncompensated. (*b*) Bourdon uncompensated. (*c*) Pressure compensated.

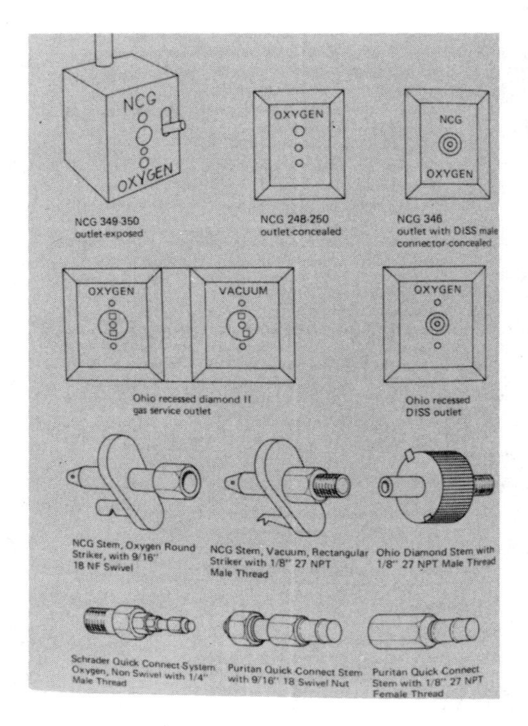

Figure 11-4 Wall outlets and adapters. (Reprinted with permission of Prentice Hall.)

ogy and machines are only aids. *Respiratory therapists* of and the *respiratory therapy department* treat patients. Record keeping provides physicians, nurses, therapists and other allied health personnel with a valuable guide for treating the patient.

Since city and industrial area air is polluted, respiratory irritation frequently occurs. A surprising number of people will require artificial ventilatory machines as our modern growth expands.

11-8 Intermittent Positive Pressure Breathing (IPPB) Therapy

Intermittent positive pressure breathing (IPPB) therapy involves special procedures and equipment. IPPB is a type of assisted breathing pattern in which the lungs are inflated by positive pressure during *inspiration* and, upon release of the pressure, *expiration* occurs passively.

Intermittent positive pressure breathing is indicated for patients with the following problems: chest disorders such as bronchitis, bronchiectasis, asthma, pulmonary emphysema, edema; central nervous system disorders (i.e., drug overdose); chronic bronchopulmonary diseases (i.e., bronchial infections, respiratory acidosis); and postoperative conditions (to prevent pneumonia).

The *objectives* of IPPB are to assist and promote more uniform ventilation, facilitate better O_2 and CO_2 exchange and aspirate antibiotic drugs, relieve bronchospasm, assist in removal of bronchopulmonary secretions (drainage), and exercise respiratory muscles. Medical gases commonly used are oxygen, compressed air, and oxygen and helium.

Under emergency circumstances in which IPPB respirators are not available, *cardiopulmonary resuscitation (CPR)* may save the patient's life. This is a technique used to keep blood flow and oxygenation up when heart attack (fibrillation) or respiratory arrest is present. It consists of pumping rhythmically on the patient's chest with the palm of one hand while periodically blowing into the patient's mouth. For long-term respiratory treatment, respiratory therapy devices must be used. All patient care personnel should be familiar with CPR.

Patient orientation is extremely important for IPPB therapy success. Respiratory therapists should greet and treat patients with a cheerful attitude, explaining that this treatment is easy but requires the patient's cooperation. Should the patient fight the machine, total success is uncertain. The therapist should also explain the procedure of treatment and entertain any questions from the patient.

IPPB unit *hardware* includes a special device known as the *Bennett valve*. This valve controls an intermittent positive inspiration

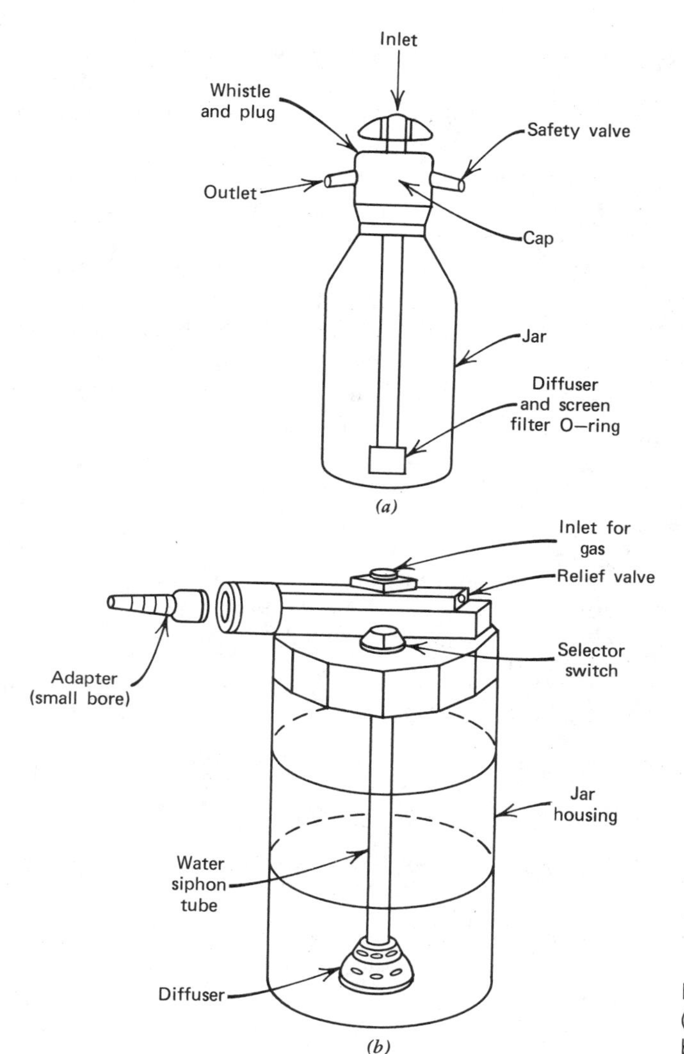

Figure 11-5 Humidifiers. (*a*) Ohio jet. (*b*) Puritan-Bennett bubble-jet.

pressure that assists in inflating the lungs. The patient's respiratory effort (rate and rhythm) controls valve cycling. Connection between the patient and gas reservoir through the valve is such that in the automatic mode, the patient can take over both inspiration and expiration.

Figure 11-6 shows a Bennett respiration unit that operates from a standard 50 psi pressure source. Supply can be through a standard hospital piped system of 40 to 70 psi or from a separate gas cylinder of 50 psi regulated.

Figure 11-7 shows a *block diagram* of the functional features of the Bennett-type IPPB unit. Air pressure at 50 psi is fed to the Bennett valve through an adjustable-diluter regulator. The control and system pressure are recorded on front panel gauges. From the main input gas line, a low-pressure regulator

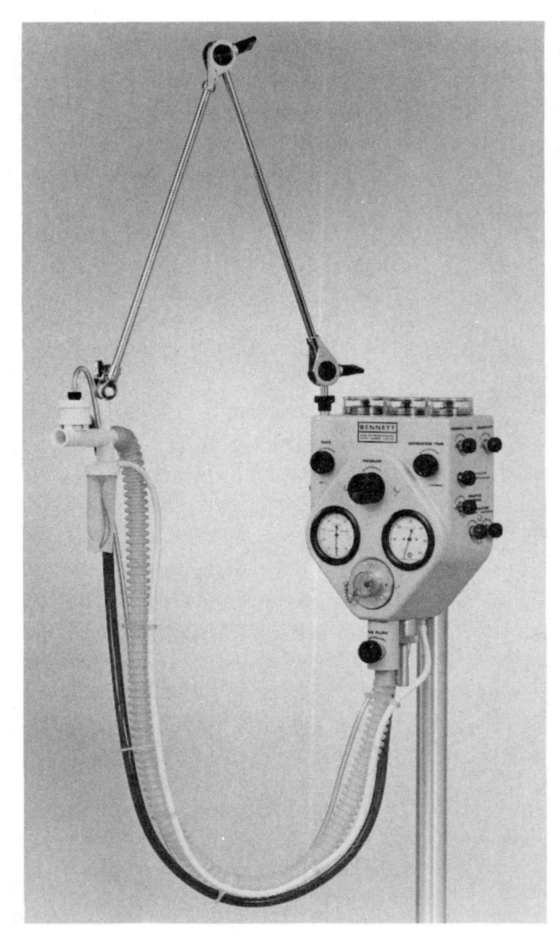

Figure 11-6 Bennett Respirator. (Photo courtesy Puritan-Bennett.)

supplies the input for sensitivity, rate, and expiration control. These lead to the Bennett valve for manually adjusted control. Terminal and negative pressures are also controlled from the main line. The *Bennett valve rotating internal ball* closes during patient expiration, and the exhalation diaphragm opens to release exhaust gases. Upon inspiration, the diaphragm closes and the Bennett valve opens to feed oxygen and water vapor (from the nebulizer) to the patient.

Accessories can be used with this unit, as shown in Figure 11-6: These include a *monitoring spirometer*—indicates each expiration volume (refer to Chapter 10); an *alarm*—set to sound below preset tidal volume; a *humidifier*—cascade type for counteracting drying effects of anhydrous gas; an *oxygen adder*—mixes oxygen from metered supply with air from the main line; *bacterial filters*—reduces infection between patients; and *volumetric ventilation control*—delivers preset tidal volume to the patient independent to lung compliance.

The assembly and operation procedures for the Bennett IPPB unit are important. During the performance of these procedures, problems are uncovered, which can cause serious injury to the patient.

An *assembly procedure* consists of the following steps.

1. Blow out the dust particles from the cylinder valve and connect it to the compressed gas cylinder (tighten with a wrench and avoid stripping the threads).

2. Screw the support arm into the threaded hole on the unit head (tighten using wing nuts and avoid forcing any section into place).

3. Attach the manifold including exhalation valve to the end of the support arm (exhalation port should face the patient).

4. Attach one end of the large main air tube to the port beneath the Bennett valve.

5. Attach the remaining end of the main air tube to the bottom manifold port for horizontal nebulizer use and attach to the back of the manifold for vertical nebulizer use.

6. Connect the smaller diameter tubing to the exhalation port on the manifold and the other end below the Bennett valve.

7. Connect the larger diameter tubing to the nebulizer tailpiece and the remaining end to the nebulizer control (top of unit head).

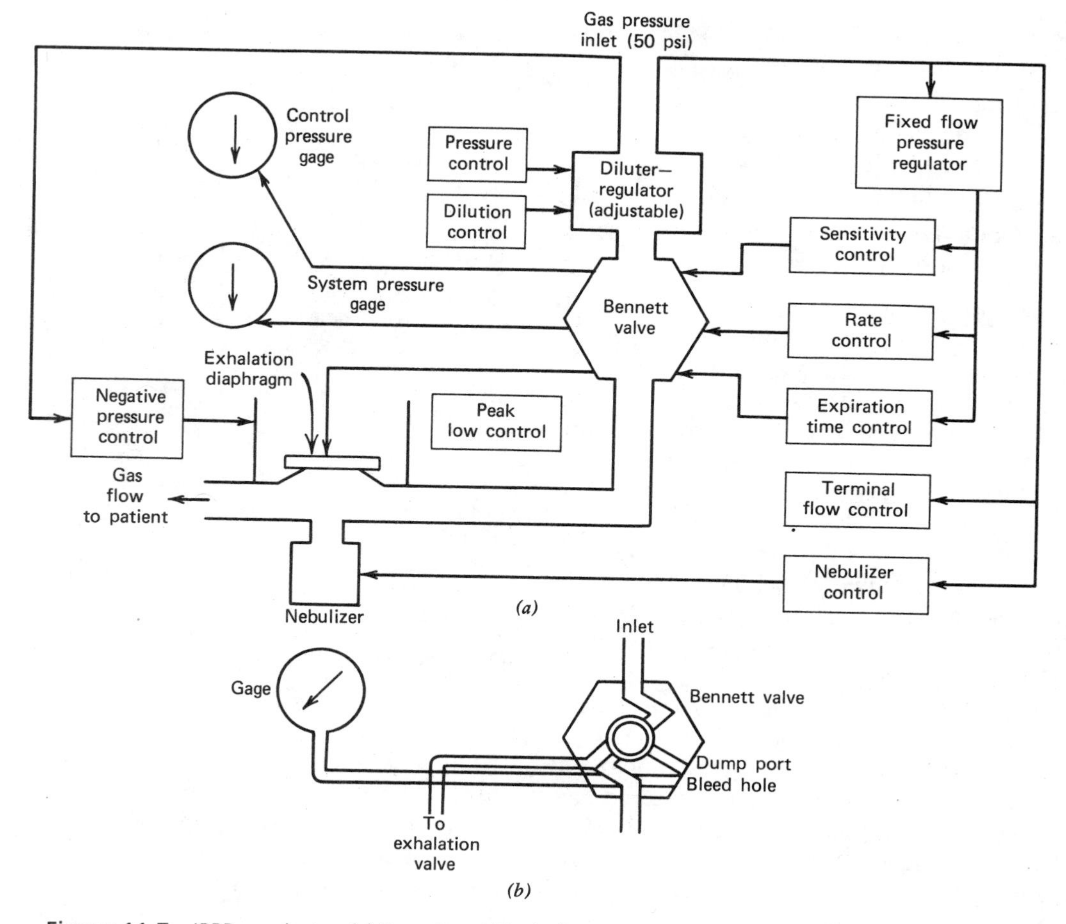

Figure 11-7 IPPB respirator. (*a*) Functional block diagram. (*b*) Internal structure of the Bennett valve.

8. Secure the flex tube and mouthpiece (or mask) to the manifold end nearest the exhalation valve.

It is important to remember that other assembly techniques yield a system that appears to be functional but may be erratic or marginal for patient therapy success.

The *operation procedure* is as follows:

1. With both hands, open the cylinder valve fully (stand in back of the unit for safety should a burst of gas dislodge any unit member).

2. Move the shutoff lever on the unit head downward to the vertical position (ON position).

3. Turn the pressure control clockwise until the gage registers the amount prescribed by the physician (typically 10 to 20 cm of water pressure).

4. Set the dilution control prescribed by

the physician to either 100 percent oxygen setting or air-oxygen setting.

5. Place the nebulizer medication in the canister (always have either medication, distilled water, or physiological saline present, as dry gas will harm the patient).

6. Open the nebulizer control and observe a slight mist or fog.

The foregoing procedure involves an *operating sequence* that involves setting inspiratory flow start (patient cycled or automatic time cycling), expiratory start (inspiratory end), system cycling (rate control), inspiration/expiration ratio, nebulizer flow, and breathing system (lung flow resistance, and lung and thoracic compliance and leaks).

It is *important* to note that the biomedical equipment technician (BMET) should not set up or administer respiratory treatment. This operation is left to a trained respiratory therapist or physician.

Automatic electronically controlled IPPB therapy units are also on the market (e.g., the Bennett AP-5). These use a motor and pump assembly that drives filtered air through a Bennett valve to the patient. Most often these are used in patient's home, physician's office, or outpatient clinics because they do not require a tank air pressure cylinder supply. They can also be fitted with accessories, such as cascaded humidifiers, nebulizers, and bacteria filters.

Some *manual IPPB units* operate by compressing room air and are equipped to provide additional oxygen, if required (e.g., the Bennett Model TA-1, shown in Figure 11-8).

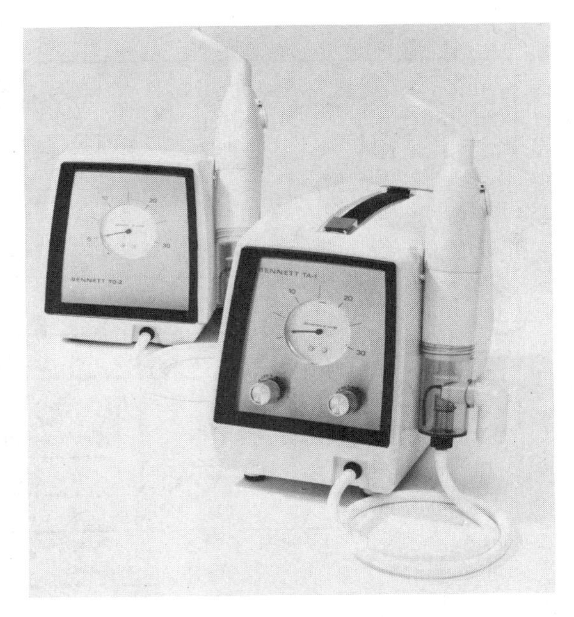

Figure 11-8 Bennett TA-1 Manual IPPB unit model. (Photo courtesy Puritan-Bennett.)

11-9 Artificial Mechanical Ventilation

Artificial lung ventilators are devices that connect to the patient's airway and are designed to augment or replace the patient's ventilation automatically. They are employed with a mask, endotracheal tube (within the trachea), or tracheostomy tube (through an artificial opening in the trachea via the throat). These ventilators consist of a controller, which operates independently of the patient's inspiratory effort, an assistor, which augments or assists inspiration of the spontaneously breathing patient, and an assist-controller, which assists and/or controls. Some lung ventilators are *pressure preset* and others are *volume preset* (tidal or minute volume). Some *cycle inspiratory to expiratory* (volume, pressure, time, or a combination). Others *cycle expiratory to inspiratory* (pressure, time, combined, patient controlled, or manual override). Safety limits are *volume pressure* or *time set*. Pressure patterns are positive atmospheric, postive-negative, or positive-positive. Most operate pneumatically and others are electrically powered.

Essentially, ventilators are *classified* into two types:

1. *Pressure-cycled (pressure-limited)*—those that continue to inflate the patient's lungs until a preset pressure is reached, at which time expiration stops and inspiration starts. These units maintain good ventilation with small system leaks but compensate poorly for obstruction.

2. *Volume-controlled (volume-limited)*—those that permit maintenance of constant minute and tidal ventilation even though changes in pulmonary compliance and airway resistance occur. Safety valves for maximum preset pressure should be included. These units maintain steady tidal volume but with partial obstruction they compensate poorly for system leaks.

Figure 11-10 Bird Mark-14 ventilator. (Photo courtesy of the Bird Corporation.)

Pressure-cycled, assistor controlled pneumatic ventilators (Bird Mark 8, shown in Figure 11-9) are used for patients with apnea, assisted ventilation, and patients with airway resistance. This system pressure is adjusted by the operator and the patient-delivered volume is variable, depending on lung compliance. The Bird Mark 14, shown in Figure 11-10, has all the capabilities of the Bird Mark 8 and, in addition, has leak-compensated, positive-phase operation, automatic flow accelerator, and provisions for handling airway pressures of up to 140 mm Hg and flow rate up to 160 l/min.

Figure 11-9 Bird Mark-8 ventilator. (Photo courtesy of the Bird Corporation.)

Volume-limited ventilators (Bennett Respiration Unit Model MA-1, shown in Figure 11-11) are used for patients who can initiate their own inspiration. The MA-1 is an electrically powered device that uses *ambient air* as the principal gas with provisions for oxygen enrichment up to *100 percent*. It can initiate inspiration by a *timing mechanism*. Inspiration may also be initiated by the patient, and under this condition the unit becomes an assistor. Adjustments are also available to permit the ventilator to take over respiratory control should the patient become apneic (cease to breathe). *Adjustable volume and pressure limits* set the end of the inspiration. Essentially, these controls are positioned to deliver a fixed volume at each inspiration that is below the selected maximum safety pressure. The *breaths per minute* setting is set on the timer dial. This is chosen based on the volume or pressure limits and calibrated maximum flow rate. Periodic deep breaths (sighs), volume and pressure limits, oxygen percentage, and nebulizer controls are also present on the front panel. There are nine *output displays:*

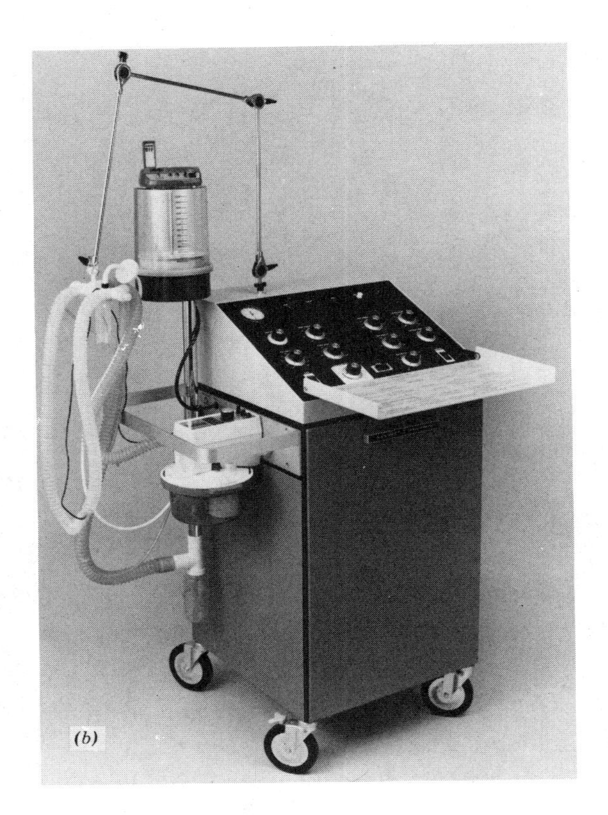

Figure 11-11 (*a*) Flow block diagram. (*b*) Bennett respirator unit Model Ma-1 solid-state IPPB unit. (Photos courtesy of Puritan-Bennett.)

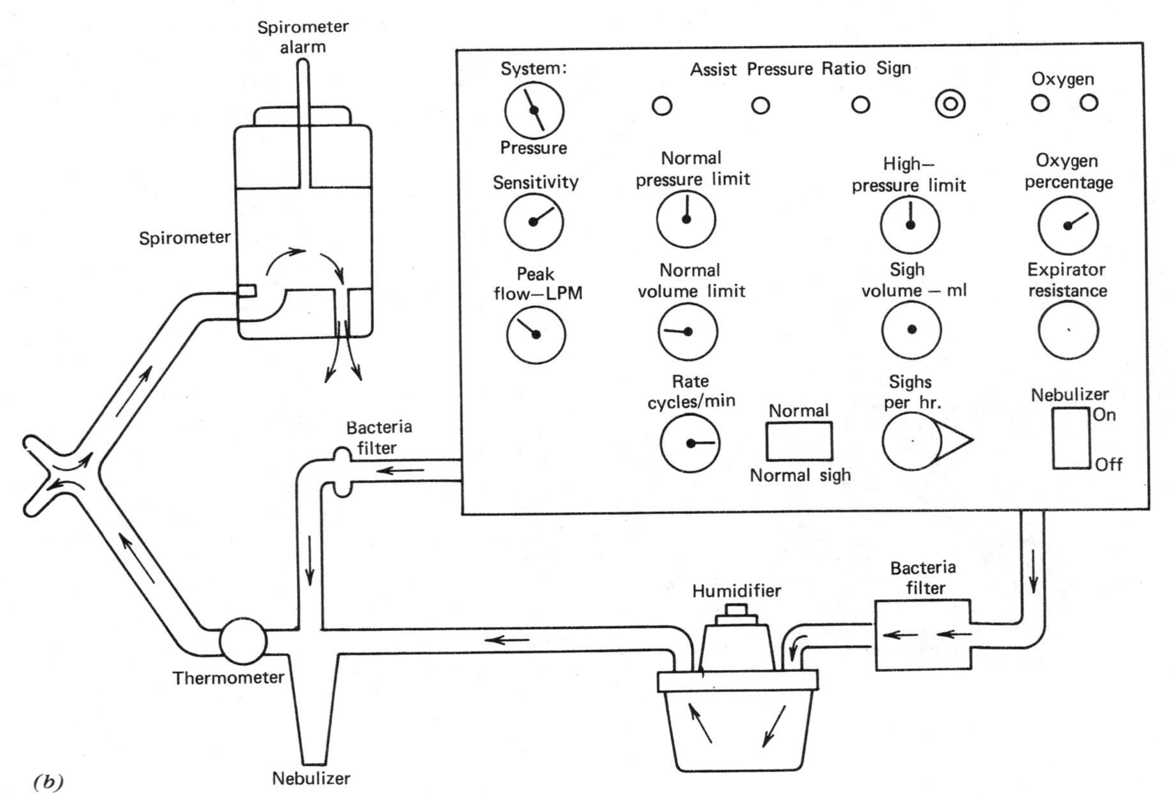

1. System pressure gage.

2. Spirometer below volume and spirometer audible alarm.

3. Adverse ratio.

4. Warning light.

5. Oxygen deficiency audible alarm.

6. Sigh inspiration warning light.

7. Pressure warning limit light/audible alarm.

8. Thermometer.

9. Elapsed time indicator.

The *MA-1 unit* operates from a 115-V, 60-Hz power line with chassis grounding for electrical safety (low leakage current). As shown in Figure 11-11, the *external controls* are as follows:

1. Adjustable calibrated normal volume (0–2200 ml)

2. Adjustable calibrated normal rate (6–60 cycles per minute).

3. Adjustable calibrated normal pressure limit (20–80 cm H_2O).

4. Manual normal inspiration start control.

5. Adjustable calibrated sigh volume (0–2200 ml).

6. Adjustable calibrated sigh intervals (4–6–8–10–15 times or deep breaths per hour to remove blockages).

7. Adjustable calibrated multiples (1–2–3 sighs per interval).

8. Adjustable calibrated sigh pressure limit (28–80 cm H_2O).

9. Controlled expiration after sigh inspiration.

10. Manual sigh inspiration start control.

11. Adjustable calibrated O_2 percentage (21–100 percent).

12. Adjustable assist sensitivity control with lockout position.

13. Adjustable calibrated maximum flow (15–100 lpm).

14. Adjustable inspiratory plateau and expiratory flow resistance.

15. Nebulizer on-off control.

16. Adjustable heated humidifier.

17. Power on-off switch.

11-10 Accessory Devices Used in Respiratory Therapy Apparatus

Accessory devices used with respiratory therapy apparatus enhance the possibilities for quality patient treatment.

Respiratory monitors (refer to Chapter 10) typically have meter displays for tidal volume, minute volume, and respiratory rate, or low/high pressure alarm settings and audible tones. These are portable and operate from 60-Hz power lines or internal rechargeable batteries. Output signals are usually provided for oscilloscopes or strip-chart recorders. Respiration transducers are thermistor probes that change temperature as the patient breathes air across its surface. *Pneumographs* plot physical respiratory excursions.

Portable oxygen analyzers are also used to give oxygen concentrations in controlled environments (refer to Chapter 10), such as tents, incubators, oxygen hoods, and ventilators.

Ventilation failure devices attach to the patient's gas delivery line and give audible warning for respiratory malfunction, supply gas loss, massive leaks, compliance-obstruction changes, and loose tracheostomy fittings.

11-11 Sterilization and Isolation Procedures in Respiratory Therapy Units

Sterilization of respiratory therapy equipment results in the *destruction of all micro-*

organisms and their spores. This is necessary because gases carrying bacteria into the lungs can cause severe lung damage due to infection. The effect of respiratory treatment is then reduced, and in the extreme the patient may die.

Sterilization processing of contaminated equipment should be done in an area separate from clean storage. All apparatus should be dismantled as completely as possible and washed with a brush in a detergent. Thorough rinsing and draining is then required. Equipment should then be packaged or arranged for sterilization by using the following:

1. *Autoclave*—steam treatment at temperatures higher than 212° for 20 min or longer.

2. *Ethylene oxide gas*—microbicidal and sporicidal agent poured over equipment parts for at least 5 min (45 min and longer are also common).

3. *Cold sterilization using Cidex*—2 percent aqueous activated dialdehyde used as a bactericidal, tuberculocidal, virucidal, and sporicidal agent poured over equipment for 10 min to 10 h.

4. *Automatic decontamination system (Cidematic)*—device that cleans, disinfects, and dries equipment automatically.

5. *Sonacide (potentiated acid glutaraldehyde)*—lemon-scented sterilizing/disinfecting solution.

These techniques are selectively designed to kill *fungi, bacteria,* and most *viruses*. However, regardless of the method used, *periodic bacteriological cultures* should be taken to evaluate sterilization effectively.

In addition to sterilization, *isolation* should be required for certain patients to prevent the transmission of infection. Infection may be caused by direct/indirect/droplet contact, through a *vehicle route* (blood, water, drugs, food), *vector route* (insects), or *airborne route* (droplet and dust). *Frequent hand washing* is always a good practice. *Wearing gloves, gowns, and masks* is an added precaution near infected patients. Isolation is required for patients with the following *infections:* staphylococcus aureus, streptococcus, diphtheria, herpes simplex, plague, rubella, smallpox, rubella (congenital syndrome), meningitis, mumps, tuberculosis, and any aspergillosis, pseudomonas, bacillus, or coccidioidomycosis (pneumonia) conditions.

11-12 Typical Faults and Maintenance Procedures for Ventilators

In general, *air tubes and their connections* cause considerable difficulty. This can be minimized by frequent inspections, especially prior to patient use. *Worn clamps and tubing* should be replaced immediately. Humidifiers and nebulizers frequently become *clogged,* but *constant cleaning* reduces this occurrence. Occasionally, someone will accidently *spill some fluid* (blood, urine, saline, antiseptic betadyne, or water) into the machine. *Extensive* disassembly and *parts replacement* (switches, relays, motors, and air filters) often becomes necessary.

Leaks in compressed gas (oxygenated helium) cylinders and ventilator systems do occur despite attentive inspections. These are tracked down by rating differential pressures in various parts of the system. Obviously, leaks reduce gas pressure and volume delivered to the patient.

Routine PM (preventive maintenance) consists of periodic inspections of system connections and functional operation. Electrical leakage current (less than 10 μA) checks should be done monthly. *Calibration* is usually performed every six months. The applicable *operation and maintenance manual* should always be consulted for calibration procedures.

Intake and patient line *air filters* must be

cleaned or replaced periodically (every few weeks depending on use and patient infection). This assures proper air flow to the patient.

Electrical components that require routine replacement include lamps, switches, actuating devices, motors, and heaters. *Electronic problems* (faulty capacitors, diodes, transistors, and integrated circuits) are relatively few.

11-13 Summary

1. Two general *respiratory diseases* that may require artificial respiratory therapy are *hypoxia* (low blood oxygen content) and *hypercapnia* (poor alveolar ventilation).
2. The *goal* of an *artificial ventilatory* control mechanism is to adjust alveolar ventilation to changing body needs of oxygen intake and carbon dioxide removal.
3. *Medical gas safety* is regulated by the NFPA, DOT, FDA, OSHA, and state/local agencies. Gas cylinders are dangerous since air, oxygen, helium, and nitrous oxide compressed gas may cause fire, explosion, or direct physical injury. Interchangeable threaded piping is controlled by specific systems and organization (DISS).
4. *Oxygen therapy* is the administration (treatment) of oxygen for treatment of conditions (pneumonia, edema, obstruction, heart failure, and thrombosis) resulting in patient oxygen deficiency. Devices such as gas regulators, flowmeters, humidifiers, nebulizers, masks, tents, and incubators are commonly used.
5. *Intermittent positive pressure breathing (IPPB)* therapy is a type of assisted breathing pattern in which the lungs are inflated with air, O_2, or O_2-helium by a positive pressure during inspiration. Upon release of the pressure, expiration occurs passively. IPPB therapy promotes more uniform breathing, facilitates better O_2 and CO_2 gas exchange, and removes secretions. Patient orientation is extremely important. Assembly and operating procedures should be followed carefully. Most IPPBs use a *Bennett*-type valve to control inspiration and expiration. Spirometer, alarms, humidifiers, and filters are common accessories used.
6. *Cardiopulmonary resuscitation (CPR)* is a mechanical manual method for maintaining respiration under emergency conditions.
7. *Artificial mechanical ventilators* are devices that connect to the patient's airway and are designed to augment or replace the patient's ventilation. They work automatically and are either *pressure-cycled* (pressure-limited) or *volume-controlled* (volume-limited). The applicable *operation and maintenance manual* should always be consulted for instructions (guessing can be dangerous to the patient).
8. Respiratory therapy apparatus *accessories* include respiratory monitors, pneumographs, O_2 analyzers, and alarms.
9. *Sterilization* of respiratory therapy equipment is a process of cleaning deposits and destroying bacteria and viruses. This is accomplished by using an autoclave, ethylene oxide gas, cold liquids (dialdehyde), and automatic decontamination chambers.
10. Typical *faults* in artificial respiratory ventilators include broken or leaky air tubes/clamps, clogged devices/filters, leaks, burned-out lamps, broken switches, and rare electronic failures (capacitors, diodes, transistors, and integrated circuits).

11-14 Recapitulation

Now return to the objectives and self-evaluation questions at the beginning of the chapter and see how well you can answer them. If you cannot answer certain questions, place a check mark next to each and review appropriate parts of the text. Next, try to answer the following questions using the same procedure. When you have answered all of the questions, solve the problems in Section 11-16.

11-15 Questions

11-1 Two general respiratory therapy diseases, ________ and ________, cause patient ventilation difficulties.

11-2 The goal of an artificial ventilatory control mechanism is to adjust ________ ventilation to body needs of ________ intake and ________ ________ removal.

11-3 The following six regulatory agencies are involved in compressed gas safety: ________, ________, ________, ________, ________, ________.

11-4 Compressed gas in an ICC-3B seamless tank at 150 psig is considered *high/low* pressure by industrial standards.

11-5 For hospital piping system safety, automatic ________ valves, as well as separate station manual ________ valves, are required.

11-6 Therapeutic gases include ________, ________, ________, ________, ________-________ mixture, and ________-________ ________ mixture.

11-7 Interchangeable threaded connections on gas cylinders are provided by the ________-________ ________ System developed by the ________ ________ Association.

11-8 The ________-________ ________ System provides the two-pin approach, which prevents incorrect ________ of medical gas cylinders with flush-type valves.

11-9 Oxygen therapy is the ________ of ________ gas for treatment of conditions resulting from ________ deficiency.

11-10 Oxygen can be introduced through nasal ________, ________, or ________.

11-11 Gas regulators are used to ________ cylinder ________ to a safer level such as ________ psig.

11-12 A flowmeter is a device that controls and ________ gas flow in units of ________ ________ ________.

11-13 Aerosol therapy units are known as ________ and may operate on the ________ principle or from ________ energy.

11-14 Infant incubators are a type of ________ therapy apparatus.

11-15 IPPB represents ________ ________ ________ ________ and is defined as a type of assisted ________ pattern in which the ________ are inflated by a ________ pressure during inspiration.

11-16 The objectives of IPPB therapy are to promote more ________ ventilation, facilitate better ________ and ________ ________ exchange, relieve ________, assist in removal of ________, and ________ respiratory muscles.

11-17 Medical gases commonly used are ________, ________, ________, and ________.

11-18 IPPB units commonly use a ________ valve, which contains a rotating internal ball. This controls an intermittent ________ ________ pressure that assists in inflating the patient's ________.

11-19 CPR refers to what procedure?

11-20 How much pressure is usually fed to a Bennett valve?

11-21 Can the Bennett valve be adjusted to respond to the patient's inspiratory initiation?

11-22 What accessories can be used with IPPB units?

11-23 Why are the assembly and operating procedures for IPPB units important?

11-24 Artificial mechanical ventilators are generally cycled by which two methods?

11-25 The Bennett MA-1 ventilator cycles by what method?

11-26 What is the maximum oxygen concentration, breaths per minute, and flow rate on the Bennett MA-1 ventilator?

11-27 Why are pressure limit settings available on the Bennett MA-1 ventilator?

11-28 The sigh control on the Bennett MA-1 ventilator serves what long-term purpose?

11-29 Name four types of accessory devices used with artificial respiratory ventilators.

11-30 What is the purpose of sterilization for respiratory therapy equipment?

11-31 How would one proceed to sterilize respiratory apparatus using available sterilizing equipment?

11-32 Should one wear gloves, gown, and mask to repair disease-contaminated respiratory therapy equipment?

11-33 Name five of the most common faults with artificial respiratory ventilators and list their solutions.

11-16 Problems

11-1 A respiratory therapist reports that a mechanical ventilator does not turn on when the power switch is toggled. What checkout procedure would you use and which system components would be suspected first?

11-2 A nurse reports that an artificial respiratory ventilator appears to be delivering insufficient gas flow to the patient with proper flow, pressure, and gas mixture front panel settings (based on the nurse's experience). What checkout procedure would you use and which system components would be suspected first?

11-17 References

1. Hunsinger, Doris L., Karl Lisnerski, Jerome J. Maurizi, and Mary Phillips, *Respiratory Technology—A Procedure Manual,* Reston Publishing Co., Inc. (Reston, Va., 1976).
2. Moore, Francis D., John H. Lyons, Jr., Ellison C. Pierce, Jr., Alfred P. Morgan, Jr., Philip A. Drinker, John D. MacArthur, and Gustave J. Kammin, *Post-Traumatic Pulmonary Insufficiency,* W. B. Saunders Co. (Philadelphia, 1969).
3. Petty, Thomas L., *Intensive and Rehabilitative Respiratory Care—A Practical Approach to the Management of Acute and Chronic Respiratory Failure,* Lea and Febiger (Philadelphia, 1971).
4. Beall, Cheryl E., Harold A. Brown, and Frederick W. Cheney, Jr., *Physiological Basis for Respiratory Care,* Mountain Press Publishing Co. (1974).
5. *Medical and Biological Engineering, 12* (2): 160–169 (1974). "A Comprehensive Volume-Cycled Ventilator Embodying Feedback Control."
6. Pace, William R., Jr., *Pulmonary Physiology in Clinical Practice,* F. A. Davis Company (Philadelphia, 1970).
7. *Operation and Service Manual* for Bennett TV-2P IPPB Pressure Breathing Therapy Unit, Puritan-Bennett Corp. (Kansas City, Mo., 1969).
8. *Operation and Service Manuals* for Bird Mark 1,7,8,10,14 Intermittent Positive Pressure Ventilator, Bird Corp. (Palm Springs, Calif., 1974).
9. *Operation and Service Manual* for Bennett PR-2 IPPB Mobile Respiratory Therapy Unit, Puritan-Bennett Corp. (Kansas City, Mo., 1969), Form 2131 B.
10. *Operation and Service Manual* for Bennett MA-1 Volume-Limited Respiration Unit, Puritan-Bennett Corp. (Kansas City, Mo., 1969, Form 5030 J.
11. *Operation and Service Manual* for Ohio 560 Volume Ventilator, Ohio Medical Products, A Division of Airco, Inc. (Madison, Wisc.) Form 1887.
12. *Operation and Service Manual* for Emerson Postoperative Constant Volume, Pressure Variable Ventilation, J. H. Emerson Co.

Chapter 12

The Human Nervous System

12-1 Objectives

1. Be able to introduce the biological principles underlying the human central nervous system (CNS), the peripheral nervous system (PNS), and the autonomic nervous system (ANS).
2. Be able to describe the structure and function of the *neuron* (single nerve cell), *CNS* (brain and spinal cord), *PNS* (nerve pairs), and *ANS* (sympathetic and parasympathetic).
3. Be able to identify specific areas of the brain concerned with bodily sensory and motor functions.

12-2 Self-Evaluation Questions

These questions test your prior knowledge of the material in this chapter. Look for the answers as you read the text. After you have finished studying the chapter, try answering these questions and those at the end of the chapter (see Section 12-12).

1. List the major divisions of the nervous system.
2. Draw a diagram and define various portions of a *neuron* (single nerve cell).
3. Describe *nerve impulse conduction* through one neuron and several neurons connected together.
4. Describe the basic structure and function of the central nervous system (*CNS*, brain and spinal cord), peripheral nervous system (*PNS*), and autonomic nervous system (*ANS*, sympathetic and parasympathetic).
5. Name the four lobes of the cerebrum.
6. Identify the areas of the brain responsible for the senses of sight, sound, touch, smell, and taste.
7. Identify the areas of the brain responsible for the function of muscle movement, memory, intelligence, judgment, imagination, creativity, and conscious thought.
8. How does blood circulate through the brain?
9. What is the purpose of the three membranes (meninges) covering the brain?
10. Does human behavior control brain function or does brain function control human behavior?

12-3 Organization of the Nervous System

The nervous system is a complex interconnection of nervous tissue that is concerned with the *integration* and *control of all bodily functions*. Knowledge of this intricate system is still quite limited. We do know, however, that this system allows an individual to *detect* internal and external environmental *changes* (stimuli) and to *interpret* (analyze) the resulting nerve impulses.

Homeostasis (constancy or stability in internal body states) is achieved through a network of *negative feedback loops* involving nervous (electrochemical) and humoral (biochemical) components. For example, blood carbonate is detected in the brain. If it is too high (indicating too much carbon dioxide in the blood), the brain initiates movement of breathing muscles. This, in turn, increases breathing rate and rids (ventilates) the body of carbon dioxide gas through the lungs.

The nervous system is generally considered the most complex bodily system and is divided into several major divisions distinguished by *anatomy* (structure and location) and *physiology* (function), including the following:

Figure 12-1 The nervous system. (From McNaught & Callander, *Illustrated Physiology.* Used by permission of Churchill Livingstone.)

NERVOUS SYSTEM

The Nervous System is concerned with the INTEGRATION and CONTROL of all bodily functions.

It has specialized in IRRITABILITY – *the ability to receive and respond to messages from the external and internal environments*

and also in CONDUCTION – *the ability to transmit messages to and from CO-ORDINATING CENTRES.*

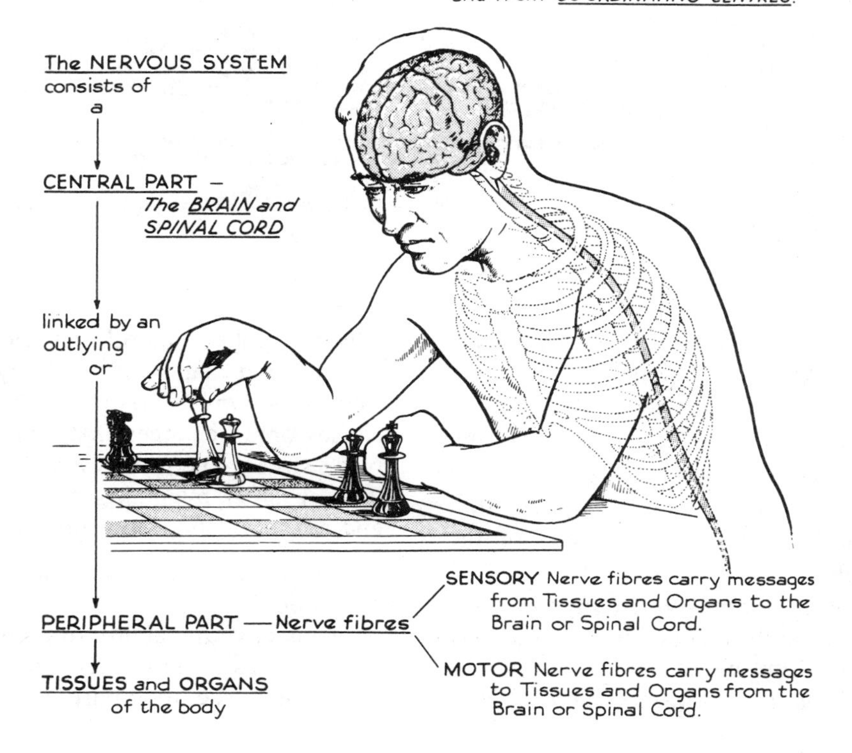

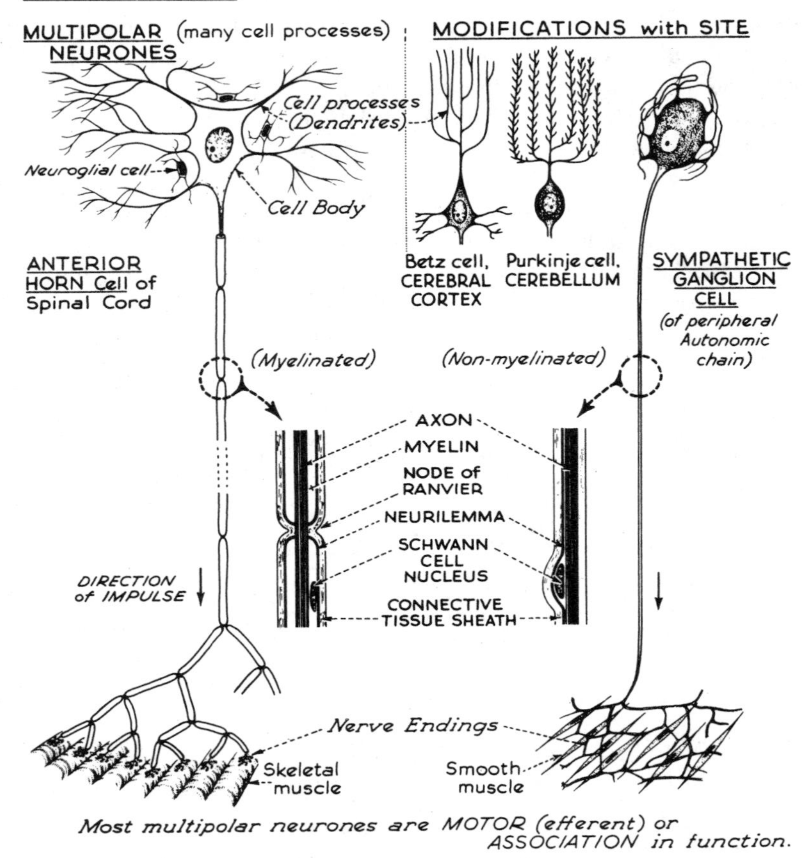

Figure 12-2 Nervous tissue. (From McNaught & Callander, *Illustrated Physiology.* Used by permission of Churchill Livingstone.)

1. *Central nervous system (CNS),* which is enclosed within the skull and vertebral column—*brain and spinal cord.*

2. *Peripheral nervous system (PNS),* which consists of nervous tissue outside the skull and vertebral column—periphery (extremity) of the body. Subdivisions of the PNS include:

(a) *Somatic system,* which supplies sensory motor and sensory fibers to the skin and skeletal muscles.

(b) *Autonomic nervous system (ANS),* which supplies smooth muscle, cardiac muscle, and glands in the body viscera. The *sympathetic* (stimulatory) *system* causes organ changes that help the body resist stress. The *parasympathetic* (inhibitory) *system* maintains normal function and conserves body resources.

Figure 12-1 shows an overview of the nervous system.

Humans have the ability to perceive their environment as well as receive it. *Reception* involves response to environment stimuli, while *perception* relates to the recognition of symbolic patterns. These patterns are abstract (not physical) and are composed of linguistic symbols. The ability of the human to think of the physical world in abstract terms accounts for the extraordinary talent to manipulate objects, construct houses, and control his or her environment.

12-4 The Neuron (Single Nerve Cell)

The *neuron* is the fundamental unit of the nervous system. It is a single cell composed of a *cell body (soma),* several short *"input" projections (dendrites),* and a *long propagation channel (axon).* The axon together with its sheath (covering) forms the nerve fiber. Figure 12-2 shows a nerve cell connected to *skeletal muscle* and to *smooth muscle*. Notice that the nerve impulse travels in one direction only from dendrites to nerve endings. The *axon* extends the entire length of the nerve cell, and some are surrounded by a *myelin sheath* (segmented insulating covering). The *neurilemma* encompasses this sheath and is composed of *Schwann cells. Nodes of Ranvier* act to speed up the nerve impulse transmission.

The *transmission of the nerve impulse* is actually a result of biochemicals that travel across the *synapse* (space between nerve cells). The change in membrane permeability is the chief reason for nerve pulse transfer across the synaptic junction. Figure 12-3 shows the synaptic ultrastructure. Observe

Figure 12-3 Synaptic ultrastructure.

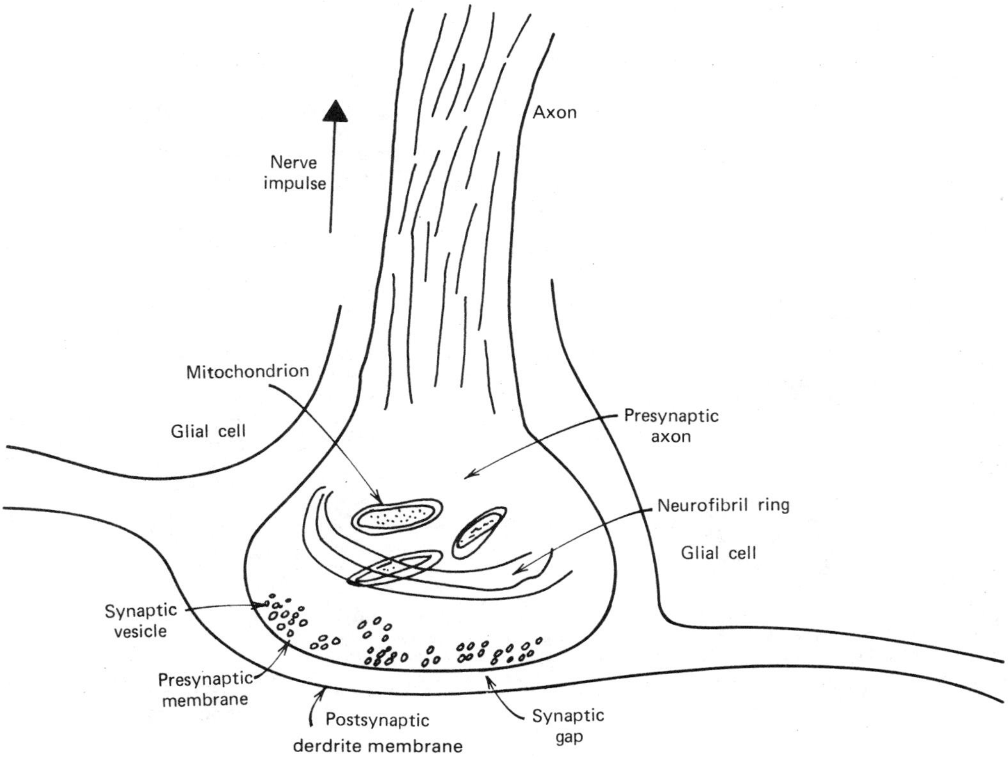

the presynaptic and postsynaptic membranes and synaptic gap. The excitatory postsynaptic potential (*EPSP*) results in *higher* levels of conduction, and the inhibitory postsynaptic potential (*IPSP*) results in *decreased* or inhibited conduction.

The conduction pulse results in a wave of depolarization (action potential) similar to that presented for heart muscle in Chapter 2, particularly Figures 2-3 and 2-4. *Nerve conduction is in one direction* only (constant speed) from dendrites through axon to nerve endings. Neurons can connect to each other in the following arrangements: *one to one, one to many, many to one, or many to many.* Figure 12-4 shows these possibilities.

The axons and dendrites (nerve fibers) bundle together to form a *nerve*. The sensory nerves carried to the brain are known as *afferent nerves,* and the ones carried away from the brain are called *efferent nerves.* Nerves switch on and off in such a manner as to cause abrupt changes in cell voltages. In effect, this nerve impulse switching is sim-

Figure 12-4 Arrangements of neurons (neurones). (From McNaught & Callander, *Illustrated Physiology.* Used by permission of Churchill Livingstone.)

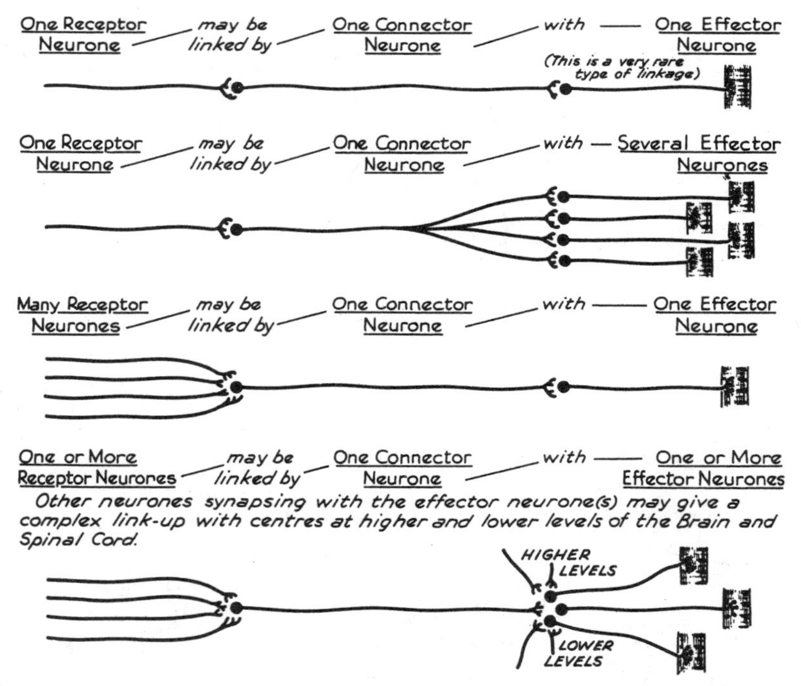

ilar to electronic digital circuit logic. Several nerve cells may be required to conduct before a triggering threshold is reached. The *AND* and *OR* functions performed by nerve cells act to control bodily coordination and reflexes.

Reflex action in the human involves many reflex arcs. A nervous reflex is an *involuntary* action response caused by stimulation of an afferent nerve ending or receptor. The knee jerk in response to the tap of a hammer is one example of a reflex arc, the components of which are shown in Figure 12-5. The components of the arc are:

1. A *receptor,* which detects change.
2. An *afferent neuron,* which conducts the nerve impulse from the sensory area to the CNS.
3. A *center or synapse,* which connects neurons together.
4. A brain processing area.
5. An *efferent neuron,* which conducts nerve impulses from the CNS to an organ for appropriate response.
6. An *effector or organ,* which responds to maintain homeostasis.

12-5 The Structure and Function of the Central Nervous System[1]

The brain is *defined*[2] as "a large soft mass of nerve tissue contained within the cranium, the *encephalon.*" Three major structures compose the *brain* (shown in Figures 12-6, 12-7, 12-8, and 12-9): the *brain stem*—automatic vital system control, the *cerebellum*—involuntary muscle control and coordination, and the *cerebrum*—voluntary movement, sensation, and intelligence.

[1] The central nervous system is composed of the brain and spinal cord. Many excellent drawings and photographs have been presented by anatomists. *Yokochi* has excellent pictures of actual brain tissue. Frank H. Netter, *Ciba Pharmaceutical Products, Inc.* has a superb 13-minute motion picture presentation from the Department of Anatomy, U.C.L.A. Medical School (Teaching Films, Houston, Texas, Clemintine and Hardwick) entitled *Guides to Dissection, The Cranial Cavity—Removal of the Brain.*

[2] *Taber's Cyclopedic Medical Dictionary,* F. A. Davis Co. (Philadelphia, 1973).

12-5.1 Brain Stem

The *brain stem* consists of the *medulla* (oblongata), *pons, midbrain,* and *diencephalon.*

The *medulla* automatically controls heart rate and breathing. (Actually, most essential life systems are controlled here.) Reflex functions such as coughing, sneezing, and vomiting are associated with the medulla. Indeed one modern definition of clinical death is the absence of lower brain EEG activity.

The *pons* is about 2.5 cm long and forms a noticeable bulge on the anterior surface of the brain stem. It *functions* as a relay station for motor respiratory and auditory fibers from the cerebrum and cerebellum. Other impulses from eye movement, head muscles, and taste sensors also pass through here.

The *midbrain* is a wedge-shaped portion of the stem. Midbrain tissues *function* as a motor relay station for fibers passing from the cerebrum to the cord and cerebellum. Integration of visual and auditory reflexes, including those concerned with avoiding objects, also occur here.

The *diencephalon* forms the superior (top) part of the brain stem. As part of the original forebrain, it develops into the *thalamus* and *hypothalamus.*

The *thalamus* (paired thalami) receives fibers from the hearing structures of the inner ear and visual system. It also provides pathways for somatic sensory systems. All this sensory information eventually reaches the *cerebrum,* where it is processed.

The *hypothalamus* responds to the properties of blood passing through nerve con-

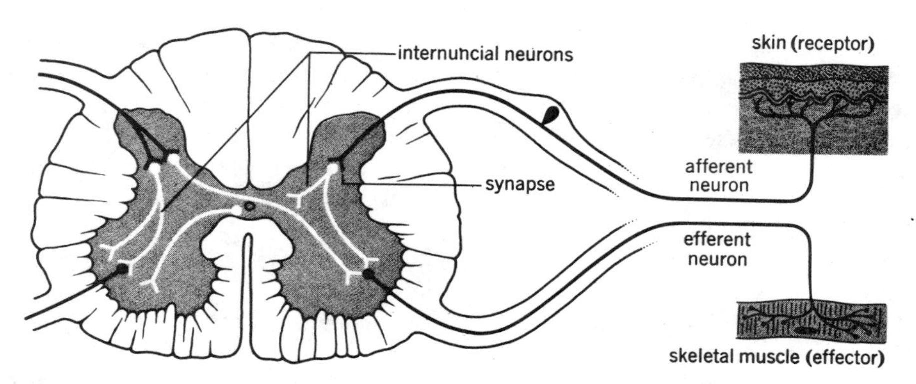

Figure 12-5 Components of reflex arc. (From *Human Anatomy and Physiology,* 2nd edition, by James E. Crouch, Ph.D. and J. Robert McClintic, Ph.D. John Wiley & Sons, New York, 1976. Used by permission.)

Figure 12-6 The cerebrum. (From McNaught & Callander, *Illustrated Physiology.* Used by permission of Churchill Livingstone.)

CEREBRUM

The largest part of the human brain is the CEREBRUM — made up of 2 CEREBRAL HEMISPHERES. Each of these is divided into LOBES.

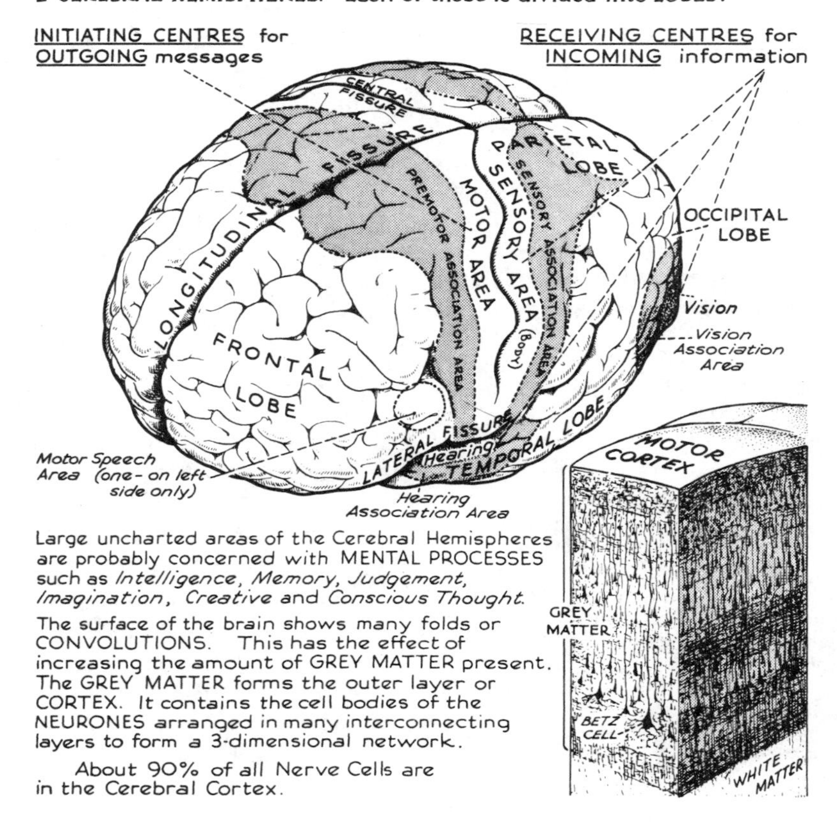

HORIZONTAL SECTION through BRAIN

This view shows surface GREY MATTER containing Nerve Cells and inner WHITE MATTER made up of Nerve Fibres.

Deep in the substance of the Cerebral Hemispheres there are additional masses of GREY MATTER:-

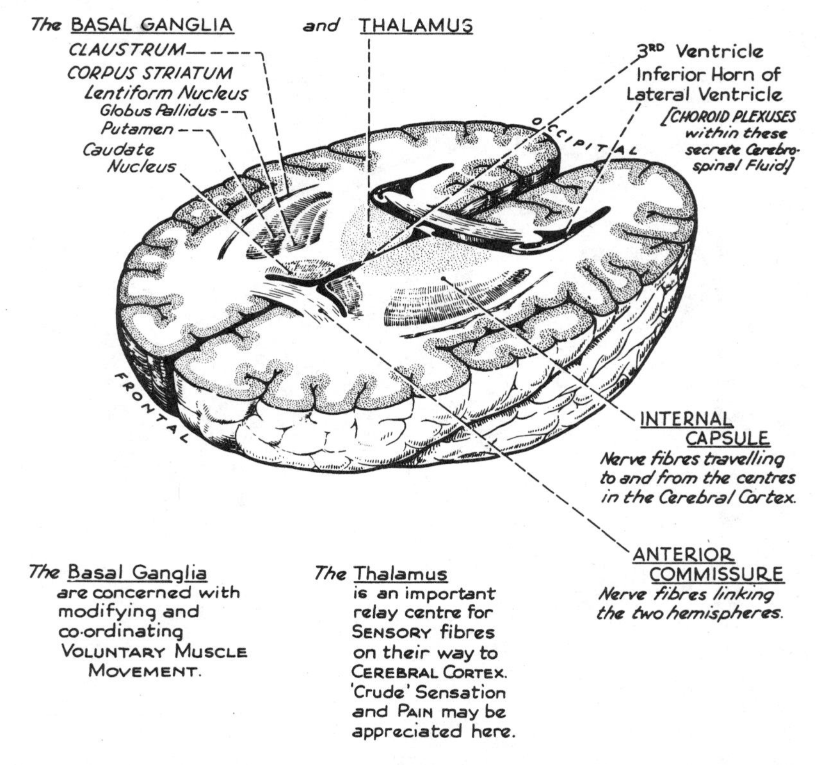

Figure 12-7 Horizontal brain section. (From McNaught & Callander, *Illustrated Physiology.* Used by permission of Churchill Livingstone.)

nections. The *endocrine system* is controlled through nerve responses, affecting emotional behavior patterns. Other functions controlled via chemical interaction with the *pituitary gland* are temperature regulation, water balance, food intake, gastric secretion, sexual behavior, and sleeping patterns.

As a general arousal mechanism, the *reticular activation system (RAS)* functions with the thalamus to prepare the cerebral cortex (higher parts of the brain) for incoming sensory stimulation data. This system is a network of gray matter placed centrally in the brain stem.

12-5.2 Cerebellum

The *cerebellum* (Figures 12-8 and 12-9) is the second largest portion of the brain (the cerebrum is the largest) and, essentially, integrates incoming sensory messages to provide smooth body *muscle movements, balance,* and *equilibrium*. This portion of the brain has an outer *cortex* of gray matter and an inner

This is a Vertical Section through the LONGITUDINAL FISSURE — a deep cleft which separates the two Cerebral Hemispheres. At the bottom of this cleft are tracts of nerve fibres which link up the different LOBES of each hemisphere and also link the two hemispheres with each other — the CORPUS CALLOSUM.

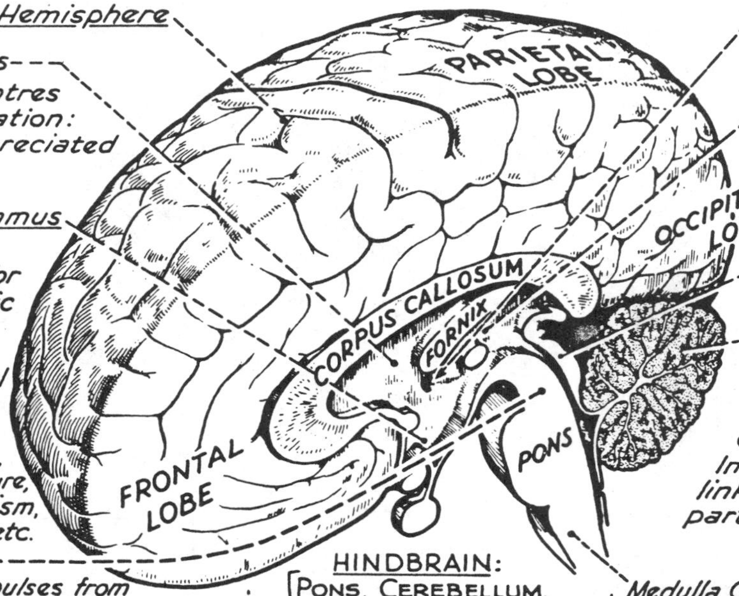

FOREBRAIN

Cerebral Hemisphere

Thalamus — relay centres for sensation: pain appreciated here.

Hypothalamus — contains centres for Autonomic Nervous System. e.g. Control of Heart, Blood pressure, Temperature, Metabolism, etc.

Opening of Lateral Ventricle

3RD Ventricle

PARIETAL LOBE

OCCIPITAL LOBE

CORPUS CALLOSUM

FORNIX

FRONTAL LOBE

PONS

Corpora Quadrigemina

Cerebellum
Centres concerned with balance and equilibrium. Important tracts link it with other parts of Brain and Spinal Cord.

MIDBRAIN
Receives impulses from Retina and Ear. Serves as a centre for Visual and Auditory Reflexes. In the Grey Matter are nerve cell bodies of III, IV Cranial nerves and the Red Nucleus which helps to control skilled muscular movements. The White Matter carries nerve fibres linking Red Nucleus with Cerebral Cortex, Thalamus, Cerebellum, Corpus Striatum and Spinal Cord. It also carries Ascending Sensory fibres in Lateral and Medial Lemnisci, and Descending Motor fibres on their way to Pons and Spinal Cord.

HINDBRAIN:
[PONS, CEREBELLUM, MEDULLA OBLONGATA]

Pons: Groups of Neurones form sensory nucleus of V and also nuclei of VI and VII Cranial nerves. Other nerve cells here relay impulses along their axons to Cerebellum and Cerebrum. Rubrospinal tract, Lateral and Medial Lemnisci pass through Pons and nerve fibres linking Cerebral Cortex with Medulla Oblongata and Spinal Cord.

Medulla Oblongata:
Groups of Neurones form Nuclei of VIII, IX, X, XI, XII Cranial nerves. Gracile and Cuneate Nuclei — second sensory neurones in cutaneous pathways. Tracts of Sensory fibres decussate and ascend to other side of Cerebral Cortex. Some fibres remain uncrossed. The larger part of each Motor pyramidal tract crosses and descends in other side of Spinal Cord.

Figure 12-8 Vertical brain section. (From McNaught & Callander, *Illustrated Physiology.* Used by permission of Churchill Livingstone.)

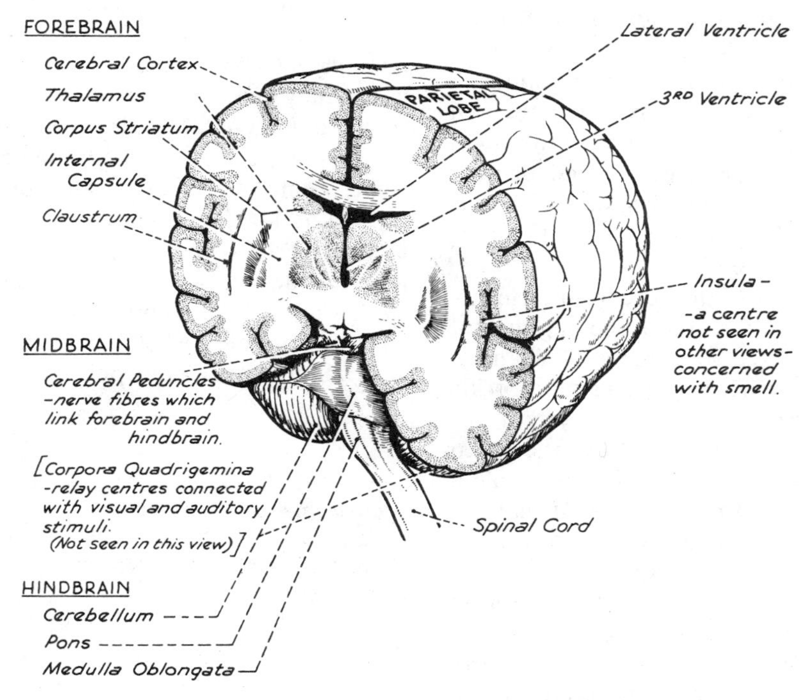

Figure 12-9 Coronal brain section. (From McNaught & Callander, *Illustrated Physiology.* Used by permission of Churchill Livingstone.)

medulla composed of white matter. The cerebellum acts at the subconscious level in coordinating reflexes that automatically establish an upright posture. Thus, *proprioception* (information on position of limbs and movements of muscles) is established through general and special body proprioceptors. Some consider the cerebellum to be a part of the "old" brain involving basic locomotion but not intellectual thought.

As an example, consider yourself driving a bicycle or car down a road that has frequent turns. When you begin drifting to the right (¼ ft), you take no action to correct your position until the drift becomes appreciable (2 ft). You then apply a force opposite in direction to that which caused the drift until a center of the lane position is obtained. This constitutes *negative feedback* from a macroscopic point of view and allows you to maintain a relatively smooth path down a winding road. Of course, each person has a different perception threshold (ability to sense appreciable drift), but we do manage (most of the time) to speed smoothly past each other at speeds of 10 mph up to even 70 mph. If this integrated brain-muscle action were not present, you would travel down the road in short, jerky paths using much more muscle energy.

12-5.3 Cerebrum

The *cerebrum* (Figure 12-6) is the largest part of the brain in humans and acts as a nerve impulse processor as well as a memory bank and controller of voluntary motor actions. A *longitudinal fissure* (a groove or natural division) divides the cerebrum into lateral halves or *hemispheres*. Each hemisphere is further subdivided by other fissures. The surface area of the brain is large due to the many folds of the cerebrum.

The outer layer of the cerebrum forms the *cerebral cortex* and is composed of gray matter. Approximately 90 percent of all nerve cells are in this cortex. Inner portions of the brain are comprised of white matter.

John Von Neumann[3] (mid-twentieth-century mathematician—Von Neumann Computer Architecture) had calculated that in a 60-year lifetime a human would store about 2.8×10^{20} bits of information in the nearly 10 billion neurons found in the cerebrum. The conclusion of this grand computer-type model of the brain reveals that every neuron must have a memory capacity equivalent to 30 billion off-on switches. This seems inconsistent with the (early) idea that one neuron stores only one bit of data. But recent evidence indicates that neuronal activity does increase the amount of the nucleic acid RNA. Although a firm conclusion is not yet available, John Von Neumann might have been correct in his estimation of the mechanism of human memory, and this is only a fraction (one-third) of the total brain capability. Humans do seem, however, to be more than just learning machines or computers.

The *structure* of the *cerebrum* is typically divided into the following *lobes* (two each, one on each side) that are named for the bones they lie beneath: frontal, temporal, parietal, and occipital. These areas of the brain are, in part, concerned with the sensation of sight, sound, smell, taste, and touch.

Frontal lobes contain motor neurons that control certain bodily muscles. Within the lobe on the left side, speech action is initiated. Figure 12-10 shows the *motor homunculus* ("diminutive of" or "little man"), which indicates various portions of the brain

Figure 12-10 Sensory locations in the cortex. (From *Human Anatomy and Physiology,* 2nd edition, by James E. Crouch, Ph.D. and J. Robert McClintic, Ph.D. John Wiley & Sons, New York, 1976. Used by permission.)

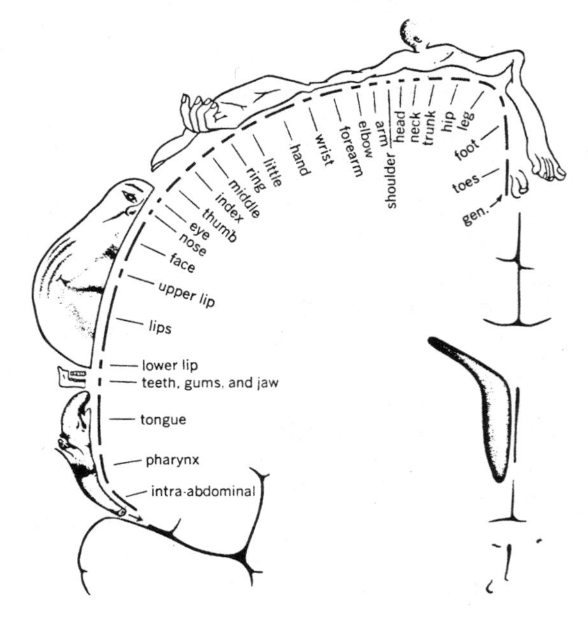

[3] Before the age of 25, Von Neumann wrote *The Theory of Games and Economic Behavior,* which characterized the theoretical precise steps and outcomes of games such as poker. He also wrote *The Computer and the Brain* (1956), in which he described the brain as having a language that involves specific activities that are related.

designated to motor action. Notice that a large portion of the frontal lobe is dedicated to the face, especially the lips and tongue. This accounts for human speech development. This is much less predominant in most animals. Figure 12-11 shows the *sensory homunculus*.

Prefrontal lobes are most likely related to mental processes such as intelligence, memory, judgment, imagination, and creative and conscious thought. Disease of these areas such as anoxia (lack of oxygen) at birth can lead to mental retardation or even behavioral disorders.

Temporal lobes house the auditory sensory areas superior (above) to the ears. High-frequency auditory sensation seems to occur in the temporal areas closest to the frontal lobes. These lobes are also concerned with learning and memory of visual and auditory images. Damages to these areas can result in epilepsy, with severe behavioral disorders and aggressive outbursts. Even the loud, excitable, noisy child (*the hyperkinetic syndrome*) may be accompanied by brain damage. Although brain and nervous tissue will not regenerate once damaged, other nearby portions may take over the intended functions.

Minimal cerebral dysfunction (minimal brain dysfunction, or MBD) has been recognized in recent years as a very subtle cerebral palsy. Mild frontal and temporal lobe damage might be more widespread than previously anticipated.

Parietal lobes relate to the interpretation of size, shape, texture, degree of head, and other qualities of touch. As seen in Figure 12-11, large sensory areas are devoted to the head, especially the lips, face, and tongue. Greater motor brain area is devoted to thumb movement than is devoted to thumb sensation.

Occipital lobes located at the back of the head contain the visual cortex. This region of the brain receives impulses from the eyes by way of the optic nerves and tracts. The left eye is controlled by the right occipital cortex and the right eye by the left. The nerve impulses cross over in the brain through the *optic chiasma* (common crossing point). In fact, much of the left half of the body is controlled by the right brain and vice versa.

The senses of *smell* and *taste* do not have specific cerebral areas associated with them, although the *olfactory bulb* (smell) is near the center of the brain.

Association areas are those areas of the cerebrum from which specific motor responses or sensations cannot be elicited when stimulated (link of many sensations). Internal brain communication occurs across bands of tissues (corpus collosum, fornix,

Figure 12-11 Motor locations in the cortex. (From *Human Anatomy and Physiology,* 2nd edition, by James E. Crouch, Ph.D. and J. Robert McClintic, Ph.D. John Wiley & Sons, New York, 1976. Used by permission.

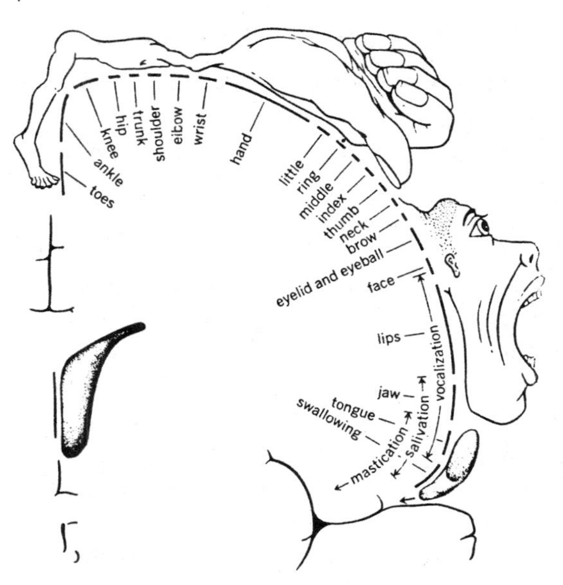

and commissures, seen in Figure 12-7 and 12-8).

The cerebrum also contains four major ventricles through which circulates *cerebrospinal fluid (CSF)*. It is similar to blood plasma but has no clotting agents. Three major functions for this fluid are known: protective covering against physical shock, constant cranial fluid volume regulation, and exchange of metabolic substances to nerve cells. *CSF circulation* occurs due to the forces of *hydrostatic pressure*. Internal hemorrhaging (bleeding) near the brain or nervous system can be detected by withdrawing a sample of CSF from the vertebral column (*spinal tap*). No red blood cells (*RBC*) should appear in the CSF.

Blood circulation is another circulatory system in the brain. Arterial blood is pumped to the head and brain via the *common carotid artery*. Venous blood is returned to the superior vena cava of the heart through the *jugular vein*.

The brain is covered by three major *membranes (meninges)*. The meninges serve to protect and transfer CSF. The common term for inflammation of these membranes is *meningitis*.

Cranial nerves supply the head, neck, and most of the viscera. The 12 pairs of nerves arise directly from the undersurface of the brain, as shown in Figure 12-12. *Sensory* and *motor* nerves are evident.

The *spinal cord* can be described as a long column of nervous tissue. The extension, *nerve cord,* is protected within the bony *spinal column*. All nerves to the trunk and limbs arise from the spinal cord, and it is the center of reflex action containing the conductive paths to and from the brain. It is bathed in cerebrospinal fluid, which also circulates through the brain. Figure 12-13 shows the spinal cord and vertebral canal. Notice the five general divisions with 31 pairs of spinal nerves (cervical, thoracic, lumbar, sacral, and coccygeal).

12-6 The Peripheral Nervous System

The *peripheral* (away from the brain) *nervous system* (PNS) includes the *craniosacral nerves* or *spinal nerves,* the organs of special sense or body skin and muscle receptors, and the sympathetic nervous system or stimulatory nerve responses.

This system acts on the extremities of the body to control muscles that permit mobility.

All of the nerves that branch from the CNS and link to the body are either *motor* (efferent), *sensory* (afferent), or a combination of the two.

12-7 The Autonomic Nervous System

The *autonomic nervous system (ANS)* is typically classified as part of the PNS. It supplies motor fibers to smooth and cardiac muscles and glands, functioning "automatically" at the reflex and subconscious levels to keep basic body systems operating (i.e., heart rate and breathing). Figure 12-14 shows the ANS including *sympathetic* (stimulatory) and *parasympathetic* (inhibitory) areas.

12-7.1 Dynamic Equilibrium

Equilibrium of the following bodily functions is obtained through the ANS:

1. Heart function—rate and volume output.

2. Blood pressure—arterial-venous vessel size.

3. Blood sugar—regulation of liver action.

4. Digestion—regulation of gastric secretions.

5. Growth—hormone secretion through the endocrine system.

6. Body temperature—sweat glands.

7. Body fluid balance—sweat and kidney functions.

8. Emotional reactions—endocrine system and the brain.

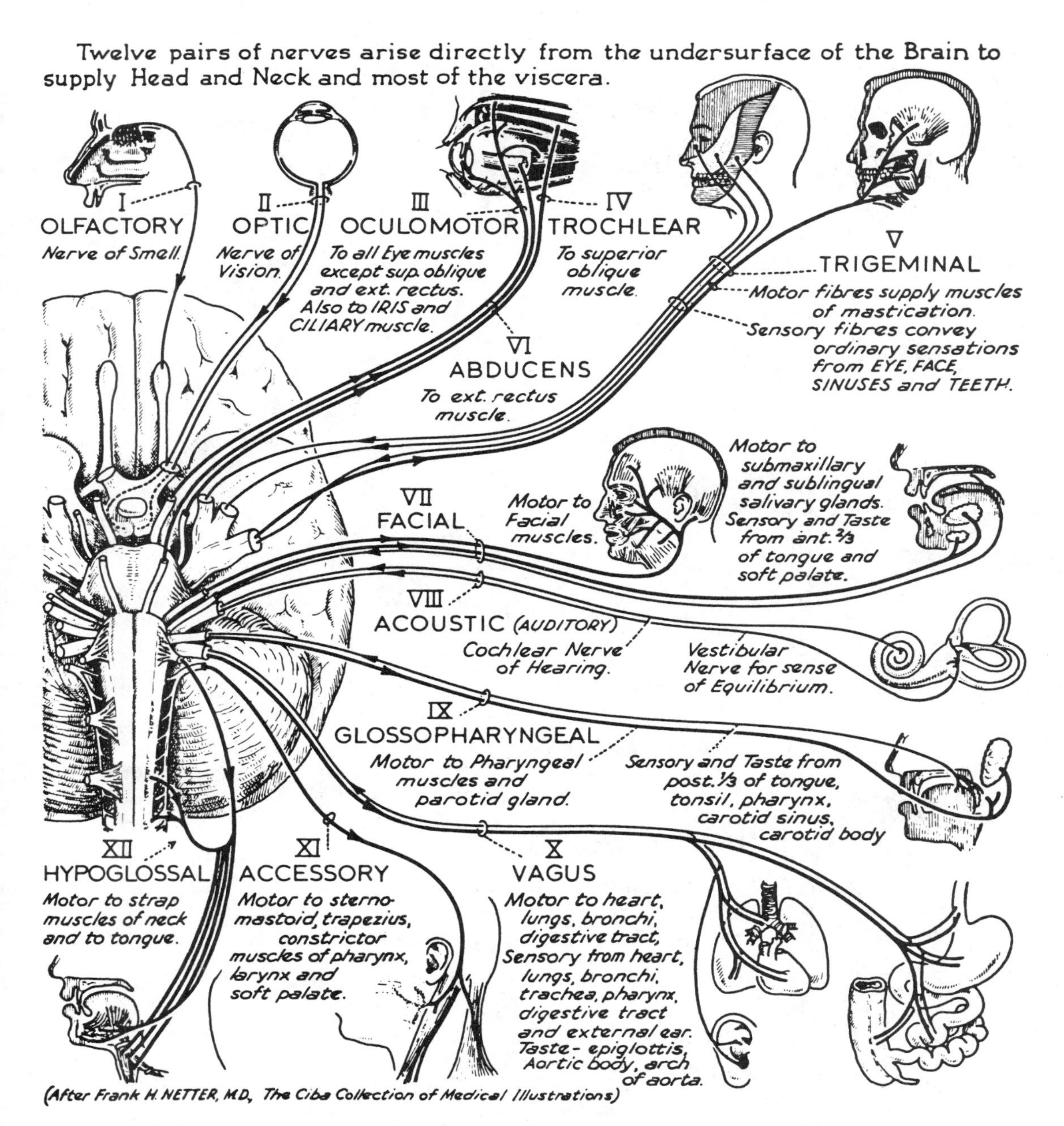

Figure 12-12 Cranial nerves. (From McNaught & Callander, *Illustrated Physiology.* Used by permission of Churchill Livingstone.)

The SPINAL CORD lies within the Vertebral Canal. It is continuous above with the Medulla Oblongata.

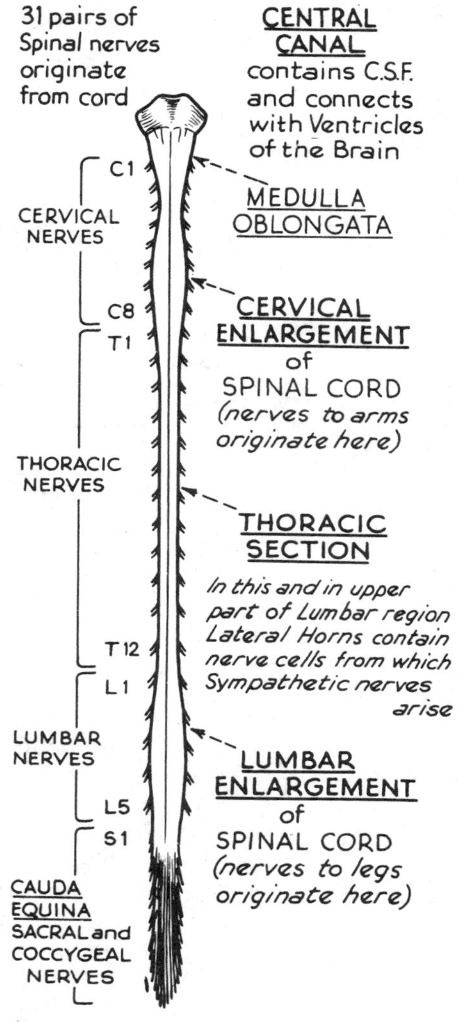

The spinal nerves travel to all parts of the trunk and limbs.

Figure 12-13 Spinal cord. (From McNaught & Callander, *Illustrated Physiology.* Used by permission of Churchill Livingstone.)

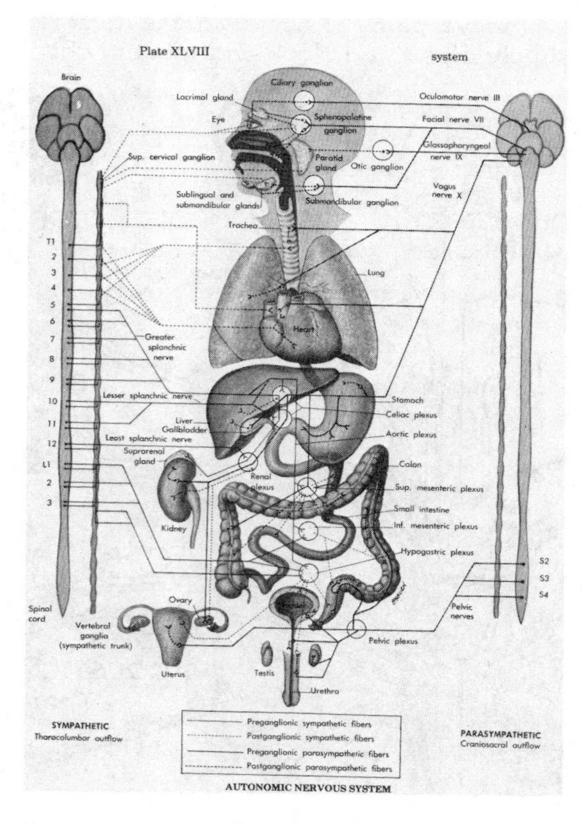

Figure 12-14 Autonomic nervous system. (From *Dorland's Illustrated Medical Dictionary,* 25th edition, W. B. Saunders Co., Philadelphia, 1974.)

12-7.2 Sympathetic Nervous System

This system is formed by thoracic and lumbar nerve outflows. Body resources are made more available through stimulation by this system. Visceral functions are involved. For example, a massive sympathetic discharge is associated with the liberation of *epinephrine (adrenalin)* from the adrenal glands (on top of the renal glands or kidneys). This increases heart rate and skeletal muscle movements that characterize the ''fight-or-flight'' state.

12-7.3 Parasympathetic Nervous System

This system arises from cranial motor nerve outflows, involving the brain stem and spinal nerves. During sleep this system acts to slow the heart rate down to conserve energy.

The *master tissues* (nervous system) control the *vegetative systems* (respiration, digestion, growth, repair, reproduction, and others). Essentially, body control mechanisms originate and are communicated through *nerve impulse* transmission and reception or through *hormonal secretions*. Thus, the body is a conglomerate of *electrochemical* and *biochemical* activity.

12-8 Behavior and the Nervous System

The most sophisticated computer system or machine that humans have ever built is but a fraction of the complexity of the human brain. The hereditary carrier, *deoxyribonucleic acid (DNA),* within a cell's nucleus stores an extensive amount of information concerning structure and function. The most complex organ in the body, the brain, is still greatly uncharted in terms of function, although its structure has been identified.

Linking human brain activity and behavior is uncertain. We do know, however, that *human behavior is adaptive,* and this trait seems to be controlled by the *prefrontal lobes*. The electrical activity of the brain (*EEG*) changes during sleep and other forms of behavior but is difficult to relate to specific patterns of behavior.

The process of *memory* appears to involve three mechanisms. First is *momentary retention* (recall of a phone number long enough to dial it), second is *short-term memory* of events occurring minutes to hours before (classroom lecture), and third is *long-term memory* of events occurring in the past (childhood joys, successes, sorrows, failures, and horrors). Most of our memory bank is hidden from us (*subconsciousness*) in a cloud of misplaced data. Nevertheless, stimulation of areas within the *temporal lobes* evoke recall of long-past experiences.

As we will see in Chapter 13, EEG waveforms can easily be obtained from the electroencephalography machine. The biomedical equipment technology is well developed, even if the interpretation of the recording is not.

12-9 Summary

1. The human nervous system is the *most complex* system in the body.
2. The major divisions of the nervous system distinguished by *anatomy* (structure and location) and *physiology* (function) are the *central nervous system* (brain and spinal cord), *peripheral nervous system* (periphery or extremities), *somatic system* (body sensory and motor areas), and *autonomic nervous system* (sympathetic stimulatory), and *parasympathetic* (inhibitory nervous systems).
3. The *neuron* (a single nerve cell) is the fundamental unit of the nervous system. It consists of a cell body (*soma*), several short "input" projections (*dendrites*), and a long propagation "output" channel (*axon*).
4. Neurons may be connected *one to one, one to many, many to one,* or *many to many.* These combinations may form *afferent nerves* (sensory, toward the brain) and *efferent nerves* (motor, away from the brain). The multiconnection off/on action is similar to that of digital logic circuits (AND/OR).
5. *Reflex arc action* is an involuntary response caused by stimulation of an afferent nerve ending or receptor. The

knee jerk in response to the tap of a hammer is one example.

6. The *central nervous system* is composed of the *brain* (soft mass of nerve tissue within the skull) and *brain stem* (long column of nervous tissue).
7. The *cerebellum* is the second largest part of the brain and integrates incoming sensory messages to provide smooth body muscle movements, balance, and equilibrium (*proprioception*). It acts below the level of consciousness to control functions as *breathing* through *negative feedback loops*.
8. The *cerebrum* is the largest part of the brain and acts as a nerve impulse processor as well as a memory bank and controller of voluntary motor actions. The *longitudinal fissure* separates the cerebrum into two hemispheres. The *cerebral cortex* is composed of *gray matter.* Inner portions consist of *white matter.* Four pairs of lobes are noticeable in the cerebrum:
   (a) *Frontal lobes* (voluntary movements, intelligence, memory, thought).
   (b) *Temporal lobes* (auditory and visual sensation, learning, memory).
   (c) *Parietal lobes* (touch, object size/shape/texture sensation, thumb).
   (d) *Occipital lobes* (visual sensation, crossover in optic chiasma).
9. The cerebrospinal fluid circulatory system operates on hydrostatic pressure, and the CSF circulates between the cerebrum and spinal cord (vertebral column).
10. Blood circulates through the brain via the *common carotid artery* (arterial flow) and *jugular vein* (venous flow).
11. The *peripheral nervous system* includes nerves that supply the extremities of the body.
12. The *autonomic nervous system* includes nerves that supply motor fibers, smooth and cardiac muscles, and glands. It functions *"automatically"* at the reflex and subconscious levels to keep basic body systems operating. The *sympathetic* (stimulatory) and *parasympathetic* (inhibitory) nervous systems function as antagonists to regulate these body systems.
13. The body is composed of *master tissues* (nervous system) and *vegetative systems* (respiration, digestion, growth, repair, reproduction, and others).
14. *Human behavior and brain activity* are difficult to relate.
15. Electrical activity of the brain can be measured and recorded on *EEG* machines. The biomedical equipment technology for EEG measurement is well developed.

12-10 Recapitulation

Now return to the objectives and self-evaluation questions at the beginning of the chapter and see how well you can answer them. If you cannot answer certain questions, place a check next to each and review appropriate parts of the text. Next, try to answer the following questions using the same procedure. When you have answered all of the questions, solve the problems in Section 12-12.

12-11 Questions

12-1 The most complex system in the body is the ______________ ______________.

12-2 The major divisions of the nervous system are the central nervous system, the ____________ ____________ ____________, and the ____________ ____________ ____________.

12-3 The CNS consists of ______________ and ______________ ______________.

12-4 The neuron is composed of a soma, many ______________ and an ______________, and has a nerve impulse traveling in ______________ direction only.

12-5 A synapse is a ______________ between neurons across which electrochemical substances pass.

12-6 Afferent nerves carry nerve impulses *to/away from* the brain.

12-7 Efferent nerves carry nerve impulses *to/away from* the brain.

12-8 A reflex arc is an ______________ action caused by stimulation of an afferent nerve ending.

12-9 The CNS is composed of the __________ and ______________.

12-10 The medulla acts to ______________ control basic physiological systems such as ______________ ______________ and ______________.

12-11 The cerebellum (cortex and medulla) provides the body with smooth __________ __________ and balance or __________.

12-12 Biological/physiological negative feedback loops act to *increase/decrease* the output when an *increase/decrease* in the input occurs.

12-13 Four lobes make up the ______________ according to the ______________ structures they lie beneath.

12-14 Frontal lobes control voluntary ________ ______________, ______________, and ______________.

12-15 Temporal lobes control ______________ sensation and ______________.

12-16 Parietal lobes control the qualities of ______________.

12-17 Occipital lobes control ______________ sensation.

12-18 Cerebrospinal fluid (CSF) circulates through ______________ in the brain and functions to provide the brain with ___________, constant ___________ ______________ ______________, and exchange of ______________________ ______________.

12-19 Blood circulates through the brain by the ______________ ______________ artery and returns through the __________ vein.

12-20 Meningitis refers to ______________ of brain membranes.

12-21 The PNS consists partly of __________ nerves.

12-22 The ANS consists of the ______________ or stimulatory and ______________ or inhibitory nervous systems.

12-23 Dynamic body function equilibrium control through the ANS involves __________ rate, __________ pressure, __________ sugar, ______________ temperature, and ______________ reactions, among others.

12-24 Modern understanding of the brain *does/does not* include knowledge of which brain areas control which body sensory and motor functions.

12-25 Electroencephalography (EEG) refers to ______________ activity of the brain.

12-26 EEG patterns change with the behavior of ______________.

12-27 Visual or auditory sensory stimulation *does/does not* change the EEG pattern.

12-12 Problems

The human nervous system is *more complex* than any device invented by the human.

12-1 (a) Some scientists, including John Von Neumann, have estimated that the human brain has *10 billion neurons* each capable of storing 30 billion bits of data (one bit equals one off-on switch) over a *60-year lifetime*. From this postulated information, *how many total bits* of data can the human store in an average lifetime?

(b) Given a large modern electronic computer with 12 computer memory banks (internal and external) each capable of storing *200,000,000 bits* of data, *how many total bits* can the computer store?

(c) Approximately (numbers on the brain are difficult to prove exactly) *how many times more data* can the brain store over the electronic computer?

(d) What can you say about the relative physical size of the biological versus the electronic device?

12-2 (a) The *human eye* passes nerve impulses to the brain through the optic nerve and cranial nerve II. The band of frequencies passed by the eye range from *red* (0.7μm wavelength, 4.3×10^{14} Hz) to *violet* (.4μm wavelength, 7.5×10^{14} Hz). *What is the approximate bandwidth* of the *human visual system?*

(b) Commercial VHF and UHF electronic television operates on frequencies (granted by the FCC) that range from *Channel 2* (54 MHz) to *Channel 83* (890 MHz) neglecting the break in the band between Channels 6 and 7 and Channels 13 and 14. *What is the approximate bandwidth* of the *television system?*

(c) Approximately *how many times wider* is the bandwidth of the *human visual system* over the television system?

(d) What can you say about the *relative physical* size of the biological versus the electronic device?

12-13 References

1. Crouch, James E. and Robert J. McClintic, *Human Anatomy and Physiology,* Wiley (New York, 1976).
2. Guyton, Arthur C., *Textbook of Medical Physiology,* W. B. Saunders Co. (Philadelphia, 1976).
3. McNaught, Ann B. and Robin Callander, *Illustrated Physiology,* Churchill Livingstone (Edinburgh/London/New York, 1975).
4. Yokochi, Chihiro, *Photographic Anatomy of the Human Body,* University Park Press (Baltimore/London, 1971).
5. Smith, C. U. M., *The Brain—Towards Understanding,* G. P. Putnam's Sons (New York, 1970).
6. Chusid, Joseph G. and Joseph J. McDonald, *Correlative Neuroanatomy and Functional Neurology,* Lange Medical Publications (Los Altos, Calif., 1967).
7. Karczmar, A.G. and J. C. Eccles, *Brain and Human Behavior,* Springer-Verlag (New York, 1972).
8. Bruch, Neil and H. I. Altshuler, *Behavior and Brain Electrical Activity,* Plenum Press (New York, 1973).
9. Eccles, John, *The Understanding of the Brain,* McGraw-Hill (New York, 1973).
10. Khanna, J. L., *Brain Damage and Mental Retardation—A Psychological Evaluation,* Charles C Thomas (Springfield, Ill., 1973).
11. Walton, John N. and Lord Brain, *Brain's Diseases of the Nervous System* (1933–1977), Oxford University Press (New York, 1977).
12. Avers, Charlotte, *Evolution,* Douglass College, Rutgers University, Harper and Row (New York, 1974).
13. Saparina, Y., *Cybernetics Within Us*—Foreword by Maxwell Maltz (*Psycho-Cybernetics*), Wilshire Book Co. (Calif., 1967).
14. Bronowski, Jacob, *The Ascent of Man,* Little, Brown, and Co. (Boston, 1973).
15. Cooper, R., J. W. Osselton, and J. C. Shaw, *EEG Technology,* Butterworth (London, England, 1974).
16. Strong, Peter, *Biophysical Measurements,* Tektronix, Inc. (Beaverton, Ore., 1970/1973).
17. Bachrach, Henry and Jim Mixtz, "The Wechsler Scale as a Tool for the Detection of Mild Cerebral Dysfunction," University of Pennsylvania, *J. Clinical Psychology, 30* (1):58–61 (January 1974).
18. Kaufman, Nadeen L. and Alan S. Kaufman, "Comparison of Normal and Minimally Brain Dysfunctioned Children on the McCarthy Scales of Children's Abilities (MSCA)," *J. Clinical Psychology, 30* (1):69–72 (January 1974).

Chapter 13

Instrumentation for Measuring Brain Parameters

13-1 Objectives

1. Be able to describe biomedical instrumentation for measuring anatomical and physiological parameters of the brain.
2. Be able to describe the procedure and equipment used in cerebral angiography, cranial X-rays, and brain scans.
3. Be able to describe the procedure and equipment used in echoencephalography.
4. Be able to describe the procedures and equipment used in electroencephalography (EEG).
5. Be able to describe the origin, cranial location, amplitude, and frequency bands of EEG signals.
6. Be able to describe EEG electrodes and the 10–20 electrode placement system.
7. Be able to describe diagnostic uses of EEG patterns (waking/sleeping and diseased states).
8. Be able to describe multichannel EEG recording systems, including simplified block diagram.
9. Be able to list typical external controls on EEG machines.
10. Be able to list typical characteristics of EEG preamplifiers and machine specifications.
11. Be able to describe the setup and equipment used to obtain visual and auditory evoked potential recordings.
12. Be able to describe the setup and equipment used for EEG telemetry.
13. Be able to list typical EEG system artifacts, faults, troubleshooting, and maintenance.

13-2 Self-Evaluation Questions

These questions test your prior knowledge of the material in this chapter. Look for the answers as you read the text. After you have finished studying the chapter, try answering these questions and those at the end of the chapter (see Section 13-21).

1. What is the origin of the physiological parameter (EEG signal) measured by the electroencephalography (EEG) machine?
2. Give the frequency bands usually specified for EEG signals.
3. What is meant by the 10–20 EEG electrode placement system? Describe it.
4. Name five common EEG machine malfunctions.

13-3 Instrumentation for Measuring Anatomical and Physiological Parameters of the Brain

Instrumentation used to measure *anatomical* (structure) and *physiological* (function) parameters of the brain includes X-ray equipment, ultrasonic equipment, and electrophysiological equipment.

X-ray equipment transmits high-energy electromagnetic (light) waves (0.05–100 Å units) that pass through the body and indicate relative tissue density on a photographic plate. X-ray brain measurements include cerebral angiography, cranial X-rays, and brain scans.

Ultrasonic equipment transmits high-frequency sound waves (1-ms bursts at 2.5 MHz) that pass into the body and indicate tissue location by reflecting waves. This measurement involves *echoencephalography* (echo ranging from the brain).

Electrophysiological equipment detects low-voltage (1–100 μV) and low-frequency (0.1–100 Hz) bioelectric signals that are *picked up* by electrodes, signal *conditioned* by amplifiers/filters, and *displayed* on graphic recorders/CRTs. This measurement involves *electroencephalography,* EEG (electrical activity produced in the brain).

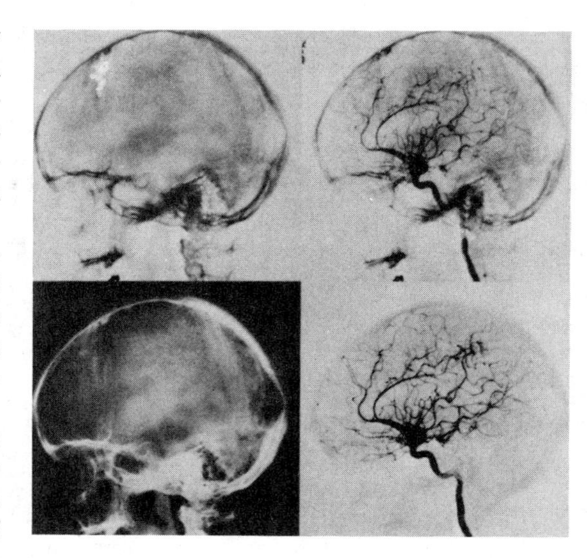

Figure 13-1 Cerebral angiography.

13-4 Cerebral Angiography

Cerebral angiography is an X-ray technique used to display brain structures and blood vessels with the aid of a contrast medium. Radio-opaque dyes that block X-rays are injected in an artery (e.g., common carotid) and disperse throughout the cerebrovascular tree. X-ray images are taken at 1-s intervals and can reveal blockages or tumors, as shown in Figure 13-1.

Recent techniques in *nuclear medicine* are somewhat different than angiography and considerably safer. Small amounts of short-lived *radioactive isotopes* (e.g., iodine 131 taken up by the thyroid gland) are introduced into the cardiovascular system, accumulating in various organs throughout the body. The concentrated radioactivity is measured with a *scintillation counter* that responds to impinging alpha, beta, or gamma rays, depending on and given off by the radioactive material used. The amount of substance taken up by a specific gland indicates the physiological function of the gland. Since very little radioactive substance is required and its emission life is very short, the danger is minimal.

13-5 Cranial X-Rays

Cranial X-ray "pictures" are simply two-dimensional X-ray exposures taken of the cranium. They may be used to indicate fractures in cranial bones and, on occasion, blood clots or tumors. The problem in cranial diagnosis, as in chest X-ray exposures, is the difficulty in reading (evaluating) the exposure. Only when the contrast is high, indicating relatively large tissue density differences, can positive diagnosis be made. Therefore, in recent times (early 1970s), the practice of X-ray scans have become more popular.

13-6 Brain Scans

Brain scans are radiographs that are taken through successive scanning by highly collimated X-ray beams. Small-contrast differences in selected planes can be seen on the "picture" and thus provide considerably more information than simple cranial X-ray exposures.

13-6.1 Computerized Axial Tomography (CAT)

CAT is a technique of recording and processing a set of image projections that represent a reconstruction of the object scanned. A *thin layer* of the object is *tomographically* (sectioned) scanned with a pencil beam. The X-ray attenuation is then recorded with a *scintillation counter* in groups of parallel scans. Many angle X-ray exposures are obtained every degree for 180°, as shown in Figure 13-2. Many projection scans (three dimensional) are stored in a computer and, through a complex program, the original object is redrawn (two dimensional) on a CRT (cathode ray tube) screen by the computer. The CAT scanner is, then, capable of producing a tomogram of the skull/brain (Figure 13-3), which shows the ventricles of the brain.

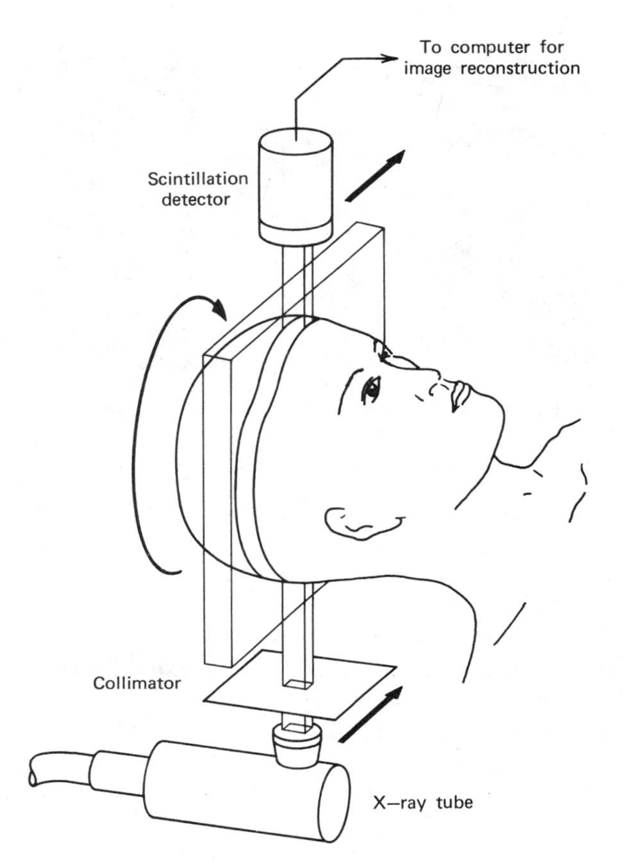

Figure 13-2 Pencil beam multiprojection cerebral scanning. (From *Medicine and Clinical Engineering*, by Bertil Jacobson and John Webster, Prentice Hall, Englewood Cliffs, N.J. 1977. Used by permission.)

13-6.2 Whole Body Scanners

These scanners present views of whole body areas. Such tomographic X-ray results are more costly than regular X-ray patterns. However, they show more detail, more resolution, and have much higher diagnostic quality. This is especially true for color presentations in which different colors represent fine differences of tissue density. One such device is the *ACTA scanner* (automatic computerized transaxial X-ray tomography) developed by Robert S. Ledley and his colleagues at Georgetown University in Washington, D.C.

13-7 Ultrasonic Equipment

Diagnostic ultrasound may prove to be as great a medical discovery as the X-ray and diagnostic radiology. Diagnostic ultrasound is described in detail in Chapter 17.

In this chapter, we are concerned with some diagnostic qualities of *echoencephalography* (echo ranging from the brain to reveal abnormal brain structures such as tumors and dilated brain ventricles). Diagnostic success in such ultrasound neurological examinations depends as much on the operator as the equipment. The principle underlying diagnostic ultrasound is similar to that of radar and sonar. All ultrasonic equipment emits short bursts of sound at pulse

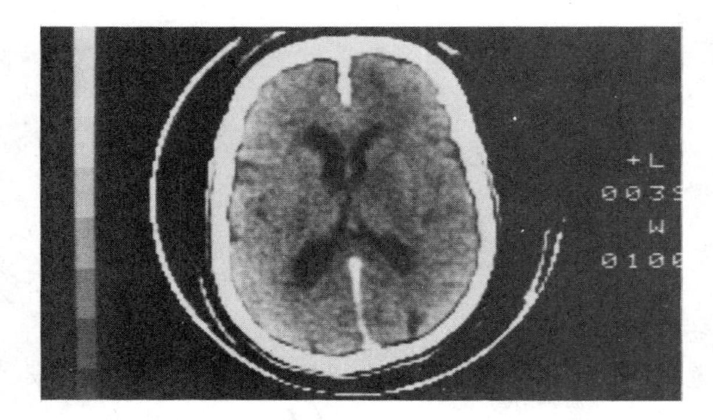

Figure 13-3 Data tomogram of normal skull.

frequencies at or near 2.5 MHz. The sound energy is formed into a beam by a transducer (usually piezoelectric) and directed into the body. The echo received back indicates the distance of the reflecting tissues. The *reflections* are displayed as *peaks versus distance* and show symmetry of brain structures. Asymmetrical peaks may indicate tumors or swellings in the brain. Figure 13-4 shows a *normal* sonogram with reflected echo from the Sylvian fissure (see Chapter 12 for brain anatomy). Figure 13-5 shows an abnormal echo resulting from a subarachnoid (second of three brain meninges) hemorrhage.

13-8 Electroencephalography (EEG)

Electroencephalography is a representation (writing on paper or display on CRT) of the electrical activity of the brain. The technique involves the following:

1. Biopotential pickup—cranial or cerebral surface transducer electrodes.
2. EEG signal conditioning—transducer output amplification and filtering (0.1–100 Hz).
3. EEG signal recording—signal displayed on graphic recorder or CRT.
4. EEG signal analysis—visual or computer interpretation of resulting EEG.

The record obtained is called the electroencephalogram.

The *EEG record* obtained in (3) in the preceding list is used primarily for diagnosis, including the following:

1. Help detect and localize cerebral brain lesions (asymmetry/irregularity in EEG tracings).
2. Aid in studying epilepsy (recurrent, transient attacks of disturbed brain function

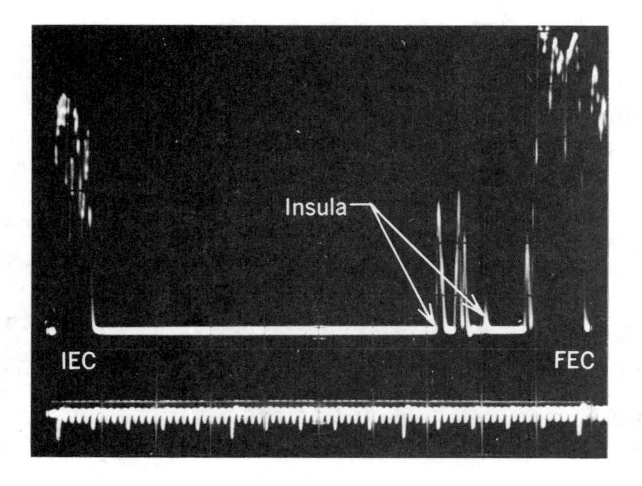

Figure 13-4 Cerebral sonogram reflection from Sylvian fissure. IEC: initial echo complex; FEC: far echo complex. (From M. S. Tenner and G. M. Wodraska, *Diagnostic Ultrasound in Neurology: Methods and Techniques,* John Wiley & Sons, New York, 1975. Used by permission.)

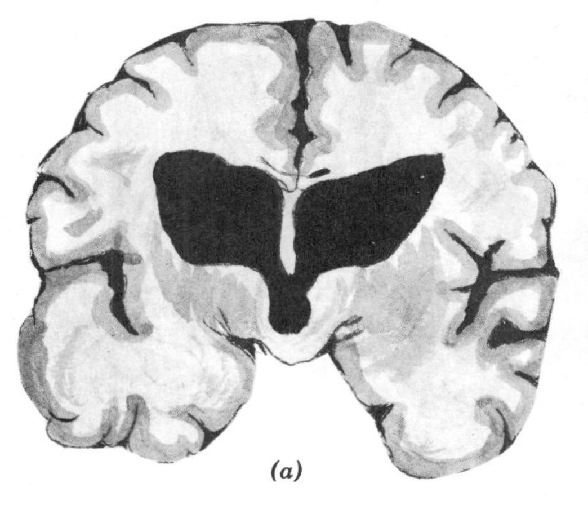

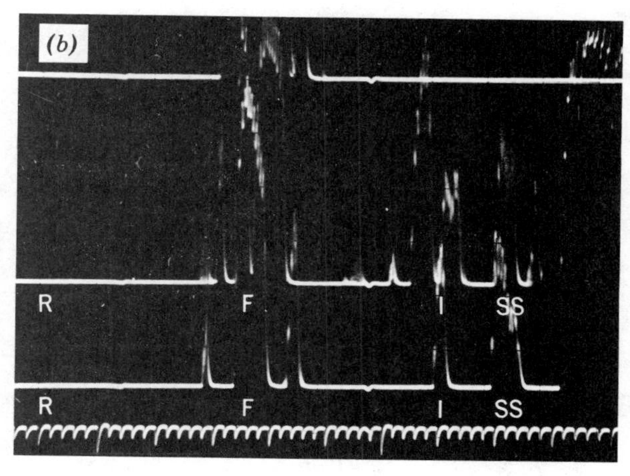

Figure 13-5 Subarachnoid hemorrhage. (*a*) Brain structure. (*b*) Prominent echoes and hemisphere symmetrical reflections indicate an enlarged subarachnoid space (SS). (From M. S. Tenner and G. M. Wodraska, *Diagnostic Ultrasound in Neurology: Methods and Techniques,* John Wiley & Sons, New York, 1975. Used by permission.)

with irregular sensory and motor activity such as convulsions).

3. Assist in diagnosing mental disorders.

4. Assist in studying sleep patterns.

5. Allow observation and analysis of brain responses to sensory stimuli.

Notice the words *help, aid,* and *assist* in the preceding listings. They indicate that the EEG is a *collaborative tool* in diagnosing brain function and disease. Many physicians and neurologists view EEG signals as interesting artifacts but confess that they are not certain of the signal origins. In fact, until recently, EEG waveforms were originally thought to be a summation of action potentials of neurons as they made their way to the cranial surface. Later ideas reflect stimulation associated by diverse neurons.

Modern interpretation of EEG origin rests with knowledge of basic *neuronal electrochemical processes*. The *action potential* (AP) from neurons has been recorded with microelectrodes at the cellular level. Essentially, the synaptic fibers, terminal boutons, neuronal membrane, and axon contribute the distinguishable response characteristics. Electrical reaction of neurons includes the following potentials:

1. *Presynaptic spike* potential (rapid 1-ms positive event resulting from presynaptic depolarization).

2. *Excitatory postsynaptic* potential (EPSP) (prolonged 2-ms graded *positive* potential).

3. *Spike* potential (high-voltage, sudden 2-ms *positive* discharge of 10 to 30 mV).

4. *After hyperpolarization* (prolonged *positive* potential).

5. *Inhibitory postsynaptic* potential (IPSP) (*negative* potential associated with neuronal inhibition).

Figure 13-6 shows the various neuron membrane potentials. Because of the short time durations, a high-frequency oscilloscope (500 Hz bandwidth) and Polaroid camera are used in place of chart recorders (150 Hz bandwidth) to display the neuron potentials.

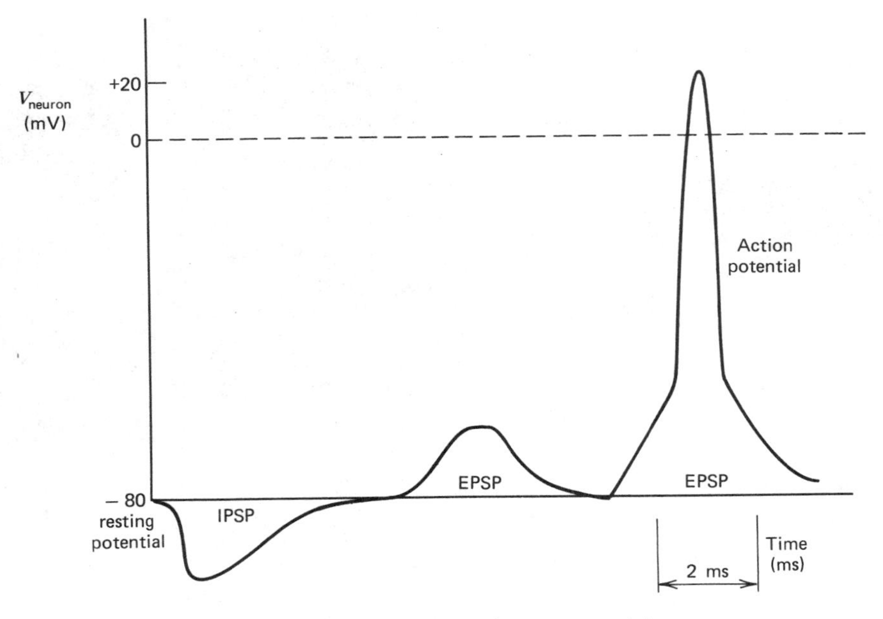

Figure 13-6 Neuron membrane potentials.

The EEG is composed of *electrical rhythms* and transient discharges which are distinguished by *location, frequency, amplitude, form, periodicity,* and *functional properties*. *Synchronization* appears in the EEG and the resulting slow activity is evident. In fact, some EEG investigators have discovered an "*EEG pacemaker*" located just above the brain stem. Among other animals, cats have displayed this synchronism from their unanesthetized cerebrum.

EEG measurements from the cranial surface can be considered by the following *analogy*. Imagine a Martian who visits Earth and lands on the Astrodome (stadium in Houston, Texas) during a football game. If he places a microphone on the Astrodome surface (similar to cranial surface), he hears many sounds (similar to EEG patterns). The complex sound patterns are difficult to interpret on a gross scale. The Martian may hear loud sounds from one side of the dome (cheering) but cannot distinguish its origin. He may also lower a microphone and listen to one person (similar to one neuron), who is selling "hot dogs." Again, this one set of sounds is hardly representative of the entire situation and what is happening in the stadium. EEG signals are, in many respects, just as foreign to us as diverse football game noises are to an extraterrestrial visitor.

13-9 EEG Electrodes and the 10–20 System

EEG electrodes transform ionic currents from cerebral tissue into electrical currents used in EEG preamplifiers. The electrical characteristics are determined primarily by the type of metal used. Silver-silver chloride (Ag-Ag Cl) is commonly found in electrode discs. (The equivalent circuit of such an electrode is shown in Figure 3-1.)

Essentially, five types of electrodes are typically used.

1. *Scalp* electrodes—silver pads/discs/cups, stainless steel rods, and chlorided silver wires.

2. *Sphenoidal* electrodes—alternating insulated silver and bare wire and chlorided tip inserted through muscle tissue by a needle.

3. *Nasopharyngeal* electrodes—silver rod with silver ball at the tip inserted through the nostrils.

4. *Electrocorticographic* electrodes—cotton wicks soaked in saline solution that rests on the brain surface (removes artifacts generated in the cerebrum by each heartbeat).

5. *Intracerebral* electrodes—sheaves of Teflon-coated gold or platinum wires cut at various distances from the sheaf tip used to electrically stimulate the brain.

Reusable scalp disc or cup electrodes (most common in the clinic) are placed on the head using a *conductive cream* (similar consistency to body fluids/electrolytes). The area is first *cleaned* with *alcohol* or *acetone* to remove skin oils. It is *good practice* (using conductive paste) to lower this contact resistance *below 10 kΩ* to insure good EEG signal recording. A test of this resistance can be made with a dc ohmmeter, but electrode polarization results after a few seconds. A better approach is to use an *ac ohmmeter,* which applies an ac signal between two electrodes, avoiding polarization.

The amplitude, phase, and frequency of EEG signals depend on electrode *placement*. This placement is based on the frontal, parietal, temporal, and occipital cranial areas described in Section 12-5.3. One of the most popular schemes is the *10–20 EEG electrode placement system* (Figure 13-7) established by the International Federation of EEG Societies. In this setup, the head is mapped by four standard points: the *nasion* (nose), the *inion* (external occipital protuberance or projection), and the *left* and *right* preauricular points (ears). Lead F_{p1}, for example, is in the frontal area and lies on a circle with other leads. Nineteen electrodes, plus one for grounding the subject, are used. Electrodes are placed using a flexible tape measure by measuring the nasion-inion distance and marking points on the (shaved) head *10, 20, 20, 20,* and *10 percent* of this length. The *vertex,* C_2 electrode is the midpoint. Figure 13-7 shows the complete 10–20 electrode placement system. Here, 19 *electrodes* are used on the scalp, plus one additional for grounding the subject.

Electrode arrangements may be either *unipolar* or *bipolar,* as shown in Figure 13-8. A *unipolar arrangement* (Figure 13-8*a*) is composed of a number of scalp leads connected to a *common indifference point* such as an earlobe. Hence, one electrode is common to all channels. For example, F_{p2} may be measured with respect to two ear electrodes connected together (Figure 13-8*a*) or summation of scalp electrodes (Figure 13-8*b*). A *bipolar arrangement* (Figure 13-8*c*) is achieved by the interconnection of scalp electrodes. The difference voltage between F_{p2} and F_{p8} may also be measured (Figure 13-8*c*). *Montages* are patterns of connections between electrodes and recording channels. All of these combinations have inputs to a three-lead differential amplifier and use a third connection for the reference (two ears, forehead, or nose).

13-10 EEG Amplitude and Frequency Bands

EEG signal voltage amplitudes range from about *1* to *100 μV* peak to peak at low frequencies (0.5–100 Hz) at the *cranial* surface. At the surface of the *cerebrum,* signals may be 10 times larger. Also, brain stem signals measured at the cranial surface are often no larger than 0.25 μV peak to peak (100–3000 Hz). In contrast, ECG chest surface signals are about 500 to 100,000 μV peak to peak. Weak EEG signals require input preampli-

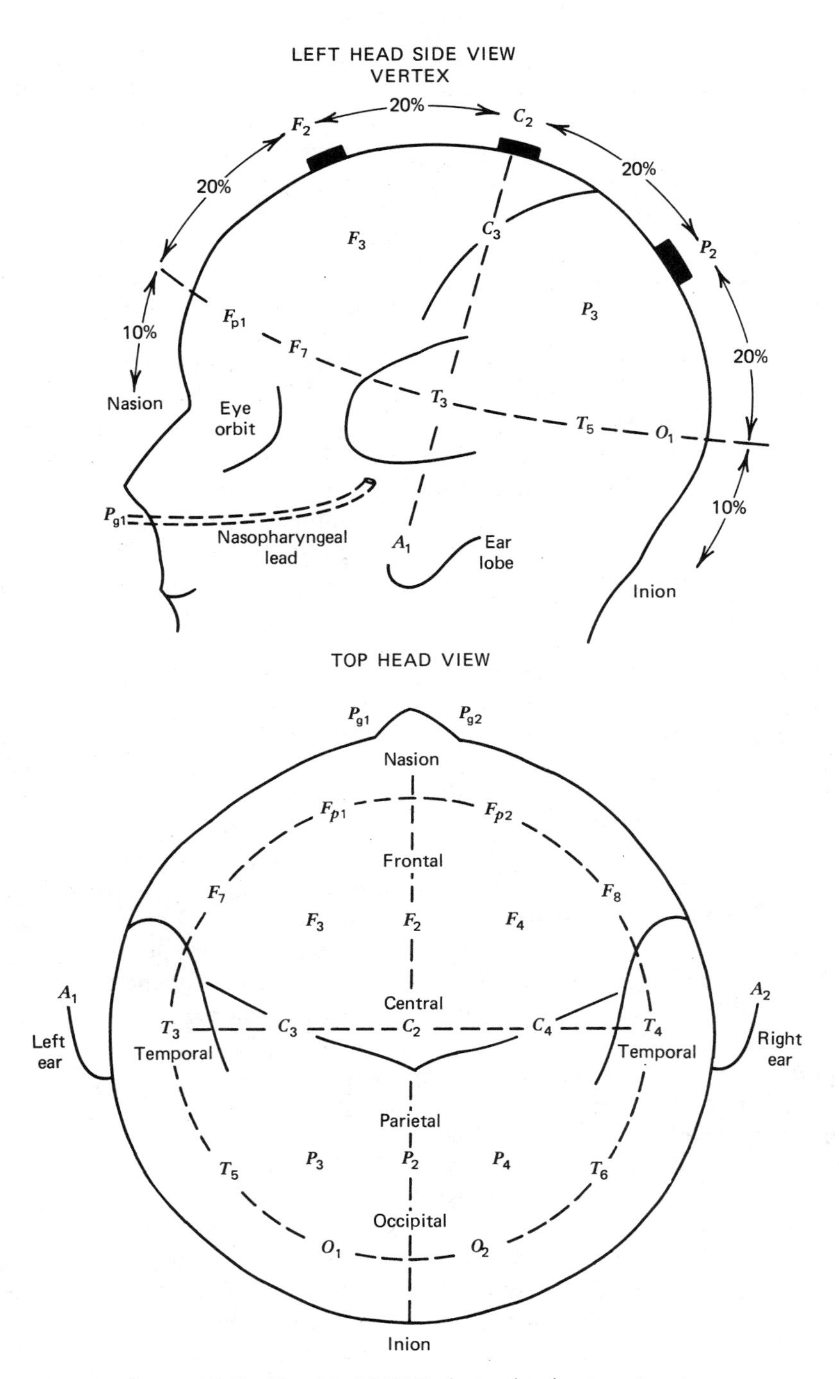

Figure 13-7 The 10–20 EEG electrode placement system.

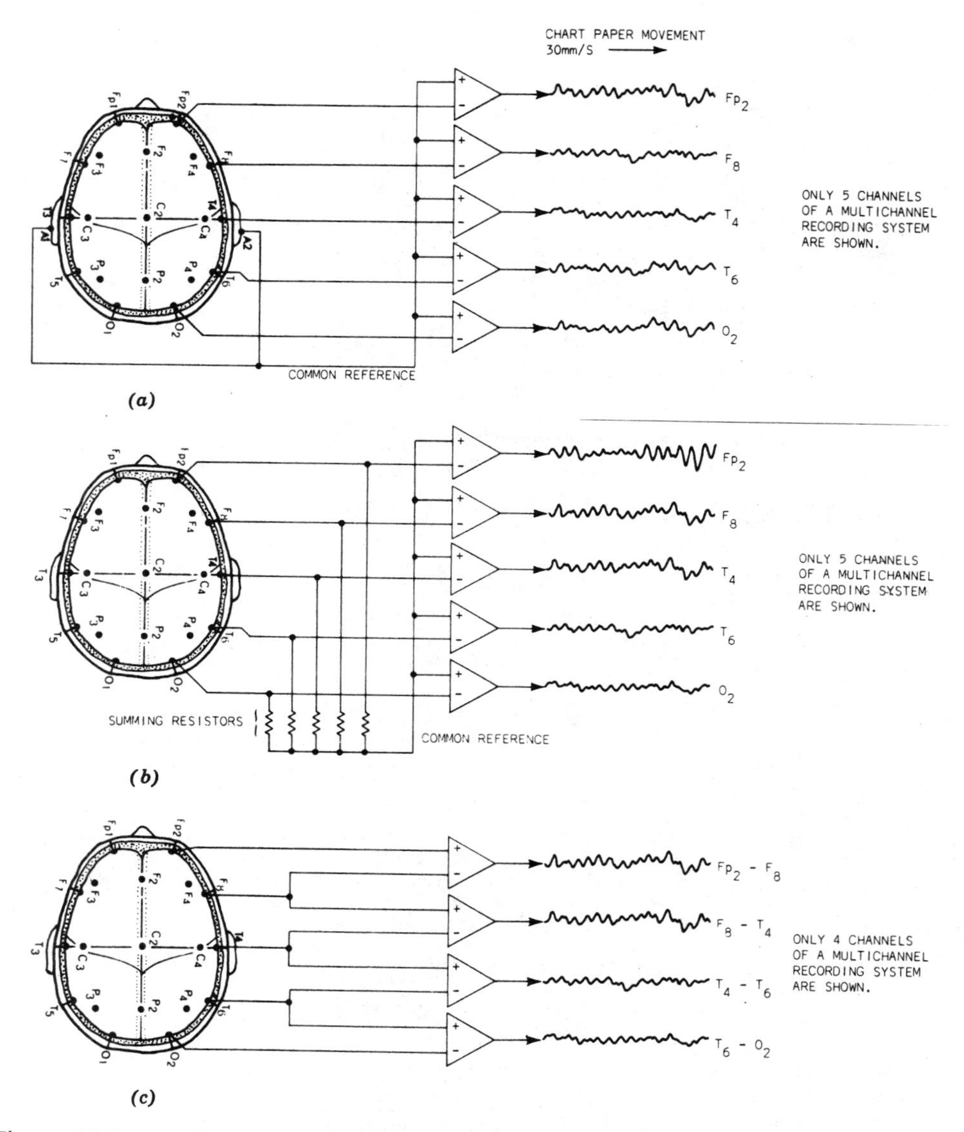

Figure 13-8 EEG recording modes. (*a*) Unipolar. (*b*) Average. (*c*) Bipolar. (Courtesy of Tektronix, Inc. Copyright 1970. All rights reserved.)

fiers (differential type) that have high gain and internal/external noise rejection. Typical EEG signals for *wakefulness* and *sleep* in a normal adult are shown in Figure 13-9.

EEG *frequency* bands are normally classified into five categories:

Delta (δ)	0.5– 4 Hz
Theta (θ)	4– 8 Hz
Alpha (α)	8–13 Hz
Beta (β)	13–22 Hz
Gamma (γ)	22–30 Hz and higher

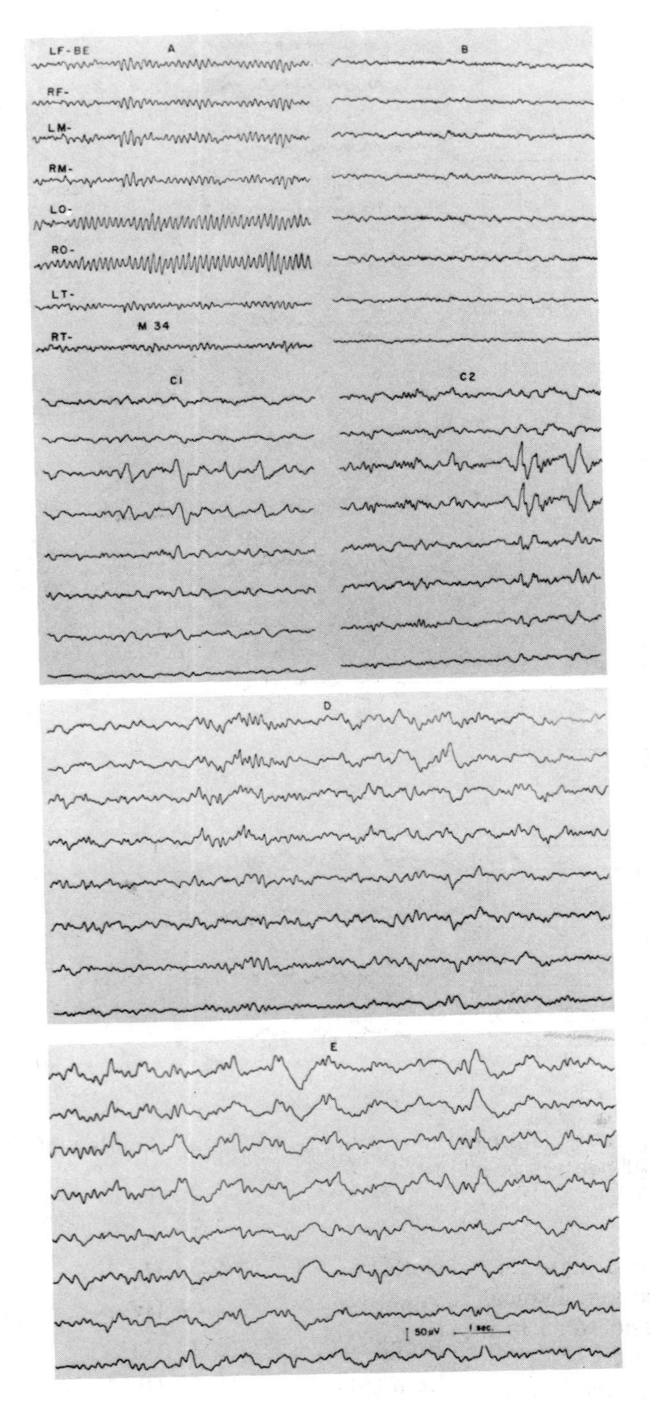

Figure 13-9 EEG patterns. (*a*) Alert state. (*b*) Drowsiness. (*c*) Theta and beta waves. (*d*) Moderately deep sleep. (*e*) Deep sleep. (From *Fundamentals of Electroencephalography,* by Kenneth A. Kooi, M.D., Harper & Row, New York, 1971. Used by permission.)

The meaning of these different frequencies is not completely known. However, *alpha activity* is less than 10 μV peak to peak, and reasonably stable (deviating less than 0.5 Hz). These signals arise from the posterior brain in the waking person with eyes closed. Opening the eyes and focusing attention greatly reduces alpha waves.

Beta activity is less than 20 μV peak to peak arise over the entire brain but is most predominant over the central region at rest. High states of wakefulness and desynchronized alpha patterns produce beta waves.

Gamma activity is less than 2 μV peak to peak and are low-amplitude, high-frequency waves resulting from attention or sensory stimulation.

Theta and delta activity (less than 100 μV peak to peak) are strongest over the central region and are indications of sleep. The waveforms of Figure 13-9 represent *adult* EEG patterns; those of infants are almost nonexistent (lack of brain development). Children show increasingly stronger signals as their brain matures (a 12-year-old child has an adult-looking EEG).

The *frequency spectrum* of the EEG is shown for the normal adult in Figure 13-10. This reveals alpha (10 Hz) and beta (18 Hz) peaks with eyes opened and closed. The usable bandwidth is not much beyond 50 Hz.

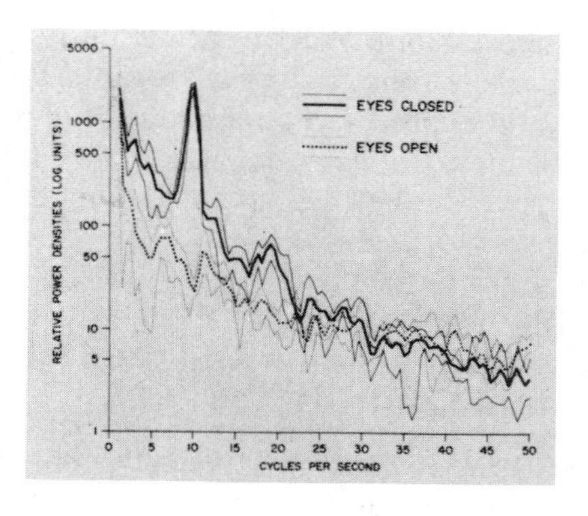

Figure 13-10 EEG frequency spectrum of normal adult. (From *Fundamentals of Electroencephalography,* by Kenneth A. Kooi, M.D., Harper & Row, New York, 1971. Used by permission.)

13-11 EEG Diagnostic Uses and Sleep Patterns

EEG waveforms show remarkable changes just prior to an *epileptic seizure. Grand mal* seizures are associated with wild, uncontrollable muscle contractions (convulsions) and may be accompanied with coma (unconscious state in which the patient cannot be aroused by external stimuli). Changes in EEG patterns are very predominant and usually reflect a large-amplitude, random, low- to high-frequency EEG oscillation, especially near brain motor areas. *Petite mal* seizures are associated with small muscle movements and, occasionally, temporary loss of consciousness. Some symptoms in young children involve simply a few moments staring into space (hardly noticeable). Evoked cerebral potentials to visual or auditory stimuli are also useful in diagnosing brain disorders (Section 13-15).

EEG sleep patterns show dramatic changes to the four *stages of sleep,* as shown in Figure 13-9. These are *drowsiness, light sleep, moderately deep sleep, deep sleep* and *rapid eye movement (REM)*—sleep usually follows deep sleep. Notice the progressively higher amplitude, lower frequency as sleep takes place. EEG changes are also apparent in patients with sleep disorders such as *insomnia* (most prevalent—complaints of lack of adequate sleep), *narcolepsy* (recurring, uncontrollable sleep episodes), *chronic hypersomnia* (excessive sleep or sleepiness), *sleep paralysis* (characterized by an inability to move during apparent full consciousness), and *nightmares* (night terrors revealed by sudden scream and arousal).

EEG pattern changes are also present with changes in behavior. Examples are relative depressions of EEG peaks in *alcoholics,* sporadic runs of slow waves in *drug addicts,* and depth EEG abnormalities in those who display *violence and aggression* (more conclusive studies are still under investigation).

13-12 Multichannel EEG Recording Systems and Typical External Controls

Clinical EEG machines typically consist of 8, 16, or 32 channels, as shown in Figures 13-11 and 13-12. *Eight-channel* devices are most common and record 8 switch-selectable signals from the 20 cranial electrodes (10–20 systems). The older EEG machine (Figure 13-11) is rather large but tends to be very reliable. Newer solid-state EEG devices have very stable circuitry; their weakest section is often the ink-pen graphic display.

A large selector-switch box usually accompanies the EEG machine proper. These switches allow the selection of particular *montages* (specific electrodes connected to EEG input amplifiers for eventual graphic display). The International Federation of EEG Societies suggests the following guidelines for setting up montages:

1. Recording channels should be connected, in sequence, to rows of electrodes along the anteroposterior or traverse lines of the head.

2. The sequences should run from the front to the back of the head and from right to left.

3. For bipolar recordings, channels should be connected so that the black lead of an amplifier (grid 1—right side of head near nose) is anterior to or to the right of the white lead (grid 2—right side of head near ear).

The *EEG technician,* psychologist, physiologist, researcher, or physician selects these montages to obtain desired recordings. The

Figure 13-11 EEG machine.

Figure 13-12 EEG machine (newer solid-state type).

BMET sometimes encounters switch problems, which can be found using an ohmmeter or by viewing problems in the EEG recording.

External controls on EEG machines usually include the following: (see Figures 13-11 and 13-12):

1. Gain or sensitivity multiplier switch—selects sensitivity ranges usually ×20, ×4, ×1, ×500, and ×250.

2. Gain control or sensitivity potentiometer—sets the overall system gain (must be high enough to give good pen deflection but not high enough to clip EEG peaks). It is useful to plot gain in μV/cm for specific machines since most EEGs have all channel master control and individual channel control.

3. Low-frequency (time constant) filter attenuator or high-pass filter switch—selects low-frequency cutoff, usually 0.16 Hz, 0.53 Hz, 1 Hz, and 5.3 Hz.

4. High-frequency filter attenuator or low-pass filter switch—selects high-frequency cutoff, usually 15 Hz, 35 Hz, 50 Hz, 70 Hz, and 100 Hz.

5. Sixty hertzes, notch filter switch—connects or removes 60-Hz filtering (reduces 60 ± 0.5 Hz by −60 dB typically but does cause some signal phase distortion).

6. Calibration push button—sets 5 to 1000 μV peak to peak for rectangular-wave calibration pen deflection.

7. Baseline (position) potentiometer—sets graphic display baseline.

8. Individual electrode selection switches—select specific electrodes.

9. Event marker push button—places a graphic display mark to identify desired events.

10. Chart speed switch—selects speed of graphic display chart paper, usually 10, 15, 30, and 60 mm/s.

13-13 The EEG System—A Simplified Block Diagram

System operation can be understood by studying a *simplified block diagram*. For specific EEGs, detailed manufacturer's block diagrams and schematic must be consulted. If a block is not available, an industrious investigator could draw one from observation of the schematics. Figure 13-13 shows a simplified block diagram for an EEG. Twenty electrodes are placed on the patient's scalp, and these are switch selected to the input of eight differential preamplifiers (differential input–single-ended output). The eight outputs are further amplified and presented to eight driver/power amplifiers that supply sufficient current to drive the pen deflectors. A calibration signal usually in the form of a pulse is generated by a separate circuit and applied to the *diff-amp inputs*. It is more advantageous to connect the calibration signal to the electrode switch selector box to check the system operation. *Calibration signal amplitude* gives an indication of correct sensitivity settings. If the reading is not within specifications, the amplifier system must be adjusted. *Calibration signal waveshape* gives an indication of frequency response. As with the ECG machine (see Chapter 5), *pulse ringing* will occur when underdamping exists and *rounded pulse corners* will be evident when overdamping occurs.

The low-voltage power supply (Figure 13-13) design and operation is very important in

Figure 13-13 Simplified block diagram of an eight-channel EEG system.

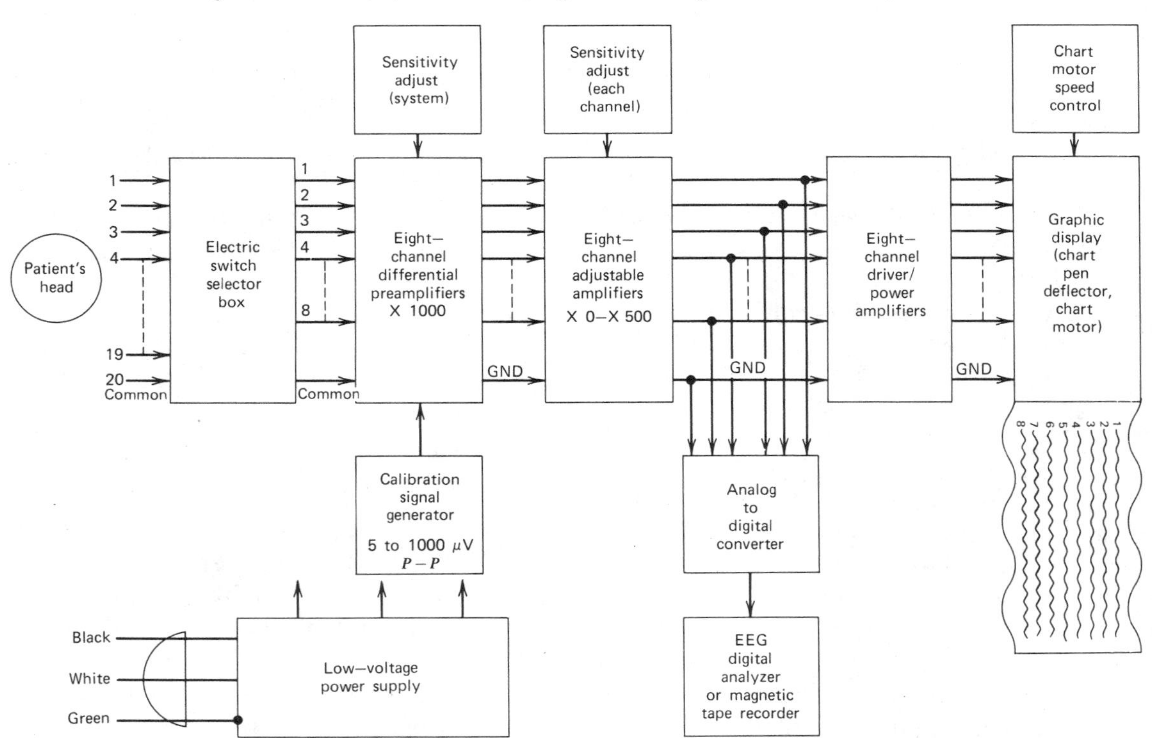

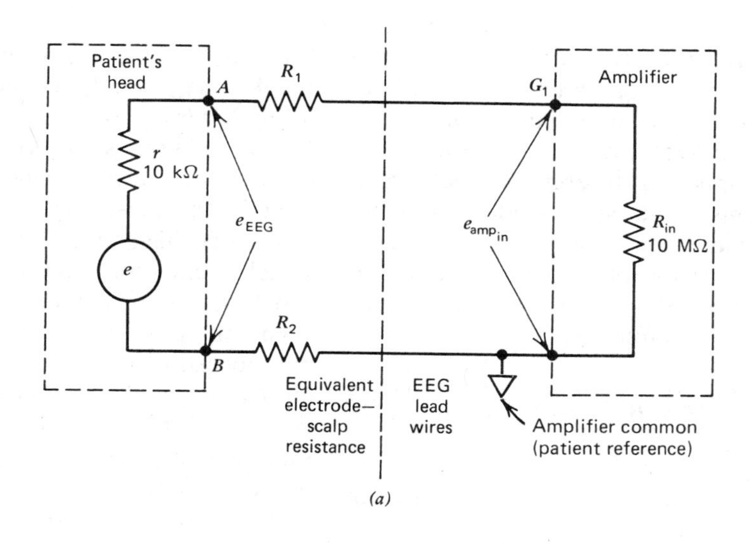

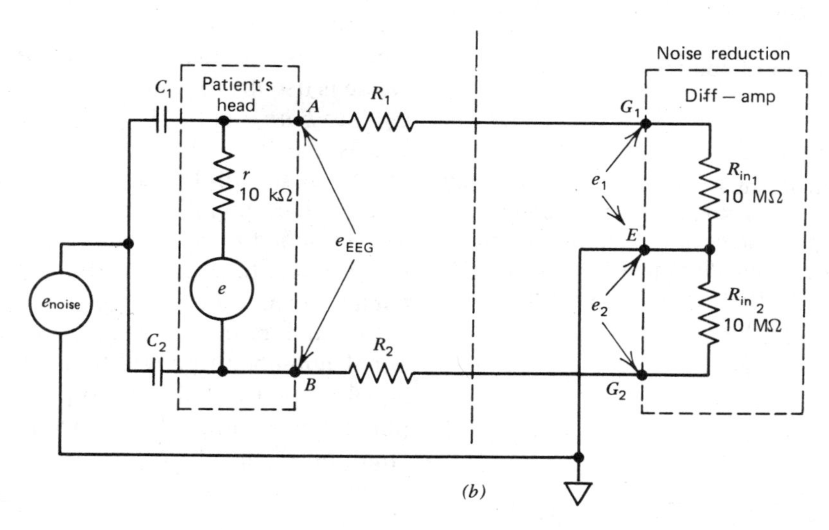

Figure 13-14 EEG amplifiers. (*a*) EEG input circuit—single ended. (*b*) EEG input circuit—differential amplifier.

EEG systems. This is due to the low-level input signals (as small as 5 μV peak to peak), which can easily pick up extraneous 60-Hz internal noise as well as external noise.

EEG output signals can be digitized in an analog-to-digital converter and then analyzed in a digital analyzer (computer) or stored on digital magnetic tape.

13-14 Preamplifiers and EEG System Specifications

EEG preamplifiers are perhaps the most important link in the EEG system. They are usually differential amplifiers and have the following characteristics: low internal noise, high gain (X5k to X10k), high common-mode rejection ratio (CMRR of 100 dB), low-frequency ac-coupled operation (1 Hz and below), low dc drift, and high input impedance (10 mΩ and above). Figure 13-14*a* shows single-ended input and differential input amplifier diagrams. The single-ended amplifier simply provides a ground for one scalp electrode and uses the other as an active site. The current resulting from the ''cranial voltage source'' is as follows:

$$i = \frac{e}{r + R_1 + R_2 + R_{\text{in}}} \tag{13-3}$$

where

e is the cranial voltage source acting through cranial impedance, r

R_1 and R_2 represent equivalent electrode scalp resistance

R_{in} is the input impedance of the electronic amplifier

EXAMPLE 13-2

Given the cranial generator voltage (e) to be 100 μV_{p-p}, impedance (r) to be 10 kΩ, the equivalent electrode-scalp resistances (R_1 and R_2) to be 10 kΩ each, and the amplifier input impedance (R_{in}) to be 10 MΩ, calculate the amplifier input voltage ($e_{\text{amp}_{\text{in}}}$) (refer to Figure 13-14*a*).

SOLUTION

$$i = \frac{e}{r + R_1 + R_2 + R_{\text{in}}} = \frac{100\ \mu V_{p-p}}{10\ \text{k}\Omega + 10\ \text{k}\Omega + 10\ \text{k}\Omega + 10\ \text{m}\Omega}$$

$$i = \frac{100\ \mu V_{p-p}}{10.03\ \text{M}\Omega} = 10 \times 10^{-12}\ a_{p-p}$$

$$e_{\text{amp}_{\text{in}}} = iR_{\text{in}} = 10 \times 10^{-12}\ a_{p-p} \times 10\ \text{M}\Omega$$

$$e_{\text{amp}_{\text{in}}} = \mathbf{100\ \mu V_{p-p}}$$

The amplifier input voltage is calculated as:

$$e_{\text{am}_{\text{in}}} = iR_{\text{in}} = \frac{eR_{\text{in}}}{r + R_1 + R_2 + R_{\text{in}}} = \frac{e}{1 + (r + R_1 + R_2)/R_{\text{in}}} \tag{13-4}$$

When $(r + R_1 + R_2)/R_{\text{in}}$ in Example 13-2 is small, $e_{\text{amp}_{\text{in}}}$ nearly equals e_{EEG}. That is, when R_{in} is large compared to $(r + R_1 + R_2)$, say 100 times greater, then the amplifier receives the *cranial-generated* EEG signal. That is one reason that EEG preamplifiers have such high input impedance—to avoid signal attenuation (and also reduce possible shock hazard). Figure 13-15*a* shows one such amplifier. This circuit (single-ended input operational amplifier) has one serious disadvantage. It will amplify noise voltages induced from lights/power equipment by the same amount as the signal. If the noise amplitude is larger than the EEG signal, the EEG recordings will be obscured.

The balance or differential input amplifier, shown in Figures 13-4*b* and 13-15*b*, cures most of the noise pickup problem. Noise is usually capacitively coupled (C_1 and C_2) into *both* inputs. If the gain of path 1 (G_1) equals the gain of path 2 (G_2), then equal noise signals will be canceled (ideally). This results since amplifier gain equals $G_1 - G_2$ and amplified noise equals $(e_{\text{noise}} - e_{2\ \text{noise}})\ G$. G equals total gain, and $e_{1\ \text{noise}} - e_{2\ \text{noise}}$ equals zero. Since $e_{1\ \text{EEG}}$ is not equal to $e_{2\ \text{EEG}}$, the difference is not zero. The differential amplifier then subtracts the two unequal input EEG signals to produce an amplified EEG output. But it also subtracts the *equal* noise signals to produce zero (or very small) noise output.

EXAMPLE 13-3

Given $R_1 = 1$ kΩ and $R_2 = 500$ kΩ in Figure 13-15*a*, calculate the gain of the op-amp (see Chapter 4 for op-amp details).

SOLUTION

$$A_v = 1 + \frac{R_2}{R_1} = 1 + \frac{500\ \text{k}\Omega}{1\ \text{k}\Omega} \tag{13-5}$$

$$A_v = 501$$

The input impedance is on the order of 10 MΩ (typical for a noninverting op-amp). This circuit

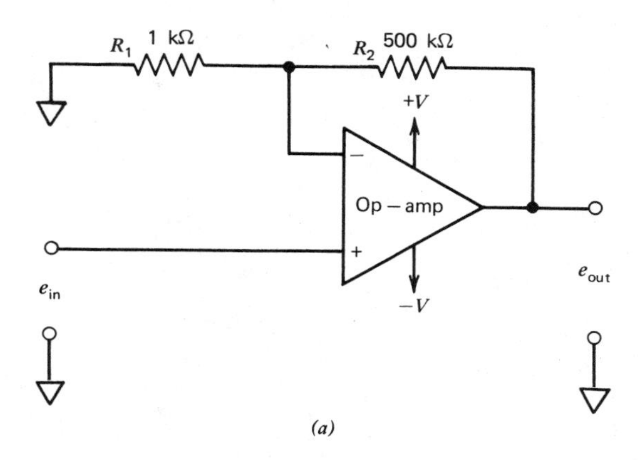

(a)

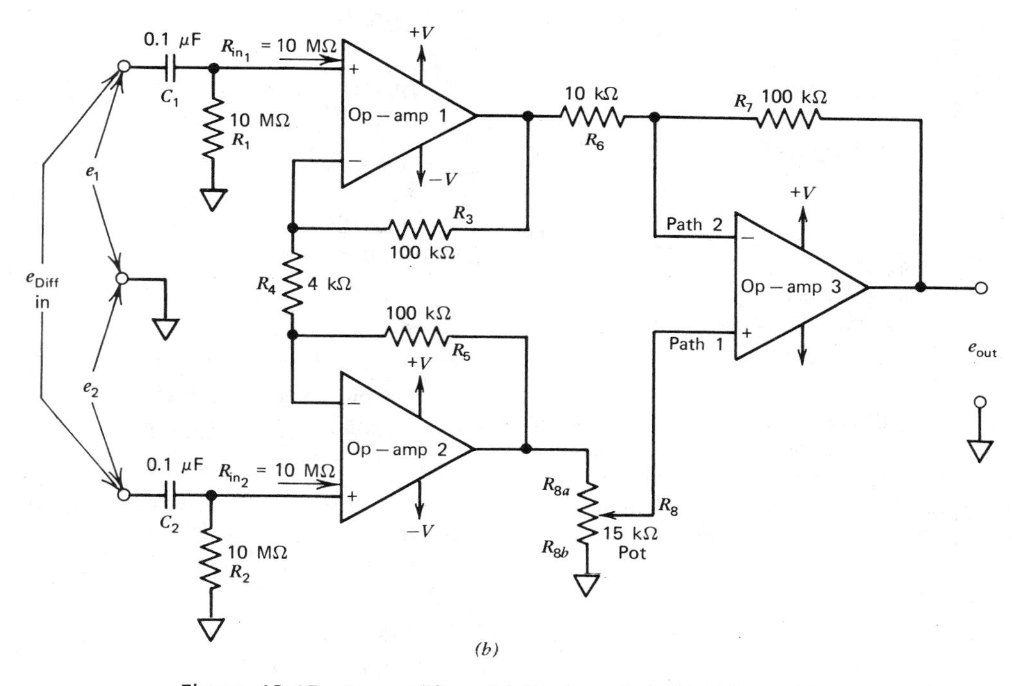

(b)

Figure 13-15 Preamplifiers. (*a*) Single ended. (*b*) Differential.

arrangement is *not* particularly suitable for EEG preamplifiers.

EXAMPLE 13-4

Given $R_3 = R_4 = 100$ kΩ, $R_4 = 4$ kΩ, $R_6 = 10$ kΩ, and $R_7 = 100$ kΩ in Figure 13-15*b*, calculate the output voltage for the *differential EEG input signal*.

SOLUTION

$$e_{o\ \text{EEG}} = e_{\text{in EEG}} A_v$$

$$= (e_{2\ \text{EEG}} - e_{1\ \text{EEG}}) \left(1 + \frac{2R_3}{R_4}\right)\left(\frac{R_7}{R_6}\right) \qquad (13\text{-}6)$$

$$= e_{\substack{\text{diff}\\ \text{in EEG}}} \left(1 + 2\frac{100\ \text{k}\Omega}{4\ \text{k}\Omega}\right)\left(\frac{100\ \text{k}\Omega}{10\ \text{k}\Omega}\right)$$

$$e_{o\,\text{EEG}} = e^{\text{diff}}_{\text{in EEG}}\,510$$

If $e^{\text{diff}}_{\text{in EEG}} = 100\ \mu V_{p-p}$, then $e_{o\,\text{EEG}} = 51\ \text{mV}_{p-p}$

EXAMPLE 13-5
Given the same values in Example 13-4, calculate the *output voltage for the noise input.*

SOLUTION

$$e_{o\,\text{noise}} = e_{\text{in noise}} A_v$$

$$= (e_{2\,\text{noise}} - e_{1\,\text{noise}})\left(1 + 2\frac{R_3}{R_4}\right)\left(\frac{R_7}{R_6}\right) \quad (13\text{-}7)$$

$$= e^{\text{diff}}_{\text{in EEG}}\left(1 + 2\frac{100\ \text{k}\Omega}{4\ \text{k}\Omega}\right)\left(\frac{100\ \text{k}\Omega}{10\ \text{k}\Omega}\right)$$

$$e_{o\,\text{noise}} = e^{\text{diff}}_{\text{in noise}}\,510$$

If $e^{\text{diff}}_{\text{in noise}} = 0$, that is, input noise signals on two EEG leads are equal, then $e_{o\ \text{noise}} = 0\ V_{p-p}$ ideally.

R_8 is adjusted to give equal gain in both paths 1 and 2 (balancing the diff-amp). Even with careful adjustment, some gain differences are present and, thus, some residue noise voltage is present in the output.

EXAMPLE 13-6
Given the EEG output is 51 mV_{p-p} (to a 100 μV_{p-p} differential EEG input) and the noise output is 0.005mV_{p-p} (to a 100 μV_{p-p} common-mode noise input), calculate the *common-mode rejection ratio* (CMRR). Refer to Figure 13-15*b*.

SOLUTION

$$\text{CMRR}_{\text{ratio}} = \frac{e_{o\,\text{EEG}}}{e_{o\,\text{noise}}} = \frac{51\ \text{mV}_{p-p}}{0.005\text{m}V_{p-p}} = 10{,}200 \quad (13\text{-}8)$$

$$\text{CMRR}_{\text{dB}} = 20\log_{10}\text{CMRR}_{\text{ratio}}$$

$$= 20\log_{10}10{,}200 = 80\ \text{dB}$$

The CMRR is a number that indicates, in this case, that the *differential input gain* is 10,200 times the *single-ended input gain*. Most EEG machines have 60 to 120 dB of CMRR.

EXAMPLE 13-7
Given the preamplifiers shown in Figure 13-15*b*, which is also ac coupled to reduce slow baseline drift, calculate the *low cutoff frequency* (*half power* or − 3 dB down frequency).

SOLUTION

$$f_{\text{low}} = \frac{1}{2\pi\,(R_1 \,\|\, R_{\text{in1}})(C_1)} \quad (13\text{-}9)$$

$$= \frac{1}{2\pi(5\ \text{M}\Omega)(0.1\ \mu\text{F})}$$

$$= 0.318\ \text{Hz}$$

The *single-ended input impedance* in Figure 13-15*b* is approximately 5 MΩ.

EEG preamplifiers are the predominant stage that influences EEG machine specifications.

EEG machine specifications typically include:

1. Input impedance: 12 MΩ min at 10 Hz.
2. Sensitivity: 0.5 μV/mm maximum.
3. Sensitivity controls: 10 position master (2 to 75 μV/mm, six-position individual channel (X20 to X0.25), and individual channel gain equalizer.
4. Calibration voltages: 5 to 1000 μV.
5. Common-mode rejection ratio (CMRR): 2000 or 66 dB min at 60 Hz and 10,000 or 80 dB min at 10 Hz.
6. Noise: 1 μV_{rms} (equivalent referred to input) with input shorted.
7. Low frequency (time constant): 30 percent attenuation—0.16 Hz through 5.3 Hz at time constants of one through 0.03 s, respectively.
8. High-frequency response: 30 percent attenuation at 1 Hz to 1000 Hz.
9. 60-Hz filter: 50 dB down at 60 Hz.
10. Chart speeds: 10 to 60 mm/s.

13-15 Visual and Auditory Evoked Potential Recordings

Early EEG investigators discovered that *cortical potentials* could be evoked from sensory stimuli. EEG changes were noted when loud sounds were present. Although wave and spike resources, as well as K-complexes, are evident in the primary record (real time), most evoked responses on the scalp are too small to be recorded by typical EEG machines.

The technique *of averaging repetitive EEG signals* allows the tiny evoked potential to be separated from the background EEG. Each EEG response (5 μV_{p-p}) to a sensory stimulus is added to produce a readable signal. Along with this evoked signal enhancement, the ongoing EEG signal (100 μV_{p-p}) is reduced (average out due to its random nature). Some information is lost in this process and, therefore, the number of stimulations should be kept to a minimum. Unfortunately, each response is unique, and the end result can only represent the addition of all 100 different potentials.

The *block diagram* in Figure 13-16 shows a *visual and auditory evoked potential system*. Three electrodes pick up a one-channel EEG signal. This signal is then filtered through a 60-Hz notch filter and presented to the *evoked potential averaging computer*. The raw or ongoing EEG signal is available at the 60-Hz filter output. The computer can be set, for example, to give 100 trigger pulses to the visual or auditory stimulator. Each brain response is added synchronously with each trigger in the computer. The computer output is then the *summed response potentials* divided by the number of responses (an *average* of 100 signals). This information can be on magnetic tape or hard copied onto a strip-chart recorder. A 5 μV_{p-p} *calibration signal* can be used to test the averaging process accuracy.

Figure 13-17 shows an *auditory* (*tone burst*) *evoked potential* from a 22-year-old male. Notice the positive response at 5 ms and the negative peak at 15 ms. Early responses probably indicate lower-level brain reactions and later higher-level (cerebral cortex) reactions. Waveform origin is still being

Figure 13-16 Block diagram of a visual and auditory evoked potential system.

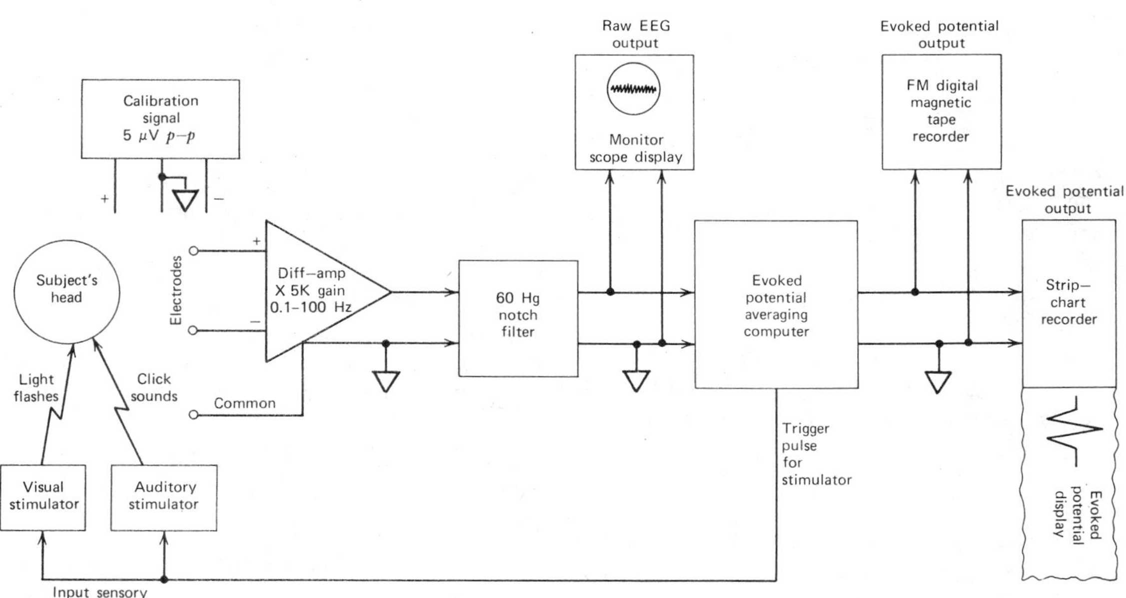

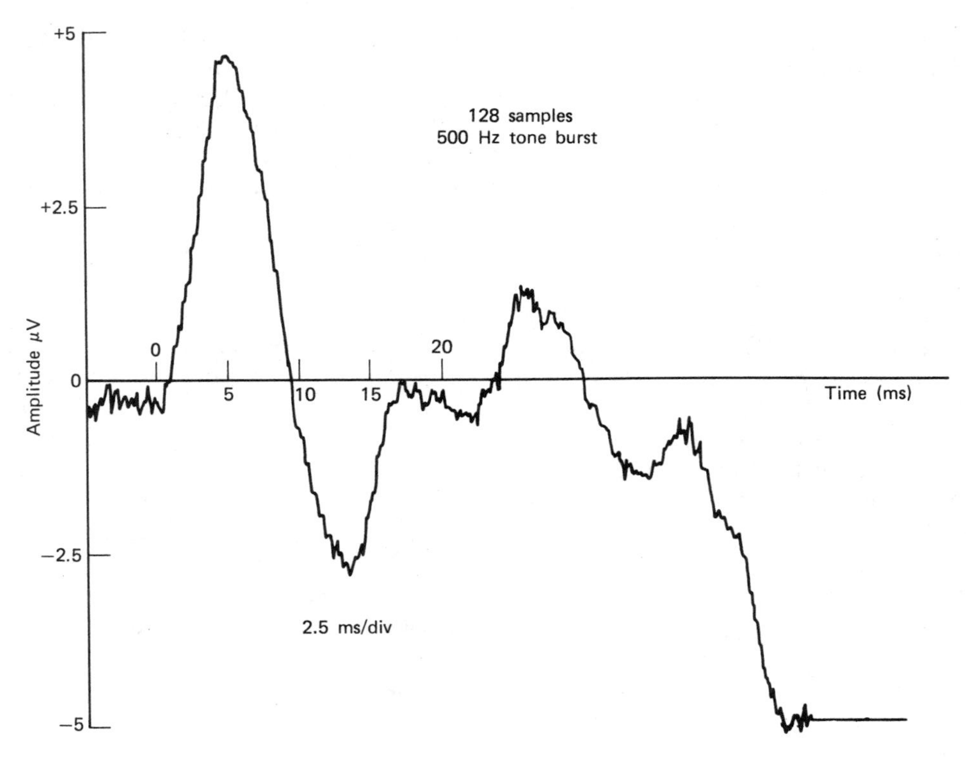

Figure 13-17 Auditory (tone burst) evoked potential from a 22-year-old male.

investigated. Evoked potentials can be used to evaluate sensory action and impairment.

13-16 EEG Telemetry System

EEG telemetry systems are used to transfer the EEG from the patient's cranium to a remote site without encumbering wires. This technique is useful for children or other persons (mentally disturbed) who may be uncooperative in the data-gathering process. Young children can be monitored while they play, or epileptic patients can be monitored while active or just prior to an attack.

Figure 13-18 shows a block diagram for a two-way telemetry system. From this system, evoked responses can be studied in free-roaming patients. The EEG is amplified and modulated (AM, FM, or PM) and transmitted to a remote site. There, the EEG is demodulated and presented to an EEG machine and evoked-response computer. Tone bursts or clicks can also be transmitted via radio. Radio transmission is, however, plagued with noise interference. Control of the frequency band and power output is required by the *Federal Communication Commission* (FCC) in the United States. Furthermore, it is difficult to transmit multichannel as bandwidth problems become important. Transmission media other than radio may prove more suitable for line-of-sight EEG telemetry.

13-17 Typical EEG System Artifacts, Faults, Troubleshooting, and Maintenance

EEG recording systems suffer from artifacts that can obscure the signals of interest and

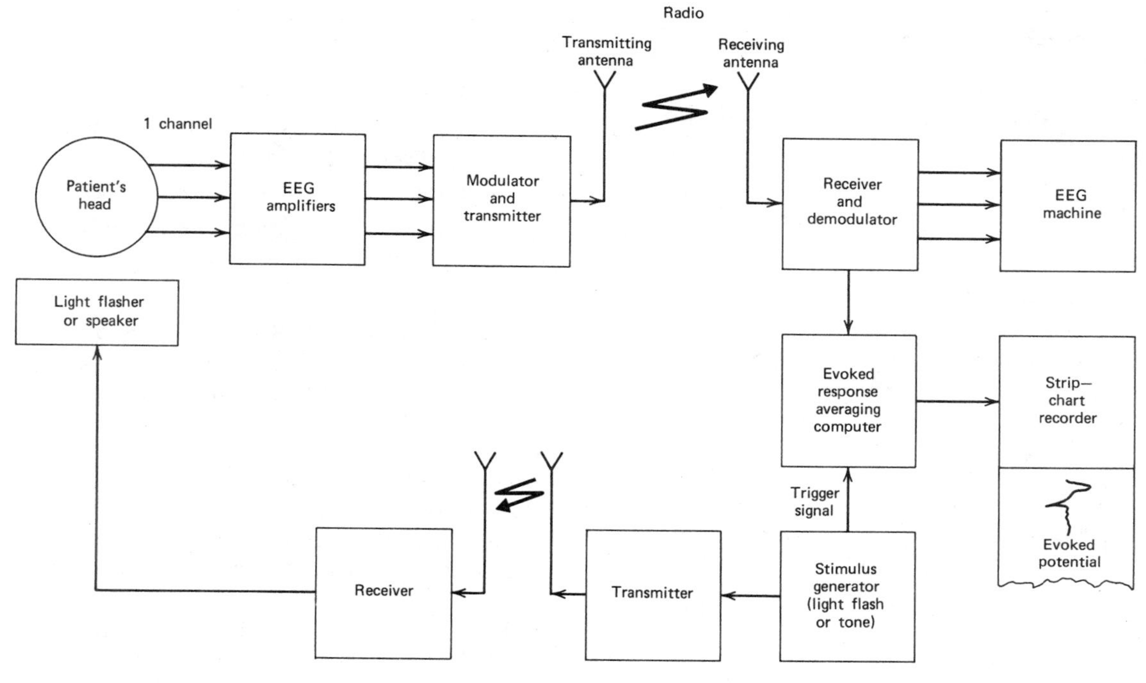

Figure 13-18 EEG telemetry system block diagram.

render diagnosis impossible. Figure 13-19 shows typical artifacts. Aside from 60-Hz interference and *eye blinks* that result in spikes, *muscle activity* from the scalp causes significant interference (Figure 13-19*a*). This can be reduced by adjusting external EEG machine control to give 30 percent high-frequency attenuation at 25 Hz (higher frequencies are effectively removed in Figure 13-19*b*). *Cardiac activity* may be evident as arterial pulse (Figure 13-19*c*) or *R*-wave peaks (Figure 13-19*d*). *Perspiration* under the common electrode, for example, can cause a progressive reduction in time constant (Figure 13-19*a*).

Typical faults fall into the following categories:

1. Patient electrode connection problems—high impedance connections to the scalp or broken electrode wires.

2. Cable connection problems—broken wires and connector pins bent.

3. Incorrect switch position—operator error or broken knob indicators.

4. Broken switches—faulty switch contacts.

5. Graphic recorder malfunctions—drive roller slipping, ink pens clogged or unseated.

6. Electronic malfunctions—circuit faults in individual channels, system control, or power supply.

Troubleshooting, maintenance, and repair of EEG machines is similar to that of ECG machines (see Chapter 5, Sections 5-9 and 5-10, particularly examples 5-1, 5-2, and 5-3). EEG machines are typically less rugged than clinical ECG machines but need not be inspected by the BMET on a daily basis. However, the EEG technician should per-

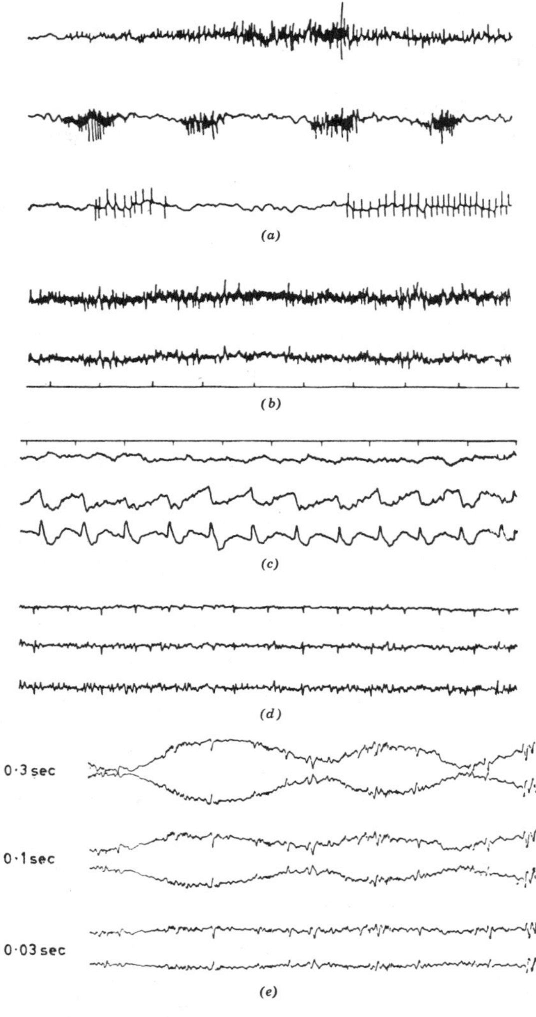

Figure 13-19 Typical EEG recording artifacts. (From *Fundamentals of Electroencephalography,* by Kenneth A. Kooi, M.D. Harper & Row, New York, 1971. Used by permission.) (*a*) Examples of muscle activity recorded from scalp electrodes. (*b*) The effect of high-frequency attenuation (30 percent cut at 25 Hz) on the muscle activity shown in the upper trace. Artifacts due to cardiac activity. (*c*) Pulse artifacts from the same subject. (*d*) ECG artifacts from the same subject. (*e*) Effects of progressively reducing the time constant in a pair of channels affected by perspiration under the common electrode.

form a routine inspection procedure prior to daily use. This includes:

1. Machine turn on warm up.

2. Calibration set, usually 100 μV and observation of rectangular pulse on all channels (pen recorder).

3. Sensitivity set (system and individual channel) for proper deflection corresponding to 100 μV.

4. Pressing and holding calibration switch to observe time constant decay.

5. Grounding all inputs to observe zero signal on all channels.

Individual manufacturer's *operation and maintenance manuals* give specific calibration and maintenance procedures.

Electrical/electronic and mechanical faults are actually rare occurrences in clinical EEG machines. ECG machine faults occur much more frequently due to higher usage, but even these are relatively few. Manufacturer's troubleshooting trees (charts) are often helpful. The following examples are representative of typical problems:

EXAMPLE 13-10

Symptom: Machine runs, but the tracing on one or more channels is missing.

Possible Causes:

1. Ink reservoirs for pens are dry [on missing channel(s)].
2. Ink tubes are clogged.
3. Pen not touching.

Troubleshooting (machine off):

1. Check ink reservoirs.
2. Check ink tube for clogging.
3. Check for upwardly bent pens—gently push pen onto paper with finger or pencil to observe any tracing.

SOLUTIONS

1. For dry ink reservoirs, fill to level suggested by manufacturer (usually just below top rim). To overfill causes messy operation and

can damage circuitry and mechanisms if allowed to drip into the machine.

2. For clogged ink tubes, remove the tube and pen and soak in warm water. Use a fine wire to gently push the clog through. Be certain not to punch a hole in the tube.
3. For bent pens, remove the pen in question and gently bend the pen downward. Be careful not to bend at right angles, as these pens are delicate and will crack.

EXAMPLE 13-11

Symptoms: Spotty recordings (light/dark).

Possible Causes: Worn pens or incorrectly loaded paper.

Troubleshooting: Check paper loading, and if proper, then check pen for worn tip (ink not feeding properly).

SOLUTIONS

1. For paper loading, perform manufacturer's procedure.
2. For worn pen tip, replace with manufacturer's part or equivalent.

EXAMPLE 13-12

Symptoms: Noisy or poor recording.

Possible Causes: Lead connection or electronic/mechanical problems.

Troubleshooting:

1. Place selector switches to standard calibration position and check for noise and improper operation.
2. If calibration operation is normal, the problem is probably the patient connection.
3. Ground all EEG leads and check for straight line tracing (noiseless) and, if good, connect an EEG simulator if available. Check for good tracings. If noise appears on the trace, the problem is probably inside the machine. Refer to the service manual for troubleshooting.

Interference (60 Hz) in EEG machines is the most common problem next to chart-recorder difficulties. Since EEG signals are very low amplitude (5 to 100 μV_{p-p}), great care must be taken to shield and connect the patient leads. Open power supply filter capacitors and shorted voltage regulators cause symptoms that often appear to be similar to lead problems.

Muscle jitter (Figure 13-19*a*) is different in both *amplitude* and *frequency* than 60-Hz interference. *Filtering* may be used, but it is almost imperative that the patient be helped to relax.

Erratic or wandering baseline can be caused by poor electrode connections or long-term patient connection (Figure 13-19*e*). Clean, well-secured electrodes (low electrode resistance below 10 KΩ) are the only sure cure for this phenomenon. Also, junction box leads may have intermittent connections. Occasionally, amplifiers may have excessive dc drift.

13-18 Summary

1. Instrumentation for measuring anatomical and physiological parameters of the brain include *X-ray equipment* (cerebral angiography, cranial X-rays, and brain scans), *ultrasonic equipment* (echo encephalography), and *electrophysiological equipment* (electroencephalography—EEG and brain stimulators).
2. X-rays pass more readily through soft tissue than bone and reveal *differences in tissue density*. They are *dangerous* and can cause mutations, physical illness, and death.
3. *Cerebral angiography* is an X-ray technique used to display brain structures (tumors) and blood vessels (blockages) by circulating a radio-opaque dye contrast medium.
4. *Cranial X-rays* are two-dimensional exposures taken of the cranium and can reveal cranial fractures and tumors.
5. *Brain scans* (computer tomography) are radiographs that are taken through successive scanning by highly collimated

X-ray beams. They show small differences in tissue densities and reveal obscure tumors and swellings.

6. *Echo encephalography* is an ultrasound pulsing Doppler shift technique that shows differences in acoustic properties of various tissues. It shows brain structures and blood flow.
7. *Electroencephalography (EEG)* is a representation (writing on paper or display on CRT) of the electrical activity of the brain. It is used to help detect and localize cerebral brain lesions and study epilepsy, mental disorders, sleep patterns, and brain responses to sensory stimuli.
8. *Neuronal electrochemical processes* form the basis for EEG activity.
9. The *10–20 EEG electrode placement system* establishes 10 and 20 percent distances across the cranium (*nasion* to *inion*) for electrode positions. Nineteen electrodes are used with one as a reference. EEG patterns are distinguished by location, frequency, amplitude, form, periodicity, and functional properties. A common or indifference point (earlobe) is used to supply the differential amplifier input with an inactive reference. *Unipolar* recordings indicate the difference between an active and inactive site, while *bipolar* is between two active sites.
10. *EEG amplitude* ranges from 1 μV to 100 μV_{p-p} at 0.5 to 100 Hz. The following signal waves are defined: delta (0.5 Hz–4 Hz), theta (4 Hz–8 Hz), alpha (8 Hz–13 Hz), beta (13 Hz–22 Hz), and gamma (22 Hz–30 Hz and above).

    Alpha activity (the most predominant) occurs awake with eyes closed. *Beta waves* begin with eyes open. *Delta* and *theta waves* appear at various stages of drowsiness and sleep.
11. *Clinical EEG machines* typically consist of 8, 16, or 32 channels (8 is most common). Switch-selectable electrode signals make up the montage.
12. External EEG machine *controls* consist of gain or sensitivity control (system and individual channel), low- and high-frequency attenuation filter adjusts, 60-Hz notch filter switch, calibration push button, baseline (position) potentiometer, electrode selection switch box, event marker push button, and chart speed switch.
13. The *basic EEG block diagram* consists of electrode switch-selector box, differential amplifiers (system), adjustable gain amplifier (individual channel), driver amplifiers, graphic display, and power supply.
14. *EEG preamplifiers* (differential amplifiers) compose the most important section of the EEG machine. Their characteristics are low internal noise, high gain (X5 to 10), high external noise rejection (CMRR of 100 dB), low-freqnency ac-coupled operation (1 Hz), low dc drift, and high input impedance.
15. EEG machine *specifications* typically include 12 MΩ input impedance, 0.5 μV/mm sensitivity, 5 to 1000 μV calibration voltages, common-mode rejection ratio (CMRR) of 60 to 120 dB, low frequency attenuation (30 percent cut-off) at 0.16 through 5.3 Hz, high frequency attenuation at 1 to 100 Hz, 60-Hz notch filtering (50 dB down), and chart speeds of 10 to 60 mm/s.
16. Sensory (visual and auditory) *evoked potential* recordings are obtained by repetitive sensory stimulation and synchronous computer averaging of the low-amplitude evoked signal (5 μV_{p-p}).
17. *EEG telemetry systems* are used to transmit and receive EEG signals via radio (AM, FM, PM) and other media for mobile or uncooperative patients.
18. EEG recording systems suffer from *artifacts* (60-Hz noise, eye blinks, muscle

tremor/activity, cardiac activity, and perspiration). These may obscure the recording but can be corrected by good electrode technique, high-quality preamplifiers, and filtering.

19. *Typical EEG machine faults* are poor patient-electrode connection, broken cable wires, bent connector pins, broken switches, graphic recorder paper drive and ink problems, electronic/mechanical malfunction, and operator error.
20. *Troubleshooting and maintenance* takes the form of an inspection procedure (calibration, control check, tracing evaluation). Individual equipment manufacturer's operation and maintenance manuals must be consulted for proper maintenance. Typical trouble symptoms include missing trace, spotty, noisy, or poor recordings, 60-Hz noise interference, muscle jitter, and erratic or wandering baseline.

13-19 Recapitulation

Now return to the objectives and self-evaluation questions at the beginning of the chapter and see how well you can answer them. If you cannot answer certain questions, place a check mark next to each and review appropriate parts of the text. Next, try to answer the following questions using the same procedure. When you have answered all the questions, solve the problems in Section 13-21.

13-20 Questions

13-1 Three classes of instrumentation used to measure anatomical and physiological parameters of the brain are __________, __________, and __________.

13-2 Cerebral angiography and brain scans both use __________.

13-3 Mutations, physical illness, and death are hazards that can result from overexposure to __________.

13-4 During cerebral angiography, __________ or contrast media are injected into the circulatory system of the brain.

13-5 In brain scans, CAT refers to __________ __________ __________.

13-6 The ACTA scanner utilizes a __________ counter to record X-ray beam attenuation.

13-7 Diagnostic ultrasonic brain instrumentation is known as __________ and uses frequencies of __________ MHz.

13-8 Brain *symmetry/asymmetry* indicated by ultrasonic reflections in the cranium indicate normal conditions.

13-9 Electroencephalography (EEG) results from __________ activity of the __________.

13-10 EEG systems include __________ pickup, signal __________, signal __________, and signal __________.

13-11 EEG records are used to detect cerebral __________, study convulsive problems of patients with __________, assist in diagnosing mental __________, study __________ patterns, and observe __________ brain responses to __________ stimuli.

13-12 Neuronal action potentials can be measured with __________ electrodes.

13-13 EEG patterns are distinguished by __________, __________, __________, and __________.

13-14 The 10–20 EEG electrode placement system derives its name from what measurements?

13-15 How many EEG electrodes are typically used?

13-16 The EEG vertex electrode is half of what distance?

13-17 Why is a common or indifference point used for EEG recordings?

13-18 What is the difference between unipolar and biopolar recordings?

13-19 EEG electrode resistance should be below what level?

13-20 State the typical amplitude range for EEG signals.

13-21 List the names and frequency bands of EEG waveforms.

13-22 EEG multichannel recording systems typically use how many channels?

13-23 Montage refers to the selection of what connections?

13-24 List 10 typical EEG machine external controls and describe each.

13-25 List the main sections of a typical EEG machine block diagram and describe the flow of signals.

13-26 List six typical EEG preamplifier (differential amplifier) characteristics.

13-27 List 10 typical EEG machine specifications.

13-28 In sensory evoked potential systems, the averaging computer performs what two functions?

13-29 List two uses for evoked potential studies.

13-30 EEG telemetry systems transmit signals (wireless) and are used for which type of patients?

13-31 List four types of EEG artifacts.

13-32 List six typical EEG machine/system faults.

13-33 Troubleshooting and maintenance of EEG machines includes a routine inspection procedure typically consisting of which steps?

13-34 Name two EEG machine problems (symptoms, causes, troubleshooting, and solutions).

13-21 Problems

13-1 What peak-to-peak output voltage level will be present from an EEG input level of 50 μV_{p-p} and EEG system gain of 5000?

13-2 From Figure 13-15*b*, what voltage gain will result if $R_3 = R_5 = 150$ kΩ, $R_4 = 10$ kΩ, $R_6 = 5$ kΩ, $R_7 = 100$ kΩ, and R_8 is set to balance the diff-amp?

13-3 From Figure 13-15*b*, will the low-frequency response be satisfactory if capacitor C_1 leaks and becomes 0.005 μF?

13-4 What common-mode rejection ratio (CMRR), number ratio and dB, exists if the EEG output is 1 V_{p-p} (to a 100-μV_{p-p} EEG input) and the noise output is 0.01 mV_{p-p} (to a 100 μV_{p-p} noise input)?

13-22 References

1. Scott, Donald, *Understanding the EEG, An Introduction to Electroencephalography*, J. B. Lippincott Co. (Philadelphia, 1975).
2. Kooi, Kenneth A., *Fundamentals of Electroencephalography*, Medical Department, Harper and Row Publishers, (New York, 1971).
3. Kiloh, L. G., A. J. McComas, and J. W. Osselton, *Clinical Electroencephalography*, Appleton-Century-Crofts Educational Division/Meredith Corp. (New York, 1972).
4. Cooper, R., J. W. Osselton, and J. C. Shaw, *EEG Technology*, Butterworths (London, 1974).
5. Strong, Peter, *Biophysical Measurements*, Tektronix, Inc., Measurement Series (Beaverton, Ore., 1970).
6. Williams, Robert L. and Carolyn J. Hursch, *Electroencephalography (EEG) of Human Sleep: Clinical Applications*. Wiley (New York, 1974).
7. Burch, Neil and H. I. Altshuler, *Behavior and Brain Electrical Activity*, Plenum Press (New York, 1975).
8. Mishkin, Fred S. and John Mealey, Jr., *Use and Interpretation of the Brain Scan*, Charles C Thomas (Springfield, Ill., 1969).
9. Tenner, Michael S. and Georgina M. Wodraska, *Diagnostic Ultrasound in Neurology*, Wiley (New York, 1975).
10. Jacobson, Bertil and John Webster, *Medicine and Clinical Engineering*, Prentice Hall, Inc. (Englewood Cliffs, New Jersey, 1977).
11. Geddes, L. A. and L. E. Baker, *Principles of Applied Biomedical Instrumentation*, Wiley (New York, 1968/1975).
12. Ledley, Robert S., "Computerized Transaxial X-Ray Tomography of the Human Body," *Science, 186*:207–212 (1974).

Chapter 14

Intensive and Coronary Care Units

14-1 Objectives

1. Be able to describe the functions and purpose of special care units in the hospital.
2. Be able to list the types of instrumentation systems used in the ICU/CCU.
3. Be able to identify and troubleshoot common ICU/CCU instruments.

14-2 Self-Evaluation Questions

These questions test your prior knowledge of the material in this chapter. Look for the answers as you read the text. After you have finished studying the chapter, try answering these questions and those at the end of the chapter (Section 14-11).

1. List the types of monitoring equipment normally found in the ICU/CCU of a hospital.
2. What is the function of a *cardiac memory* unit?
3. What is an *arrhythmia monitor?*
4. Describe the principal functions of a *bedside monitor.*

14-3 Special Care Units

Special care units go by a variety of names, some of which are descriptive of function (i.e., intensive care unit [ICU], critical or coronary care unit [CCU], etc.). But all of these units are designed to offer the advantages of a low nurse-patient ratio and a concentration of the equipment and resources needed to take care of critically ill or seriously injured patients.

The situation "on the floor"—on the regular medical and surgical wards of the hospital—is quite different. It is not unusual in some hospitals on the night shift to find one or two nurses caring for 50 to 80 patients, with the ratios going as high as 50 patients per nurse when crowded. In the not-too-distant past a critical patient would be assigned to such a ward, possibly with a one-on-one private duty nurse. But in the event of a crisis, few resources were available in short enough time to do any good. Even today in some rural hospitals, there might not be a physician in the building at odd hours, especially the night shift from midnight to morning.

But in the ICU/CCUs of modern hospitals, physicians are available, and the nurses are specially trained to recognize and deal with signs of impending disaster. Rapid action can be taken to counter the event. In fact, the critical care nurse is no longer an emerging skilled specialty but an established fact. In medical and surgical ICUs, there are usually one or two patients per nurse, and in many cases the ratio is one to one. Additionally, the staff have the physical resources, supplementing their skills and abilities, necessary to handle emergencies.

Most hospitals have medical-surgical emergency teams that rush to the scene of an emergency that could be life threatening. Most hospitals call the team by radio pager (i.e., "beeper") and by voice page over the public address system. It is common to find the team called using a cryptic phrase such as "code 2 MICU," which might be interpreted by appropriate personnel as "respond immediately to an emergency in the medical intensive care unit, do not stop at the vending room for coffee!" The use of a cryptic code for an emergency may stem from a desire to keep from alarming the visitors to the hospital but also serves a communications role in that it provides an unmistakable command to the team. Since emergencies rarely follow a prepublished time schedule, it is not likely that team members will be expecting trouble. The use of a code allows them to be trained to respond immediately, even though they were otherwise engaged at the moment. The emergency team will respond to emergency situations anywhere in the hospital, including the ICU/CCU areas that will need additional help in such cases.

A key factor in the success of the ICU/CCU is the information that is available to the staff. Providing that information is the role of the technology and equipment.

14-4 ICU/CCU Equipment

The information used by physicians and nurses in the ICU/CCU comes from several sources, including their own skilled observations, the laboratory, and an array of electronic monitoring equipment at each bedside. The particular parameters measured on any given patient vary from one hospital to another and from one physician to another. There is some disagreement over certain types of monitoring, but in all cases the ECG waveform is monitored.

Arterial blood pressures are routinely monitored in some units. Less often, but increasingly frequent, we also find monitors for temperature, respiration rate/apnea, venous pressure, right arterial pressure, cardiac output, and EEG. Almost all monitors include a heart rate meter that is triggered from the ECG waveform, and in some cases the arterial pressure waveform.

Additionally, other equipment may be present for use as required. Examples are respirators/ventilators, hypo/hyperthermia machines, defibrillators, rotating tourniquets, aortic balloon pumps, etc.

Most ICU/CCU facilities provide a bedside monitor near the patient, complete with an oscilloscope and sometimes its own dedicated strip-chart recorders. The bedside monitor will have auxiliary outputs that are connected by wire to remote readouts at a central console (Figure 14-1). On computer-based systems, which are common today, a *local area network* (LAN) may be used for intercommunications.

The console will display physiological waveforms and numerical data on one or more patients; a popular unit size seems to be eight patients, although that is not a general rule.

Several philosophies regarding ICU/CCU layout seem to be prevalent. In some, all of the beds can be monitored from a single console. In some cases, the console is at the nurses' station, while in others it is at a nearby location. Monitor watching may be the responsibility of a designated person (i.e., a nurse or a specially trained "monitoring technician") or it may be a general responsibility shared by everyone on duty at the time. A variation on this theme is to have several consoles, each carrying part of the load, no more than a few patients. This design concept minimizes the possibility that a

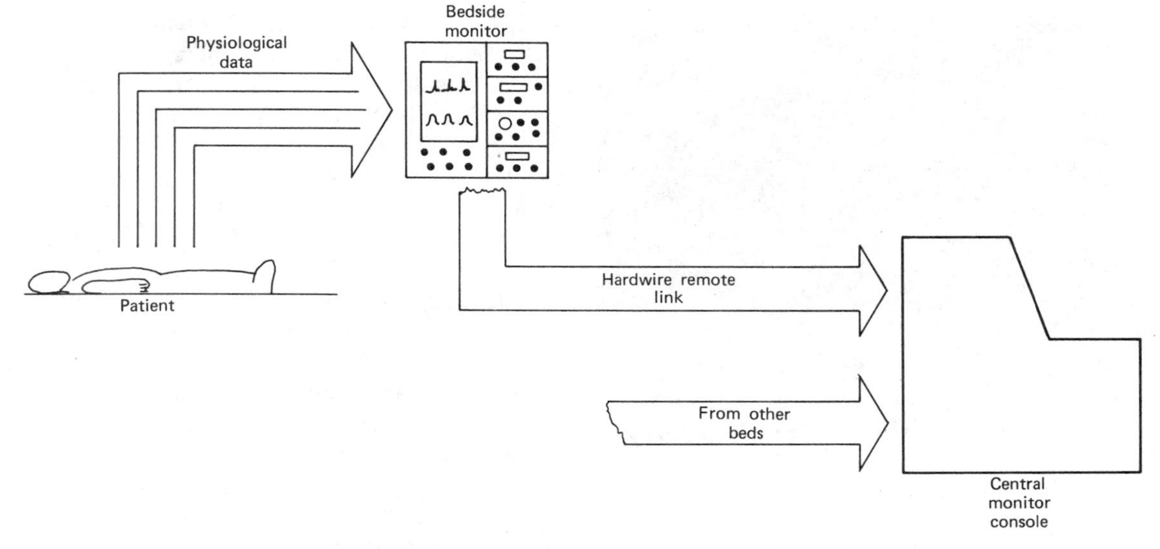

Figure 14-1 ICU/CCU central monitoring system.

single equipment failure will cost the unit all of its monitoring capability.

14-5 Bedside Monitors

The bedside monitor in the ICU/CCU may consist of a simple ECG amplifier, oscilloscope, and heart rate meter package, or a complex array of physiological instruments. Despite differences in features, however, all bedside monitors will be equipped with alarms, at least for heart rate, so that the staff is warned if the patient gets into trouble. Alarm circuitry will be discussed more fully in a later section.

Figure 14-2 shows several types of bedside monitoring equipment. The instrument shown in Figure 14-2*a* is an older style self-contained basic monitor, the Hewlett-Packard Model 78330A. The oscilloscope in this model is nonfade, that is, the waveform is stored on the CRT screen until it is erased by an updated waveform on the next sweep. The oscilloscope sweep rate is selectable between 25 and 50 mm/s.

The ECG waveform displayed may be any of the standard leads. Additionally, the monitor can be used to display EEG or the output of a photoplethysmograph. The display selector in the upper left-hand corner shows both *ECG-DIAG* and *ECG-MON* positions, which refer to *diagnostic* and *monitoring* grade displays, respectively. These switch positions select different amplifier bandwidths in the ECG channel (i.e., 0.05 to 100 Hz in diagnostic mode, and 0.05 to 45 Hz in monitoring mode). The *ECG-MON* mode provides for reducing the muscle artifact and gross patient movement artifact, without sacrificing the ability to display and recognize life-threatening cardiac arrhythmias.

The alarms are set by front panel controls. A switch allows them to be turned on and off and to be reset once the condition that caused the alarm ceases to exist. The alarm limits are displayed on the CRT screen along with the heart rate. The display is in the form of a band of light at the bottom of the screen. The low limit is indicated by the end of the thick band from the left side of the screen,

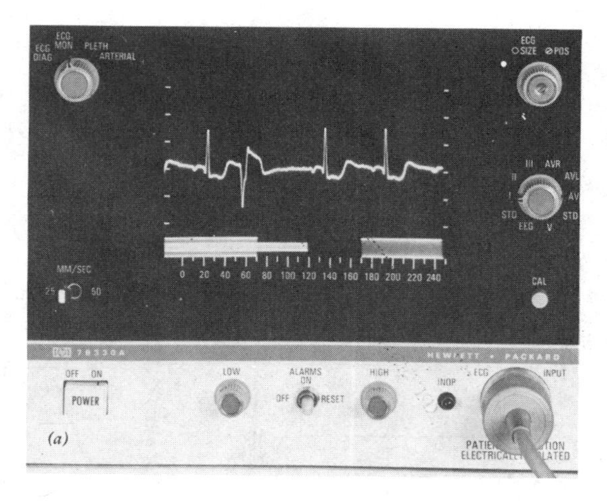

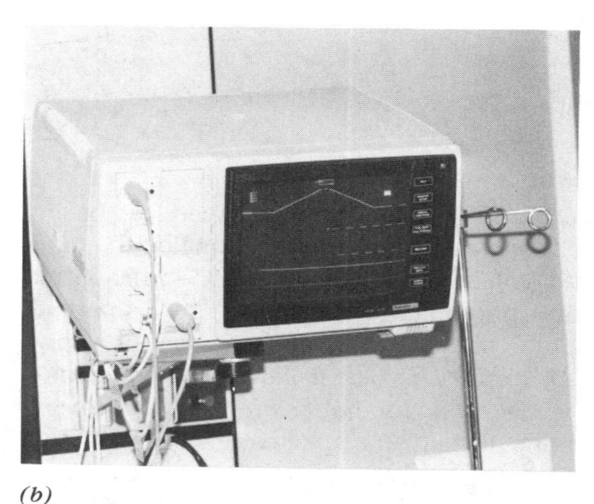

(b)

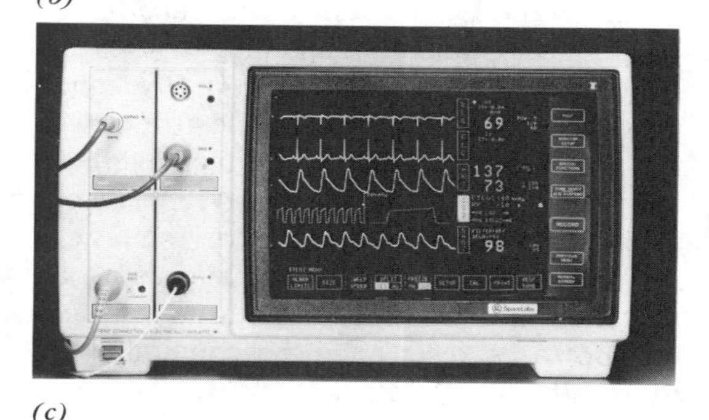

(c)

Figure 14-2 Bedside monitors. (*a*) H-P Model 7833OA. (Photo courtesy Hewlett-Packard.) (*b and* c) Integrated bedside monitor. (Photo courtesy Hewlett-Packard.)

in this case 70 BPM. The *high* alarm limit is indicated by the end of the band coming from the right side of the CRT, in this example 170 BPM. The patient's actual heart rate is determined by the narrow band of light from the left side of the CRT, in this case 120 BPM.

The patient monitor unit depicted in Figure 14-2*b* represents a trend away from rigidly defined "black box" models that offered only what the manufacturer thought was needed. While that type of design is still common among portable monitors, it has long since departed from the formal bedside monitor scene. The modular design still survives because it allows the manufacturer, the hospital, and the individual nursing and physician staffs to custom configure a monitor for a patient's needs. For example, it may be policy to provide continuous arterial pressure and ECG monitoring for all patients. All of the bedside monitors in that unit would be equipped with these modules. Other measurements, such as venous pressure or continuous rectal temperature or EEG, on the other hand, would typically be ordered for fewer patients. The physician could order these facilities, and they could then be provided by a monitoring technician or biomedical equipment technician. The hospital would thus be spared the cost of completely equipping each bed with all functions, while still providing the functions whenever needed. A related bedside monitor unit is shown in Figure 14-2*c*. This model shows the various physiological traces that can be displayed.

These monitors have a built-in self-test mode. Modern microprocessor electronic monitors can conduct self-tests and display the results. Thus, the unit will self-test whenever it is first turned on and whenever a self-test is initiated by the medical, nursing, or biomedical equipment staffs. With proper design, self-test results can be reported over the data lines to the central computer and be used there for record-keeping purposes. Some self-test capability is able to locate the problems in the system to the printed circuit board level and therefore tell the repair technician which subassembly to replace.

The Hewlett-Packard instrument in Figure 14-2*d* is an integrated bedside monitor that makes extensive use of digital electronic circuitry. The unit contains ECG, heart rate, two pressure channels, respiration, temperature, and (optionally) a second temperature channel with ΔT capability. An internal character display generator, such as those used in computer video terminals, displays the numerical data along the right-hand vertical edge of the CRT.

14-6 Bedside Monitor Circuits

Most of the circuits in the bedside monitor are ordinary ECG or pressure amplifiers, covered in detail elsewhere in this text. In this section, we will discuss those circuits that are unique to, or commonly employed in, bedside monitors.

14-6.1 Cardiotachometers

The *cardiotachometer* is a heart rate meter. It provides an analog or digital display of the heart rate, usually developed from the patient's ECG signal but sometimes derived from the arterial pressure waveform.

The vast majority of all cardiotachometers are analog circuits that produce a *dc voltage* proportional to the patient's heart rate. This dc voltage will be displayed on an analog or digital voltmeter. The block diagram of a cardiotachometer is shown in Figure 14-3*a* and consists of four sections: *R-wave discriminator, monostable multivibrator* (i.e., *one-shot*), *integrator,* and *readout*.

The *R*-wave discriminator is necessary so that the circuit will count only once for each heart beat, and the *R*-wave is the most easily identified feature of the ECG waveform. All *R*-wave discriminators use a *level detector* (i.e., a circuit that will produce an output change only when the predetermined input voltage level is exceeded). A simple voltage comparator will perform this function.

But a simple voltage comparison circuit is not sufficient to prevent false counting. Equipment that uses only level detection often suffers from double counting because, in some patients, the *T*-wave amplitude also exceeds the level detector threshold.

A *differentiator* stage (dE/dt) preceding the level detector is used by some manufacturers to reduce the double counting artifact. This circuit is called a "high-pass filter" in some service manuals, but it is the same circuit under a different name. The high-pass filter produces an output voltage that is proportional to the *slope* (i.e., *rate of change* of the input signal). The fast-changing *R*-wave, therefore, will produce a much larger output voltage than will the slow-rising, low-frequency *T*-wave and *P*-wave features.

The output of the *R*-wave discriminator triggers a monostable multivibrator. This "one-shot" stage will generate one output pulse for each *R*-wave. These pulses have *constant duration* and *amplitude;* only the pulse repetition rate varies with the patient's heart rate.

The reason why the one-shot stage is required is that the dc output voltage, which must be proportional to the heart rate, is obtained from an *integrator* (called a low-pass filter in some service manuals) stage. The integrator *averages* the pulses applied to its input and produces an output voltage proportional to the total area under the pulses. This area is determined by the number of pulses received and the area of each pulse (i.e., *duration* $\times$ *height*). The one-shot stage, however, produces constant-area pulses, and only one output pulse is generated for each *R*-wave. The dc output of the integrator, therefore, is proportional only to the time average of the heart rate (i.e., number of *R*-waves per unit of time).

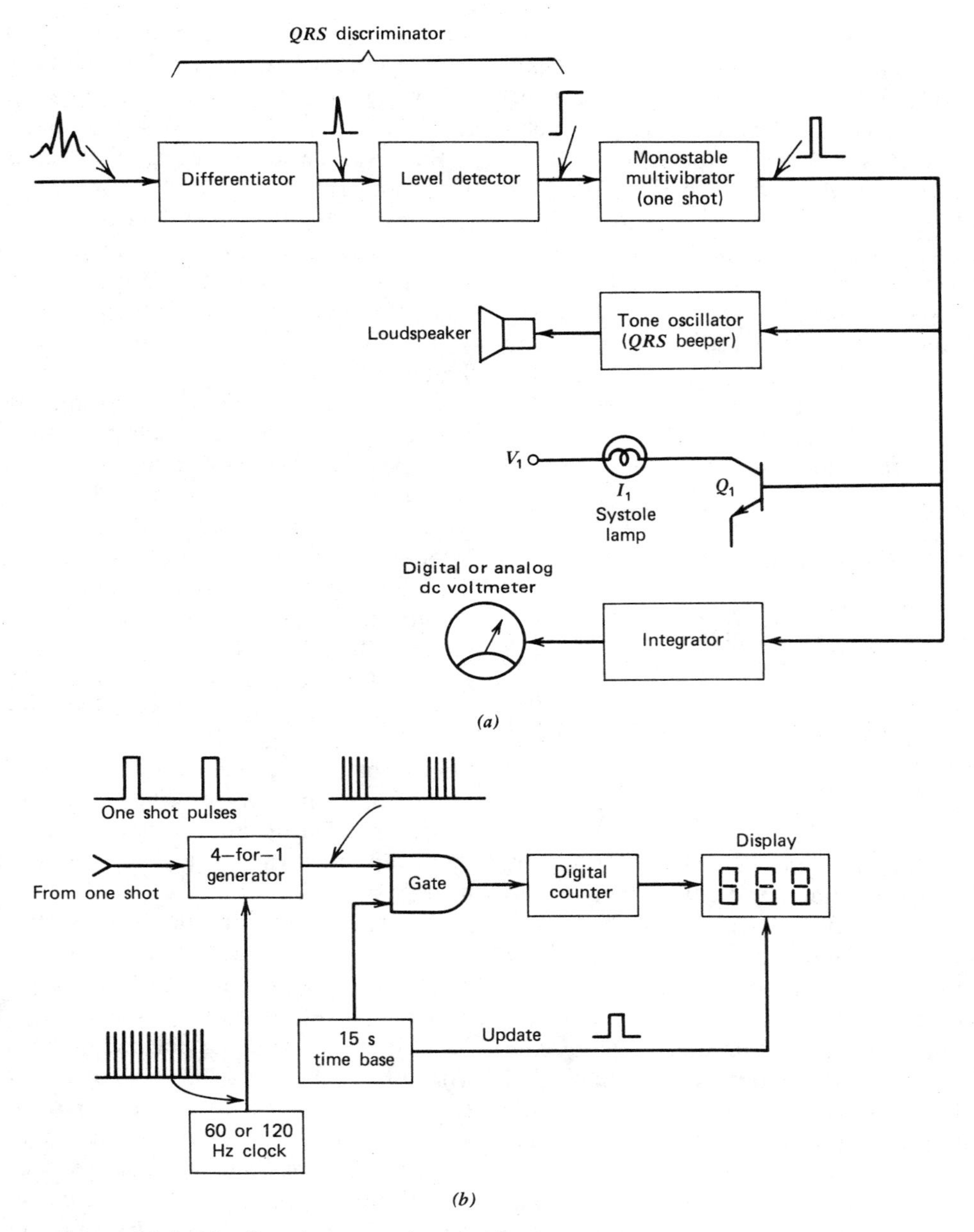

Figure 14-3 Cardiotachometer circuits. (*a*) Integrating analog type. (*b*) Digital type.

The readout device will be a simple voltmeter. Even many "digital" heart rate meters use analog circuitry and then display the result on a digital voltmeter.

A *systole lamp* and a *systole "beeper"* may also be driven by the output of the one-shot stage. The systole lamp or LED is turned on by a drive transistor that becomes forward biased only when the output of the one-shot is high. The systole beeper is a tone oscillator that is turned on by the output pulse from the one-shot stage. The beeper is especially useful when transporting the patient on a stretcher, and in other cases where the oscilloscope screen is not always in view, or the alarms are inoperative. At most other times the beeper is an annoyance and so will remain turned off.

Figure 14-3*b* shows a digital cardiotachometer. This circuit, or similar circuits, is found only in a few instruments. The input section of the circuit is the same as the analog version, up until the output of the one-shot stage.

The one-shot pulses occur at the same rate at the patient's *R*-waves and are used to trigger a *four-for-one* circuit, which outputs four clock pulses for each pulse received at the input. In most cases, the clock signal is derived from either a free-running *R-C* oscillator or a 120-Hz wave train from the full wave rectified power supply.

The digital counter circuit, then, sees four input pulses for each *R*-wave and is gated on for 15 s. After the 15-s count period, the accumulated data is equal to the patient's heart rate in BPM, so the time base issues an *update* command to the display and then initiates a new count cycle. This allows the digital counter to measure the averaged heart rate in 15 s (i.e., one-fourth of a minute).

Both of these cardiotachometer circuits are basically averaging types. It is not generally deemed prudent to use a short time constant in these circuits because many patients' heart rates may normally change slightly from beat-to-beat (i.e., the *period* of *R-R interval* changes). This is one reason why few instruments use the period to derive heart rate.

14-6.2 Alarms

Alarm circuits are provided on bedside monitors to warn the staff of an emergency condition. On heart rate, for example, it is normal practice to bracket the patient's indicated heart rate with high and low alarm limits. If the patient's rate either speeds up or slows down significantly, then an alarm sounds.

Two common approaches are taken in the design of alarm circuits, and these are shown in Figures 14-4 and 14-5. In Figure 14-4 we see the photocell type of alarm circuit used in analog readout meters. Inside of the meter housing are two lamp-photocell assemblies (Figure 14-4*a*), one each for high and low alarms. These assemblies are positioned by the *alarm set* tabs on the front of the meter housing. A metal vane attached to the rear of the meter pointer is used to trigger the alarm condition. The photocell assembly is built such that the light will shine on the pc element, keeping its resistance low all of the time unless the limit is exceeded. If that situation should occur, then the vane of the meter pointer will blind the photocell assembly, causing its resistance to increase substantially. In one model, the dark resistance is over 1 MΩ, while lighted resistance is under 20 kΩ.

The alarm circuit is shown in Figure 14-4*b*. Only one is shown here, but two will be in each monitor; one each for high and low.

Alarm driver transistor Q_1 is normally reverse biased by the potential across capacitor C_1. This voltage is the sum of two sources, $V-$ and $V+$. Under normal conditions the resistance of the photocell is low, so the voltage across C_1 is negative, thereby reverse biasing Q_1. If the photocell is *blinded* when an alarm situation occurs, then its resistance goes very high, and the voltage across C_1 goes positive to forward bias Q_1 and turn on the alarm circuit.

Figure 14-5 shows two alarm circuits that use voltage comparators to detect the alarm condition. In the version of Figure 14-5*a* there is a resistance element inside of the

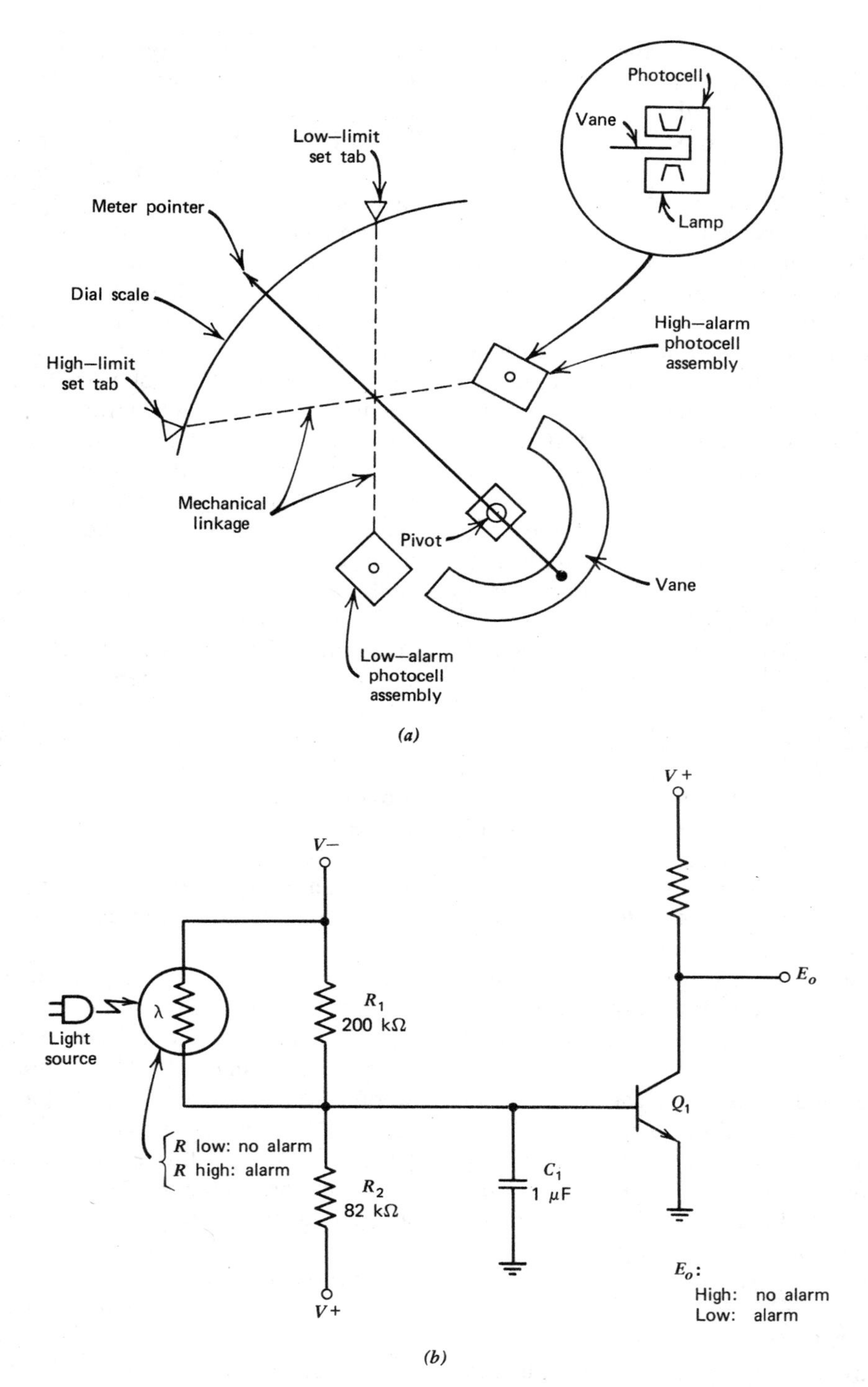

Figure 14-4 Photocell alarm. (*a*) Mechanical arrangement (top view). (*b*) Circuit.

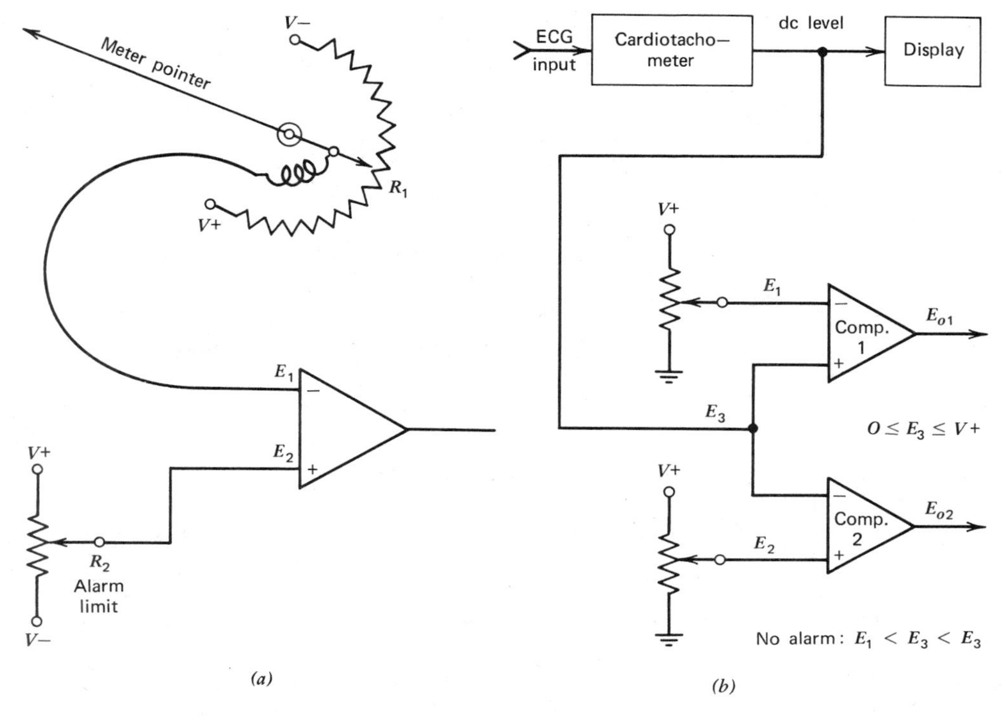

Figure 14-5 Alarm circuits. (*a*) Comparator alarm with special meter movement. (*b*) Alarm circuit using cardiotachometer output.

analog meter movement. The output voltage, E_1, indicates the heart rate, while the alarm limit voltage E_2 is produced by a front panel potentiometer. As long as the set limit is not exceeded by E_1, no alarm occurs. But if the limit is exceeded, then the alarm triggers.

A similar, and more popular, circuit is shown in Figure 14-5*b*. In this case the cardiotachometer output voltage, a dc level, is fed to the input of a standard dual-limit window comparator. No alarm occurs if E_3 remains between E_1 and E_2.

14-6.3 Lead Fault Indicator

When a monitor electrode, or its lead wire, comes loose, the appearance of the display will be either 60-Hz interference or a flat baseline (i.e., no signal at all). In the latter case the flat baseline may be mistaken for asystole and emergency resuscitation procedures initiated inappropriately. A *lead fault indicator* will help prevent this occurrence.

A simplified version of the lead fault circuit used in the Tektronix Model 414 monitor is shown in Figure 14-6*a*. Two very high-value resistors (R_1 and R_2) place a 10-nA current on each lead from the patient. Voltages E_1 and E_2 are formed by voltage divider action of the patient-electrode resistance and resistors R_1/R_2, respectively. Ordinarily, E_1 and E_2 are very low and nearly equal to each other and so do not affect the ECG preamplifier output. For example, E_1 is

$$E_1 = \frac{ER}{R + R_1} \qquad (14\text{-}1)$$

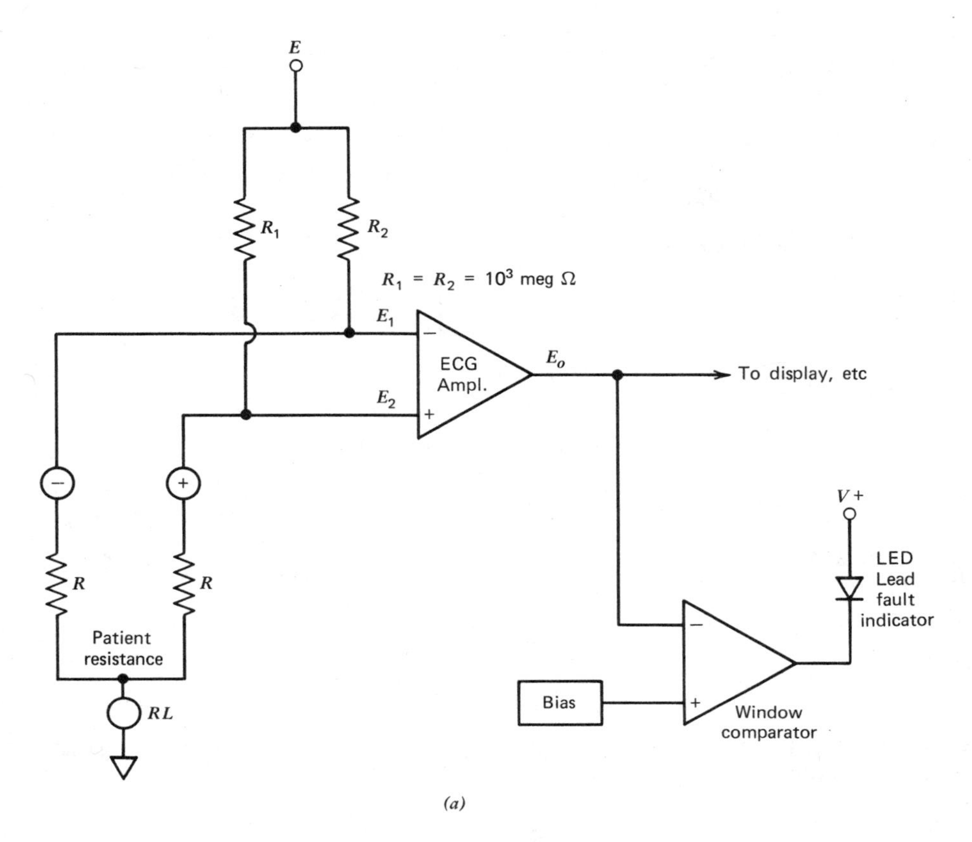

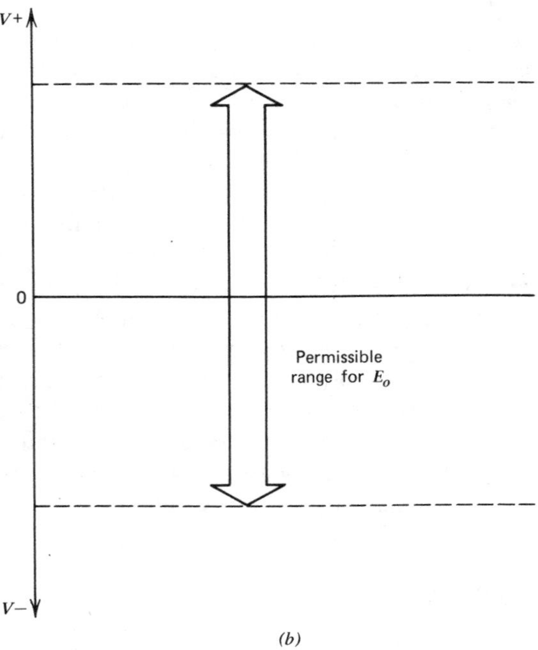

Figure 14-6 Lead fault indicator. (*a*) Circuit. (*b*) Range of E_0.

where

E_1 is the dc voltage at the inverting input of the ECG preamplifier
E is the reference voltage
R is the patient resistance in ohms
R_1 is the value of resistor R_1 (i.e., $10^9\ \Omega$)

EXAMPLE 14-1

Calculate the value of E_1 in Figure 14-6*a* if $E = +10$ V dc, and R is 10 Ω.

SOLUTION

$$E_1 = ER/(R + R_1) \qquad (14\text{-}1)$$
$$E_1 = (10\ \text{V})(10^4\ \Omega)/10^4 + 10^9)\Omega$$
$$E_1 = (10^5\ \text{V})/(10^4 + 10^9) \approx \mathbf{100\ \mu V}$$

If a lead should come off of the patient's body, however, the voltage at the input of the ECG preamplifier rises to almost +10V, saturating the output. A level detector turns on the *lead fault* lamp if E_o exceeds the limits due to the offset caused by the loss of R, the patient resistance. Figure 14-6*b* shows the permissible limits of E_o.

14-7 Central Monitoring Consoles

The central monitoring console is credited with allowing each critical care nurse to attend more patients. Without central monitoring, critical care would be more one to one, and total number of nursing personnel is greater. But life-threatening events do not usually occur on all patients in the unit at the same time, so such a costly nurse-patient ratio is not always cost effective.

The central station serves several functions in the intensive care environment. One thing that is immediately apparent is that it amplifies the abilities of the staff to keep track of the situation and so reduces the number of nurses and doctors needed to staff a unit. All of the analog signals, plus numerical data and the alarm status signals, are routed from each bedside monitor to the central nurses' station unit. The console will provide an array of multichannel oscilloscopes, heart rate meters (plus occasionally blood pressure meters), a computer terminal, an alarm status annunciator panel, and a communications system. On modern systems, the entire information display may be part of a video terminal controlled by a personal computer, with a strip-chart recorder for hard-copy readout (Figure 14-7). With this information a single operator can keep track of the condition of several patients at once, relieving the need to station a nurse in each patient room all of the time.

Figure 14-8 shows the block diagram of a typical central monitoring system. Electrodes and transducers (sensors) attached to the patient provide signals through an array of input amplifiers to the local bedside monitor. These signals are locally displayed on a monitor oscilloscope, various numerical (digital) readouts, and sometimes a local strip-chart (paper) recorder. Local alarms are also provided to alert the staff if any parameter (e.g., heart rate) goes outside of set limits.

Figure 14-7 Modern central monitoring system.

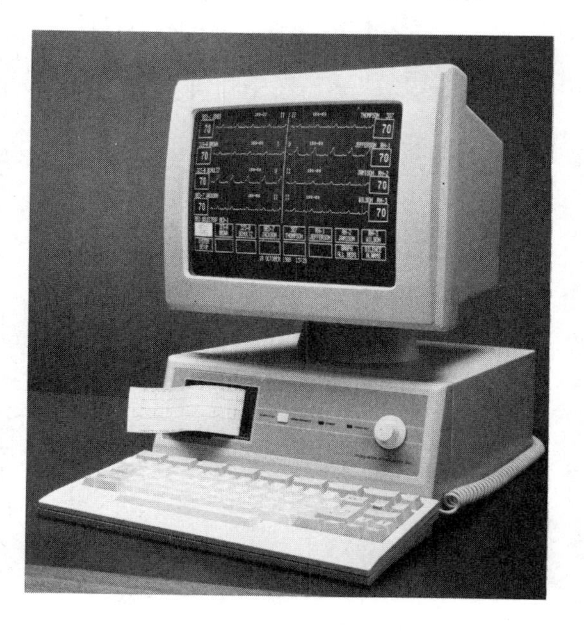

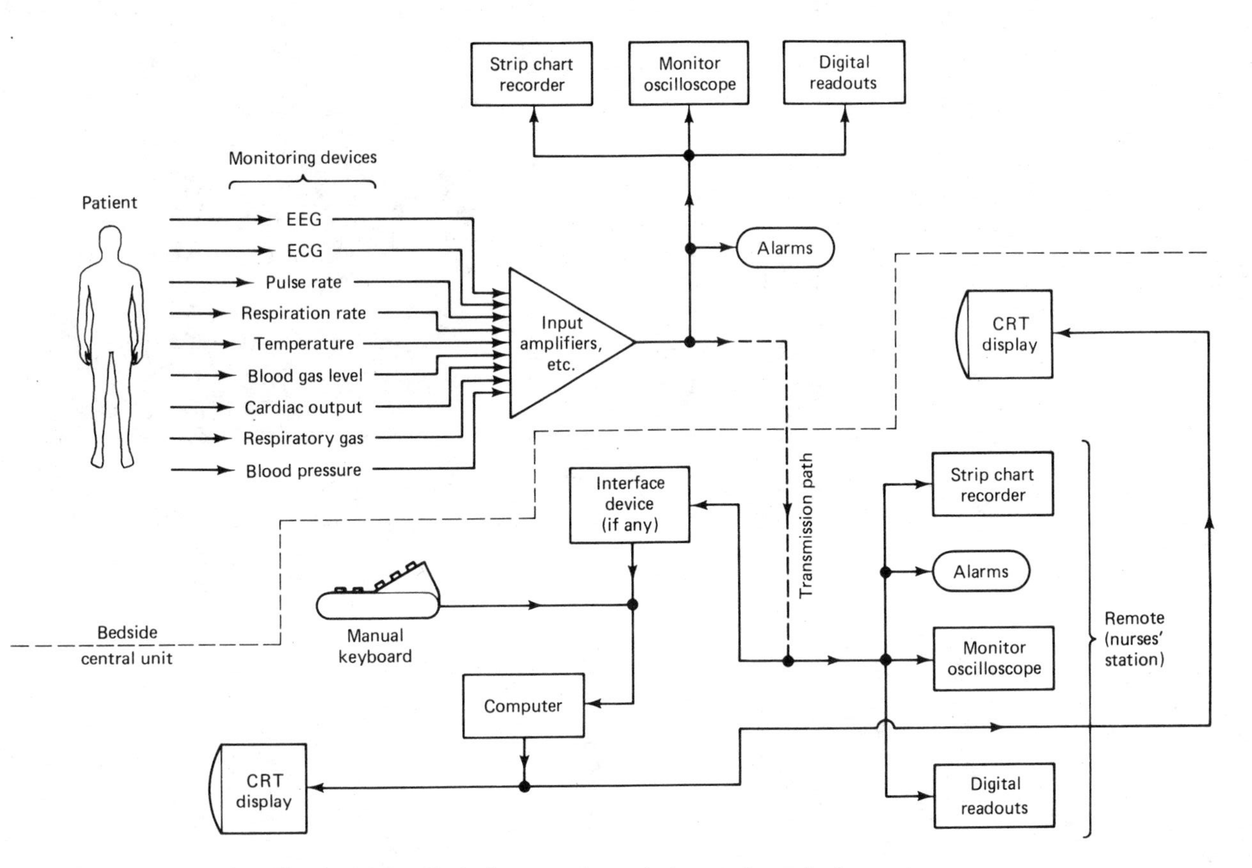

Figure 14-8 Block diagram of a typical central monitoring system.

There will be a transmission path between the bedside monitors and the central station. The transmission path might be analog or digital, or a mixture of both. We will discuss these options later.

The signals from the bedside units are routed to the central nurses' station console, where they are displayed on slave units of the bedside monitors. Shown in the system of Fig. 14-8 are the following units: oscilloscope (usually multichannel), digital readouts for numerical data, alarms, and a strip-chart recorder. In many modern designs the numerical data and analog data are both displayed simultaneously on the screen of a video terminal.

Although older central station units (some of which are still in use) were simple analog instruments that were slaved to the bedside monitors, modern design uses a microcomputer to keep track of things and record data. The typical system will contain a computer (often an IBM-XT or AT-class machine), a manual data entry keyboard, a video/CRT display, and an interface device (not always used). The *architecture* of the central monitoring system determines how the various units relate to one another and how the system is interconnected. Figure 14-8 showed one example of system architecture.

Older systems were strictly analog and so used the type of system shown in Figure 14-9. The bedside monitor is equipped with an output interface that consists of a large multiconductor cable that carries two types of signal: analog waveforms and alarm/control *discretes*. A "discrete" is a single wire or wire-pair that is either open circuited or closed circuited when a certain condition exists. The ECG alarm discrete, for example, is "open" when there is no alarm present and is shorted to ground when the alarm occurs. These conditions signal the central station circuits as to the alarm condition. This type of architecture requires a large number of wires to implement. A 24-bed ICU system, for example, has as many as 20 discrete and analog signal lines per bed, for a total of 480 analog lines.

Note in Figure 14-9 that the output of the ECG amplifier (labeled "analog ECG signal") is not fed directly to the multiconductor cable, but rather is directed to a buffer amplifier ("BUFF"). This stage is used to isolate the bedside monitor from faults in either the multiconductor cable, the interconnections, or the central station itself. If, for example, a short circuit occurs in the transmission path, the output of the buffer will be shorted to ground. But, because of the buffer amplifier stage, the local oscilloscope and alarm system remains working—protecting the patient. If the external fault does not affect the alarm discrete, then even the central station alarm function will remain (even though the analog signal disappears).

Buffering in analog signals has its parallel in digital systems and represents a feature that should be required in specifying any new systems or modifications to existing systems. Some things to include in the purchase order and/or request for quotation are as follows:

1. No single-point fault will remove more than one single patient unit from service (i.e., a fault on, say, bedside monitor no. 2 only affects that unit and does not affect all other units).

2. No single-point fault, such as an output short, on a bedside monitor unit shall remove all functions of that unit either locally or at the nurses' station.

3. It shall be possible to disconnect any bedside monitor unit from the system without adversely affecting any other unit or function either in the local rooms or at the central station.

Most modern bedside monitor (BSM) units are digitized to allow their use with a computer. Early computerized monitoring systems were configured like any analog sys-

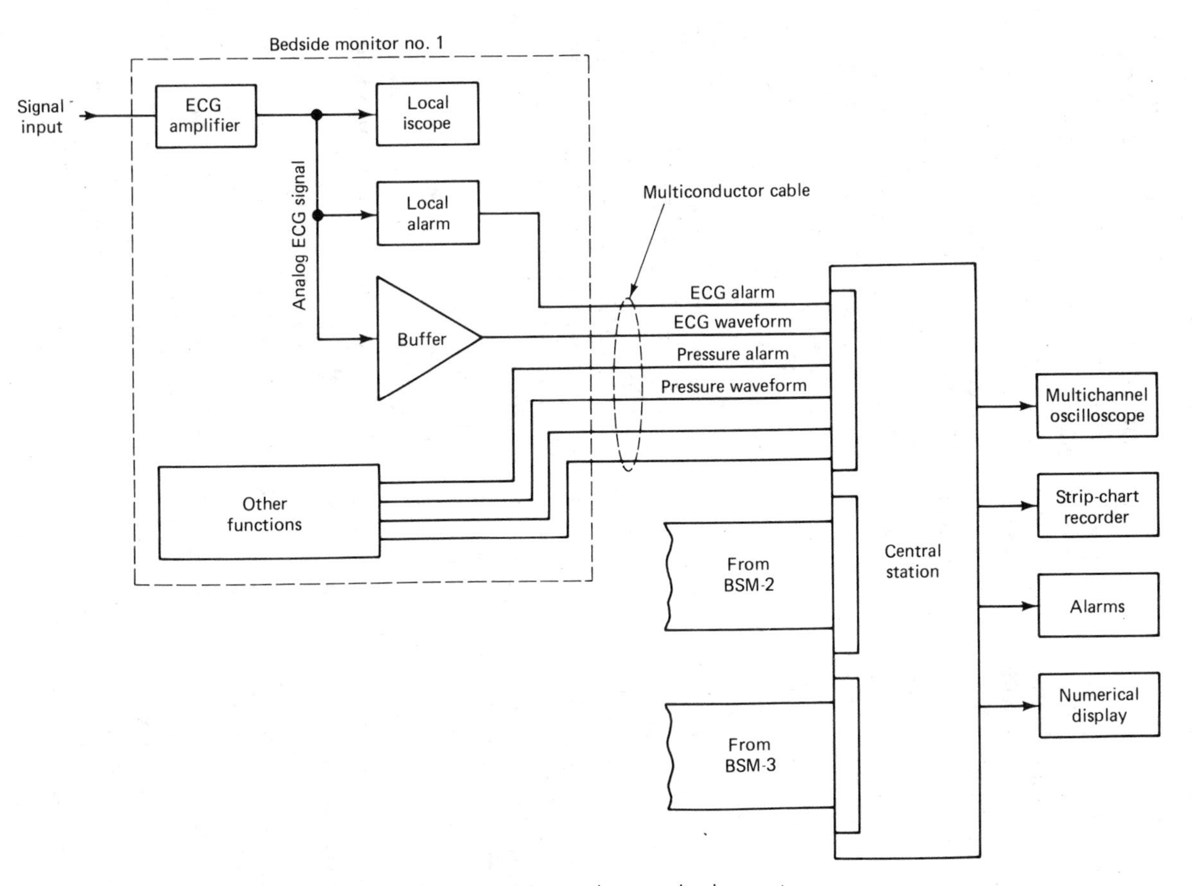

Figure 14-9 Older analog monitoring system.

tem, with the exception that *digitization* took place at the central nurses' station computer. In modern systems, however, the BSM itself may include a microprocessor to perform many of the chores once performed elsewhere (in addition to chores not offered before).

Figure 14-10 shows the block diagram for a simple digitized ECG monitor or BSM. The ECG amplifier, local alarm, and local oscilloscope are similar (often identical) to those of the strictly analog system. The difference is that no analog signals are passed to the central station. An *analog-to-digital converter* (A/D) is used to convert the analog voltage that represents the ECG signal into binary "words" that are transmitted to the central computer over a data bus. Because the voltage levels of the binary word do not transmit well over distances greater than about 20 yards, it is often the practice to use a modulator/demodulator (MODEM) unit to convert the binary signals to a series of audio tones. These tones are transmitted to the central station computer, where another MODEM unit reconverts them to data words.

The alarms can either be sent along the bus as a tone by way of the modem or via a separate discrete line. Both systems are known. In some cases, both modem and discrete alarms might be used as a safety feature.

The BSM controller module usually contains a microprocessor or simple digital computer to control the operation of the bedside monitor, run self-tests, and perform the alarm functions. It can communicate with the main computer unit in order to synchronize operations. In some systems "handshaking" between the central computer and BSMs indicates when data is ready or may be transmitted.

A function of the controller is to respond to the central computer when it is being polled for data. Each BSM is given a unique *address*. The BSM will not respond to traffic on the bus unless it "hears" either its own address or an "all units listen" broadcast address.

Although there are many variations on the following themes, and variations also exist in the specifics of implementation, the systems shown in Figure 14-11 represent a large number of data connection schemes between BSMs and central stations. In Figure 14-11*a* we see the system in which a local area controller receives the data lines from each BSM, prioritizes the signals, and then transmits the data to the central station computer. In some implementations this system is called a "star" connection.

A parallel connection is shown in Figure 14-11*b*. In this case, a common main data bus connects all of the BSMs and the central station computer. The controller is located inside the computer. Be sure to avoid connections that are truly "daisy chained."

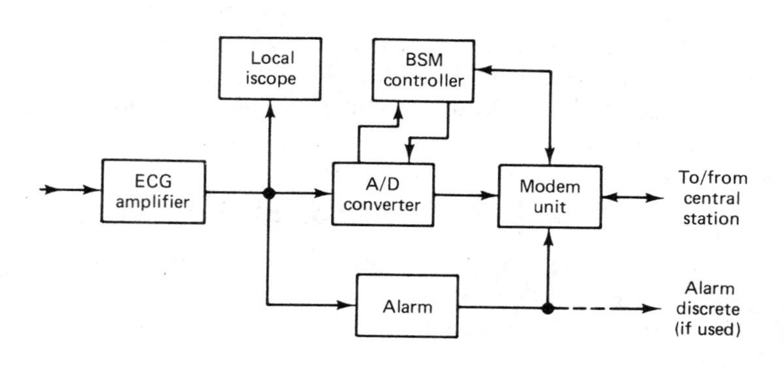

Figure 14-10 Digitized ECG/bedside monitor system.

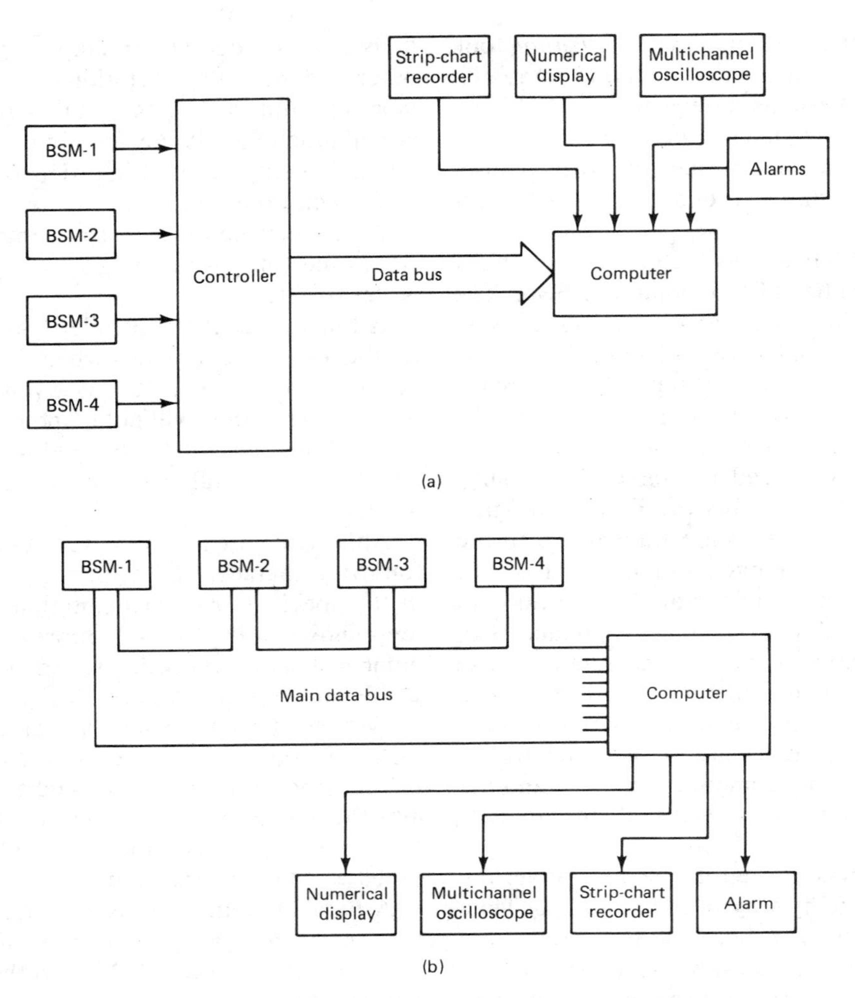

Figure 14-11 (*a*) Local area controller configuration; (*b*) parallel connection configuration.

These modules should be paralleled (and isolated to prevent a single failure from taking out the whole system). In a daisy chain system, data is passed from BSM-1 to BSM-2 to BSM-3 to BSM-4 and then to the computer. In that type of system, which is analogous to a series of strung Christmas tree lights, a single failure in BSM-4 blocks all data from the other three units. In other words, the entire ICU monitoring system goes down because of a single failure.

The actual data bus between the bedside units might be either of the following: twisted pair wires, multiconductor wires, coaxial cables (like TV antenna cables) for a computer local area network (LAN), or special wires.

14-8 ECG/Physiological Telemetry

Telemetry systems are special cases of CCU patient monitoring systems in which the monitoring function is done from a remote location. There are two basic forms of telemetry systems: *radio telemetry* and *landline telemetry.* The radio form, which is used in CCU "step-down" units, uses a small radio transmitter attached to the patient that picks up the electrocardiograph (ECG)—or other physiological data—waveform and transmits it via radio waves to a receiver at a central monitoring station. The landline form of telemetry uses an audio tone, or computer-like modem, to represent the analog ECG signal and then transmit it over telephone lines to a central office. We will discuss both forms.

14-8.1 Radio Telemetry Systems

Many hospitals use radio telemetry systems to monitor certain patients. The most common use of radio telemetry is to keep track of improving cardiac patients and at the same time keep them ambulatory. These units are sometimes called *post-coronary care units* (PCCU), or *step-down CCU,* or some other name that indicates a less rigorous monitoring regime than the full CCU, where new heart patients are treated. The telemetry unit is short of a "half-way house" between the full-up CCU and either the general medical floor patient population or home care.

The telemetry unit uses a tiny VHF or UHF radio transmitter (Figure 14-12) that is attached to the patient by either a belt clip or small sack hung around the patient's neck. Most transmitters contain an analog ECG section that acquires the signal and uses it to frequency modulate (FM) the radio transmitter. The nurses' station is equipped with a bank of radio receivers tuned to the same frequencies as the transmitters. The receiver demodulates the FM signal to recover ECG waveform. The waveform is then displayed on an oscilloscope and/or strip-chart recorder, as in any other patient monitoring system. The signal may also be input to a computerized monitoring system (indeed, today it probably will be so processed).

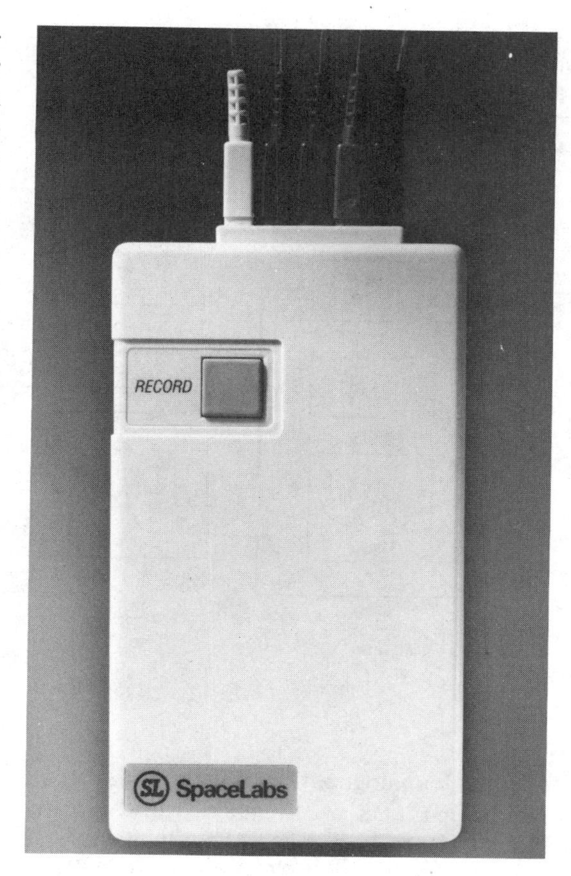

Figure 14-12 UHF ECG telemetry unit.

It is common practice to use either specialized radio frequencies set aside by the Federal Communications Commission (FCC) for medical telemetry or unused television channel frequencies. It is not unusual for a telemetry transmitter to operate in the quieted "guard band" between the sound and video carrier of a television signal. These VHF and UHF frequency allocations allow the telemetry system designer to use hardware that was originally designed for the master antenna television (MATV) or cable TV markets to process signals for medical systems.

Figure 14-13 shows the block diagram of

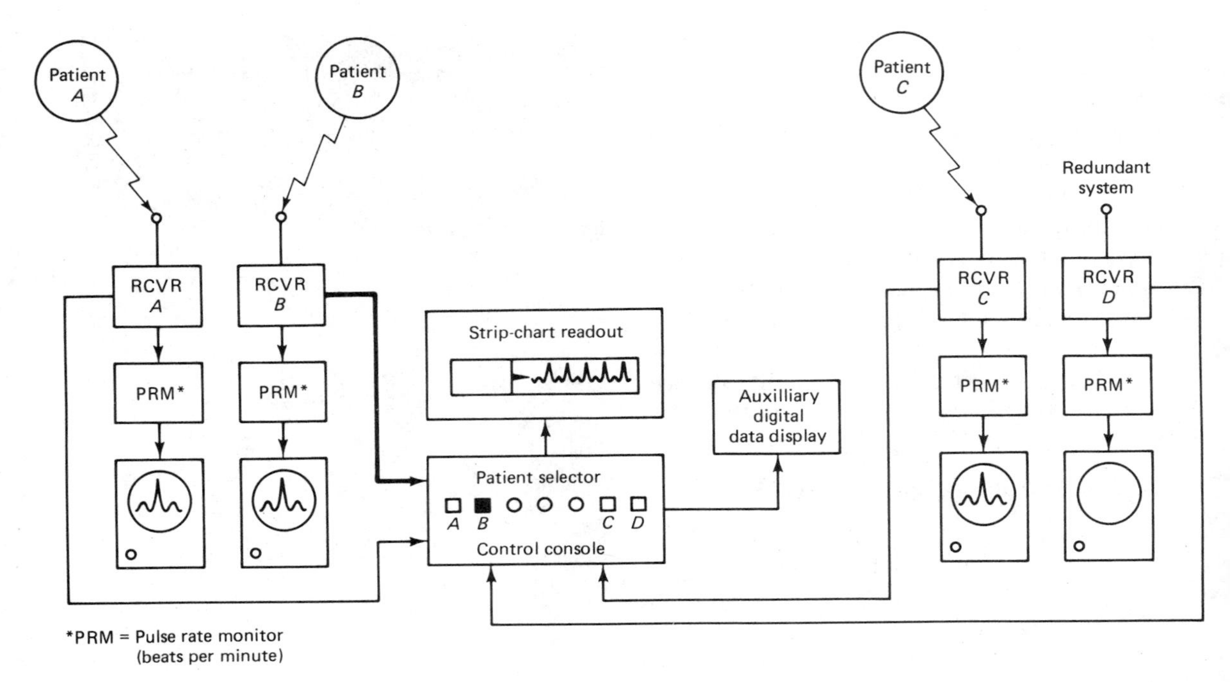

Figure 14-13 Block diagram for an analog ECG telemetry system.

a typical analog patient telemetry system. Several patients (A, B, and C) are wearing the miniature transmitters that pick up and then transmit the ECG waveform over a VHF or UHF radio frequency. In each case, a radio receiver picks up and demodulates the signal, recovering the analog waveform. This waveform is output to an oscilloscope display and a pulse rate meter. The pulse rate meter will also have a high rate (tachycardia) alarm and a low rate (bradycardia) alarm built in. In addition, the analog signal is also sent through a patient selection switch to a strip-chart recorder that can provide a "hard copy" of the waveform for the nurses and doctors who care for the patient.

Some telemetry systems are now using an analog-to-digital converter (A/D) inside the transmitter to digitize the analog waveform prior to transmission. The transmitter then sends a series of tones that represents the ones and zeros of the binary numbers recognized by the computer attached to the receivers.

A practical analog telemetry system is shown in Figure 14-14*a*. The corridors around the nurses' station are set aside for use of ambulatory patients. However, there are defined limits to the patients' permitted zone of travel. Some hospitals paint the walls of the permitted area either a different color from the rest of the building or use a color-coded stripe on the walls. Still other hospitals make no modifications of the paint scheme at all but rather depend on telling the patient where to stop walking, or rely on a prominent landmark such as a fire door or elevator lobby to limit the zone. The nurses' station is usually located to permit surveillance of all the permitted area in case a patient has problems.

The transmitters carried by the patient are very low-power units, on the order of a few milliwatts at best. Therefore, it is possible

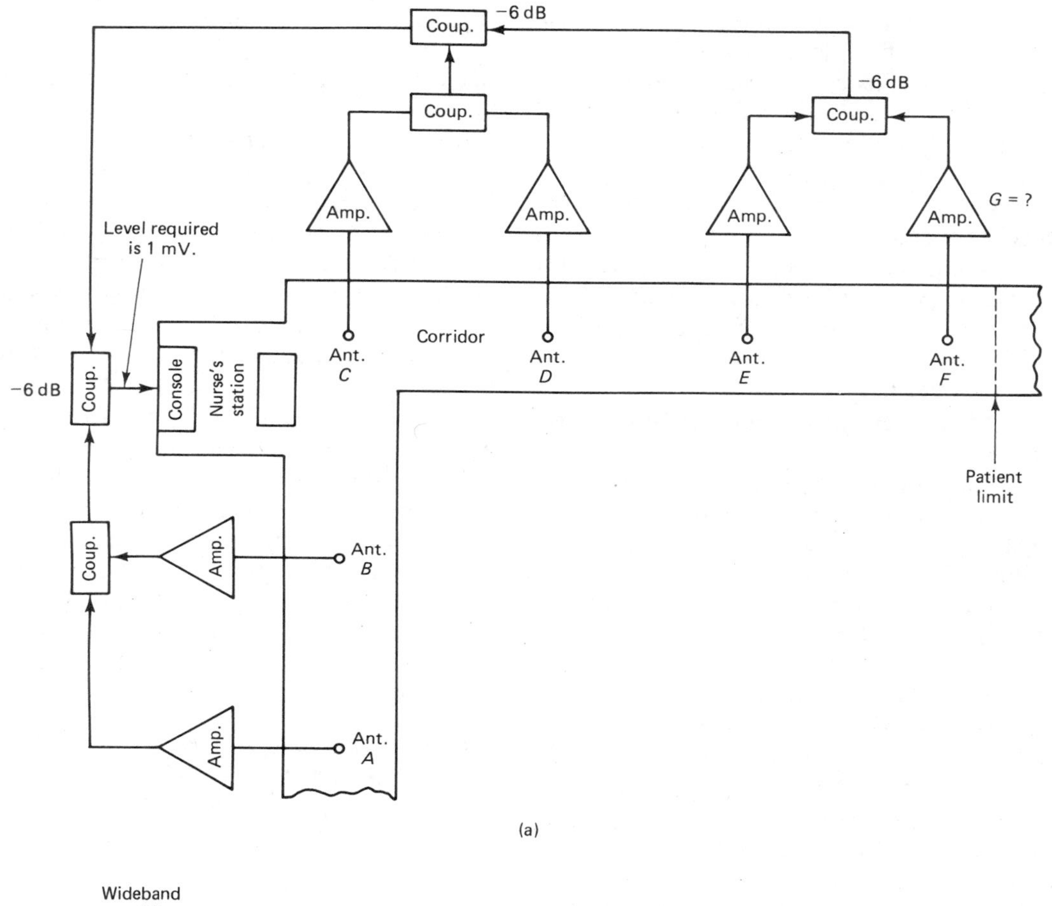

(a)

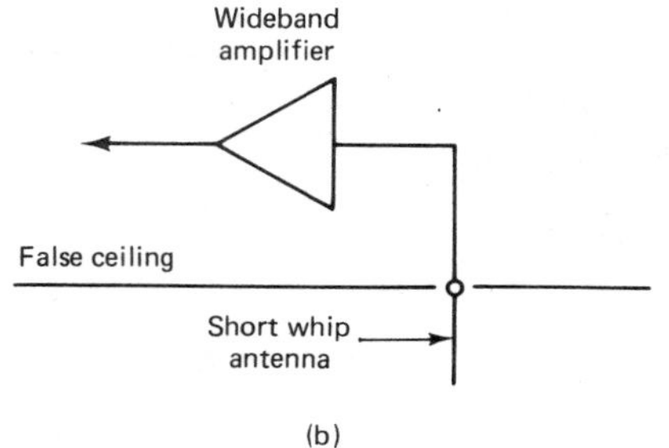

(b)

Figure 14-14 (*a*) Practical analog telemetry system; (*b*) antenna installation in false ceiling.

for the signal levels to be too low for the receiver, even at a short distance, even though the area permitted for travel is not extensive. As a result, a series of small whip antennas are placed at strategic locations throughout the unit. These antennas usually hang from the false ceilings (Figure 14-14*b*) in a manner that does not interfere with pedestrian traffic in the corridor.

Even with several antennas in place, however, the losses in the system added to the low power of the original signal conspire to prevent adequate reception. As a result, booster amplifiers (labelled "Amp" in Figure 14-14*a*) are placed at each antenna site. Gains up to 60 dB are required. These amplifiers are usually selected from MATV and cable TV equipment.

The outputs of the various amplifiers are mixed together in standard VHF/UHF two-set TV couplers ("Coup" in Figure 14-14*a*). It is common to find ordinary television receiver couplers in this application. These devices are passive and so will offer losses of −2 to −9 dB.

Some modern ECG telemetry systems are classified as *digital telemetry.* These systems are more sophisticated than analog systems and are also more free of fading and signal artifacts than analog systems. Digital technology allows the use of frequency synthesis for the transmitter and receiver, which means that the frequency can be entered for each unit (making them more interchangeable). In addition, complex error correction and checking schemes can be incorporated to ensure proper transmission of the waveform and other data. Some systems incorporate a zone pattern scheme like cellular radio (i.e., a unit will alter its operation as it travels from one antenna cell to another), a scheme that reduces—some claim eliminates—fading or mutual interference.

14-8.2 Troubleshooting Telemetry Systems

ECG telemetry systems are like all other hardware systems, so from time to time they will malfunction. Some of the faults can be handled directly by the user, while others must be referred to various grades or levels of service shop.

Nurses, emergency medical technicians (EMTs), and other medical personnel can perform several minor troubleshooting tests. First, the patient electrodes and the wires that connect them to the transmitter are usually replaceable by the user, so they can be checked before taking a unit out of service. Second, the battery (which is a common fault) can be replaced on most units by the user (alternatively, a battery charger is used). Finally, it is permissible for users to swap telemetry transmitters and receivers on most systems and therefore restore operation (if on a different channel) by the simple process of elimination. The problem on single-channel systems is that this process does not reveal whether it is the receiver or the transmitter that is at fault.

For routine troubleshooting of the telemetry system by a service technician, much can be said to owning either a television field strength meter (FSM), which can also be used for making site surveys prior to installation or ordering, or a continuously tunable VHF/UHF receiver that covers all of the operating frequencies that might normally be expected to be covered (Fig. 14-15). The re-

Figure 14-15 Using a field strength meter to check ECG telemetry transmitter.

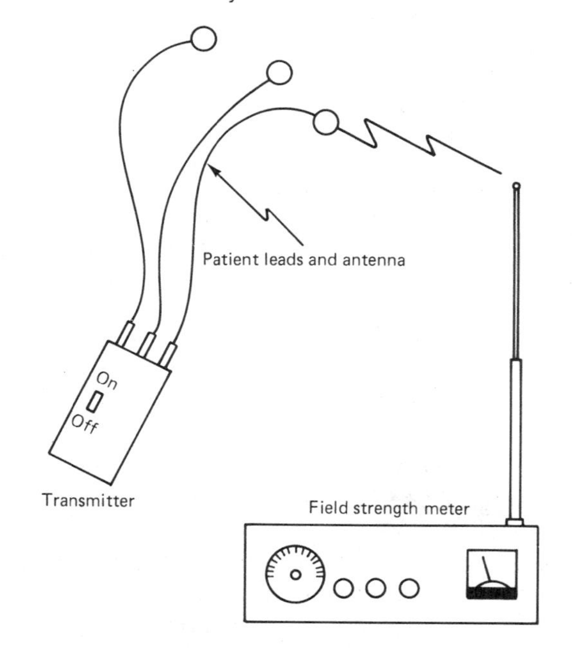

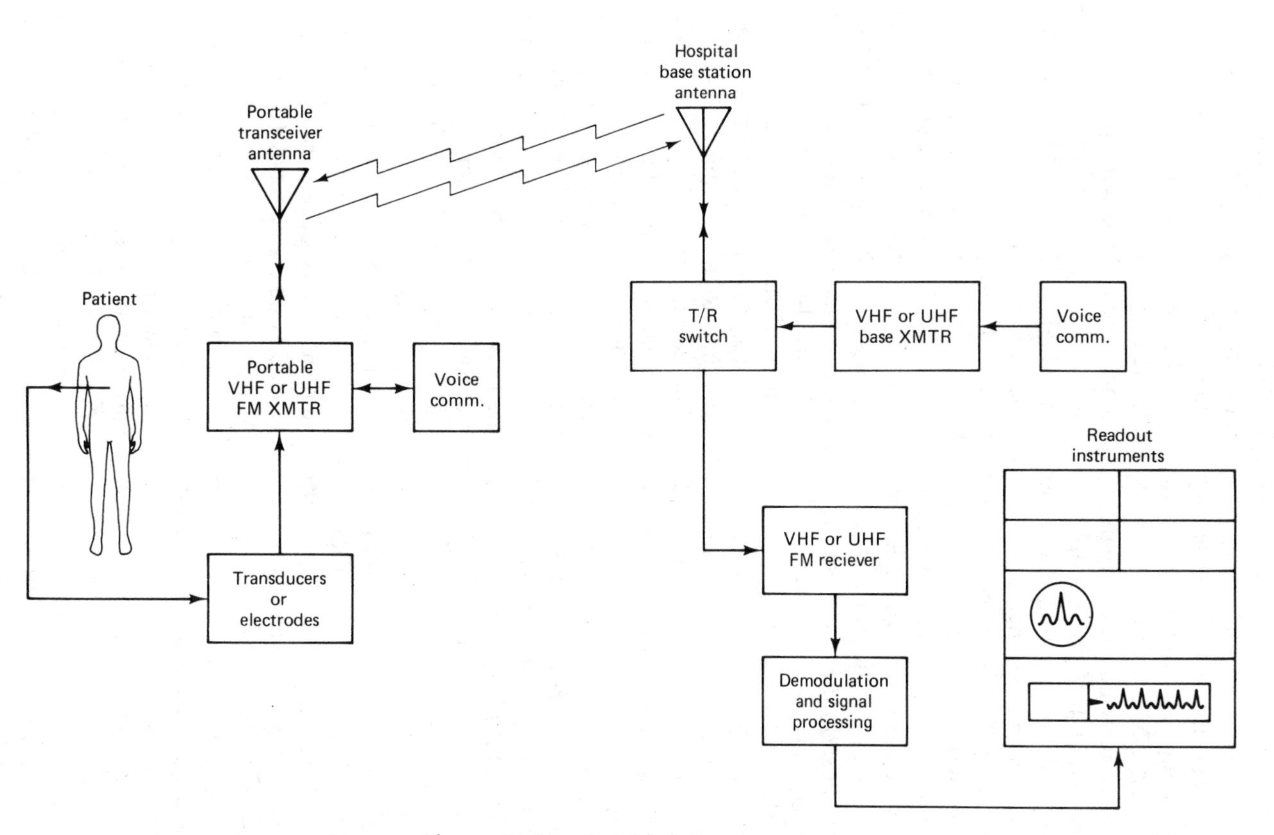

Figure 14-16 Portable telemetry system.

ceiver or FSM can be used to monitor the output of the transmitter to determine whether or not the transmitter is putting out a signal.

Perhaps the most useful form of service instrument for telemetry equipment is the spectrum analyzer. This device is a swept frequency receiver in which the signals received are displayed in the form of amplitude versus frequency. The spectrum analyzer can show output signal strength, its harmonic content, and interactions between two units or a unit and some external signal source. Spectrum analyzers are less expensive than they were only a few years ago and so are appearing in more and more shops.

Receiver troubleshooting requires a signal generator that covers the frequency of operation. The selected instrument should be an FM signal generator that is capable of external modulation so that a low-frequency square wave, simulated ECG from a "chicken heart" generator, or other signal source can be used to modulate the FM output of the generator. The FM signal generator should be capable of deviating at least 25 KHz and preferably the entire range of the deviation expected in the system.

If no FM generator is available, then a common continuous wave (CW) generator can be used by the technician who knows what he or she is doing. The telemetry transmitter can also be used as a signal source, but this approach is fraught with difficulty when there is an ambiguity over whether the receiver or transmitter is at fault.

Also useful for troubleshooting telemetry systems is the usual collection of dc multimeters and oscilloscopes that are needed for all forms of complex electronic service work. However, be aware that most of the faults are "trauma" items, such as broken battery connectors, open switches, and other components that are subject to abuse in normal service.

14-8.3 Portable Telemetry Units

The increase in emergency medical technicians in the rescue services of local communities gives us an immensely useful tool in dealing with trauma and coronary victims outside of the hospital. Although very highly trained, the EMT is not a physician, so some means is often required to communicate physiological data to the local hospital emergency room, where they are interpreted by a trained physician. In addition, two-way voice communications must be established for the EMT team to converse with, and receive instructions from, the physician at the hospital. Specialized communications equipment is needed for these requirements.

Figure 14-16 shows a portable telemetry system that has the range and power needed for the EMT/ambulance crew to establish a data link to the hospital. The transmitter might be a special unit, or a modified version of the standard handheld transceivers normally used by fire, police, and other services. The modulating signal, however, is either the analog ECG or a digitized version of the same. The signal is transmitted over the airwaves to a base station transceiver at the hospital. From there, the demodulation and display is similar to that of other telemetry systems.

Because the size of handheld radio transceivers used for telemetry and voice communications is necessarily very small, the available radio frequency (RF) power is low. As a result, the range is short for these units. Where the required range is greater, however, a *repeater* system can be used. At critical locations around the city, receiver sites can pick up a small signal from handheld units. This scheme is commonly used in police and fire communications systems and so is no great leap in technology to extend it to the EMT communications. Another method, shown in Figure 14-17, is to install the re-

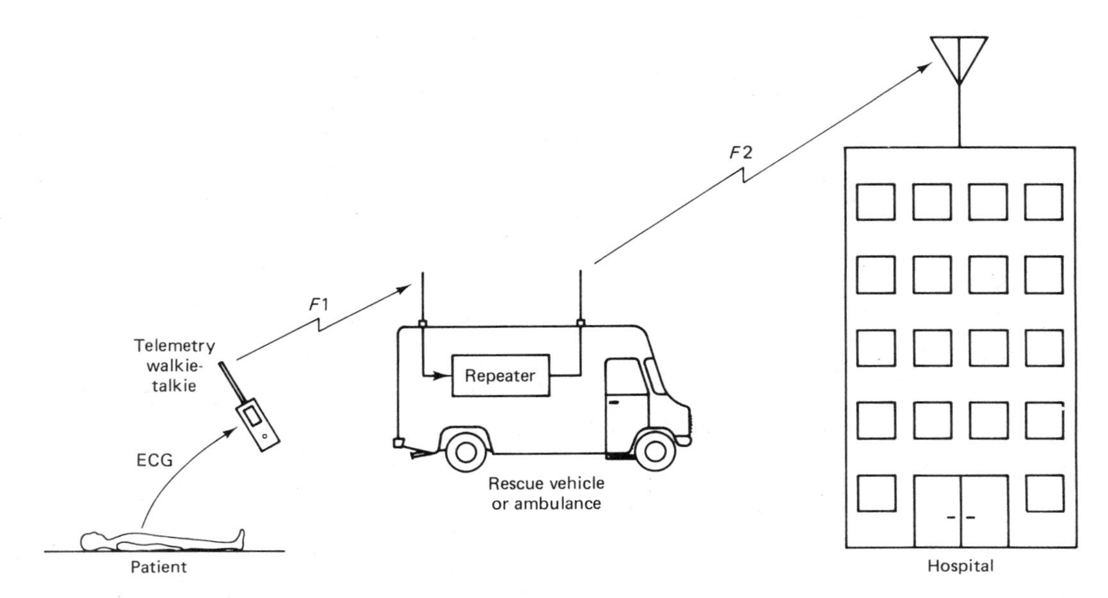

Figure 14-17 Repeater ECG telemetry system.

peater on the ambulance or rescue vehicle itself. The handheld unit only has to transmit on frequency F_1 as far as the vehicle, where the signal is picked up and retransmitted at a higher power level on frequency F_2 to the hospital site.

14-8.4 Landline Telemetry

It is possible to transmit ECG signals over the telephone lines (called ''landlines'' in the trade). Both analog and digitized ECG signals can be transmitted. This form of telemetry is used for several purposes. For example, the converted ECG waveforms can be transmitted to a distant location for interpretation either by a computer or a specialist physician.

In other cases, some patients are asked to send in their ECG waveforms on a periodic basis. For example, some cardiac pacemaker clinics ask the patient to start sending in the waveform after so many months, because it is known that an imminent failure of the battery is usually preceded by a sudden shift of heart rate. Because the approximate life expectancy of the battery is known statistically, the monitoring can commence when the patient is in the dangerous period between the onset of permanent failures and the routine replacement cycle.

Figure 14-18 shows the block diagram for a typical landline telemetry unit. The basis for this system is a voltage-controlled oscillator (VCO), or digital modem, as is used in radio telemetry. The signal will be acquired from ECG electrodes or a photoplethysmograph (PPG) sensor. The waveform is applied to the input of a signal processor (which includes filtering) and the modulator, where it is converted to the audio FM or digital signal. This signal can be recorded on an ordinary audiotape recorder or a digital recorder, or it can be transmitted over telephone lines to the central station.

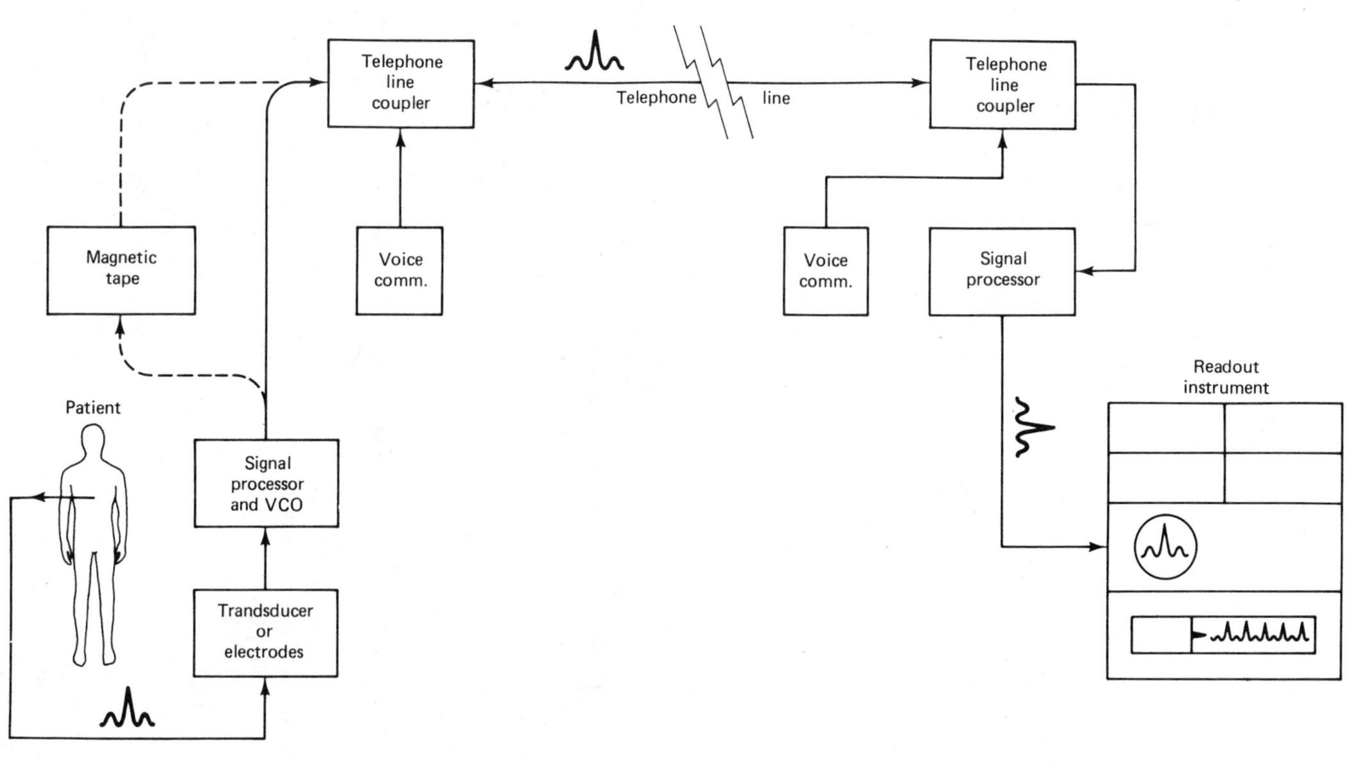

Figure 14-18 Block diagram of a landline telemetry system.

14-9 Summary

1. In the ICU and CCU, patients are continuously monitored. All units monitor ECG, and some also monitor pressures, temperatures, etc.
2. The principal instrument is a bedside monitor, which may be able to monitor the ECG only, or a combination of several physiological parameters.
3. In many units, the ECG and other waveforms, in addition to certain physiological data, is remotely displayed at a central console.
4. Computerized monitoring systems are used to automatically classify and record heart rate, life-threatening or premonitory arrhythmias, and alarm conditions.

14-10 Recapitulation

Now return to the objectives and self-evaluation questions at the beginning of the chapter and see how well you can answer them. If you cannot answer certain questions, place a check mark next to each and review appropriate parts of the text. Next, try to answer the following questions, using the same procedure.

14-11 Questions

14-1 Discuss in your own words the purpose and function of an intensive care unit.

14-2 Each patient's ECG is continuously monitored for ________________ and ________________ cardiac arrhythmias.

14-3 List five physiological parameters, other than ECG and heart rate, that might be measured in an ICU or CCU.

14-4 List four pieces of nonmonitoring equipment often found in the ICU or CCU.

14-5 Describe a typical bedside monitor.

14-6 What is the principal difference between *monitoring* and *diagnostic* ECG modes in a bedside monitor?

14-7 Why are two ECG monitoring modes needed?

14-8 What is the frequency range of a typical ECG radio telemetry transmitter?

14-9 List two different types of ECG telemetry transmitter.

14-10 What is the function of a *cardiotachometer?*

14-11 Draw a block diagram for, and discuss the operation of, an integrating cardiotachometer.

14-12 What is a *systole* lamp? What does it indicate?

14-13 Describe the operation of a digital cardiotachometer.

14-14 List two types of alarm circuit used in bedside monitors.

14-15 What is the function of a cardiomemory unit?

14-16 List two types of cardiac memory unit.

14-17 Describe in your own words, using diagrams if needed, the operation of a low-frequency tape recorder using FM. Why is it necessary to use an FM recording system?

14-18 Describe the operation of a lead fault indicator circuit.

Chapter 15

Operating Rooms

15-1 Objectives

1. Be able to describe the protocols for working in the operating room (OR) suite.
2. Be able to list sterilization techniques.
3. Be able to describe the different types of personnel employed in the OR.
4. Be able to list the different types of surgical specialties.
5. Be able to list and describe special equipment used in the OR.

15-2 Self-Evaluation Questions

These questions test your prior knowledge of the material in this chapter. Look for the answers as you read the text. After you have finished studying the chapter, try answering these questions and those at the end of the chapter (Section 15-10).

1. Why are aseptic working procedures and techniques required in surgery?
2. List four methods for sterilizing surgical instruments, etc.
3. List four types of worker employed in the OR other than the surgeon.
4. Anesthesia is used to manage ________ in surgical procedures.
5. True or false? You may enter the OR suite for brief periods only wearing street clothes.

15-3 Surgery

The surgeon is a physician trained to perform operations and to use other techniques to treat diseases. Three aspects of surgery are pain associated with cutting into the body, the tendency to bleed, and the problem of infection.

Pain is managed by the use of anesthesia agents that put the patient to sleep or, alternatively, deaden the pain sensation in the region where the surgeon is working.

The bleeding is managed mainly by the practices of the surgeon, who must continually stop bleeding as it occurs during the procedure. The surgeon will tie off cut blood vessels with suture thread or cauterize them with an electrical device called an *electrosurgery machine* (Chapter 18).

Infection is caused by microorganisms that exist almost everywhere on earth. In past centuries, before the role of these microorganisms in infection was discovered and prior to the introduction of drugs that fight infection, surgery was considerably more dangerous than it is now. Many patients survived the operation, only to succumb a little while later due to an infection. Today, however, the threat of infection is reduced by the use of antibiotic drugs and aseptic techniques in the operating room.

Aseptic technique requires that a zone be created surrounding the operation site that is essentially free of microorganisms. The surgeon and any assistants will wash their hands extensively and then be dressed in a sterile gown and wear sterile gloves. All instruments used during the procedure will have been sterilized. Those instruments capable of withstanding high temperatures and humidity will be steam sterilized in an *autoclave,* while more sensitive devices are gas sterilized in an ethylene oxide atmosphere (see Section 15-6).

All personnel in the OR must wear special clothing provided by the hospital (i.e., scrub suits or dresses). These garments are not sterile but are well laundered. Their use is to limit the introduction of microorganisms from outside. Although the specific dress codes vary slightly from one hospital to another, it is *never* proper to wear street clothes into an OR suite, even if only for a few seconds duration. Before you enter the OR suite, you must go to an appropriate locker room and change clothes. When we say "never," we mean exactly that! No matter how "short" the visit, it is always necessary to change clothes. If you are not sure of the dress code in any particular hospital, then ask somebody to explain it to you.

15-4 Types of Surgery

The use of surgical techniques is found in general surgery, several surgical specialties, and in other areas of medicine where surgery is but one of several different treatment options. Dentists also perform *oral surgery* in hospital operating rooms. Several specialty areas are as follows:

1. *Ophthalmology*—treats diseases of the eye by various methods, including surgery.

2. *Otolaryngology*—treats diseases of the ear, nose, and throat using various techniques including surgery.

3. *Orthopedics*—treats diseases of the bones, joints, and other locomotor organs and structures.

4. *Neurosurgery*—treats certain diseases of the brain and nervous system using surgical techniques.

5. *Thoracic surgery*—surgical specialty that performs operations to treat diseases of the organs in the chest cavity.

6. *Urology*—treats diseases of the urinary system using various techniques, including surgery.

7. *Obstetrics*—manages pregnancy and delivers babies.

8. *Gynecology*—treats diseases of the female reproductive system using various techniques, including surgery. Many physicians in this specialty are also obstetricians.

15-5 OR Personnel

The *surgeon* is a physician who has been trained beyond medical school in the art of performing operations and the postsurgical management of the patient's recovery. In many, perhaps most, states any licensed physician may legally perform surgery. But most hospitals of any size require additional training in a residency program, if not board certification of the physician as a surgeon. There may also be another physician assisting the surgeon, very often a resident or intern learning the trade.

The anesthesia agent is administered and managed either by a specially trained physician called an *anesthesiolgist,* or a nurse trained in anesthesia called a *nurse anesthetist* or, simply, *anesthetist*. Some nurse anesthetists who pass a certification procedure are called *certified registered nurse anesthetists* (CRNA).

At least two other persons will be assigned to each case. A *circulating nurse* (usually an RN) is outside of the sterile zone and is used for various purposes including (but certainly not limited to) keeping records, obtaining supplies, preparing drugs, etc. Because of the matter of drugs, and certain other medical-legal problems, most hospitals require an RN in the circulator's position.

The *scrub nurse* may be an RN, licensed practical nurse (LPN), or an operating room technician. This person works in the sterile zone and must follow the same antiseptic rules as the surgeon. The scrub nurse or technician is responsible for keeping the instruments, tools, and supplies straight. This job is not similar to the so-called mechanics helper, who merely hands tools to the mechanic but is a skilled job in its own right and is often done by registered nurses.

Anesthesia technicians assist the anesthesiologists and anesthetists by bringing supplies, running errands during the procedure, and cleaning or maintaining the anesthesia machines and related equipment after the day's case load is finished.

Monitoring technicians operate the physiological monitoring equipment such as pressures monitors, ECG, etc. These people may be of a minimal skill level or may be somewhat more extensively trained as to be indistinguishable from cardiovascular technicians.

The *cardiovascular technician* is trained to operate and perform elementary maintenance on a wide variety of physiological monitoring and measurement equipment, intraaortic balloon pumps, heart-lung machines, blood gas machines, and so forth. CV technicians who mainly operate the heart-lung machine are sometimes called *perfusionists* or *perfusion technicians*.

Orderlies and *nursing assistants* perform menial to semiskilled work in the OR, usually under the direct supervision of an RN or other person.

15-6 Sterilization

Sterilization is the process of killing microorganisms on tools, instruments, and other objects used in surgery. It is also applied to linen and clothing. In some cases, sterilization is also applied to objects and tools not normally used in surgery or other sterile fields, so as to prevent the spread of disease from one patient to another as the object is used. Blood pressure transducers (those not using sterile disposable domes) and certain items in respiratory therapy, for example, are not used in a sterile environment but could possibly spread pathogenic (i.e., disease causing) microorganisms between patients.

Several different methods are used for sterilization: *steam heat, dry heat, gas, liquid,* and *radiation*.

Steam sterilization is done in a device called a *steam autoclave,* which is a pressure lock chamber of rather substantial construction. Temperatures inside of an autoclave are in excess of 100°C (212°F), so the pressure must also be elevated. Never attempt to operate or service a steam autoclave without first receiving instructions in that particular model. In some localities, a steam "engineer's" license is required of anyone who services the autoclave.

The objects that are being sterilized are wrapped inside of a double thickness of linen or certain other suitable material that is porous to pressurized steam. They are then loosely packed with other such bundles inside the autoclave chamber. There are three periods of the steam sterilization cycle:

1. *Vacuum period*—the air is withdrawn from the sealed autoclave chamber, producing a vacuum.

2. *Sterilization period*—superheated (i.e., $T > 100°C$) steam is introduced into the chamber. The temperature will be either 120°C for 15 min or 144°C for 2 min. These temperatures correspond to pressures of 10^5 Pa (i.e., 1 atm) and 3×10^5 Pa (i.e., 3 atm), respectively.

3. *Poststerilization period*—the steam is withdrawn, allowing the wrapped packs to dry before readmitting room air and unlocking the door.

A special tape is applied to the wrapped packs, and special indicator strips are placed inside of the pack to give the user an indication of whether the sterilization process was completed. There are marks on these indicators that become visible or change color when sterilized. Most autoclaved bundles can be stored for a considerable length of time provided that they are not opened or become damp. The efficacy of the autoclave is checked periodically by sterilizing a *spore strip*. After the process, the strip is sent to the laboratory in a sterile container. It is placed in an incubator that allows any live spores, those not killed in the autoclave, to grow and reproduce.

The use of dry heat allows sterilization in room air, at room pressures. The dry heat process requires that objects or bundles be in an environment with temperatures of 160°C for 120 min, 170°C for 60 min, or 180°C for 30 min. Dry heat is considered to be less effective than steam sterilization.

Gas sterilization uses *formalin* gas, or more commonly, *ethylene oxide* gas. The wrapped bundles are exposed to the gas for 10 to 12 hours and then must be allowed to vent for 24 hours or more to get rid of gas residue. Gas sterilizers are typically more difficult to use than are steam or dry heat models. As a result, many hospitals place steam and dry heat sterilizers in the OR suite and send objects requiring gas sterilization to a central supply department.

Gas methods are used to sterilize objects that cannot withstand the high temperatures of the steam and dry heat methods but can tolerate the 30 to 60°C temperatures used in the gas autoclave. Typically, plastics, rubbers, synthetics, fabrics, and certain sharp-edged metal objects that become dull in high-temperature environments are candidates for gas sterilization. Occasionally, a question will arise of whether certain objects are immune to reaction with the gas used in the process because they seem to deteriorate rapidly under gas sterilization.

There are several antiseptic liquids that can be used to sterilize objects. The object is left immersed in the liquid for periods ranging from 30 min to several hours, depending upon the type of solution and the nature of the object being sterilized.

Ionizing radiation (i.e., gamma and X-rays from a cobalt-60 source) can also be used for sterilizing objects. The object must be exposed to the radiation for as long as 24 hours, so the process is used primarily in industrial situations where the high degree of sterilization of medical products being manufactured requires and justifies the radiation equipment.

15-7 OR Equipment

The range of medical equipment found in various operating rooms depends on several factors (i.e., the types of surgery performed, physician's preferences, level of activity, etc.). There are, however, certain items that are found in all operating rooms.

Electrosurgery (ES) *machines* (Chapter 18) are radio frequency ac generators that produce currents of the intensity needed to cut tissue and cauterize bleeding blood vessels. These machines are commonly, if erroneously, called *Bovies* after the inventor. But the word *Bovie* is the brand name of one of the oldest producers of ES equipment. Liebel-Flarshiem. The word has become, however, a generic term in jargon.

Light sources are used with fiber-optic endoscopes to view inside the patient's body through certain openings or surgical incisions. They are also used to illuminate the surgeon's head lamp. Most light sources are quartz-halogen lamps, either 110 V or 21 V.

Suction apparatus are used to remove blood, mucous, and other material from the patient's body, mouth, or the surgical wound. The suction device usually has a bottle to hold the material that is removed but may be powered by either its own compressor, or a vacuum port in the OR wall.

Anesthesia machines are used to control and deliver oxygen and the anesthesia gas to the patient. It is not wise to attempt to work on these machines without training in that area. While they are not terribly complicated, they can be an issue in a malpractice case.

Many hospitals now routinely use ECG and blood pressure monitors during surgical procedures. These instruments are essentially the same as those devices discussed in Chapters 3 to 8.

15-8 Summary

1. Three aspects or problems in all surgery are pain, bleeding, and infection.
2. Aseptic techniques and working practices are designed to reduce the incidence of infection.
3. The most commonly used forms of sterilization are steam heat, dry heat, gas, liquid, and radiation.
4. Anesthesia agents are administered by either a physician (called an anesthesiologist) or a nurse anesthetist.

15-9 Recapitulation

Now return to the objectives and self-evaluation questions at the beginning of the chapter and see how well you can answer them. If you cannot answer certain questions, place a check mark next to each and review appropriate parts of the text. Next, try to answer the following questions, using the same procedure.

15-10 Questions

15-1 Three problems common to all surgery are ________, ________, and ________.

15-2 Pain is managed by ________ agents administered by an ________ (MD) or ________ (nurse).

15-3 List two practices used to control bleeding during surgery.

15-4 Infection is controlled through the use of ________ drugs and ________ techniques.

15-5 List five medical/surgical specialties that use surgical treatment as all or part of their repertoire.

15-6 The ________ nurse works in the sterile zone, while the ________ nurse works outside of the zone.

15-7 List four types of sterilization for objects used in the OR.

15-8 In a steam autoclave the temperature is raised to ________°C for 15 min or ________°C for 2 min.

15-9 In a dry-heat autoclave the temperature is maintained at ________°C for two hours, ________°C for one hour, or ________°C for one-half hour.

15-10 Ethylene oxide autoclaves operate at a temperature that is between ________ °C and ________°C.

Chapter 16

Medical Laboratory Instrumentation

16-1 Objectives

1. Be able to state the purpose of blood.
2. Be able to list the components and describe the composition of blood.
3. Be able to list and describe blood tests (cells and chemistry).
4. Be able to state the purpose, uses, principle of operation, and maintenance of the following blood instrumentation: colorimeter/densitometer, flame photometer, spectrophotometer, blood cell counter, blood gas analyzers (pH, P_{O2}, and P_{CO2}), chromatograph, and autoanalyzers.

16-2 Self-Evaluation Questions

These questions test your prior knowledge of the material in this chapter. Look for the answers as you read the text. After you have finished studying the chapter, try answering these questions and those at the end of the chapter (see Section 16-16).

1. Define *blood* and state its purposes.
2. Name the components of blood (cells and plasma).
3. What is the difference between blood cell tests and blood chemistry analysis?
4. Why is the medical laboratory department of critical importance in diagnosing and treating the patient?
5. Describe the principle of operation and maintenance of the colorimeter, flame photometer, spectrophotometer, blood cell analyzer, blood gas analyzer (pH, P_{O2} and P_{CO2}), chromatograph, and autoanalyzer.

16-3 Blood (Purpose and Components)

Blood is the fluid that circulates through the heart, arteries, veins, and capillaries carrying nourishment, electrolytes, hormones, vitamins, antibodies, heat, and oxygen to body tissues and taking away waste matter and carbon dioxide.

Whole blood, as shown in Figure 16-1, is composed of *cells* and *plasma* (fluid containing dissolved and suspended substances). The *blood cell portion* consists of the following elements:

1. *Red blood cells (RBC) or erythrocytes*—these are concaved disc-shaped cells (8 μ length, 3 μ width) that contain no nucleus and live about 120 days before being replaced by the bone marrow. Their number is 4.5 to 5.5 × 10^6 cells/mm^3. Internally, each RBC contains four iron atoms in a structure known as the *hemoglobin molecule*. Oxygen

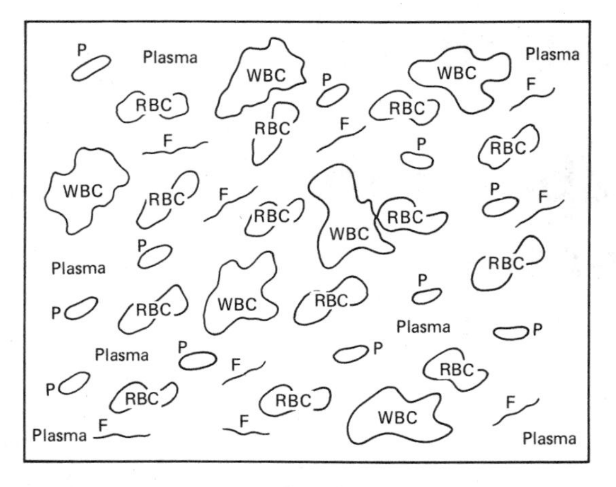

Figure 16-1 Whole blood composed of cells and plasma. RBC—red blood cells. 4.5–5.5 × 10^6 cells/mm^3, 8 × 3 μ, no nucleus, 120-day life. WBC—white blood cells. 6–10 × 10^3 cells/mm^3 10 μ diam., nucleus, 15-day life. P—platelet cells. 200–800 × 10^3 cells/mm^3 3 μ diam., no nucleus. F—fibrinogen protein. Plasma—fluid portion. Inorganic and organic substances dissolved in H_2O. Whole blood = RBC + WBC + P + F + plasma. Plasma = whole blood − (RBC + WBC + P). Serum = plasma − fibrinogen.

from the lung alveoli enter the bloodstream and chemically combines with hemoglobin to form *oxyhemoglobin*. RBCs transport oxygen to the tissues and pick up carbon dioxide to form *carbaminohemoglobin*.

2. *White blood cells (WBC) or leucocytes*—these are amoebalike cells (10 μ in diameter) that contain a nucleus and live from 13 to 20 days. Their number is 6 to 10 × 10^3 cells/mm^3. They are also present in the lymph fluid and engulf invading bacteria and foreign substances to destroy the invaders' effect. For example, bacteria invading the leg are encapsulated by WBCs in the lymph fluid, transported to the inferior vena cava, circulated through the right atrium/ventricle through the lungs to the left atrium/ventricle, and pumped to the kidneys, where they are extracted in the urine. They are then excreted from the body as harmless cell fragments. Specific *antibodies* are also produced to kill the invaders' *antigen* (toxin).

3. *Platelets*—these are cell fragments (3 μ in diameter) that contain no nucleus. Their number is 200 to 800 × 10^3 cells/mm^3. These form a repair substance that initiates blood coagulation and clotting. A protein *thrombin* also acts on *fibrinogen* (soluble protein formed in the liver) to generate insoluble *fibrin*. Fibrin deposits as fine threads to form the framework of the blood clot. *Platelets* cling to intersections of fibrin threads. As fibrinogen is used up, serum is secreted. *Serum* will not clot, as it contains no fibrinogen.

Blood plasma consists of the following elements:

1. *Plasma proteins*—organic *repair* substances. These are *albumins* (synthesized in the liver) that help regulate plasma/tissue cell osmotic pressure. *Fibrinogen* and *prothrombin* is used in the clotting process. *Globulin* substances (alpha, beta, and gamma) are catalysts and aid in the immunizing (disease protection) process.

2. *Plasma nutriments—energy-storing* substances. These are *glucose* (blood sugar), *lipids* (fats), and *amino acids* (make up proteins for tissue growth).

3. *Regulatory and protective substances*—these are *enzymes* (catalysts for digestion and cell metabolism), *hormones* (stimulatory/inhibitory function to target organs), and *antibodies* (providing immunity against infection).

4. *Plasma electrolytes*—acid-base and *nerve-impulse* transmission substances. These are inorganic salts (metal and nonmetal combination) and pure chemical substances (Na^+, K^+, Cl^-).

5. *Metabolic waste substances*—these include *urea* and *uric acid* waste from the kid-

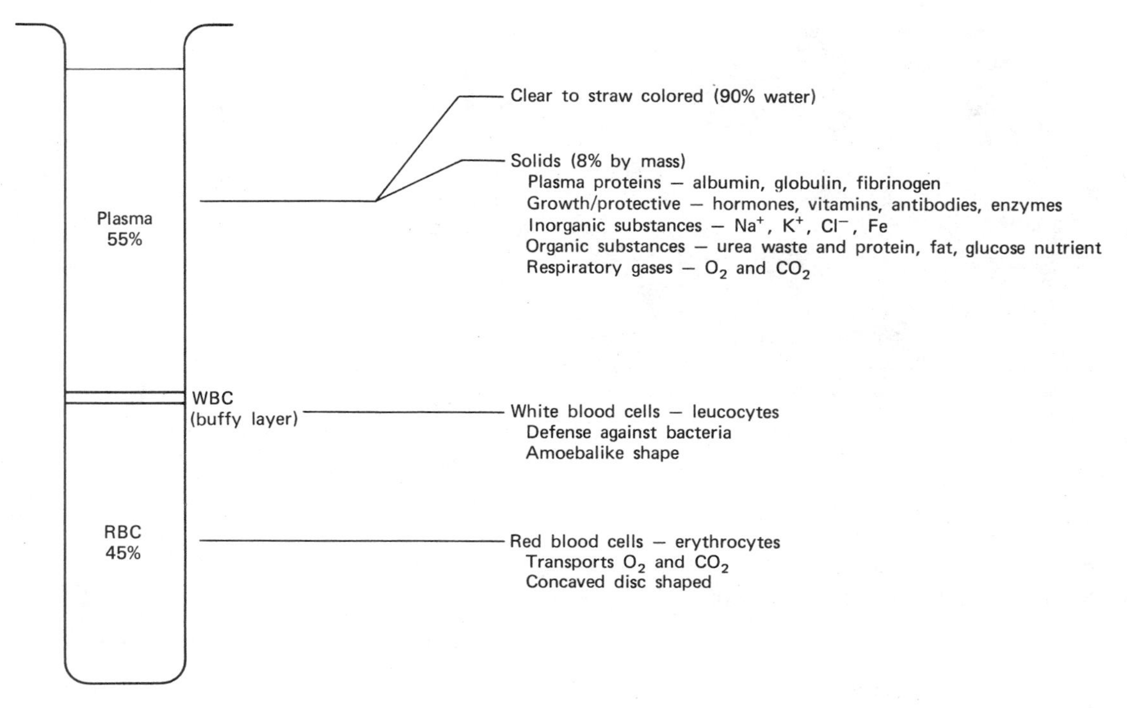

Figure 16-2 Blood that has been spun in a centrifuge.

neys and carbon dioxide waste from cellular metabolism.

Figure 16-2 shows blood that has been spun in a centrifuge (motor-driven mechanical device that generates a circular motion). A centrifuge is shown in Figure 16-3. Since the RBCs are the heaviest, they sink to the bottom and form *45 percent* of the total by volume. Plasma occupies *55 percent* and contains substances as indicated. Blood plasma does contain some dissolved oxygen, but 97 percent of the transported oxygen is carried in the RBC hemoglobin molecules. During its passage through the body, blood hemoglobin still remains 70 percent oxygen saturated (see oxygen dissociation curve, Figure 9-2). The total carbon dioxide carried by the blood is 30 percent in the RBCs and 60 percent in the blood plasma.

The body functions as a biological machine that receives life-giving input substances of oxygen and food nutriments and gives off waste substances. All living creatures typically display the characteristics of *organization* (life process control), *irritability* (response to change), *contractility* (movement), *nutrition* (ingestion and digestion of food), *metabolism/growth* (liberation of stored chemical energy), *respiration* (intake of O_2 and ventilation of CO_2), *excretion* (elimination of waste), and *reproduction* (generation of a new structure).

Metabolism is the sum total of all chemical and biochemical processes in the body. It involves *catabolism* (breaking down of complex protein/sugar substances to simpler ones) and *anabolism* (building up of complex substances for body use). Waste products from the digestive process are eliminated in the feces. Toxic substances that result from

Figure 16-3 Refrigerated centrifuge.

metabolic processes are removed from the blood by the kidneys and excreted in the urine.

The purpose of *medical laboratory instrumentation* is to provide a means of measuring required substances and metabolic waste products in urine and blood.

16-4 Blood Tests (Cells and Chemistry)

Blood cell tests include the following elements:

1. *Red blood cell count (RBC)*—accomplished manually (diluting a blood sample 100 to 1 and counting the cells per mm^3 by use of a microscope) or automatically (blood cell counting analyzers).

2. *White blood cell count (WBC)*—accomplished manually (10 to 1 dilution) or automatically (blood cell analyzer).

3. *Platelet count*—accomplished automatically by a blood cell analyzer.

4. *Hematocrit (Ht)*—percentage of total blood volume that is solid (WBC volume is negligible). This is measured by spinning a blood sample in a test tube and optically observing the percentage of packed RBCs (see Figure 16-2). It normally ranges from 45 to 55 percent.

5. *Mean cell (corpuscular) volume (MCV)—average volume* of an RBC measured by a value based on the RBC count (number per mm^3). This volume is measured in femtoliters (10^{-15} l).

$$\text{MCV} = \frac{\text{Ht}}{\text{RBC}} \tag{16-1}$$

6. *Mean cell hemoglobin (MCH)*—the *proportional mass* of RBC/100 ml to the total number of RBCs is expressed as:

$$\text{MCH} = \frac{\text{hemoglobin in grams/100 ml}}{\text{RBC}} \tag{16-2}$$

This value is indicated in picograms (10^{-12} g).

7. *Mean cell hemoglobin concentration (MCHC)*—hemoglobin *color concentration* measured by lysing RBCs (breaking their membranes) to release hemoglobin. Acid hematin or cyanmethemoglobin can be generated by hemoglobin chemical reaction. The resultant value is measured by a colorimeter and normally indicates 32 to 36 percent color index.

$$\text{MCHC} = \frac{\text{hemoglobin in grams/100 ml}}{\text{Ht}} \tag{16-3}$$

Blood chemical tests check for amounts of *acidity*—pH normally 7.36 to 7.44 (blood is normally slightly alkaline); *glucose*—lactic acid, lactose sugar; *nonprotein nitrogen substances*—amino acids, peptides, urea waste, and uric acid; *lipids*—fatty acids of cholesterol and triglycerides; *proteins*—plasma albumin, globulins, and fibrinogen; and *enzymes and steroids,* among other elements.

Blood serological tests involve testing for

agglutination (clumping) of cells due to the addition of *antigens* (bacterial toxins) to blood serum. This occurs following the reaction of a specific *antibody* produced by WBCs in response to the specific invader.

Blood bacteriological tests include growth of blood bacteria in a petri dish with appropriate nutriments.

Histological tests are studies of small thin tissue samples under the microscope. Specimens are obtained by cutting tissues with a precision slicer known as a microtome.

16-5 Medical Laboratory Department

The *medical laboratory department* includes facilities, personnel, and equipment within the hospital or public/private location. The facilities must include a clean, safe surrounding with a special area for sterilization of contaminated blood/urine samples and equipment. Since high-volume blood testing occurs in this department, sufficient storage and cleaning areas must be designated. In such a situation, the chance of error (misreading or a patient record mixup) is high.

Medical laboratory *personnel* includes equipment operators (medical technologists), supervisors, and physicians. The director of these facilities is usually a physician.

Equipment contained in the laboratory includes glassware, centrifuges, suction devices, and sophisticated instrumentation such as colorimeters, spectrophotometers, blood/cell and gas analyzers, chromatographs, autoanalyzers, and computer-based record and operation systems.

Record keeping is extremely important. This information is used by physicians as an aid in diagnosing disease and inbalanced physiological states. Standard cards with printouts of RBC/WBC count, Ht, MCV, MCHC, and blood chemistry are presented by most clinical instrumentation.

16-6 Overview of Clinical Instrumentation

Clinical instrumentation in the early 1900s was almost nonexistent. Since the 1950s, sophisticated apparatus have been developed to measure blood parameters, as described in Section 16-4. The complex substances appearing in blood serum can be evaluated for concentration based on chemical color reactions. Blood cells can be counted by electrical conductivity changes as they pass through a fixed diameter aperture. The following types of instrumentation are used to analyze blood:

1. *Colorimeter or filter photometer* is an optical electronic device that measures the color concentration of a substance in solution (following the reaction between the original substance and a reagent). The results are displayed in percent optical color transmittance or absorbance to indicate hemoglobin concentration, for example. The *densitometer* is a device similar to the colorimeter and measures optical transmittance (density) of particles in fluid suspension.

2. *Flame photometer* is an optical electronic device that measures the color intensity of substances (i.e., sodium or potassium) which have been aspirated into a flame.

3. *Spectrophotometer* is an optical electronic device that measures light absorption at various wavelengths for a given liquid sample. This is a type of sophisticated colorimeter.

4. *Blood cell analyzer* is an electromechanical device that measures the number of red and white blood cells per scaled volume. This is accomplished by noting the changes in electrical conductivity as the cells pass an aperture of fixed diameter.

5. *pH/Blood gas analyzer* is an electromechanical device which measures blood pH (acid-base balance), P_{O_2} (partial pressure of

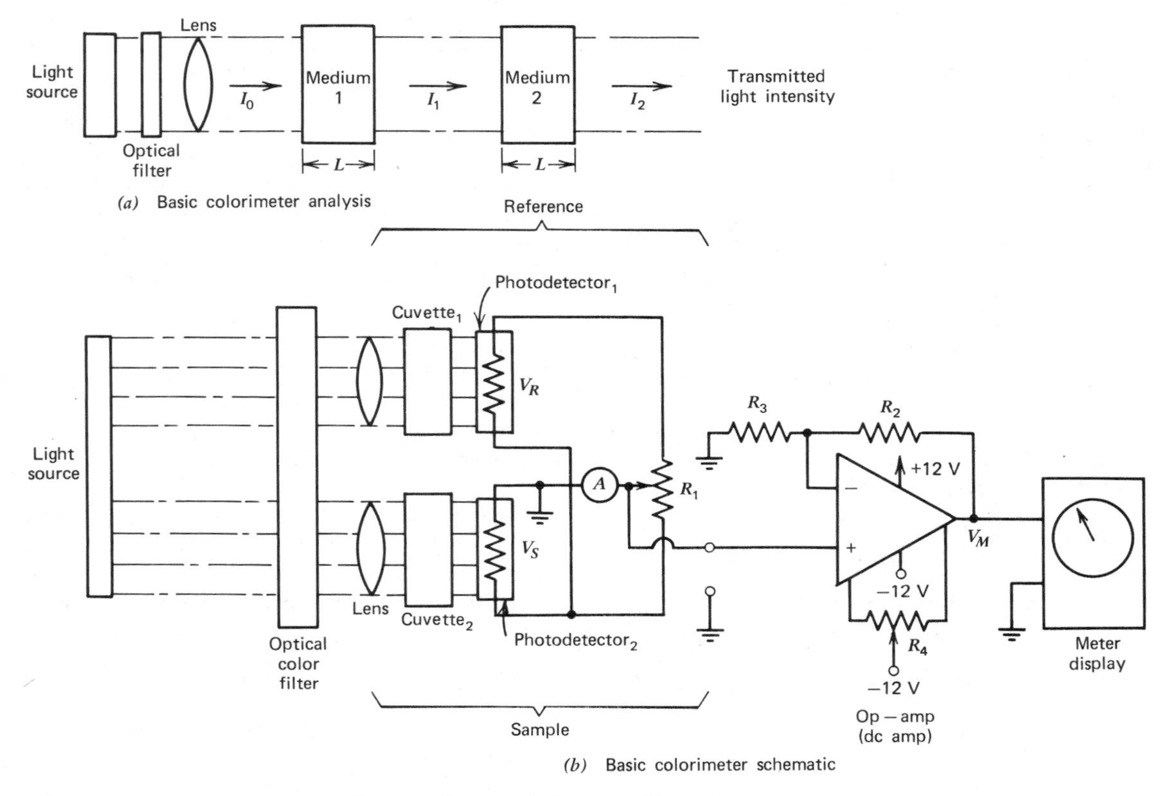

Figure 16-4 Colorimeter—filter photometer.

blood oxygen), and P_{CO_2} (partial pressure of blood carbon dioxide). This is accomplished through use of glass electrode transducers.

6. *Chromatograph* is an electromechanical device used to separate, identify, and measure the concentration of substances in a liquid medium. Results are displayed as colored bands in a liquid column or as colored strips on paper.

7. *Autoanalyzer* is an electromechanical-electronic device that sequentially measures and displays blood chemistry analysis. This is accomplished by using mixing tubes and colorimeters arranged in a serial system connection.

16-7 Colorimeter

The *colorimeter,* shown in Figure 16-4, is a filter photometer that measures the color concentration of a substance in solution. This is accomplished electronically by detecting the color light intensity passing through a sample containing the reaction products of the original substance and a reagent. A yellow-colored urine sample, for example, passes yellow light and absorbs blue and green. For this reason and to obtain purity in measurement, optical color filters are used to select a narrow wavelength spread (bandwidth) of light that shines on the photodetectors.

Basic colorimeter analysis (Figure 16-4*a*) involves the precise measurement of light intensity. Transmittance is defined as:

$$T = \frac{I_1}{I_0} \times 100 \text{ percent} \quad (16\text{-}4a)$$

or

$$I_2 = TI_1 \quad (16\text{-}4b)$$

where

$$I_2 = \mathrm{T}^2 I_0 \quad (16\text{-}4c)$$

I_0 is initial light intensity
I_1 is first attenuated light intensity
I_2 is second attenuated light intensity
T is transmittance in percent

Absorbance (optical density) is defined as:

$$A = \log \frac{I_1}{I_0} = \log \frac{1}{T} \quad (16\text{-}5)$$

where

A is absorbance
I_1 and I_2 are as before

If the path length or concentration increases, the transmittance decreases and the absorbance increases. Essentially, this phenomenon can be expressed by *Beer's law:*

$$A = aCL \quad (16\text{-}6)$$

where

A is absorbance
L is cuvette path length
C is concentration of absorbing substance
a is absorbtivity related to the nature of the absorbing substance and optical wavelength (known for a standard solution concentration)

Therefore, the concentration of the unknown solution can be found from the following relationship:

$$C\mu = C_s \frac{A_u}{A_s} \quad (16\text{-}7)$$

where

C_u is unknown concentration
C_s is standard concentration (for calibration)
A_u is unknown absorbance
A_s is standard absorbance

A *basic colorimeter schematic* is shown in Figure 16-4*b*. Observe that light passes through an optical color filter, is focused by lenses on the reference and sample cuvettes, and falls on the reference and sample photodetectors. The difference in voltage between the two detectors is increased by a dc amplifier and applied to a meter. A *calibration procedure* is as follows:

1. Ground the amplifier input (V_1) and adjust potentiometer (R_4) for 0 V ± 5 mV at the amplifier output.

2. Remove the ground and place reference concentrations in cuvettes 1 and 2 (empty cuvettes or open spaces may also be used).

3. Adjust potentiometer (R_1) for 0 V ± 10 mV at the amplifier output.

4. Leave the reference concentration in cuvette 1 and replace cuvette 2 with a cuvette containing the sample.

5. Read the unbalanced voltage on the meter in percent transmittance or absorbance units.

A photo of a colorimeter is shown in Figure 16-5.

EXAMPLE 16-1 (Refer to Figure 16-4*b*)
Given $V_1 = +1$ mV with reference cuvettes (step 3 in calibration procedure), $V_1 = +25$ mV with reference and sample cuvettes (step 5 in calibration procedure), $R_2 = 2\ \text{k}\Omega$, $R_3 = 1\ \text{k}\Omega$, *calculate* the voltage read on the meter display for both conditions of V_1.

(Noninverting op-amp voltage gain.) *See Chapter 4.*

$$A_r = 1 + \frac{R_2}{R_3} \quad (16\text{-}8)$$

$$A_v = 3 = 1 + \frac{2\ \text{k}\Omega}{1\ \text{k}\Omega} = 3$$

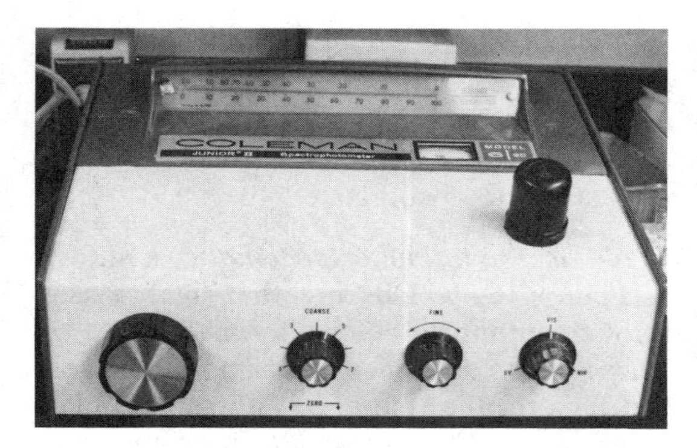

Figure 16-5 Colorimeter.

Condition 1:

$$V_1 = +1 \text{ mV}$$

$$A_v = \frac{V\text{m}}{V_1} \quad (16\text{-}9)$$

$$V_m = A_v V_1 \quad (16\text{-}10)$$
$$= 3\,(+1 \text{ mV})$$

$$V_m = \mathbf{+3\ mV}$$

Condition 2:

$$V_1 = +25 \text{ mV}$$
$$= A_v V_1$$
$$V_m = 3\,(+25 \text{ mV})$$
$$V_m = \mathbf{+75\ mV}$$

Note that the sample measurement (balanced) voltage is 25 times larger than the reference measurement (unbalanced) voltage. This is a desirable low-error situation.

Precipitating reagents are usually mixed with samples to remove substances from the sample. Table 16-1 shows common chemical tests and their normal ranges.

Maintenance includes calibration adjustment and replacement of burned-out lamps and photodetectors. Colorimeters are very reliable and usually do not have frequent electronic problems.

16-8 Flame Photometer

The *flame photometer,* shown in Figure 16-6, measures the color intensity of a flame that is supported by oxygen and a specific substance. The basic schematic shows that a reference gas containing a *lithium salt* causes a *red color* to shine on the *reference photodetector* through the reference optical filter. A yellow or violet light from sample sodium or potassium falls on the *sample photodetector.* Basically, the flame photometer is *calibrated* in a manner similar to the colorimeter. However, continuous calibration can be accomplished by inspiration of air (oxygen to support combustion) and lithiums. The output is read in units of sodium or potassium concentration. A photo of a flame photometer is shown in Figure 16-7.

Maintenance includes calibration adjustment and replacement of bulbs and photodetectors. Aspiration devices and flame chambers occasionally require cleaning. Electronic failures are usually infrequent.

Table 16-1
Common Chemical Blood Tests and Normal Ranges

Test	Normal Ranges	Units
1. Sodium	135–145	m Eq/l
2. Potassium	3.5–5	"
3. Chloride	95–105	"
4. Total CO_2	24–32	"
5. Blood urea nitrogen	8–16	mgN/100 ml
6. Glucose	70–90	mg/100 ml
7. Inorganic phosphate	3–4.5	"
8. Calcium	9–11.5	"
9. Creatine	0.6–1.1	"
10. Uric acid	3–6	"
11. Total protein	6–8	g/100 ml
12. Albumin	4–6	"
13. Cholesterol	160–200	mg/100 ml
14. Bilirubin	0.2–1	"
15. SGOT	20–50	mU/ml

$$1 \text{ milli equivalent per liter} = \frac{1\text{m Eq}}{1} = \frac{\text{concentration}\left(\frac{\text{mg}}{1}\right)}{\text{molecular weight}}.$$

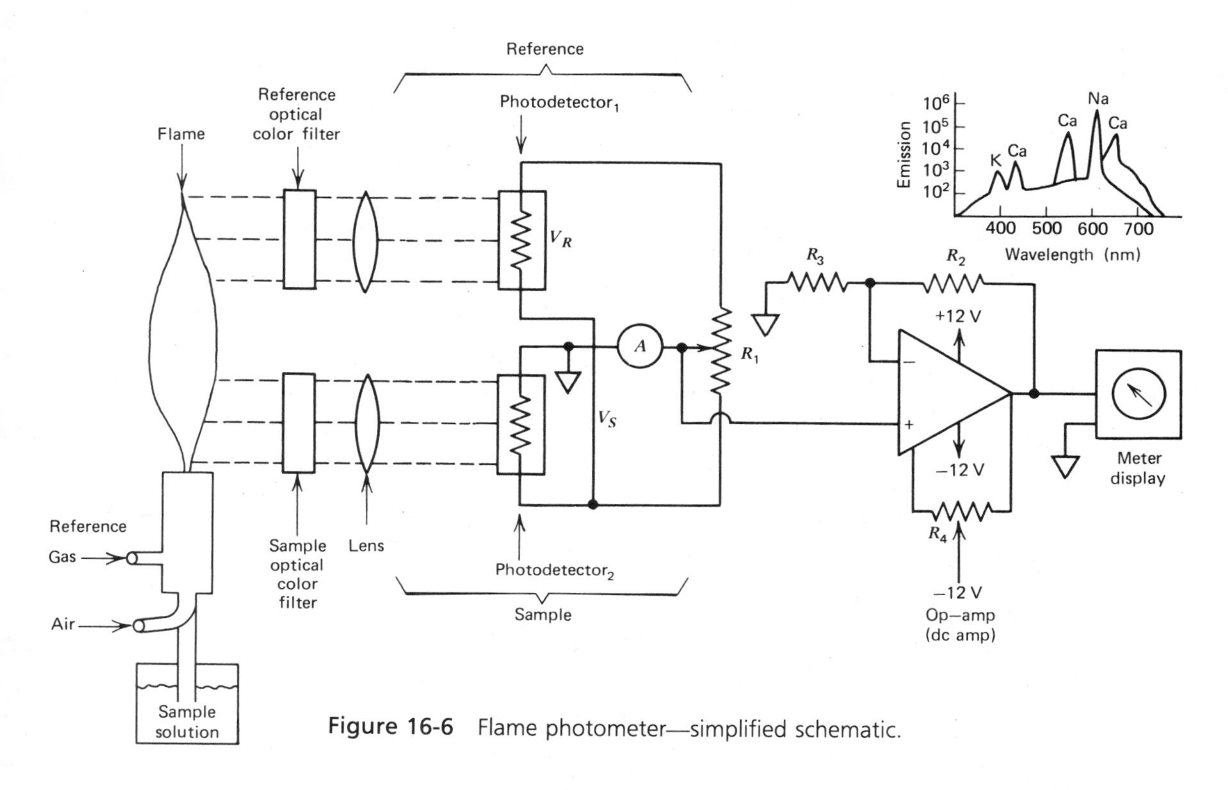

Figure 16-6 Flame photometer—simplified schematic.

16-9 Spectrophotometer

The *spectrophotometer,* as shown in Figure 16-8, measures light absorption by a liquid substance at various wavelengths. From this, the components of an unknown material can be determined or the concentration of a number of known substances can be measured. A *monochromator* uses a diffraction grating or prism to disperse the light from the lamp (slit S_1). The light is broken into its spectral components as it arises from slit S_2 and falls on the sample in the cuvette. Narrower slits give rise to shorter wavelengths. The angle of the diffraction grating determines light wavelength if all other parameters are fixed and the mirror reduces equipment size. Light output, photodetector

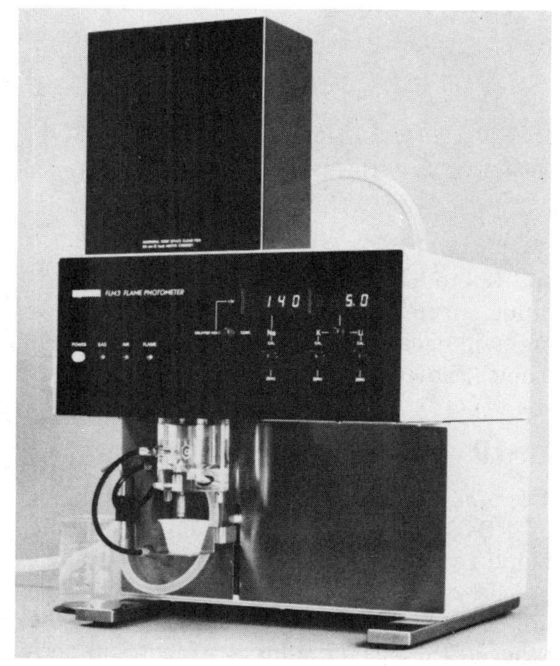

Figure 16-7 Flame photometer. (Courtesy The London Co.)

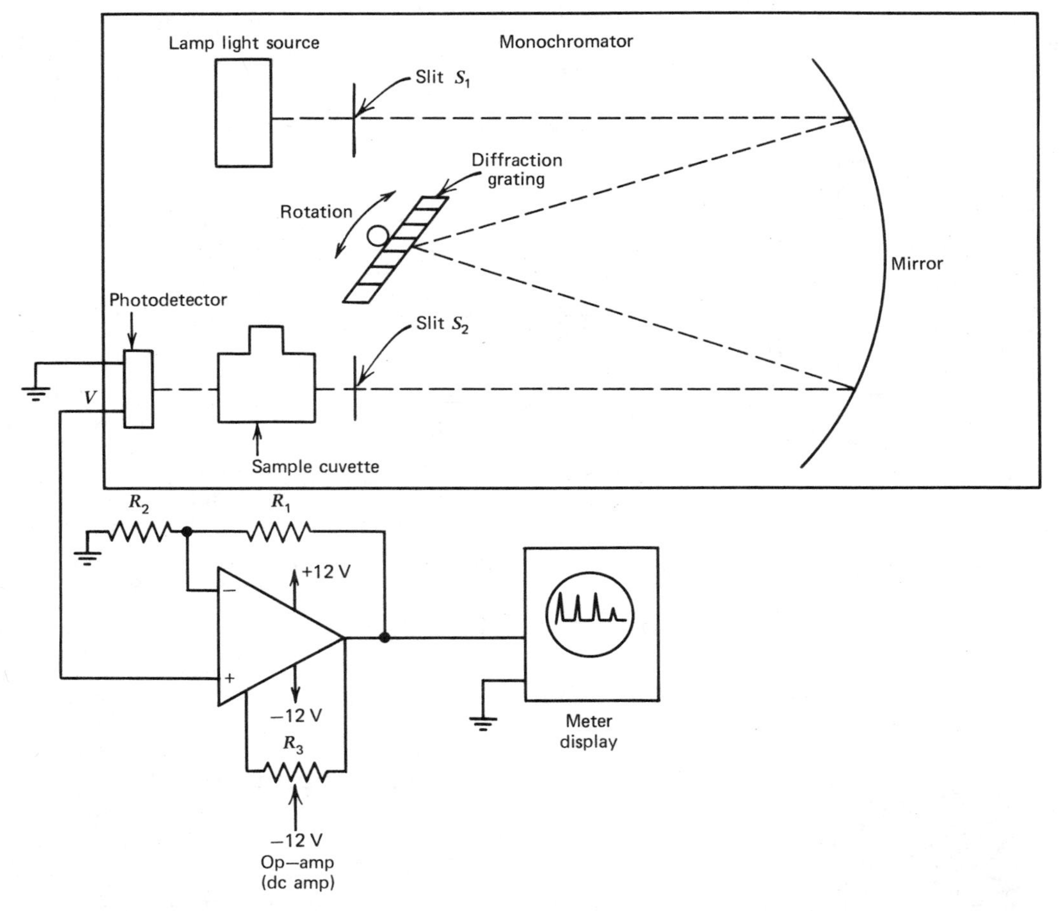

Figure 16-8 Spectrophotometer—simplified schematic.

sensitivity, and sample substance absorption change with wavelength, and this necessitates zero calibration for each wavelength measurement. The *double-beam* spectrophotometer accomplishes this automatically by beam path switching (sample to reference) via a mechanical shutter or rotating mirror. The ratio of path absorbances can then be computed. Figure 16-9 shows a photograph of a spectrophotometer with input sample to the left and output graph to the right.

Maintenance includes calibration adjustment and replacement of light source bulbs and photodetectors. Also, mechanically rotating assemblies (mirrors, diffraction grating) will occasionally malfunction. The electronics, however, are very reliable.

16-10 Blood Cell Counter

The *blood cell counter* measures the number of RBCs or WBCs per volume of whole blood by placing diluted blood in a beaker, as shown in Figure 16-10. Electrical plates are placed, one inside a test tube and the other in the beaker solution. Electrical con-

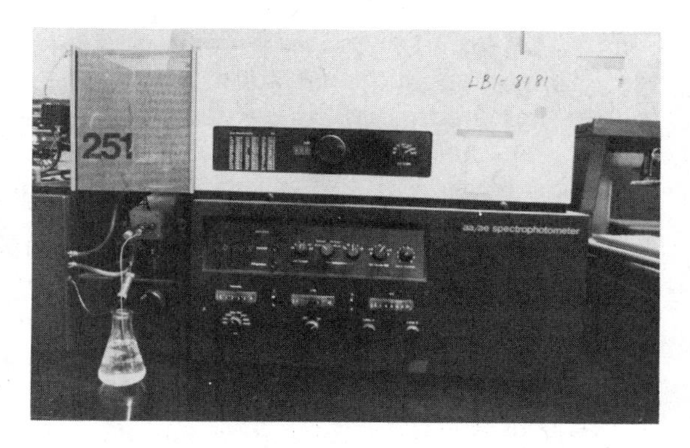

Figure 16-9 Spectrophotometer.

ductivity is measured between the plates. When RBCs or WBCs traverse the test tube aperture (100 μ fixed diameter) due to suction pump action, electrical conductivity drops. This results from the relatively lower conductance of the cells as compared to the surrounding salt solution. Each cell that crosses through the aperture produces a pulse. The *number of pulses* indicates the *concentration* of RBCs since a fixed volume is drawn (start-stop mercury contact/pulse gate control circuit). A gated electronic counter counts pulses above the threshold amplitude and presents the result to the dispay panel. The *area* under the pulse curve (or peak) indicates the *volume* of the cell passing through.

A complete count is made in 20 s and is displayed as a digital count and CRT peak versus time display. This conductivity count-

Figure 16-10 Blood cell counter—basic block diagram.

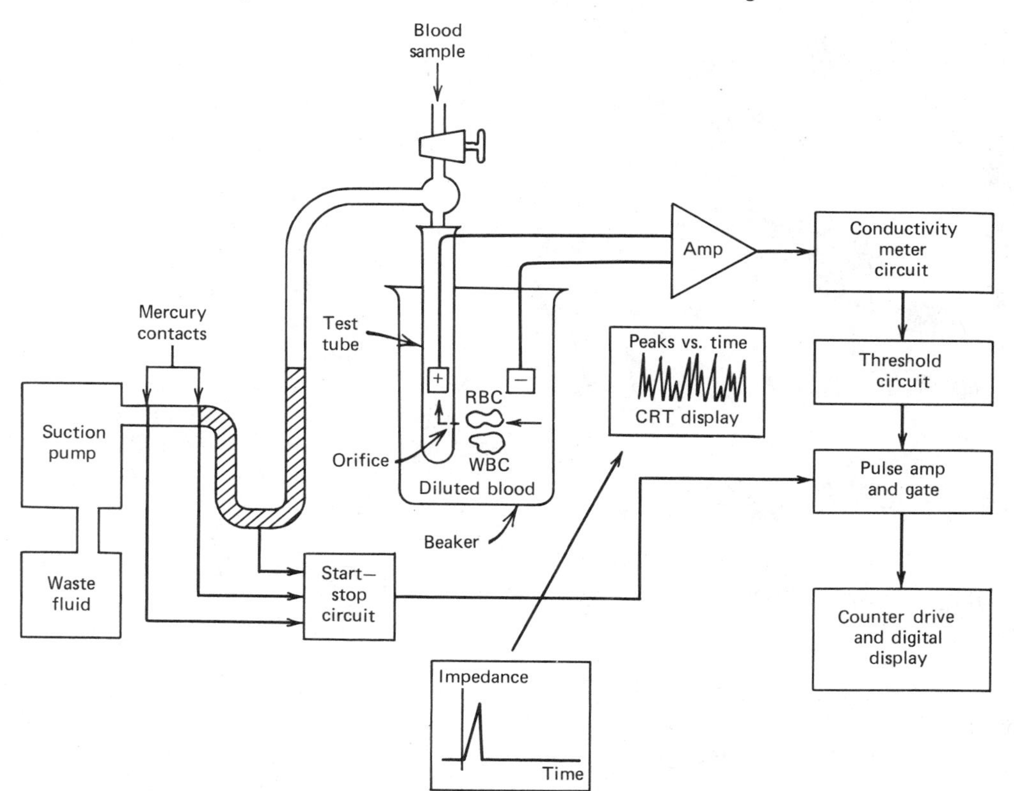

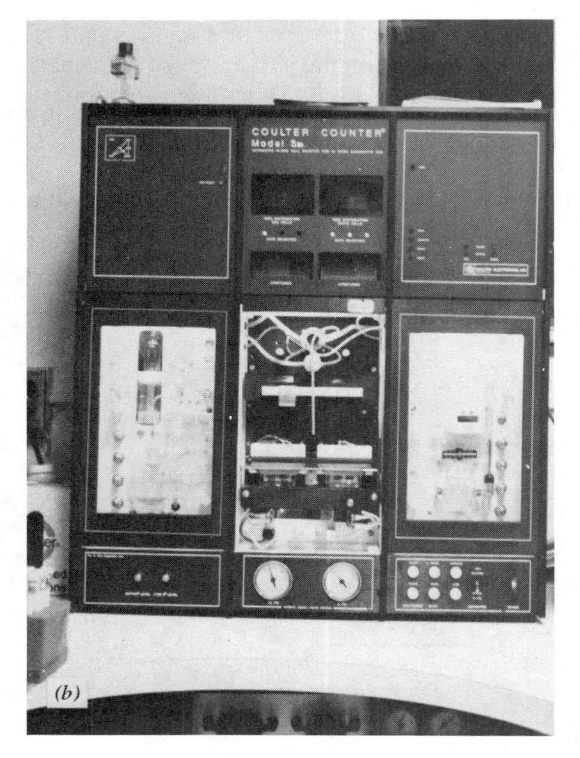

Figure 16-11 Blood cell counters. (*a*) Coulter model F. (*b*) Coulter model senior.

ing method is known as the Coulter Counter Method. Two *Coulter Counters* are shown in Figure 16-11. These devices *measure* (see Section 16-4) the RBC count, the WBC count, the MCV (mean cell volume), and Hgb (Hemoglobin concentration) and *calculate* Ht (hematocrit), MCH (mean cell hemoglobin), and MCHC (mean cell hemoglobin concentration).

Results are printed out on a specially prepared report form (multicopy card), as shown in Figure 16-12. The printout represents a statistical weighting of counts up to 100,000.

Maintenance includes frequent calibration checks and replacement of burned-out lamps. Coulter counters have more mechanical problems than electronic. For example, the pipette system that draws the sample into the machine must be frequently cleaned. Tubing also becomes dislodged or obstructed.

16-11 pH/Blood Gas Analyzers

Acid-base balance of the blood is generated by body electrolytes and measured by a pH meter. The *respiratory system* provides an immediate buffer to sudden blood pH changes and the *renal system* provides a slower, more long-range balance adjustment (protection).

An early *pH meter* is shown in Figure 16-13. An electrode is dipped into the solution under test (bottom) and read on a meter nulled by a precision dial. The pH reading is taken from the dial.

The *glass pH electrode* is the heart of the pH meter. Acidity or alkalinity is indicated by the concentration of hydronium ions (H_3O^+) in solution. This gives rise to hydrogen ions (H^+). pH is a measure of this ion concentration and is defined as:

$$\text{pH} = \log_{10}\frac{1}{(\text{H}^+)} = \log_{10}(\text{H}^+) \quad (16\text{-}11)$$

where

pH is acidity

(H^+) is hydrogen ion concentration

TEST REQUESTED

☐ PROFILE ☐ DIFF.

SPECIMEN TAKEN

DATE	TIME A.M. P.M.

JULIAN DATE

TEST NO.	TEST
IF 99.9 RE-DILUTE	WBC $\times 10^3$
	RBC $\times 10^6$
	Hgb gm
	Hct %
	MCV μ^3
	MCH $\mu\mu g$
	MCHC %

	IMMATURE	DIFFERENTIAL
	NEUTRO. BANDS	
	NEUTRO. SEGS.	
	LYMPHOCYTES	
	EOSINOPHILS	
	BASOPHILS	
	MONOCYTES	
	PLATELETS	
	SED. RATE	
	PLATELET COUNT	
	RETICULOCYTE COUNT	
	CLOTTING TIME	
	BLEEDING TIME	
PTT	CONTROL	
	PATIENT	
PROTIME	CONTROL	
	PATIENT	
	% ACTIVITY	

Enter in above space: PATIENT IDENTIFICATION – TREATING FACILITY – WARD NO. – DATE

REQUESTING PHYSICIAN'S SIGNATURE

REPORTED BY

TECH

MD

DATE

LAB ID. NO.

HEMATOLOGY II

URGENCY: ☐ ROUTINE ☐ TODAY ☐ PRE-OP. ☐ STAT.

SPECIMEN/LAB REPORT NO.

PATIENT STATUS: ☐ BED ☐ AMB. OUTPATIENT ☐ NP ☐ DOM.

SPECIMEN SOURCE: ☐ VEIN ☐ CAP. ☐ OTHER

REMARKS

VA FORM MAR 1973 10-1429 HEMATOLOGY II

Figure 16-12 Hematology output card.

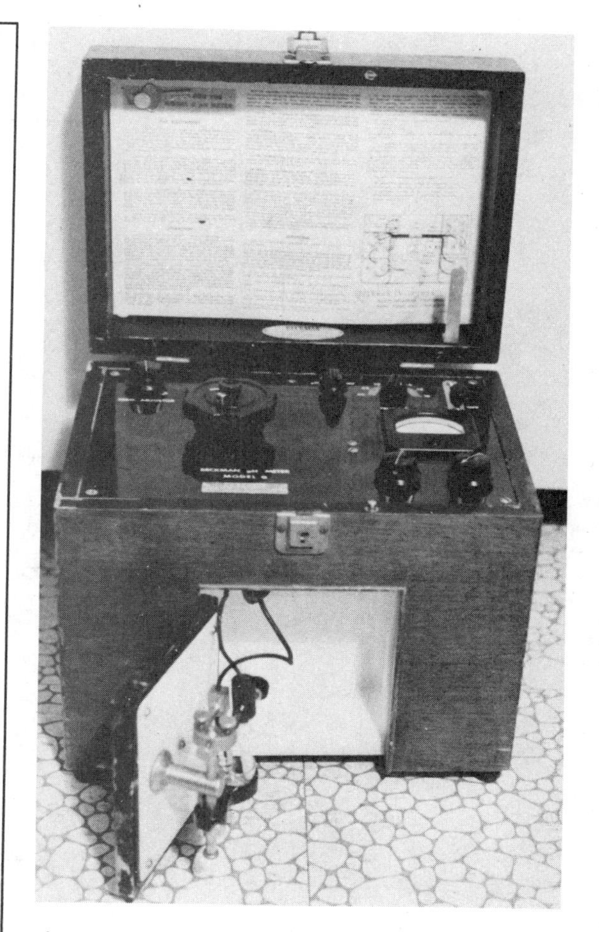

Figure 16-13 Early 1950s pH meter.

Pure water has a pH of 7 (neutral, $[H^+] = 10^{-7}$). *Stomach acid* has a pH of 1.5 (greatly acid, $[H^+] = 10^{-1.5}$). *Blood* has a pH of 7.36 to 7.44 (slightly alkaline, $[H^+] = 10^{-7.4}$ average).

The glass electrode, shown in Figure 16-14, consists of a platinum wire immersed into a highly acidic buffer solution contained within a thin *glass bulb*. This bulb wall is 0.1 mm thick and has an electrical resistance of *1000 MΩ*. It passes hydrogen ions only and thus acts as a *membrane* for separating out these ions. Also immersed into the test solution is a *calomel reference cell* (mercurous chloride). The platinum wire electrode generates a *half-cell* electrical potential that acts

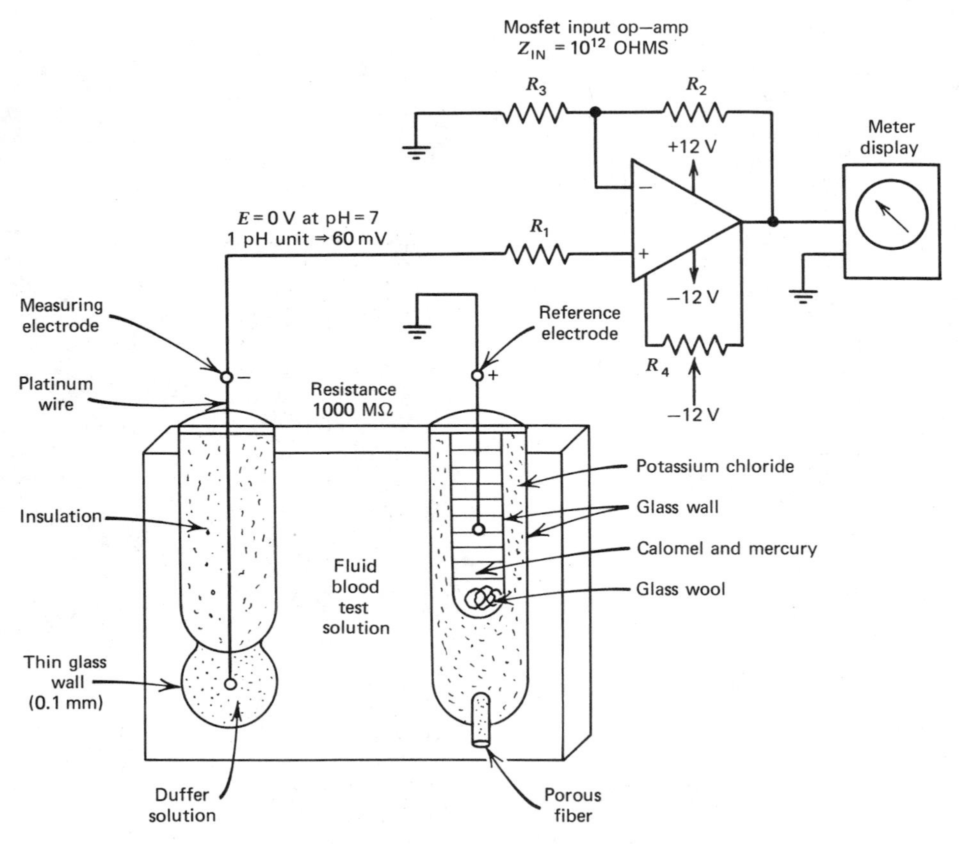

Figure 16-14 Basic pH glass electrode.

in combination with the stable reference calomel halfcell. The test solution is in common with both half-cells. The resultant voltage is amplified by a high input impedance amplifier such as a MOSFET input op-amp (10^{12} Ω differential input impedance). Op-amp baseline drift (dc voltage stability) is important since the glass electrode signal is low-level dc (50 mV). These amplifiers are often chopper stabilized, which returns the amplifier to ground potential at a relatively high rate (1 KHz).

Blood gas analyzers measure the partial pressure of oxygen and carbon dioxide (P_{O_2} and P_{CO_2}) in solution. This gives an indication of respiratory function. The *carbon dioxide electrode* is known as the *Severinghaus* electrode. It has a thin teflon CO_2 permeable membrane over the glass bulb of the pH electrode. Final readings are reached quickly. pH values are measured and compared to pH values of standard calibration solutions with partial pressures of 60 and 30 Torr (mm Hg). This is done on a *nomograph* and is called the *Astrup method*. Essentially, partial pressures correspond to specific pH values and, thus, P_{CO_2} electrodes are actually modified pH electrodes.

The *oxygen electrode* is composed of a thin platinum wire and a reference silver-silver chloride (Ag-Ag Cl) electrode. The magnitude of a small current generated by a

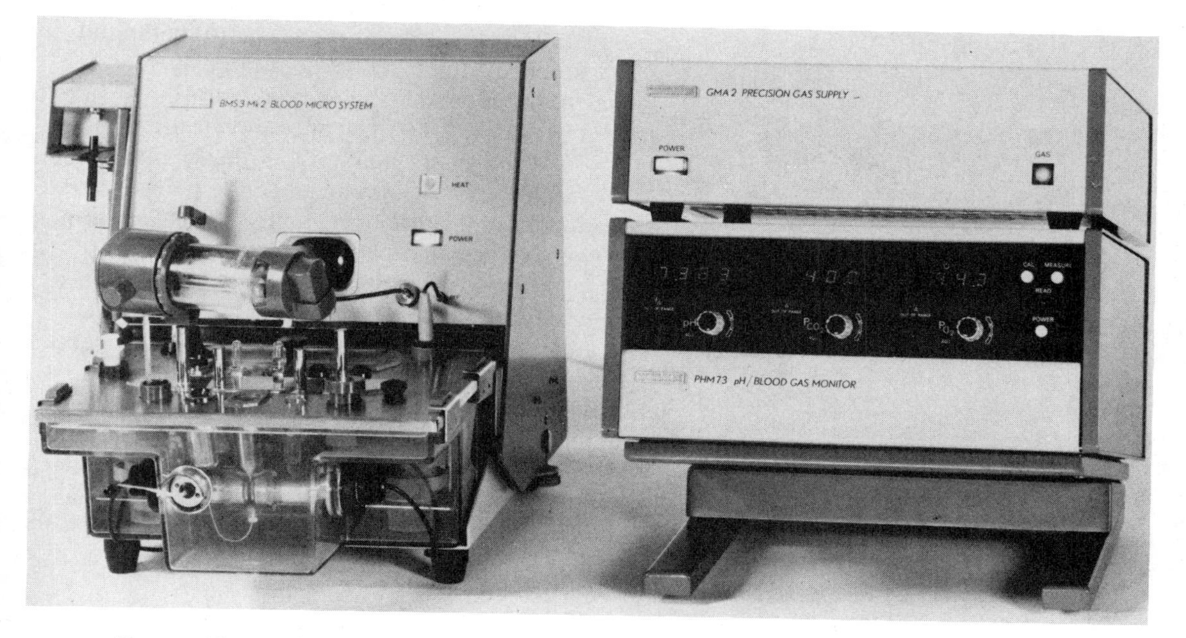

Figure 16-15 Blood gases/blood pH analysis system. [Photo Courtesy The London Co. (Radiometer).]

battery connected across the electrodes is proportional to the oxygen concentration of the solution. Modern measurement techniques use the *Clark* P_{O_2} *electrode*. This consists of a platinum wire and Ag-Ag Cl electrode mounted inside a glass housing containing a saturated potassium-chloride (K-Cl) solution. A polythene membrane, semipermeable to oxygen molecules, covers an opening at the bottom. A current driven by a battery indicates P_{O_2}.

A combination electrode can also be designed which measures blood pH and gases. This is known as a *Clark-Severinghaus* electrode assembly.

Modern blood gas analyzers are precision devices that typically measure blood *pH*, P_{CO_2}, *and* P_{O_2}. One is shown in Figure 16-15. The blood micro system is a thermostated unit incorporating a micro pH electrode (40 μl sample), a microtonometer for four disposable equilibrium tubes, and a high-value suction pump. The glass electrode is seen at the middle and the suction nozzle at the upper left. The *pH/blood gas monitor* gives accurate pH, P_{CO_2}, and P_{O_2} determinations on the blood sample or other body fluids. It is designed such that less frequently used controls are covered by a hinged front panel. Essentially, the blood sample can be measured for a preset (delay function) time following the push of the measure button. The unit will measure pH (15 s typically), P_{CO_2} (30 s typically), and P_{O_2} (50 s typically) and then display the values. The values shown in the photograph are within normal ranges. Calibration can be accomplished easily. This unit contains a combination of stable analog amplifiers and many digital control and storage circuits.

A *fully automated photometric analyzer* is shown in Figure 16-16. This unit measures hemoglobin concentration and oxygen saturation. Whole blood or erythrocyte concen-

Figure 16-16 Hemoximeter photometric analyzer. [Photo courtesy The London Co. (Radiometer).]

trate is aspirated into the system, where hemolysis (breaking open of RBCs by 40-kHz ultrasonic energy), measurement, calculation, and rinsing are performed automatically. A precision colorimeter is used to measure hemoglobin concentration.

Figure 16-17 Gaseous oxygen analyzer. (Photo courtesy Hewlett-Packard.)

Neonatal oxygen monitors are commonly found in nurseries to continuously monitor environmental oxygen concentration inside the incubator. The one shown in Figure 16-17 monitors infant ECG, heart rate, and incubator P_{O2}. A transducer is placed next to the child and a manually selected alarm setting permits warning of low or high P_{O2}.

Maintenance of pH meters and blood gas analyzers includes frequent calibration adjustment and replacement of glass electrodes. These electrodes age and require increasing times to produce accurate readings. For continuous P_{O2} monitors, electrodes must be periodically cleaned and occasionally replaced. Electronic failures are relatively few.

16-12 Chromatograph

The *chromatograph* provides a means for separation and assay of complex substances. Material identification and concentration can be determined by any one of the following methods:

1. *Liquid column chromatography*—liquid is percolated down a tube, and the bands formed along the tube, as well as the time to travel through, indicate the type of substance present.

2. *Paper partition chromatography*—solid and liquid phases are separated out on strips or hollow cylinders of filter paper. The solvent movement is timed.

3. *Gas chromatography*—movement of gas instead of liquid is used to separate out samples. Gas passing through a solid is known as *gas solid (adsorption) chromatography (GSC)* and gas through liquid as *gas liquid chromatography* (GLC).

Figure 16-18 shows a gas chromatograph with input to the left and graph output to the upper right.

Figure 16-18 Gas chromatograph.

16-13 Autoanalyzer

The *autoanalyzer* sequentially measures blood chemistry and displays this on a graphical readout. As shown in Figure 16-19, this is accomplished by *mixing, reagent reaction,* and *colorimetric measurement* in a continuous stream. The system includes the following elements:

1. *Sampler*—aspirates samples, standards, and wash solutions to the autoanalyzer system.

2. *Proportioning pump and manifold*—introduces (mixes) samples with reagents to effect the proper chemical color reaction to be read by the colorimeter. It also pumps fluids at precise flow rates to other modules, as proper color development depends on reaction time and temperature.

3. *Dialyzer*—separates interfacing substances from the sample material by permitting selective passage of sample components through a semipermeable membrane.

4. *Heating bath*—heats fluids continuously to exact temperature (typically 37°C incubation equivalent to body temperature). Temperature is critical to color development.

5. *Colorimeter*—monitors the changes in optical density of the fluid stream flowing through a tubular flow cell. Color intensities (optical densities) proportional to substance concentrations, are converted to equivalent electrical voltages.

6. *Recorder*—converts optical density electrical signal from the colorimeter into a graphic display on a moving chart.

The *heart* of the autoanalyzer system is the *proportioning pump*. This consists of a *peristaltic* (occluding or roller) *pump*. Air

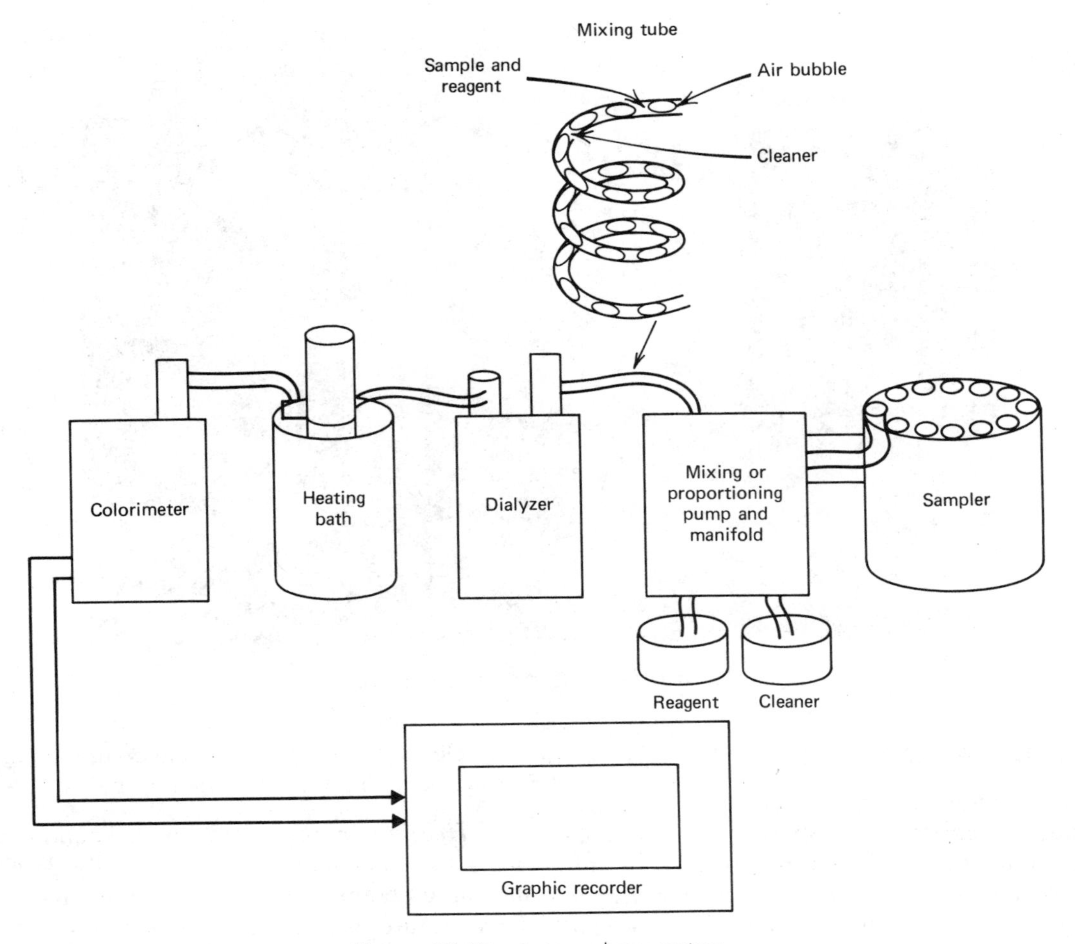

Figure 16-19 Autoanalyzer system.

segmentation in the *mixing tube* separates the sample/reagent mixture from the cleaning fluid and other samples (Figure 16-19). As these air-separated fluids traverse the coil of the mixing tube, effective mixing action is achieved.

The Technicon SMA 12/60, shown in Figure 16-20, is a sequential multiple analyzer that performs 12 different tests on 60 samples per hour. It is a continuous flow process that produces a chemical profile read on a graphic chart. Tests accomplished include most of those shown in Table 16-1.

A later computerized version is shown in Figure 16-21. This is the Technicon *SMAC*. Up to 40 different tests can be performed on an individual serum sample.

One *problem* with automatic analyzers is certain identification of samples. Patient data can be intermixed with other patients if care is not taken.

Sterilization is also needed for samples, glassware, and equipment parts that are contaminated with disease. Diseases such as hepatitis or other communicable infections can be spread to equipment operators. Fig-

Figure 16-20 Technicon SMA 12/60.

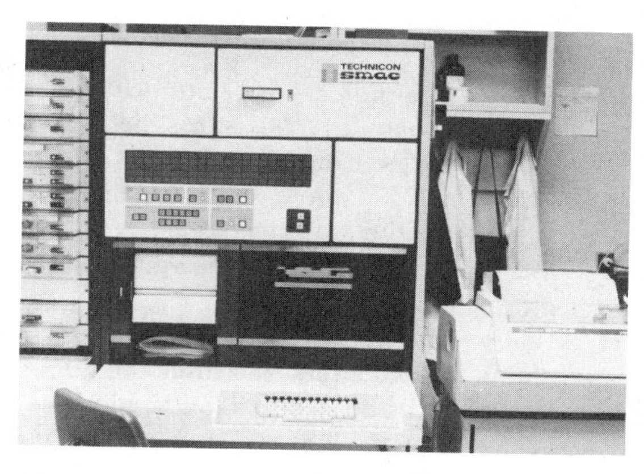

Figure 16-21 Computerized autoanalyzer.

Figure 16-22 Autoclave sterilizer.

ure 16-22 shows an *autoclave* unit used to sterilize small and large items. It operates at saturated steam pressures and temperatures of 120°C for 20 min to one hour.

Maintenance on autoanalyzers include frequent calibration adjustment. Most problems are mechanical (tubes, moving pump parts) and electrical (switches, motors). Electronic failures are few. Sophisticated autoanalyzer system maintenance and repair requires that the BMET have gone through manufacturer's schools. Operation and service manuals must always be consulted. A patient's life may hinge on accurate measurement results obtained by clinical instrumentation.

16-14 Summary

1. *Blood* is defined as the fluid that circulates through the heart, arteries, veins, and capillaries carrying nourishment, electrolytes, hormones, vitamins, antibodies, heat, and oxygen to body tissues and taking away waste matter and carbon dioxide.
2. *Whole blood* consists of *cells* (RBCs, WBCs, and platelets) and *plasma* (fluid with dissolved inorganic and organic substances such as proteins, fibrinogen, glucose, fats, enzymes, vitamins, antibodies, electrolytes, salts, and urea.
3. Toxic substances from metabolic processes are carried by the blood and excreted in the urine. These substances can be measured by clinical laboratory instrumentation.
4. *Blood cell tests* include *RBC, WBC,* and *platelet* count, hematocrit (*Ht*), mean hemoglobin concentration (*MHC*), mean corpuscular volume (*MCV*), and mean corpuscular hemoglobin concentration (*MCHC*).
5. *Blood chemistry tests* include determination of acidity (pH), glucose (sugar), nonprotein nitrogen substances (amino acids, peptides, urea), lipids (fats), proteins (plasma, albumin, globulin, and fibrinogen), and enzymes and steroids.
6. Other blood tests include *serological* (agglutination), *bacteriological,* and *histological.*
7. The *medical laboratory department* includes special facilities, personnel, and equipment. Equipment is expensive and critical to accurate diagnosis of the patient.
8. A *colorimeter* or *filter photometer* is an optical electronic device that measures the color concentration of a substance in solution. Maintenance includes routine cleaning and calibration.
9. A *flame photometer* is an optical electronic device that measures the characteristic color intensity of substances (sodium, potassium) that have been aspirated into a flame. Maintenance is similar to the colorimeter.
10. A *spectrophotometer* is an optical electronic device that measures light absorption at various wavelengths for a given liquid sample. Maintenance is similar to the colorimeter except for mechanically moving parts (diffraction grating, rotating mirror).
11. A *blood cell analyzer* is an electromechanical device that measures the number of red and white blood cells per scaled volume. This is accomplished by the electrical conductivity (Coulter Counter) method. The Coulter Counter output is in card form on which is printed RBC, WBC, Hgb, Ht, MCV, MCH, and MCHC.
12. A *pH/blood gas analyzer* is an electromechanical device that measures blood pH (acid-base balance, $pH = -\log [H^+]$), P_{O_2} (partial pressure of blood

oxygen), and P_{CO_2} (partial pressure of blood carbon dioxide). pH glass electrodes can be modified to measure P_{CO_2} and P_{O_2} (Severinghaus and Clark electrodes).

13. A *chromatograph* is an electromechanical device used to separate, identify, and measure concentration of substances. Types include liquid column, paper partition, and gas chromatographs.
14. An *autoanalyzer* is an electromechanical-electronic device that measures and displays blood chemistry. The system consists of a sampler, proportioning pump (mixing coil), dialyzer, heating bath, colorimeter, and recorder. Examples are the Technicon SMA 12/60 and SMAC sequential multiple analyzers.

16-15 Recapitulation

Now return to the objectives and self-evaluation questions at the beginning of the chapter and see how well you can answer them. If you cannot answer certain questions, place a check mark next to each and review appropriate parts of the text. Next try to answer the questions (Section 16-16) using the same procedure. When you have answered all of the questions, solve the problems in Section 16-17.

16-16 Questions

16-1 Blood is defined as that ________ which circulates through the ________, ________, ________, and capillaries carrying nourishment, ________, ________, ________, antibodies, heat, and ________ to the body tissues, and taking away ________ ________ and ________ ________.

16-2 Whole blood is composed of ________ and ________ blood cells and ________ fluid, which consists of proteins, nutriments, ________, ________, antibodies, and metabolic ________.

16-3 Blood spun in a centrifuge is composed of ________ percent RBCs and ________ percent plasma.

16-4 Blood cell tests include counts for ________ and ________, volume of ________, and color intensity of ________.

16-5 Blood chemistry includes ________ (acid-base balance), ________ (sugar), ________ (fats), ________ (protein), and enzymes.

16-6 Blood antibody agglutination evaluation is known as a ________ test and organism growths as a ________ test.

16-7 The importance of the medical laboratory department rests ultimately with the ________ of disease.

16-8 Since the medical laboratory has high volume, good record keeping helps to prevent mixup in a patient's ________.

16-9 State the purpose, uses, operation, calibration, and maintenance for colorimeters (include a block diagram/schematic).

16-10 State the purpose, uses, operation, and maintenance for a flame photometer (include a block diagram/schematic).

16-11 State the purpose, uses, operation, and maintenance for spectrophotometers (include block diagram/schematic).

16-12 State the purposes, uses, operation, and maintenance for blood cell counters (include block diagram).

16-13 State the purpose, uses, and maintenance for pH/blood gas analyzers (include a diagram and description of a pH glass electrode).

16-14 State the purpose, uses, operation, and operation for chromatographs (include the different types).

16-15 State the purpose, uses, and maintenance of autoanalyzers (include system block diagram).

16-17 Problems

16.1 Refer to Figure 16-4*b*. Given $V_1 = +1.5$ mV with reference cuvette and $V_1 = +35$ mV with reference and sample cuvettes, $R_2 = R_3 = 2.5$ kΩ, *calculate* the voltage read on the meter display for both conditions of V_1.

16-2 A nurse reports that a *colorimeter* that was previously working is now giving very low transmittance reading for all blood samples. Which portion(s) of the device would you check first?

16-3 A medical technologist reports that the *Coulter Counter* Model S in the clinical lab gives correct RBC, WBC, Ht, and Hgb readings, but incorrect MCV, MCH, and MCHC readings. What checkout procedure would you use to identify the trouble source?

16-4 A nurse in the neonatal nursery cannot obtain accurate readings with an isolette *oxygen monitor.* Which portion(s) of the device would you check first?

16-5 A medical technologist reports that a *pH meter* in the clinical lab has an output that wanders about an apparently correct pH value. What part of the device should probably be replaced?

16-18 References

1. Crouch, James E. and Robert J. McClintic, *Human Anatomy and Physiology,* Wiley (New York, 1976).
2. Henry, R. J., D. C. Cannon, and J. W. Winkelman, *Clinical Chemistry, Principles and Techniques,* Harper & Row (Hagerstown, Md., 1974).
3. Lee, W. L., *Elementary Principles of Laboratory Instruments,* C. V. Mosby Co. (St. Louis, Mo., 1970).
4. Jacobson, Bertil and John G. Webster, *Medicine and Clinical Engineering,* Prentice Hall, Inc. (Englewood Cliffs, N.J., 1977).
5. Geddes, L. A. and L. E. Baker, *Principles of Applied Biomedical Instrumentation,* Wiley (New York, 1968, 1975).
6. Laughlin, Alice, *Roe's Principles of Chemistry,* C. V. Mosby Company (St. Louis, Mo., 1976).
7. Nave, Carl R. and Brenda C. Nave, *Physics for the Health Sciences,* W. B. Saunders Co. (Philadelphia, 1976).
8. Veterans Administration, *Preventive Maintenance Guides for Select Hospital Equipment,* Vol. 1, Clinical Lab, CTR Inc. (Concord, Mass., G-29, 1974).
9. *Operation and Maintenance Manual for Coulter Counter,* Models S and Plus, Coulter Electronics, Inc. (Hialeah, Florida).
10. *Operation and Maintenance Manual for SMA 12/60 and SMAC,* Technicon Instruments (Tarrytown, N.Y.).
11. *Principles of Operation, Basic Autoanalyzer System,* Publ. No. TN1-0169-00, Technicon Instruments (Tarrytown, N.Y., 1969).
12. *Programmed Instruction for the Basic Autoanalyzer System,* Manual TPO-0169-11, Technicon Instruments (1973).
13. *Training Manual, ABL-2,* Practical Course in the Techniques, Applications and Maintenance of the Radiometer Acid-Base Laboratory, London Company (Cleveland, Ohio, 1977).
14. *Service Manual for BMS2MK2 Blood Micro System, PMH73 pH/Blood Gas Monitor,* and *OSM2 Hemoximeter Radiometer* (Copenhagen).

Chapter 17

Medical Ultrasound

17-1 Objectives

1. Be able to describe the physics of ultrasound.
2. Be able to list the types of ultrasonic transducers.
3. Be able to list and describe the different types of ultrasonic instruments.
4. Be able to inspect and evaluate the performance of ultrasound instruments.

17-2 Self-Evaluation Questions

These questions test your knowledge of the material in this chapter. Look for the answers as you read the text. After you have finished studying the chapter, try answering these questions and those at the end of the chapter (Section 17-16).

1. What is the range of frequencies designated as ultrasonic?
2. How does a 2500-kHz ultrasound wave *differ* from a 2500-kHz radio wave?
3. What is the principal ultrasonic transducer?
4. A frequency suitable for use in Doppler flowmeters is ______________ mHz.

17-3 What Is Ultrasound?

The term *ultrasound* refers to *acoustical* waves above the range of human hearing (i.e., frequencies greater than 20,000 Hz). Many electronics textbooks arbitrarily list as ultrasonic those frequencies between 20 and 100 kHz. There is, however, no reason for placing the upper limit at 100 kHz, or any other frequency. An ultrasonic wave is acoustical (i.e., a mechanical wave in a gaseous, liquid, or solid medium). The upper frequency limit presently accepted for medical ultrasound would have to be in the 10- to 15-mHz range, because instruments currently in existence operate in that range. If someone were to invent a transducer that could operate at higher frequencies, then we could reasonably claim a new ''higher'' limit.

Most textbooks list frequencies in the high kHz and mHz region as *radio* waves. But there is a distinguishing difference between radio waves and ultrasonic waves: Radio waves are *electromagnetic* in nature, while ultrasound waves are *acoustical*. If an electrical oscillation of, say, 2500 kHz, were connected to an appropriate antenna, an electromagnetic wave would be created. But if that same electrical oscillation were connected to an ultrasound transducer, then an ultrasound wave would be created.

17-4 Physics of Sound Waves

Ultrasound waves are vibrations or disturbances in a physical medium such as gas, liquid, or solid matter. Although some animals are able to hear acoustical waves above the range of human hearing (i.e., 20–30 kHz), none are believed capable of hearing waves in the range above 500 kHz at which most medical ultrasound instruments operate.

17-4.1 Wavelength and Velocity

All waves, acoustical or electromagnetic, will obey the following relationship:

$$V = F\lambda \qquad (17\text{-}1)$$

where

V is the velocity of progation through the medium in *meters per second* (m/s)
F is the frequency in *hertz* (Hz)
λ is the wavelength in *meters* (m)

Equation 17-1 is sometimes written in the alternate form of Equation 17-2, which expresses the relationship in terms of the wave's *period,* instead of frequency. Period is defined as the reciprocal of frequency, so we may write Equation 17-2 as

$$V = \frac{\lambda}{T} \qquad (17\text{-}2)$$

where

T is the period of the wave in *seconds* (s), and the other terms are as described in Equation 17-1

EXAMPLE 17-1
An ultrasonic wave in human tissue has a frequency of 2500 kHz and a wavelength of 6×10^{-4} m. Calculate its velocity of propagation.

SOLUTION

$$V = F\lambda \qquad (17\text{-}1)$$

$$V = \left(2500\ \text{kHz} \times \frac{10^3\ \text{Hz}}{\text{kHz}}\right)(6 \times 10^{-4}\ \text{m})$$

$$V = (2.5 \times 10^6\ \text{Hz})(6 \times 10^{-4}\ \text{m}) = \mathbf{1500\ m/s}$$

Table 17-1

Material	Density (g/cm^3)[a]	Sound Velocity (m/s)
Air	0.001	330
Bone	1.85	3360
Muscle	1.06	1570
Fat	0.93	1480
Blood	1.0	1560

[a] At 25°C.

The figure 1500 m/s is generally accepted as the velocity of sound waves in tissue for devices operating in the 2- to 3-mHz region. But we find that the actual velocity depends on factors that are properties of the transmission media, as well as frequency. Table 17-1 lists several materials so that the sound velocity through each can be compared.

EXAMPLE 17-2
A 2000-kHz electrical oscillator can be connected to either an ultrasonic transducer or a radio antenna. Calculate (a) the wavelength of the acoustical wave generated by the transducer in fatty tissue (V = 1500 m/s) and (b) the wavelength of the electromagnetic wave and (c) compare the two wavelengths for the two waves. [Note 2000 kHz = 2×10^6 Hz, and V for electromagnetic waves is the speed of light (i.e., 3×10^8 m/s).]

SOLUTION

$$\text{(a)}\ \lambda_1 = \frac{V}{F} \qquad (17\text{-}1)$$

$$\lambda_1 = \frac{1500\ \text{m/s}}{2 \times 10^6\ \text{Hz}} = \mathbf{7.5 \times 10^{-4}\ m}$$

$$\text{(b)}\ \lambda_2 = \frac{V}{F} \qquad (17\text{-}1)$$

$$\lambda_2 = \frac{3 \times 10^8\ \text{m/s}}{2 \times 10^6\ \text{Hz}} = \mathbf{150\ m}$$

$$\text{(c)}\ \frac{\lambda_2}{\lambda_1} = \frac{150\ \text{m}}{7.5 \times 10^{-4}\ \text{m}} = \mathbf{2 \times 10^5}$$

Note that the wavelength of a 2000-kHz *radio* wave is 200,000 times *greater* than the wavelength of a 2000-kHz *acoustical* wave. This relationship remains true despite the fact that the same electronic oscillator circuit generated both waves.

17-4.2 Reflection and Refraction Phenomena

Reflections and refraction are phenomena affecting all waves and in fact define "wave" behavior. These phenomena occur when the waves impinge on a surface, or boundary, between zones of different density materials. We are most familiar with these phenomena for *light* (i.e., we see our reflection in the mirror). Fishermen are familiar with refraction—light rays bending at the water-air boundary—because it makes a fish *appear* at a location that is somewhat different from its actual location. Rarely does one phenomenon exist without the other. In almost every case, the situation will be as in Figure 17-1; some of the energy in the incident wave will be reflected from the surface, while the rest will be refracted inside of the material.

Figure 17-1 Reflection and refraction phenomenon.

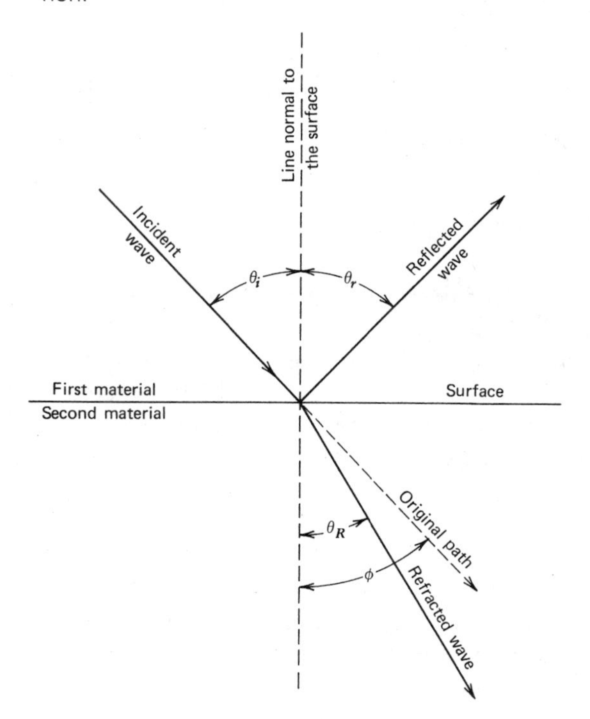

By convention, refraction and reflection angles are measured as an acute angle to a line normal (i.e., perpendicular) to the boundary surface at the point where the incident wave strikes. On curved surfaces, the normal line is perpendicular to a line tangent to the surface at the point where the wave strikes the surface.

In reflection, we know that the angle of incidence and the angle of reflection are equal to each other:

$$\theta_i = \theta_r \tag{17-3}$$

where

θ_i is the angle of incidence
θ_r is the angle of reflection

If the incident wave impinges the surface at an angle of 90° (i.e., it is coincident with the normal line), it will be reflected back on itself. But if the angle is anything other than 90°, then the reflected wave will travel away from the surface at the same angle.

Refraction phenomena affects the portion of the incident wave that enters the second medium. We may infer the behavior of ultrasound waves from the behavior of light waves in an optical medium.

$$\frac{\text{Sin } \theta_i}{\text{Sin } \theta_r} = n_{2-1} \tag{17-4}$$

where

θ_i is the angle of incidence
θ_r is the angle of refraction
n_{2-1} is the relative index of refraction of medium 2 with respect to medium 1

The *index of refraction* (n) is defined as the ratio of the wave velocity in air (or a vacuum for electromagnetic waves) to the wave velocity in the medium, that is,

$$n = \frac{V_{\text{air}}}{V_{\text{med}}} \tag{17-5}$$

Table 17-2

Material	Value of *n*
Water	1.33
Ethyl alcohol	1.36
Quartz	1.46
Flint glass	1.66
Sodium chloride	1.53
Polyethylene	1.50–1.54
Air	1.0[a]

[a] Value is 1.003 when measured against a vacuum.

Table 17-2 shows the index of refraction for several common materials.

In medical ultrasound applications it is rare to find waves impinging tissue from air, but more commonly, waves will enter the tissue from some other medium. In optical systems we know that the relative index of refraction of the second material to the first material is

$$n_{2-1} = \frac{V_1}{V_2} \tag{17-6}$$

where

V_1 is the velocity of the wave in the first material
V_2 is the velocity of the wave in the second material
n_{2-1} is the relative index of refraction

So, Equation 17-4 may be written as

$$\frac{\text{Sin } \theta_i}{\text{Sin } \theta_r} = \frac{V_1}{V_2} \tag{17-7}$$

From Equation 17-7 we may infer the behavior of ultrasound waves in the human body. We know that these waves tend to have a higher velocity in denser materials,[1] so we may conclude from Equation 17-7 that:

[1] This is only *generally* true. The velocity of sound in lead, for example, is *less* than that in aluminum. See Reference 1, p. 394.

1. θ_r is less than θ_i if the incident waves cross into a more dense material.

2. θ_r is greater than θ_i if the incident waves cross into a less dense material.

17-4.3 Acoustical Impedance

The acoustical impedance of a material is a measure of its opposition to the propagation of sound waves and is defined as

$$Z_a = \zeta V \tag{17-8}$$

where

Z_a is the acoustical impedance in *grams per square centimeter second* (g/cm²-s)
ζ is the density of the medium in *grams per cubic centimeter* (g/cm³)
V is the velocity of sound in the medium in *centimeters per second* (cm/s)

EXAMPLE 17-3

Calculate the acoustical impedance of water if the velocity of sound in water is 1450 m/s and the density of water at body temperature (i.e., 37°C) is 0.99 g/cm³.

SOLUTION

$$Z_a = \zeta V \tag{17-8}$$

$$Z_a = \left(\frac{0.99 \text{ g}}{\text{cm}^3}\right)\left(\frac{1500 \text{ m}}{\text{s}} \times \frac{10^2 \text{ cm}}{\text{m}}\right)$$

$$Z_a = \left(\frac{0.99 \text{ g}}{\text{cm}^2}\right)\left(\frac{1.5 \times 10^5}{\text{s}}\right) = \frac{148{,}500 \text{ g}}{\text{cm}^2\text{-s}}$$

17-5 Absorption of Ultrasound Energy

In any system where both reflection and refraction occur, the energy of the incident wave is divided between the reflected and refracted components, that is,

$$U_i = U_r + U_R \tag{17-9}$$

where

U_i is the energy of the incident wave
U_r is the energy of the reflected wave
U_R is the energy of the refracted wave

The percentage of incident energy U_i that is reflected is determined by the *coefficient of reflection:*

$$\Gamma = \left(\frac{Z_1 - Z_2}{Z_1 + Z_2}\right)^2 \qquad (17\text{-}10)$$

where

Γ is the coefficient of reflection
Z_1 is the impedance of material number 1
Z_2 is the impedance of material number 2

To express the reflected energy in *percent,* multiply Equation 17-10 by the factor 100 percent.

EXAMPLE 17-4

Calculate the coefficient of reflection at a boundary betwen air ($Z_1 = 50$ g/cm²-s) and tissue ($Z_2 = 150{,}000$ g/cm²-s).

SOLUTION

$$\Gamma = \left(\frac{Z_1 - Z_2}{Z_1 + Z_2}\right)^2 \times 100 \text{ percent} \qquad (17\text{-}10)$$

$$\Gamma = \left(\frac{50 - 150{,}000}{50 + 150{,}000}\right)^2 \times 100 \text{ percent}$$

$$\Gamma = -0.999^2 \times 100 \text{ percent} = \mathbf{99.9\ percent}$$

Note in Example 17-4 that almost *all* of the incident energy is reflected from the boundary between the two materials. This example illustrates a problem that exists in medical ultrasound: the vastly different impedances of air and tissue. Medical ultrasonic transducers are coupled to the patient's skin through a special jelly that provides an air-free path for the sound waves.

Let us rework Example 17-4 using the impedance of the jelly in place of the impedance of air and see what difference exists. The impedance of the jelly can be taken to be approximately 1.48×10^5 g/cm-s.

$$\Gamma = \left(\frac{Z_1 - Z_2}{Z_1 + Z_2}\right)^2 \times 100 \text{ percent}$$

$$\Gamma = \left(\frac{148{,}000 - 150{,}000}{148{,}000 + 150{,}000}\right)^2 \times 100 \text{ percent}$$

$$\Gamma = (-6.7 \times 10^{-3})(100 \text{ percent}) = \mathbf{-0.67\ percent}$$

In this case, with the impedances "matched," the reflected energy is very small (i.e., 0.67 percent, so the remaining 99.33 percent is absorbed by the tissue).

In many ultrasonic medical instruments a receiver transducer examines the *reflected wave.* If the structure causing the reflection is acoustically dense (i.e., bone), then the reflection coefficient is large. But when the reflection is caused by an object that has a density only slightly different from the surrounding tissue, then only a small amount of energy reflects back toward the transducer.

17-6 Biological Effects of Ultrasound

The increased use of ultrasound monitoring, therapeutic, and diagnostic equipment has led to some concern over the possible hazard levels. Thus far, no definitive information exists concerning safe levels. Most clinical and research instruments are designed to produce power output levels between 5 and 50 mW/cm² at the transducer. Unfortunately, the measurement of ultrasound power levels is difficult, and there is no well-established standard accepted by all for the measurement of power levels.

Several factors are believed to have importance in the biological interaction of tissue and ultrasound waves: *frequency, irradiation time, beam intensity,* and *duty cycle.*

The principal biophysical effects of ultrasound are *thermal, cavitation, shearing action,* and *intracellular motion.*

The thermal effects are caused by sonic agitation of cells and are affected by the nature of the tissue, blood flow, thermal conduction losses, etc. In general, we can state that heat created is the product of the applied power intensity (I) in W/cm^2 and the absorption coefficient per cm^{-1}.

$$U_h = \alpha I \qquad (17\text{-}11)$$

where

U_h is the heat energy in watts (W)
α is the absorption coefficient in cm^{-1}
I is the beam intensity in W/cm^2

The cavitation phenomenon is rather complex but can be described as the creation of gaseous bubbles in the fluid medium due to agitation by the ultrasonic waves. Cavitation is also the mechanism by which ultrasonic cleaners work.

Shear forces operate in media of low viscosity that are able to flow. The ultrasound waves create shear forces by generating eddy currents in the medium.

Intracellular movement is created when cell membrane vibrations take place under the influence of the ultrasound waves. These movements often take the form of twisting action.

17-7 Ultrasonic Transducers

Recall from Chapter 3 that a transducer is a device that converts energy from one form to another for purposes of measurement or control. In the case of ultrasound, two transducer functions are recognized: (1) conversion of ac electrical oscillations into acoustical vibrations and (2) conversion of acoustical vibrations into ac oscillations of the same frequency. These two functions are the *transmit* and *receive* transducers, respectively, and in many cases are merely different applications of the same device.

Several different devices can be used to transduce ultrasonic energy (i.e., dynamic microphone elements and capacitor assemblies), but none are as effective as the *piezoelectric resonator,* usually called simply a "crystal."

The piezoelectric phenomenon exists in certain crystalline structures such as natural quartz, barium titanate, and Rochelle salts. The piezoelectric element (see Figure 17-2) will create an electrical current when it is mechanically deformed. Flexing the crystal in either direction will cause an output voltage to appear across the electrodes. The crystal element, therefore, will serve as an ultrasonic receive transducer.

But the opposite effect is also noted in piezoelectric elements. That is, the application of a voltage across the electrodes causes the crystal element to deform. If an ac potential is applied to the crystal, then the crystal element will oscillate back and forth at the ac frequency. Unfortunately, most crystals exhibit a *resonance* effect. So these effects only occur at a *specific frequency.*

The exact form taken by any given piezoelectric transducer will depend on the application. Several of these are discussed in later sections of this chapter when we discuss specific instruments.

17-8 Transcutaneous Doppler Flow Detectors

Transcutaneous flow detectors are designed to use the Doppler effect to detect the flow of blood in arteries close to the surface of the body. An example of a commercial flow detector, and several different transducers, is shown in Figure 17-3.

Figure 17-2 Crystal transducers.

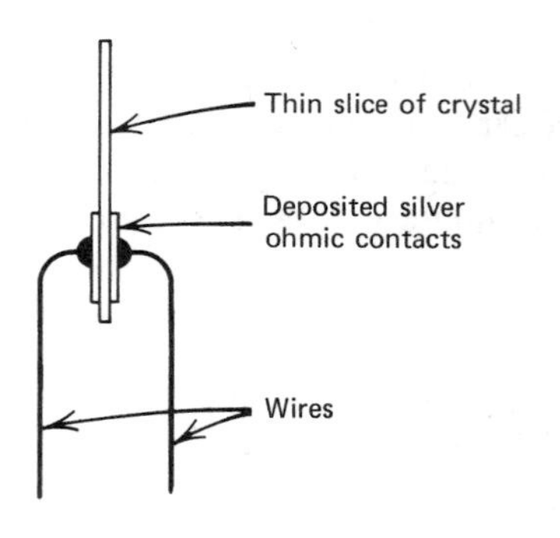

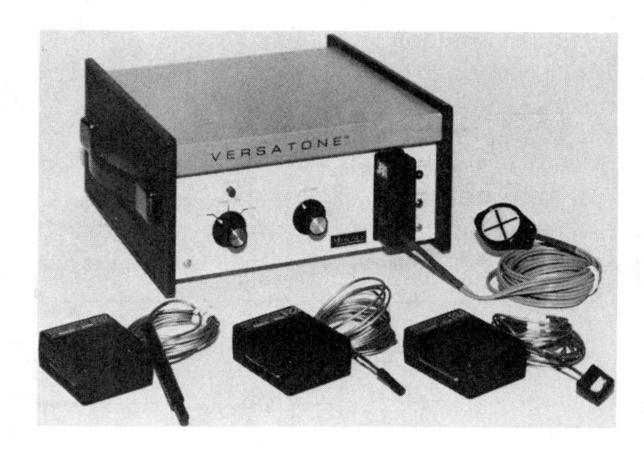

Figure 17-3 Ultrasound flow meter. (Photo courtesy Medsonics, Inc.)

Figure 17-4*a* shows the basic configuration of a Doppler flow detector. Two crystals, one each for transmit and receive functions, are placed in a plastic housing that is positioned over a peripheral artery. A frequency near 10 mHz is usually selected for these instruments to take advantage of the natural attenuation of high ultrasonic frequencies in tissue. Such frequencies also permit better focusing and a larger frequency change for any given amount of motion. Reflections of significant amplitude from underlying tissues are therefore attenuated, while reflections from blood vessels near the surface are stronger.

The back-scatter reflections from the moving blood will have a slightly different frequency than the incident wave due to Doppler shift. If the blood were motionless, the frequency shift would be zero, so the return wave frequency will be identical to the incident wave frequency. But moving blood produces a shift that is proportional to the blood velocity. A frequency shift of approximately 200 Hz at 10 mHz corresponds to a blood velocity of approximately 6 cm/s.

The 10-mHz oscillator in Figure 17-4*a* excites the transmit crystal and, in some models, one port of the detector in the receiver section of the instrument. The incident wave will be shifted $\pm\Delta F$ by the flowing blood, and this frequency excites the receive crystal. The signal at the output of the receive crystal is amplified and then fed to a detector. The output of the detector is the $\pm\Delta F$ component. In most instruments the filtered output takes the form of a low-frequency "hiss."

The audio output, through earphones or a loudspeaker, serves to give the surgeon a subjective, yet accurate, indication of blood flow. In a few models, an integrator circuit is used to produce a dc voltage that is proportional to the flow value. This voltage, however, is difficult to calibrate accurately but does serve as a relative flow indicator.

A probe-type flow transducer is shown in Figure 17-4*b*. In this assembly, two crystal elements are positioned side by side but are separated by a barrier with a high acoustical impedance.

17-9 Flowmeters

There are two general types of flowmeter. One type depends on the velocity difference between upstream and downstream sound waves. The second technique uses the Doppler shift of scattered waves from the moving medium (i.e., blood).

Figure 17-5*a* shows a transit time ultrasonic flow transducer. Two piezoelectric crystal elements are aimed at each other obliquely across the flow path. The axis between these crystals has length D and forms angle θ with the flow axis.

The crystals in Figure 17-5*a* serve both transmit and receive functions. When measuring the downstream transit time of an ultrasonic pulse, crystal A is the transmitter, while crystal B is the receiver. When measuring the upstream time, however, the roles of the crystals are reversed; A becomes the receiver and B is the transmitter.

The difference between upstream and

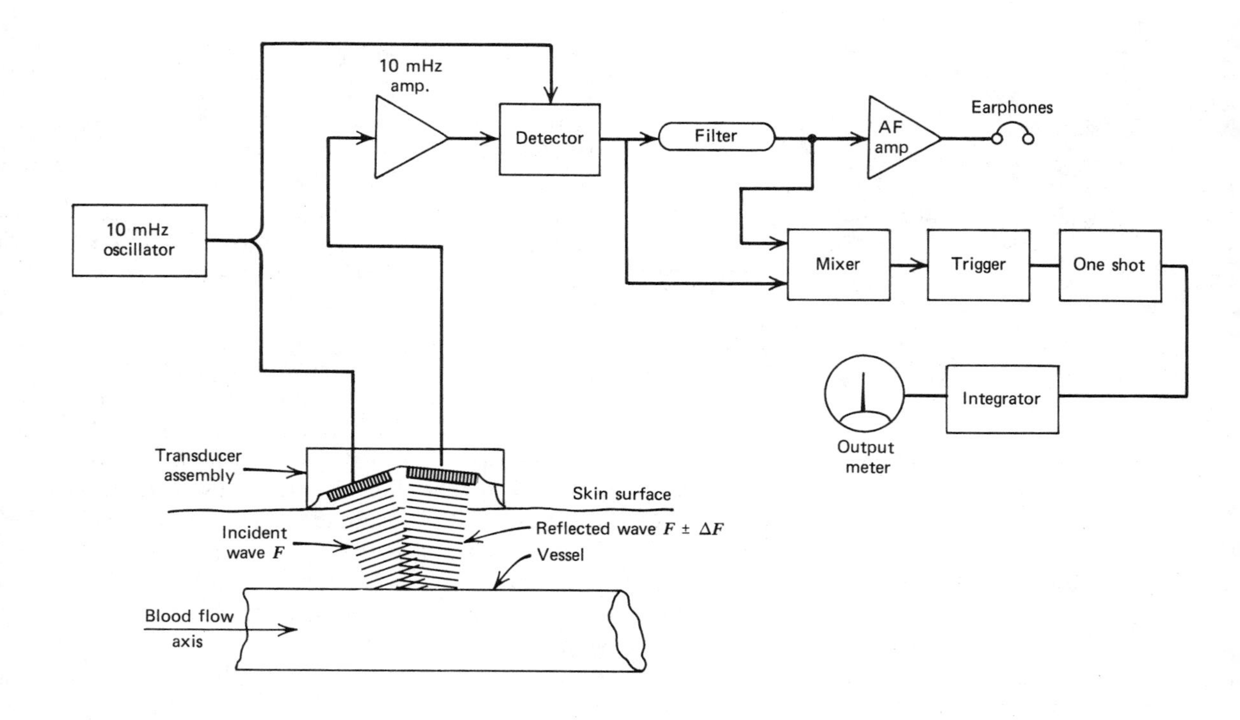

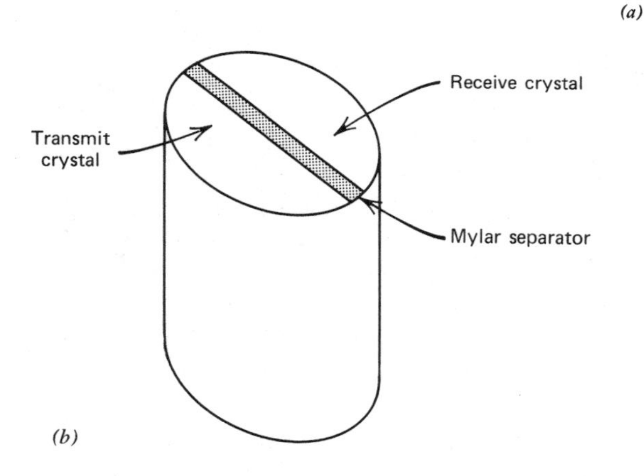

Figure 17-4 Ultrasonic flow meter transducers. (*a*) Regular. (*b*) Pencil probe.

downstream transit times (ΔT) is measured in an electronic circuit and is then used to solve the expression

$$\overline{V} = \frac{C^2 \, \Delta T}{2 \, D \cos \theta} \tag{17-12}$$

where

$\overline{V}$ is the average flow velocity of the medium

C is the speed of sound in the medium

ΔT is the difference between upstream and downstream transit times

θ is the angle between the crystal axis and the flow axis

EXAMPLE 17-5

An ultrasonic flow transducer is designed to measure the mean velocity of a gaseous medium such

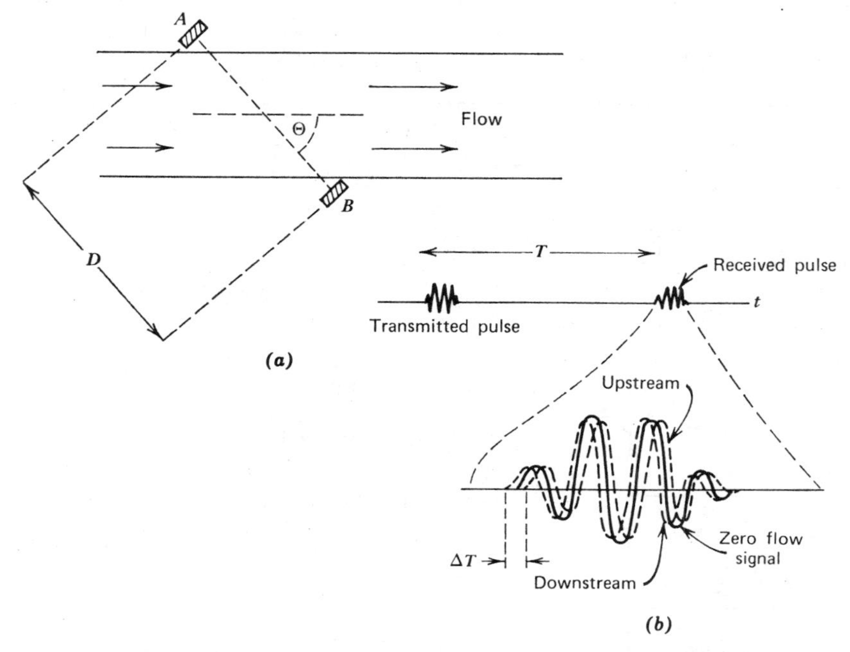

Figure 17-5 Ultrasound flowmeter capable of yielding numeric data. (From R. S. C. Cobbold, *Transducers for Biomedical Measurements,* John Wiley & Sons, New York, 1974.)

as respiratory air. The crystal axis makes a 30° angle with the flow axis, and the crystals are spaced 1.25 cm apart. Calculate the air velocity if the transit time is 8.6×10^{-9} s. (*Hint:* the speed of sound in air is 335 m/s.)

SOLUTION

$$V = \frac{C^2 \, \Delta T}{2D \cos \theta} \qquad (17\text{-}12)$$

$$V = \frac{[(335 \text{ m/s}) \times (100 \text{ cm/m})][8.6 \times 10^{-9} \text{ s})]}{(2)(1.25 \text{ cm})(\cos 30°)}$$

$$V = \frac{9.6 \text{ cm}^2\text{/s}}{2.2 \text{ cm}} = \mathbf{4.4 \text{ cm/2}}$$

Note in Figure 17-5*b* that the quantity ΔT tends to be very small. When analog circuits are used to measure ΔT, then matters such as noise and drift become very important. As a result, many instruments based on transit time differences use phase detection techniques rather than actual timing techniques.

A Doppler flow transducer is shown in Figure 17-6*a*. This transducer uses the frequency change of a wave scattered from particulate matter flowing in the vessels. It has been shown that the change in frequency, ΔF, is given by:[2]

$$\Delta F = \pm F_s (\cos \theta + \cos \phi) \frac{v}{c} \qquad (17\text{-}13)$$

where

ΔF is the frequency shift in *hertz* (Hz)
F_s is the transmitter frequency in *hertz* (Hz)
θ is the angle of the transmit crystal axis to the flow axis
ϕ is the angle of the receive crystal axis to the flow axis

[2] Richard S. C. Cobbold, *Transducer for Biomedical Measurements,* Wiley (New York, 1974), p. 280.

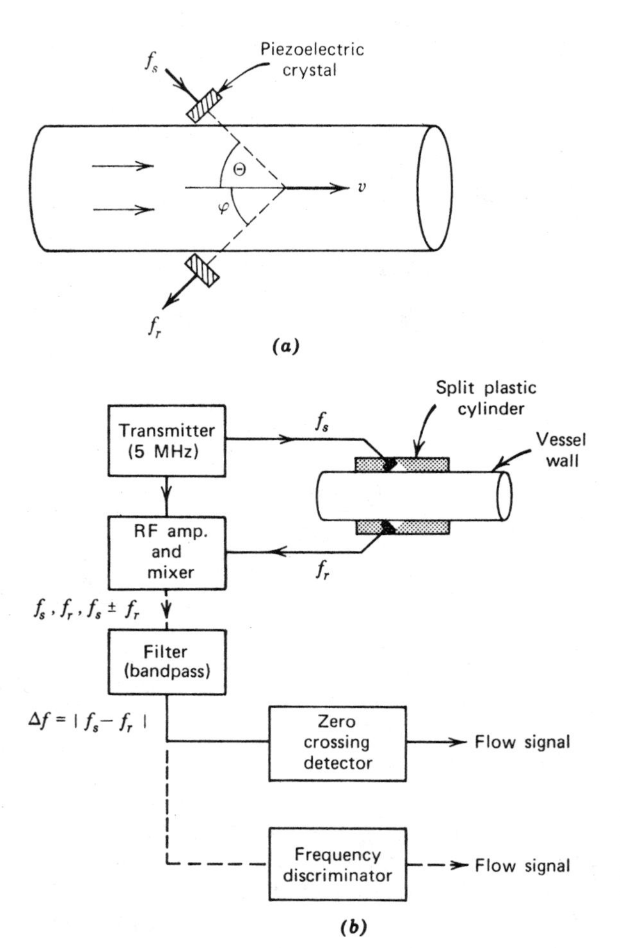

Figure 17-6 Doppler flowmeter. (*a*) Transducer. (*b*) Block diagram. (From R. S. C. Cobbold, *Transducers for Biomedical Measurements,* John Wiley & Sons, New York, 1974.)

A minus sign is used in Equation 17-13 when the flow is toward the crystal faces, and a plus sign denotes flow away from the faces.

The block diagram for an ultrasonic flow detector based on Equation 17-13 is shown in Figure 17-6*b*. In this instrument the signal from the receive crystal F_r (i.e., $F_r = F_s \pm \Delta F$) is mixed with F_s in a radio frequency (rf) amplifier/mixer stage. A bandpass filter is used to insure that only the *difference* frequency $(F_s - F_r)$ is actually used.

Two alternate output stages are shown in Figure 17-6*b*. One type uses a zero-crossing detector to produce spike pulses at each point where ΔF goes through zero. If these pulses are integrated, we obtain a dc flow signal. Alternatively, the output pulses could be processed in a digital circuit that is designed to recognize the repetition rate of the pulses.

The alternate technique is to use a frequency-sensitive discriminator to produce a flow signal.

The flow signal that is obtained from an instrument such as Figure 17-6*b* represents the *mean flow velocity* in cm/s of the fluid (i.e., blood) in the tube (i.e., vessel). Examples of both liquid and gaseous flowmeters are known using this technique.

The data from the instrument does not

represent volumetric flow (i.e., ml/s) rate. Only velocity data can be discerned from the output of the transducer unless the cross-sectional area of the vessel is known, in which case volumetric flow is proportional to the product of area and velocity.

17-10 Ultrasonic Blood Pressure Measurement

Ultrasonic techniques can be used to measure arterial blood pressure indirectly. Both manual and automatic (i.e., Figure 17-7) models have been manufactured.

Figure 17-8 shows how such a system operates. Piezoelectric crystals are placed between the patient's arm and a blood pressure cuff (Figure 17-8*a*). The ultrasonic circuits measure the Doppler shift caused when the incident wave reflects from a moving wall of the underlying brachial artery. The frequency of the reflected signal is proportional to the instantaneous velocity of the vessel wall. When the brachial artery is occluded, the Doppler shift is near zero. But when the arterial pressure is able to overcome the cuff pressure, the occlusion will snap open, causing a Doppler shift in the 200- to 500-Hz range.

There are actually two Doppler events during each cardiac cycle (Figure 17-8*b*). When the arterial pulse pressure rises to cuff pressure, we find the opening event characterized by the 200- to 500-Hz Doppler shift. But as the arterial pressure recedes back toward the diastolic, the vessel will close (i.e., reocclude) once the arterial pressure is less than the cuff pressure. This event produces a low-frequency Doppler shift in the 25- to 100-Hz range.

Figure 17-7 Ultrasonic blood pressure monitor. (Photo courtesy Roche Medical Electronics, Cranbury, N.J.)

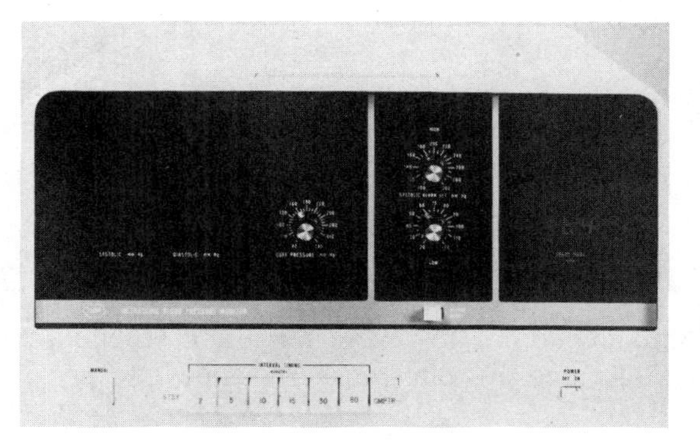

When measuring blood pressure, the cuff pressure is allowed to bleed down at a fixed rate. The systolic pressure is indicated by the onset of high-frequency Doppler shift sounds. At that time, the high- and low-frequency components are almost coincident (see Figure 17-8*b*). But as the cuff pressure drops below systolic, the occurrence of the two events becomes further apart, until the cuff pressure drops to the diastolic pressure. At that point, the low-frequency component signifying the closing events becomes almost coincident with the high-frequency component of the *next* cardiac cycle. In a manual blood pressure instrument, these events are audible in a loudspeaker, and the operator uses them as if the instrument were a stethoscope; the ultrasonic audible events being analogous to the Korotkoff sounds of sphygmomanometry.

A simplified block diagram of an automatic ultrasonic blood pressure measurement system is shown in Figure 17-9. This instrument consists of three sections: pneumatic subsystem, control system, and the ultrasonic section. The operation is as follows (the control section coordinates all events):

1. The pump turns on and inflates the cuff to the predetermined level set by the *cuff pressure* control on the front panel.

2. Next, the bleed valve (V_1) opens to allow the cuff to *slowly* deflate, while the ultrasonic section looks for the high-frequency tones indicating systolic pressure.

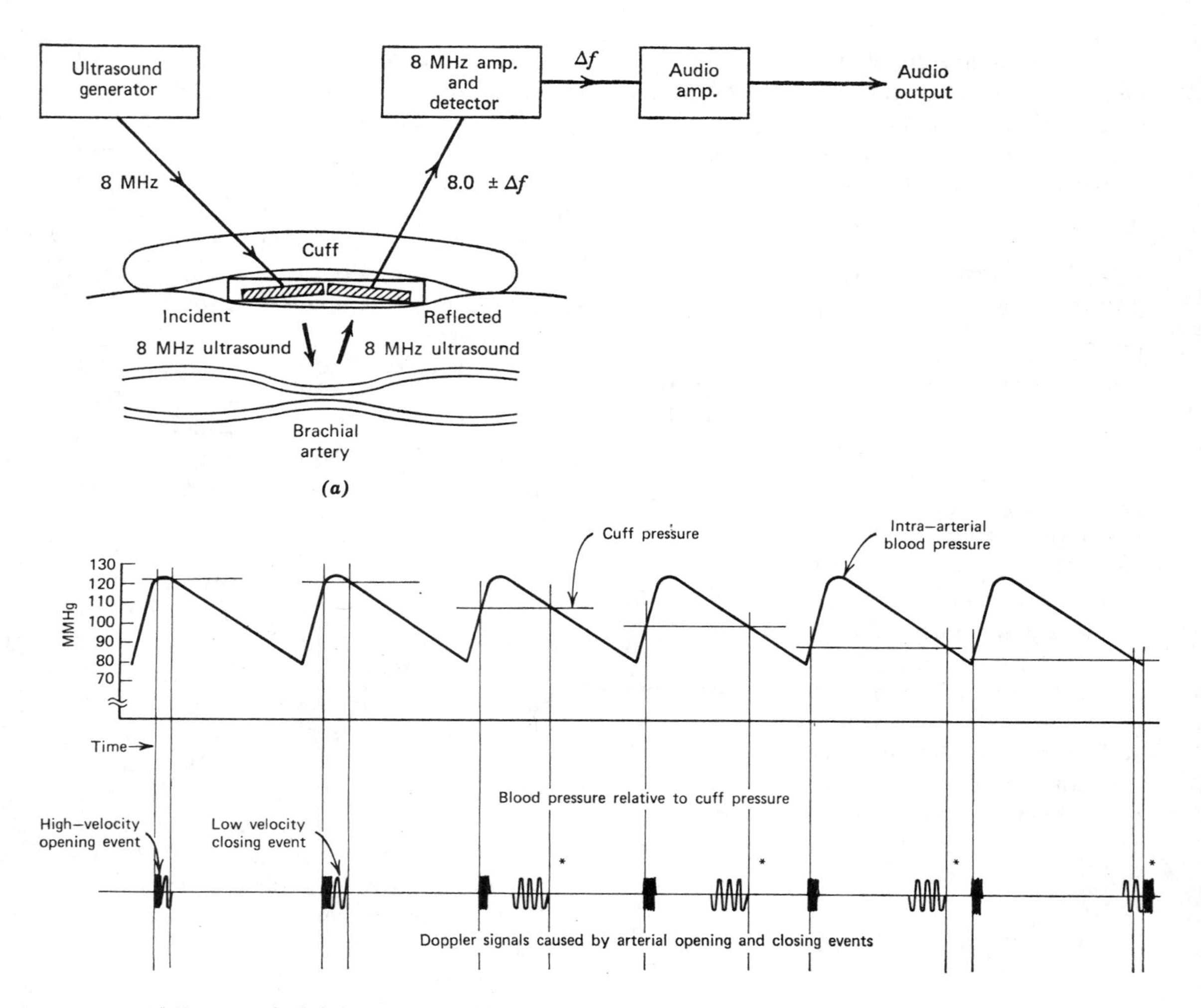

Figure 17-8 An ultrasonic pressure system. (*a*) Transducer placement and block diagram. (*b*) Timing.

3. When the cuff pressure has bled down to the systolic pressure, the electronic circuit will receive the high-frequency tones. When this occurs, the systolic hold valve (V_2) is closed, causing the systolic manometer to remain at the systolic pressure value.

4. When the cuff pressure drops to the diastolic value, the control system closes valve V_3, holding the diastolic manometer at the diastolic pressure.

5. Following determination of the diastolic pressure, the control system opens valve V_4 to vent the cuff to atmosphere. The manometers, however, remain at the diastolic and systolic pressure values, respectively.

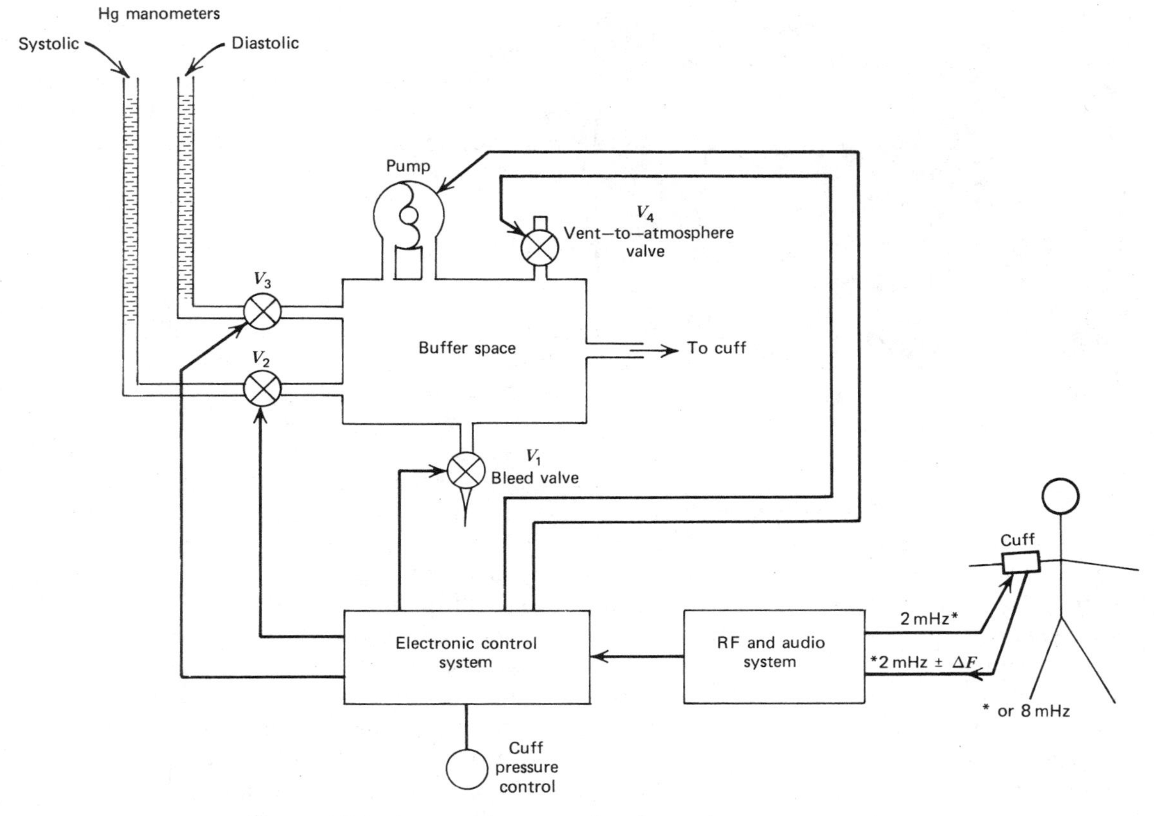

Figure 17-9 Pneumatic system for ultrasonic pressure monitor.

The ultrasonic blood pressure device has demonstrated an ability to measure the pressure of normal patients to within ± 2.5 mm Hg, and on hypotensive patients it can obtain a valid response where auscultation is difficult or impossible. The automatic version allows continuous, noninvasive monitoring of blood pressure during surgical procedures or in the ICU/CCU.

The updated version shown previously in Figure 17-6 replaces the mercury manometers with electronic measurement of the chamber pressure. Such a scheme eliminates the need for valves V_2 and V_3 in Figure 17-9. If electronic pressure measurement is used, it is possible to simply update the systolic or diastolic display at the instants dictated by the control system.

17-11 Fetal Monitors

Fetal monitors fall into two different categories. One type is essentially an electronic stethoscope and uses the ultrasonic Doppler shift to detect the fetal heart sounds. The other type is more complex and measures the fetal ECG and pressure of the amniotic fluid.

In the simplest instruments, we have a circuit similar to a Doppler flow meter, although operating near 2 mHz instead of 10 mHz. The ultrasonic transducer is placed

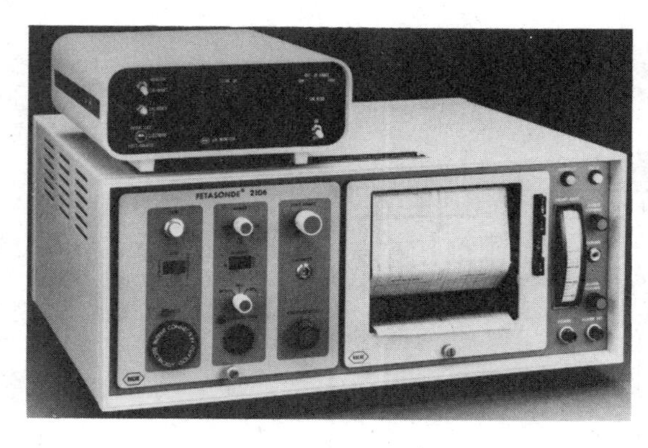

Figure 17-10 Roche Fetasonde®. (Photo courtesy Roche Medical Electronics.)

over the mother's abdomen and aimed toward the fetal heart. As the heart beats, it scatters signal back toward the transducer with a Doppler shift proportional to the instantaneous velocity of the heart wall.

Unfortunately, the signal is not pure but will contain Doppler components from other structures in the maternal abdomen. The return signal, therefore, must be filtered before being applied to the earphones or loudspeaker.

Fetal heart rate can be obtained from this type of instrument if the audio output is rectified and filtered to form a low-frequency pulse from the return waveform envelope. The pulses from the filter will have the same repetition rate as the fetal ECG.

Obtaining a fetal ECG is done by attaching scalp electrodes to the fetus's head. The fetal ECG (FECG) will have a lower amplitude than the maternal ECG (MECG). An electrode on the mother's abdomen acquires the MECG signal, while the scalp electrode acquires the signal defined as (MECG + FECG). By summing these signals in a differential amplifier, we obtain an output signal that is FECG + artifact). The artifact signal can be substantially suppressed by frequency-selective filtering.

Figure 17-10 shows a fetal monitor capable of scalp electrode ECG, ultrasonic, and measurement of the uterine pressure. The latter capability allows the obstetrician and delivery room personnel to note when contractions occur. Such monitors have seen increased use in the past few years.

17-11.1 Ultrasonic Fetal Scans

Ultrasonics are also used in diagnostic radiology to view objects normally hidden from view. The use of ultrasound on the fetus is preferred over X-ray examination for reasons of safety.

Figure 17-11 shows a simple *A-scan* depth detection scheme. In this instrument, *pulsed ultrasound* is used. A pulse generator fires a burst of energy from the transmitter and simultaneously triggers a horizontal sweep on the oscilloscope. The output of the receiver is connected to the vertical input of the oscilloscope. The A-scan display, then, is a *time-versus-amplitude* presentation. The depth of the object is measured by noting how far to the right along the horizontal time scale its reflection appears. The A-scan is also used to locate and detect foreign objects in the body.

Figure 17-12 shows how a *linear-B scan* is created. The pulse generator initiates the ultrasound energy burst and the vertical sweep of the cathode ray oscilloscope (CRO). A mechanical position transducer creates an electrical signal for the horizontal deflection of the CRO beam. The CRO beam is *intensity modulated* by the output of the ultrasound receiver section.

The fetus, or other object, is repeatedly scanned as the transducer moves back and forth across the maternal abdomen. This produces a cross-sectional view of the object known as a *tomogram*.

A similar scan technique called a *compound-B scan* is made by rocking the transducer back and forth as it undergoes its linear back-and-forth motion. This method

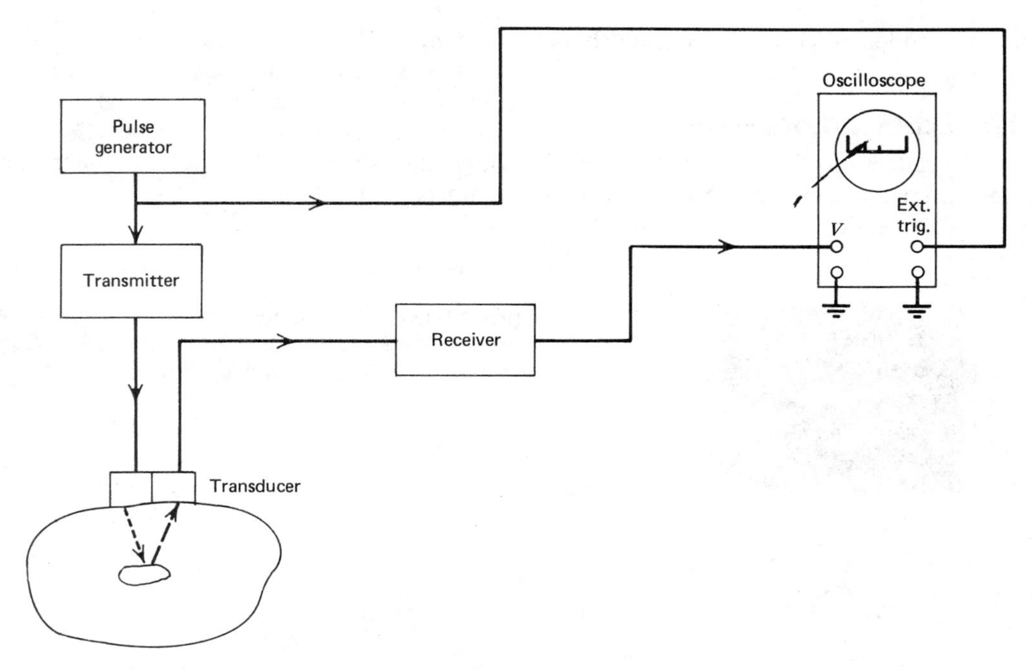

Figure 17-11 Simple ultrasonic fetal detector.

Figure 17-12 Fetal ultrasonic instrument.

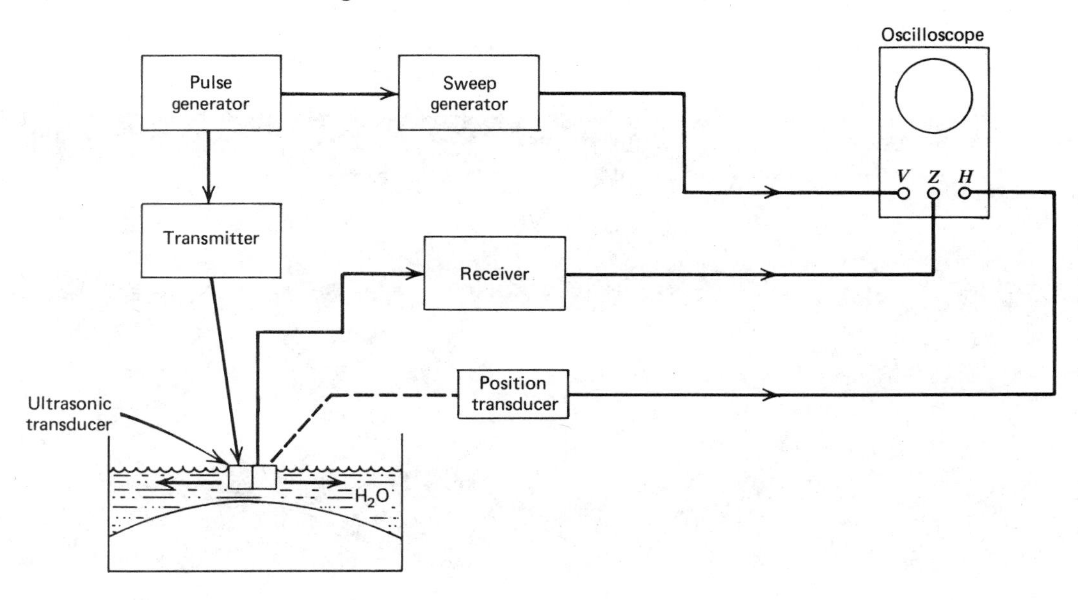

provides a superior view of the underlying structures.

17-12 Echocardiography

Ultrasonic reflections from the moving heart can be plotted against time to form a *time-motion*, or *T-M*, plot. Figure 17-13*a* shows the T-M scan schematically, while Figure 17-13*b* shows an actual echocardiogram. This system detects motion of the heart's mitral valve and is particularly useful in diagnosing stenosis.

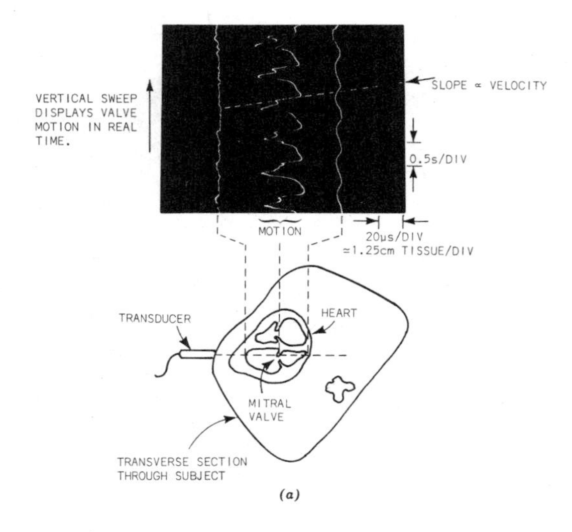

Figure 17-13 Echocardiography. (*a*) T-M mode cardiography. Reprinted courtesy of Tektronix, Inc. Copyright 1970. All rights reserved. (*b*) Echocardiogram.

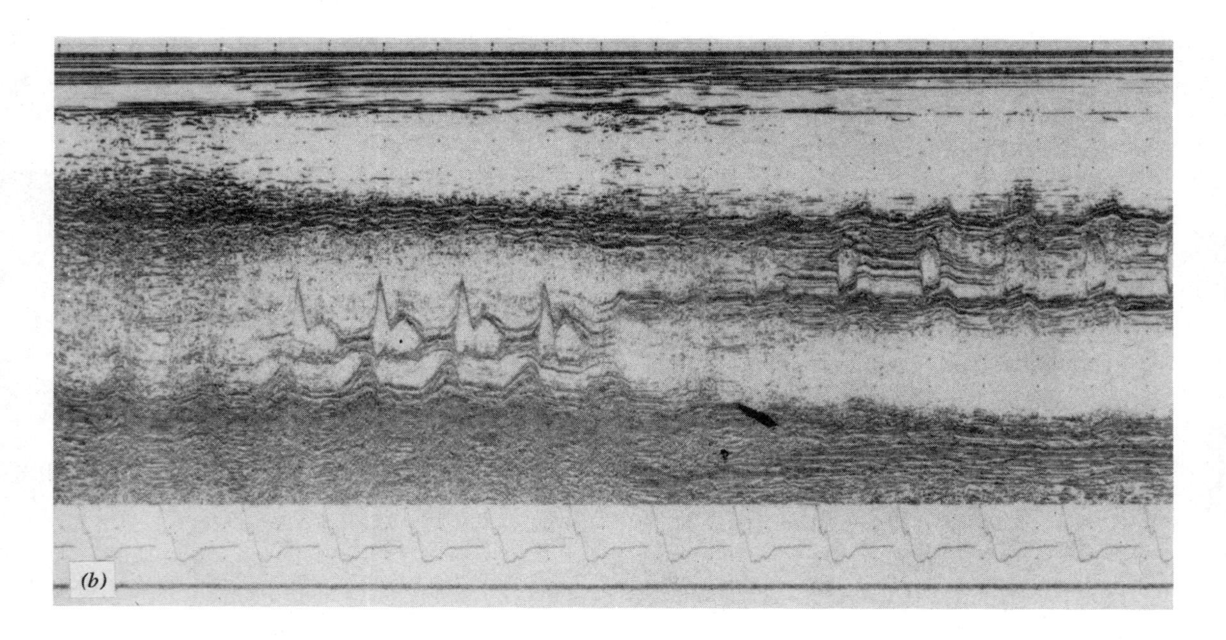

A commercial echo machine is the *Electronics for Medicine* model shown in Figure 17-14*a*. Echo instruments are usually equipped with a method for compensating for differing signal attenuation in tissue at differing depths. This instrument provides selectable compensation between 0 and −40 dB at increments of 2 cm, from 0 to 24 cm below the surface.

Depth and gain information is displayed separate from the T-M information in the form shown in Figure 17-14*b*. This display is called a *time-compensated gain* (TCG) curve.

Figure 17-14 Echocardiograph machine. (*a*) An E-for-M instrument. (Photo courtesy Electronics for Medicine, Pleasantville, N.Y.) (*b*) Time-compensated gain curve.

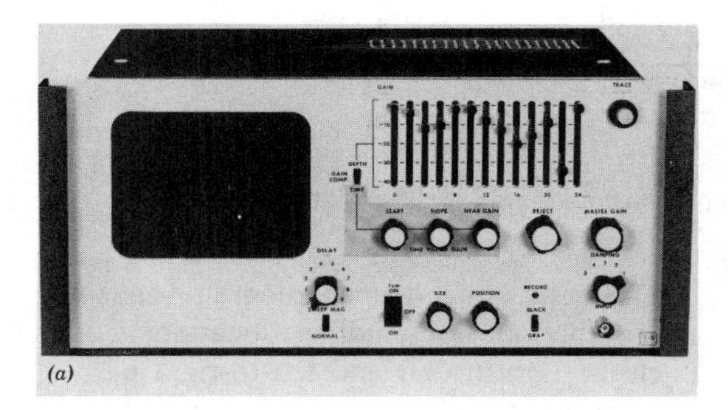

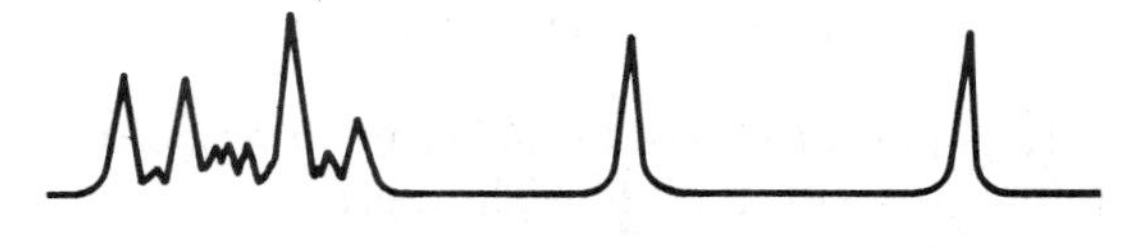

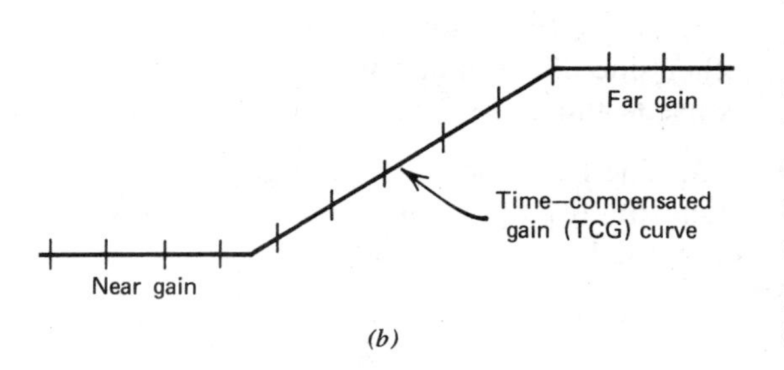

In a T-M display such as Figure 17-13*a*, there are actually two time bases used. The typical recorder is an optical oscillograph in which photosensitive paper is pulled across a cathode ray tube screen at approximately 0.5 s/cm. The CRO beam is swept left to right at 20 μs/cm. Reflections will intensity modulate the CRO and thus appear as bright spots on the CRO screen. As the object being scanned moves, the location of the bright spots on the CRO sweep changes, creating the patterns shown in Figure 17-13*b*.

17-13 Echoencephalography

Echo ranging can also be used to probe the brain, and it forms the basis for an ultrasonic device called the *echoencephalograph*. Figure 17-15 shows the basic principle of an A-scan echoencephalograph. This illustration shows two transducer locations for the comparison of results.

The echoencephalograph will fire 1-μs bursts of 2 to 3 mHz ultrasound energy at a repetition rate of 500 bursts per second. An oscilloscope connected for A-scan will show traces such as Figure 17-15. The left-most reflection corresponds to the skull wall nearest the transducer. The right-most feature in this type of scan is the reflection from the far side of the patient's skull. The features between the skull wall reflections depend upon transducer placement and certain other factors.

The midline feature is caused by reflections from the lateral ventricles, the third ventricle, and the septum pellucidum between the lateral ventricles. Ordinarily, the septum lies within ±2 mm of the center line of the skull, and its reflection on the CRO will be located exactly midway between the spikes representing the third ventricle in Figure 17-15. A shift of more than ±3 mm from the correct position is often considered to be pathologic and may indicate the presence of a tumor or other lesion.

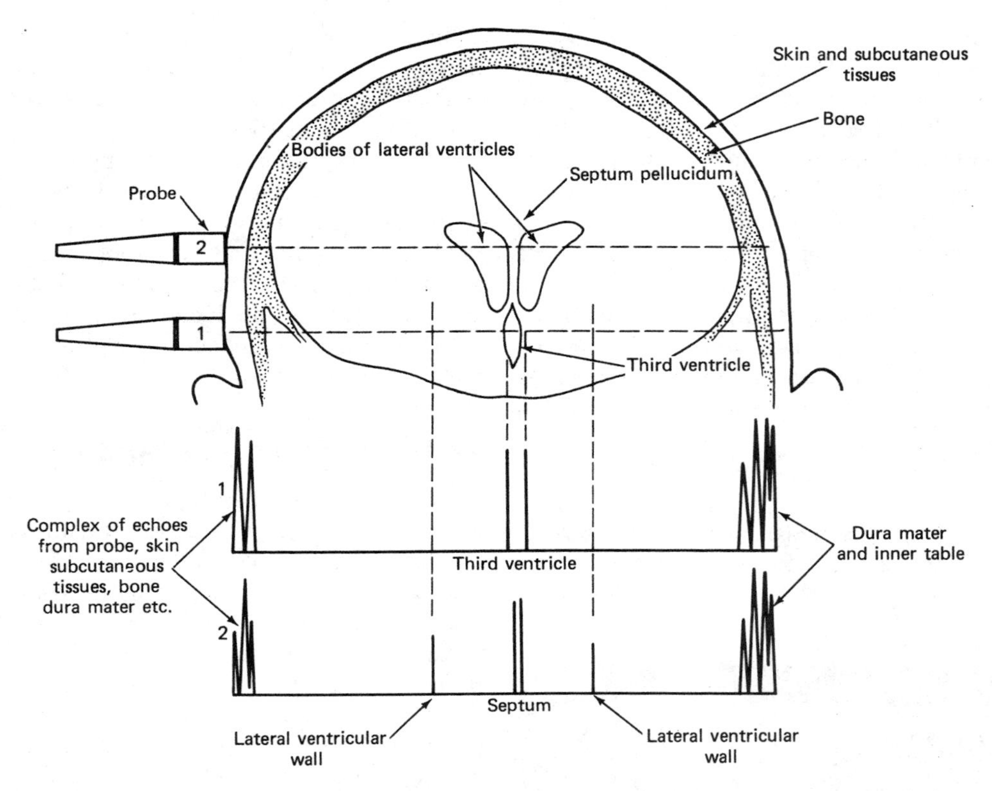

Figure 17-15 Echoencephalograph transducer placement and patterns generated. (Reprinted courtesy of Hewlett-Packard.)

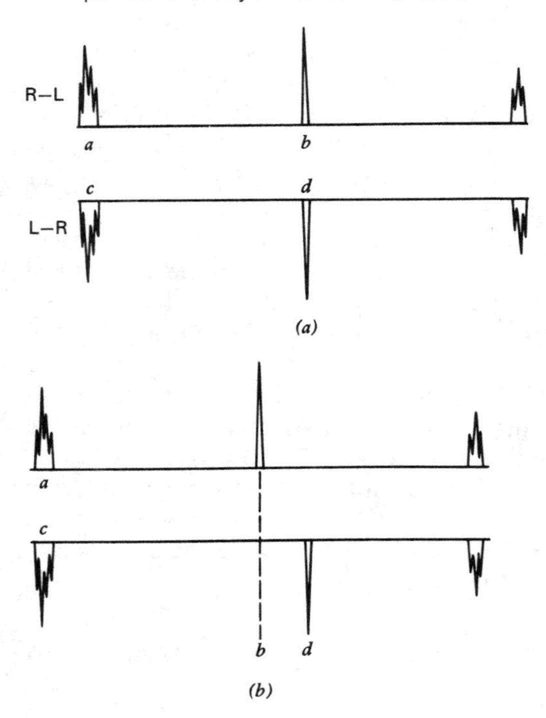

One form of echoencephalogram uses two determinations of midline locations, one each for right-to-left and left-to-right paths. Figures 17-16*a* and 17-16*b* show such a presentation. The procedure is as follows:

1. A good R-L path display is obtained on an oscilloscope and then photographed on an oscilloscope camera. The film is *not* advanced at this time.

2. Next, the transducer is moved to the other side of the skull (or another transducer is used) and an L-R display is obtained.

3. The L-R display is photographed on the same photographic plate so that it may be compared to the R-L display.

Figure 17-16 R-L and L-R patterns. (*a*) Normal. (*b*) Displacement of midline. (Reprinted courtesy of Hewlett-Packard.)

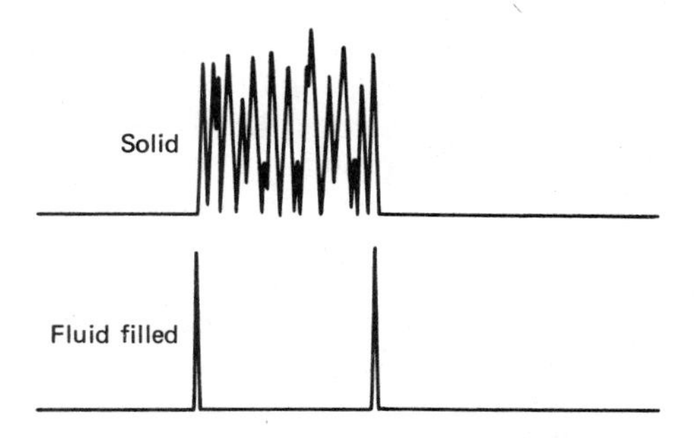

Figure 17-17 Different displays of solid and fluid-filled masses. (Reprinted courtesy of Hewlett-Packard.)

Figure 17-16*a* shows a normal display in which the R-L and L-R presentations of the midline structures coincide. But in Figure 17-16*b* the structures do not coincide because some midline shift is present.

The locations of tumors, aneurysms, and other lesions can be determined by multiple scans to find the distance from the skull walls at different angles.

Note that solid and soft (i.e., fluid-filled) tumors produce radically different echo responses, as shown in Figure 17-17.

17-14 Summary

1. Ultrasound uses acoustical waves in the frequency range above 20 kHz.
2. Ultrasound waves obey the reflection and refraction rules ordinary to wave behavior.
3. Several factors affect biological interaction with ultrasound: frequency, irradiation time, beam intensity, and duty cycle.
4. Most ultrasound transducers are piezoelectric crystals.
5. Transcutaneous flow detectors use Doppler shift to detect the flow of blood.
6. Two types of flowmeters are commonly employed: transit time and Doppler shift.
7. Ultrasonic blood pressure monitors use Doppler-shifted reflections from vessel walls and a sphygmomanometer to determine the systolic and diastolic values.

17-15 Recapitulation

Now return to the objectives and self-evaluation questions at the beginning of the chapter and see how well you can answer them. If you cannot answer certain questions, place a check mark next to each and reread appropriate sections of the text. Next, try to answer the following questions, using the same procedure. When you have answered all of the questions, solve the problems in Section 17-17.

17-16 Questions

17-1 Define in your own words the meaning of "ultrasound."

17-2 A 2-mHz signal will be ______________ if radiated by an antenna and ______________ if radiated by a piezoelectric crystal.

17-3 Ultrasonic waves are ______________ in a liquid, gas, or solid medium.

17-4 Ultrasound waves exhibit both ______________ and ______________ phenomena, characteristic of all wave behavior.

17-5 A measure of a medium's opposition to the propagation of ultrasound waves is called the ______________ ______________.

17-6 The percentage of energy *reflected* at any interface between media is given by the ______________ of ______________.

17-7 Impedance matching of the path between an ultrasonic transducer and the patient's skin surface requires the use of a ______________.

17-8 List four factors affecting the interaction between ultrasound waves and biological tissue.

17-9 Most diagnostic ultrasound instruments limit output power at the transducer to the range ______________ mW/cm^2 to ______________ mW/cm^2.

17-10 List four biophysical effects of ultrasound waves in tissue.

17-11 Most ultrasound transducers used in medical instruments are ______________ ______________.

17-12 A transcutaneous blood flow detector uses the ______________ effect to detect the presence of blood flowing in arteries.

17-13 A transcutaneous blood flow detector is a(n) (invasive) (noninvasive) technique.

17-14 Most transcutaneous flow detectors use frequencies in the ______________-mHz range to take advantage of the natural attenuation of tissues.

17-15 Draw a block diagram of a transcutaneous flow detector.

17-16 List two methods for making flow measurements using ultrasound.

17-17 Most flow meters measure ______________ flow velocity.

17-18 Ultrasonic blood pressure monitors use Doppler shift components to detect ______________ and ______________ points.

17-19 The systolic pressure causes Doppler shifts in the ______________-to ______________-Hz range.

17-20 Ultrasonic fetal monitors use Doppler shift to detect fetal ______________ ______________.

17-21 What is an *A-scan display* in ultrasonic terminology?

17-22 What is a *linear B-scan?* Compound B-scan?

17-23 An echocardiograph uses a ______________- ______________ display to detect motion of the mitral valve.

17-24 Depth and gain information in an echocardiograph are displayed on ______________ ______________ ______________ curve.

17-25 The echoencephalograph probes the ______________ using an ultrasonic A-scan.

17-17 Problems

17-1 Calculate the velocity of a 5-mHz ultrasound signal in water if its wavelength is 6.8×10^{-3} cm.

17-2 Calculate the frequency of an ultrasound signal in tissue if its velocity is 1500 m/s and the wavelength is 7.6×10^{-3} cm.

17-3 Calculate the wavelength of a 7 mHz (a) *ultrasound* wave in tissue (V = 1500 m/s) and (b) electromagnetic wave ($V = 3 \times 10^8$ m/s).

17-4 An ultrasound wave impinges a boundary between media at an angle of 28°. Find the angle of reflection.

17-5 Calculate the angle of refraction if the incident angle is 32° and the relative index of refraction is 1.33.

17-6 Calculate the index of refraction for tissue if the ultrasound wave travels at 1500 m/s. (*Hint:* V_{air} = 335 m/s.)

17-7 Find the relative index of refraction at the interface between two mediums if sound travels at 1500 m/s in one and 970 m/s in the other.

17-8 An ultrasound wave impinges the interface in Problem 17-7 at 29°. Calculate the angle of refraction.

17-9 Find the acoustical impedance if sound travels at 1250 m/s in a medium with a density of 1.6 g/cm^3.

17-10 The density of material A is 1.2 g/cm^3, while the density of material B is 0.88 g.cm^3. The velocity of sound in material A is 1500 m/s, and in B it is 1000 m/s. Calculate the coefficient of reflection.

17-11 A transit time blood flow meter uses crystal faces 1.8 cm apart, forming an angle with the flow axis of 32°. Find the mean flow velocity of the blood if the velocity of sound is 1100 m/s and the transit time is 9×10^{-9} s.

17-12 A Doppler flow *meter* uses two crystals, one at 40° to the flow axis and the other at 30° from the flow axis. Calculate the Doppler shift in hertzes of a 5-mHz signal if the flow velocity of blood is 6 cm/s and the speed of sound is 1000 m/s.

17-18 References

1. Halliday, David and Robert Resnick, *Physics Parts I and II*, Wiley (New York, 1966).
2. Nave, Carl R. and Brenda C. Nave, *Physics for the Health Care Sciences*, W. B. Saunders Co. (Philadelphia, 1975).

3. Veluchamy, V. ''Medical Ultrasound and its Biological Effects,'' *Clinical Engineering, 3* (2): 162–166 (April-June 1978).
4. Schwartz, Morton D. and Dominic De Cristofaro, ''Review and Evaluation of Range-Gated, Pulsed, Echo-Doppler,'' *J. Clinical Engineering, 3* (2): 153–161 (April–June 1978).
5. Hewlett-Packard applications note AN708 ''7215A Echoencephalograph Applications Notes'' (Waltham, Mass., 1969).
6. Jacobson, Bertil and John G. Webster, *Medicine and Clinical Engineering*, Prentice-Hall (Englewood Cliffs, N.J., 1977).
7. Strong, Peter, *Biophysical Measurements*. Measurement Concepts Series, Tektronix, Inc. (Beaverton, Ore., 1970).
8. Cobbold, Richard S. C. *Transducers for Biomedical Measurements*, Wiley (New York, 1978).

Chapter 18

Electrosurgery Generators

18-1 Objectives

1. Be able to describe the principles behind the electrosurgery machine.
2. Be able to draw the principal electrosurgery machine waveforms and the circuits used to generate them.
3. Be able to describe the correct procedure for the safe handling of electrosurgery machines.
4. Be able to state how to test electrosurgery machines.

18-2 Self-Evaluation Questions

These questions test your prior knowledge of the material in this chapter. Look for the answers as you read the text. After you have finished studying the chapter, try answering these questions and those at the end of the chapter (Section 18-9).

1. What is the range of frequencies used in electrosurgery?

2. The *patient plate* is considered the ______________ electrode.

3. A ______________ rf ammeter can be used to make rms power measurements of electrosurgery machine output.

4. A dummy load is a resistor used to test electrosurgery machines. This resistor must be a ______________ type so that the impedance is purely resistive.

18-3 Electrosurgery Machines

An electrosurgery machine is an alternating current source that operates at radio frequencies (rf). Typical electrosurgery devices operate in the range 300 to 3000 kHz. The surgeon uses the electrosurgery machine to cut tissue and cauterize bleeding vessels.

Figure 18-1 shows the basic principle behind the electrosurgery machine. There are two electrodes connected to the rf power generator. One electrode is said to be *active* and has a cross-sectional area that is very small (i.e., a few square millimeters) with respect to the other electrode. The active electrode is usually fashioned into the form of a tool or probe and is manipulated by the surgeon.

The *passive electrode* has a much larger area than the active electrode, on the order of 100 cm^2 or larger. In the past, the passive electrode was a metal surface called a *patient plate*. It was positioned beneath the buttocks or thigh. More recently, however, many hospitals have switched to a disposable electrode pad that attaches to the patient's thigh by adhesive.

Regardless of the type of passive electrode, the operating principle (Figure 18-1)

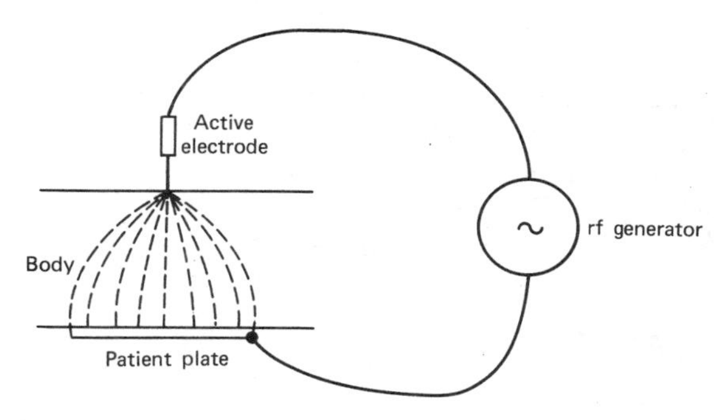

Figure 18-1 Basic principle behind the electrosurgical machine.

remains the same. The current flowing into the patient plate is the same as the current flowing into the active electrode. But since the active electrode has a far smaller cross-sectional area than the passive electrode, the *current density* in amperes per square meter (A/m^2) is far *greater*. As a result of the difference in current density between the two electrodes, the tissue underneath the passive electrode heats up slightly, while the tissue underneath the active electrode is heated to destruction.

The heating of tissue is due to the power dissipated in the tissue, which is found from the expression:

$$P = \rho V I_d^2 \qquad (18\text{-}1)$$

where

P is the power in *watts* (W)
ρ is the resistivity of the tissue in *ohm-meters* (Ω-m).
V is the volume in *cubic meters* (m^3)
I_d is the *current density* in *amperes per square meter* (A/m^2)

EXAMPLE 18-1

Calculate the power dissipated in 0.2 m^3 of tissue that has a resistivity of 1.6×10^3 Ω-m if the current density is 0.36 A/m^2.

SOLUTION

$$P = \rho V I_d^2 \qquad (18\text{-}1)$$

$$P = (1.6 \times 10^3\ \Omega\text{-m})(0.2\ \text{m}^3)\frac{0.36\ \text{A}^2}{\text{m}^2}$$

$$P = (1.6 \times 10^3\ \Omega\text{-m})(0.2\text{m}^3)(0.13\ \text{A}^2/\text{m}^4)$$

$$P = 41.6\ \Omega\text{A}^2 = \mathbf{41.6\ W}$$

Sometimes the surgeon may elect to use a *bipolar electrode* that does not require a passive electrode. This designation is actually something of a misnomer because all rf generators, indeed all electrical current sources, require two poles. In the so-called *unipolar* systems the two electrodes are the active and passive electrodes described earlier. In the bipolar system both output terminals of the rf generator are connected to the hand piece used by the surgeon. The rf current flows between the two electrodes in the handpiece. The current density at each electrode, therefore, is quite high.

18-4 Electrosurgery Circuits

The first electrosurgery machines were developed in an era when the electrical spark gap was the only means for generating significant amounts of rf power. Ships, radiotelegraph stations, and amateur radio operators used spark gap transmitters for radio communications. The first electrosurgery machines, therefore, used spark gap technology. Interestingly enough, although modern solid-state machines have largely supplanted the spark gap and vacuum tube models, it is still possible as of this writing to buy a *new* spark gap machine. There are many spark gap machines still in service, and these are likely to remain in service for several more years.

Figure 18-2 shows the circuit to a basic spark gap machine. It consists of a high-voltage 60-HZ ac transformer (T_1), the spark

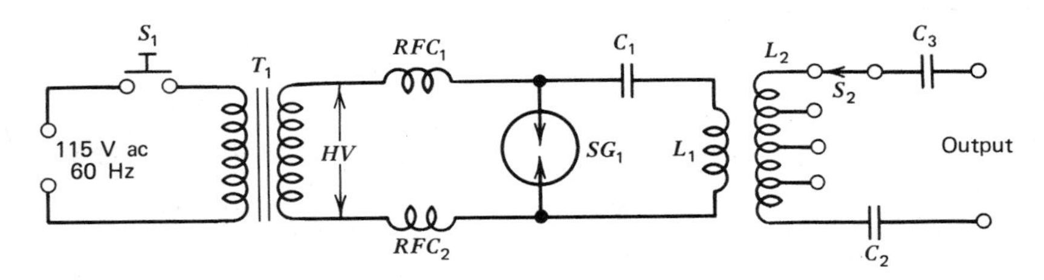

Figure 18-2 Spark gap rf generator.

gap (SG1), and a series resonant tank (C_1/L_1).

Transformer T_1 increases the line voltage from approximately 115 V ac to a potential that is capable of ionizing the air in the space between the tungsten points of the spark gap (i.e., about 0.006 in. of air). Most generators use a potential of about 2000 to 3000 V ac for this job.

When the gap begins to arc, it does so in an oscillatory manner that is rich in rf components. These currents set up an oscillation in the tank circuit (C_1/L_1), which is coupled to the output circuit by induction (L_1/L_2).

The power output level applied to the patient is selected by switch S_2, which connects the active electrode to different taps on inductor L_2. Depending on design and applied voltage, spark gaps produce output power levels between 25 and several hundred watts.

The frequency spectrum produced by the spark gap generator is centered about the resonant frequency of L_1/C_1 but contains significant components at frequencies far removed from the center frequency. A 500-kHz spark gap signal is audible on radio receivers tuned to frequencies well into the VHF region.

The radio frequency chokes (RFC_1 and RFC_2) are used to prevent rf energy from getting into the power supply, thereby being radiated through the power lines. Some models delete the rf chokes but use instead a capacitor in parallel with the secondary of T_1.

The principal use of the spark gap generator is in *coagulation* or *cauterization* of bleeding vessels. Most machines offer two modes, *cut* and *coagulate,* selectable by a footswitch. The *cut* waveform is a pure, continuous sinewave (Figure 18-3*a*), while the coagulation mode will produce a spark gap waveform, damped oscillation (Figure 18-3*b*), or chopped sinewave (Figure 18-3*c*), depending on which technology is used to generate the signal.

The waveform in Figure 18-3*c* is a *cut* waveform produced on a machine that does not use a smoothing filter in the high-voltage dc power supply circuit, a common practice. The use of pulsating dc produces the amplitude modulated waveform shown in the illustration.

A damped coagulation waveform such as Figure 18-3*b* is generated in a circuit such as Figure 18-4. Vacuum tube V_1 is a *power thyratron* that acts like a voltage-level triggered switch. A thyratron is a gas-filled triode. The plate-cathode path remains a high impedance until the voltage between the grid and cathode exceeds a certain threshold point. At that point, the gas inside of the tube becomes ionized, creating a low impedance between plate and cathode. A thyratron, then, is analogous to a unijunction transistor. It behaves as a switch that turns on in an all-or-nothing manner when the grid-cathode potential exceeds a given threshold.

The circuit of Figure 18-4 uses the thyratron as a switch to *ring* an rf tank circuit by

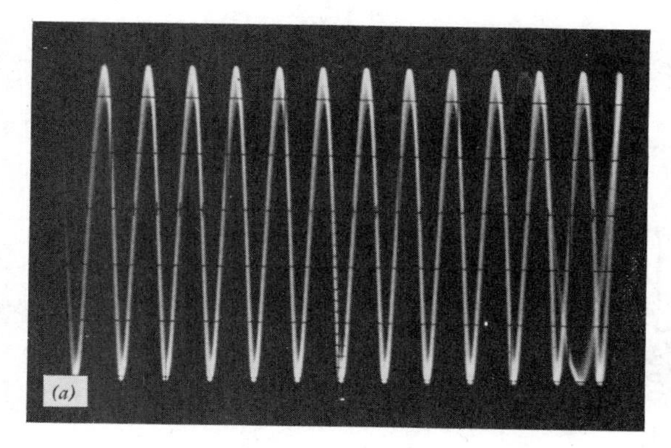

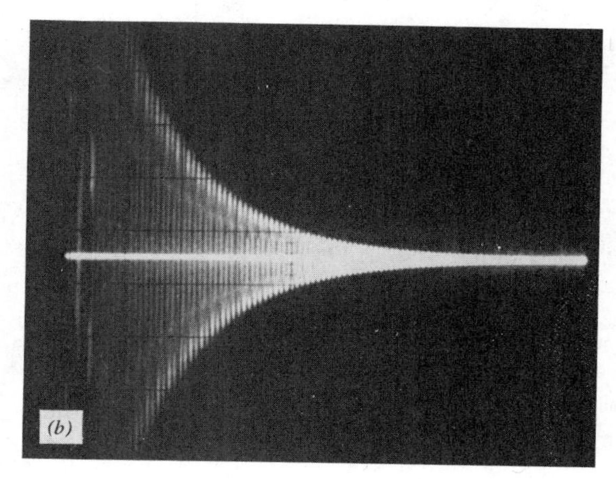

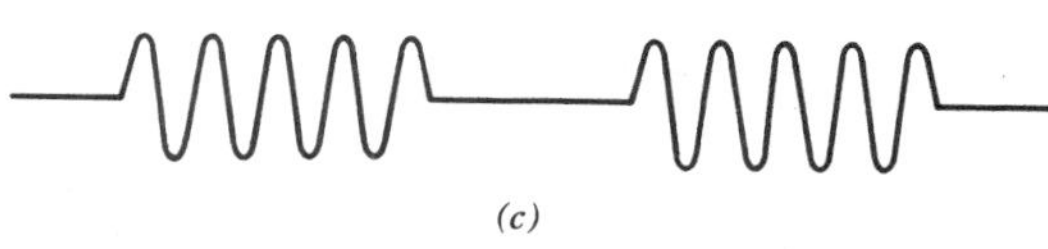

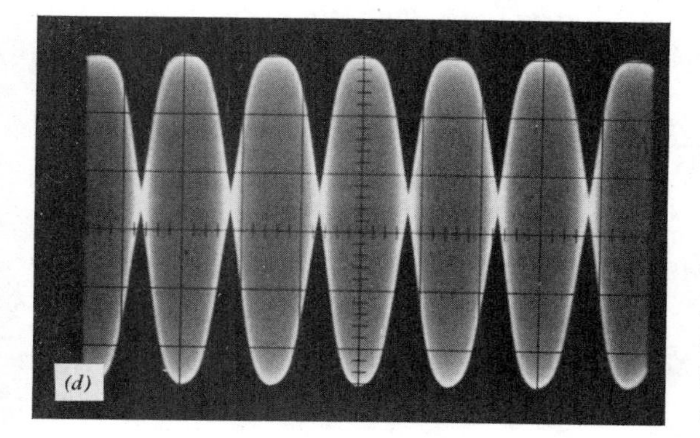

Figure 18-3 Electrosurgery waveforms. (*a*) Sinewave cut waveform. (*b*) Damped coagulation waveform. (*c*) Chopped cut waveform. (*d*) Modulated (60 Hz) cut waveform.

applying a step function potential, operating as follows:

1. When power is initially applied, thyratron V_1 is turned off. Capacitor C_3 charges through rf choke L_3 and inductor L_1.

2. A pulse generator circuit applies bursts of pulses to the grid of V_1 at a 25-kHz rate.

3. When the pulse amplitude applied to the grid exceeds the threshold potential for V_1, then the thyratron ionizes and rapidly discharges C_3, applying the step function that rings L_1/C_3.

4. The energy from L_1/C_3 is coupled to the output. Diode D_1 serves to damp the waveform, producing the shape shown in Figure 18-3*b*.

5. The thyratron then deionizes, allowing C_3 to charge.

6. On the next pulse burst from the pulse generator, the cycle occurs all over again.

The *cut* oscillator is usually connected to the output terminal of the electrosurgery machine in parallel with the coagulation oscillator. To prevent interaction between the two oscillators, circuits such as Figure 18-4 use a parallel resonant *wavetrap* (L_2/C_4). This tank circuit presents a high impedance to the cut oscillator sinewave, reducing its effect on the coagulation circuit. A similar wavetrap would also be used in the cut oscillator circuit to block the coagulation oscillator signal.

18-4.1 Vacuum Tube Sinewave Circuits

Radio transmitters typically use two or more stages to generate high rf power levels: a low-power (lightly loaded) oscillator that drives a power amplifier, or chain of power amplifiers. But most vacuum tube electrosurgery machines use a power oscillator such as Figure 18-5. The triodes tend to be universally UXCV-11s in older models, although a newer design uses the type 7986.

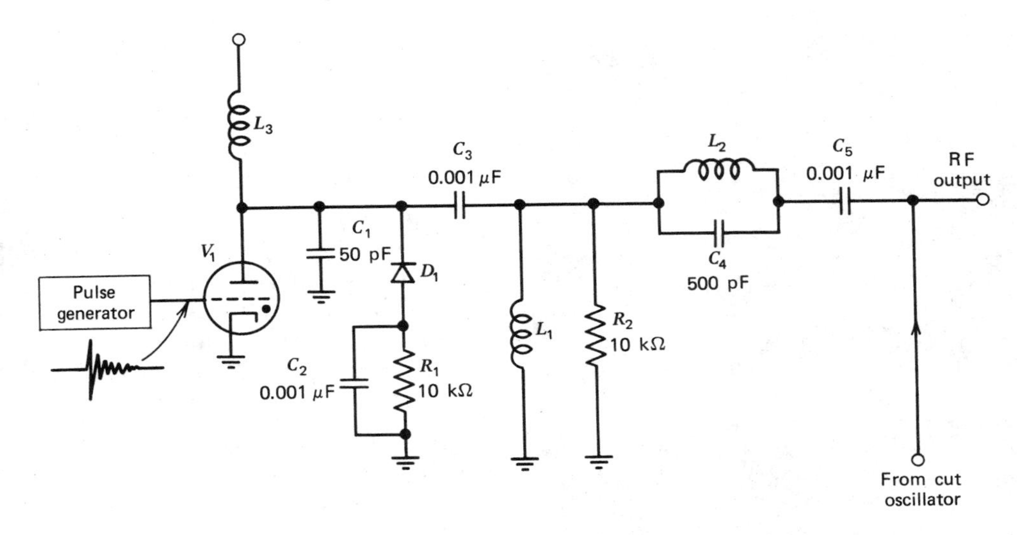

Figure 18-4 Thyratron coagulation current generator.

The operation of this circuit is straightforward. When power is applied to the circuit, tank circuit $C_2/C_3/L_2$ begins ringing at its resonant frequency. Rf energy from the tank circuit is coupled to the V_1/V_2 grids by the tube's own interelectrode capacitance. The signal is kept above ground, thereby forming a grid signal, by L_1. Dc bias is obtained from C_1/R_1 by the grid-leak method.

18-4.2 Solid-State Circuits

Solid-state technology progressed to the point several years ago where transistorized

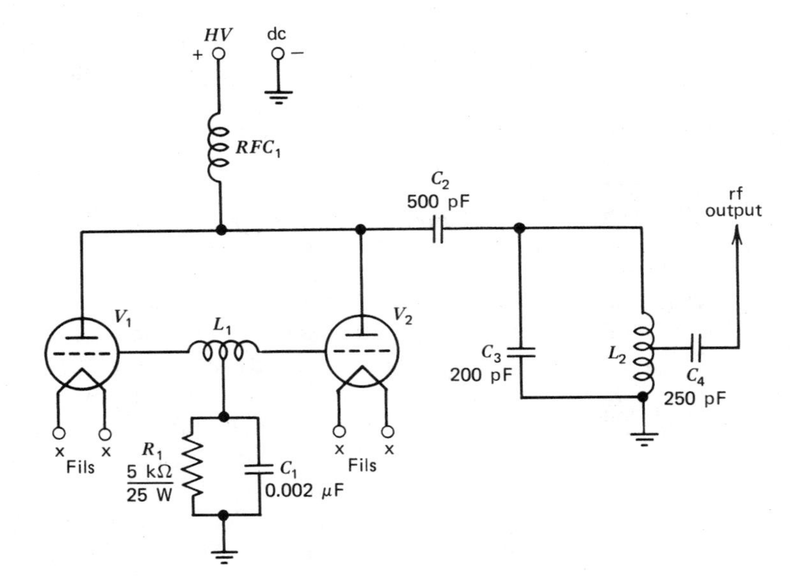

Figure 18-5 Dual triode sine-wave oscillator for cut current.

electrosurgery machines could be produced at a reasonable cost. Today, most newly purchased instruments are solid-state models. For comparison of relative sizes, examine Figures 18-6*a* and 18-6*b*. The instrument in Figure 18-6*a* uses a spark gap to generate the coagulation signal and a pair of vacuum tubes to generate the cut signal. The solid-state machine in Figure 18-6*b* is only a few inches tall and rests easily on a small "typing table."

The circuit to a solid-state rf power amplifier is shown in Figure 18-7. This basic type of design is similar to the power amplifier stages in a number of different models. The circuit is push-pull/parallel. There are two banks of three transistors each (i.e., Q_1–Q_3 and Q_4–Q_6). The transistors within each bank are parallel connected, while the two banks are connected in push-pull with each other. The transformers used for input and output coupling are usually wound on a toroid-shaped ferrite core. This technique is preferred over other types of construction because the toroid shape limits the magnetic field of the coil to the immediate vicinity of the winding, thereby limiting the types of problem that are caused by mutual induction and other forms of coupling.

The rf signal is created in an oscillator circuit such as Figure 18-8. When transistor Q_1 is forward biased, the stage will oscillate at an rf frequency determined by the circuit constants.

Transistor Q_2 operates as a switch to control Q_1. When Q_2 is turned on, then the bias network for Q_1 (i.e., R_3) is grounded, causing Q_1 to be forward biased into oscillation. In the *cut* mode Q_2 remains forward biased, so the output of the oscillator is a sinewave. But in the *coagulation* mode, a squarewave is applied to the base of Q_2, causing it to switch into and out of conduction. This causes a *chopped sinewave* output from the oscillator (see inset at Figure 18-8).

Other manufacturers use different techniques, but most are variations on the chopped sinewave technique shown in Figure 18-8.

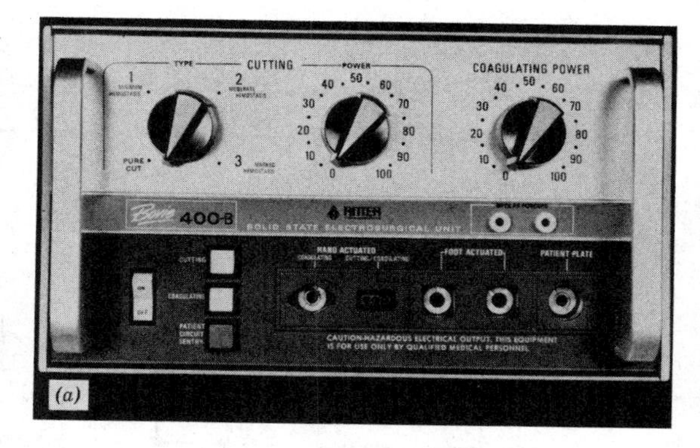

Figure 18-6 Electrosurgery machines. (*a*) Solid-state machine. (*b*) Older vacuum tube model. (Photos courtesy The Ritter Co.)

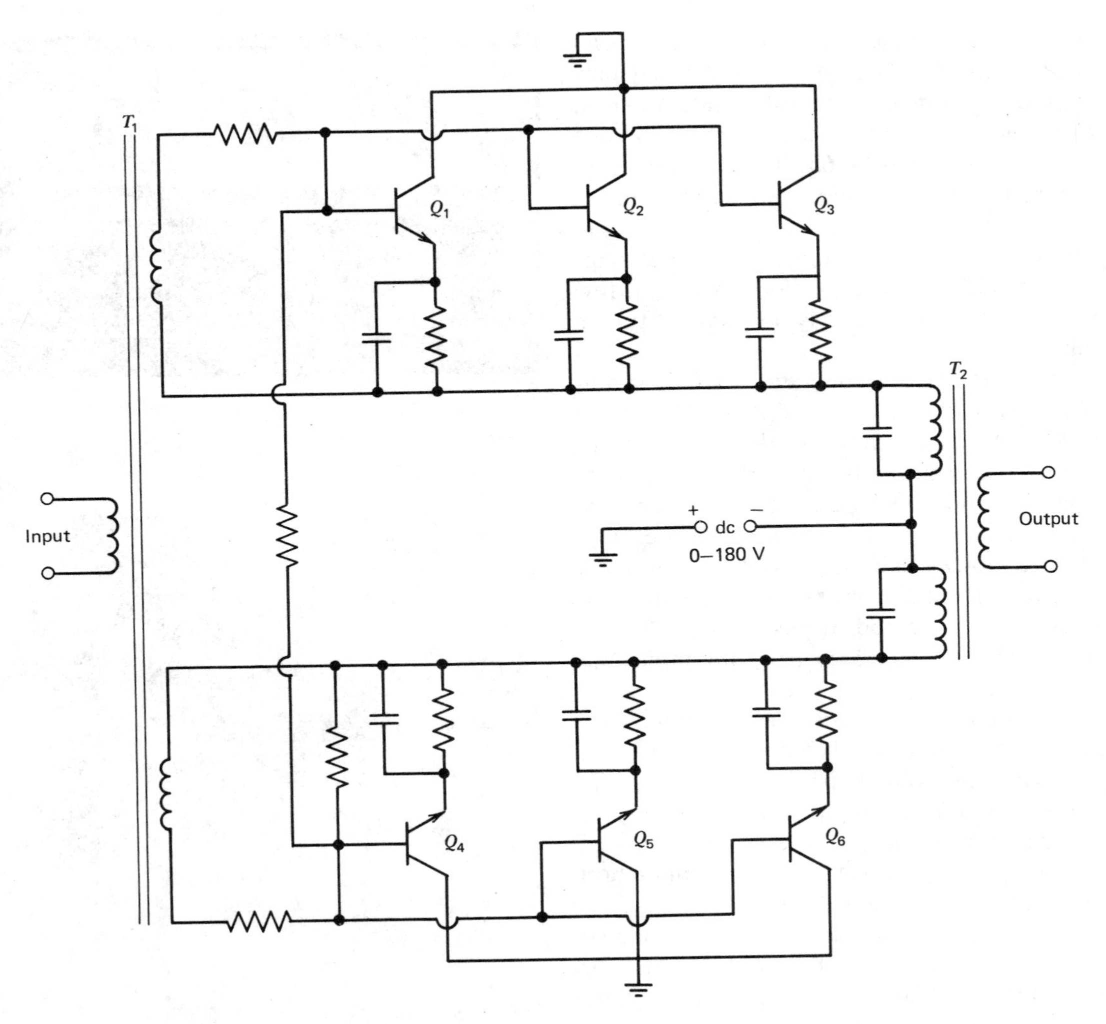

Figure 18-7 RF power amplifier—solid state.

18-5 Electrosurgery Safety

The electrosurgery machine is a high-powered rf generator and, as such, can pose a threat to users and patients alike if misused or abused. When used in the manner intended by the manufacturers, however, the machine is considered safe, but misuse is potentially dangerous.

The most common injury to patients involves burns at inappropriate points on the body. Perhaps the most likely problem in this respect is a damaged or misapplied patient plate. Recall from Section 18-3 that burning at the patient plate site is eliminated by keeping the current density low. Local hotspots of high-current density can be accidentally formed, however, by placing the plate over

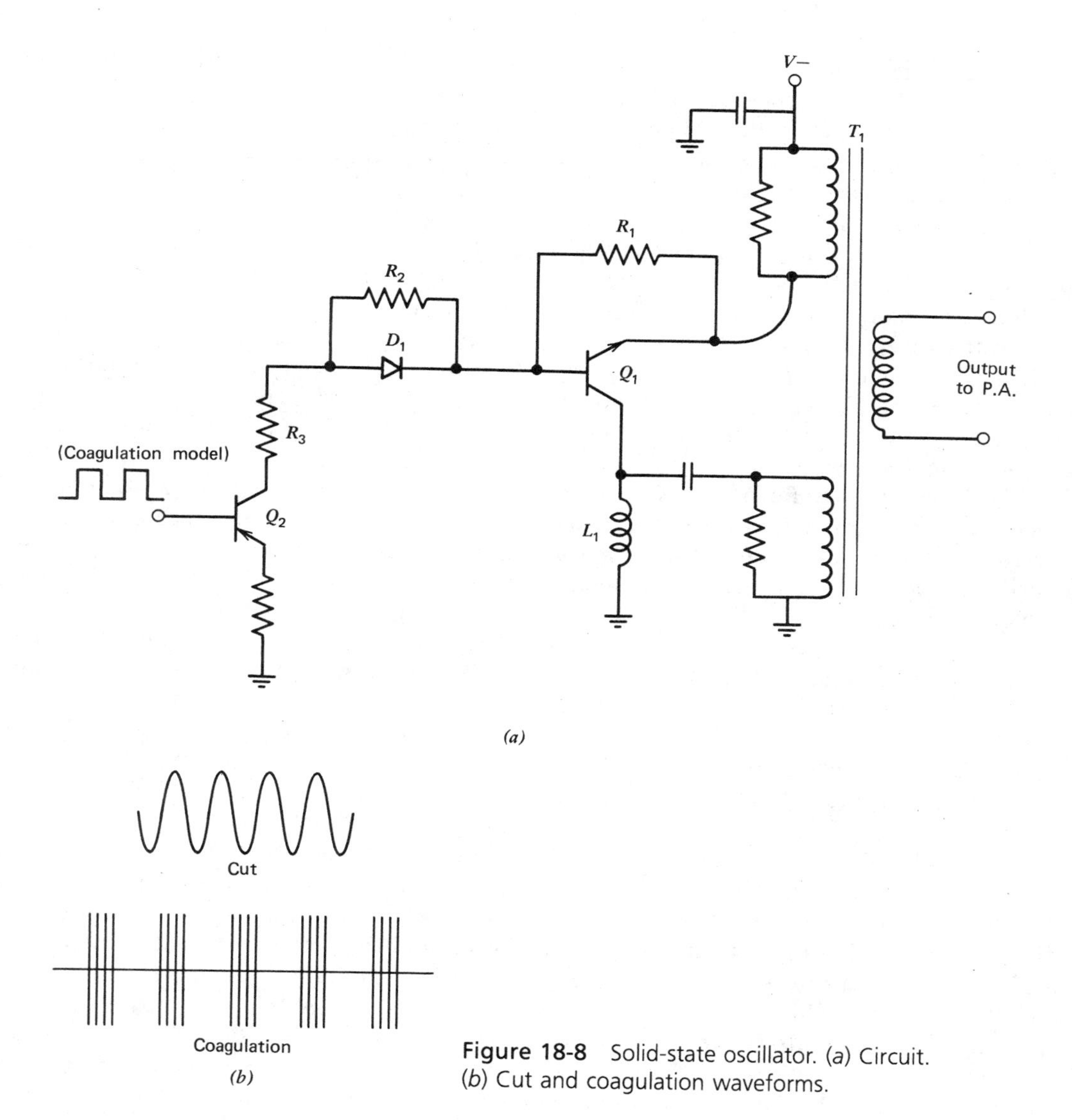

Figure 18-8 Solid-state oscillator. (*a*) Circuit. (*b*) Cut and coagulation waveforms.

a bony prominence instead of the fleshy portions of the buttocks or thigh. Hard bony regions place a more firm pressure point on the plate, thereby decreasing the contact impedance, and increasing the current density.

The problem of patient burns can also be caused if there are any dents, creases, or bends in the surface of the patient plate; the plate must be perfectly flat. Anomalies in the surface can create the local hotspots of increased current density that cause burns. In most cases, damaged patient plates should be discarded instead of repaired, so that the efficacy of the repair cannot be challenged in a malpractice suit.

An electroconductive gel or paste is used

to decrease the likelihood of burns to the patient. This gel is smeared over the surface of the plate that contacts the patient's skin and is similar to the gels and pastes used in ECG and defibrillator electrodes.

A third mechanism for patient burns is the inadvertent ground path, fortunately more common on older machines than on present machines. In older designs we find that one side of the output circuit is grounded. The patient plate on those models was usually connected to the chassis, which was in turn connected to the ac power mains ground. If a point on the patient's body became grounded with a significantly lower resistance, then a major current would be diverted to the spurious ground. If the current density through the point that is grounded is high enough, then burns will occur.

Danger to the surgeon using an electrosurgery machine is minimal. But if there are even microscopic holes in the surgeon's gloves, then there is a possible shock hazard. Surgeons from time to time request that a machine be inspected after they have experienced a "tingling" sensation.

It is possible for severe damage and electrical shock to occur if any liquids are spilled into the machine. Operating rooms are usually very crowded, so it is tempting for the staff to use the flat top of an electrosurgery machine as a table. This practice often leads to spillage of liquids into the machine, often with spectacular results, and so should be strongly discouraged. Many of the solutions used (i.e., blood, saline, betadine, etc.) in operating rooms are quite conductive.

When testing the electrosurgery machine in the shop or lab, be sure to observe practices that prevent your injury. In Section 18-6 we discuss some testing procedures and the criteria for test equipment that reduce the possibility of electrical shock or burns. In general, however, common sense will dictate that you use only secure fixtures that are well insulated and are isolated from ground.

18-6 Testing Electrosurgery Units

A traditional, but no longer seriously considered, method for testing electrosurgery machine output was to buy a piece of beef and then cut it with the machine. This technique is now used mostly by salespeople who wish to dramatize the effect of their particular product. But engineers and technicians generally prefer a more objective and quantitative testing procedure.

Almost all manufacturers of medium- to high-power electrosurgery machines recommend the test unit of Figure 18-9*a* for checking the rf output. This circuit uses a *dummy load* resistor (R_1) to simulate the patient and an rf ammeter to measure the current applied to R_1 by the machine. The resistor should have a resistance between 200 and 500 Ω, as specified by the manufacturer of the electrosurgery machine. A "universal" test box should use a 500-Ω resistor, although one will not be able to use the same output current figures as supplied by the manufacturer. In that case, establish normal values when the machine is known to be in good working order, and then attempt to use these to test the unit in the future. The meter should have a 0 to 1.5- or 2-A full-scale range for high-power machines, and 0 to 500 mA for low power levels. In all cases, however, the meter must be a *thermocouple rf ammeter.*

When constructing the testing in Figure 18-9*a*, or any of the other circuits, care must be taken to prevent rf coupling to the metal shielded enclosure. It is the usual practice to mount the components inside of a box or prefabricated enclosure. If the enclosure is metallic, then the meter should be mounted on a piece of plastic, which is in turn mounted on the front panel. Allow at least a half inch all around the meter, or there will

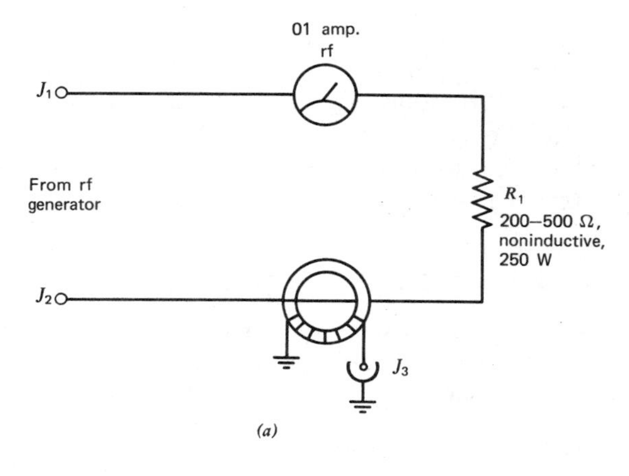

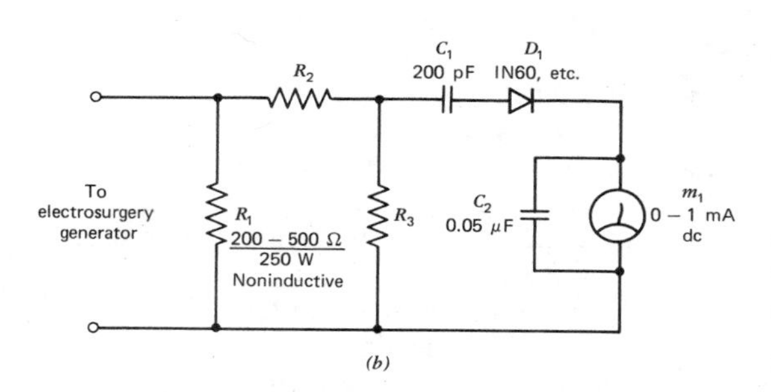

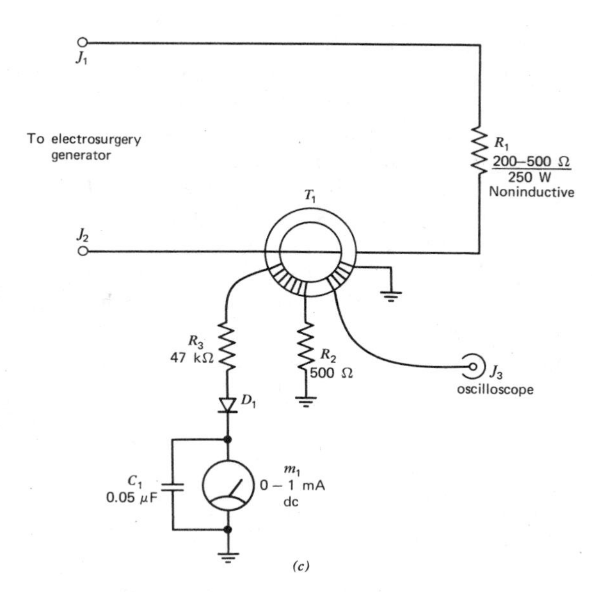

be insufficient isolation from the chassis. Similarly, we often find corona arcing and other problems if the resistor, input jacks, and wiring are not also isolated from the chassis.

The thermocouple rf ammeter is inherently an rms-reading device, so it provides a true picture of the actual output level. Peak-reading devices are also sometimes used, but these give a true picture of output level *only* when sinewave cut mode is used, or when the levels have been previously calibrated against an rms-reading device.

On the disadvantage side, however, we find two major problems with the thermocouple rf ammeter. They are, for one thing, more expensive than other types of meter movement. The other problem is the matter of linearity. The thermocouple rf ammeter does not have a linear scale; it is more crowded on the low end of the range than at the top. As a result, it is often very difficult to read low power levels on the 1.5- or 2-A movement, necessitating two meters with different ranges.

Transformer T_1 in Figure 18-9*a* is used to sample the rf waveform for display on an oscilloscope. It is dangerous to try connecting the oscilloscope directly to the load, unless the correct probes are used, so a toroid current transformer is used. Viewing the waveform did not matter on older machines, but many of the solid-state models require adjustment of timing circuits using a CRO display of the output waveform.

A slightly different test box is shown in Figure 18-9*b*. This circuit uses resistor voltage divider R_2/R_3 and a diode rectifier (D_1) to develop a dc level to drive a 0- to 1-mA

Figure 18-9 Simple electrosurgery testers. (*a*) Toroid transformer oscilloscope output. (*b*) Voltage divider. (*c*) Toroid current transformer.

dc meter movement. The circuit here is essentially a peak-reading device and so can be calibrated in *watts* only by comparison to an rms device such as the thermocouple rf ammeters.

Another peak-reading circuit that has found popularity is shown in Figure 18-9*c*. This circuit employs a toroidal current transformer, but with two windings. One winding is connected to J_3 for display of the waveform on an oscilloscope, while the other drives a dc meter movement through a rectifier diode. The circuit in Figure 18-9*c* is particularly useful when testing electrosurgery machines of low power rating (i.e., microcoagulators, ophthalmic, and laparoscopy machines).

There are several commercial electrosurgery testers available. Specific brand names often appear as "recommended" models in the service manuals of the electrosurgery machines.

18-7 Summary

1. An electrosurgery machine is an rf power generator used to cut and cauterize tissue during surgical procedures.
2. Two output electrodes are used, a small cross-sectional area *active electrode* and a larger *passive electrode*.
3. The electrosurgery machine cuts at the active electrode, but not at the passive electrode, due to the difference in the cross-sectional areas.
4. Burns can occur to the patient if any pressure point against the passive electrode, or patient plate, has a substantially lower contact impedance than surrounding points.

18-8 Recapitulation

Now return to the objectives and self-evaluation questions at the beginning of the chapter and see how well you can answer them. If you cannot answer certain questions, place a check mark next to each and review appropriate parts of the text. Next, try to answer the following questions using the same procedure. When you have answered all of the questions, solve the problems in Section 18-10.

18-9 Questions

18-1 An electrosurgery machine is an ac current source operating at rf frequencies between ________ and ________ kHz.

18-2 The passive electrode is also called the ________ ________ and must have an area not less than ________ cm^2.

18-3 Describe in your own words how an electrosurgery machine cuts tissue.

18-4 Early electrosurgery machines, some of which are still in use, used a ________ ________ to create the rf oscillations.

18-5 Describe the principal difference between the cut and coagulate waveforms.

18-6 Many solid-state electrosurgery units produce a ________ sinewave in the coagulate mode.

18-7 List three mechanisms by which accidental burning of a patient can occur. Also discuss the methods for preventing these burns in each case.

18-8 What instrument should be used to measure the rms power produced by an electrosurgery machine in the coagulate mode?

18-9 List a principal design feature of an electrosurgery machine tester that enhances *your* safety.

18-10 What range ammeters are suitable to test (a) low-power, (b) high-power electrosurgery machines?

18-10 Problems

18-1 Find the resistivity in ohm-meters of an object whose volume is 0.6 m^3 if an electrosurgery unit delivers 56 W and a line current of 600 mA.

18-2 Find the output power in watts of an electrosurgery machine that produces 1.2 A of rf current in a 300-Ω noninductive dummy load.

18-3 Find the rms power of an electrosurgery machine that delivers 500 mA into a 200-Ω load.

18-4 Calculate the peak rf voltage produced by the electrosurgery machine if 450 W is dissipated in a 500-Ω load while using the cut mode.

Chapter 19

Electrical Safety in the Medical Environment

19-1 Objectives

1. Be able to define electrical safety as applied to medical institutions.
2. Be able to describe the scope (electrical equipment and specific environmental areas) of electrical safety in medical institutions.
3. Be able to list major publications and organizations concerned with electrical safety.
4. Be able to identify degree of involvement and nature of responsibilities of various hospital personnel for electrical safety.
5. Be able to describe how preventive maintenance (equipment inspection and documentation) can reduce electrical hazards.
6. Be able to list legal and insurance requirements relating to electrical safety in medical institutions.
7. Be able to list the major areas of concern in setting up a hospital electrical safety program.
8. Know how to describe the physiological effects of electricity on the human (i.e., the theory of *macroshock* and *microshock*).
9. Be able to list macroshock (arm to arm) and microshock (through the heart) 60-Hz shock current levels that produce adverse effects.
10. Be able to define leakage current.
11. Be able to describe the subtle hazards and cautions of microshock in hospitals.
12. Know how to describe monitoring instrument design considerations for reducing electrical shock hazards.
13. Be able to describe a power isolation transformer system/line isolation monitor and their use in reducing electrical shock hazards.
14. Be able to describe an equipotential grounding system and its use in reducing electrical shock hazards.
15. Be able to describe a ground fault interrupter (GFI).
16. Know how to draw a diagram for a proper power wiring, distribution, and ground system for providing a safe patient environment.
17. Be able to list specialized electrical safety test equipment and its use in electrical safety testing programs.

19-2 Self-Evaluation Questions

These questions test your knowledge of the material in this chapter. Look for the answers as you read the text. After you have finished studying the chapter, try answering these questions and those at the end of the chapter (Section 19-21).

1. What is the definition of *electrical safety* as applied to medical institutions?
2. What specific medical equipment causes electrical hazards?
3. List the major publications and organizations concerned with electrical safety.
4. Who is responsible for electrical safety in medical institutions?
5. State specific legal and insurance problems that can arise from electrically hazardous environments.
6. Define the terms *macroshock* and *microshock*.
7. Explain why leakage current is potentially dangerous.
8. What is a line isolation transformer and monitor?
9. Explain what is meant by an equipotential ground system.
10. How does a ground fault interrupter (GFI) operate?
11. How is a proper power wiring, distribution, and ground system used to provide a safe patient environment?
12. What special test equipment is used in hospital electrical safety testing programs?

19-3 The Definition of Electrical Safety

Electrical *safety* is the *containment* or limitation of hazardous electrical shock, explosion, fire, or damage to equipment and buildings.

Electrical shock refers to both *macroshock* (high-value arm-to-arm current ultimately passing through the heart) and *microshock* (low-value current passing directly through the heart). A difference of potential must be present (two points of contact in either case). Shock may occur to patients, employees, and visitors to a hospital or health care facility. Shock results from improperly wired or maintained electrical equipment or power systems.

Explosion may result from electrical contact sparks that ignite a variety of explosive gases, such as ether or cyclopropane anesthetic.

Fire may result from heat produced by overloaded, incorrectly wired, or improperly maintained equipment or power systems.

Damage to equipment and buildings may result from explosion, fire, or electrical overload.

Safety may be defined as the condition of being safe from hurt, injury, or loss. Actually, safety is often referred to as a situation that is harmless. However, in reality, no situation can be rendered *completely* safe. As such, electrical *safety* in the *medical* environment refers only to the *limitation* of hazardous situations.

In the practical daily routine of hospital life, it is important to remember that electrical safety is not so much a static state as it is a dynamic, continuous course of action involving hazard *detection* and *correction*.

19-4 The Scope of Electrical Safety in Medical Institutions

The *scope* of the electrical safety problem in medical institutions has actually never been clarified. A multitude of articles appearing in health care-oriented publications in the early 1970s represented an attempt to define the vague hazard of electrical microshock. *Microshock* is defined as a low-value current (microamperes) passing *directly* through the heart of a *catheterized* patient, causing *ventricular fibrillation* and possible death.

Let us define each of these terms. Microshock current usually results from *leakage current*. Leakage current passes from an equipment metal chassis to earth ground.

A *catheter* is a plastic tube inserted into an artery or vein to measure blood pressure/

flow, to inject substances into the bloodstream, or to electrically pace the heart. *Ventricular fibrillation* (V-FIB) is a condition of the heart in which the myocardium *vibrates* and *quivers* instead of pumping steadily and rhythmically. Unless corrected naturally by the patient or artificially by a defibrillator unit, V-FIB results in *death* due to inadequate blood circulation.

This potentially lethal microshock hazard causes physicians, nurses, biomedical engineers, technicians, administrators, lawyers, patients, and others to be closely concerned with hospital electrical safety.

Dr. Carl Walter of Peter Bent Brigham Hospital, Boston, stated in the early 1970s that 1200 patients were electrocuted in hospitals annually. In 1971, Friedlander referred to the Food and Drug Administration estimation of 1600 injuries and 100 deaths caused annually by electrical equipment. Since the problem of microshock is subtle, no one seems to know the precise extent of the problem. It is probably somewhere between the two above-mentioned estimates. But this problem is real. It is a result of the many new electrically operated biomedical instruments and the introduction of advanced medical techniques involving catheterization.

In essence, the scope of electrical safety in medical institutions predominantly involves potential microshock from any electrically operated device that can come in contact with the patient in any way.

Hazards exist in public areas (hallways, etc.), patient care areas (patient rooms, etc.), and critical care areas (CCU, ICU, etc.).

19-5 Major Organizations Producing Publications Pertinent to Electrical Safety

Electrical safety in medical institutions is now entangled in a vast array of codes and standards. Some of these involve *national law* such as the Occupational Safety and Health Administration (*OSHA*), Health, Education and Welfare (*HEW*), Public Health Service (*PHS*), and Food and Drug Administration Medical Devices Safety Group (*FDA*). Others involve *voluntary consensus standards* (nationally agreed-on specifications) such as the National Fire Protection Association (*NFPA*), National Electrical Code (*NEC*), American National Standards Institute (*ANSI*), Underwriters Laboratories (*UL*), Association for the Advancement of Medical Instrumentation (*AAMI*), and Joint Commission on the Accreditation of Hospitals (*JCAH*). Furthermore, local codes and standards are also influential in defining electrical safety specifications.

Interpreting the many pertinent publications is a matter for the medical institution safety officer or medical safety technician. But the biomedical engineer (BME) and biomedical equipment technician (BMET) must know the accepted specifications to be used in their electrical safety test procedures.

19-6 Responsibilities of Hospital Personnel

The responsibility for electrical safety in medical institutions rests with *every* employee and patient. However, the scope of responsibilities differ. The *patient* should report any suspected electrical hazard to the attending physician or nurse, who in turn should report this to the safety officer/technician, if the condition is questionable. Even though the patient should be alert, he or she will probably be unaware of most of the potential electrical hazards. Therefore, the *bulk* of responsibility rests with the following hospital personnel:

1. *Medical staff*—includes physicians, nurses, and medical technicians who should constantly inspect all electrical equipment connected to the patient and report any suspicious electrical hazards.

2. *Support staff*—includes BMEs, BMETs, safety officer/technicians, plant operations personnel, and medical technicians, who must be able to recognize, test for, and correct all electrical hazards as well as educate other hospital personnel.

3. *Administration staff*—includes administrators, managers, and supervisors, who must hire competent employees and sponsor electrical safety educational programs.

19-7 Preventive Maintenance Programs to Reduce Electrical Hazards

One description of *preventive maintenance* is[1] "the performance of nonfunctional repairs, component replacements, cleaning, and general service in order to prevent improper and/or inadequate operation, thus, extending the period between possible malfunctions and also extending the service life of the instrument by months and years." *Calibration* is "the assessment and correlation of instrument performance against standards traceable to the National Bureau of Standards." *Repairs* are "effected either during the scheduled preventive maintenance interval or on an unscheduled immediate need basis."

Preventive maintenance (PM) differs from *corrective maintenance* (CM) in that PM involves *routine* inspection and testing while CM involves *total* calibration or replacement of defective parts. Electrical safety test measurements made during PM can reduce electrical hazards by uncovering early signs of degradation. Faults can then be corrected during CM. By replacing broken plugs, faulty power receptacles, and poor ground connections, the medical environment can be made safer for the patient.

[1] See reference 18 at the end of the chapter.

PM involves *testing* to specific *standards*. These standards or specifications are stated in the publications listed in Section 19-5. For example, AAMI publishes *Safety Standards for Electromedical Apparatus, Safe Current Limits*. Its objectives are to provide techniques for measuring risk currents and to provide the user-authority with basic guidelines for proper use and care of equipment. This standard sets risk current limits for *all categories of electrical apparatus* used on or in the vicinity of any patient.

Following the functional and electrical safety testing of any biomedical instrument, a *sticker* should be affixed to the equipment. This denotes the date on which the equipment has been tested. Another sticker gives the location of the *manufacturer's operation and service manual*. This manual typically describes general equipment setup procedures, installation, operational conditions and procedures, underlying theory of operation, preventive maintenance and test procedures, and spare parts. For specific details on particular biomedical instrumentation, consult the AAMI manual.

Special electrical safety test equipment is required to perform the specific tests. One such unit is the Neurodyne-Dempsey Model 431F *Safety Analyzer* shown in Figure 19-1*a* and Figure 19-1*b*. This comprehensive electrical safety unit, designed especially for hospital use, measures leakage current (chassis to ground and ECG machine leads to ground), receptacle polarity, power ground resistance, and other pertinent parameters. Another is the Dempsey Model 449 *Microspector* shown in Figure 19-2. This is used to make leakage current measurements. High-resistance values can be made with the Dempsey Model 447 *Cordohmeter* shown in Figure 19-3. This is a megohmmeter used to inspect electrical power cables. These are discussed in further detail in Section 19-18.

All of this testing must ultimately be *documented*, and such data recording schemes

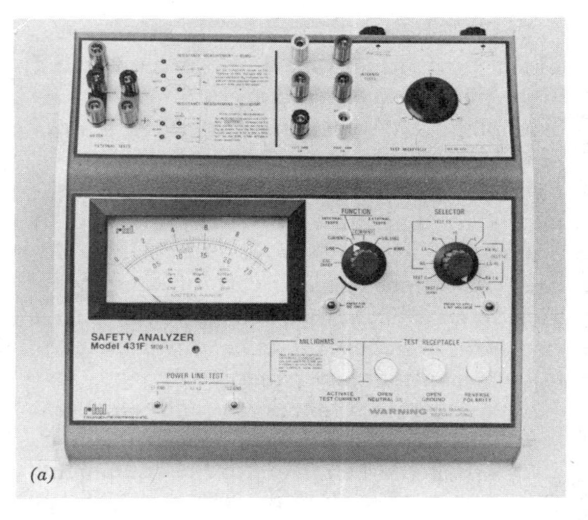

(a)

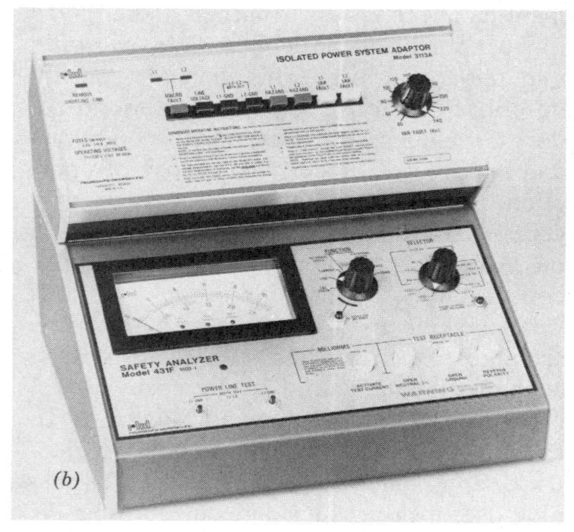

(b)

Figure 19-1 Safety analyzers. (*a*) Neurodyne-Dempsey Model 431F. (*b*) Model 431F with isolated power system analyzer. (Photo courtesy of Neurodyne-Dempsey, Inc.)

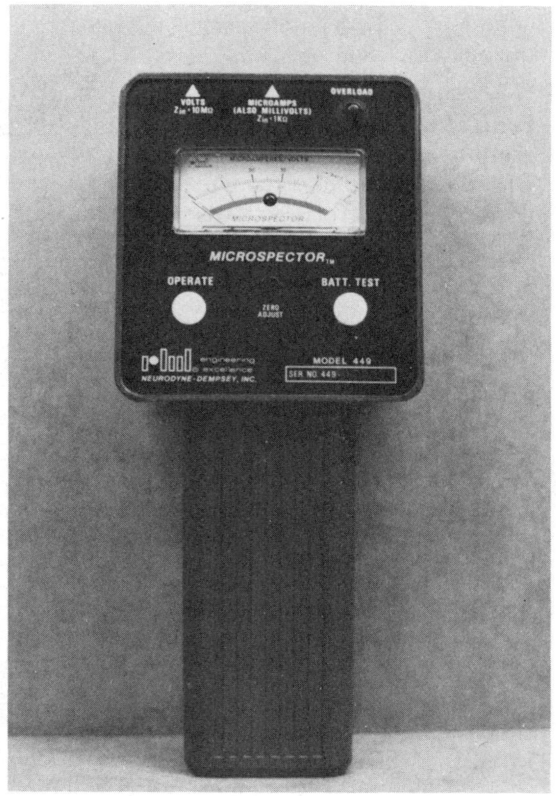

Figure 19-2 Leakage current meter. (Photo courtesy of Neurodyne-Dempsey, Inc.)

are numerous. Although the *JCAH* inspection team reviews individual documentation techniques, they do not recommend nor designate a specific documentation structure. However, they do require a fixed logical documentation system and proof that the hospital is following it. They point to the AAMI standards and the procedures in the operation and service manuals. In addition, they may wish to review data recorded on a standard form, shown in Figure 19-4 (The Emergency Care Research Institute, Plymouth Meeting, Pennsylvania). This record is part of a testing system that insures patient safety.

19-8 Legal and Insurance Requirements of Electrical Safety

Legal requirements of electrical safety programs are varied and confusing. Dr. Martin Lloyd Norton, Associate Professor of Anesthesiology and Adjunct Professor of Law at

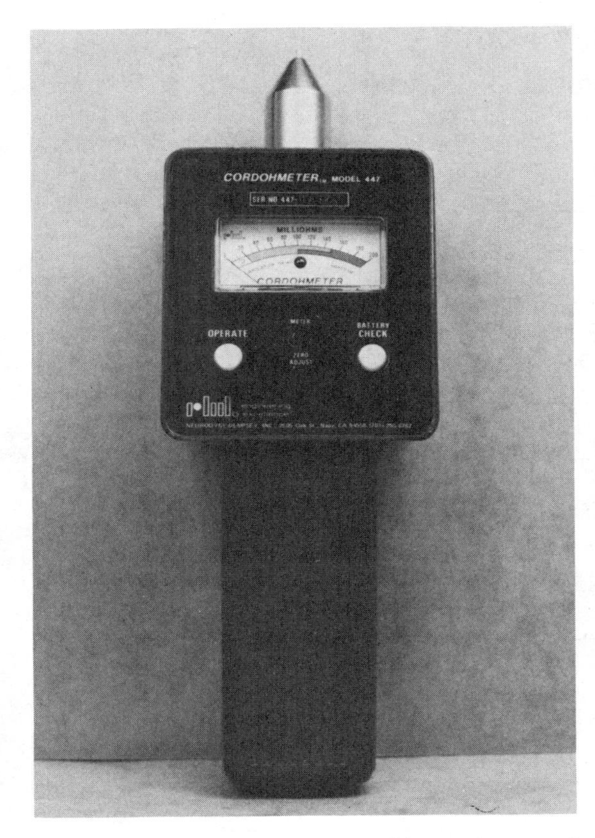

Figure 19-3 Resistance tester. (Photo courtesy of Neurodyne-Dempsey, Inc.)

Wayne State University, gave some clarification regarding negligence. He stated that failure of a hospital to have a PM program and/or use of adequately trained personnel could be considered *negligence* under "reasonable man" standards. Since electrical safety refers to limitation and not elimination of hazards, personnel carrying out PM programs must perform and document tests that are consistent with current acceptable practices (reasonable). In the case of Butler *v.* N. W. Hospital of Minnesota, the court opinion was that "general rule is that equipment furnished by a hospital for a patient's use should be reasonably fit for the uses and purposes intended under the circumstances, and injury suffered because of failure of this duty leads to liability based on negligence." Any person may initiate a lawsuit if he or she believes to have suffered a wrongdoing. Cases of negligence involve *standard of conduct, damages,* and *appropriate relationship.* If hospital personnel are adequately trained, take reasonable care to test equipment, and document the test results, negligence and malpractice lawsuits can be reduced.

Insurance of the hospital is obviously a functional requirement. *Malpractice insurance* is unfortunately costly and often nebulous. The escalated cost of malpractice insurance premiums for anesthesiologists ($20,000 to $50,000 per year) are well documented in current hospital-oriented periodicals. Even the BMET may be sued under modern guidelines. Therefore, blanket policies taken by hospitals cover support as well as medical personnel.

In any event, the authors of this text believe that *the patient comes first*, and that legal liability can be reduced by proper and adequate documentation from a PM program.

19-9 Setting Up an Electrical Safety Program in the Hospital

Every BME and BMET should be familiar with procedures for setting up or running an electrical safety program in the hospital. BMETs may be called on to give electrical safety *instruction* to nurses, physicians, and other medical care personnel. A film presentation should not exceed 20 min. Lecture and group discussions should not exceed 45 min per session. The instruction program is, briefly, as follows.

1. Introduction—purpose and objectives of safety instruction.
2. Basic concepts of electricity.

General-Purpose Inspection Form

AC Line-Powered Equipment

ACTION
Not Needed ☐ Needed ☐ Taken ☐
Control No.
Date of Inspection:
Next Inspection Due:

TYPE OF DEVICE ____________________

LOCATION ____________________ SERIAL NO. ____________________

MODEL ____________________ MANUFACTURER ____________________

	OK	ACTION NEEDED	ACTION TAKEN (Date & Initials)
1. ATTACHMENT PLUG ____________________			
2. LINE CORD & STRAIN RELIEFS ____________________			
3. GROUNDING RESISTANCE: ________ Ohms ____________________			

4. LEAKAGE CURRENT (Circle unacceptable values)

	OFF	OPERATING (List Modes) ________	________	OK	ACTION NEEDED	ACTION TAKEN (Date & Initials)
PROPERLY GROUNDED	________ uA	________ uA	________ uA			
UNGROUNDED, CORRECT POLARITY	________	________	________			
UNGROUNDED, REVERSED POLARITY	________	________	________			

	OK	ACTION NEEDED	ACTION TAKEN (Date & Initials)
5. ____________________			
6. ____________________			
7. ____________________			
8. ____________________			
9. ____________________			
10. ____________________			

COMMENTS & DESCRIPTION OF DEFICIENCIES (Refer to item numbers) ____________________

____

____

____

☐ TYPE A INSPECTION TAG AFFIXED
☐ TYPE B INSPECTION TAG AFFIXED
☐ GROUND WARNING TAG AFFIXED
☐ EQUIPMENT SERVICE REQUEST FORM COMPLETED

INSPECTED BY ____________________

Health Devices Evaluation Service
Form HD-303-1171

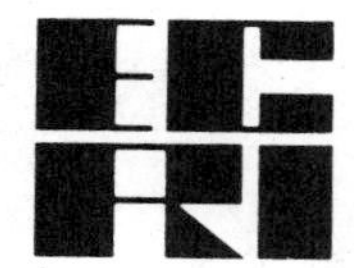

THE EMERGENCY CARE RESEARCH INSTITUTE
5200 BUTLER PIKE, PLYMOUTH MEETING, PENNSYLVANIA 19462

Figure 19-4 Electrical safety documentation form for ac-powered equipment. (Reprinted by permission, Emergency Care Research Institute.)

3. Physiological nature of electrical shock.
4. Macroshock and microshock hazardous situations.
5. Identifying electrically hazardous equipment and situations.
6. Reporting all suspicious conditions.
7. Avoiding electrically hazardous situations.
8. Establishing responsibility.
9. Relating to pertinent publications.
10. Making a friend of the BME/BMET, safety officer/technician.
11. Demonstration.
12. Conclusion, discussion.

BMETs in the hospital usually give such presentations periodically, but they may also be involved in setting up an entire electrical safety program. This is coordinated with the *hospital safety committee* and consists of the following:

1. Purposes and objectives—patient safety at a cost.
2. Scope—hospital size/type, personnel available, safety committee.
3. Pertinent publications—law, codes, standards.
4. Test procedures—in-house, nationally published.
5. Inspections—specific operations, time schedule, preventive maintenance.
6. Documentation—in-house, commercially available.
7. Education—medical staff, users.
8. Assessment of effectiveness—feedback.
9. Cost of operation—dollars versus benefit.

19-10 Physiological Effects of Electricity on the Human (Theory of Macroshock and Microshock)

Electrical current passing through the human body has *three primary effects:* injury to tissues, uncontrollable muscle contraction or unconsciousness, and fibrillation of the heart. Appreciation of these effects rests with knowledge of cellular action potentials. *Muscle and nerve cells* in the body act as tiny batteries or *polarized units*. Polarized, depolarized, or reversed-polarized cellular potentials arise from sodium, potassium, and chlorine ion concentration differences across semipermeable cell membranes. At rest, cells polarized at resting potentials of −70 MV can be stimulated to depolarization by any of the following means: mechanical, chemical, thermal, optical, and electrical.

It is the electrical stimulation of muscle and nerve cells that is of primary interest in electrical safety. For a more complex discussion of traveling waves depolarization in heart muscle see Chapter 2, particularly Figures 2-3 and 2-4. For a more complete discussion of nerve cell action, see Chapters 12 and 13.

Electrical shock involves electrical stimulation of tissue, and its effects range from a *tingling sensation* to the violent reactions of *muscle tetanus* to ventricular fibrillation. Thus, electrical shock is measured in terms of current *intensity* at specific *frequencies*. *Macroshock* is defined as a *high-value current* level (mA), which passes arm to arm through the body by (skin) contact with a voltage source. There must be two points of body contact. The resulting current eventually passes through the heart and may cause ventricular fibrillation or death.

Microschock is defined as a *low-value current* (μA), which passes *directly through the heart* via a needle or catheter in an artery or vein. The catheter may touch the interior surface of the heart where blood pressure is measured or cardiac pacing is effected.

Larger currents are required to cause death from macroshock because the skin is a relatively good insulator. Table 19-1 shows the effect of 60-Hz arm-to-arm electric current shock. These currents range from *1 mA* (threshold of sensation) to *10 mA* (can't let go) to *100 mA* (respiratory failure and ventricular fibrillation—death) to *1 A* and larger (tissue damage—burns). In contrast, microshock currents of *10* to *100 μA* can cause ventricular fibrillation and death.

The *frequency of the current* is also important when considering the shock phenomenon. For example, an arm-to-arm shock at 50 to 60 Hz is particularly potent as compared to higher or lower frequencies. A 1-mA, 60-Hz current establishes the threshold of perception for most people, and a 60-Hz, 100-mA current may cause respiratory difficulty, ventricular fibrillation, and/or death. However, if the frequency is raised above 1 kHz, these current levels no longer produce such sensations or *life-threatening phenomenon*. High frequencies in the megahertz region, for example, will not cause shock at all. This relates to the function of the *electrosurgical* unit described in Chapter 18. This unit cuts, burns, and cauterizes tissue but does not induce electrical shock.

Table 19-1
Effects of 60-Hz Electric Shock (Current) On an Average Human Through the Body

Current Intensity—1 S Contact	Effect
1 mA	Threshold of perception.
5 mA	Accepted as maximum harmless current intensity.
10–20 mA	"Let-go" current before sustained muscular contraction.
50 mA	Pain. Possible fainting, exhaustion, mechanical injury; heart and respiratory functions continue.
100–300 mA	Ventricular fibrillation will start but respiratory center remains intact.
6 A	Sustained myocardial contraction followed by normal heart rhythm. Temporary respiratory paralysis. Burns if current density is high.

Actually most danger from macroshock or microshock is *ventricular fibrillation*. This is defined as a condition of the heart in which the *myocardium quivers* instead of pumping rhythmically. The result is an ineffective heart pump. Heart tissue is one of the most sensitive tissues in the body. Normal internal electrical stimulation of the heart, beginning with the sinoatrial node in the right atrium, initiates *synchronous* cardiac activity. This electrical activity of the heart gives rise to the electrocardiogram (ECG) as described in Chapter 2, particularly Figures 2-5 and 2-6.

Ventricular fibrillation may result when the heart is *externally* stimulated. A few heart cells become deranged and can trigger off a "chain reaction" of chaotic activity. The *ectopic foci* of electrical impulses spread and cause *asynchronous* cardiac function. Death follows in minutes if not treated by *cardiopulmonary resuscitation* (CPR) or defibrillation using a *defibrillator unit* as described in Chapter 8. Defibrillation requires large currents of 6 A or more passing through the chest to act as a "reset to the heart."

Since 10 mA or less is the let-go shock level, a safe 5-mA leakage current has become the standard. In fact, for a manufacturer to receive an *Underwriters Laboratories listing* (approval), the equipment must have a 60-Hz leakage current from power line to equipment metal case of less than 5 mA.

The safe microshock current levels originate from studies on dogs. In the 1960s, experiments showed that 20 μA at 50 to 60 Hz directly through the heart could cause *ventricular fibrillation in dogs*. This is due to the high-current density in the myocardium. Since experiments on humans are limited, the 10-μA safe microshock current level was extrapolated from the dog studies. These studies were inspired in the early 1960s by those who felt that little attention was given to patient safety in the modern hospital environment. This setting includes fluid-filled catheters, different manufacturers' grounding and isolation systems, inadequate hospital power wiring, electrically unaware hospital maintenance personnel, and electrically untrained medical staff.

19-11 Leakage Current

Leakage current is defined as the low-value electrical current (μA) that inherently flows (leaks) from the energized electrical portions of an appliance or instrument to the metal chassis. All electrically operated equipment has some leakage current. This current is not a result of a fault but is a natural consequence of electrical wiring and components.

Leakage current has two major parts: capacitive and resistive. *Capacitive leakage current* results from distributive capacitance between two wires or a wire and a metal chassis/component case. For example, the "hot" copper wire (black for power systems) forms one plate, the wire insulation forms the dielectric, and the metal chassis (ground) forms the other plate of a capacitor. This capacitor is actually distributed over the entire length of the power cord, and the longer the cord the greater the capacitance. A capacitance of 2500 pF at 60 Hz on a 120-V power system gives approximately 1 MΩ of capacitive reactance and 120 μA of leakage current. Components that cause capacitive leakage current are rf filters, power transformers, power wires, and any device that has stray capacitance.

Resistive leakage current arises from the resistance of the insulation surrounding the power wires and transformer primary windings. Modern thermoplastic dielectrics or power lines and cords are of such high resistance that the resultant leakage current is negligible compared to capacitive leakage. Figure 19-5 shows the *origin* of leakage current and Figure 19-6 the increased leakage with use of rf filters.

The *classical remedy* for excessive leakage current is the third or safety ground wire. Understanding of electrical power *wiring distribution* and *grounding* is a prerequisite to understanding the leakage current phenomena. The *hot wire* in U.S. systems is *black*

Figure 19-5 Origins of leakage current (stray capacitance). (Reprinted courtesy Hewlett-Packard.)

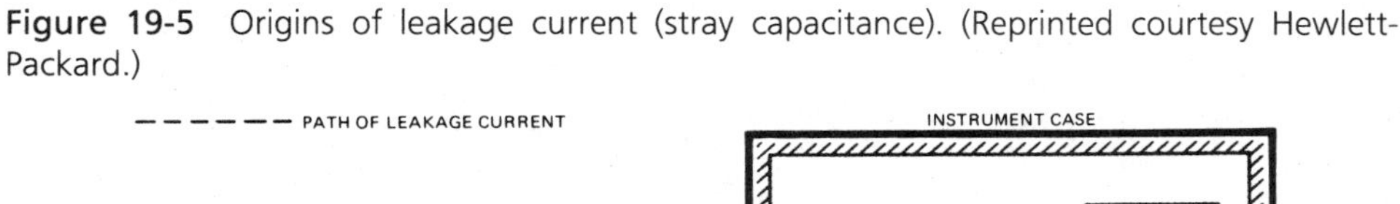

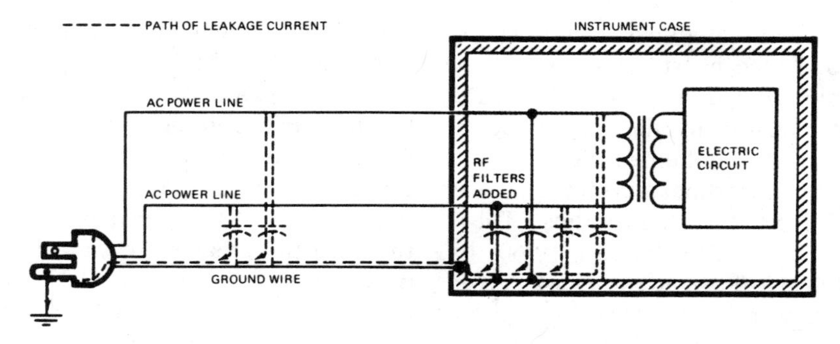

Figure 19-6 Rf filters increase leakage current. (Reprinted courtesy Hewlett-Packard.)

and is the *ungrounded* wire. The *neutral ground wire* is *white* and is the return wire that is connected to earth ground in the main power/fuse panel. The *safety ground wire* (which normally carries no current) is *green* and is the ground current return only under

Figure 19-7 Receptacles connected to fuse panel.

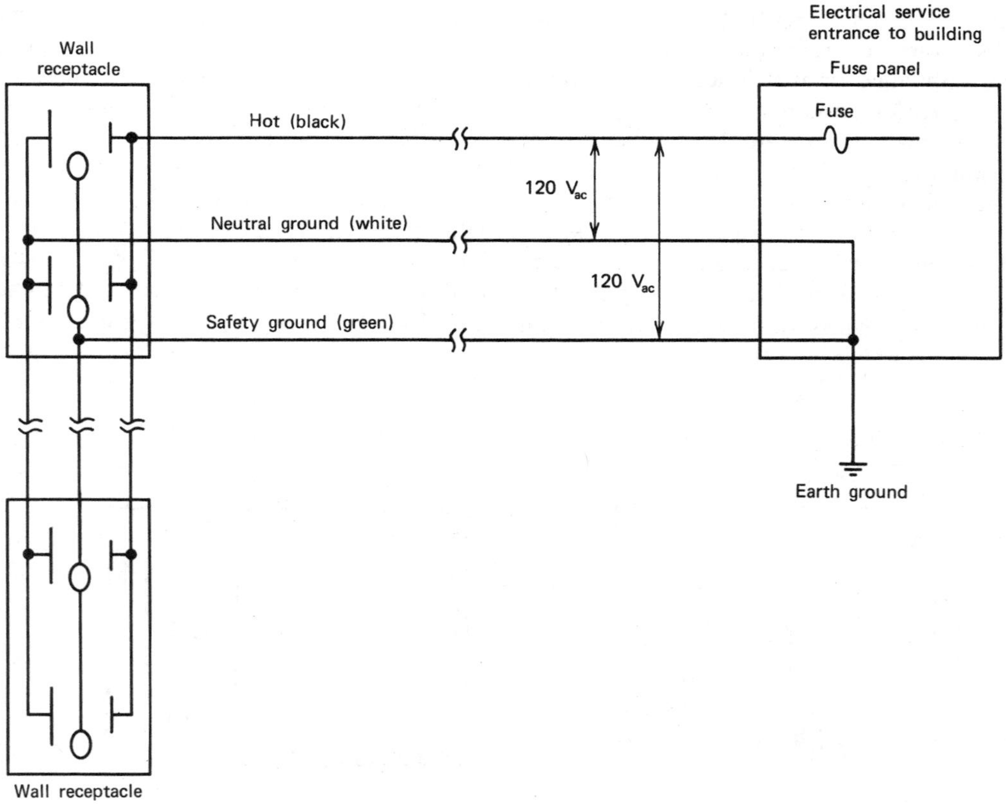

leakage and fault conditions. Actually, two purposes of safety grounding are to drain off leakage current and blow the fuse in the hot line in case of catastrophic fault (such as hot wire shorts to grounded metal case or overload). The National Electrical Code (NEC) clearly specifies wiring distribution and grounding techniques. Figure 19-7 shows two electrical receptacle outlets connected to a power/fuse panel.

As an example of the effect of safety ground, consider an electrical instrument connected to a power system in which the leakage current through a 1-Ω ground resistance is assumed to be 100 μA (no patient connection). If a patient of 500-Ω resistance touches the instrument metal case, 0.2 μA of leakage current flows through the patient, and 99.8 μA flows through the safety ground. Clearly, the safety ground is a much lower resistance connected in parallel with the patient. Hence, most of the leakage current flows through safety ground. Figure 19-8 shows the normal ground system. If the safety ground connection should become broken or defeated by using a 3 to 2 adapter or two-wire extension cord, *all* leakage current flows through the patient. Figure 19-9 shows a broken ground system that is potentially very hazardous to the patient.

In the case of three-wire (safety ground) failure, the following measures could prevent patient shock or electrocution:

1. Reduce internal equipment leakage current to below 10 μA.
2. Continuously monitor ground wire continuity.
3. Add an additional ground wire in parallel with the power cord safety ground.
4. Periodically inspect ground-connection integrity.
5. Use a power-isolated system that isolates the equipment and, hence, the patient from neutral ground.

Figure 19-8 Normal path for leakage current. (Reprinted courtesy Hewlett-Packard.)

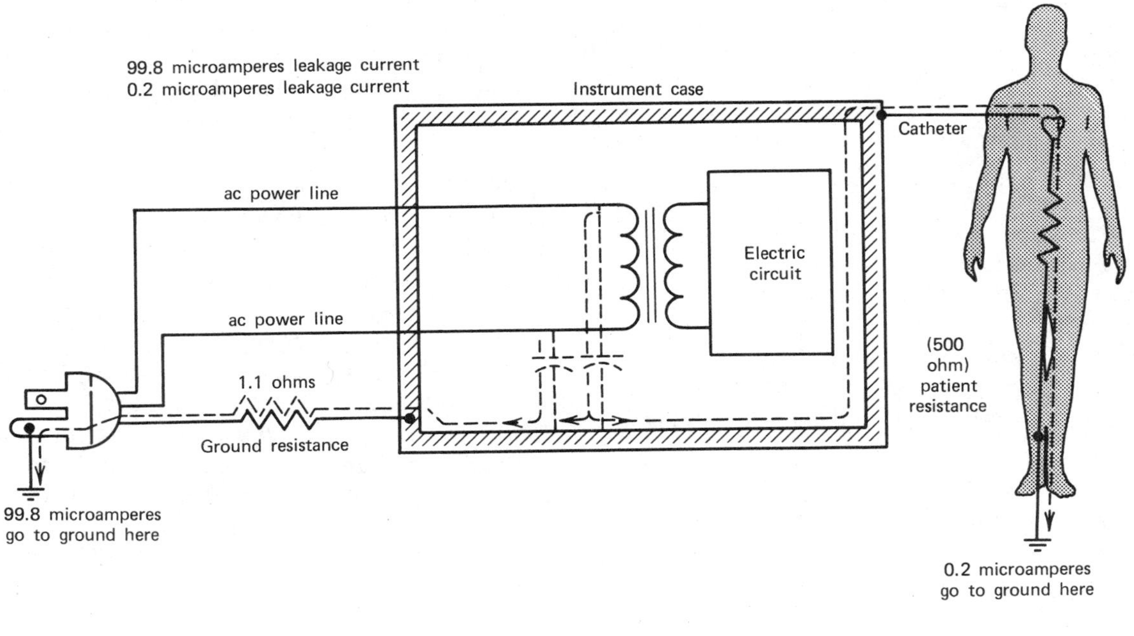

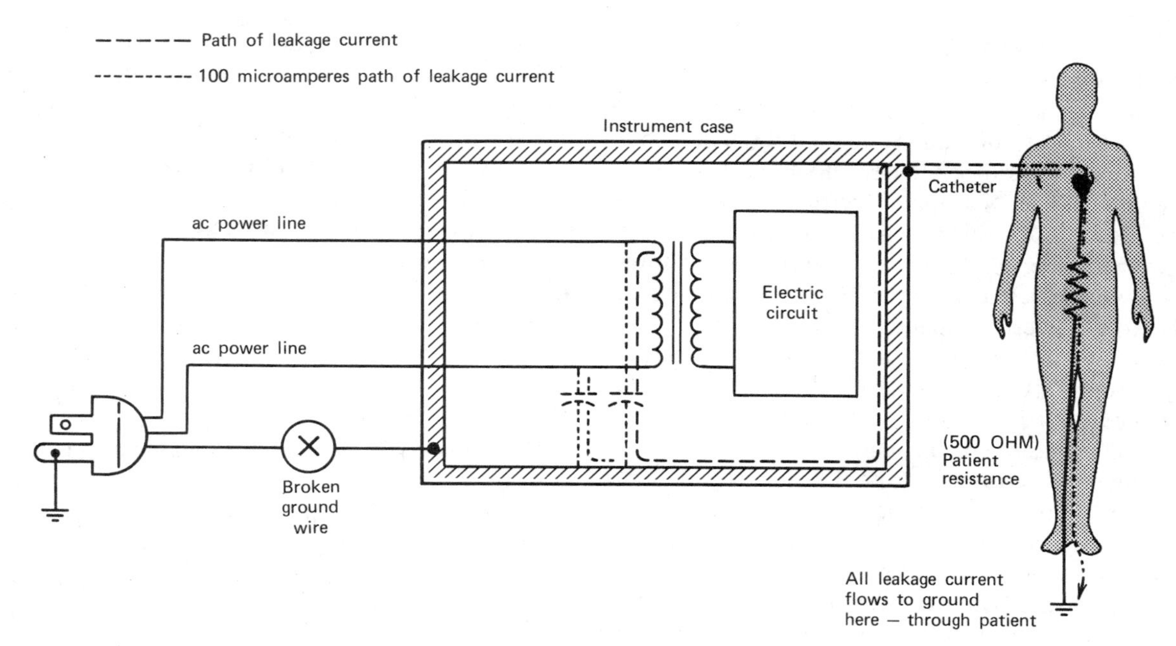

Figure 19-9 Path of leakage current with defective ground wire. (Reprinted courtesy Hewlett-Packard.)

19-12 Subtle Hazards and Cautions for Microshock in Hospitals

Now that we understand the physiological effect of small currents on catheterized patients and the elusive nature of leakage currents, hazards can be pinpointed and contained.

The following three cases each include *parameters or situations present, analysis, recommendations*, and *summary*.

The first case illustrates the subtle electrical hazards that result from inadequate or nonexisting grounds but that do not produce sufficient hazardous current to be felt or seen by the staff members using the equipment:

Case 1

Parameters

(1) The patient is lying on an electrically operated bed. (2) The ground connection from the wall plug is faulty. (3) The patient is equipped with a transvenous pacing catheter connected to a small, battery-operated pacemaker. (4) The patient is connected to an ECG monitor. The right-leg ECG electrode is connected to the hospital grounding system through the monitor (see Figure 19-10).

Analysis

The faulty ground connection on the electric bed allows a voltage to exist on the bed frame due to capacitive coupling between the bed frame and the primary wiring in the bed. Normally this voltage produces a current that is conducted harmlessly to ground. But if this ground wire breaks, the current can follow other paths. In this example, assume that an attendant comes to the bedside to adjust the pacing catheter connections and, without thinking, simultaneously touches the pacemaker terminals and the bedrail. Assume the attendant supplies a 100,000-Ω con-

CASE 1

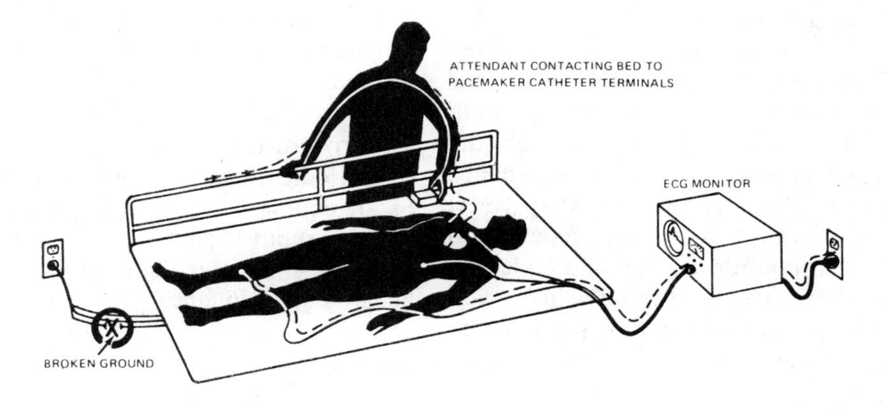

CASE 1. Analysis

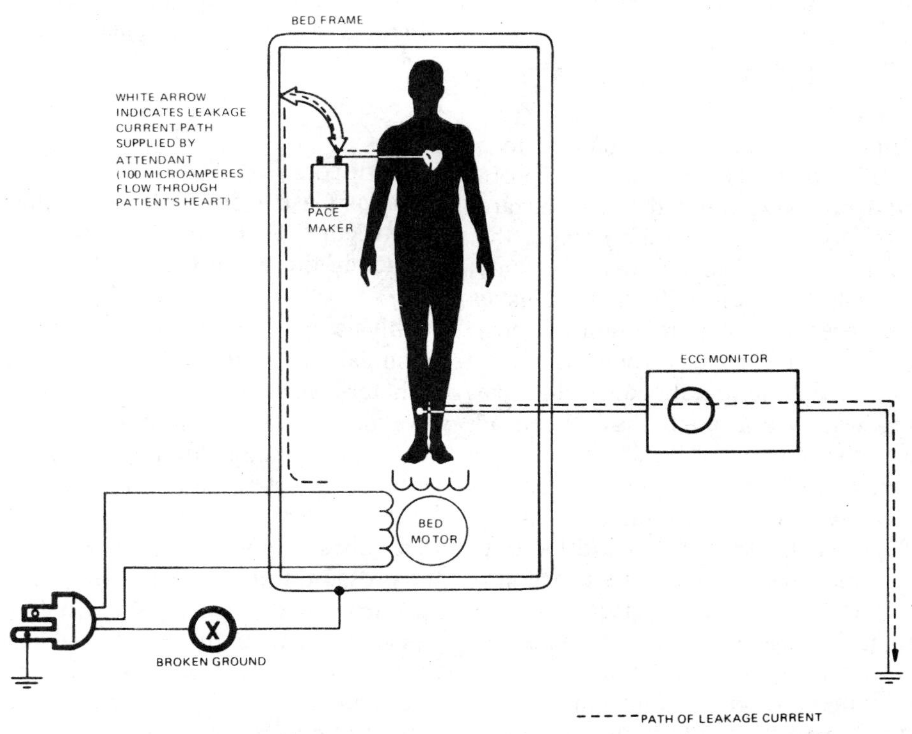

Figure 19-10 Case 1 setup and analysis. (Reprinted courtesy Hewlett-Packard.)

nection between these two points, and that the resistance between catheter terminals and patient is 500 Ω. We can see from the analysis that the attendant completes the path between the power line and ground, with the path going directly through the heart. If we assume the leakage impedance of the electrical bed is approximately 1 MΩ (assuming 2500 pF of capacitance from power line to bedframe), then a simple calculation shows that over 100 μA pass through the patient's heart (200 μA if a 240-V supply is assumed).

$$I = \frac{V}{Z} = \frac{V}{R - jXc} = \frac{V}{\sqrt{R^2 - Xc^2}}$$

$$V = 120\ V_{rms},\ R_{attendant} = 100{,}000\ \Omega$$

$$R_{patient} = 500\ \Omega,\ X_c = 1{,}000{,}000\ \Omega$$

$$I_L - \frac{120\ V}{\sqrt{(100{,}500\ \Omega)^2 + (1{,}000{,}000\ \Omega)^2}}$$

This is almost certain to be a hazard to a catheterized patient. The medical staff probably would not have noticed that a hazard existed. If one of them were to touch the bedframe and ground simultaneously, only 100 μA would pass through them. This is below the threshold of perception of most adults for current passing through the skin and would have gone unnoticed. A clue that something was wrong might have been an increase in the amount of line frequency interference on the ECG trace on the monitor. The natural reaction of the nurse would be to see if the electrode creme had dried out, requiring replacement. Since this procedure would fail to reduce the interference, the nurse might then assume that something was wrong with the monitor and call the monitoring equipment service technician. During all this time the bed would continue to operate, so that a fault in it probably would not be suspected. Although the fault in this case was due to a faulty ground connection from the bed, the same kind of hazard would exist if the ground connection in this right-leg-grounded type of ECG monitor was broken instead.

Recommendations

(1) Periodic check of ground wire continuity of all equipment in vicinity of patient, (2) isolated input circuits on ECG monitor, (3) training staff to recognize potential shock hazards and remedies.

Summary

Fault: Broken ground wire in electric bed power cord. *Hazard:* Leakage current from the bed that would normally be conducted to ground now can flow through the patient grounded through the right-leg electrode of the ECG monitor. *Indication of Hazard:* Possible increases in interference on ECG monitor.

Case 2

Parameters

Same as Case 1 but with saline-filled catheters (Figure 19-11) and a two-wired cord appliance in the vicinity of the patient.

Analysis

A similar situation could occur if saline-filled catheters were used to monitor pressures or take blood samples in the vicinity of the heart. The saline column in the catheter is a sufficiently good conductor to provide a path for hazardous currents to reach the heart. Often these catheters are grounded through the pressure transducer to the monitoring instrument case. This presents a hazard because the patient or an intermediary could touch improperly grounded equipment. The intermediary would inadvertently provide a source of current that flows into the patient, through the catheter, and to ground via the pressure transducer and monitor.

The source of current could be any device

CASE 2

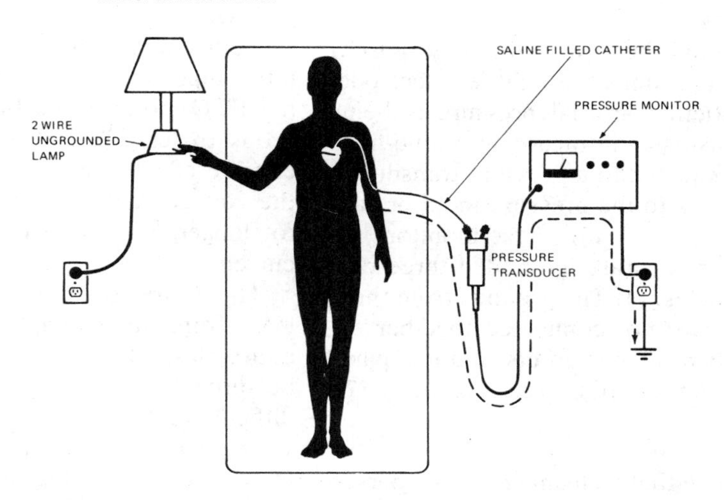

Figure 19-11 Case 2. (Reprinted courtesy Hewlett-Packard.)

with a two-wire power cord as well as improperly grounded equipment. Many such devices that connect to the power outlet with only a two-wire cord can present a hazard to the patient even though their power cords and insulation are in good condition. Sufficient capacitive coupling often exists in the power wires to allow leakage currents greater than 20 μA to flow if the patient just touches the outer case of some of these devices. In some equipment this current flow can be as high as 500 μA.

The leakage current available from ungrounded television sets, radios, electric shavers, and lamps is usually so feeble that it is not felt by the attending staff. But that feeble current is sufficient to be a hazard to the patient with electrodes implanted in the vicinity of the heart, where 20-μA currents are considered hazardous.

Recommendations

(1) Within 15 ft of the patient, use only apparatus with three-wire power cords and proper grounds. (2) Train staff in recognition of hazards. (3) Eliminate as many permanent ground paths from the patient as possible through the use of isolated input monitoring devices.

Summary

Fault: Devices, with two-wire power cords.
Hazard: Leakage currents present on outer surface of bedside devices. Ground path through saline column in indwelling catheter.
Indications of Hazard: None, unless leakage current can be felt by attending staff.

In the situations discussed thus far, the sources of current have been due to leakage current from properly functioning equipment which was disconnected from ground, either through ground connection failure or because a three-wire power cord was not used.

Unfortunately, a hazardous situation can still occur when equipment appears to be properly grounded. As an example, assume that a patient is being monitored in an ICU under the following conditions.

Case 3

Parameters

(1) The patient is being monitored by an ECG monitor that grounds the right-leg electrode. (2) The patient's arterial pressure is being monitored using an intracardiac, saline-filled catheter connected to a pressure transducer, which attaches to the pressure monitor case and then to ground. (3) These monitors are connected to separate, grounded three-wire wall receptacles. (4) The grounds from these two outlets are not connected together except at a central power distribution panel many feet from the ICU area (Figure 19-12).

Analysis

Let us assume that a cleaning service person now plugs a vacuum cleaner into a wall outlet on the same circuit as the ECG monitor. The cleaner has a three-wire power cord with the third wire grounding its outer case. This is a necessary safety feature for vacuum cleaners, as they are notoriously hazardous devices from an electrical safety point of view. The windings of the motor are continually exposed to dust, often damp, which provides a good path for an eventual "winding-to-outer-case" short. Because this kind of short makes the case rise to full-line voltage, the case is grounded to protect the operator. In this example the vacuum cleaner hasn't completely failed but has developed a fault sufficient to allow 1 A to flow down the ground wire, back to the power distribution panel. If we assume that the power distribution panel is 50 ft away and that the power wiring is #12 AWG, the 50 ft of ground wire has 0.08 Ω of resistance.

One ampere of current flowing in the ground wire common to the ECG monitor results in a voltage drop of 80 mV. Since a very small current is flowing in the ground wire from the pressure monitor, its case remains very close to the ground. We see that this potential difference appears directly across the patient, between the ECG monitor and the pressure monitor.

If we assume that 10 μA is the maximum safe current, this amount of current will flow if the impedance through the patient between the ECG monitor and the pressure monitor drops to less than 8000 Ω. This might be considered a low resistance for this path, but the voltage could also have been higher due to longer ground wires or higher fault currents.

There are several important points to learn from this example, as it relates to an entire class of low-voltage hazards that can be difficult to detect and the causes more difficult to find. In the example cited, the wiring would have met the provisions of most existing wiring codes. Such wiring could be in an older hospital, where additional power outlets and circuits were added as part of a modernization program without abandoning existing outlets and wiring.

These low-voltage hazards would not be detected by the medical staff, since the resulting current through them would be too feeble to be felt. There is a remote possibility of an increase in the amount of interference on the ECG monitor trace, but if it occurs it may be interpreted as a fault in the monitor, not in the wiring. Also, the hazard may exist only for short periods of time, such as when the vacuum cleaner is in use, so that the staff may be unable to find the cause. If the voltage were sufficient to cause fibrillation in the patient in this example, it is unlikely that the medical staff would associate the patient's difficulty with a cleaning-service vacuum cleaner.

Summary

Fault: Two devices connected to the patient are plugged into outlets with grounds connected together by excessively long wire. *Hazard:* Faulty appliance causes difference in ground potential between two devices and allows current to flow through the patient.

CASE 3

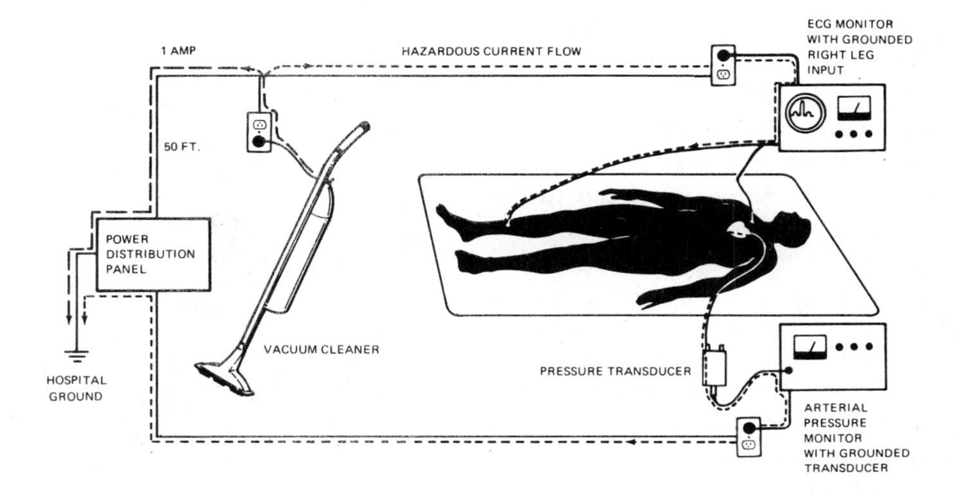

CASE 3. Analysis

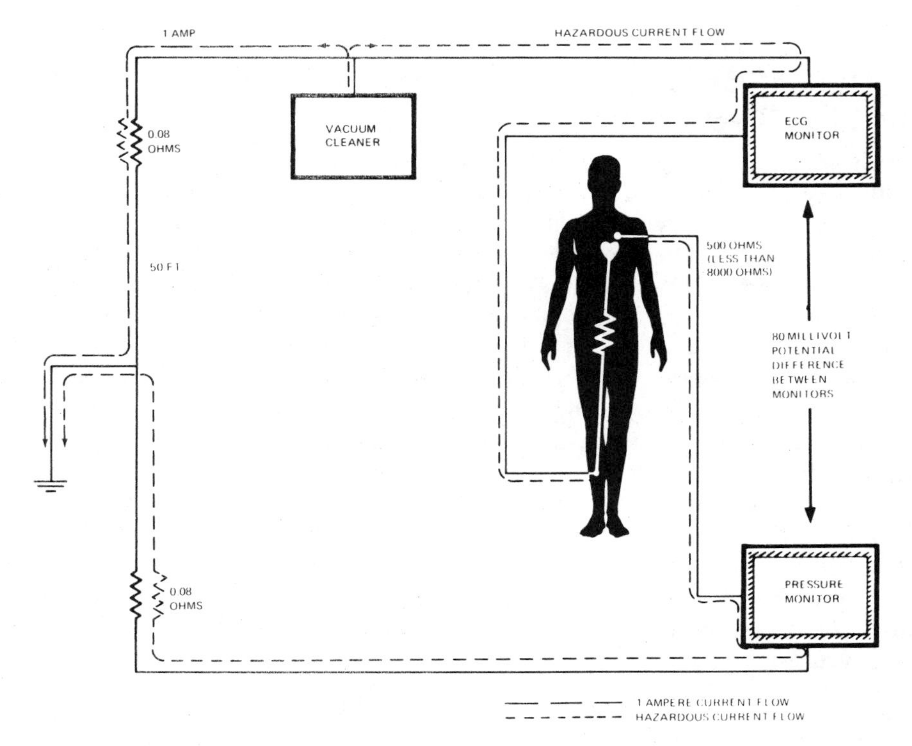

Figure 19-12 Case 3. (Reprinted courtesy Hewlett-Packard.)

Indication of Hazard: None likely; possible increase in ECG interference.

Recommendations

1. Place all power outlets in vicinity of patient on a common panel, with ground connections strongly bonded together.

2. Assign a specific power circuit to operate patient care equipment and *prohibit its use for any other purpose.*

3. Routinely check potential on ground terminal of outlets to be used for operating patient care equipment, with respect to all other conductive surfaces within 15 ft of the patient.

4. Provide isolated input monitoring equipment to eliminate possible paths for, and sources of, hazardous current.

5. Training staff to recognize potentially hazardous conditions and provide procedures for having them investigated and corrected promptly.

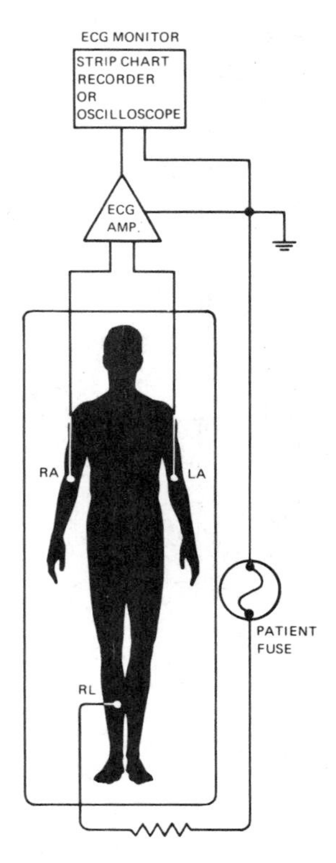

Figure 19-13 Ground-referenced ECG amplifier (1946). (Reprinted courtesy Hewlett-Packard.)

19-13 Design Considerations for Reducing Electrical Hazard

The hazard of shock is always present when electrically operated biomedical instrumentation is connected to a patient. This is especially true for long-term connection to patient monitors. Early ECG machines (late 1940s) that employed the *"ground-referenced differential amplifier"* posed a particular problem in this regard. Since the right leg (RL) of the patient was wired directly to ground via an electrode (good electrical connection), a difference of potential across the patient's body was easy to establish. Old machines had a fuse in this RL lead to limit current to the 5-mA UL standard. Newer machines, however, do not use this fuse but are still widely used. In any case, such RL grounded systems are only hazardous if large differential voltages (10–100 mV) are within reach of the patient and are especially dangerous to catheterized patients. Figure 19-13 shows the RL grounded system. Essentially, two systems are utilized to minimize this potential shock hazard in addition to constant safety checks by maintenance personnel: a driven right-leg ECG preamplifier and a ground-isolated input ECG preamplifier.

Driven right-leg ECG amplifier systems became a reality following the emergence of transistor circuits in the early 1960s. Figure 19-14 shows the RL driven system. This amplifier samples the 50- or 60-Hz power line noise interference pickup by the patient and presents a signal back to the patient, which cancels the interference. The *current feedback* never exceeds the original noise current already flowing through the patient. Thus, the patient is effectively isolated from

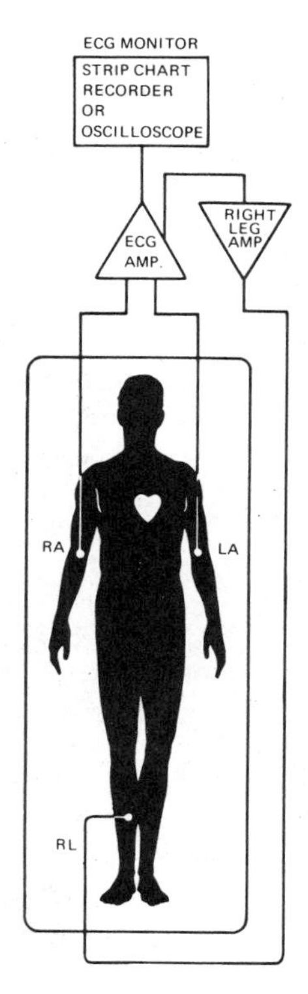

Figure 19-14 Driven right-leg ECG amplifier (1962). (Reprinted courtesy Hewlett-Packard.)

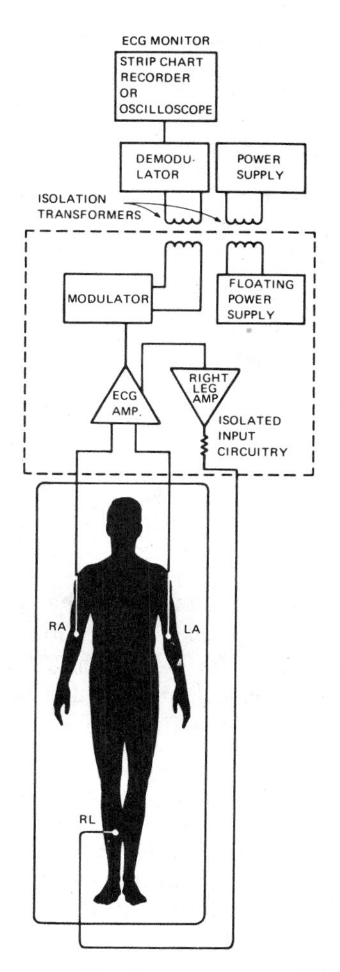

Figure 19-15 Isolated input ECG amplifier (1967). (Reprinted courtesy Hewlett-Packard.)

ground with very low leakage current, and the ECG recording is much cleaner (reduced hum). This *common mode amplifier* is also shown in Figure 5-3*a*.

Isolated patient circuits are achieved by providing an *isolation transformer* in the signal path as well as in the ac power circuit. The effect is to isolate the patient from ground and other portions of the equipment such as the graphic recorder. Figure 19-15 shows the isolated input ECG system. The engineering objective of low distributed capacitance is a reality today. Isolation impedances of over 10 MΩ at 50 Hz or 60 Hz between input terminals and ground are typical.

Table 19-2 compares three different types of ECG monitor amplifiers: grounded input ECG amplifier with 1000 Ω patient impedance; driven right-leg circuit with 1000 Ω patient impedance; and isolated input circuit with effective isolation impedance of 15 MΩ.

Clearly, the driven right-leg circuit provides better leakage current protection, especially from large differential potentials. Isolated input circuits provide better protection. Since no single approach provides the complete answer to patient safety, the entire patient environment must be considered.

Table 19-2
Current Flow versus Potential Difference (Voltage) Between Patient and ECG Amplifier

Voltage Between Patient and Monitor (60 Hz)	ECG Monitor with Grounded Input μA	ECG Monitor with Driven Right-Leg Circuit μA	ECG Monitor with Isolated Input Circuit μA
0	0	0	0
2.5 mV	2.5	0.72	0.0002
5.0 mV	5.0	1.4	0.0003
80 mV	80	2.3	0.0054
150 mV	150	3.2	0.01
1.0 V	1000	38	0.067
50 V	50,000	1000	3.3
120 V	120,000	2400	8.0
240 V (50 Hz)	240,000	4800	16.0

The important points are proper grounding of equipment, regular inspection to verify grounding integrity, instruments that isolate the patient from ground, and the value of the power isolation transformer.

19-14 Line Isolation Systems

Power isolation transformers produce isolated systems by breaking direct electrical connection to neutral ground. (*Note: Autotransformers* will *not* produce isolated systems.) The power used inside the electrical equipment is not effectively referenced to ground. This amounts to *reducing if not eliminating low-voltage hazards* since contact with either side of the isolation transformer secondary and ground produces no shock. Patient isolation through a driven RL amplifier amounts to *reducing possible current paths through the patient*. While these approaches are different, they both aim at preventing electric currents greater than 10 μA from flowing through the myocardium via an indwelling catheter.

Power isolation transformer design is geared toward reducing differential voltages to 5 mV between the catheter and equipment chassis or earth ground. These devices are used in the operating room. If the standard patient resistance of 500 Ω is used, no more than 10 μA will flow. This design goal is not easy to achieve, and such transformers can add $1000 to $3000 per bed cost for the monitoring installation. Furthermore, if the isolated hot wire should short to the grounded metal case, the fuse or circuit breaker will not blow since this isolated power is not ground referenced. In this case excessive leakage current will flow and only be detected by the line isolation monitor buzzer. Figure 19-16 shows a short circuit with a power isolation transformer.

In terms of *historical perspective,* the original isolation transformer systems were designed to *prevent heating and sparking* due to the hot wire shorting to the metal case. This caused explosion of flammable anesthetic gas such as ether. To understand this phenomenon, consider a *ground-referenced (nonisolated) system* and an ECG monitor. If the *hot wire shorts to the metal case* in a *two-wire system,* the floating case (not grounded) becomes 120 V with respect to ground. The fuse does not blow, and this is a macroshock and microshock hazard of a

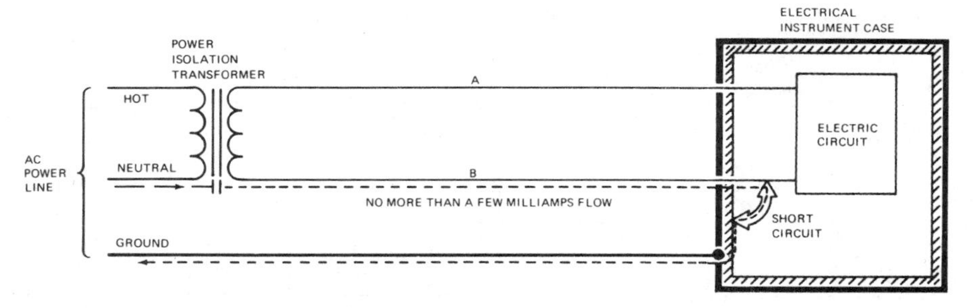

Figure 19-16 Result of short circuit with power isolation transformer. (Reprinted courtesy Hewlett-Packard.)

very serious nature. For this reason, *two-wire systems are never used in the hospital.* If the *hot wire shorts to the metal case* in a *three-wire system,* the grounded case passes a large current (15–30 A) until the fuse blows. This disconnects the power from the equipment and removes the shock hazard. However, this large current causes heating and may jump small gaps to cause sparking. At levels near the floor, where anesthetic gases collect, this can cause explosion. *In a non-grounded (isolated) system,* if the *hot wire (either isolation transformer secondary wire) shorts to the grounded metal case,* no large currents flow through the case. Only leakage current flows (a few milliamperes) and *sparking is nonexistent.* This system protects against explosion, and corrective action is taken when the line isolation monitor sounds (loud buzzer) or the monitor leakage meter reads in the red (2–3 mA).

The *line isolation monitor (LIM)* is a device that continuously monitors the impedance of either isolated power line to ground. The modern *dynamic LIM* monitors this impedance several times per second. The effect is to monitor leakage current. This device is used with power isolation transformer systems. Figure 19-17 shows a power LIM system with fault detector.

Essentially, there are several ways one could be shocked. These are by touching one power line and ground, one metal chassis and ground, and two metal chassis. For example, if the patient touches two metal chassis (devices A and B), both of which are grounded, and the *insulation on device B breaks down,* the *patient can be protected by the LIM system* (shown in Figure 19-18). Only 2 μA will flow through the 500-Ω patient and 998 μA through the ground wire as a result of the LIM (alarm) limit of 1 mA total fault current.

Unfortunately, one class of *faults is not eliminated* by the LIM system. This is the *open ground.* The fuse does not blow and the LIM alarm does not sound. This is a relatively difficult hazard to solve, and the leakage currents depend on relative leakage capacitance values that generate a total leakage current. A separate ground-wire, equipotential ground system presents added security in which all equipment chassis are connected through a separate wire to the same ground terminal.

19-15 Equipotential Grounding System in Reducing Electrical Shock Hazards

An *equipotential ground system* simply consists of separate connections from each equipment chassis to a common ground ter-

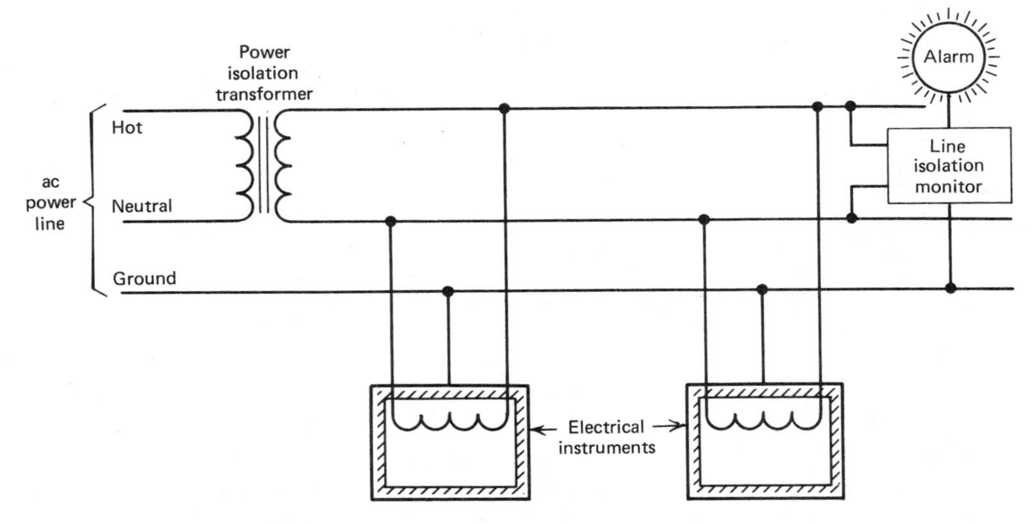

Figure 19-17 Power line isolation transformer with fault detector. (Reprinted courtesy Hewlett-Packard.)

minal. This is achieved by adding another grounding wire from each chassis to a central point that is in parallel with the third wire in the power cord. These ground wires have nearly the same length, and as such each metal chassis is at or near the same potential with respect to the other. Also, all metal surfaces are connected to the common ter-

Figure 19-18 Single insulation failure. (Reprinted courtesy Hewlett-Packard.)

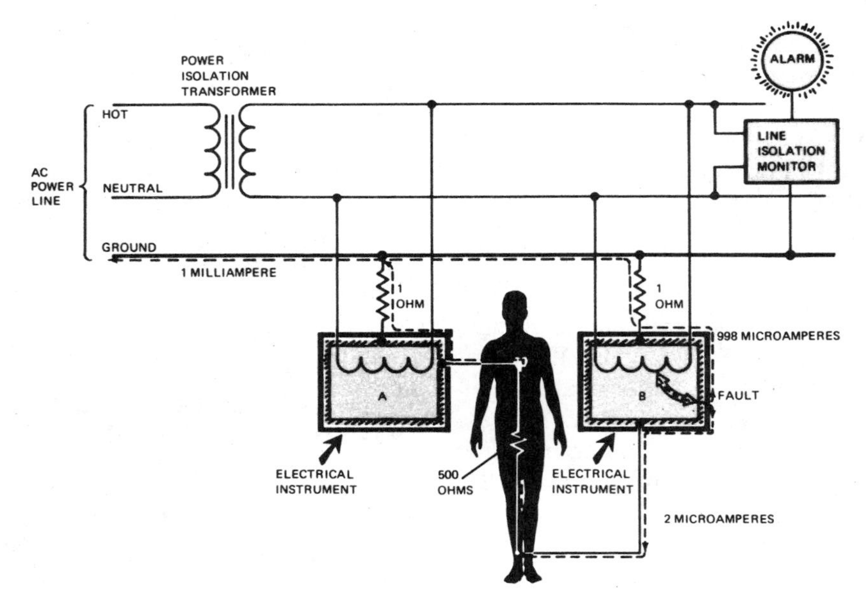

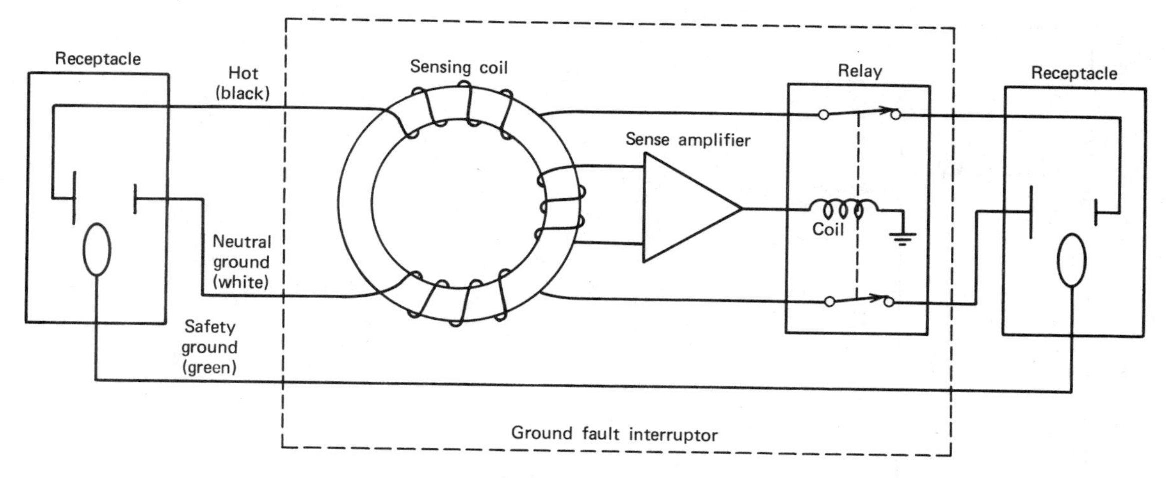

Figure 19-19 Ground fault interrupter.

minal. If the maximum differential potential between any two metal surfaces is held to under 5 mV, then no more than 10 μA will flow through a 500-Ω patient. These systems can be recognized by the large bulky ground wires (AWG #8) drooping from the equipment. Such systems are used in the OR, ICU, and CCU.

19-16 Ground Fault Interrupters in Reducing Electrical Shock Hazards

A *ground fault interrupter (GFI)* is an *automatic switch* that disconnects power if excessive leakage current is present. These devices utilize a toroidal coil on which several turns of the hot and neutral conductors are wrapped. Figure 19-19 shows a block diagram of a GFI. When the current in the hot and neutral wires is equal, no net magnetic flux is present, which indicates no leakage current. The relay remains closed. When these currents are unequal a net magnetic flux is present, which indicates the presence of leakage current. The sensing winding presents a signal to the relay coil via the sensing amplifier, and the relay contacts open, removing power from the wall receptacle. The sensitivity can be set to detect up to 5 mA of leakage current. The GFI is usually used in *wet areas of the hospital* such as the hemodialysis ward. However, it would be hazardous to use it in the operating room on all biomedical equipment because electrical power interruption could remove life support equipment function from the critically ill patient.

19-17 Proper Power Wiring, Distribution, and Ground System in Reducing Electrical Shock Hazards

Electrical power wiring, distribution, and grounding is as important in electrical safety as the electrically operated equipment. Four diagrams serve to illustrate proper wire distribution and grounding. The general guidelines are to distribute power from a central junction box connection and to keep wires to all outlets approximately equal length, es-

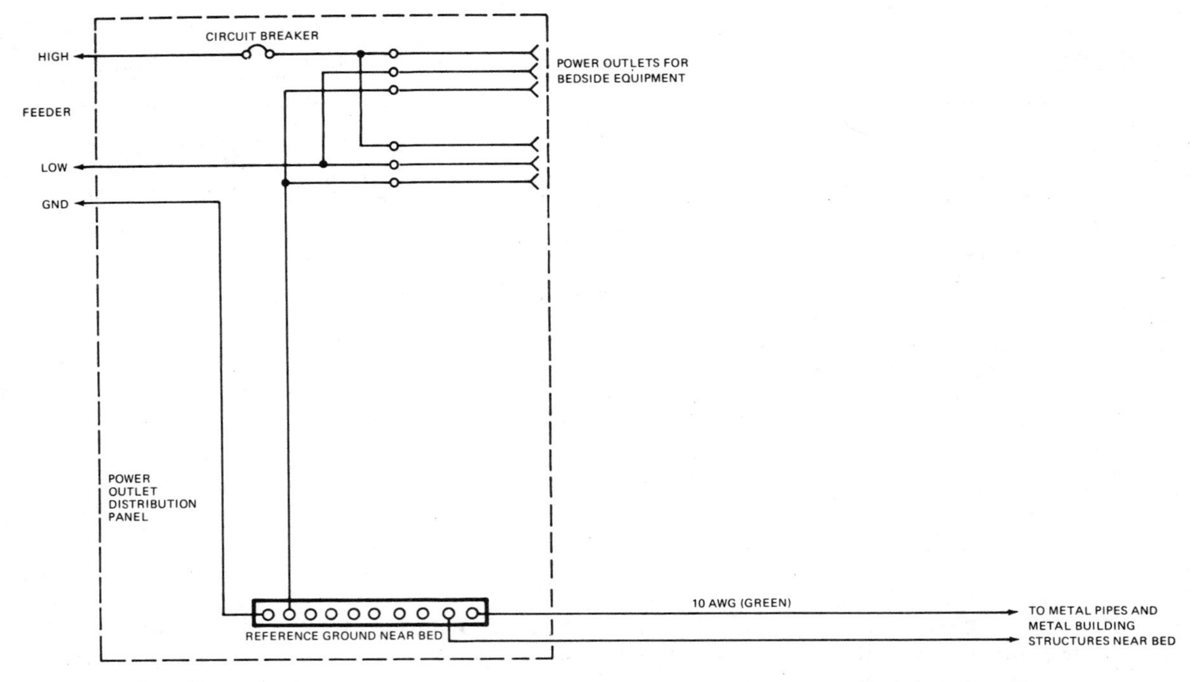

Figure 19-20 Wiring installation for grounded line outlet cluster near bed. (Reprinted courtesy Hewlett-Packard.)

pecially ground wires. Ground wires between outlets should be less than 15 ft long. The National Electrical Code (NEC) specifies wiring standards. Figures 19-20 and 19-21 show isolated input systems with a reference ground near the bed to reduce voltages between equipment chassis. Figure 19-22 shows a single bed system with isolation transformer wired in the common-ground point configuration. Figure 19-23 shows an isolated wiring system for multiple beds in which differential voltages are reduced by short grounding wires.

19-18 Specialized Electrical Safety Test Equipment

To conclude our discussion on electrical safety, it will be instructive to reread Section 19-7. To be knowledgeable in the electrical safety, a BMET should know the material presented in this chapter. However, to be functional in the hospital, a BMET should be able to perform equipment/system PM and document the results of the electrical safety inspection and tests. These tests are geared to uncover such hazards as shown in Figure 19-24. The following special test equipment will be required to insure proper measurement and tests:

1. *Tension tester* to test the spring tension on the hot, neutral, and ground lugs of the wall receptacle. Tension should be 8 oz of pull or over to insure good physical contact of the plug in the receptacle.

2. *Ground wire loop resistance tester* for

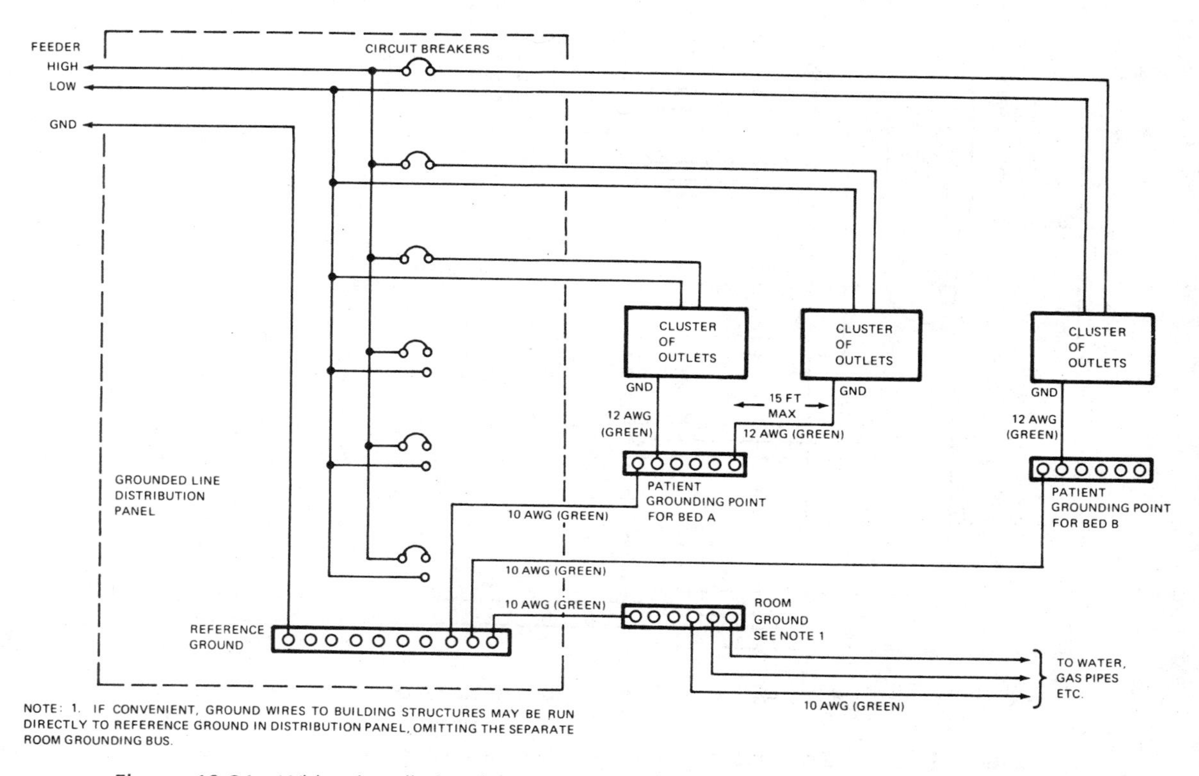

Figure 19-21 Wiring installation for grounded line (beds remote from distribution panel). (Reprinted courtesy Hewlett-Packard.)

resistance measurement between safety (green) and neutral (white) wires from the power system. This can be accomplished using the electrical Safety Analyzer 431F shown in Figure 19-1 and should be *less* than *1* Ω.

3. *Receptacle polarity tester* for correct wiring. This can be accomplished with a separate hand tester or the Neurodyne-Dempsey Safety Analyzer 431F shown in Figure 19-1.

4. *Resistance* between *third wire prong* on the plug to equipment metal *chassis* tester. This can be accomplished using the Neurodyne-Dempsey Cordohmeter 447 shown in Figure 19-3 and should be *less* than 0.1 Ω.

5. *Resistance* from *hot* (black) wire to *chassis* and *neutral* (white) wire to chassis insulation resistance tester. This can be accomplished using the Neurodyne-Dempsey Cordohmeter 447 shown in Figure 19-3 and should be *greater* than *1 M*Ω.

6. *Leakage current tester* for chassis to ground and separate ECG leads to ground. This can be accomplished using the Neurodyne-Dempsey Microspector 449 shown in Figure 19-2 or the Safety Analyzer 431F shown in Figure 19-1 and should be *less* than *10* μ*A* for critical care areas.

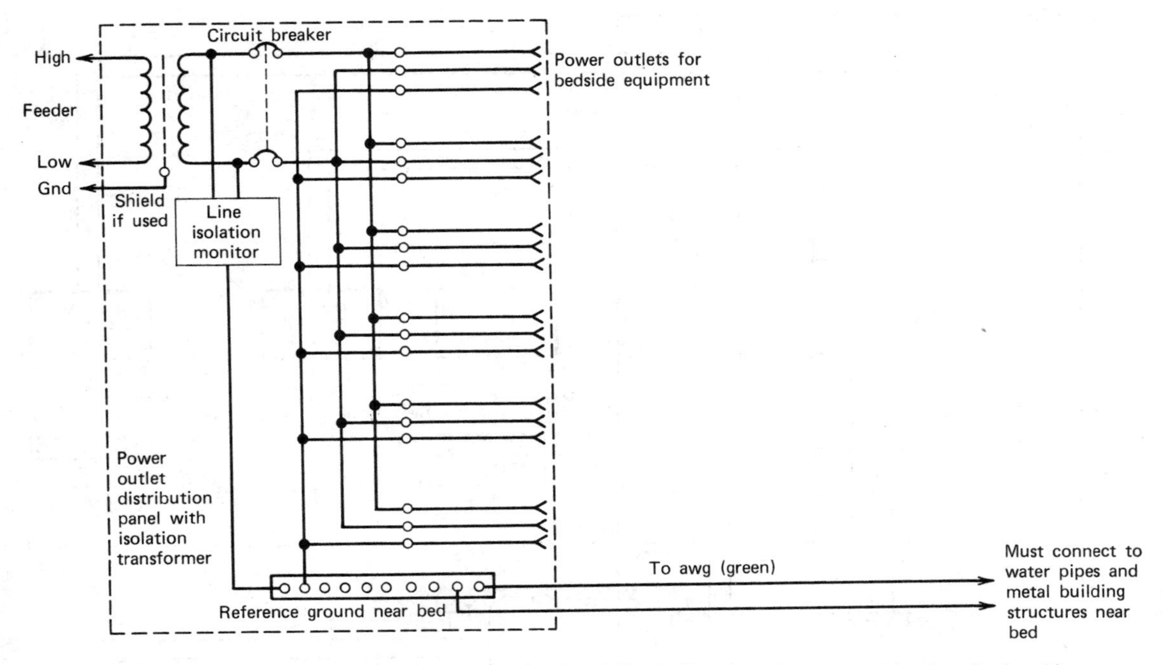

Figure 19-22 Isolated line system near patient's bed. (Reprinted courtesy Hewlett-Packard.)

Figure 19-23 Isolated line system remote. (Reprinted courtesy Hewlett-Packard.)

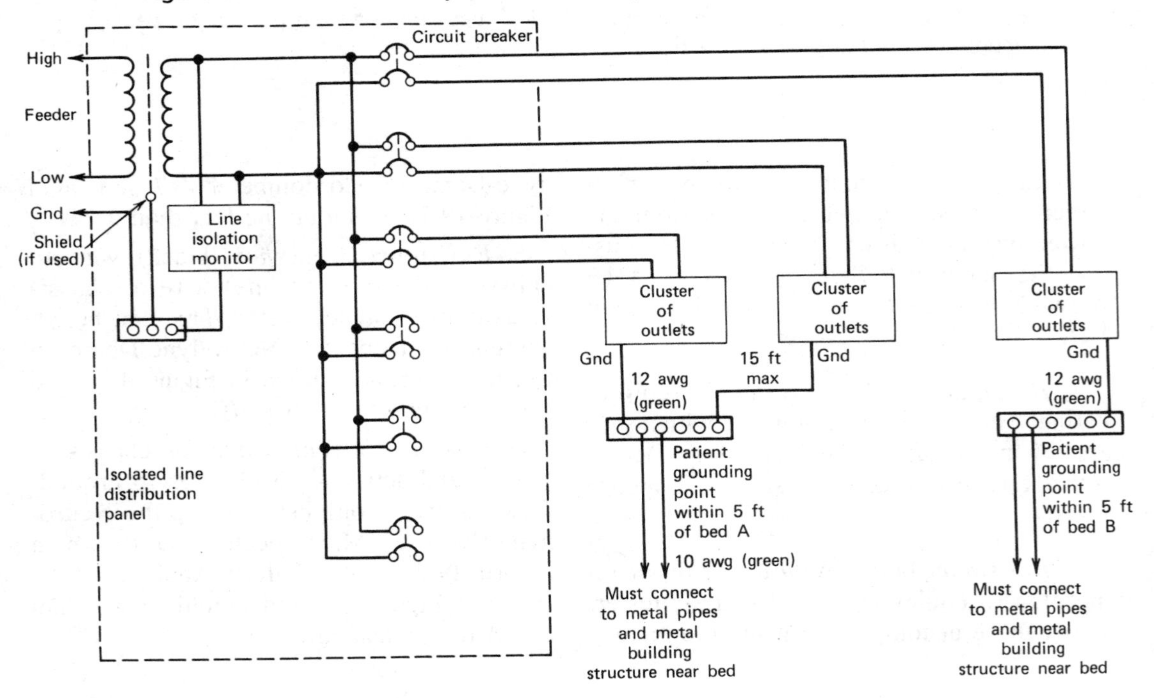

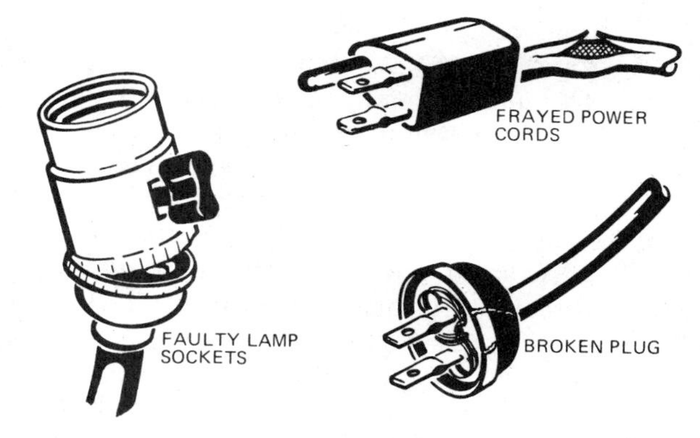

Figure 19-24 Common, very lethal electrical hazards. (Reprinted courtesy Hewlett-Packard.)

19-19 Summary

1. *Electrical safety* in medical institutions is defined as the containment or limitation of hazards: (a) electrical shock to patients, employees, and visitors (macroshock and microshock), (b) explosions, (c) fire, and (d) damage to equipment and buildings. Hazards can be minimized but not eliminated.
2. The *scope* of electrical safety in medical institutions involves any electrically operated equipment used in (a) public, (b) general care, and (c) critical care areas of the hospital.
3. The following organizations produce *pertinent documents* for hospital electrical safety: (a) OSHA, (b) FDA, (c) NFPA, (d) NEC, (e) ANSI, (f) UL, (g) AAMI, and (h) JCAH.
4. *Responsibility* for electrical safety in the medical institutions rests with all personnel and more specifically with the *medical staff* (physicians, nurses), *support staff* (BMET, safety technicians, plant operation personnel), and *administrative staff* (administrators, managers, supervisors).
5. *Preventive maintenance* involving frequent equipment inspections and safety checks can reduce electrical hazards by uncovering early signs of degradation and allowing for correction of faults, such as broken plugs and poor electrical ground contacts.
6. *Legal and insurance requirements* of electrical safety involve possible negligence under the "reasonable man" standards. It is possible that BMETs in the hospital could be sued. Insurance requirements revolve around claims cost to insurance companies, but proper PM and documentation can reduce premiums through reduced hazards. *The patient should come first.*
7. Electrical safety *instruction sessions* may be delivered by BMETs to nurse/medical staff and should include such topics as purposes/objectives, basic electricity, nature of shock, identification and avoidance of shock hazard, reporting of hazardous situations, responsibilities, pertinent publications, demonstration, conclusion, and discussion.
8. Hospital electrical safety *programs* should cover areas of objectives, scope, pertinent publications, PM, test procedures, inspections, documentation, education, assessment of effectiveness, and cost versus benefit.
9. The *physiological effects* of electricity on the human body involve injury to tissues, uncontrollable muscle contractions, and fibrillation of the heart.
10. *Macroshock* is a large value electrical current (mA) that passes arm to arm and eventually through the heart. It may be lethal. Accepted values are (a) 1 mA, sensation (b) 10 mA, can't let go, (c) 100 mA, respiratory failure/fibrillation of the heart, and (d) 1 A and above, burns.
11. *Microshock* is a small value electrical

current (μA) that passes directly through the heart. It may be lethal. The safe limit is established at 10 μA or less.

12. *Leakage current* is a naturally occurring current that results primarily from distributed capacitance within equipment or power cords and leaks from the hot side (black wire) to equipment metal chassis to safety ground (green wire). Safe limits have been established at 10 μA or less for critical care areas, 100 μA or less for general patient areas, and 500 μA and less for public areas.
13. The *subtle hazards of microshock* have caused the primary interest in hospital electrical safety. On a case-by-case basis, the description of parameters and technical analysis pave the way for recommendations of hazard limitation. Hazards include faulty ac power cord ground connections, use of two-wire cord on appliances, and long ground wire connections between ac power outlets. Hazard containment is accomplished through personnel education and PM programs.
14. *Monitoring instrumentation design*, such as in the ECG machine, originally included a differential preamplifier that required a ground on the patient's right leg (RL). Better designs with lower leakage currents include the *driven RL amplifier* (patient current path limitation), *ground isolated amplifier* (patient lead isolation), and *ac power isolation transformer* (ac power ground to patient isolation).
15. *A power isolation transformer* produces an isolated system by breaking direct electrical connection to neutral ground. Hence, equipment internal electrical power is not ground referenced. Low-voltage hazards are reduced (5-mV limit) if not eliminated. Explosion from sparks are effectively nonexistent in these systems. The transformer is used in the operating room.
16. The *line isolation monitor* (*LIM*) is a device that monitors the impedance of either isolated power line to ground. In effect, the modern dynamic LIM continuously monitors leakage current in an isolated power system. It is used with power isolation transformer systems.
17. An *equipotential ground system* consists of separate additional ground wire connections from each equipment chassis and metal surface to a central ground terminal. This reduces differential potentials between metal surfaces to near zero. Consequently, leakage current hazards are greatly reduced. Such systems are used in the OR, ICU, and CCU.
18. The *ground fault interrupter* (*GFI*) is an automatic switch that disconnects power if excessive leakage current is present. It is used in wet areas of the hospital such as the hemodialysis ward.
19. Proper *power wiring, distribution, and grounding* is as important in electrical safety as electrically operated equipment. Correct polarity and ground wire integrity are the underlying safe features of the medical equipment system. The National Electrical Code (NEC) specifies wiring standards.
20. *Special electrical safety test equipment* must be included in the hospital electrical safety inspection, testing, and PM program. This includes an electrical output tension tester, outlet and line cord ground wire resistance tester, outlet polarity tester, and leakage current tester.

19-20 Recapitulation

Now return to the objectives and self-evaluation questions at the beginning of the chapter and see how well you can answer them. If you cannot answer certain questions, place a check mark next to each and review appropriate parts of the text. Next, try to answer the following questions using the same procedure. When you have answered all questions, solve the problems in Section 19-22.

19-21 Questions

19-1 Electrical safety in medical institutions involves the limitation of ______, ______, and ______ to ______ as well as electrical shock hazards.

19-2 Electrical shock hazards exist in ______ ______, ______ ______, and ______ hospital areas.

19-3 Eight major organizations producing pertinent publications related to electrical safety in medical institutions are ______, ______, ______, ______, ______, ______, ______, and ______.

19-4 The medical staff, administrative staff, and ______ staff are responsible for hospital electrical safety.

19-5 Electrical safety test measurements made during ______ ______ can reduce electrical hazards by uncovering early signs of ______.

19-6 Faults are corrected during ______ ______.

19-7 AAMI publishes safety standards for ______ ______ ______ ______ ______.

19-8 All electrical safety tests must ultimately be ______.

19-9 Typically, the manufacturer's operation and service manual has six sections describing ______ ______, ______, ______ ______, ______ of ______, ______ ______, and ______ ______.

19-10 Electrical safety test equipment measures outlet polarity, ground wire resistance, and ______ ______.

19-11 The JCAH inspection team reviews hospital ______ techniques.

19-12 Legal requirements for hospital electrical safety PM programs relate to "______ ______" standards.

19-13 Lawsuits may be avoided if hospital personnel are ______ trained.

19-14 Poor electrical safety PM programs may result in ______ insurance premiums.

19-15 One ultimate goal of an electrical safety PM program is to protect the *patient/hospital*.

19-16 BMETs are often required to give electrical safety presentations to ______ and ______.

19-17 A hospital electrical safety program should include objectives, scope, pertinent ______, test ______, inspection, practices, documentation, education, assessment of ______, and cost of ______.

19-18 The effect of electrical current passing through the body includes injury to tissues, ______ muscle contractions, and ______ of the heart.

19-19 Fibrillation refers to (a) asynchronous skeletal contractions, (b) synchronous heart waves, or (c) asynchronous heart contractions.

19-20 Macroshock results from *high/low* current passing arm to arm.

19-21 Microshock results from *high/low* current passing directly through the heart.

19-22 Leakage current results primarily from *inductive/capacitive* effects of ac power cords and electrical transformers.

19-23 Leakage current standards are ______ μA or less for critical care areas, ______ μA or less for patient care areas, and ______ μA or less for public areas of the hospital.

19-24 Leakage current can be reduced by adding a ______ wire from equipment metal chassis to a common ______ terminal.

19-25 In power systems, the black wire is

__________, the white wire is ________ ______________, and the green wire is ______________ ______________.

19-26 The third wire in power systems serves to blow the ____________ ____________ if the hot wire shorts to equipment metal case and to drain off ______________ ______________ ______________ to earth ground.

19-27 Subtle hazards exist in biomedical equipment systems as a result of *open/shorted* ground connections and the use of two/three-wire extension cords.

19-28 In the ECG "ground referenced differential amplifier," the patient's ____________ leg is connected to safety ground.

19-29 The UL standard for leakage current is ______________ milliamps.

19-30 The driven right-leg ECG amplifier system *increases/decreases* ECG recording noise.

19-31 A signal isolation transformer (ECG am modulation scheme) isolates the patient from *neutral/safety* ground and reduces ______________ ______________.

19-32 A power isolation transformer breaks direct electrical connection from *neutral/safety* ground and reduces ____________ ______________.

19-33 The modern equivalent patient resistance is ______________ Ω.

19-34 The maximum differential voltage standard in critical care areas is __________ mV.

19-35 The line isolation monitor (LIM) monitors ______________ current and is used with the *signal/power* isolation transformer.

19-36 The maximum differential voltage between any two ______________ surfaces in an equipotential ground system is ______________ mV.

19-37 A ground fault interrupter (GFI) disconnects ______________ power if excessive ______________ current exists.

19-38 GFI devices are usually used in the (a) operating room, (b) EEG laboratory, or (c) hemodialysis ward.

19-39 Electrical power wiring, distribution, and grounding is detailed in the ____________ ______________ Code.

19-40 Specialized hospital electrical safety test equipment measures ______________ ______________ resistance, __________ polarity, ______________ spring tension, and ______________ current.

19-22 Problems

19-1 Refer to Figure 19-10 and case 1. Given the *attendant's resistance* as 500 Ω, the *patient's resistance* as 500 Ω, and the *differential voltage* between the *attendant's right hand* and the patient's *right leg* to be 200 mV, answer the following questions: (a) Does current pass through the attendant? (b) If so, how much? (c) Is this a macroshock or microshock situation for the attendant? (d) Is the attendant most probably electrically shocked (above sensation)? (e) Is the attendant most probably electrocuted (killed)? (f) Does current pass through the patient? (g) If so, how much? (h) Is this a macroshock or microshock situation for the patient? (i) Is the patient most probably shocked (above sensation)? (j) Is the patient most probably electrocuted (killed) by modern standards?

19-2 Refer to Figure 19-11 and case 2. Given the *patient's resistance* as 500 Ω and the *differential voltage* between the *patient's right hand* and blood pressure saline-filled *catheter* as 4 mV, answer the following questions: (a) Does current pass through the patient? (b) If so, how much? (c) Is the patient most probably shocked (above sensation)? (d) Is the patient most probably electrocuted (killed) by modern standards?

19-3 Refer to Figure 19-12 and case 3. Given the *patient's resistance* to be 500 Ω and the *differential voltage* between the *patient's right leg* and blood pressure saline-filled *catheter* to be 40 mV, answer the following questions: (a) Does current pass through the patient? (b) If so, how much? (c) Is the patient most probably shocked (above sensation)? (d) Is the patient most probably electrocuted (killed) by modern standards?

19-23 References

1. *Patient Safety—Electrical Safety*, Hewlett-Packard applications note AN718 (Waltham, Mass., 1971).
2. *Using Electrically-Operated Equipment Safely with the Monitored Cardiac Patient*, Hewlett-Packard applications note AN735 (Waltham, Mass., 1970).
3. *AAMI Safety Standards for Electromedical Apparatus, Safe Current Limits*, Association for the Advancement of Medical Instrumentation (Arlington, Va., 1977).
4. National Fire Protection Association, *National Electrical Code* (Boston, Mass., 1977).
5. *Joint Commission on the Accreditation of Hospitals* Manual (Chicago, Ill., 1978).
6. Walter, Carl W., "Green Wire Spells Electrical Safety in Hospitals," *Hospital Topics, 50* (2): 25–29 (February 1972).
7. Friendlander, G. D., "Electricity in Hospitals—Elimination of Lethal Hazards," *IEEE Spectrum* (September 1971).
8. Butler *v.* Northwestern Hospital of Minneapolis, 202 MINN. 282 278 N.W. 37, legal case.
9. Norton, M. L., "Biomedical Instrumentation and Liabilities," *J. AAMI Mag* (June 1971).
10. *Operating and Service Manual*, Safety Analyzer for Model 431A/431F, Neurodyne-Dempsey, Inc. (Napa, Calif., 1975).
11. *Operating and Service Manual*, Microspector, Model 449, Neurodyne-Dempsey, Inc. (Napa, Calif., 1975).
12. *Operating and Service Manual*, Cordometer, Model 447, Neurodyne-Dempsey, Inc. (Napa, Calif., 1974).
13. Strong, Peter, *Biophysical Measurements*, Measurement Concepts Series, Tektronix, Inc. (Beavertown, Ore., 1970).
14. Leeming, Michael N. and Eric Perron, *Electrical Safety Program Guide*, Biotek Instruments, Inc. (Shelburne, Vt., 1975).
15. Roth, Herbert H., Erwin S. Teltscher, and Irwin M. Kane, *Electrical Safety in Health Care Facilities*, Academic Press (New York, 1975).
16. *National Safety News*—Periodical, October 1976, Data Sheet, "Electrical Safety in Health Care Facilities."
17. Hoenig, Stuart A. and Scott H. Daphne, *Medical Instrumentation and Electrical Safety—The View from the Nursing Station*, Wiley (New York, 1977).
18. Simmons, David A., *Medical and Hospital Control Systems, The Critical Difference,* Little, Brown Co. (Boston, 1972).
19. Spooner, Robert B., *Hospital Instrumentation Care and Servicing for Critical Care Units*, Instrument Society of America (Pittsburgh, Pa., 1977).
20. Caceres, Cesar A. and Albert Zara, *The Practice of Clinical Engineering*, Academic Press (New York, 1977).

Chapter 20

Waveform Display Devices

20-1 Objectives

1. Be able to describe the principles behind the servomechanism recorder and recording potentiometer.
2. Be able to describe the operation of the permanent magnet moving coil (PMMC) galvanometer mechanism.
3. Be able to list and describe the different types of writing system used in mechanical recorders.
4. Know how to describe the principles behind the medical cathode ray oscilloscope (CRO).
5. Be able to list the different types of CROs available.
6. Be able to describe how a *nonfade* medical CRO operates.
7. Know how to describe the principal differences between medical and engineering CROs.

20-2 Self-Evaluation Questions

These questions test your prior knowledge of the material in this chapter. Look for the answers as you read the text. After you have finished studying the chapter, try answering these questions and those at the end of the chapter (Section 20-18).

1. Describe in your own words a recording potentiometer.
2. Describe in your own words the operation of a PMMC galvanometer.
3. What is a "deadband"?
4. Describe in your own words the type of paper used in thermal recorders.
5. Medical CROs use ______________ persistence phosphors on the viewing screen.
6. List two methods for creating two or more channels from a single CRO electron beam.
7. Typical sweep speeds for medical oscilloscopes are ______________ mm/s and ______________ mm/s.
8. List two different forms of sweep on a nonfade medical CRO.

20-3 Permanent Magnet Moving Coil Instruments

The chart recorder is used in medical instrumentation to make permanent recordings (i.e., "hard copies") of the analog waveforms produced by the physiological measurement instruments. One of the most common types of mechanism is the *permanent magnet moving coil* (PMMC) *galvanometer*. The PMMC is very similar to the D'Arsonval meter movement. A *writing pen* replaces the meter point and is attached to the moving coil assembly that is in the field of a strong permanent magnet. Current flowing in the moving coil creates a magnetic field that interacts with the magnetic field of the permanent magnet. This will cause deflection of the pen assembly in the same manner as the meter needle will deflect in the D'Arsonval mechanism.

The tip of the pen in a PMMC assembly is positioned over a strip of chart paper that is pulled under the pen tip at a constant speed. This mechanism, therefore, will make *Y-time* recordings in which the *Y*-axis is the deflection of the pen and the *X*-axis is the time base established by moving the chart paper at a constant speed past the pen tip. The PMMC assembly, then, will produce a chart recording of the *waveshape* of the applied waveform. These instruments are sometimes called *recording oscillographs*.

The PMMC pen assembly sweeps an arced path and so will write in a curvilinear manner, as in Figure 20-1*a*. The tip of the pen travels in an arc because of the rotary motion of the moving coil assembly. In some cases this may be tolerable, especially if the user will record on paper that has a special curved grid. This type of recorder is used in student instruments and where amplitude rather than waveshape is the important parameter.

The pivoted pen motor assembly of Figure 20-1*b* is a solution to the problem of curvilinear action that is used on some higher-priced ink machines. The PMMC is not connected to the pen directly, but through a mechanical link that translates the curvilinear motion of the PMMC to the *rectilinear* motion at the pen tip.

A *pseudorectilinear* writing system is shown in Figure 20-1*c*. In this type of recorder the pen assembly is very long compared with the width of the chart paper. The pen tip, therefore, travels in an arc whose length is very small compared with the radius (i.e., pen length). The trace will appear to be nearly linear in this type of recorder.

There are several different writing methods used in strip-chart recorders, but two are amenable to a special type of pseudorectilinear recording: *thermal* and *direct contact*. An example of the thermal recorder is shown in Figure 20-1*d*. Both of these types of recorder use a special writing stylus (rather than a pen) and a *knife edge*, also called a *writing edge*. The mark is made on the paper by the contact of the stylus on the paper along the knife edge. The stylus tip still travels in a curvilinear path, but the resulting trace is rectilinear because the knife edge is straight. This technique works because these recording methods work by contact pressure between the stylus and the knife edge. The stylus can write anywhere along its length, so by keeping the knife edge straight under the paper, we obtain the rectilinear recording that clearly shows waveshape as well as amplitude.

20-4 PMMC Writing Systems

There are several different writing systems commonly used on PMMC recorders: *direct contact, thermal, ink pen, ink jet,* and *optical*.

Recorders that use any type of pen (i.e., ink) or stylus (i.e., direct contact and thermal) have a relatively low-frequency response due to the inertia of the pen or stylus assembly. Most such recorders have an upper −3 dB frequency response point of 100 to 200 Hz. Ink jet and optical types, on the other hand, use lighter-weight fixtures in the writing system and so have a higher natural frequency response. An upper −3 dB point in the range 1000 to 3000 Hz is possible in some of these instruments.

The direct contact writer uses a special type of chart paper that is chemically treated to have a carbonized back. When a pressure

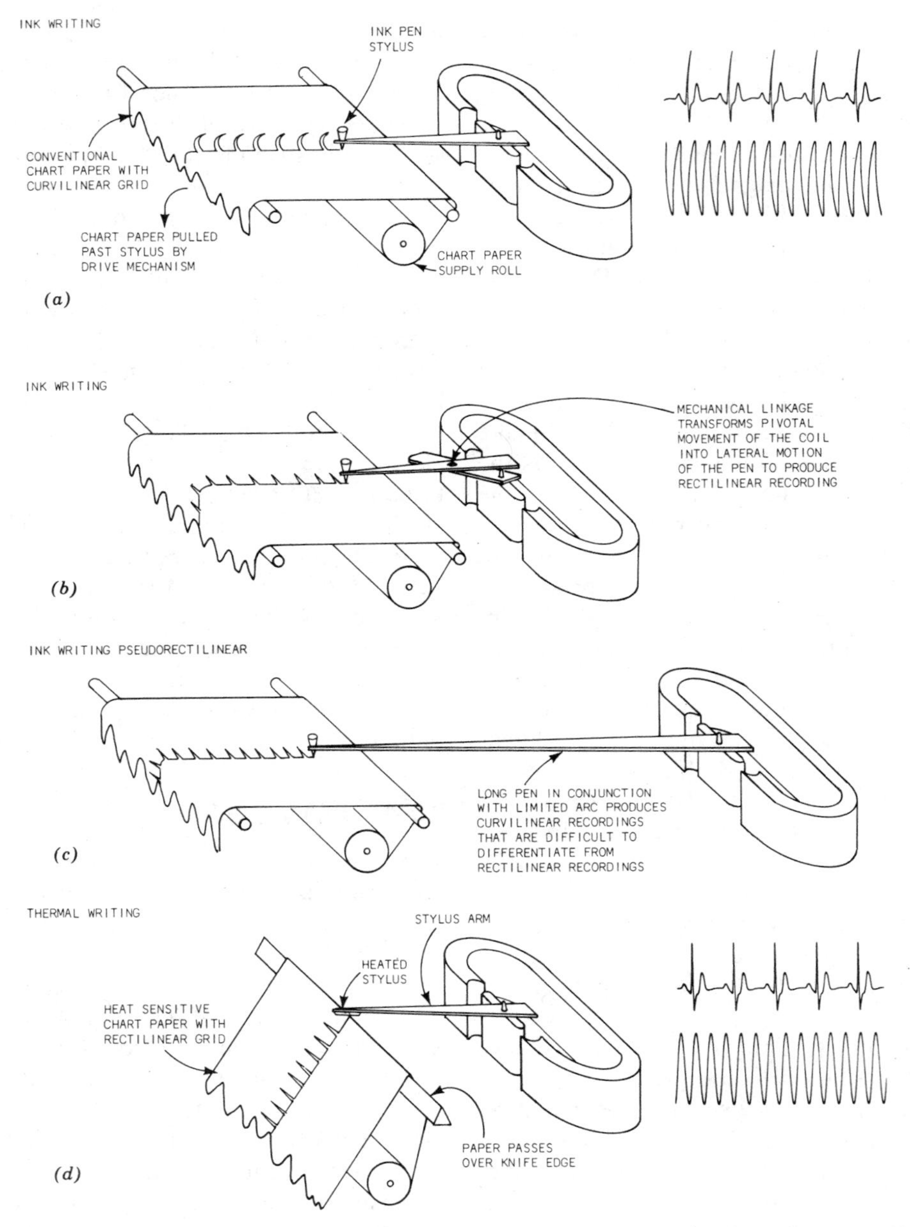

Figure 20-1 Curvilinear and rectilinear galvanometric recorder mechanisms. (Courtesy Tektronix, Inc., Beaverton, Ore. Copyright 1970. All rights reserved.)

is applied to the front of the paper, a black mark will appear through the paper. Most of these instruments have a frequency response of less than 25 Hz and so are not commonly used in medical instrumentation.

The thermal recorder also uses a special paper, but in this case it is *waxed* or *paraffin* treated so that it will turn black when *heated*. The thermal recorder is, by far, the most commonly employed in medical instrumentation, especially cardiovascular instruments such as the ECG or pressure monitors.

The stylus in a thermal system is little more than a heated resistance wire connected to a low-voltage ac or dc power supply. Early models formed the stylus tip from a U-shaped electrical resistance element, while modern models use a resistance wire inside of a cylindrical metal stylus. In both cases a low-voltage electrical power supply energizes the element, causing the tip to become heated almost to incandescence. The black mark is made at the points where the heated stylus touches the paper. Note that it takes less time to write at high temperatures than at low temperatures, so multispeed instruments typically apply a higher voltage to the heater at high speeds than at low speeds.

Ink pen writers use a hollow pen and an ink supply to write on the chart paper. In some machines the ink is pressurized by an atomizerlike (i.e., squeeze ball) hand pump. Many polygraph machines used in EEG recording use this type of writing system. Other instruments, however, are somewhat more automatic and use a thick, viscous ink in a special cartridge that is placed under pressure by a spring driven piston. In multichannel instruments (Figure 20-2), the ink may be distributed to all pens from the same cartridge by connecting the line from the cartridge to the input side of a special *ink manifold* and the lines to the pens to the output sides of the manifold. In many multichannel machines additional pressure is applied to the ink by the solenoid-operated manifold bladder.

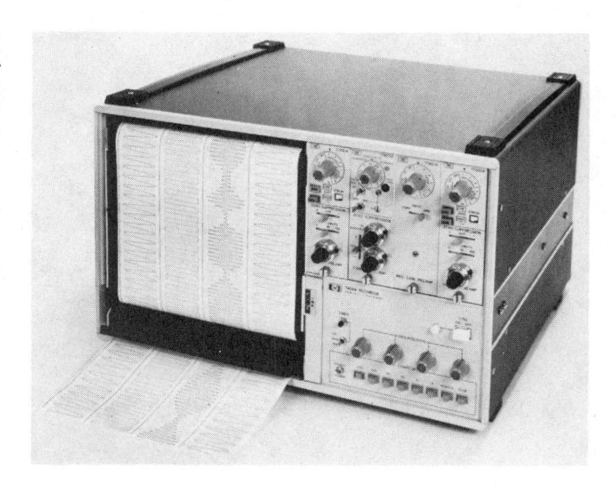

Figure 20-2 Four-channel recorder. (Photo courtesy Hewlett-Packard.)

The high-velocity ink jet recorder is capable of higher-frequency responses than are the types mentioned previously. This type of recorder is popular on European instruments, of which many have found their way to the United States. In the ink jet recorder, low-viscosity ink is directed to a *nozzle* mounted on a PMMC galvanometer in place of the pen assembly. The ink jet produced by the nozzle is directed at the paper, and, when the system is properly adjusted, it will produce a recording that is very nearly rectilinear. Only a small amount of trace fuzziness due to ink splattering is apparent.

There are actually two types of optical recorder in common use. One is a PMMC type and uses a mirror in place of the pen or stylus. The other optical recorder uses photographic paper that is pulled across a cathode ray tube screen and is called a CRT camera recorder.

An example of the PMMC optical recorder is shown in Figure 20-3. A small, low-inertia mirror is mounted on the PMMC galvanometer in place of the pen assembly. The

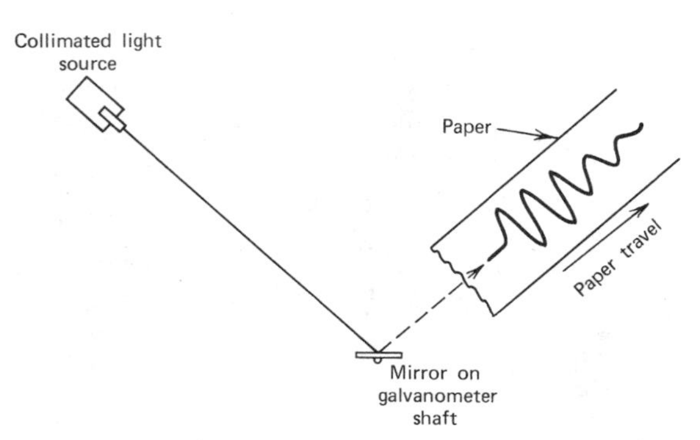

Figure 20-3 Light-beam galvanometer.

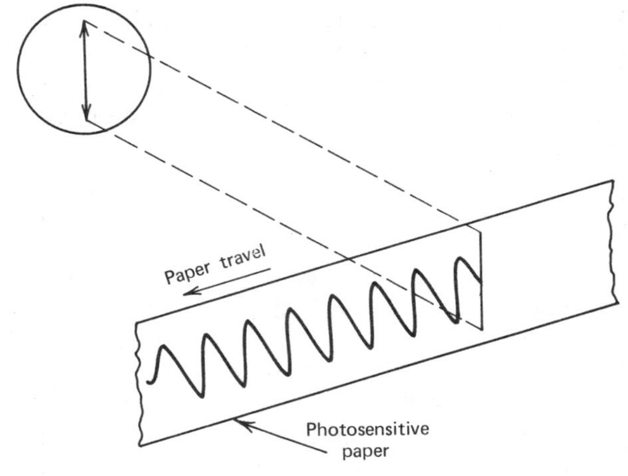

mirror reflects a pencil-thin beam of collimated light onto the photosensitive paper. Most of these recorders use wide paper (i.e., 6 in. wide or wider) so the resolution is better than on the smaller paper widths used in pen and thermal recorders. Additionally, on a multichannel optical recorder it is possible to examine the time relationships between different traces more easily because the traces can be allowed to *overlap* each other.

The paper in the optical recorder is often developed by exposure to an ultraviolet lamp as the paper comes out of the recorder, an example of *postfogging*. The trace will fade over a long period of time, however, unless the paper is either wet-developed following the recording session or is stored in a light-tight box.

The other type of optical recorder is the CRT type shown in Figure 20-4*a*; a commercial example (*Electronics-for-Medicine* VR-12) is shown in Figure 20-4*b*. In the CRT camera type of recorder, the CRT sweeps only the vertical axis (as many as there are channels). The time base is provided by pulling the photosensitive paper in front of the CRT screen.

The frequency response of the CRT camera recorder is better than that of any of the other types, being limited mostly by the writing speed of the photosensitive paper.

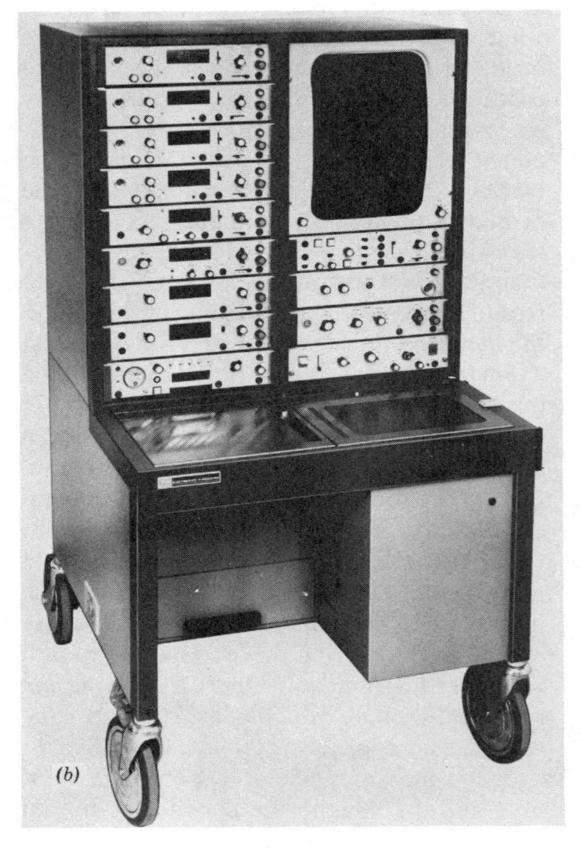

Figure 20-4 CRT camera optical recorder. (*a*) Writing system. (*b*) E-for-M VR-6. (Photo courtesy Electronics for Medicine.)

20-5 Servorecorders and Recording Potentiometers

In potentiometric measurements a three-terminal variable resistor (i.e., a potentiometer) is connected to produce an output voltage that is a function of both a reference potential and the position of the variable resistor's wiper arm. A galvanometer such as a zero-center D'Arsonval or taut-band meter movement will read *zero* when the unknown voltage and the potentiometer voltage are equal. In normal operation the operator will manually null the voltage displayed on the meter and then read the value of the unknown voltage from the voltage calibration on the potentiometer dial. Figure 20-5*a* shows a simplified schematic for a *recording potentiometer* that is self-nulling. A *servorecorder* is a self-nulling potentiometer that records the waveshape of the applied signal on graph paper. A commercial example of a servorecorder is shown in Figure 20-5*b*.

Figure 20-5*a* shows the dc potentiometer servorecorder mechanism. The pen is attached to a string that is wound around a pair of *idler pulleys* and a *drive pulley* that is on the shaft of a dc servomotor. The pen assembly is also linked to a potentiometer (R_1) in such a way that the position of the wiper arm on the resistance element is proportional to the pen position.

The potentiometer element is connected across a precision reference potential, E_{ref}, so potential E will represent the position of the pen; that is, E is the electrical analog of pen position. When the pen is at the left-hand side of the paper in Figure 20-5*a*, E will be zero, and when the pen is full scale at the right-hand side of the paper, E is equal to E_{ref}.

The pen position is controlled by the dc servomotor, which is in turn driven by the output of the servo amplifier. The amplifier has *differential* inputs; E_{in} (the unknown) is connected to one input and the position signal E is connected to the other input. The difference signal ($E_{in} - E$) represents the *error* between the *actual* pen position and the position the pen *should* be in for the applied voltage. If the error signal is zero, meaning that the amplifier output is also zero and the pen is correctly positioned, then the motor remains turned off. But if $E \neq E_{in}$, then the amplifier sees a nonzero input signal and therefore creates an output signal that turns on the motor.

The motor drives the pen and potentiometer in such a direction as to *cancel* the error signal. When the input signal and position signal are equal, then the motor turns off and the pen remains at rest.

A paper drive motor forms a time base because it pulls the paper underneath the pen at a constant rate. In most servorecorders a sprocket drive is used instead of the friction rollers used in most PMMC machines. The paper used in these machines (see Figure 20-5*b*) has holes along the margins to accept the sprocket teeth.

Most high-quality servorecorders use a *stepper motor* to drive the paper supply. Such a motor will rotate only a few degrees every time a pulse is applied to its windings. A few models use a continuously running motor that drives the sprocket through a speed-reducing transmission gear box.

The stepper motor system is actually capable of very good time base accuracy because a crystal oscillator or the ac power mains (i.e., 60 Hz) is used to drive the pulses used to advance the motor. Digital integrated circuit frequency dividers (i.e., countercircuits) can be used to reduce the clock frequency to the frequency required to drive the motor at the desired speed.

The reference potentiometer used to measure the position of the pen may be any of the following devices: slide wire, rectilinear, or rotary models. The slide-wire system is often used because it can be built with less friction and no mechanical linkage. An ex-

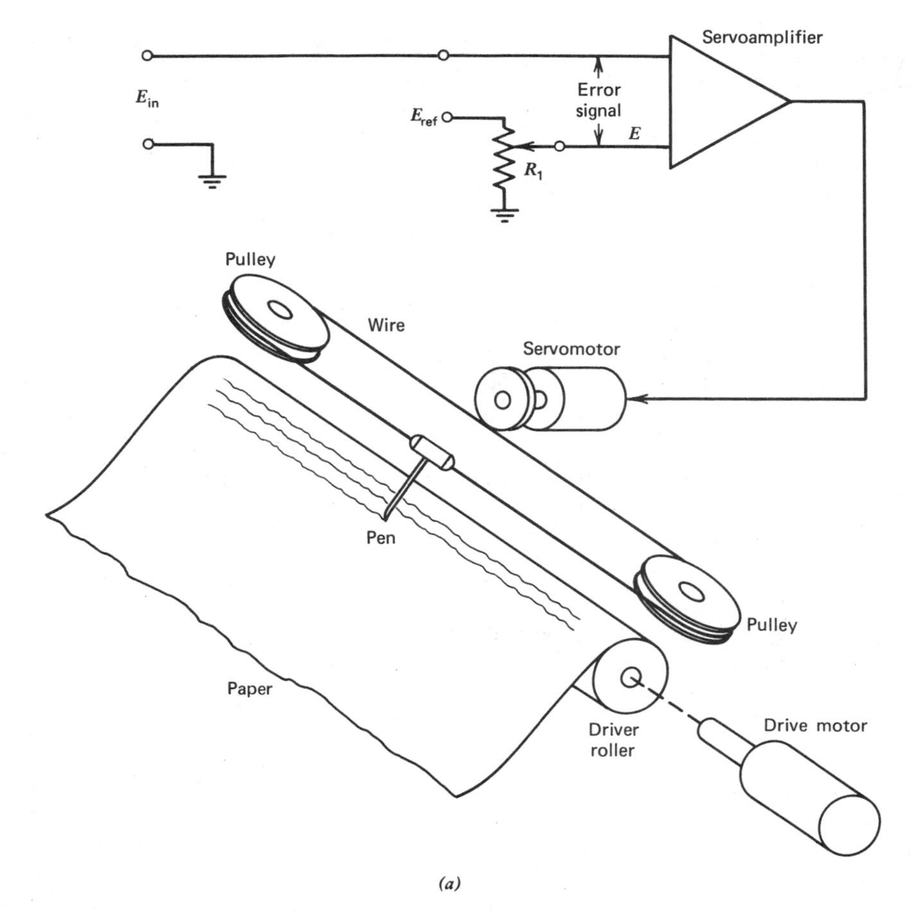

(a)

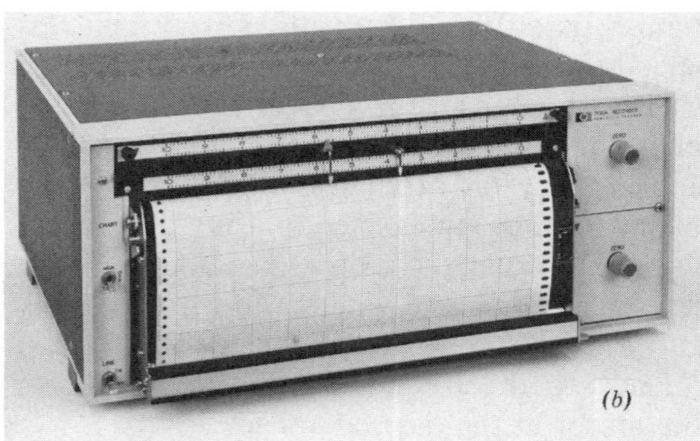

(b)

Figure 20-5 Servorecorder. (*a*) Simplified schematic. (*b*) Commercial example. (Photo courtesy Hewlett-Packard.)

ample of a slide-wire model is shown in Figure 20-6*a*, while the equivalent points are shown in Figure 20-6*b*. A *resistance wire* and a shorting wire are stretched taut parallel to each other and the direction of pen travel. A shorting bar on the pen assembly serves as a wiper on the resistance element and also connects the shorting wire. *A*, *B*, and *C* in Figure 20-6*a* refer to the potentiometer terminals shown in Figure 20-6*b*. The shorting wire serves as terminal *B* of the potentiometer (i.e., the wiper).

20-6 *X-Y* Recorders

An *X-Y* recorder (see Figure 20-7) uses *two* servomechanisms connected to the same

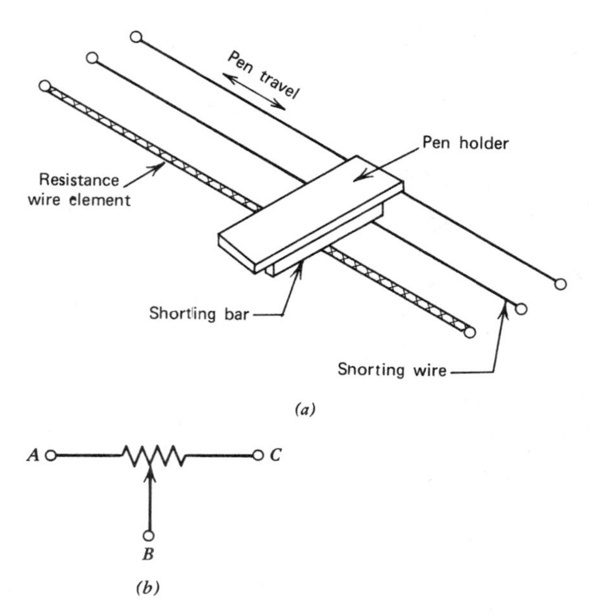

Figure 20-6 Slide-wire potentiometer. (*a*) Mechanism. (*b*) Equivalent circuit.

writing assembly, but at right angles to each other. The *X*-axis servomechanism moves the pen-bar assembly back and forth across the paper in the horizontal plane, while the *Y*-axis servomechanism moves the pen vertically up and down along the bar.

Figure 20-7 *X-Y* recorder. (Reprinted courtesy Hewlett-Packard.)

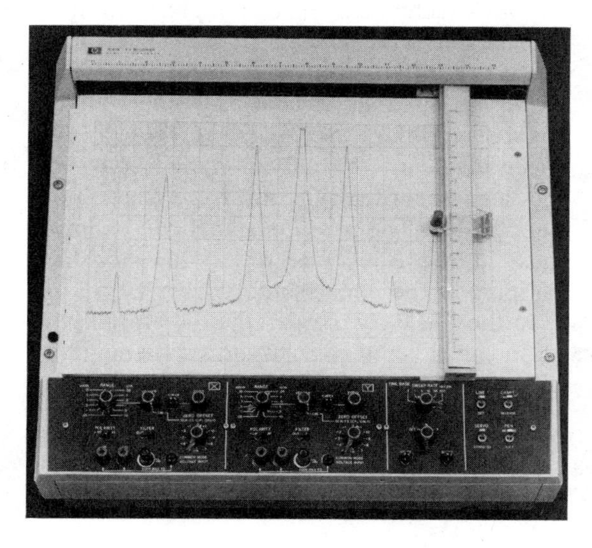

The paper itself does not move. It is held in place either by clamps or, in high-quality instruments, by a vacuum pump that is used to evacuate a hollow chamber below the paper platform. Holes in the platform create the negative pressure needed to keep the paper in place.

One advantage to the *X-Y* recorder over almost any other type of instrument is that almost any type of paper may be used. One could, for example, use specially printed paper that gives units proper to, say, pulmonary function tests (Chapter 10).

The *X-Y* recorder can plot Lissajous patterns because both *X* and *Y* axes can be driven by external signals. If a *Y*-time display is required, then we can drive the *X*-axis with a linear ramp function.

20-7 Problems in Recorder Design

All pen assemblies will have mass, regardless of which type of drive mechanism is used. Because of the inertia produced by this mass, the pen will not begin moving from being at rest until a certain minimum signal voltage level is provided. This phenomenon, called the *deadband,* is shown in Figure 20-8. A deadband signal is the largest signal to which the recorder will *not* respond. In most recorders the deadband is approximately 0.05 to 0.1 percent of full scale.

The deadband can create severe distortion in low-amplitude signals (i.e., those whose amplitude approximates the deadband voltage). The solution to the deadband is to slew through it as rapidly as possible. This means that sufficient preamplification of the signal is required so that as much of the recorder's span as possible is used.

Another problem that can distort waveforms is *overshoot* and *undershoot* of the

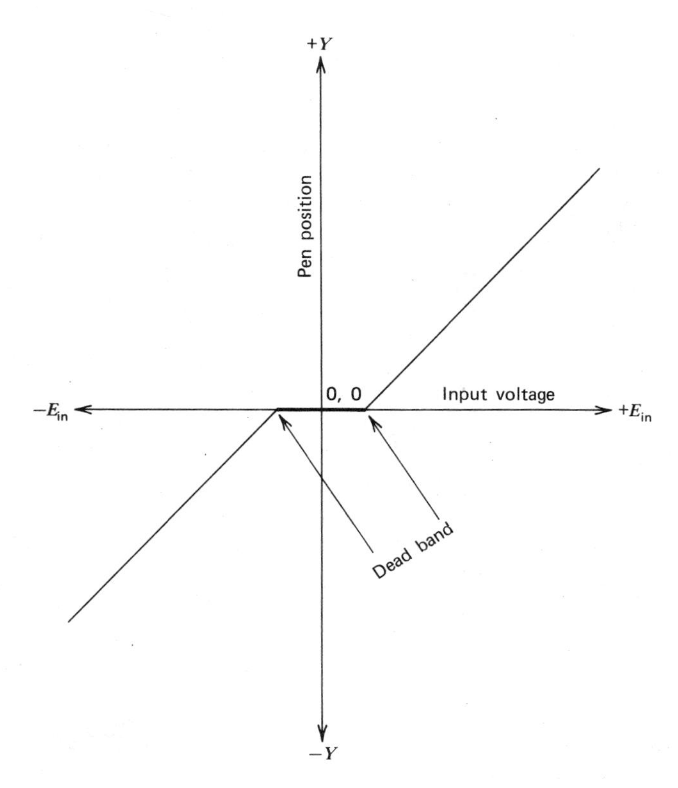

Figure 20-8 Deadband.

recorded trace in response to a *step-function* input signal. This problem affects all mechanical writing systems. Figure 20-9 shows the three types of responses found in recorders. In the *critically damped* case the signal rises smoothly and quickly to the proper value with little or no curvature as it approaches the proper point.

An *undercritically damped* recorder will *overshoot* the correct point and then hunt back and forth across the correct point for a few cycles until it hones in and settles properly. If a squarewave is applied to such a recorder, the trace will show *ringing* on the waveform.

An *overcritically damped* recorder is sluggish; the pen approaches the correct position very slowly. A squarewave applied to such a system will have a rounded corner characteristic of a squarewave that has been passed through a low-pass filter circuit.

The three different types of damping property are mechanical analogs of electrical frequency-selective filtering. In fact, electronic filters are often used to correct improper (i.e., over or undercritical) damping. Some PMMC recorders use a *pen position transducer* inside of the galvanometer housing, and servorecorders already have a pen position signal. The *position* signal can be low-pass filtered (i.e., integrated) to vary the damping of the pen assembly.

In the PMMC recorder the integrated position signal is fed back to the servoamplifier to be summed with the error signal.

Pen assemblies are often damaged if they strike the limits-of-travel stops at a high speed. If a signal that is very large relative

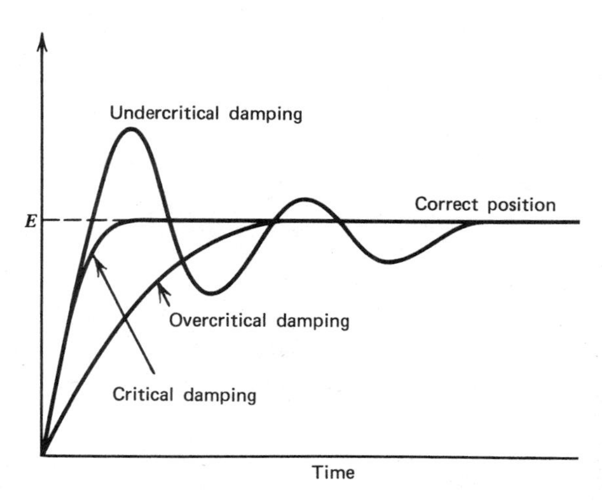

Figure 20-9 Critical, overcritical, and subcritical damping.)

to the full-scale signal is applied to the input, then the pen may hit the mechanical stops and break.

Some PMMC recorder amplifiers use output limiting to guard against such damage. Alternatively, a pair of Zener diodes connected back to back across the PMMC coil are sometimes used to accomplish the same job. The Zener potential of the diodes is selected so that the diodes break over and conduct current only when a voltage greater than the normal full-scale potential is applied to the input of the amplifier.

If a PMMC *position* signal is available, then it can be *differentiated* to produce a *velocity* signal (i.e., $v = dx/dt$), and the position signal may be differentiated *twice* to form an *acceleration* signal (i.e., d^2x/dt^2). These signals can be used to apply a hard *braking* signal to the PMMC coil if either pen velocity or acceleration exceeds certain limits. The pen will still strike the stops, but with considerably reduced force.

Most recorders have a *damping control* that is adjusted when a squarewave (i.e., 1 Hz) is applied to the amplifier input. The damping control is adjusted for best ''squareness'' of the tracing.

20-8 Maintenance of PMMC Writing Styluses and Pens

There are several common faults of medical paper recorders. These are sometimes amenable to adjustment, while in other cases repair is necessary. Let's consider separately the ink pen and heated stylus types of recorder. Because they find only limited use in medical equipment, we will not consider some of the other types (even though you need to be aware of them conceptually). In this section we will discuss some common maintenance actions that do not always require trained engineering technicians.

Figure 20-10 shows how to remove an ink blockage from an ink pen recorder that has been allowed to stand too long without being used. As a general rule, such recorders should be run for about five minutes or so once a week when not in regular service. Otherwise, the ink will dry up at the tip and prevent the recorder from writing. Fill a 3- to 10-cc syringe with water (or acetone, if certain types of ink are used), and insert the needle end into the ink inlet on the rear end of the pen assembly. On most recorders the pen will have to be removed from the machine for this operation. The needle should be inserted up to the Luer lock hub in order to make a good fluid seal. Quickly, and with a single sharp motion, drive the plunger ''home'' so that a high-pressure jet of water or acetone is forced into the pen. The ink clot should be forced out the other end.

This procedure usually works well. Some precautions are in order, however. First, always wear protective goggles over your eyes when doing this operation. Also, wear protective clothing, as ink and water will splatter everywhere. Second, make sure that the pen tip is aimed downward into a sink. Fi-

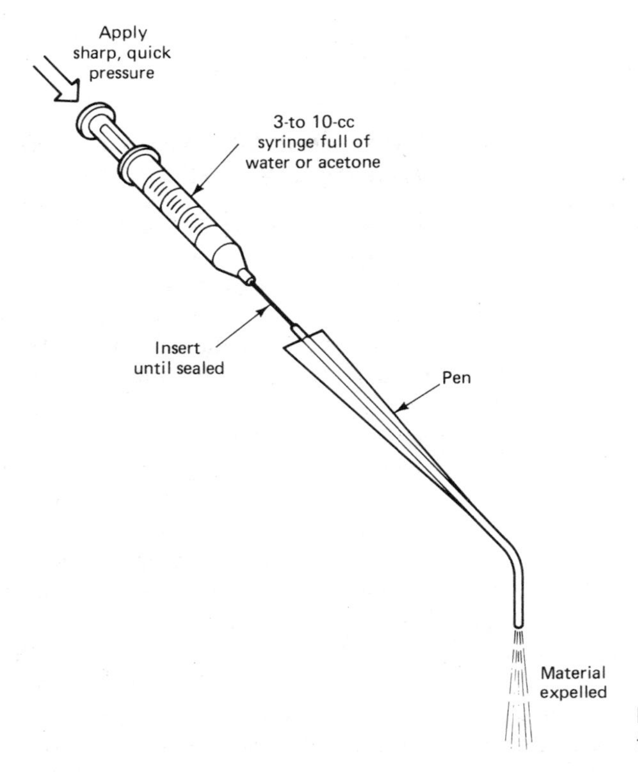

Figure 20-10 Removing ink clots from a pen recorder ink tip.

nally, as always when dealing with needles, be careful not to stick yourself.

Keep in mind that the thick, high-viscosity ink used in these machines stains everything it touches and is nearly impossible to clean.

Ink pen tips are designed to operate parallel to the paper surface (see inset in Figure 20-11). If the pen is worn, or when a new pen is installed, it is necessary to ''lap'' the tip in order to reestablish the parallelism. The symptom that lapping is needed will be either (or both) of the following: (a) a ''blob'' of ink when the machine first starts recording a waveform, or (b) a too-thick trace. In most such machines the ink should be dry before the paper leaves the paper platform at the drive roller end of the surface. If it is not dry, then lapping may be needed. To lap the pen, place a piece of very fine emery cloth (a sandpaper-like material available at any hardware store) under the tip. The pen tip is worked back and forth 5 to 10 times to ''sand'' the tip parallel to the paper.

The pressure of the stylus or pen is also important. If the pressure is not correct, then the waveform may be distorted. On medical equipment it is possible to make a normally healthy Lead-I ECG signal look like it has either a heart block or a recent myocardial infarction because of improperly adjusted stylus pressure. The manufacturer will specify a pressure in grams. These numbers vary from 1 to 10 grams depending upon the machine model (and its year of manufacture—older model machines tended to use heavier stylus pressures).

A stylus pressure gage (see Figure 20-12) is used to lift the pen or stylus from the paper as the machine is running until the trace just

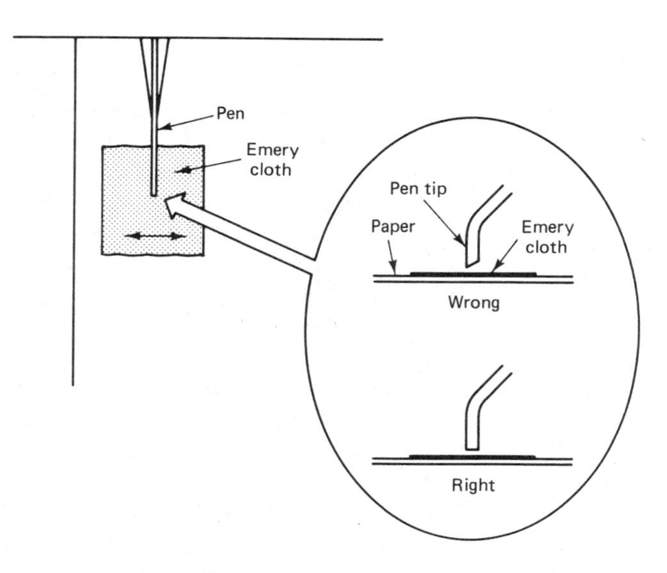

Figure 20-11 Lapping a pen tip to make it parallel to the paper.

disappears. The pressure reading is then made from the barrel of the gage. Suitable stylus pressure gages can be purchased from ECG machine manufacturers' service or parts departments. Alternatively, the stylus pressure gage used for record player "tone arms" is also useful if the specified pressure is within their relatively limited range (usually 0 to 4 grams). The stylus pressure adjustment is made using a screw that is usually located on the rear of the stylus (or pen), or the assembly that holds it in place.

Figure 20-12 Measuring stylus or pen pressure in analog recorder.

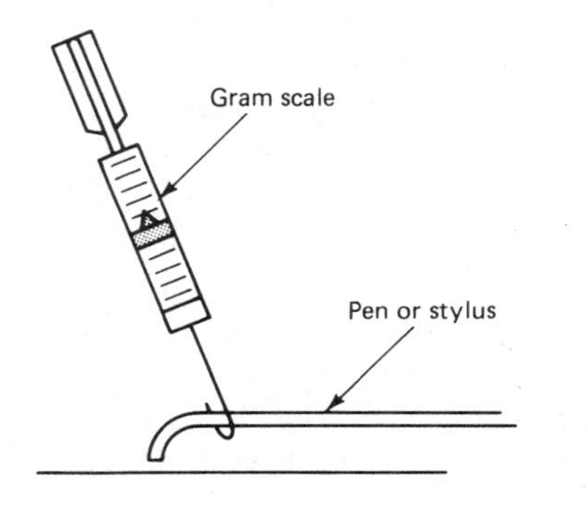

Figure 20-13 shows several different 1-mV calibration pulses from an ECG machine. These traces can be made by pressing the "1-MV" or "CAL" button on the front panel of the machine. The ideal shape is perfectly square, as shown in Figure 20-13*a*. But this ideal is almost never achieved in practical machines because of the inertia of the pen or stylus assembly. Usually we see the slightly rounded features of the pulse shown in Figure 20-13*b*. This waveform is usually acceptable. What is not acceptable, however, are the overdamped and underdamped waveforms of Figures 20-13*c* and 20-13*d*, respectively.

On all recorders, there may be a damping control available for adjustment by a properly trained technician. This control is adjusted (usually internally to the machine) in order to compensate for problems. On ink pen machines the stylus pressure can affect the waveshape, especially if it is set to too high a value (it produces the overdamped waveform). On heated stylus machines, both the stylus pressure and the heat can affect the waveform.

On some heated stylus machines, the

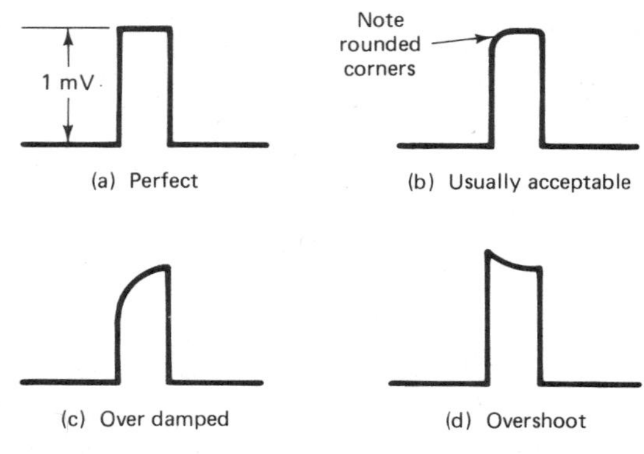

Figure 20-13 One millivolt (1 mV) calibration pulses: (*a*) ideal shape; (*b*) slightly rounded figures normally seen; (*c*) overdamped pulse; (*d*) underdamped pulse.

standard procedure is to set the pressure to a specified value, set the voltage applied to the stylus heating element to a specified value (usually either 5.00 or 7.00 volts), and then adjust the internal damping control to produce the waveform of Figure 20-13*b* in response to either a squarewave input or successive presses of the "1-MV CAL" button.

If the manufacturer of the machine did not provide a knob on the stylus heat control, then do not adjust it without the correct equipment (usually stylus pressure gage and voltmeter) *and* the manufacturer's service manual.

20-9 Dot Matrix Analog Recorders

The dot matrix printer is long familiar to users of computer equipment. The dot matrix printer was developed in response to the high cost of traditional computer printers. When they became popular, dot matrix printers cost about one-fifth to one-third the cost of daisy wheel, Selectric, or similar printer mechanisms. The original dot matrix printers (mid-1970s) used a 5 × 7 matrix of dots (Figure 20-14*a*) to form alphanumeric characters.

The dot matrix machine used a print head (Figure 20-14*b*) to cause the correct dot elements to be energized to make a mark on the paper. Two different methods were once popular, although one has since faded almost to obscurity. Some of the earliest machines were thermally based. The dots were thermally connected to heating coils and could be heated when needed. Special temperature-sensitive paper was used to receive the text. This method is no longer used widely. The second method used an array of seven print hammers (actually pins). The pins would either extend or retract depending upon whether or not that particular dot was active for the character being printed. An advantage of the pin method is that ordinary paper can be used. The pins impact an inked ribbon to leave the impression. Although the original low-resolution printers were seven-pin models (as shown in Figures 20-14*a* and *b*), higher-resolution models are now available with 9-pin, 18-pin, and 24-pin print heads.

The computer world rapidly discovered that clever programmers could make a dot matrix print head do graphics as well as alphanumerics. The spate of newsletters, church and school bulletins, and other low-cost (but often creatively done) publications are a testimony to the graphics capability of modern dot matrix technology.

Dot matrix printing can be used to make analog recorders that outperform most older mechanical analog recorders. Figure 20-14*c* shows the concept schematically. A dot matrix print head with a very large number of pins is arrayed over a platen. The action depends upon whether thermal, electro-arc, or plain strip-chart paper is used. Regardless of the particulars, however, the result is a strip-chart recording that mixes analog and digital data on the same chart.

Figure 20-15 shows dot matrix analog recordings. The digital computer backing up this system can print the appropriate grid (notice the difference between Figures

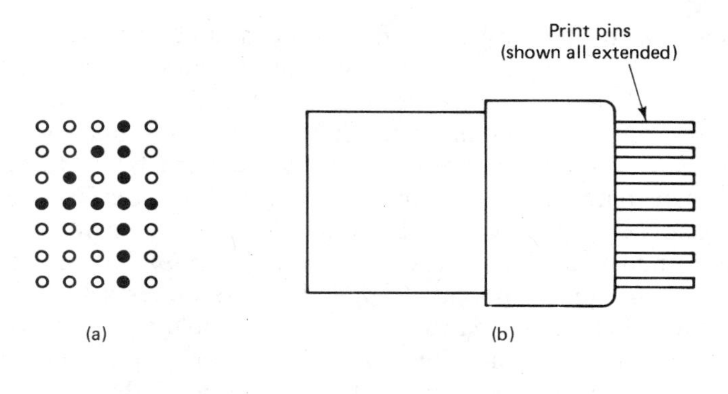

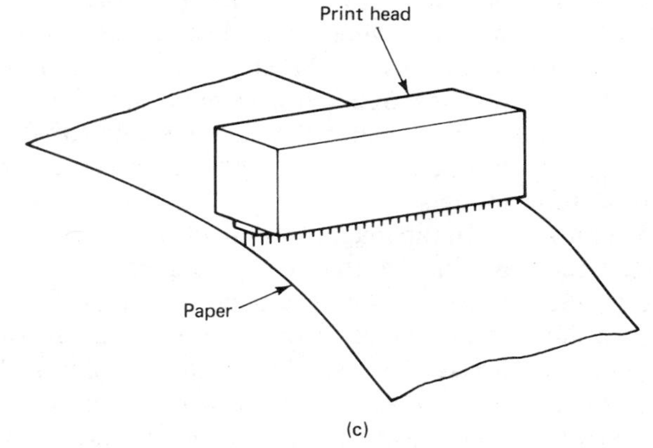

Figure 20-14 (*a*) 5 × 7 dot matrix; (*b*) dot matrix print head; (*c*) dot matrix recorder system.

20-15*a* and 20-15*b*; these recordings were made sequentially on the same machine without changing paper). Printed as alphanumeric characters along the top of the strip are data including ICU bed number, time, date, ECG lead number, heart rate, and blood pressure values. Figure 20-15*b* is similar to Figure 20-15*a* but includes an arterial

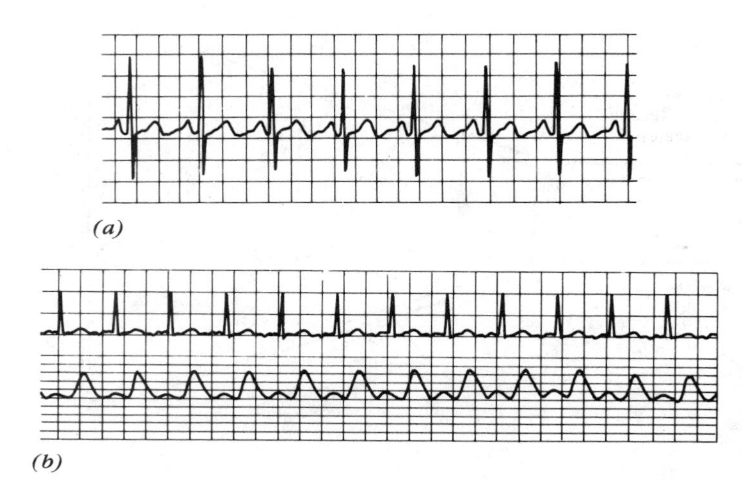

Figure 20-15 Dot matrix analog recordings: (*a*) ECG waveform; (*b*) ECG and pressure waveform.

blood pressure waveform along with the ECG waveform.

20-10 Oscilloscopes

The cathode ray oscilloscope has become an elementary instrument in the sciences, engineering, medicine, and several industries. Many measurement processes can take place, or are made easier, by the CRO because it will display not only amplitude, but also time and waveshape relationships. Many medical instruments use the CRO to display physiological waveforms as an alternative to paper-consuming strip-chart recorders.

The heart of any oscilloscope is the *cathode ray tube* (CRT) shown in Figure 20-16. An electron gun at the rear of the tube emits a beam of electrons that are accelerated and focused by special electrodes beyond the gun. When the accelerated electrons strike the phosphor-coated screen, they give up their kinetic energy in the form of light. Without any other external influences, the beam will impact exactly in the center of the screen.

Patterns can be drawn on the CRT screen by deflecting the beam up and down and left and right of its normal path. There are two basic types of CRT deflection systems in common use in medical CROs: *magnetic* and *electrostatic*. The electrostatic form is shown in Figure 20-16 and consists of two pair of deflection plates: one each for *horizontal* and *vertical deflection*. An electrical potential applied across either set of plates creates an electrostatic field that deflects the electron beam. The *polarity* of the potential determines the *direction* of the deflection, while its *magnitude* determines the *amount* of deflection. Most laboratory and service oscilloscopes use electrostatic deflection CRTs because they can operate to very high frequencies—up to 200–300 mHz in the best models.

In the magnetic deflection system, vertical and horizontal electromagnet coils are positioned around the neck of the CRT, concentric to the electron beam path. Both coils are housed in a single assembly called a *deflection yoke*. Current flowing in the deflection coils creates magnetic fields that deflect the electron beam. The inductance of the coils, plus their distributed capacitance, reduces the frequency response of the CRT system

Figure 20-16 Basic cathode ray tube structures.

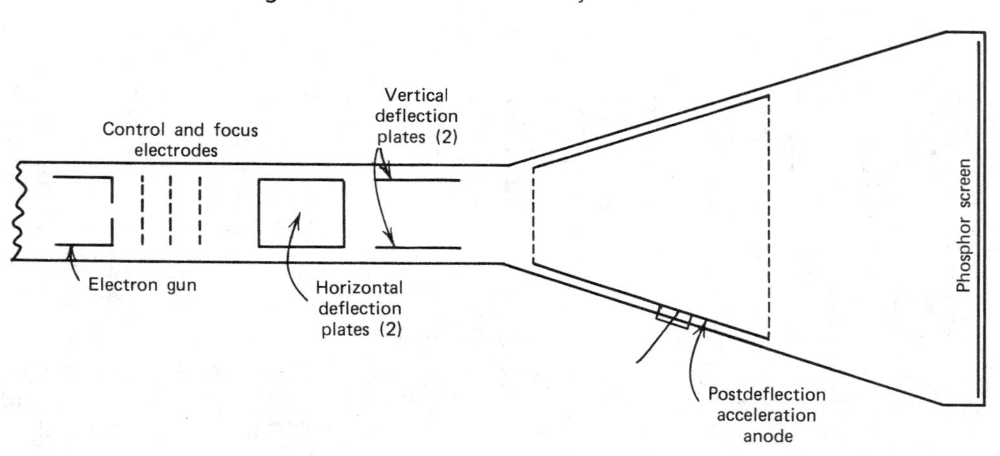

to less than 20 to 25 kHz. The frequency limitations of magnetic deflection systems prohibit their use in laboratory and service oscilloscopes. But in medical oscilloscopes, the frequency response required of the CRT is usually in the dc to 1000 Hz range, or less. Magnetic deflection, therefore, is also suitable for use in medical CROs. In fact, the magnetic deflection CRT usually has a shorter neck length than its electrostatic counterpart, and that is an advantage in that it allows the oscilloscope to be packaged in a shorter cabinet.

Most medical oscilloscopes are of the *Y-time* type, meaning that the signal, a time-varying voltage, is applied to the Y-axis (i.e., vertical channel), while a sawtooth time base signal is applied to the X-axis (i.e., horizontal channel). This type of sweep allows us to view the waveshape of the time domain signals such as the ECG and arterial pressure waveform. The horizontal sweep speed for most medical oscilloscopes is 25 mm/s or 50 mm/s, with some offering 100 mm/s as an option. On a standard 10-cm graticule, then, a 25-mm/s sweep speed requires a sweep time of

$$T = 10 \text{ cm} \times 10 \text{ mm/cm} \times 1 \text{ s}/25 \text{ mm}$$

$$T = ((10 \times 10)/25)\text{s} = \mathbf{4\ s}$$

which corresponds to a sweep frequency of 0.25 Hz, a distinct contrast to the megahertz sweep speeds of some laboratory oscilloscopes.

Some CRTs use only the electrodes prior to the deflection system to accelerate the electron beam to the point where it can produce light when it strikes the phosphor coating. But most medical CROs use CRTs that have a postdeflection acceleration electrode. This electrode is given a high (i.e., over 1 kV) positive potential to attract and accelerate the electron beam. The use of a postdeflection accelerator anode improves the linearity of the deflection system. Such tubes can usually be identified by the high-voltage "nipple" electrode on the glass bell.

20-11 Medical Oscilloscopes

The medical oscilloscope differs from service and laboratory models in several principal ways: horizontal sweep speed, vertical amplifier bandwidth, and CRT phosphor persistence.

The horizontal sweep speed difference has already been mentioned. The medical CRO sweeps in the subhertz range instead of kilohertz or megahertz. This low-frequency range is consistent with the fact that physiological signals tend to have fundamental frequencies in the range of the human heart rate.

In service and laboratory oscilloscopes, we look for as much vertical bandwidth as possible, within the limitations of budget. We pay a heavy premium for wide vertical amplifier bandwidth. But in a medical oscilloscope, that bandwidth is totally wasted and, in fact, may be detrimental if there is a lot of high-frequency noise on the input signal. The amplifiers driving the vertical deflection system of the medical oscilloscope are purposely limited in frequency response in order to eliminate or reduce its response to artifacts. An upper -3 dB point for medical oscilloscopes is typically found in the range of 200 to 2000 Hz, depending on application.

The other principal difference is the nature of the phosphor material used to coat the viewing screen. Service and laboratory CROs tend to have *short persistence* CRTs, while medical CROs use *long persistence* CRTs. Persistence is the property where the light remains on the screen for a short time after the electron beam has swept past the point. Once generated, the light *decays* after the electron beam moves to another point; it does not turn off instantly. On the medical CRO we use long persistence CRTs because the waveforms being viewed have such low fundamental frequencies; such long sweep times are needed because if they were not used, the features of the waveform on the leading edge of the waveform would fade before the trailing edge is written onto the screen. But a long persistence CRT (i.e., P_7 phosphors) allows the entire waveform to be viewed at once.

The long persistence phosphors have an interesting light spectrum—it is a mixture of yellow and violet-blue. Without a filter over the viewing screen, it appears blue to violet. Yellow and green displays can be created by the use of an appropriate color filter. Both yellow and green displays, however, use the blue-violet CRT!

20-12 Multibeam Oscilloscopes

There are no true multibeam cathode ray tubes, but a multitrace display can be created by certain *switching* techniques and through the use of a *gating amplifier*. Various versions of these methods are used in multichannel medical oscilloscopes.

An example of a switching system is shown in Figure 20-17. In this example, the switching function is performed by a pair of CMOS analog switches such as the 4016 and 4066 devices. One side of both switches is connected to the input of the vertical amplifier. The other side of each switch is connected to the outputs of the preamplifiers, one for each input channel. The control terminals for the two switches are connected to out-of-phase squarewave drive signals, so that one is turned on while the other is off. This permits only one of the two channels to be connected to the input of the vertical amplifier at one time. In most models, the switching rate is 2000 to 100,000 Hz.

Figure 20-18 shows another switching circuit, this one used by Hewlett-Packard in the model 7803C two-channel medical CRO. This circuit uses bipolar transistors Q_1 through Q_4 as an electronic switch; Q_1 and Q_2 operate channel 2, while Q_3 and Q_4 operate channel 1. Switching action is controlled by the signal from a 2-kHz multivibrator. The multivibrator is designed such that the signal applied to CR_1 is exactly 180° out of phase with the signal applied to CR_2. On one-half of the 2-kHz squarewave, CR_1 will be forward biased and CR_2 will be reverse biased. Under this condition Q_3/Q_4 will be turned off, and channel 2 will be connected to the vertical amplifier through Q_1/Q_2. On the opposite half of the 2-kHz squarewave, however, exactly the opposite situation occurs: Q_3/Q_4 are turned on, connecting channel 2 to the vertical amplifier, while Q_1/Q_2 are turned off. As was also true in the previous example, we alternately apply the two signal channels to the signal vertical amplifier at a 2-kHz rate.

In a gating amplifier system, the vertical axis of the CRT is scanned at a fixed rate, usually in the 15- to 25-kHz range, while the horizontal axis is swept left-to-right at a 25 mm/s rate. This means that each successive vertical sweep will scan the CRT at a slightly different point, moving left to right with the horizontal sweep. Ordinarily, this action would produce a *raster* on the CRT screen, just as is used on the CRT screen of television receivers. But in medical CROs, the electron beam is kept *blanked* (i.e., turned off) most of the time. The screen is unblanked at specific *times* by pulses from the *gating amplifier* (Figure 20-19*a*), which is actually a voltage-controlled pulse generator. The pulse repetition rate is controlled by the input signal voltage. An input voltage of zero volts produces a certain fixed pulse rate, while positive and negative input voltages produce lower and higher rates, respectively.

Figure 20-19*b* shows the operation of the gating amplifier. When the input signal is zero (i.e., upper trace), then the pulses occur at a given, fixed rate. This causes each vertical sweep to be unblanked at precisely the

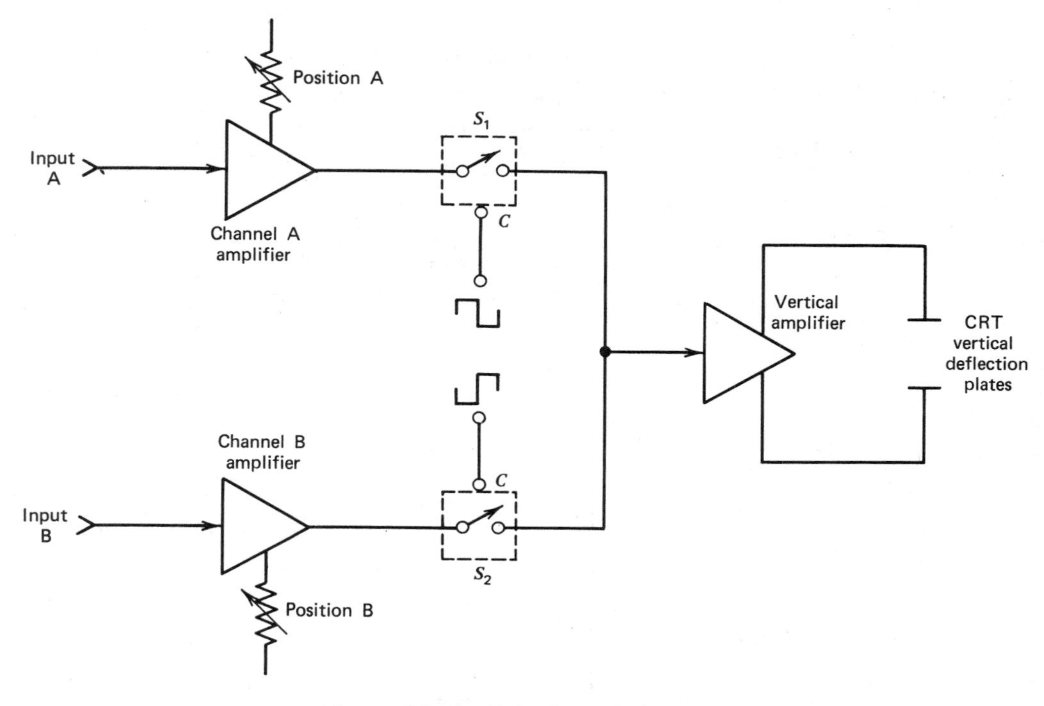

Figure 20-17 Two-channel chopper.

Figure 20-18 Two-channel chopper from the H-P 7803B. (Reprinted courtesy Hewlett-Packard.)

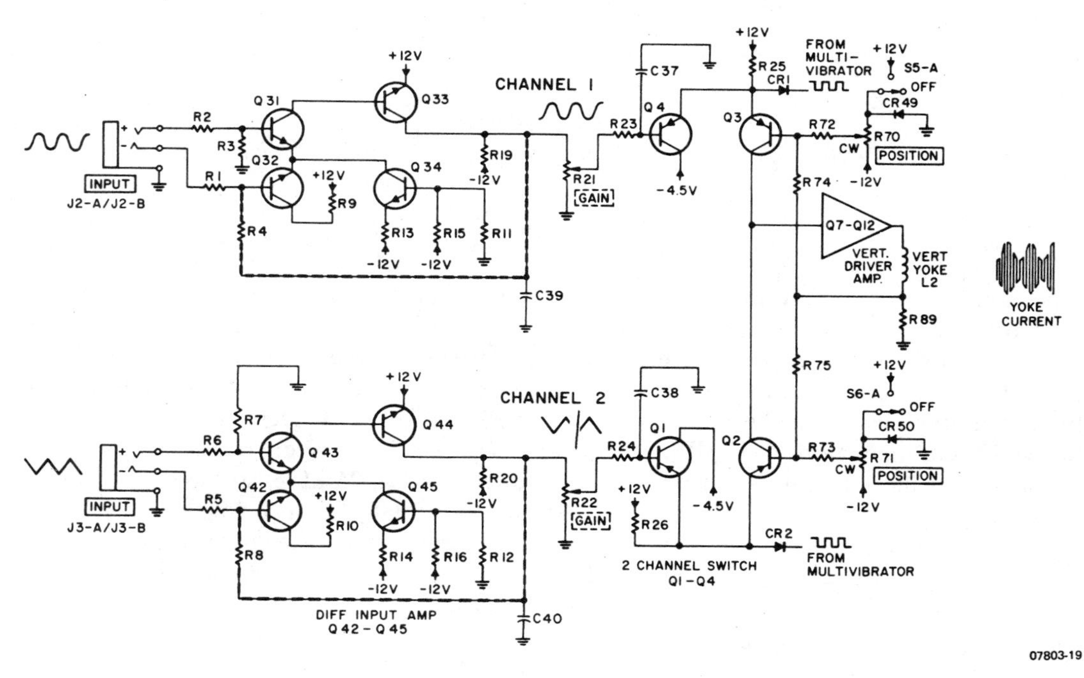

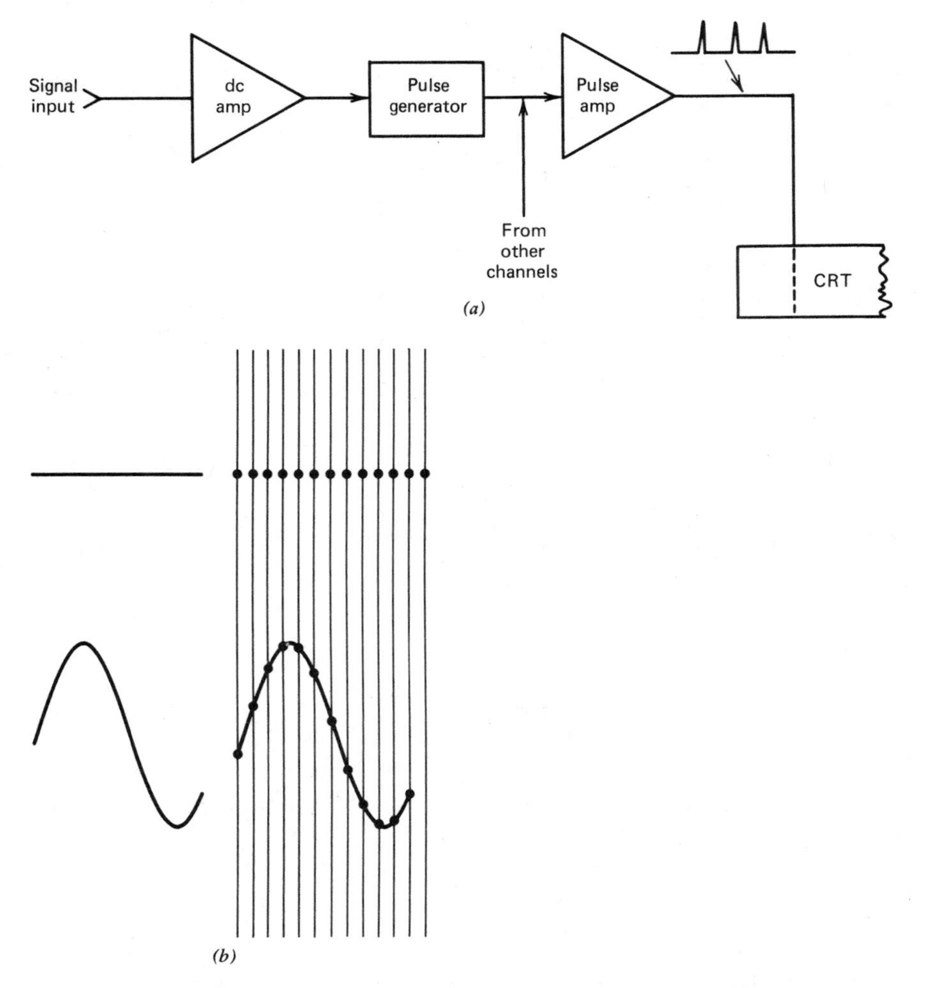

Figure 20-19 Gating amplifier system. (*a*) Block diagram. (*b*) Operation.

same point approximately in the middle of the screen. The result is a series of lighted spots that move across the CRT screen at 25 mm/s. If a dc level is applied to the input of the gating amplifier, then the line of lighted dots will occur higher or lower on the screen, depending on the amplitude and polarity of the voltage.

Applying a voltage waveform to the gating amplifier input causes the timing of the pulses to vary according to the amplitude and polarity of the waveform. This causes the vertical sweeps to be unblanked at different times, scanning out the waveshape of the signal on the screen.

The vertical sweep frequency is on the order of 15 to 25 kHz, so the dots of Figure 20-19*a* actually appear to be a continuous line, an effect improved by the persistence of the phosphor and human eye.

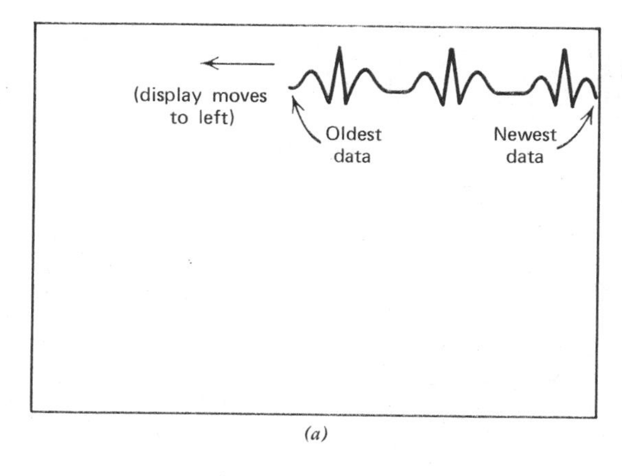

(a)

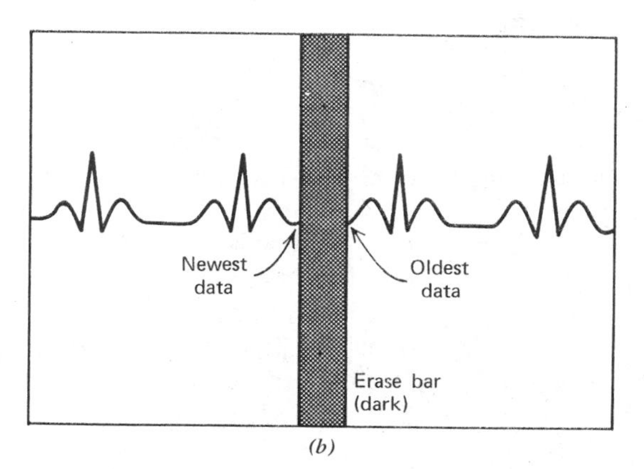

(b)

Figure 20-20 Types of nonfade display. (*a*) Waveform parade. (*b*) Erase bar or "cursor."

20-13 Nonfade Oscilloscopes

The traditional oscilloscope uses a beam of electrons to sweep the screen, writing the waveform as it is deflected. Even with long persistence phosphors, however, we find that the trace vanishes shortly after it is written onto the screen. This type of CRO is sometimes called a *bouncing-ball* display in the jargon. To the medical personnel using the bouncing-ball display, it is very difficult to evaluate waveform anomalies because the trace fades too rapidly.

The solution to this problem is the *digital storage oscilloscope*, which is called a *nonfade display* in the jargon. CRT storage systems are not used in medical oscilloscopes because, at the low frequencies involved, the digital types offer a better display at competitive prices. Also, the digital type of storage oscilloscope does not "bloom" when the display is erased.

Two different nonfade formats are commonly used: *parade* and *erase bar* (see Figure 20-20). The parade display is shown in Figure 20-20*a*. The newest data, that being written in real time, appear in the upper right-hand corner of the screen. The light beam bounces up and down at a fixed (horizontal) point in response to the vertical waveform; it does not move along the time base as in normal CROs. The old data (i.e., immediate past waveforms) moves to the left to disappear off the left-hand side of the screen. This type of display has been named the "parade" because the oldest waveform appears to lead the succeeding waveforms "marching" off the screen.

The erase bar format display is shown in Figure 20-20*b*. On this type of nonfade CRO, the beam of light travels left to right (it is not stationary as on parade models). There is an erase bar (i.e., dark region) traveling ahead of the beam that obliterates the oldest data so that new data can be written onto the CRT screen.

Figure 20-21 shows a block diagram for a nonfade oscilloscope. The principal sections are *input amplifier, analog-to-digital converter, scratch pad memory, main memory, digital-to-analog converter, output amplifier,* and *control logic section*. Some models also include a second D/A converter to create a horizontal time base signal that is synchronized with the memory. In many other models, however, more traditional analog methods are used to generate the horizontal sawtooth, but it is triggered by the control logic section.

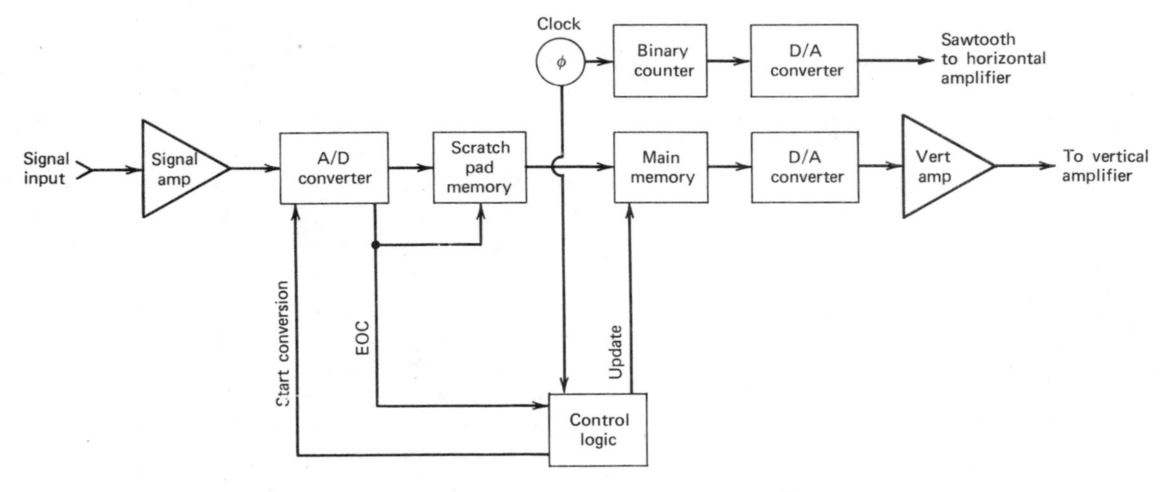

Figure 20-21 Block diagram for nonfade oscilloscope.

The input amplifier serves both to *scale* the amplitude of the input signal to the range of the oscilloscope and to *buffer* the oscilloscope from the outside world. This stage tends to be a low-gain (i.e., A_v is usually less than 10) transistor or IC operational amplifier. The gain is usually variable so that the input signal amplitude may be scaled properly.

The A/D converter serves to create a digital binary *word* that is proportional to the applied signal amplitude. Eight and 10-bit A/D converters are very common, although the 8-bit are probably most common. As an example, consider an 8-bit system. It might represent zero signal amplitude as 00000000_2, and a full-scale signal as 11111111_2. If +2.55 V represents full scale, then each 10-mV change in the input signal will represent a 1 least significant bit (i.e., 1 LSB) change in the digital word used to represent the voltage.

In most A/D converter designs, the operation must be synchronized with a series of pulses. The control logic section will generate a *start* pulse to initiate a conversion, and the A/D will generate an *end-of-conversion* (EOC) pulse to let the rest of the circuits know when it is finished with the conversion cycle.

The scratch pad memory is a shift register that holds one to four of the *most recent* data produced by the A/D converter.

The main memory contains all of the data appearing on the CRT screen (scratch pad memory data do not appear on the CRT screen until it is transferred to the main memory). Most medical nonfade oscilloscopes use either 256, 512, or 1024 eight-bit memory locations. Each location contains a binary word proportional to some instantaneous signal amplitude. Both scratch pad and main memory are shift registers.

The D/A converter produces an analog voltage level from the binary word applied to its inputs. The output of the D/A converter is applied to the vertical channel of the oscilloscope through an output buffer amplifier.

In some models horizontal sweep is generated by a second D/A converter. A binary counter used to sequentially address memory locations also drives the horizontal D/A converter. The binary counter is driven by a clock signal, so the output lines increment by one bit for each clock pulse. The result of this action at the output of the D/A converter is that a linear ramp voltage is generated that rises a few millivolts for every clock pulse received. When the counter overflows, its output word goes from full scale (i.e., 11111111_2) back to zero (00000000_2). The output of the D/A converter, then, will drop from full scale back to zero at this time, completing the sawtooth waveform needed to drive the horizontal deflection system of the oscilloscope.

In certain other nonfade models, analog circuitry is used to generate the sawtooth. Many of these models use the *Miller* integrator circuit, driven from a constant reference voltage. The integrator capacitor is kept discharged by an electronic switch until a trigger pulse from the control logic section tells it to initiate a sweep. At that same instant the binary address counter is reset and then begins counting from 00000000_2.

Regardless of the sweep system used, the binary counter steps through the 1024 memory addresses as the beam is swept left to right across the CRT screen. Since this action is synchronized, each *location* in memory represents a *point* along the horizontal axis of the CRT screen. The amplitude of the reproduced signal at that point is proportional to the digital word stored in the memory location defining that point.

The waveform on the CRT screen does not fade because it is *refreshed* 64 or 128 times per second by rapidly incrementing the counter. The refresh rate is selected to allow little fading of the CRT beam between refreshes; data are constantly being rewritten onto the screen before they can be lost.

The control circuit coordinates the action of the A/D converter, scratch pad memory, main memory, and the binary address counter. In most nonfade oscilloscopes the data in the scratch pad memory will replace the oldest data in memory every 4, 8, or 16 refresh cycles. In other words, the binary address counter increments through its range 4 to 16 times to refresh the CRT for every new data point entered. The sequence of events is as follows:

1. The control logic section issues a *start* pulse to the A/D converter, which then initiates a conversion cycle.

2. When the A/D converter is completed, the A/D converter issues an *EOC* pulse that causes the A/D data to be input to the scratch pad memory.

3. When the scratch pad memory is full (i.e., one to four words), its data are transferred to the main memory. This data replaces the four oldest data in the memory.

4. The control logic section scans the memory to output display data 64 or 128 times per second, between those times when the scratch pad memory data are being transferred to main memory.

There are a large number of nonfade oscilloscopes that use a simpler design, as shown in Figure 20-22. All of the other stages

Figure 20-22 Recirculating Shift Register Memory System.

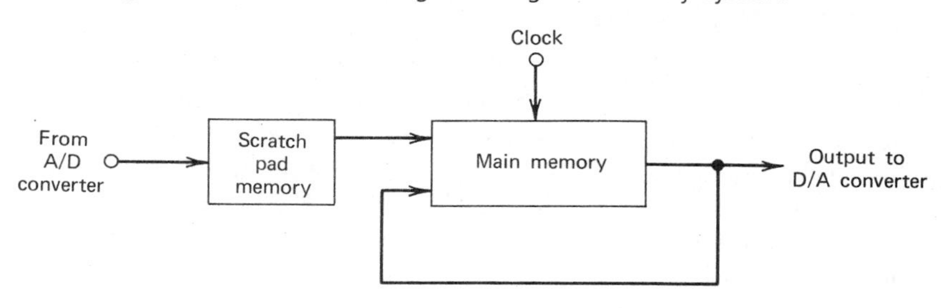

in this type of oscilloscope are the same as in the model shown previously, so only the memory is shown here. The memory is a recirculating serial shift register. Data are input on the left and will shift to the right one position (i.e., one bit at a time) each time a clock pulse is received on the shift line. As the data are output, they are also written into the input, hence the term *recirculating*. Again, data in the short-term scratch pad memory will be written into the main memory, replacing the oldest data every 4, 8, or 16 circulations.

Many nonfade models include a *freeze* capability. Such models will have a front panel switch that allows the operator to transfer all of the data in the memory to an auxiliary memory of the same bit length. The freeze memory data continues to recirculate but is never updated. As a result, the waveform on the CRT screen remains unchanged (i.e., frozen in place).

Another optional feature found on many models is the ability to *cascade* channels. A multichannel oscilloscope would require one memory chain, and possibly one A/D-D/A pair, for *each channel*. An eight-channel nonfade oscilloscope, therefore, uses eight separate memories. But when fewer than eight channels are in use, then the unused memory is wasted. If the operator selects the cascade mode, however, the output of one memory will be fed into the input of another. So when the memory for, say, channel 1 is filled, it will overflow and begin filling up channel 2.

20-14 Modern Oscilloscope Designs

Modern medical oscilloscopes are based on a variety of digital and analog technologies and often include a microprocessor for signals processing and control functions. Figure 20-23 shows a typical medical monitor scope used in an intensive care unit. The pattern on the screen is physiological data. This type of monitor is used in a computer-based system that actually forms the image.

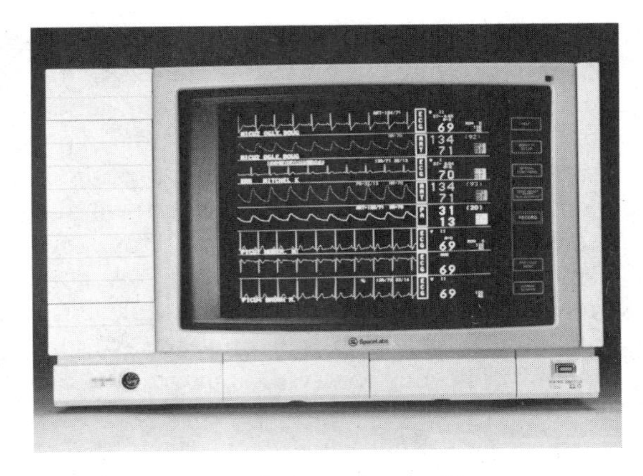

Figure 20-23 Medical monitor oscilloscope.

The monitor of Figure 20-23, like certain other displays, uses a "touch screen" method for the selector "switches." Figure 20-24 shows how most such scopes operate. Positioned along the edges of the display are a series of infrared sources (light-emitting diodes, LEDs, operating in the IR region), and infrared detectors. The IR light is invisible to the naked eye and so is not seen by the operator. The function labels are either painted onto the CRT by the computer (as in Figure 20-24) or affixed to the edge. When the operator touches the screen over any label, his or her finger interrupts one vertical and one horizontal beam, causing a unique pattern. For example, suppose all detector outputs are at binary low when the IR reaches the detector. The X outputs X_1-X_2 are L-L, and the Y outputs Y_1-Y_2-Y_3-Y_4-Y_5-Y_6 are L-L-L-L-L-L. And then someone touches the "self-test" label. In this case, X_1 and Y_1 are both interrupted and their respective outputs go high. Thus, the patterns become H-L and H-L-L-L-L-L. The internal computer recognizes this as an operator command to branch to the self-test software stored in program memory.

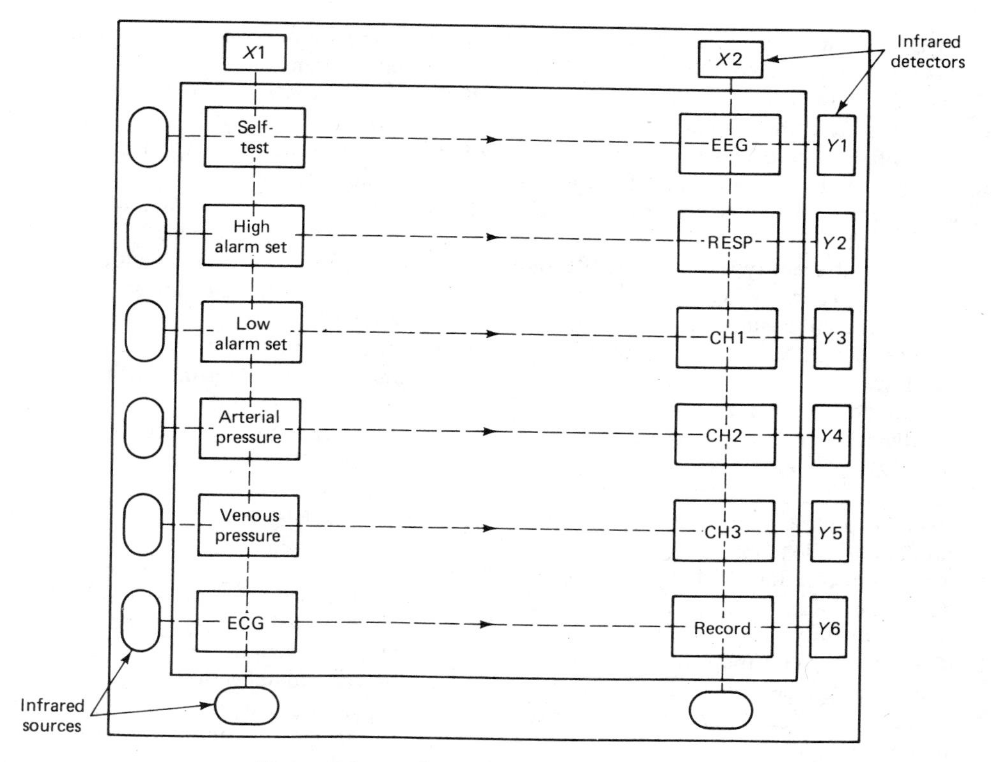

Figure 20-24 "Touch screen" oscilloscope.

20-15 Summary

1. Medical oscilloscopes can be classified as the *bouncing* ball design for normal *Y-T* models and *nonfade* for digital storage models.
2. Medical oscilloscopes use very low-frequency horizontal sweep (i.e., 25, 50, or 100 mm/s).
3. The CRTs used in medical oscilloscopes have *long persistence* phosphors.
4. Two principal techniques are used to make multichannel oscilloscopes from single-beam CRTs: *switching* and *gating amplifier*.

20-16 Recapitulation

Now return to the objectives and self-evaluation questions at the beginning of the chapter and see how well you can answer them. If you cannot answer certain questions, place a check mark next to each and reread appropriate parts of the text. Next, try to answer the questions below, using the same procedure. When you have answered all of the questions, solve the problems in Section 20-18.

20-17 Questions

20-1 List the principal *differences* between medical and service/laboratory oscilloscopes.

20-2 What property is required of the CRT in a medical oscilloscope?

20-3 List two general types of medical oscilloscopes, as classified by their display.

20-4 What is the normal sweep speed for a medical oscilloscope designed to display

ECG and arterial pressure waveforms? Also list two other optional speeds often encountered.

20-5 List two different display formats found on nonfade oscilloscopes.

20-6 Give two different techniques for creating a two-channel oscilloscope from a single-beam CRT.

20-7 List two different types of CRT deflection systems used in medical oscilloscopes.

20-8 The use of a postaccelerator anode improves the ______________ of the CRT display.

20-9 The upper −3 dB point in the vertical bandwidth for medical oscilloscopes is usually between ______________ Hz and ______________ Hz.

20-10 List the principal stages found in typical nonfade oscilloscopes.

20-11 The output of the A/D converter represents the ______________ value of the input waveform at that time.

20-12 The waveform on the screen of a nonfade oscilloscope does not fade because it is ______________ 64 or 128 times per second.

20-13 Most medical ECG and arterial pressure recorders use the ______________ writing system.

20-14 Describe in your own words the operation of a PMMC galvanometer. Draw a comparison between the PMMC and the D'Arsonval meter movement.

20-15 What is meant by "curvilinear recording"?

20-16 Describe a method by which a PMMC galvanometer can be constructed to render a *rectilinear* recording.

20-17 Describe in your own words, or using a picture, a pseudorectilinear PMMC writing system.

20-18 How does a knife edge (i.e., writing edge) render a trace rectilinear even though the stylus tip describes angular or curvilinear motion?

20-19 Describe a thermal writing system. Mention any special paper needed.

20-20 Compare the pen-and-ink and ink-jet writing systems.

20-21 Describe the optical PMMC writing system.

20-22 Describe an optical CRT camera.

20-23 What is a deadband?

20-24 Describe undershoot and overshoot. How is it related to damping?

20-25 How can a position signal be used to adjust the damping of the stylus or pen in a PMMC system?

20-26 *X-Y* recorders use two ______________ recorder mechanisms positioned at right angles to each other.

20-27 How can an *X-Y* recorder be made to make *Y*-time recordings?

20-18 Problems

20-1 Calculate the sweep frequency of a medical oscilloscope that has a 10-cm wide CRT graticule if the sweep speed is 25 mm/s.

20-2 An ECG oscilloscope is calibrated so that the vertical deflection factor is 1 V/cm, and an external preamplifier has a gain of X1000. How much deflection occurs if a 1- to 4-mV ECG signal is applied to the input?

20-3 An arterial pressure monitor is calibrated to produce an output of 10 mV/mm Hg, which is to be displayed on an oscilloscope that has a 10-cm vertical graticule. It is decided that the oscilloscope should be calibrated such that it has a scale factor of 25 mm Hg/cm. Find (a) the maximum pressure to be displayed and (b) the number of volts per division.

20-19 References

1. Diefenderfer, James, *Principles of Electronic Instrumentation*, W. B. Saunders Co. (Philadelphia, Pa., 1972).
2. Strong, Peter, *Biophysical Measurements*, Tektronix, Inc. (Beaverton, Ore., 1973).

Chapter 21

Electro-Optics (Fiber Optics and Lasers)

21-1 Objectives

1. Be able to list the advantages of fiber-optic data/signal transmission.
2. Be able to describe the propagation modes in fiber optics.
3. Be able to describe the different types of laser.
4. Be able to describe the physical basis for lasers.

21-2 Self-Evaluation Questions

These questions test your prior knowledge of the material in this chapter. Look for the answers as you read the text. After you have finished studying the chapter, try answering these questions and those at the end of the chapter (Section 21-26).

1. List the basic advantages of fiber-optic video, data, and voice communications links.
2. Define *intermodal dispersion* as applied to fiber optics.
3. ____________ is a synthetic gem material used to make red-light lasers.
4. Define an *insulating crystal laser*.

21-3 Fiber-Optic Technology

Fiber optics are the technology in which light is passed through a plastic or glass fiber so that it can be directed to a specific location. If the light is encoded (i.e., *modulated*) with an information signal, then that signal is transmitted over the fiber-optical path. There are many advantages to the fiber-optical communications or data link including:

- Very high bandwidth (accommodates video signals, many voice channels, or high data rates in computer communications).
- Very low weight and small size.
- Low loss compared with other media.
- Freedom from electromagnetic interference (EMI).
- High degree of electrical isolation.
- Explosion proof.
- Good data security.
- Improved "fail-safe" capability.

The utility of the high-bandwidth capability of the fiber-optical data link is that it can handle a tremendous amount of electronically transmitted information simultaneously. For example, it can handle more than one video signal (which typically require 500 KHz to 6 MHz of bandwidth, depending on resolution). Alternatively, it can handle a tremendous number of voice communication telephone channels. A high-speed computer data communications capability is also obtained. Either a few channels can be operated at extremely high speeds, or a larger number of low-speed parallel data channels are available. Fiber optics are so significant that one can expect to see them proliferate in the communications industry for years to come.

The light weight and small size, coupled with relatively low loss, makes the fiber-optic communications link a very good economic advantage when large numbers of channels are contemplated. To obtain the same number of channels using coaxial cables or "paired wires," the system would require a considerably larger, and heavier, infrastructure.

Electromagnetic interference (EMI) has been an annoying factor in electronics since Marconi and DeForest interfered with each other in radio trials for the Newport Yacht Races prior to the turn of the twentieth century. Today, EMI can be more than merely annoying and can cause tragic accidents. For example, airliners are operated more and more from digital computers. If a radio transmitter, radar, or electrical motor is near one of the data lines, then it is possible either to introduce false data or corrupt existing data—with potentially disastrous results. Because the EMI is caused by electrical or magnetic fields coupling between electrical cables, fiber optics (free of such fields) produce dramatic freedom from EMI (see Chapter 24).

21-4 Fiber-Optic Isolation

Electrical isolation is required in many instrumentation systems either for the safety of the user or the health of the electronic circuits connected to the system. For example, in some industrial processes, high electrical voltages are used, but the electronic instruments used to monitor the process are both low voltage and ground referenced. As a result, the high voltage can damage the instruments. In fiber-optical systems, it is possible to use an electrically floating sensor and then transmit the data over a fiber link to an electrically grounded, low-voltage computer, instrument, or control system.

The fact that fiber optics use light beams, and these are generated in noncontacting electronic circuits, makes the fiber-optic system ideal for use around flammable gases or fumes; for example, in monitoring systems in which natural gas or medical anesthetic agents (such as ether or cyclopropane)[1] are used. Regular mechanical switches or relays arc either on contact or when decontacting, and those sparks can create an explosion if flammable gases or fumes are present. A number of operating room explosions in hospitals occurred prior to 1980. Similarly, gasoline stations have exploded because of electrical arcs in switches.

System security is enhanced because fiber optics are difficult to tap. An actual physical connection must be made to the system. In wire systems, capacitive or inductive pick-ups can acquire signals with less than total physical connection (i.e., no splice is needed). Similarly, a system is more secure in another sense of the word because the fiber-optical transmitters and receivers can be designed to fail safe so that one fault does not take down the system. There was once a hospital coronary care unit data system that used parallel wire connections between the data output ports on bedside monitors and the central monitoring computer at the nurses' station. A single short circuit would reduce the system to chaos. That is less likely to happen in a fiber-optic system.

[1] These agents are only rarely used today. Since the late 1970s, when nonflammable substitutes came into widespread use, they have been considered too dangerous.

21-5 History of Fiber Optics

The basic fact of fiber optics—that is, the propagation of light beams in a transparent glass conductor (Figure 21-1)—was noted in the early 1870s when a man named John Tyndall introduced members of Britain's Royal Society to his experimental apparatus. An early, but not very practical, color television system patented by J. L. Baird used glass rods to carry the color information. By 1966, G. Hockham and C. Kao (Great Britain) demonstrated a system in which light beams carried data communications via glass fibers. The significant fact that made the Hockham/Kao system work was the reduction of loss in the glass dielectric material to a reasonable level. By 1970, practical fiber-optic communications were possible.

Medicine has made use of fiber optics for more than two decades. Fiber-optic *endoscopes* can be passed into various orifices of the body, either natural or surgically made, to inspect the interior of a patient's body. Typically there are two bundles, one for viewing and one for passing a light from a (misnamed) "cold" light source into the body. For example, gynecologists can inspect and operate on certain internal organs in females using a *laprascope* introduced through a "band-aid" incision in the abdomen. Knee surgeons can use a fiber-optic arthroscope to perform nearly miraculous operations on the human knee with far less trauma than previous procedures. Other physicians use fiber-optic endoscopes to inspect the stomach and gastric tract. A probe is passed through the mouth or nose, down the esophagus, and into the stomach so that tumors and ulcers can be inspected without resort to surgery. In more recent times, miniature TV cameras using charge coupled diode (CCD) arrays have been made available, with the fiber optics carrying the light into the stomach.

Fiber-optic inspection is used elsewhere than in medicine. A septic tank service company may use fiber optics and television to inspect septic tanks; plumbers may also use fiber optics. Other industrial and residential services also use fiber optics to inspect areas that are either inaccessible or too dangerous for direct viewing.

Before examining fiber-optic technology, it might be useful to review some of the basics of optical systems as applied to the fiber-optic system.

Figure 21-1 Light transmission through a glass or plastic rod.

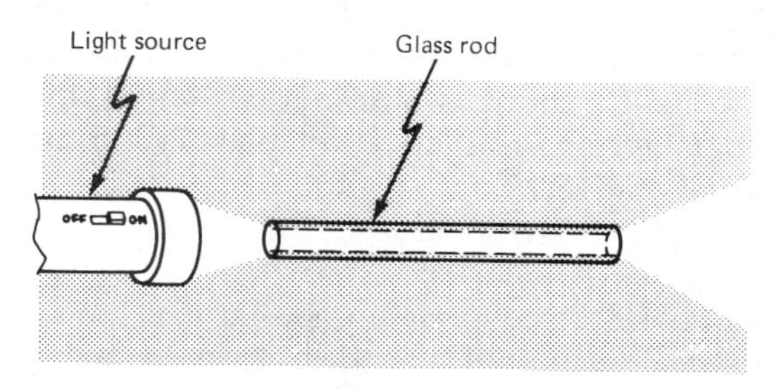

21-6 Review of Some Basics

The *index of refraction* (n), or *refractive index*, is the ratio of the speed of a light wave in a vacuum to the speed of a light wave in a medium (e.g., glass, plastic, water); for practical purposes, the speed of light in air is close enough to the speed in a vacuum to be considered the same. Mathematically, the index of refraction (n) is:

$$n = \frac{c}{v_m} \tag{21-1}$$

where

c is the speed of light in a vacuum ($\sim 3 \times 10^8$ m/s)

v_m is the speed of light in the medium

Refraction is the phenomenon in which a light ray changes direction as it passes across the boundary surface (or "interface") between two mediums of differing indices of refraction. Consider Figure 21-2, in which two materials have indices of refraction n_1 and n_2, respectively. In this illustration, n_1 is optically less dense than n_2. Consider incident light ray A, approaching the interface from the less dense side ($n_1 \rightarrow n_2$). As it crosses the interface, it changes direction toward a line normal (i.e., at right angles) to the surface. In the opposite case, ray B, the light ray approaches the interface from the more dense side ($n_2 \rightarrow n_1$). In this case, the light ray is similarly refracted from its original path, but the direction of refraction is away from the normal line.

In refractive systems the angle of refraction is a function of the ratio of the two indices of refraction (i.e., it obeys *Snell's law*):

$$n_1 \sin \theta_{ia} = n_2 \sin \theta_{ra} \qquad (21\text{-}2)$$

or

$$\frac{n_1}{n_2} = \frac{\sin \theta_{ra}}{\sin \theta_{ia}} \qquad (21\text{-}3)$$

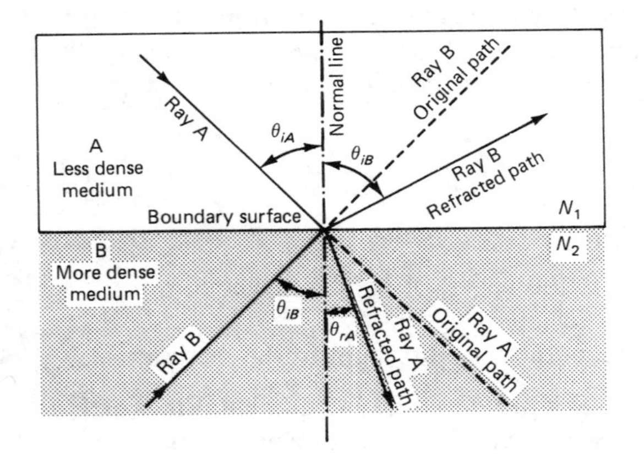

Figure 21-2 Refraction under two different circumstances: (*a*) from less dense to more dense, (*b*) from more dense to less dense.

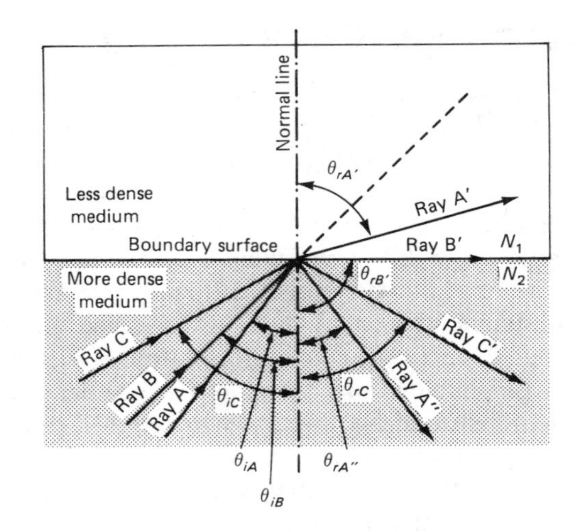

Figure 21-3 Refraction phenomenon from more dense to less dense material showing subcritical (Ray A), critical (Ray B) and supercritical (Ray C) refraction.

The particular case which concerns fiber optics is where the light ray passes from a more dense medium to a less dense medium. We can use either a water-to-air system, or a system in which two different glasses, with different indices of refraction, are interfaced. This type of system was addressed in Figure 21-2 by ray B. Figure 21-3 shows a similar system with three different light rays (ray A, ray B, and ray C) approaching the same point on the interface from three different angles (θ_{ia}, θ_{ib}, and θ_{ic}, respectively). Ray A approaches at a *subcritical angle*, so it will split into two portions (A′ and A″). The reflected portion contains a relatively small amount of the original light energy and may indeed be nearly indiscernible. The major portion of the light energy is transmitted across the boundary and refracts at an angle $\theta_{ra'}$ in the usual manner.

Light ray B, on the other hand, approaches the interface at a *critical angle*, $\theta_{rb'}$, and is refracted along a line that is orthogonal to the normal line (i.e., it travels along

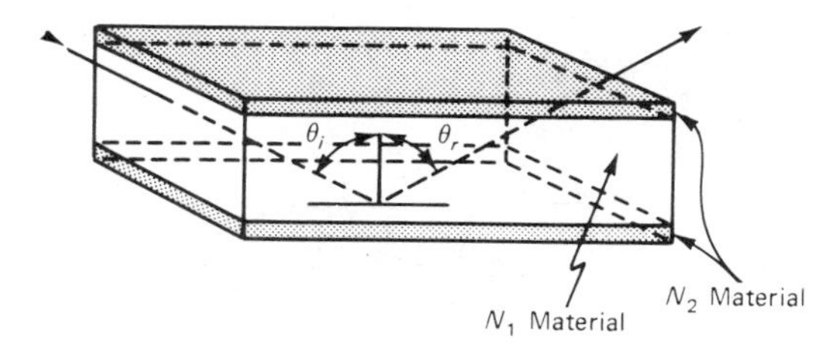

Figure 21-4 Waveguide analogy for fiber optics.

the interface boundary surface). This angle is sometimes labelled θ_c in optical textbooks.

Finally, ray C approaches the interface at an angle greater than the critical angle (i.e., a *supercritical angle*). None of this ray is transmitted, but rather it is turned back into the original media—that is, it is subject to *total internal reflection* (TIR).[2] It is the phenomenon of total internal reflection that allows fiber optics to work.

21-7 Fiber Optics

The optical fiber is essentially similar to a microwave *waveguide*, and an understanding of waveguide action is useful in understanding fiber optics. A schematic model of a fiber optic is shown in Figure 21-4. A slab of denser material (n_1) is sandwiched between two slabs of a less dense material (n_2). Light rays that approach from a supercritical angle are totally internally reflected from the two interfaces ($n_2 \rightarrow n_1$ and $n_1 \rightarrow n_2$). Although only one "bounce" is shown in our illustration, the ray will be subjected to successive TIR reflections as it propagates through the n_1 material. The amount of light energy that is reflected through the TIR mechanism is on the order of 99.9 percent, which compares quite favorably with the 85 to 96 percent typically found in plane mirrors.

Fiber-optic lines are not rectangular but

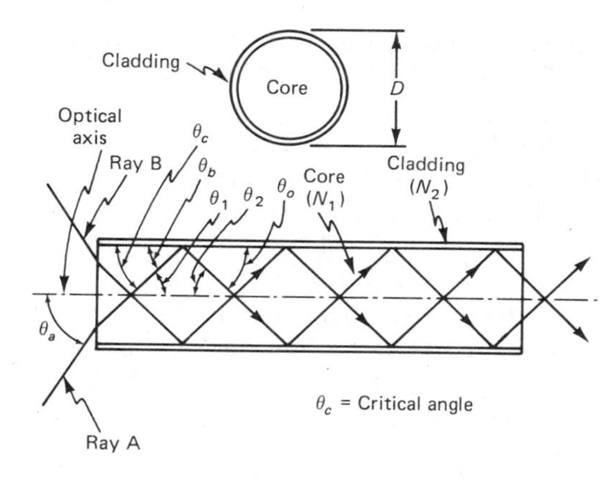

Figure 21-5 Total internal reflection forms basis for propagation in cylindrical fiber optics.

rather are cylindrical, as shown in Figure 21-5. These components are called *clad fiber optics* because the denser inner core is surrounded by a less dense layer called *cladding*. Figure 21-5 shows two rays, each of which is propagated into the system such that the critical angles are exceeded. These rays will propagate down the cylindrical optical fiber with very little loss of energy. There are actually two forms of propagation. The minority form (Figure 21-6*a*), called *meridional rays*, are easier to understand and mathematically model in textbooks because all rays lie in a plane with the optical axis.

Figure 21-6 Two types of fiber optic light propagation: (*a*) meridional propagation, (*b*) skew propagation.

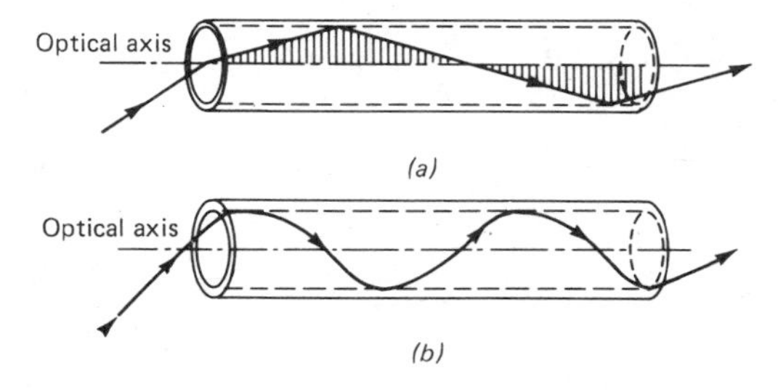

[2] TIR is called total internal *reflection*, but it is actually a refraction phenomena.

The more numerous *skew rays* (Figure 21-6*b*) follow a helical path and so are somewhat more difficult to discuss.

The light acceptance of the fiber optic (Figure 21-7) is a cone-shaped region centered on the optical axis. The *acceptance angle* θ_a is the critical angle for the transition from air ($n = n_a$) to the core material ($n = n_a$). The ability to collect light is directly related to the size of the acceptance cone and is expressed in terms of the *numerical aperture* (NA), which is:

$$NA = \sin\theta_a \quad (21\text{-}4)$$

The refraction angle of the rays internally, across the air-n_1 interface, is given by Snell's law:

$$\theta_{b1} = \arcsin\left(\frac{n_a \sin\theta_a}{n_1}\right) \quad (21\text{-}5)$$

It can be shown that:

$$\theta_{a1} = \theta_{a2} \quad (21\text{-}6)$$

$$\theta_{b1} = \theta_{b2} \quad (21\text{-}7)$$

$$\theta_{a1} = \frac{\theta_a}{n_1} \quad (21\text{-}8)$$

In terms of the relative indices of refraction between the ambient environment outside the fiber, the core of the fiber, and the cladding material, the numerical aperture is given by:

$$NA = \sin\theta_a = \frac{1}{n_a} = \sqrt{(n_1)^2 - (n_2)^2} \quad (21\text{-}9)$$

If the ambient material is air, then the numerical aperture equation reduces to:

$$NA = \sqrt{(n_1)^2 - (n_2)^2} \quad (21\text{-}10)$$

Internally, the angles of reflection (θ_{a1} and θ_{a2}), at the critical angle, are determined by

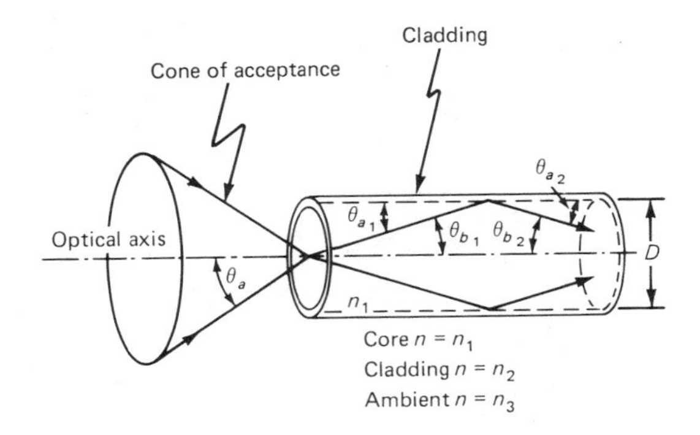

Figure 21-7 Cone of acceptance of fiber optic.

the relationship between the indices of refraction of the two materials, n_1 and n_2:

$$\theta_{a1} = \frac{\arcsin\sqrt{(n_1)^2 - (n_2)^2}}{n_1} \quad (21\text{-}11)$$

Typical fiber-optic components have numerical apertures of 0.1 to 0.5; typical fibers have a diameter D of 25 μm to 650 μm. The ability of the device to collect light is proportional to the square of the numerical aperture:

$$\zeta \; \alpha \; (NA \times D)^2 \quad (21\text{-}12)$$

where

ζ is the relative light collection ability
NA is the numerical aperture
D is the fiber diameter

21-8 Intermodal Dispersion

When a light ray is launched in a fiber optic, it can take any of a number of different paths, depending in part on its angle of arrival (Figure 21-8). These paths are known as *transmission modes* and vary from very low-order modes parallel to the optical axis of the fiber (ray A in Figure 21-8) to the highest-order mode close to the critical angle (ray C); in addition, there are a very large number of rays in between these two limits.

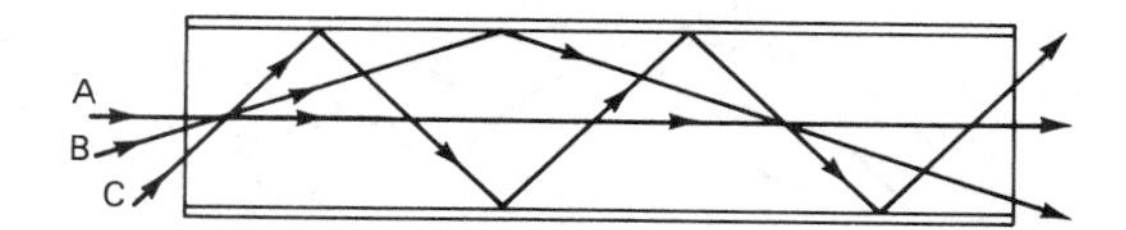

Figure 21-8 Multimode light propagation.

An important feature of the different modes is that the respective path lengths vary tremendously, being shortest with the low-order modes and longest with high-order modes. If a fiber optic has only a single core and single layer of cladding, it is called a *step index* fiber because the index of refraction changes abruptly from the core to the cladding. The number of modes (N) that can be supported are given by:

$$N = \frac{\left(\dfrac{\pi D\,[\mathrm{NA}]}{\lambda}\right)^2}{2} \qquad (21\text{-}13)$$

Any fiber with a core diameter (D) greater than about 10 wavelengths (10λ) will support a very large number of modes and so is typically called a *multimode fiber*. A typical light beam launched into such a step index fiber optic will simultaneously find a large number of modes available to it. This may or may not affect analog signals but has a deleterious effect on digital signals called *intermodal dispersion*.

Figure 21-9 illustrates the effect of intermodal dispersion on a digital signal. When a short-duration light pulse (Figure 21-9*a*) is applied to a fiber optic that exhibits a high degree of intermodal dispersion, the received signal (Figure 21-9*b*) is smeared, or "dispersed," over a wider area. At slow data rates this effect may prove negligible because the dispersed signal can die out before the next pulse arrives. But at high speeds, the pulses may overrun each other (Figure 21-10), producing an ambigous situation that exhibits a high data error rate.

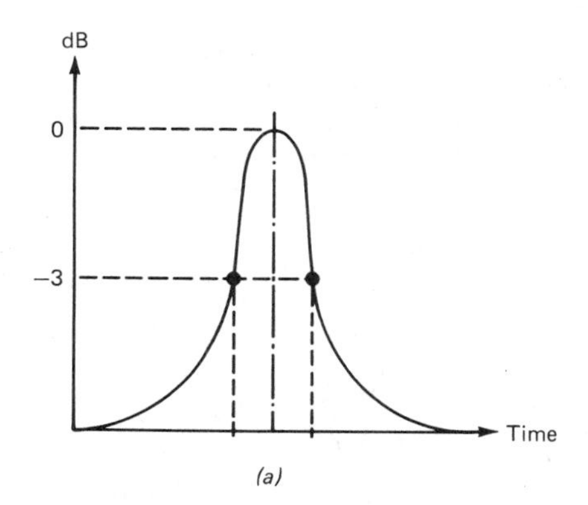

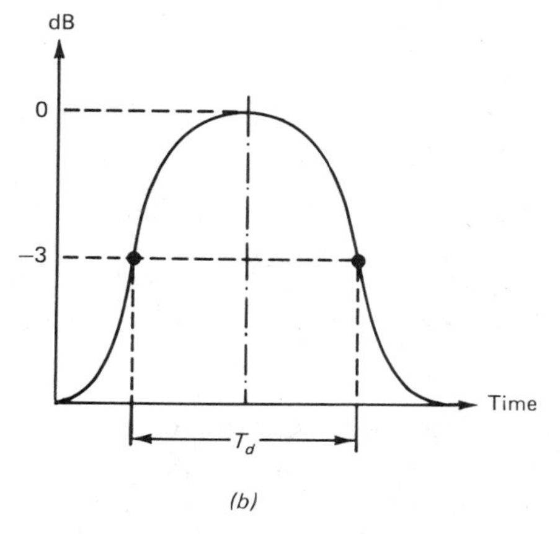

Figure 21-9 Light signal transmodal dispersion: (*a*) input pulse; (*b*) dispersed pulse.

Intermodal dispersion is usually measured relative to the widths of the pulses at the −3 dB (i.e., halfpower) points. In Figure 21-9, the −3 dB point on the incident pulse transmitted into the fiber optic is T, while in the received pulse the time between −3 dB points is T_d. The dispersion is expressed as the difference, or:

$$\text{Dispersion} = T - T_d \qquad (21\text{-}14)$$

A means for measuring the dispersion for any given fiber-optic element is to measure the dispersion of a Gaussian (normal distribution) pulse at those −3 dB points. The

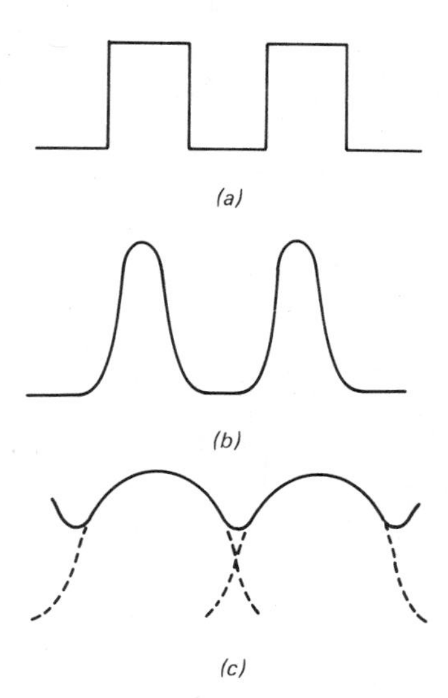

Figure 21-10 Transmodal dispersion creates problems in data fiber optic systems: (*a*) input data square wave; (*b*) resultant light pulse signal; (*c*) dispersed signal causes overriding of pulses.

cable is then rated in terms of nanoseconds dispersion per kilometer of fiber (ns/km).

The bandwidth of the fiber, in megahertz per kilometer (MHz/km), can be specified from knowledge of the dispersion, using the expression:

$$\text{BW (MHz/km)} = \frac{310}{\text{Disp. (ns/km)}} \quad (21\text{-}15)$$

21-9 Graded Index Fibers

A solution to the dispersion problem is to build an optical fiber with a continuously varying index of refraction such that n decreases at distances away from the optical axis. While such smoothly varying fibers are not easy to build, it is possible to produce a fiber optic with layers of differing index of refraction (Figure 21-11). The relationship of the respective values of n for each layer is:

$$n_1 > n_2 > n_3 > n_4 > n_5 > \ldots n_i \quad (21\text{-}16)$$

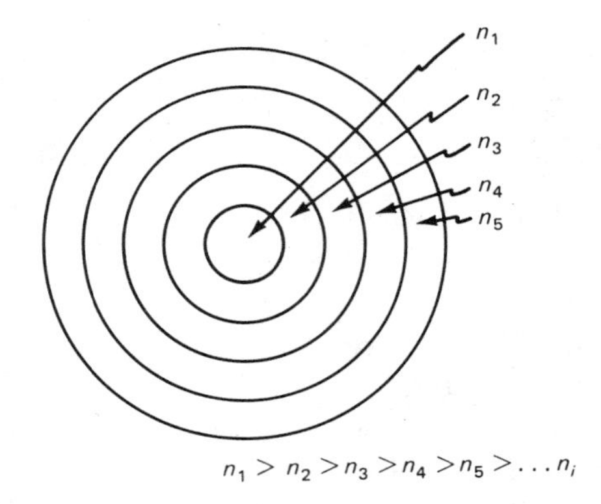

Figure 21-11 Graded index fiber optic.

The overall index of refraction determines the numerical aperture and is taken as an average of the different layers.

With graded fibers, the velocity of propagation of the light ray in the material is faster in the layers away from the optical axis than in the lower layers. As a result, a higher-order mode wave will travel faster than a wave in a lower order. The number of modes available to the graded index fiber is:

$$N = \frac{\left(\dfrac{\pi D[\text{NA}]^2}{\lambda}\right)}{4} \quad (21\text{-}17)$$

Some cables operate in a *critical mode*, designated HE_{11} (to mimmick microwave terminology), in which the cable is very thin compared with multimodal cables. As a diameter of the core decreases, so does the number of available modes, and eventually the cable becomes *monomodal*; if the core gets down to 3 to 5 microns, then only the HE_{11} mode becomes available. The critical diameter required for monomodal operation is:

$$D_{\text{crit}} = \frac{2.4\lambda}{\pi[\text{NA}]} \quad (21\text{-}18)$$

Because the monomodal cable potentially reduces the number of available modes, it also reduces intermodal dispersion. Thus, the monomode fiber is capable of extremely high data rates or analog bandwidths.

21-10 Losses in Fiber-Optic Systems

Understanding and controlling losses in fiber-optic systems is integral to making the system work properly. Before examining the sources of such losses, however, let's take a quick look at the notation for losses in the system, as well as the gains of the electronics systems used to process the signals applied to, or derived from, the fiber-optic system. This notation uses the *decibel* as the system of measurement.

The subject of decibels almost always confuses the newcomer to electronics, and even many an oldtimer seems to have occassional memory lapses regarding the subject. For the benefit of both groups, and because the subject is so vitally important to understanding electronics systems, we will examine the decibel.

The decibel measurement originated with the telephone industry and was named after telephone inventer Alexander Graham Bell. The original unit was the *bel*. The prefix *deci* means "1/10," so the *decibel* is one-tenth of a bel. The bel is too large for most common applications and so it is rarely if ever used. Thus, we will concentrate only the more familiar decibel (usually abbreviated *dB*).

The decibel is nothing more than a means of expressing a *ratio* between two signal levels (e.g., the "output-over-input" ratio of an amplifier). Because the decibel is a ratio, it is also dimensionless, despite the fact that "dB" looks like a dimension. Consider the voltage amplifier as an example of dimensionless gain; its gain is expressed as the output voltage over the input voltage (V_o/V_{in}).

EXAMPLE 21-1

A voltage amplifier outputs 6 volts when the input signal has a potential of 0.5 volts. Find the gain (A_V).

SOLUTION

$$A_V = V_o/V_{in}$$

$$A_V = (6 \text{ volts})/(0.5 \text{ volts})$$

$$A_V = 12$$

Note in the preceding example that the "volts" units appeared in both numerator and denominator and so "cancelled out," leaving only a dimensionless "12" behind.

To analyze systems using simple addition and subtraction, rather than multiplication and division, a little math trick is used on the ratio. We take the base-10 logarithm of the ratio and then multiply it by a scaling factor (either 10 or 20). For voltage systems, such as our voltage amplifier, the expression becomes:

$$\text{dB} = 20 \log \left(\frac{V_o}{V_{in}}\right) \tag{21-19}$$

EXAMPLE 21-2

In Example 21-1 we had a voltage amplifier with a gain of 12 because 0.5 volts input produced a 6-volt output. How is this same gain (i.e., V_o/V_{in} ratio) expressed in decibels?

SOLUTION

$$\text{dB} = 20 \log(V_o/V_{in})$$

$$\text{dB} = 20 \log(6 \text{ volts}/0.5 \text{ volts})$$

$$\text{dB} = 20 \log(12)$$

$$\text{dB} = +21.6 \text{ decibels}$$

The fact that the quantity represented in Example 21-2 is a gain is indicated by the plus sign. If the quantity represented a loss (e.g., if $V_o < V_{in}$), then the sign of the result would be negative. You can see this by working the problem in Example 21-2 for the ratio 0.5/6, which results in a loss of -21.6 dB. Note that the numerical result for a loss using the same voltages is the same as for a gain, but the sign is reversed.

Despite the fact that we have massaged the ratio by converting it to a logarithm, *the decibel is nonetheless nothing more than a means for expressing a ratio*. Thus, a voltage gain of 12 can also be expressed as a gain of 21.6 dB. A similar expression can be used for current amplifiers, where the gain ratio is I_o/I_{in}:

$$dB = 20 \log \left(\frac{I_o}{I_{in}}\right) \qquad (21\text{-}20)$$

For power measurements, which are what is important in light and fiber-optical systems, a modified expression is needed to account for the fact that power is proportional to the square of the voltage or current:

$$dB = 10 \log \left(\frac{P_o}{P_{in}}\right) \qquad (21\text{-}21)$$

We now have three basic equations for calculating decibels (i.e., one each for current ratios, voltage ratios, and power ratios). The usefulness of decibel notation is that it can make nonlinear power and gain equations into linear additions and subtractions. Gains (+dB) and losses (−dB) can be added to find the total gain or loss of the system.

The light power at the output end of a fiber optic (P_o) is reduced compared with the input light power (P_{in}) because of losses in the system. As in many natural systems, light loss in the fiber material tends to be exponentially decaying (Figure 21-12*a*) and so obeys an equation of the form:

$$P_o = P_{in} e^{(-\Lambda/L)} \qquad (21\text{-}23)$$

where

Λ is the length of the fiber optic being considered

L is the unit length (i.e., length for which $e^{-\Lambda/L} = e^{-1}$)

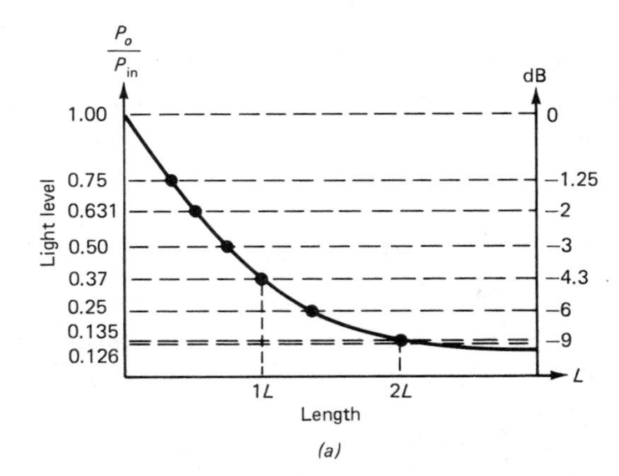

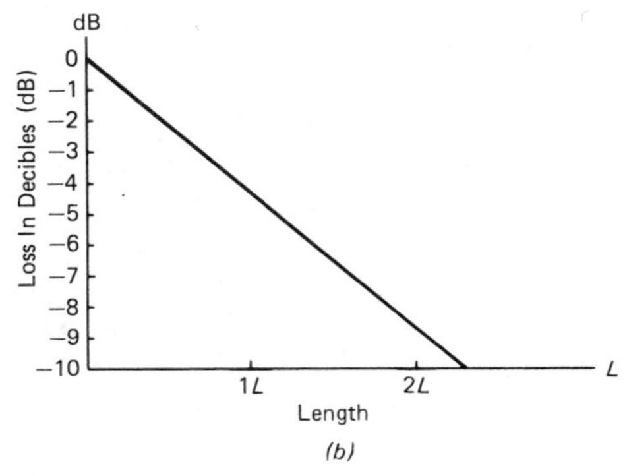

Figure 21-12 Length dependent signal attenuation in fiber optic: (*a*) log-scale graph of attenuation; (*b*) decibel scale of attenuation.

From Equation 21-22, and by comparing Figures 21-12*a* and 21-12*b*, you can see why decibels are used. The decibel notation eliminates the exponential notation, allowing us to add and subtract dB in order to calculate losses in any given system.

21-11 Losses in Fiber-Optic Circuits

There are several mechanisms for loss in fiber-optic systems. Some of these are inherent in any light-based system, while others are a function of the design of the specific system being considered.

Defect Losses

Figure 21-13 shows several possible sources of loss due to defects in the fiber itself. First, in unclad fibers, *surface defects* (nicks and scratches) that breach the integrity of the surface will allow light to escape. In other words, not all of the light is propagated along the fiber. Second (also in unclad fibers), grease, oil, or other *contaminants* on the surface of the fiber may form an area with an index of refraction different from what is expected and cause the light direction to change. And if the contaminant has an index of refraction similar to glass, then it may act as if it were glass and cause loss of light to the outside world. Finally, there is always the possibility of *inclusions* (i.e., objects, specks, or voids in the material making up the fiber optic). Inclusions can affect both clad and unclad fibers. When light hits the inclusion, it tends to scatter in all directions, causing a loss. Some of the light rays scattered from the inclusion may recombine either destructively or constructively with the main ray, but most do not.

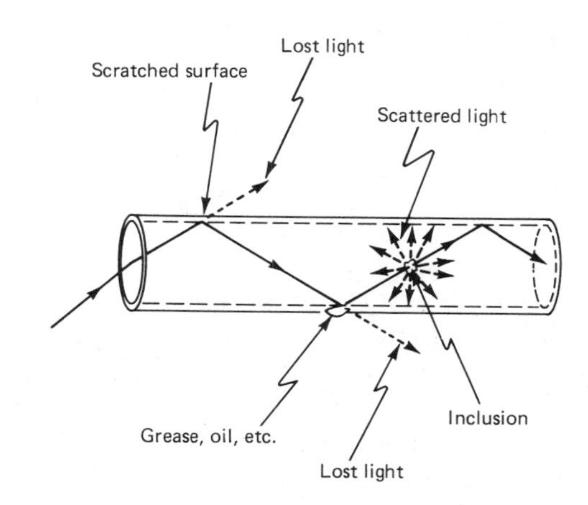

Figure 21-13 Causes of loss in fiber optics include scratches, contaminants and inclusions.

Inverse Square Law Losses

In all light systems, there is the possibility of losses due to spreading of the beam. If you take a flashlight and point it at a wall, and measure the power per unit area of the wall at a distance of, say, 1 meter, and then back off to twice the distance (2 meters) and then measure again, you will find that the power dropped to one-fourth. In other words, the power per unit area is inversely proportional to the square of the distance ($1/D^2$).

Transmission Losses

These losses are due to light that is caught in the cladding material of clad fiber optics. This light is either lost to the outside, or is trapped in the cladding layer and is thus not available to be propagated in the core.

Absorption Losses

This form of loss is due to the nature of the core material and is inversely proportional to the transparency of the material. In addition, in some materials absorption losses may not be uniform across the light spectrum but may well be wavelength sensitive.

Coupling Losses

Another form of loss is due to coupling systems. All couplings (discussed in more detail later) have loss associated with them. Several different losses of this sort are identified.

Mismatched Fiber Diameters

This form of loss is due to coupling a large-diameter fiber (D_L) to a small-diameter fiber (D_S); the larger-diameter fiber transmits to the lesser-diameter fiber. In decibel form, this loss is expressed by:

$$\text{dB} = -10\log\left(\frac{D_S}{D_L}\right) \qquad (21\text{-}24)$$

Numerical Aperture Coupling Losses

Another form of coupling loss occurs when the numerical apertures of the two fibers are mismatched. If NA_r is the receiving fiber, and NA_t is the transmitting fiber, then the loss is expressed as:

$$dB = -10 \log\left(\frac{NA_r}{NA_t}\right) \qquad (21\text{-}25)$$

Fresnel Reflection Losses

These losses occur at the fiber-optic interface with air (Figure 21-14*a*) and are due to the fact that there is a large change of index of refraction between the glass and the air. There are actually two losses to consider. First is the loss caused by internal reflection from the inner surface of the interface; second is the reflection from the opposite surface across the air gap in the coupling. Typically the internal reflection loss is on the order of 4 percent, while the external reflection is about 8 percent.

We may model any form of reflection in a transmission system similarly to reflections in a radio transmission line. Studying *standing waves* and related subjects in books on rf systems can yield some understanding of these problems. The amount of reflection in coupled optical systems uses similar arithmetic:

$$\Gamma = \left(\frac{n_1 - n_2}{n_1 + n_2}\right)^2 \qquad (21\text{-}26)$$

where

Γ is the coefficient of reflection

n_1 is the index of reflection for the receiving material

n_2 is the index of reflection for the transmitting material

Mismatched indices of refraction are analogous to the mismatch of impedances problem seen in transmission line systems, and the cure is also analogous. Where the transmission line uses an impedance matching coupling device, the fiber optic will use a coupler that matches the "optical impedances" (i.e., the indices of refraction). Figure 21-14*b* shows a coupling between the two ends of fibers (lenses may or may not be used depending on the system) that uses a liquid or gel material with a similar index of refraction to the fiber. The reflection losses are thereby reduced or even eliminated.

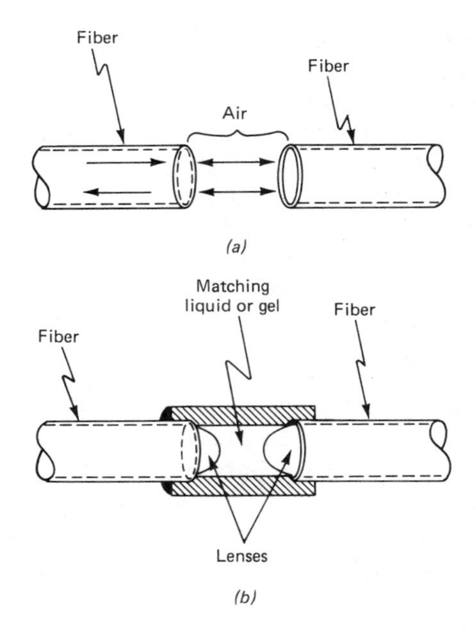

Figure 21-14 Fiber optic coupling: (*a*) in air, reflections occur; (*b*) "impedance matched" coupler.

21-12 Fiber-Optic Communications Systems

A communications system requires an information signal source (e.g., voice, music, digital data, or an analog voltage representing a physical parameter), a transmitter, a propagation media (in this case fiber optics), a receiver and an output. In addition, the transmitter may include any of several different forms of *encoder* or *modulator*, and the receiver may contain a *decoder* or *demodulator*.

Figure 21-15 shows two main forms of

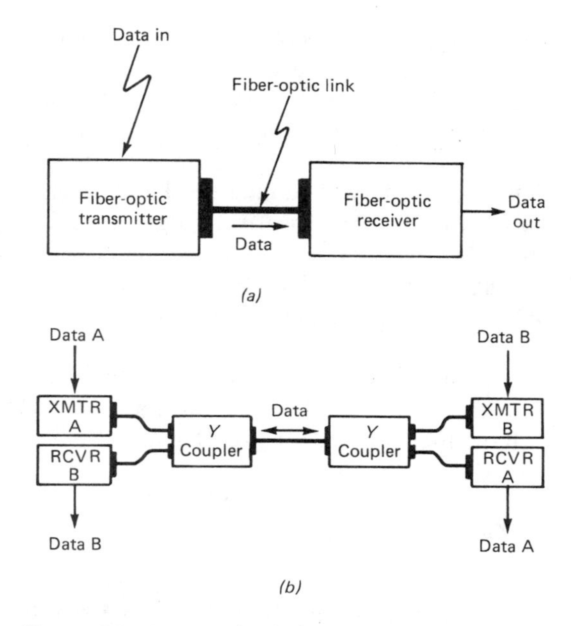

Figure 21-15 Two basic forms of communication systems: (*a*) simplex system provides one-way communication; (*b*) duplex system provides two-way communications.

communications link. The *simplex* system is shown in Figure 21-15*a*. In this system, a single transmitter sends light (information) over the path in only one direction to a receiver set at the other end. The receiver end cannot reply or otherwise send data back the other way. The simplex system requires only a single transmitter and a single receiver module per channel.

A *duplex* system (Figure 21-15*b*) is able to simultaneously send data in both directions, allowing both send and receive capability at both ends. The duplex system requires both a receiver and a transmitter module at both ends, plus two-way beam splitting *Y*-couplers at each end.

There is also a *half-duplex* system known in communications, but it is of little interest here. A half-duplex system can transmit in both directions, but not at the same time.

21-13 Receiver Amplifier and Transmitter Driver Circuits

Before the fiber-optical system is useful for communications, a means must be provided to convert electrical (analog or digital) signals into light beams. Also necessary is a means for converting the light beams from the fiber optics back into electrical signals. These jobs are done by *driver* and *receiver preamplifier* circuits, respectively.

Figure 21-16 shows two possible driver circuits. Both circuits use light-emitting diodes (LEDs) as the light source. The circuit in Figure 21-16*a* is useful for digital data communications. These signals are characterized by on/off (HIGH/LOW or 1/0) states in which the light-emitting diode is either ON or OFF, indicating which of the two possible binary digits is required at the moment.

The driver circuit consists of an open-collector digital inverter device in a light-tight container. These devices obey a very simple rule: If the input (A) is HIGH, then the output (B) is LOW; if the input (A) is LOW, then the output (B) is HIGH. Thus, when the input data signal is HIGH, point B is LOW, so the cathode of the LED is grounded. The LED turns on and sends a light beam down the fiber-optic line. But when the input data line is LOW, point B is HIGH, so the LED is now ungrounded (and therefore turned OFF)—no light enters the fiber. The resistor (R_1) is used to limit the current flowing in the LED to a safe value. Its resistance is found from Ohm's law and the maximum allowable LED current:

$$R_1 = \frac{(V^+) - 0.7}{I_{max}} \qquad (21\text{-}27)$$

An analog driver circuit suitable for voice and instrumentation signals is shown in Figure 21-16*b*. This circuit is based on the operational amplifier (see Chapter 10). There are two aspects to this circuit: the *signal*

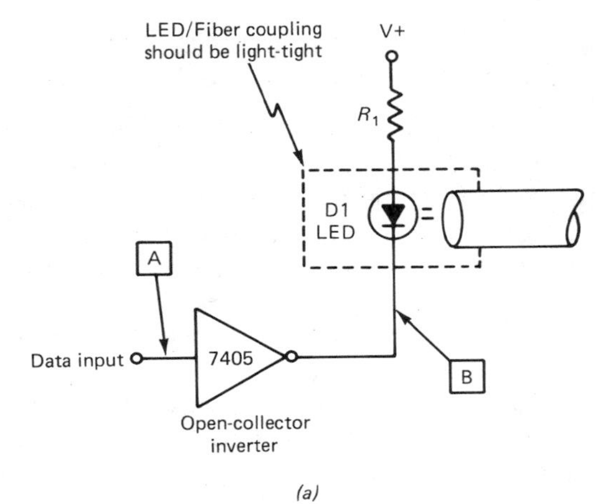

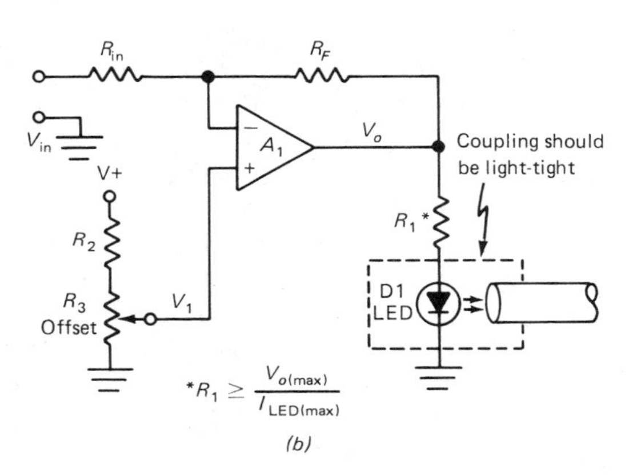

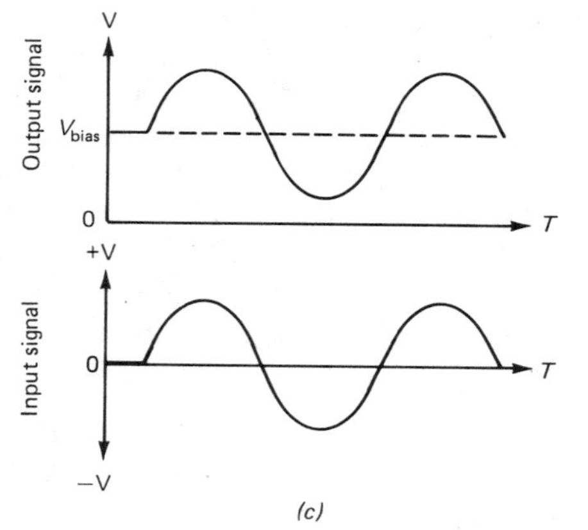

Figure 21-16 Fiber optic communications driver circuits: (*a*) Digital fiber optic driver circuit; (*b*) analog fiber optic driver circuit; (*c*) effect of DC offset in preventing signal distortion due to clipping in driver circuit.

path and the *dc offset bias*. The latter feature is needed in order to place the output voltage at a point where the LED is lighted at about one-half of its maximum brilliance when the input voltage V_{in} is zero. This way, negative polarity signals will reduce the LED brightness, but will not turn it off (see Figure 21-16*c*). In other words, biasing avoids clipping off the negative peaks. If the expected signals are monopolar, then set V_1 to barely turn on the LED when the input signal is zero.

The signal V_{in} sees an inverting follower with a gain of $-R_f/R_{in}$, so the total output voltage (accounting for the dc bias) is:

$$V_o = \left(\frac{-V_{in}R_f}{R_{in}}\right) + V_1\left(\frac{R_f}{R_{in}} + 1\right) \quad (21\text{-}27)$$

Because the network R_2/R_3 is a resistor voltage divider, the value of V_1 will vary from 0 volts to a maximum of:

$$V_1 = \frac{(V^+)\,R_3}{R_2 + R_3} \quad (21\text{-}28)$$

Therefore, we may conclude that $V_{o(max)}$ is:

$$V_{o(max)} = \left(\frac{-V_{in}R_f}{R_{in}}\right) + \left(\frac{(V^+)\,R_3}{R_2 + R_3}\right)\left(\frac{R_f}{R_{in}} + 1\right) \quad (21\text{-}29)$$

Three different receiver preamplifier circuits are shown in Figure 21-17; analog versions are shown in Figures 21-17*a* and 21-17*b*, while a digital version is shown in Figure 21-17*c*. The analog versions of the receiver preamplifiers are based on operational amplifiers (Chapter 10). Both analog receiver preamplifiers use a photodiode as the sensor. These PN or PIN junction diodes (see Chapters 7 and 8) produce an output current I_o that is proportional to the light shining on the diode junction.

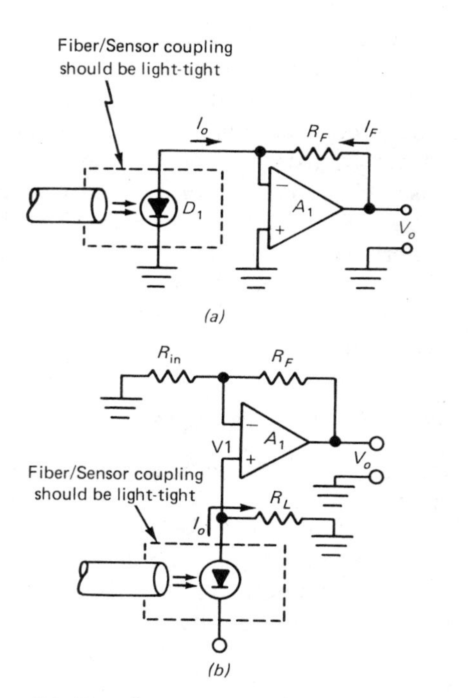

Figure 21-17 Fiber optic receiver circuits: (*a*) inverting, (*b*) noninverting.

The version shown in Figure 21-17*a* is based on the inverting follower circuit. The diode is connected such that its noninverting input is grounded, thereby set to zero volts potential, and the diode current is applied to the inverting input. The feedback current (I_f) exactly balances the diode current, so the output voltage will be:

$$V_o = -I_o R_f \tag{21-30}$$

The noninverting follower version shown in Figure 21-17*b* uses the diode current to produce a proportional voltage drop (V_1) across a load resistance, R_L. The output voltage for this circuit is:

$$V_o = I_o R_L \left(\frac{R_f}{R_{in}} + 1 \right) \tag{21-31}$$

Both analog circuits will respond to digital circuits, but they are not optimum for that type of signal. Digital signals will have to be reconstructed because of sloppiness caused by dispersion. A better circuit would be one in which the sensor is a phototransistor connected in the common emitter configuration. When light shines on the base region, the transistor conducts, causing its collector to be at a potential only a few tenths of a volt above ground potential. Conversely, when there is no light shining on the base, the collector of the transistor is at a potential close to V^+, the power supply potential.

The clean-up action occurs because the following stage is a digital Schmitt trigger. The output of such a device will snap HIGH when the input voltage exceeds a certain minimum threshold, and it remains HIGH until the input voltage drops below another threshold (snap-HIGH and snap-LOW thresholds are not equal). Thus, the output of the Schmitt trigger is a clean digital signal, while the sensed signal is a lot more sloppy.

21-14 Lasers

Lasers can be extremely dangerous. The light produced by some lasers can cause severe, sometimes permanent, damage to eyes and skin. Even low-power lasers are dangerous and must be treated with respect. Never look directly into a laser. Even reflections can be dangerous, so be very careful when working with lasers. Become familiar with the basic safety rules provided by the manufacturers of lasers. The American National Standards Institute (ANSI), 1430 Broadway, New York, NY, 10018, publishes a laser safety standard.

Lasers are optical versions of MASERs, a concept that is used in the amplification of microwave radio signals. *MASERs* (*microwave amplification by stimulated emission of radiation*) were developed for amplification of very low-level signals that would normally be obscured by noise. LASERs (*light amplification by stimulated emission of radiation*)[3] are simply optical wavelength variants on the same theme. Laser light differs from other light in several important ways. The principal characteristics of laser light are *coherency, monochromiticity,* and *low dispersion*. We will deal with these factors shortly.

21-15 Laser Classification

Lasers can be classified according to different schemes. A common system is to use the form and material of the laser (i.e., gas, solid crystal, pn junctions, and so forth). Another method of classifying lasers is according to the safety categories published by the American National Standards Institute (ANSI), and the Laser Institute of America, in ANSI Z136.1-1986 (consult ANSI for any current updates).

The ANSI classes are *Class I, Class II, Class III,* and *Class IV.* A Class I laser is one that is not capable of producing biological damage to the eye or skin during intended uses, by either direct or reflected exposure. These lasers may emit light in the 400- to 1400-nm region. The Class II lasers emit in the visible portion of the spectrum in the 400- to 700-nm region. In the Class II laser, normal aversion responses such as blinking (~250 ms) are believed to be sufficient to afford protection against biological damage from direct or reflected light. For Class II continuous wave (CW) lasers, the point source power used in exposure calculations is 2.5 mW/cm^2. Class III lasers "may be hazardous under direct and specular reflection viewing conditions, but diffuse reflection is usually not a hazard." The Class III laser is typically not a fire hazard. The Class IV laser is a dangerous, high-powered device and may be both a fire hazard and hazard to eyes and skin from both direct and specular reflected beams, and sometimes also from a diffuse reflected beam.

Laser classes are determined by authorized Laser Safety Officers (LSOs), according to the criteria established in the ANSI standard or its successor. Consult the latest edition of the ANSI standard (or its replacement) for current guidelines on laser safety.

21-16 Basic Concepts

The concept of *coherence* is very important in understanding lasers and is responsible for much of the observed behavior of lasers. According to a dictionary, coherency infers "having a definite relationship to each other." That criterion means, in lasers, that points on two or more waves traveling together maintain their relationships regarding amplitude, polarity, and time phasing. In a noncoherent light source, the various emitted waves have a random relationship, so their mutual interference is also random. In lasers, there are two types of coherence: *longitudinal* and *transverse*.

Longitudinal coherence refers to a situation similar to Figure 21-18 in which two waves, of identical frequency and wavelength, are propagated along parallel paths. Note that the time phasing of the two waves is the same (e.g., they both go through the zero line in the same direction and at the same time). In a laser, the phasing of such waves is so close that a laser beam split into two paths can still exhibit constructive interference when the two portions differ in overall path length by as much as 250 to 300 kilometers. The comparable limit for noncoherent light sources is only 50 to 100 millimeters path length difference.

Transverse coherency, which lasers also possess, refers to the properties of adjacent waves retaining their definite relationship

[3] Originally, *LASER* was an acronym. According to more recent dictionaries, however, by common usage *LASER* is now considered a noun and so is properly spelled *laser*.

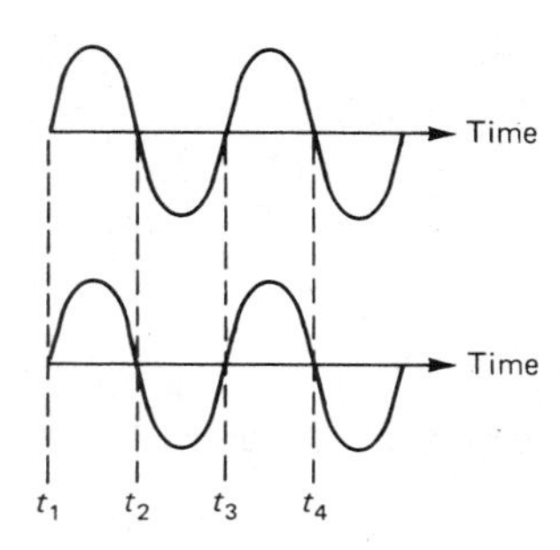

Figure 21-18 Longitudinal coherency is consistency in time phasing.

when viewed orthogonal (i.e., at right angles) to the direction of travel. For an electromagnetic wave, the transverse cut would reveal similar amplitudes and alignments of the electric and magnetic fields (Figure 21-19).

The basis of laser operation is the law of conservation of energy applied to quantum systems. When an external source of energy impinges on an atom, it will raise the energy level of the associated electrons, causing them to rise to a higher, but unstable, orbit. When the electron decays back to its ground energy state, the acquired energy is given up as light photons. The amount of change of energy state is a function of Planck's constant and the frequency of the emitted photon:

$$E = nh\nu = \frac{nhc}{\lambda} \qquad (21\text{-}32)$$

Figure 21-19 Transverse coherency is consistency across the two ways, orthogonal to the line of travel.

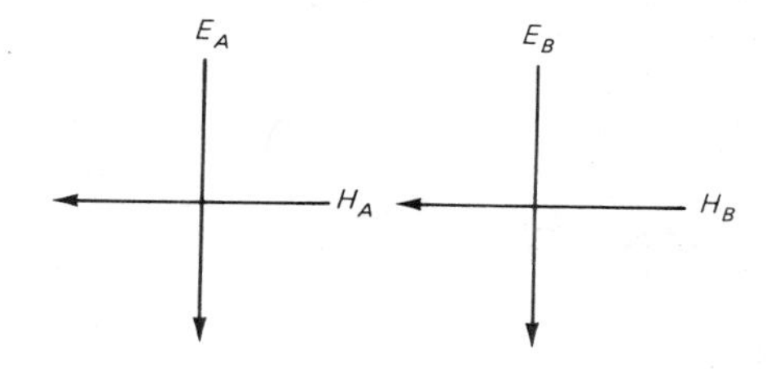

where

E is the energy level
n is an integer (1, 2, 3, . . .)
c is the speed of light (3×10^8 meters per second)
ν is the frequency of the emitted photon (Hz)
λ is the wavelength of the emitted photon (m)
h is Planck's constant (6.64×10^{-34} J-s)

Note in Equation 21-32 that the variables are energy level (E) and wavelength (γ). It can be inferred from this fact that laser light is monochromatic (i.e., of one color only). For any given material combination, there is only a very narrow band of possible wavelengths. For example, the common helium-neon (He-Ne) gas laser can produce a light output that is only a few kilohertz wide in the frequency domain, even though the center light frequency (analogous to the "carrier" frequency in rf systems) is 5×10^{14} Hz (or about 600 nm wavelength). Monochromoticity is one of the primary characteristics of laser light.

Stimulated emission of radiation was predicted as early as 1917 by Albert Einstein as part of his explanation of Planck's theory of energy quanta. According to Einstein's prediction, which proved out in practice, an atom can be stimulated into producing radiation when it is excited by an electromagnetic field that has the same frequency that would normally be emitted if the atom collapsed from the excited state to a lower state (usually ground state). Thus, a small excitation light source can create a larger stimulated light emission, and hence the light is amplified.

Laser action requires a process where there will be more atoms in the excited upper state than in the lower energy state, so that the probability of stimulated emission is increased. In normal situations, only a small portion of the atoms are in the excited state at any time, with the majority being in the lower or ground state. In good laser materials (or material combinations), it is possible to produce a *population inversion* in which more atoms are excited than are not. Such materials tend to have long-lasting excited higher energy states.

The quantum level action in typical lasers is a three-stage or four-stage transition, as shown in Figures 21-20*a* and 21-20*b*. In Figure 21-20*a*, an atom at ground state (energy level U_1) is excited by an external energy source, so it raises to a pumping level state, U_3. It then undergoes fast spontaneous decay to a metastable state (U_2), and then a slower decay back to ground state (U_1). In this second decay, light photons are emitted, so this transition is sometimes termed a *stimulated emission region*. The frequency of the photons emitted is a function of the pumping energy level (Equation 21-32). In four-stage emission (Figure 21-20*b*), there are two spontaneous decay regions (U_{4-3} and U_{2-1}), separated by the stimulated emission region (U_{3-2}).

Figure 21-20 Basis for laser operation: (*a*) three-step emission, (*b*) four-step emission.

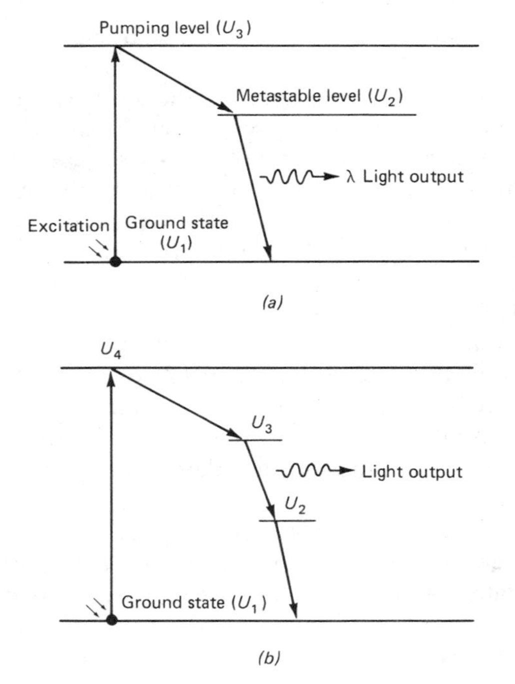

21-17 Types of Lasers

There are several different forms of lasers: insulating solid (e.g., ruby); insulating crystal; excimer, gas, injection (pn junction solid-state); chemical; and liquid (organic dye) lasers.

21-17.1 Insulating Solid Lasers

The insulating solid laser consists of a semi-transparent, solid material, such as ruby, that is optically pumped into an excited state by a pulse of light from a xenon (or similar) flash tube (Figure 21-21). The ruby laser is, perhaps, the most commonly known insulating solid type of laser. Because of the xenon tube excitation system, ruby lasers are called *optically pumped, solid crystal lasers*. The ruby laser was announced by T. Maiman of Hughes Laboratories (California) in July 1960, although patent preeminence became a matter of a legal dispute. In the ruby laser,

Figure 21-21 Ruby laser structure.

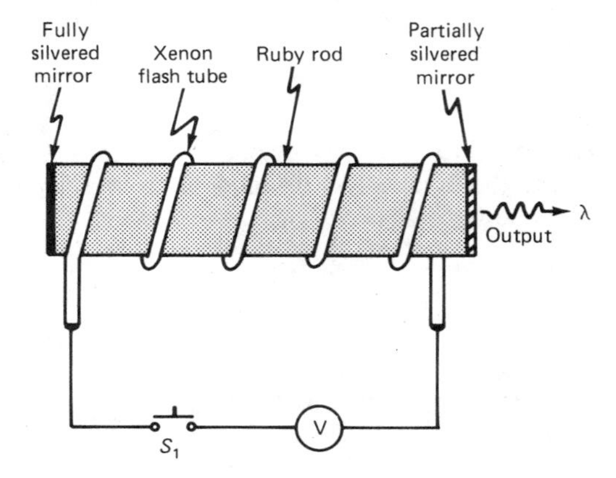

a rod of ruby material is fashioned with a fully silvered (100 percent) mirror at one end and a partially silvered (few percent) mirror at the other. Surrounding the ruby rod is the xenon high-energy flash tube, which is similar to those used in photographic flash guns. Other pumping mechanisms include krypton-filled arcs and tungsten-iodine lamps. Some lasers will "lase" on sunlight or room light, but these are not covered in detail here.

Rubies are a naturally occurring, but also synthetic, variety of the material *corundum* (Al_2O_3), which also includes sapphires (both of which are precious gemstones but can be produced synthetically); nongem/nonlaser uses of corundum include the grit used to form sandpaper. In rubies, the red coloring is provided by a small amount of chromium ions (Cr^{3+}) that replace some of the aluminum ions. When the xenon tube fires, it produces a short-duration, high-intensity blast of white light. The chromium ions absorb violet, blue, and green wavelengths from the white light, causing them to become pumped up to a higher energy state. The chromium ions then decay to a metastable state for a few milliseconds, and then undergo the stimulated decay back to a lower energy level, emitting a photon of red wavelength (6934 Å or 693.4 nm) in the process.

The initial decay is omnidirectional, so a small fraction of the photons are directed toward the fully silvered mirror at the rear of the assembly, where they are reflected back toward the partially silvered mirror. Some photons escape from the partially silvered end, but others are reflected back toward the fully silvered end. In a short time, a number of wave fronts are reflecting back and forth between the mirrors, building energy by mutual reinforcement; some waves escape on each reflection, forming an in-phase wave front.

Because the waves escape from the ruby rod in phase with each other, they are called *coherent,* and therein lays both the usefulness and the source of the intensity of laser light. Coherency is another of the principal characteristics of laser light. When the waves are emitted in random phase, there is a great deal of cancellation due to mutual interference, so a lot of the light energy is lost.

As a result of transverse coherence, the laser beam is very narrow and does not disperse the way random phased light does (as from a flashlight). Dispersion of less than 0.1 milliradians (mrad) is typical in lasers. The laser light beam remains very narrow over very long distances. Low dispersion is still another of the principal characteristics of laser light.

One way to designate the ruby laser is that it is an *optically pumped, crystalline solid-state laser.*

21-17.2 Insulating Crystal Lasers

This category of laser is very similar to the insulating solid type. Indeed, both are examples of optically pumped, crystalline solid-state lasers. Many texts place the insulating crystal lasers in the same category as ruby lasers. The insulating crystal devices differ from ruby lasers in that they use a rare earth element, such as Neodymium (Nd^{3+}), as a dopant to a material such as *yytrium aluminum garnet* (YAG)—Nd:YAG—or certain glasses—Nd:Glass.

The Nd:YAG laser typically emits a beam at infrared (1060 nm), but some are in the visible region. Energy outputs of 100 joules for several milliseconds are easily obtained. Power levels on the order of 1 watt CW are achieved using a krypton arc lamp for optical pumping. A multistage Nd:YAG experimental laser produced 1000 watts when pumped with a tungsten-iodine lamp. In the Nd:Glass configuration, up to 10 kilojoules have been produced in nanosecond-length pulses. Efficiencies are typically 3 percent, on a 2 to 4 percent range.

21-17.3 Excimer Lasers

An *excimer* is a molecule that does not normally exist except in the state where the constituent atoms are excited to a higher energy state. When the extra energy of the excitation state is given up as laser output, the de-excited molecules will revert at ground state to the original constituent atoms. Typical excimer lasers mix a rare gas (e.g., argon, krypton, xenon) with active elements such as chlorine, fluorine, iodine, and bromine.

21-17.4 Gas Lasers

Another common form of laser is the *gas laser;* included in this class are neutral atom lasers, ionic gas lasers, and molecular glass lasers. Gas lasers consist of a glass tube filled with a small partial pressure (~0.3 Torr) of a gas such as helium-neon (He-Ne), carbon dioxide (CO_2), argon, etc. (Figure 21-22). The end mirrors needed for oscillator behavior can be internal, but in most cases they are external, as shown in Figure 21-22, for ease of manufacture and adjustment. When external mirrors are used, the ends of the glass envelope of the tube are canted at a critical angle (α), called *Brewster's angle*. At this angle, light waves that are polarized in the correct manner will suffer no reflection losses at the two ends. Thus, reflection can take place between the two ends at a specific wavelength and phase relationship—forming the lasing action and retaining coherency and monochromiticity. In most gas lasers, the end mirrors will be spherical or parabolic, similar to the case of the crystal laser, in order to repeatedly reflect the light in oscillator fashion.

Figure 21-22 Gas laser structure.

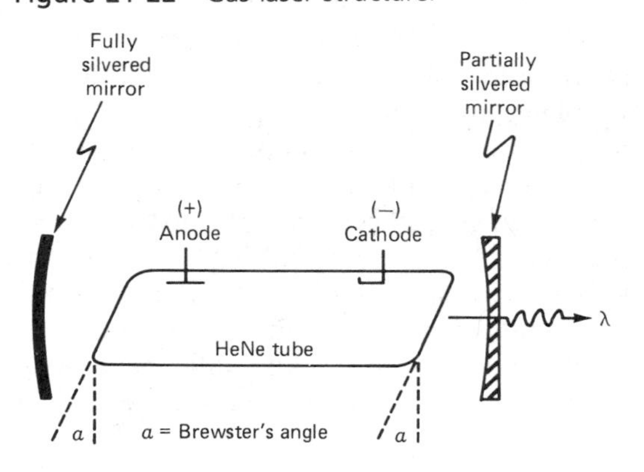

The gas laser works because electrons are injected into the gas chamber from a cathode electrode and are accelerated toward the positively charged anode electrode. Along the way, some of the electrons collide with gas ions (Figure 21-23), causing some of them to increase their energy state by absorbing the electron's kinetic energy. The excited state is not stable, so decay can be expected through several different mechanisms: collisions between the excited atom or molecule and another free electron, collision between excited and unexcited atoms, collision between excited atoms and the glass walls, and through the process of spontaneous emission.

Helium-neon (HeNe) gas lasers are usually low powered (typically 0.1 to 500 milliwatts of output power); typically drawing $\leq$10 mA, at a potential of several kilovolts, from the dc power supply. The principal light

Figure 21-23 Mechanism for gas laser action: kinetic electron excites an atom. When the atom loses its excited state, it gives up a light photon.

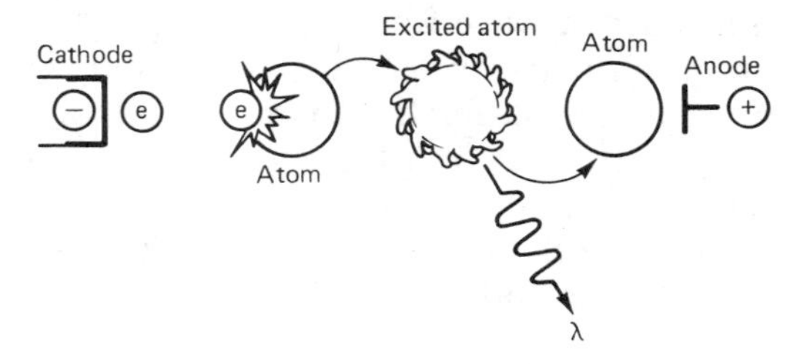

wavelength produced by the HeNe laser is at 632.8 nm (red region). The HeNe laser can also produce output light lines at wavelengths of 594, 604, 612, 1150, and 3390 nm by correct design of the end mirrors. The HeNe lasers are frequently used for classroom or hobby science demonstration projects. Some of them are in the very low milliwatt region, which are the types usually recommended for amateur experimenter use. (*Note:* Even very low-powered lasers can be dangerous to eyesight.) Divergence of the HeNe laser is typically ≤ 0.1 mrad. The glass tube on HeNe lasers is typically less than 100 cm long and has a bore diameter of 1 to 2 mm. The HeNe laser is also very low cost compared with other lasers.

Argon gas lasers produce up to 20 watts of light power at wavelengths of approximately 488 and 515 nm, in the blue and green portions of the spectrum. The typical argon laser has more gain than similar HeNe lasers but operates at efficiencies of less than 0.1 percent. These lasers often require very high current densities ($\sim$100 A/cm^2) for proper operation. Special glass is needed for the outer envelope because of heating problems.

Carbon dioxide (CO_2) lasers (which also contain small amounts of nitrogen and helium in addition to carbon dioxide) are higher powered and, indeed, can reach a very high power level (to kilowatts); CO_2 lasers can therefore be used for metal cutting and other industrial chores. The light output of the CO_2 laser has a wavelength of 10.6 μM, or about 106,000 Å, which places it in the far infrared region.

Argon and krypton gas lasers produce light in the blue and green regions and are capable of producing coherent light in two or more regions of the spectrum. These lasers find substantial application in the medical and scientific areas. Blue lasers, for example, are used for ophthalmic applications, such as "spot welding" detached retinas in the human eye.

21-17.5 Solid-State pn Junction Lasers

The final form of laser that we will consider is the solid-state pn junction diode laser (Figure 21-24), a.k.a. the *injection laser.* The laser diode is very similar to ordinary pn diodes and light-emitting diodes (LEDs), except for the very thin pn heterojunction material (e.g., GaAs) forming the pn junction, which is sandwiched between AlGaAs sections that serve as internal resonating mirrors. Population inversion, which is necessary for laser action, occurs because holes from the *p*-side, and electrons from the *n*-side, are forced by the applied electrical field into the junction region. The created conduction band becomes the upper laser energy level, while the valance band becomes the lower energy level.

The most common low-cost laser diodes use GaAs material sandwiched between AlGaAs and produce output lines in the 760- to 905-nm region, depending on the proportion of materials and whether a single or double heterojunction is used. In addition to GaAs, there are other material combinations. One such is InGaAsP (iodine gallium arsenide phosphorous), which emits in the 1200 to 1550 nm range.

In both types of injection laser, room tem-

Figure 21-24 PN diode laser structure.

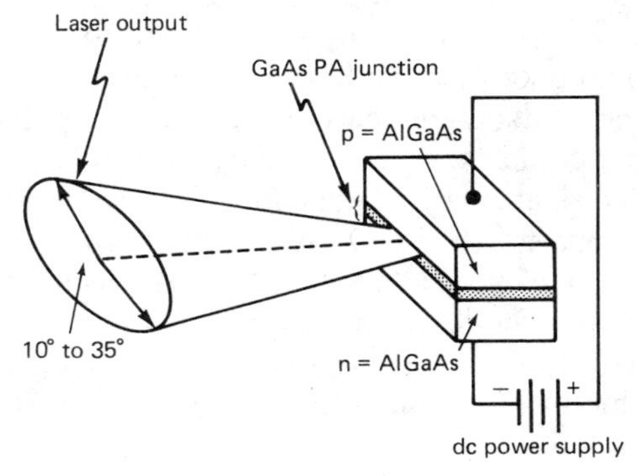

perature lasing is possible, although only at low power levels and low efficiencies. For example, one AlGaAs laser diode produces 0.02 watts (20 mw) CW, at 7 percent efficiency, at room temperature. Cooling can produce a tremendous increase in both power level and efficiency. A GaAs laser diode cooled to −253°C can be as much as 50 percent efficient. In the pulsed mode, commonly available laser diodes can produce 3 to 10 watts, drawing 10 amperes or so from the dc power supply. Noncooled GaAs devices produce 3 to 5 mw in some configurations, and 500 mw in others.

At low current levels, the diode acts like other pn junction diodes, and at higher currents will emit a broad spectrum noncoherent light much like other LEDs. But if a certain threshold current (I_{th}) is exceeded, the laser diode will begin to lase and emits light in the direction shown. Threshold currents of 1000 or more amperes per square centimeters (1000 A/cm^2) are possible, even at moderate average forward current levels, because of the very small area of the laser diode die (100 mils by 100 mils).

A stud-mounted form of laser diode package is shown in Figure 21-25*a*. This diode uses a threaded stud for mounting and to make the negative electrical connection. The positive electrical connection (anode) is made to an electrical terminal protruding from the rear of the mounting stud. Because laser diodes tend to be low-efficiency devices and therefore produce larger amounts of waste heat, it is often necessary to mount them on a metallic heatsink (Figure 21-25*b*). The heatsink will carry away the excess heat that could otherwise destroy the diode. Any electrical energy that is not converted to light output goes off as heat, so the thermal condition of the diode must be a serious concern to laser diode users. A variant shown in Figure 21-26 is a stud-mounted laser diode in which the output is coupled to a fiber-optic pig tail.

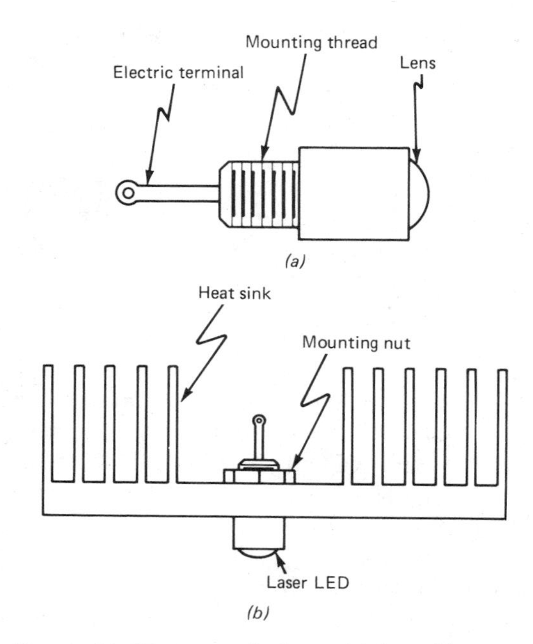

Figure 21-25 Laser diode packaging. (*a*) Typical stud-mounted laser diode; (*b*) stud-mounted laser diode mounted on heatsink to improve heat dissipation.

21-18 Driver Circuits for Solid-State Laser Diodes

The laser diode is very much like the light-emitting diode (LED) and so uses similar circuits for driving the device. Theoretically, one could use the LED-type circuit shown in Figure 21-27. In this circuit, the laser diode is connected between V$^+$ and ground, with a resistor network in series to limit cur-

Figure 21-26 Laser diode with fiber optic output.

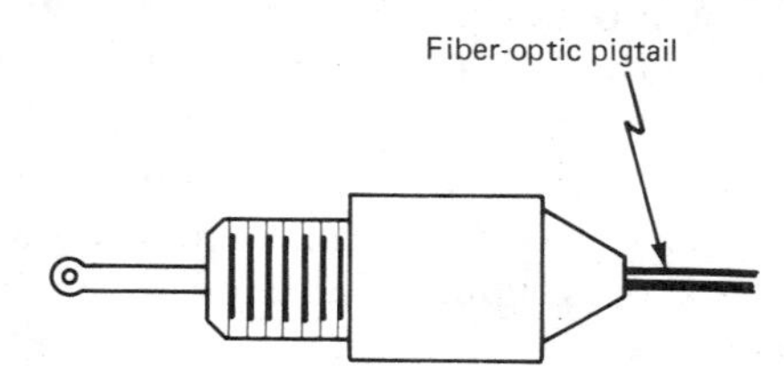

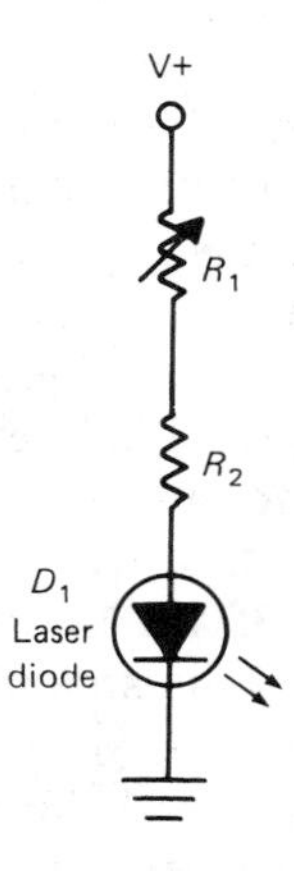

Figure 21-27 Simple laser diode driver circuit.

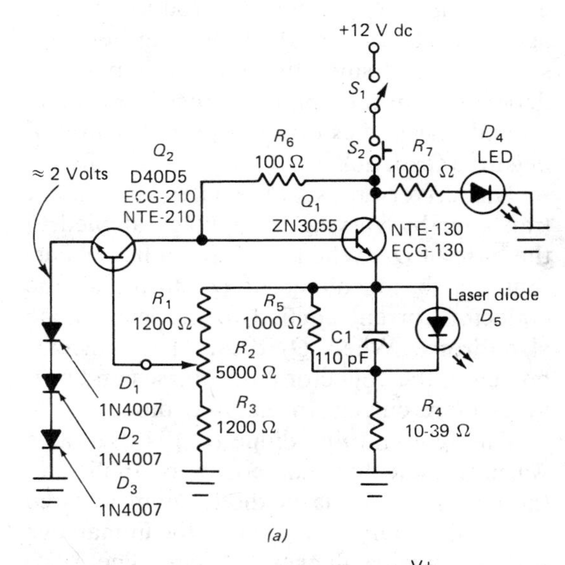

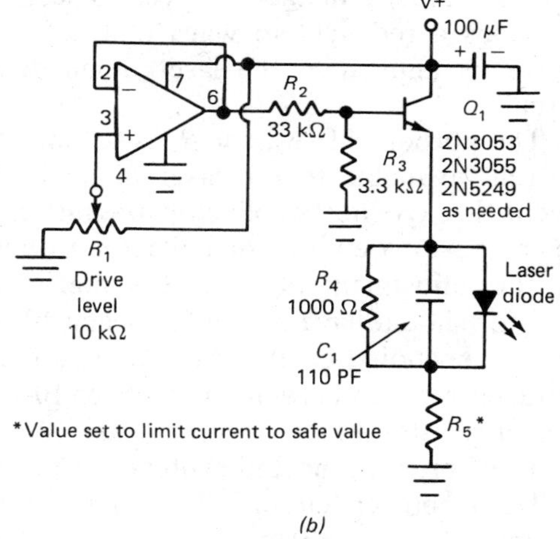

Figure 21-28 Better laser diode driver circuits: (*a*) all-transistor version; (*b*) op-amp/transistor version.

rent. According to the standard wisdom, resistor R_1 can be adjusted to vary laser diode current, and hence laser output, while R_2 is used to set the maximum current (for safety's sake) that can flow in the diode. The problem, however, is that these laser diodes tend to draw large amounts of current, so the resistors have to be 5- to 20-watt types, depending on the diode (and associated voltage and current levels). As a result, it is more common on laser diodes to use a high-power transistor driver circuit to control the laser diode current.

Figure 21-28*a* shows a two-stage transistor direct-coupled amplifier that is used as a laser diode driver. The output transistor, Q_1, is a large-power transistor such as the 2N3055. This transistor can dissipate 110 watts collector power and has a collector current rating of 15 amperes. The beta gain of the 2N3055 is about 45. The driver transistor (Q_2) is a smaller NPN power transistor in a plastic package. It will dissipate up to 5+ watts when heatsinked, and 1.33 watts in free air, at a collector current of 1 ampere; the beta gain is >120.

The emitter of Q_2 is kept at a potential close to +2 volts by three silicon 1-ampere rectifier diodes in series (although 1N4007 are shown, any of the 1N400x from 1N4001 to 1N4007 can be used). Each of these diodes has a forward voltage drop of 0.60 to 0.70 volts, resulting in a total of about 2 volts at the emitter of Q_2. The base voltage for Q_2 is derived from the emitter of Q_1, which is at a potential near 5 volts. The voltage divider consisting of R_1, R_2, and R_3 produces a voltage near 2.6 volts for the base of driver transistor Q_2, and since this voltage is partially dependent on the voltage across laser diode D_5 and series resistor R_4, it provides a bit of negative feedback.

The driver transistor excites the output transistor by varying the voltage applied to the base of Q_1. When Q_2 draws a heavy current, the base voltage of Q_1 drops, so the collector current applied to the laser diode also drops. When Q_2 draws less current, however, the collector current rises and provides more current to the laser diode.

The light-emitting diode (D_4) is used as a warning indicator that power is applied to the circuit. Some laser diodes emit infrared light which cannot be seen by the human eye and are thus a danger to eyes. The LED provides a red light to warn that collector power is applied to the laser diode driver circuit.

The purpose of resistor R_4 is to limit the current available to the laser diode in the event that Q_1 shorts collector to emitter, Q_2 opens base to emitter or collector to base, or the adjustment of R_2 is too high. This resistor should have a value between 10 and 39 Ω, depending on the laser diode's maximum current, and is generally a 5- or 10-watt power resistor.

Two series-connected switches are in the collector power circuit. The main power switch, S_1, is used to turn the circuit off entirely. Some users prefer to make this switch a key-operated model in order to limit access to the laser to those who are authorized (remember, lasers are dangerous devices). The second switch, S_2, is a pushbutton type that is used to turn on the laser when a burst of light is needed.

A variation on the circuit is shown in Figure 21-28*b*. This circuit uses the same type of power transistor output stage as the previous circuit, but the driver is replaced by an operational amplifier. The op-amp is connected in the unity gain noninverting follower configuration. The noninverting input (+IN) is connected to a potentiometer that sets the positive voltage applied to the +IN terminal. The op-amp output terminal will be at a potential that reflects the setting of R_1, and in turn drives the bias network for the output transistor (R_2 and R_3). Good values for R_2 and R_3 to start with in finding the correct setting are 33 kΩ for R_2 and 3.3 kΩ for R_3 (the correct values can be found empirically, or calculated if you know the beta gain of the transistor and the current range needed by the laser diode).

Another driver circuit is the *amplitude modulator* in Figure 21-29. In this circuit, the laser diode driver is a two-stage direct-coupled amplifier consisting of Q_1 and Q_2. The quiescent (i.e., static, no-signal) current level in the laser diode is set by adjusting potentiometer R_2. This current is modulated by applying the audio signal voltage to the base of Q_2 through dc-blocking capacitor C_4.

The audio preamplifier portion of the circuit of Figure 21-29 is an operational amplifier used as a noninverting ac-coupled amplifier. Almost any op-amp can be used for communications grade voice audio, including the 741, CA-3140, and others. In order to use just one dc power supply for the op-amp (instead of bipolar +V supplies), the noninverting input is biased to one-half the supply voltage by voltage divider network R_8/R_9. The audio gain of this circuit is set by the ratio of the two feedback resistors according to the equation:

$$A_v = \frac{R_5}{R_6} + 1 \qquad (21\text{-}33)$$

Although the gain is set to 101 in this example, it should be adjusted according to the amplitude of the audio source. The idea is to linearly modulate the current in the laser diode and so make the gain match the requirement.

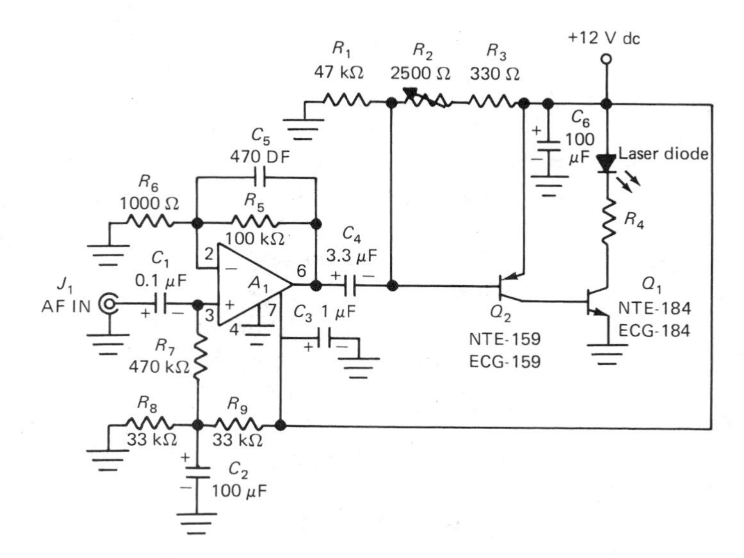

Figure 21-29 Laser diode current modulator circuit.

The frequency response of the audio amplifier is measured between the points where the gain drops off −3 dB relative to the mid-band gain. The upper −3 dB point is set by capacitor C_6 shunted across R_5, while the lower-end −3 dB point is set by capacitor C_1 and resistor R_7. These points are:

$$F = \frac{1}{2\pi R_5 C_5} \tag{21-34}$$

and

$$F = \frac{1}{2\pi R_7 C_1} \tag{21-35}$$

where the resistances are in ohms, the capacitances are in farads, and the frequencies are in Hertz.

A laser diode pulser circuit is shown in Figure 21-30. This circuit will drive only low-power laser diodes because the 555 timer (IC_1) is limited to sinking 200 mA of current. One might "get away" with more because of the short duty cycle of most laser diodes, but that is not recommended practice. It would be better to use this circuit to generate a pulse that drives a circuit similar to the previous drivers shown. The pulse repetition rate (PRR) of this circuit is set by adjusting R_1, R_2, and C_1 according to the equation:

$$PRR = \frac{1.44}{(R_1 + 2R_2)C_1} \tag{21-36}$$

The *duty cycle* is the ratio of the laser diode on-time to the off-time and can be set according to:

$$D = \frac{R_2}{R_1 + 2R_2} \tag{21-37}$$

With the values shown, the PRR is about 270 pulses per second, while the duty factor (D) is about 19 percent. At a PRR of 270 PPS, the period of each cycle, which includes pulse on-time and pulse off-time, is 0.0037 seconds (3.7 mS). The pulse on-time is thus 19 percent of 3.7 mS, or 0.7 seconds (700 nanoseconds).

A *data driver* circuit is shown in Figure 21-31. This diode uses a transistor laser diode driver similar to previous circuits, but the driver is excited by a B-series CMOS

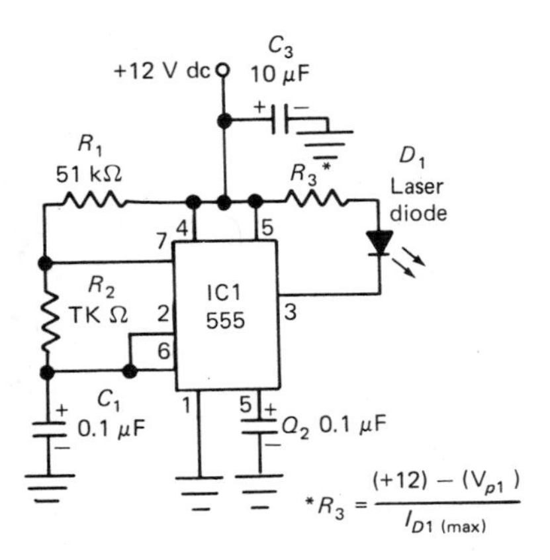

Figure 21-30 Laser diode pulse circuit.

digital logic integrated circuit (IC) device. In the specific example shown, the exciter is a NAND gate, but others could be used instead. The B-series is preferred over the A-series CMOS because the B-series has higher drive capacity.

The terms *HIGH* and *LOW* used in Figure 21-31 are from digital electronics terminology and refer to the binary bits 1 and 0, respectively (i.e., 1 = HIGH, 0 = LOW). In electronic digital circuits, it is common

Figure 21-31 Laser diode digital driver circuit.

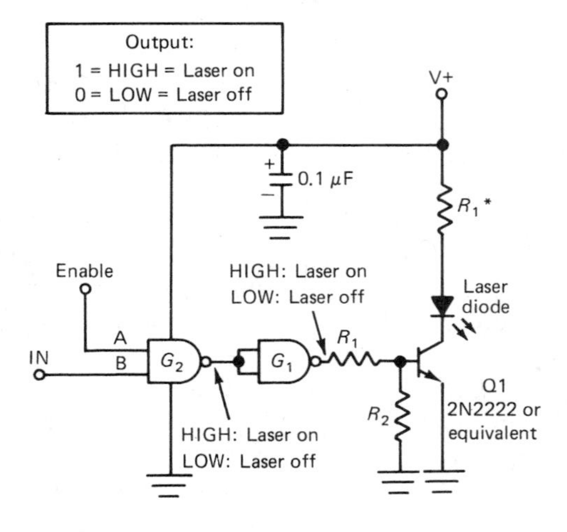

for LOW to be zero volts and HIGH to be a positive voltage. If the NAND gates were TTL (which would require additional circuitry), then +5 volts is mandatory for V+. With CMOS devices, V+ can be anything from +4.5 volts to +15 volts. Resistors R_1 and R_2 are adjusted accordingly.

The NAND gate operates according to the following rules:

1. If both inputs are HIGH, then the output is LOW.
2. If either input is LOW, then the output is HIGH.

These rules allow us to strap both inputs together and use the NAND gate as an inverter circuit. In Figure 21-31, NAND gate G_1 is used in this manner. When the inputs of G_1 are LOW, then the output is HIGH; when the inputs of G_2 are HIGH, then the output is LOW. The laser diode is turned on when the output of G_2 is HIGH, providing a bias to Q_1 through R_1/R_2.

The data stream of HIGH/LOW bits is controlled by NAND gate G_1. Input A of G_2 is used as an active-HIGH *enable* line. In other words, when A = LOW, the output of G_2 is HIGH, disabling the laser diode output. Similarly, when A = HIGH, the output of G_2 is controlled by the other input (B). When IN = LOW, then the output of G_2 is HIGH and the laser is off; when IN = HIGH, then the output of G_2 is LOW and the laser is on.

21-19 Laser Diode Receiver Circuits

The laser diode cannot be used in a communications circuit unless there is a receiver at the other end to recover the information transmitted. Figure 21-32 shows two laser preamplifier circuits that will pick up the varying pulse or AM-modulated laser signal and convert it into an ac or pulse signal. In both circuits the sensor is an NPN phototransistor connected to an operational amplifier.

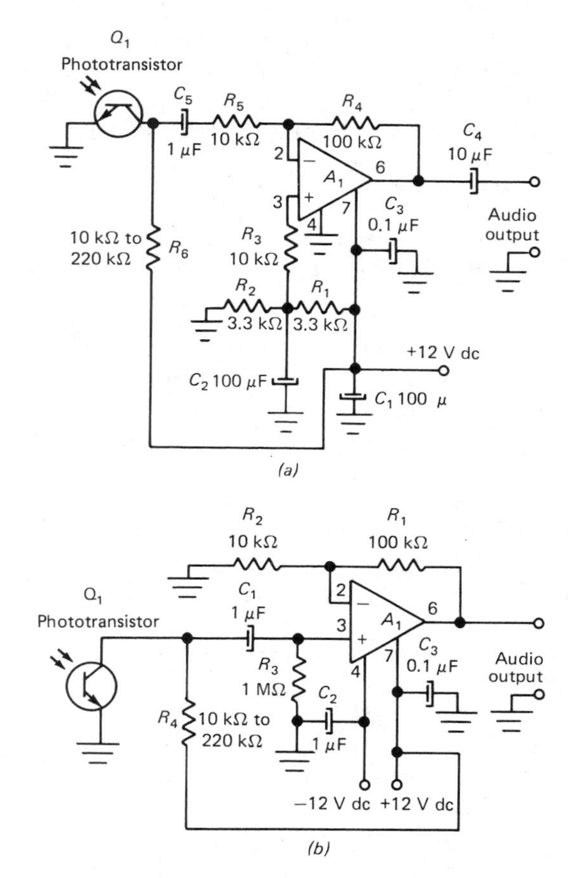

Figure 21-32 Laser diode receiver circuits: (*a*) inverting amplifier version, (*b*) noninverting amplifier version.

The circuit in Figure 21-32*a* uses an operational amplifier operated from a single dc power supply. This circuit can be used when only a single +9- to +15-volt dc power supply (e.g., battery) is available. The noninverting input is biased to one-half V+, or in this case about +6 volts dc, by the action of resistor voltage divider R_1/R_2. Resistor R_3 is used to prevent charge stored in the decoupling capacitor, C_2, from damaging the op-amp when the power is removed.

The signal gain of the operational amplifier is set by the ratio of R_4 and R_5. Because this circuit is essentially an inverting follower, the gain is:

$$A_v = -\frac{R_4}{R_5} \qquad (21\text{-}38)$$

The minus sign indicates that the output signal is 180° out of phase (i.e., it is inverted) with respect to the input signal.

The collector resistor supply voltage to the phototransistor can be set as low as 10 kΩ for high-frequency, high-speed operation, or as high as 220 kΩ for high-sensitivity operation. In most similar applications, the goal is sensitivity, so the resistor is set to some value between 150 kΩ and 220 kΩ.

A version of this circuit with a bipolar dc power supply is shown in Figure 21-32*b*. It uses V− and V+ dc power supplies rather than a single V+ dc supply (as in Figure 21-32*a*). The phototransistor circuit is very similar, but the output signal from the transistor is coupled through C_1 to the noninverting input of the operational amplifier.

A current mode receiver is shown in Figure 21-33. In this circuit, the photosensor is a photodiode that produces a current, I_d, that is a function of the light level. The output voltage, V_o, is found by:

$$V_o = -I_d R_1 \qquad (21\text{-}39)$$

where R_1 is in ohms and I_d is in amperes.

It is common practice to use an operational with a very high input impedance in this type of circuit, in order to reduce the

Figure 21-33 Simple laser diode receiver circuit.

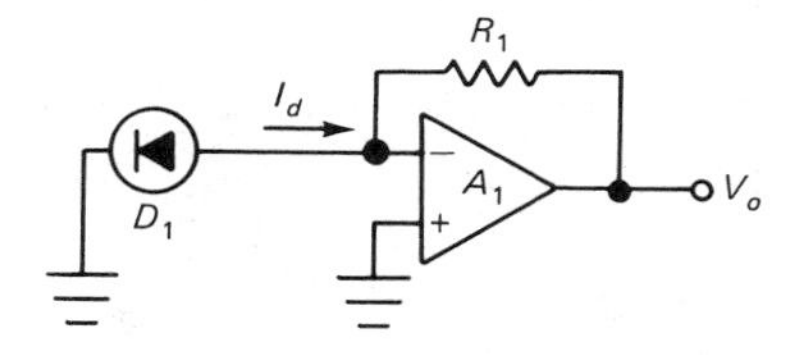

effects of input bias currents. Therefore, use a CA-3140 or its equivalent for A_1.

For those readers interested in experimenting with laser diode communications, the circuit in Figure 21-34 provides an audio amplifier stage. It will receive the outputs from any of these receivers and amplify them to a level that will drive either a small 8-Ω loudspeaker or a pair of earphones. The basis for this circuit is the LM-386 single-chip audio amplifier. The gain of this stage can be set by how you treat terminals A and B (which correspond to pins 1 and 8, respectively). For a gain of 10, leave A-B open (no connection or external circuitry required). For a gain of 50, however, connect a series *R-C* network consisting of a 1.2-Ω resistor and a 10-μF capacitor. The highest gain is 200, which is accommodated by connecting a 10-μF capacitor between A-B.

Before leaving the subject of lasers, it is prudent to reiterate the hazard warning about lasers. The coherent light from a laser, even at very low levels, is dangerous to eyes and skin. Take every step to avoid viewing laser light either directly or through either specular or diffuse reflections. Read, understand, and follow all safety instructions provided by the manufacturer of any laser device that you put into operation. Before beginning to use lasers, read and understand the laser safety guidelines in the ANSI standard.

Figure 21-34 Laser diode receiver audio stage for voice communications.

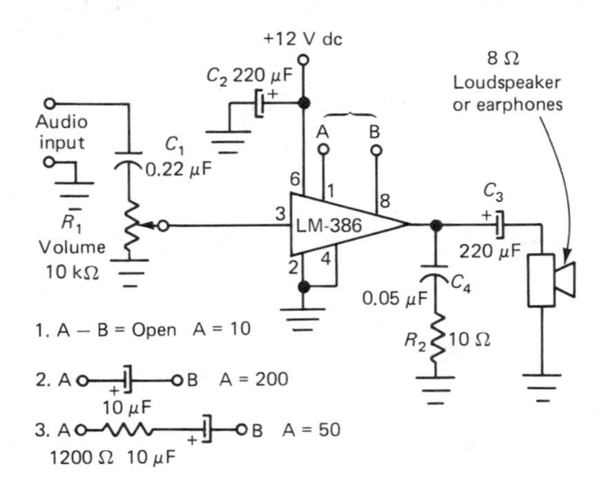

21-20 Summary

1. Fiber optics are the technology in which light waves are passed through a plastic or glass fiber so that they can be directed to a specific location.
2. Fiber optics can carry light-modulated signals and are essentially free of electromagnetic radiation interference. They provide a high degree of electrical isolation and are explosion proof.
3. Clad fiber optics are built with a dense inner core surrounded by a less optically dense cladding layer.
4. In meridional ray propagation, all rays lie in a plane with the optical axis. In skew ray propagation, the rays follow a spiral or helical path along the fiber optic.
5. The *acceptance angle* is the critical angle for the transition from air to fiber optic.
6. There are a number of transmission modes in a fiber-optic path. When a light beam attempts to find more than one mode, the resultant error is called *intermodal dispersion*, and it can have an adverse effect on digital signals. Dispersion problems can be overcome by using graded index fibers.
7. Loss modes in fiber optics include surface defects (nicks, scratches, etc.), contaminants, inverse square law losses, transmission losses, absorption losses, coupling losses, mismatched fiber diameter losses, numerical aperture coupling losses, and Fresnel reflection losses.
8. Fiber-optic communications systems

can be built by adding an encoder and decoder to opposite ends of the fiber.

9. Lasers produce a light beam that possesses the properties of coherency, monochromiticity, and low dispersion.
10. Lasers are classified by the ANSI and LIA standards for safety (Class I, II, and III).
11. *Coherence* means that the light waves are in phase with each other. Lasers can offer both longitudinal and transverse coherence.
12. Types of lasers include insulating solids, insulating crystal, excimer, gas, injection, chemical, and liquid organic dye types.

21-21 Recapitulation

Now return to the objectives and self-evaluation questions at the beginning of the chapter and see how well you can answer them. If you cannot answer certain questions, place a check mark next to each and reread appropriate parts of the text. Next, try to answer the following questions using the same procedure. When you have answered all of the questions, solve the problems in Section 21-27.

21-22 Questions

21-1 Because fiber-optic links are electrically nonconductive, they are essentially free of __________ __________ interference problems.

21-2 List several advantages of fiber optics.

21-3 Define *optical refraction*.

21-4 Define the differences between *critical, subcritical,* and *supercritical* angles.

21-5 Express Snell's law as an equation.

21-6 In a __________ fiber optic, there is an optically dense inner core overlaid with a less optically dense surface.

21-7 __________ rays lay in a single plane with the optical axis, while __________ rays follow a helical path.

21-8 Express the acceptance angle in terms of a *numerical aperture*.

21-9 In __________ __________, the light waves attempt to enter more than one transmission mode.

21-10 Any fiber with a core diameter of more than 10λ is typically called a __________ fiber.

21-11 A __________ index fiber consists of several layers of differing indices of refraction.

21-12 List several types of loss in fiber-optics systems.

21-13 List the elements of a fiber-optic communications system.

21-14 An ANSI Class __________ laser is not capable of producing biological damage to the eye or skin during intended use, by either direct or reflected exposure.

21-15 An ANSI Class __________ laser may be hazardous under direct or reflected situations but is not usually dangerous under diffuse reflection situations.

21-16 ANSI Class __________ lasers typically emit light in the 400- to 700-nm region of the spectrum.

21-17 Define *coherence* in your own words.

21-18 In __________ coherency, adjacent waves retain their definite relationship when viewed orthogonally to the direction of travel.

21-19 A __________ __________ is said to exist in a material when more electrons are in an excited state than are not excited.

21-20 List five different types of lasers.

21-21 Refraction occurs at a(n) __________ angle when total internal reflection (TIR) occurs.

21-23 Problems

21-1 Light travels in a material at 2.3×10^8 m/s. What is the index of refraction of that material?

21-2 A light ray in air approaches a material boundary at an angle of 18 degrees. If the index of refraction of the material is 1.75, the angle of refraction in the material is __________ degrees.

21-3 The acceptance angle of a fiber is 11 degrees. What is the numerical aperture?

21-4 What is the index of refraction of a material when the numerical aperture relative to air is 0.31?

21-5 A 0.51-mm fiber has a numerical aperture of 0.34. What is the relative light collection ability?

21-6 A 700-nm light is launched onto a 0.05-mm fiber. How many modes are possible if the numerical aperture is 0.122?

Chapter 22

Computers in Biomedical Equipment

22-1 Objectives

1. Be able to define basic terms (computer, input/output, data bus, hardware, software, interfacing, interrupt, ALU, CPU, controller, memory, microprocessor, microcomputer).
2. Be able to draw a simplified block diagram of a computer (organization) and describe its operation.
3. Be able to list and describe programming languages (machine, assembly, and high-level such as BASIC, FORTRAN, and COBOL).
4. Be able to draw a simplified block diagram of a microprocessor and microcomputer and describe its architecture, including word length, complexity, application, cost, memory size, program, speed constraints, and input/output.
5. Be able to describe interfacing units such as digital-to-analog and analog-to-digital converters.
6. Be able to list and describe types of large computers and microprocessors/microcomputers in biomedical instrumentation.

22-2 Self-Evaluation Questions

These questions test your prior knowledge of the material in this chapter. Look for the answers as you read the text. After you have finished studying the chapter, try answering these questions and those at the end of the chapter (see Section 22-11).

1. What is the definition of a computer?
2. What is the distinction between hardware and software?
3. What are the major blocks involved in computer organization?
4. What is the difference between low-level and high-level computer languages?
5. What is the difference between a microprocessor and microcomputer?
6. How are computers interfaced to biomedical instrumentation?
7. How are computers used in biomedical instrumentation?

22-3 Introduction

In the past 30 years, the computer revolution has unfolded. The *major factors* contributing to this *growth* are curiosity, the discovery of a powerful mathematical tool, and the enormous capital invested by Western civilization. Modern times have witnessed its use in nearly every area of human life. Computers have become commonplace in transportation systems (automatic fare collection and scheduling), department stores and supermarkets (electronic computerized cash registers and inventory control), automobiles (energy control), military (peacetime and war machines), small business (record-keeping devices), and medicine (diagnostic, therapeutic, and record systems). Hand-held calculators continue to permeate the public scene. The future will bring even greater uses for computers, as their diversity is expanding almost daily.

The *history* of computers is as interesting as it is fast and furious. Within the past 30 years, the digital computer has been structured, expanded, reduced in size, and developed for amusement, business, and scientific endeavors. Essentially, computers have transitioned from the ancient mechanical type to the gear type through the vacuum tube and transistor systems to the small-, medium-, and large-scale integration (SSI, MSI, LSI) microprocessor and microcomputer. The first known computer was the *abacus*, invented between 4000 to 3000 B.C. It is a mechanical device consisting of beads that can be moved along a wire to add, subtract, multiply, and divide. A skilled operator can perform calculations with amazing speed. Other mechanical machines include the rotating wheel mechanical calculator, invented in 1642 by Blaise Pascal, and a more elaborate one invented in 1671 by Baron Von Leibnitz. These gave rise to the mechanical desk calculator used by large numbers of people in the seventeenth and eighteenth centuries. Charles Babbage, a Cambridge University mathematics professor, proposed his "analytical engine" in the early nineteenth century, and this eventually led to computational machines with stored programs.

Herman Hollerith conceived the idea of the punched card in 1890 to store data, and manufacturers soon adopted this technique. While calculators were becoming motor driven in the 1930s, George Stibitz was developing the complex calculator at Bell Telephone Laboratories finalized in 1937. Howard Aiken of Harvard University generated the Mark I Automatic Sequence Controlled Calculator in 1937, and in 1944 it became operational as a general-purpose mechanical digital computer. The Bell V followed in 1946, along with the Harvard Mark II and IBM 604 in 1948.

Following the age of mechanical computers, *four generations* appeared. These include *vacuum tube, transistor, integrated circuit,* and *large-scale integrated microprocessor/microcomputer*. In the 1940s, John Von Neuman proposed a general-purpose computer. UNIVAC and IBM designed and built numerous versions of vacuum tube models, but in 1958 the National Cash Register Corporation delivered the first commercial digital computer, followed by the CDC 7604 and IBM 7090 in 1960 and the CDC 6600 in 1964. In 1964, IBM announced the new 360 series and by the mid-1970s the advanced 370 series. Since 1977 the "computer on a chip" has been produced by such companies as Intel (8080 A and Z 80) and Motorola (6800).

The early computer pioneers were divided into two camps. On the one hand, some of them wanted to build special-purpose computers that were optimized for a specific job (like artillery). But the other faction, in one of the greatest insights and most far-reaching visions of the twentieth century, decided to build a *programmable* digital computer. The concept is simple, but at the time it was revolutionary: a single processor that would obey instructions that were stored in a memory bank (along with data). By writing a *program* (i.e., a set of instructions that tell the computer what to do), the user could make the same machine do a lot of different chores. Thus, the same IBM-AT class machine that controls the patient monitoring system in an intensive care unit can also be used by the administrator's secretary as a word processor, by the administrator as a management information system, by central supply as an inventory and processes controller, and by the biomedical engineer to keep preventive maintenance records. When the work day is over, one can run games like *Flight Simulator®* on the same machine. All of this is possible because the computing machine is now a kind of "universal analytic engine." That is, it can be reprogrammed to perform a wide variety of different chores.

It was only three decades ago that the computer was a large mainframe unit sold by industrial giants such as IBM and Honeywell. These machines took up entire rooms (sometimes more than one) and were housed in a series of large metal cabinets. They were used for large-scale data processing chores only and were not well suited for smaller tasks. In fact, at one hospital where I worked—a university medical center—the only computers it used in the 1960s were the IBM-1401 and later an IBM-360 down the street at main university data processing. The same institution is now overrun with computers of several sizes and employs its own in-house programmers.

In the 1960s a new form of computer appeared on the market. The *minicomputer* was housed in a single cabinet that was 19 inches wide and about 7 feet high. The unit still required special facilities such as a separate high-current electrical circuit and a large-capacity air conditioner to carry off the large amount of heat generated by the machine.

In 1972 a quiet revolution took place in California's now-famous Silicon Valley. The Intel Corporation was contracted to build a small digital processor integraged circuit ("chip"). That simple unit was the start of the *microprocessor* industry. A microprocessor is a chip that contains all of the arithmetic and logic elements required for a programmable digital computer. All the designer needs to do to make a real computer is add external memory and input/output circuits. A *microcomputer* is a small desktop computer that is based on a microprocessor chip. Today, one can buy microcomputers that sit on a desktop—with plenty of room left over—and still have substantially more computing power than the roomful of equipment that I used in engineering school in 1967. In fact, the abilities of the IBM-AT that the authors used to write this chapter are considerably greater than the computer they used as engineers in past decades.

The microcomputer has literally revolutionized medical (and other) instrument and control system design. Whereas designers were once strictly analog engineers, the instrument designer today has to be a synergist who can integrate the principles of sensor selection, analog circuit design, computer hardware selection and/or design, and software. Today, even small instruments are based on microcomputer chips, and for that reason we are going to consider these devices in some detail.

Computers in medicine began with the processing of hospital business and billing information. In the 1960s, ECG and EEG signals were analyzed. In addition, respiratory function was evaluated. Computerized medical screening began during this time, and automated analysis of blood and urine was emerging. Dedicated minicomputers of modest cost allowed the generation of computerized patient monitoring, and fixed microprocessors gave rise to instantaneous analysis within the bedside unit. Currently, automation and instant analysis of medical events are requiring that computers be integrated into modern medical instrumentation.

22-4 Computer Hardware and Software

Computers can be classified as analog (continuous time signal) or digital (discrete step-time signal). An *analog computer* is an electronic (or mechanical) device that can perform continuous mathematical operations such as addition, subtraction, multiplication, division, scaling, integration, and differentiation. Its accuracy is on the order of 0.01 percent of full scale, and while it can perform many parallel operations at high speed, it is very limited in making logical decisions. Programming is accomplished by drawing block diagrams with feedback loops and then patching the front panel with cables. However, most computers in use today, especially in biomedical instrumentation, are digital.

A *digital computer* is an electronic and electomechanical device that can perform a sequence of *arithmetic* (addition, subtraction, multiplication, division, and combinations of these) and *logical* (conditional) operations. The device can solve mathematical problems or can search, sort, and arrange information according to stored instructions. The computer program is the sequence of instructions and is known as *software*. The electronic and electromechanical portion comprises the physical part and is known as *hardware*.

The hardware is organized into various sections, as shown in Figure 22-1. These include the following components:

1. *Input unit*—keyboard, teletype punched tape or cards, or magnetic tape.

Figure 22-1 Simplified block diagram of a digital computer.

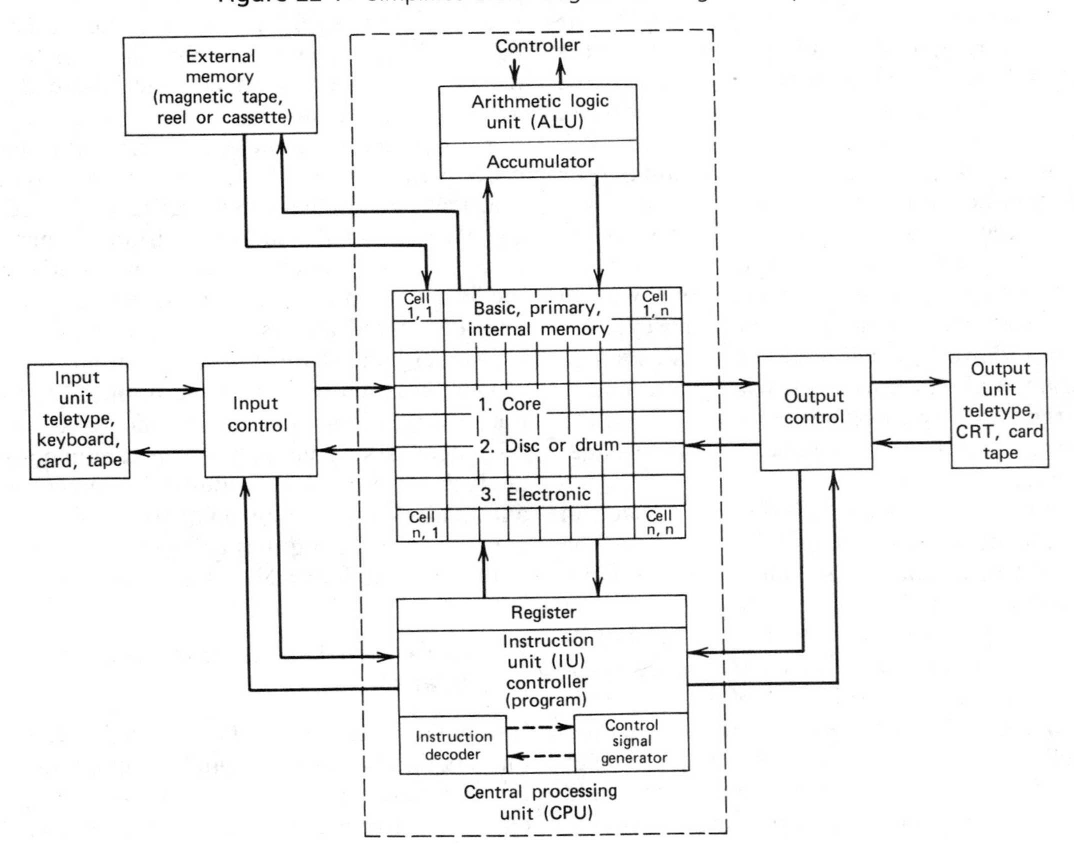

2. *Input control*—to control input data on its way to the memory.

3. *Memory unit*—internal memory (composed of cells) such as core, disc, drum, or electronic registers to store data, instructions, and results of arithmetic or logic manipulation. External memory such as magnetic tape stores large programs and data. Magnetic tape has the longest access time but is the least expensive in terms of data quantity per size of device. Core is next in speed but more expensive than tape. Disc and drum are faster still and more expensive, and electronic register is the fastest and the most costly. Magnetic tape, disc, and core are "nonvolatile" or will hold information after the power has been removed.

4. *Instruction unit and controller*—to interpret or decode computer program instructions and control operations of the computer via the control signal generator.

5. *Arithmetic logic unit (ALU)*—to perform addition, subtraction, multiplication, and division and return the calculation results to the memory. An *accumulator* temporarily holds intermediate results during calculations. Essentially, all calculations are sequential in a digital computer (i.e., multiplication is repeated addition) and can require considerable time for completion. The binary (two-state) logic makes the hardware simple (and repetitive) but is time consuming for arriving at final answers. On today's market this trade-off is very successful.

6. *Output control*—to control data on its way to the output unit.

7. *Output unit*—CRT, teletype, printers, punched tape or cards, or magnetic tape. High-speed printers are available that can produce greater than 1200 lines per minute.

The *central processing unit (CPU)* or central processor is that part of a computer system that contains the main or internal memory, arithmetic logic unit, and special registers to provide instruction processing, timing, and other housekeeping operations.

Digital computers operate in binary. Binary numbers are represented only by ones and zeros. That is, a one can be a switch that is turned on and a zero can be a switch that is off. This allows the hardware to be simple (a two-state switch) as compared to a decimal number system in which a 10-position switch is required. The term *bit* is derived from *binary digit* and indicates a single character of the binary number system.

The *signal flow* through a computer as seen from Figure 22-1 takes many paths. First, a program is written (software) and instruction and data are keyed in through the input device to be processed by the computer machine (hardware). Some keyboards contain an ASCII (American Standard Code for Information Interchange) code generator that produces a 7-bit binary number for each key pressed. Therefore, the code for the equation of a straight line could be placed into the memory via the input control as shown at the bottom of this page (* means multiply). Hardware accomplishes the software and as such they are linked together.

The ASCII code will typically be converted into a 16-bit or 32-bit binary word in large computers or an 8-bit word in mini- and microcomputers. Instructions and data share the same signal path.

After the linear equation is written, data can be placed in the memory as well. For example, if M is assigned as 2 and X as 2 and B as 1, Y can be calculated as 5 (calculate instruction) in the arithmetic logic unit (ALU). The result, through the accumulator register, can then be stored in the memory. Later, data can be taken from memory and printed (write instruction) on the output CRT. The following shows a simple program:

	Y	=	M	*	X	+	B
Character	1	2	3	4	5	6	7
ASCII Code	101 1001	011 1101	100 1101	010 1010	101 1000	010 1011	100 0010

Instructions

Read	Y = M * X + B	Input equation
Read	M = 2, X = 2, B = 1	Input data
Calculate	Y = M * X + B	Perform arithmetic calculation
Write	M, X, B, Y	Print data on CRT

The output on the CRT is as follows:

2 2 1 5

As part of the software, a flowchart is usually generated first, and lines of code are written to accomplish the desired result. An IBM standard for flowcharts is ellipse for start/stop, parallelogram for read/write, rectangle for calculate, and diamond for decision. A flowchart for the linear equation problem is shown in opposite column at top.

The loop is activated only if Y is not less than 4. Hence, M is assigned 1, X is 0, and B is 0 and the new Y will be 0, which is less than 4. Thus, the programs stops. This simple example shows how data is manipulated inside a computer and how equations can be solved. Digital computers can do this and much more. Those in biomedical instrumentation are constantly receiving updated information (data) and calculating results to be displayed for diagnostic purposes. A minicomputer and large computer system are shown in Figure 22-2*a* and *b*.

22-5 Computer Programming Languages

Computer programming languages represent the sequence of instructions by which the computer operates. They are classified as machine, assembly, and high level. *Machine language* is the actual binary number

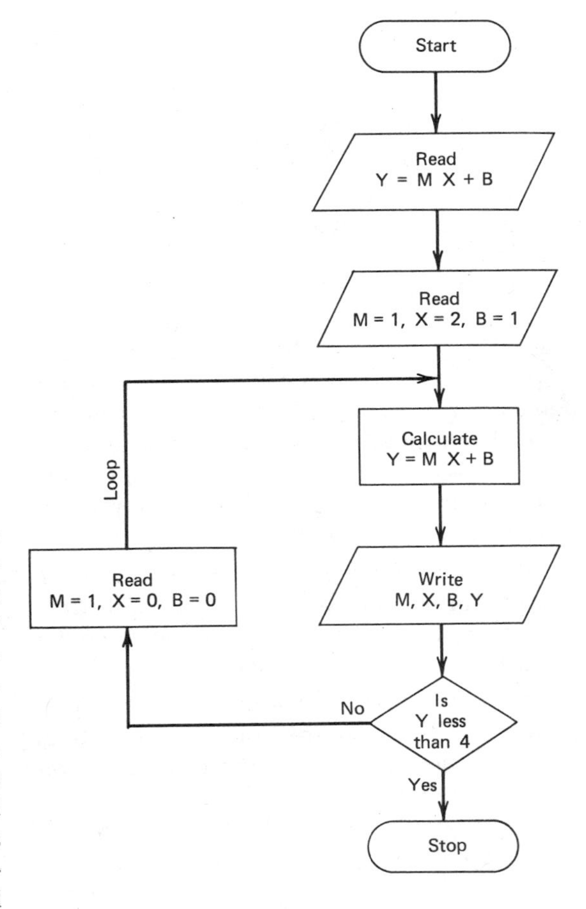

string that is keyed in using switches on the computer front panel. In a 32-bit computer, 32 switches would have to be set for data and instruction input along with another set of address switches. The process is very cumbersome and time consuming since the entire instruction set of the computer must be known. Numerous instruction books exist (see references).

Assembly language is a symbolic code of alphabet and number characters, called mnemonics, that replace the numeric instructions of machine language. It has a one-to-one correspondence with an assembly program, which directs the computer to operate or perform its designated operations. Programming

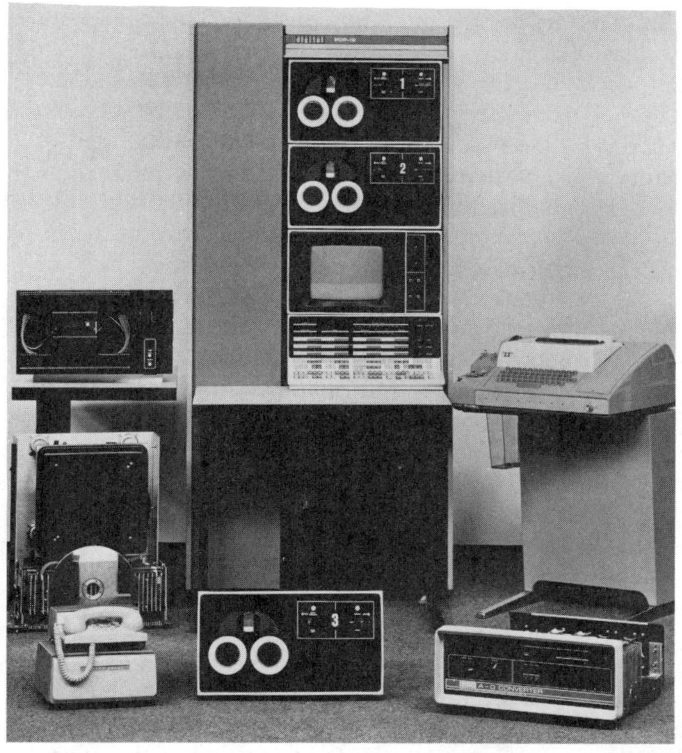

Figure 22-2 Computer systems. (*a*) Minicomputer. (*b*) Large computer.

in this language is extensive, but a knowledge of the specific instruction set of a machine is not necessary.

High-level languages are symbolic representations of alphabet, number, and special characters that are problem or function oriented. A single statement, close to the needs of the problems, may translate into a series of instructions or subroutines in machine language. Programming is relatively easy since a *compiler* converts symbolic instructions to machine language. Three commonly used languages are BASIC (Beginners All Purpose Symbolic Instruction Code), FORTRAN (Formula Translation), and COBOL (Common Business Oriented Language). Numerous texts and manuals exist explaining these languages (see references).

22-6 Microprocessor/Microcomputer System

A *microprocessor* is a large-scale integrated (LSI) electronic digital circuit that contains the control, processing, and holding registers of a microcomputer (small computer system). Metal oxide semiconductor (MOS) circuits are used for low power consumption. The microprocessor unit (MPU) handles arithmetic as well as logic data in bit-parallel fashion under control of a program.

The microprocessor is used in a system as shown in Figure 22-3*a*. Essentially, input signals are received (sense stage), processed (decide stage), placed in memory (store stage), and presented as output lines (act stage). The processor is surrounded by I/O (Input/Output) and memory. The microprocessor, as shown in Figure 22-3*b*, must handle address and data lines as well as clock, timing, and control lines. The entire processor is contained in a single integrated circuit package called a chip.

A *microcomputer* (an example is shown in Figure 22-3*c*) is a general-purpose computer composed of standard LSI components built around a central processing unit (CPU). The CPU or microprocessor is program-controlled with arithmetic and logic instructions and a common or parallel I/O bus. A microcomputer contains a microprocessor unit (MPU) as well as memory circuits and interface adapters for I/O devices. The MPU has a fixed *instruction set*, which is used during programming.

Data and address enters and leaves the MPU via a common bus line. These lines are connected together through *tristate driver/receivers*. They take on three states: logic one, logic zero, and a high impedance. In this way data or address can be targeted to a particular destination. For example, by disabling (high impedance) the tristate circuits in the memory and enabling those in the interface adapter, data or address can be steered to the I/O device. Although this method is slow, only one set of lines is required, thus reducing the hardware.

Storage of information is in either random access or read only memory. *Random access memory* (*RAM*) is a read/write memory in which data can be stored (written) at a particular address and then retrieved (read) out from the same address. This allows temporary storage of data or instructions and easy access when needed for manipulation. *Read only memory* (*ROM*) contains previously stored information impressed at the time of manufacture and can only be retrieved or read out. This device contains fixed program or control details for operation of the microcomputer. It can be called upon but cannot be altered. Special programmable read only memories (PROMs) are available, however, but must be erased with ultraviolet light and reprogrammed in a special device. All of these memories contain an array of flip-flops that are enabled during write and held during readout.

Newer types of memories include *magnetic bubble memory* (*MBM*) and *charged coupled devices* (*CCD*). *MBM* contains microscopic magnetized areas located on a thin magnetic film. A pulsating signal current generates a moving bubble or logic one. Stored data is read out by a resistive element array that changes value when a bubble passes below a particular detector. Access time can be as long as one millisecond but is many times faster than disc and less expensive considering total storage capacity versus space. *CCD* memories are faster still but cost

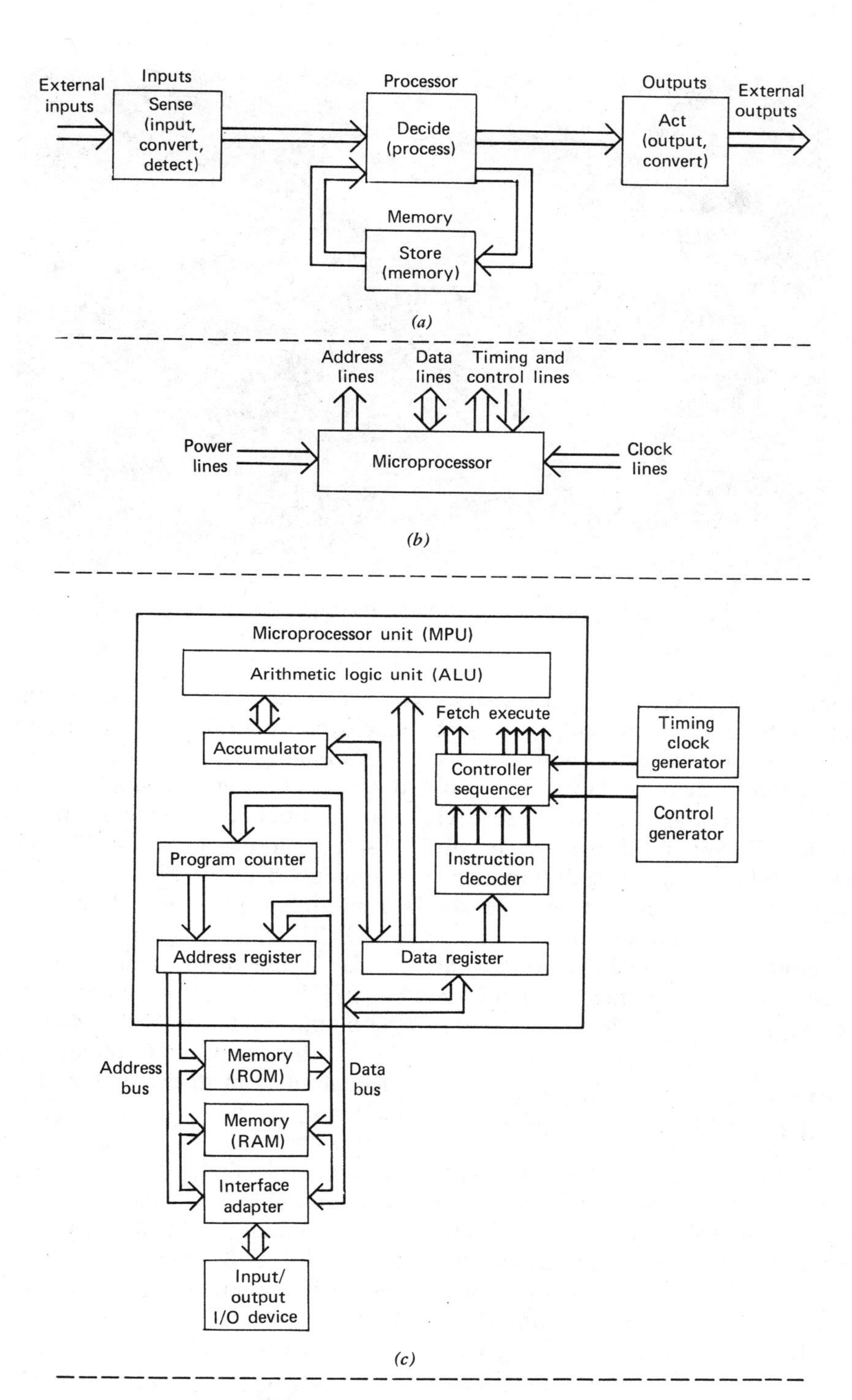

Figure 22-3 Simplified block diagram of a microprocessor/microcomputer system. (*a*) Basic system. (*b*) Processor signals. (*c*) Microcomputer system block diagram.

Figure 22-4 Microcomputer-based medical monitoring system.

much more. Small capacitive charges store the bits of information and are circulated back to the input of each path. These charges change or are refreshed as they recirculate and can be read out in parallel in several microseconds. However, this memory is volatile. Newer technologies bring different and unique memories. This is but one area of growth in the age of microcomputers.

An example of a biomedical instrument system that is based on a microcomputer is shown in Figure 22-4.

22-7 Interface Between Analog Signals and Digital Computers

Biomedical instrumentation usually transduces, amplifies, filters, and displays *analog signals* obtained from the human body. These signals are variable amplitude and frequency waveforms continuous in time. Since digital computers require binary numbers, a converter device is required before analog signals can be accepted and processed in digital machines. Similarly, *digital signals* must be converted to analog before they can be displayed on biomedical instrumentation.

An *analog-to-digital converter* (*A*/*D*) is a device that accepts an analog signal (continuous in time) and produces a binary number (discrete in time). A 4-bit number (more commonly 8 and 16 bit) can represent one of 16 possible voltage levels. For example, in Figure 22-5*a*, a 3-bit number represents eight levels. If the input signal were a constant +0.571 V (with a maximum of +1.000 V), the output would be 0100, which represents the fourth voltage level excluding zero. Since the output is in discrete steps, some information is lost and the resolution is 14.3 percent with 0.143 V per step. The resultant sequence of changing 3-bit binary numbers with time can be used as data input to a digital computer. This data can be presented in *parallel* (all 3 bits at once) or *serial* (1 bit for each computer clock time). Data can, thus, be processed and analyzed for diagnostic purposes.

A *digital-to-analog converter* (*D*/*A*), shown in Figure 22-5*b*, is the reverse of the

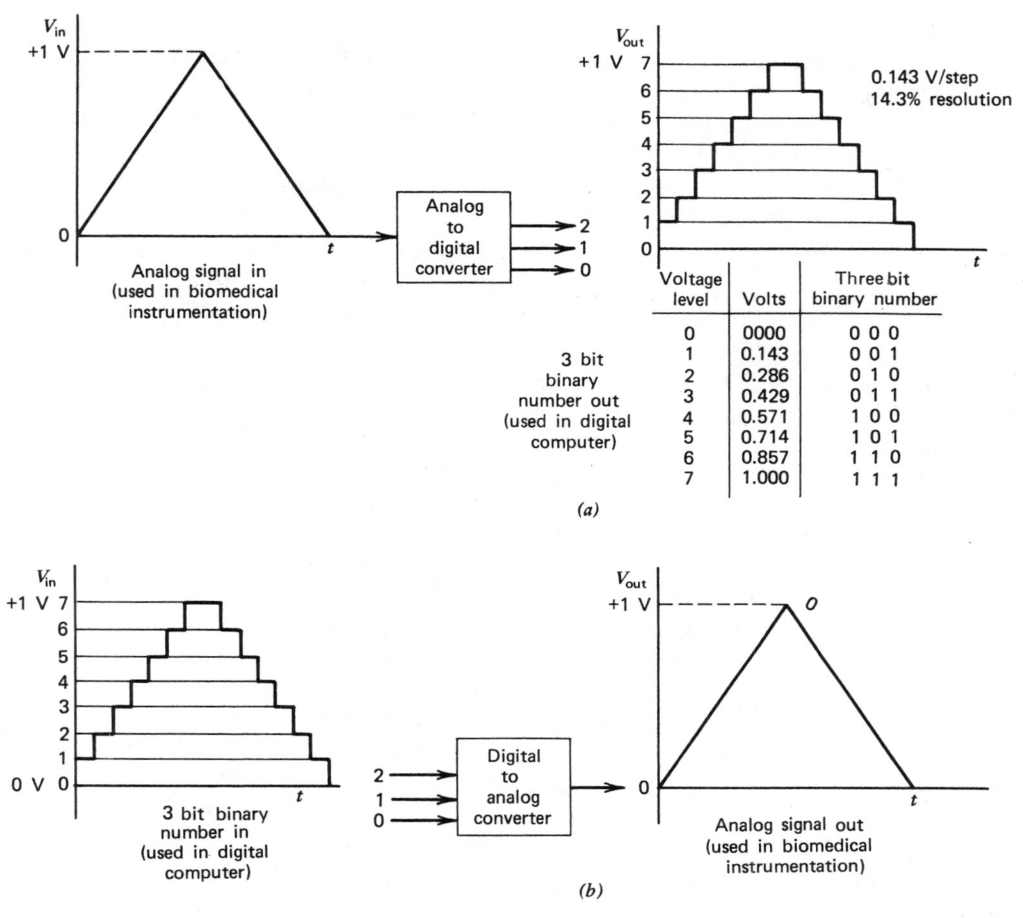

Voltage level	Volts	Three bit binary number
0	0000	0 0 0
1	0.143	0 0 1
2	0.286	0 1 0
3	0.429	0 1 1
4	0.571	1 0 0
5	0.714	1 0 1
6	0.857	1 1 0
7	1.000	1 1 1

Figure 22-5 (*a*) Analog-to-digital converter (A/D). (*b*) Digital-to-analog converter (D/A).

A/D converter. It accepts, in this case, a sequence of 3-bit binary numbers and produces a continuous analog waveform in time. This device must smooth out the discrete steps in the input signal and, hence, some information is lost. For this D/A converter (A/D as well), the 14.3 percent resolution is rather course. An 8-bit D/A has 256 levels and approximately 0.4 percent resolution, and a 12-bit D/A has 4096 steps and approximately 0.025 percent resolution. This is more acceptable for biophysical signals. For example, the accuracy of a blood transducer is 1 percent.

22-8 Computers in Biomedical Equipment

Computers are widely used in modern biomedical equipment. The following is a brief description of some of the major applications. All applications must consider techniques for *data acquisition*, ability to *store and retrieve* data, a process of *data reduc-*

tion and/or transformation to a usable form, *calculation* of variables, *recognition of patterns* in the endless waveforms, establishment of *boundaries or limits, statistical analysis* of variable signals, and *presentation formats* for the data.

The following is a list of large-scale and microcomputer applications.

1. *Computers in automated medical information systems* (MIS)—computers have been playing a large role in automated medical information systems. Financial record keeping as well as patient billing and medical history are stored in large memory banks. Pharmaceutical inventory is also computerized in larger hospitals. However, computerized patient records have caused controversy involving confidentiality of medical diagnosis. Nevertheless, computers and information go hand in hand, and more computer hardware is likely to evolve in health care facilities and organizations.

2. *Computer analysis of the electrocardiogram*—algorithms or fixed analysis sequence programs have been used in recent times to interpret ECG parameters (see Chapters 2 and 3). These include detection of all peaks (*P,Q,R,S,T*), establishment of baselines, measurement of abnormalities such as *P-R* segment depression, and determination of arrhythmias. Although sophisticated large-scale computers have been used in statistical analysis, many cardiologists seem to distrust the results. They usually read patient ECGs by eye even if a computer is used since computer-read ECGs are questionable about 20 percent of all runs.

3. *Computer analysis in patient monitoring*—most large, modern hospitals use some type of computer system in the intensive and coronary care units (see Chapter 14). CCU bedside monitors use microprocessor-based systems, and the central station uses a large microcomputer to analyze and display short- and long-term trends in ECG waveforms, pressure signals, respiration rate and depth, and temperature variations. A/D and D/A converters are very common. The information derived is very useful in determining response to treatment and the severity of disease. Figure 22-6 shows a computerized ECG management system.

4. *Computer analysis of cardiac catherization parameters*—cardiac catherization in the laboratory or operating room requires measurement of intracardiac blood pressures, especially across heart valves. Wedge pressures and vascular resistance is also important. Abnormalities can more easily be detected with computer programs that calculate stroke volumes and catheter effectivity and display results instantly. Catheters can then be placed more accurately. (See Chapter 7.)

5. *Computer averaging in electroencephalographic evoked response*—evoked responses to light flashes and tone burst show very little change in the ongoing ECG waveform (see Chapter 13). Special-purpose computers are currently being used to average evoked responses to obtain a useful display of peaks and valleys. Mini- and microcomputers are then used to statistically analyze time periods and relative peak amplitudes. The diagnosis of visual and hearing impairment in young children and adults has been made using such systems.

6. *Computer analysis of pulmonary function*—recently, equipment has appeared on the market that will analyze respiratory pulmonary function (see Chapters 10 and 11). Breathing rate, depth, and variations are displayed along with arterial blood gas changes. This offers a fast and accurate indication of a patient's condition. A system appears in Figure 22-7.

7. *Computer evaluation of clinical laboratory chemical tests*—computers blend naturally into the high-volume clinical laboratory tests of blood, urine, and other body

Figure 22-6 Coronary care monitoring computer system. (Photo courtesy Hewlett-Packard.)

Figure 22-7 Microprocessor-based respiratory analyzer. (Photo courtesy Hewlett-Packard.)

fluids (see Chapter 16). Autoanalyzers and continuous analysis equipment systems use mini- and microcomputers to aid with data acquisition and processing. Test results are accepted, lists are prepared, calculations are performed, and reports issued automatically. Remote terminals make the data available to the medical staff almost instantly, allowing easier and more rapid diagnosis of disease.

8. *Computerized axial tomography (CAT)*—CAT scanners originated through the use of computers (see Chapter 23). X-ray images obtained from pencil beam scans at various angles are stored in and analyzed by a minicomputer. The computer then redraws the picture on a CRT. Fine differences in tissue density and tumor outlines can be readily shown (even enhanced by color). *Pattern recognition* is the key to accurate results. *Algorithms*, a problem-solving set of rules or processes, are used to analyze and interpret data. This aids tremendously in diagnosing disease without surgery. Figure 22-8 shows such a system.

9. *Computer analysis of nuclear medicine results*—scintillation detection and gamma ray camera displays are analyzed by computer in some larger hospitals (see Chapter 23). Counts are statistically analyzed, and data is presented on a CRT. Diagnosis of marginal organ function is made more reliable.

Figure 22-8 Computerized axial tomography (CAT) system. (Photo courtesy of Pfizer Medical Systems, Inc.)

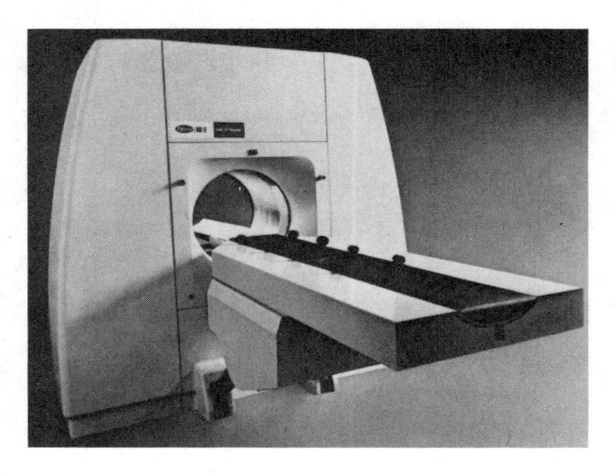

10. *Computers in other biomedical equipment*—a myriad of other uses for computers in biomedical equipment are becoming evident in the modern world of analysis and automation. Bedside ECG monitors as well as respiratory servoventilators are emerging with microprocessors embedded within them. Even the delivery of anesthesia is a target for a microprocessor-controlled patient-machine feedback system. Naturally, great distrust has arisen. Furthermore, microprocessor systems are being considered for ultrasonic scanning analysis. The future of computers in medicine will be considered on diagnostic merit, availability, and cost.

22-9 Summary

1. The *computer revolution* has unfolded in the past 30 years, and, after four generations of computer hardware, medical systems show widespread use.
2. A *digital computer* (as contrasted with analog devices) is an electronic and electromechanical device that can perform a sequence of *arithmetic* and *logical* operations according to stored instructions. It can be used to solve mathematical problems or can search, sort, and arrange information to be displayed or printed.
3. The computer program is known as the *software* and the physical portion is the *hardware*.
4. A *computer system* consists of an input unit, input control, memory, instruction control unit (IU), arithmetic logic unit (ALU), output control, and output unit. The central processing unit (CPU) contains the main or internal memory, ALU,

and IU, along with special registers, timing, and other housekeeping operations.

5. A *flowchart* is written to structure a problem, and program lines of code are written to solve the problem. The *signal path* through the computer takes many turns but always must involve the memory unit.
6. *Computer programming languages* include machine and assembly as well as high-level BASIC, FORTRAN, and COBOL, among others.
7. A *microcomputers* is a general-purpose computer composed of standard LSI components built around a central processing unit (CPU). The CPU or microprocessor is program-controlled with arithmetic and logic instructions and a common or parallel I/O tristate bus system for data and address. Memory includes RAMs, ROMs, or newer MBM and CCD.
8. *Interfacing* biomedical instrumentation to digital computers requires A/D and D/A converters.
9. *Computer applications* in biomedical equipment include automated information systems, analysis of ECG, patient monitoring, cardiac catheterization, averaging in EEG-evoked responses, pulmonary and respiratory function, clinical laboratory results, CAT scanners, nuclear medicine scintillation counting, and other bedside and automated processes.

22-10 Recapitulation

Now return to the objectives and self-evaluation questions at the beginning of the chapter and see how well you can answer them. If you cannot answer the questions, place a check mark next to each and review appropriate parts of the text. Next, try to answer the following questions using the same procedure.

22-11 Questions

24-1 Four generations of computers are ________ ________, ________, ________ ________, and ________.

24-2 A digital computer is an ________ and ________ device, which can perform a sequence of ________ and ________ operations. It can solve ________ problems or can ________, ________, and arrange information according to stored ________.

24-3 The main sections of a computer system are the ________ unit, ________, ________ unit, ________ decoder, ________ logic unit, ________ control, and ________ unit.

24-4 The central processing unit (CPU) contains the ________ memory, ________ logic unit, ________, and housekeeping ________.

24-5 The computer programs make up the ________ and the physical portions comprise the ________.

24-6 Write the ASCII code sequence for X = M + B.

24-7 What is a computer program flowchart?

24-8 Name two low-level and three high-level computer languages.

24-9 What part does the microprocessor perform in a microcomputer?

24-10 Why are tristate bus lines useful in a microcomputer?

24-11 How is a RAM different from a ROM?

24-12 What is the resolution of a 16-bit analog-to-digital converter?

24-13 What differences exist in computer systems used for medical information and records processing as compared to bedside and central station CCU monitoring?

24-14 Why is a computer-averaged, EEG-evoked response signal easier to analyze than a raw signal?

24-15 What advantage is a computer in the automated clinical laboratory?

24-16 Does CAT scanning produce better results

than a two-dimensional chest X-ray picture?

24-17 Are computer systems always applicable in biomedical equipment?

22-12 References

1. Hill, Frederick J. and Gerald R. Peterson, *Digital Systems: Hardward Organization and Design*, Wiley (New York, 1975).
2. Tanenbaum, Andrew S., *Structured Computer Organization,* Prentice Hall (Englewood Cliffs, N.J., 1976).
3. Mims, Forrest M., *Understanding Digital Computers*, Radio Shack, Tandy Corporation (Fort Worth, 1978).
4. McWhorter, Gene and Gerald Luecke, *Understanding Digital Electronics*, Texas Instruments (TI) Learning Center, Radio Shack, Tandy Corporation (Fort Worth, 1978).
5. Barna, A. and D. Porat, *Introduction to Microcomputers and Microprocessors*, Wiley (New York, 1976).
6. Cannon, Don L. and Gerald Luecke, *Understanding Microprocessors*, Texas Instruments (TI) Learning Center, Radio Shack, Tandy Corporation (Fort Worth, 1979).
7. Rony, Peter R., David G. Larson, and Jonathan A. Titus, *The 8080A Bugbook, Microcomputer Interfacing and Programming*, Howard W. Sams (Indianapolis, 1977).
8. Hausner, Arthur, *Analog and Analog/Hybrid Computer Programming,* Prentice Hall (Englewood Cliffs, N.J., 1971).
9. Bekey, George A. and Walter J. Karplus, *Hybrid Computation, Analog-Digital*, Wiley (New York, 1968).
10. Carlson, Alan, George Hannauer, Thomas Carey, and Peter J. Holsberg, eds., *Handbook of Analog Computation*, Education and Training Department of Electronic Associates, Inc. (Princeton, N.J., 1967).
11. Chapin, Ned, *360/370 Programming in Assembly Language*. McGraw-Hill (New York, 1973).
12. Keminy, John G. and Thomas E. Kurtz, *Basic Programming*, Wiley (New York, 1971).
13. McCracken, Daniel D., *A Guide to Fortran IV Programming*, Wiley (New York, 1972).
14. Haga, Enoch, Robert D. Brennen, Mario Ariet, and Josep G. Leaurodo, *Computer Techniques in Biomedicine and Medicine*, Auerbach Publishing (Philadelphia, 1973).
15. Perkins, W. J., *Biomedical Computing*, University Press (Baltimore, 1977).
16. Herbert, Ludwig, *Computer Applications and Techniques in Clinical Medicine*, Wiley (New York, 1974).
17. Payne, L. C. and P. T. S. Brown, *An Introduction to Medical Automation*, J. B. Lippincott Co. (Philadelphia, 1975).
18. *Symposium on Computer Applications in Medical Care*, IEEE Publication, November 5–9, 1978, Washington, D.C.
19. Carr, Joseph J., *Microcomputer Interfacing: A Practical Guide for Technicians, Engineers and Scientists*, Prentice Hall (Englewood Cliffs, NJ, 1991).

Chapter 23

Radiology and Nuclear Medicine Equipment

23-1 Objectives

1. Be able to list uses of diagnostic and therapeutic X-ray and nuclear medicine equipment.
2. Be able to describe the origin and list the properties and measurements of X-rays (atomic and nuclear physics).
3. Be able to describe the nature of radioactivity, including types of nuclear radiation.
4. Be able to list the dangerous effects to health from X-ray and nuclear radiation exposure.
5. Know how to draw a diagram of an X-ray tube and describe the production of X-rays.
6. Know how to draw a simplified block diagram of an X-ray machine and describe its operation, including generation, detection, and display on a photographic plate.
7. Be able to list the external controls and their use in X-ray machines (MA, kV, and exposure time settings).
8. Know how to draw a simplified block diagram of a fluoroscopic machine and describe its operation, including detection by fluorescence and display on a CRT.
9. Know how to draw a simplified block diagram of a nuclear medicine system and describe its operation, including detection by scintillation crystals, amplification by photomultiplier tubes, and display on a CRT.
10. Be able to describe computer systems used in X-ray and nuclear medicine equipment.
11. Be able to list calibration, typical faults, troubleshooting, and maintenance procedures on X-ray and nuclear medicine equipment.

23-2 Self-Evaluation Questions

These questions test your prior knowledge of the material in this chapter. Look for the answers as you read the text. After you have finished studying the chapter, try answering these questions and those at the end of the chapter (see Section 23-16).

1. What equipment is used to diagnose bone fractures, view functioning organs, and measure organ activity?
2. What type of equipment is used to treat (therapeutic function) cancer by irradiation of tumors?
3. Describe the nature of X-rays and nuclear radiation.

4. List the dangers to health from X-rays and nuclear radiation exposure.
5. How does an X-ray tube produce X-rays?
6. From a simple block diagram, describe the operation of an X-ray machine, including external controls (generation, detection, display).
7. From a simplified block diagram, describe the operation of a fluoroscopic machine (generation, detection, and display).
8. From a simplified block diagram, describe the operation of a nuclear medicine system (injected substances, detection, amplification, display).
9. What purpose do computers serve in X-ray and nuclear medicine equipment?
10. Describe general troubleshooting and maintenance procedures for X-ray and nuclear medicine equipment.

23-3 Types and Uses of X-Ray and Nuclear Medicine Equipment

X-ray machines are devices that generate exceedingly *high-frequency* (short wavelength), high-energy *electromagnetic waves* that penetrate the body during medical procedures. They are used in the radiology department. These machines serve diagnostic (measurement) and therapeutic (treatment) purposes. X-ray equipment has been used for many years to produce "pictures" of bodily tumors and skeletal fractures/deformations. While they are part of quality medicine in the modern world, they can be extremely *dangerous* (short and long term) causing nausea, vomiting, dizziness, sterility, burns, genetic mutations, cancer, and death if used incorrectly or in excess. All personnel operating the equipment should stand behind a lead wall or wear a protective powdered iron apron.

X-rays are high-energy waves (0.01 to 100 Å) that pass through the body and indicate relative tissue density on a photosensitive plate. Essentially, bones are dense and pass less X-ray than soft tissues, such as blood vessels, organs, and muscle. The X-ray that does not pass through (transmitted) is absorbed and stored within the body in accumulating doses. High accumulated doses over the long term represent a health hazard.

X-ray machines generally fall into the following *categories:*

1. *Diagnostic "still picture"* X-ray is used to examine bones and internal organ and tissue structures. The wavelengths are usually 0.01 to 1 Å, and energy levels vary with tissues to be observed. Broken bones and tumors can be detected.

2. *Diagnostic "continuous picture"* X-ray (*fluoroscopy*) is used to examine organ systems as they are functioning. Contrast substances (opaque to X-rays) fill bodily cavities and show anatomical shapes. The wavelengths are usually similar to "still picture" exposures, but the energy levels are considerably less due to the long exposure times. Tumors and blockages can be observed.

3. *Diagnostic "motion picture"* X-ray (*angiography*) is used to examine circulatory systems as they are functioning. Contrast fluids (opaque to X-rays) are introduced, for example, into the blood circulation of the heart (cardioangiography), kidney (renal angiography), or brain (cerebroangiography, Chapter 13). X-ray "still picture" exposures are then taken, one every five seconds or faster, and played back on a motion picture machine. This gives the effect of dynamic circulatory action through the blood vessels. Blockages can be visualized.

4. *Diagnostic "still picture"* X-ray scans (*tomogram*) are used to examine bones, organs, and tissues from many different angles. As in brain scans (Chapter 13), *whole body scans* are radiographs taken through successive scanning by highly collimated X-ray beams. Small contrast differences in selected planes can be seen and *provide considerably more information* than simple two-dimensional one plane or two-dimensional two plane (stereoscopy) X-ray exposures. Ultrasonic scanning (see ultrasonic devices, Chapter 17) is another technique using high-frequency sound waves instead of X-rays. X-ray *computerized axial tomography* (CAT) is a technique of recording and processing a set of image projections that represent a reconstruction of the object scanned. Essentially, a *thin layer* of the structure is *tomographically* (in sections) scanned with a pencil beam. The X-ray attenuation is then recorded with a *scintillation counter* in groups of parallel scans. Many-angle X-ray exposures are obtained every degree for 180°, and the projected scans (three-dimensional information) are stored in a computer. Through a complex program or algorithm, the original object is redrawn by the computer and displayed on a two-dimensional CRT. Colors have been used to indicate various tissue densities. A scanning layout can be seen in Chapter 13, Figure 13-2, and a tomograph of the skull (brain scan) appears in Figure 13-3.

5. *Therapeutic* X-ray is, essentially, the same as diagnostic X-ray except that the aim of therapy is to eradicate and destroy cancerous tissues and tumors. Whereas some healthy tissue is also exposed, it is hoped that it will regenerate while abnormal tissue will not. X-ray energy penetration, location, and dosage must be carefully chosen, and this can be done by using a scanning device first to pinpoint the boundaries of the tumor. Special lead forms are also used to protect surrounding healthy tissues. The wavelengths are usually long, ranging up to 100 Å, and energy levels vary according to tissue radiosensitivity.

Nuclear medicine equipment is similar to X-ray scanning machines from the standpoint of radiation detection (scintillation counters). It is somewhat different than angiography and considerably safer. Small amounts of short-lived radioactive isotopes (i.e., iodine 131 taken up by the thyroid gland) are introduced into the cardiovascular system. They accumulate in various target organs throughout the body. The concentrated radioactivity is measured with a *scintillation counter*, which responds to impinging alpha, beta, or gamma rays given off by the particular radioactive material used. The amount of substance taken up by a specific gland indicates (for diagnosis) the physiological function of the gland. Since very little radioactive substance is required and its emission life is very short, the danger to the patient is minimal. Special nuclear materials can also be used for therapy to treat tumors.

Nuclear medicine machines generally fall into the following categories:

1. *Diagnostic* low-level radiation (isotope) tracer detection devices used to measure target organ function. Radiopharmaceuticals injected and taken up by an organ are measured for concentration by gamma-ray cameras, rectilinear scanners, fixed detectors or scintillation counters, and survey instrumentation.

2. *Therapeutic* low-level localized radiation isotope source used to treat tumorous growths.

The radiology and nuclear medicine departments house the most *costly equipment* in the hospital or clinic. X-ray equipment gets high usage from almost every area of medical specialty because X-ray techniques allow visualization of internal body structures without surgery. X-ray is a noninvasive procedure when compared to other techniques, although radiation does pass through the body. Nuclear medicine equipment is being used more frequently today and represents a minor invasive process. Since equipment demand is high and total money received from services to patients is enormous, biomedical electronic servicing specialists are called frequently to repair these machines.

23-4 Origin and Nature of X-Rays

The modern era of atomic and nuclear physics (twentieth century) has paved the way for understanding the origin and nature of X-ray radiation.

X-rays are actually electromagnetic waves, as are light and radio waves. The principal difference is a matter of *frequency* or *wavelength*. Figure 23-1 shows a spectrum chart giving the relationship between radio waves, light, and X-rays. Electromagnetic waves (EM) are different from other types of waves, such as sound or water. EM waves can travel in a vacuum or in outer space. Some of the basic properties of electromagnetic waves are as follows:

1. They obey the relationship $V = F\lambda$; where V is the velocity, F is the frequency, and λ is the wavelength.

2. They propagate in a straight line.

3. They obey the *inverse square law* ($1/d^2$); their intensity falls off inversely proportionally to the square of the distance as they propagate away from the source.

4. They produce interference and diffraction patterns.

5. They are not deflected by magnetic fields.

The inverse square law can be demonstrated with the use of an ordinary flashlight. Turn on the switch and point it at a blank wall. Measure the distance between the light and the wall, and note the wall intensity with a light meter. Now move to a point exactly twice as far away and note the rapidly reducing level of illumination. Two observations are predominant. First, the intensity at any one spot has reduced to one-fourth of the previous intensity. Second, the overall area that is illuminated has increased. In fact, one will find that the increase in the illuminated area has the same factor as the reduction in spot intensity. This is because the light energy from the flashlight remained constant, while the area of illumination increased. This same phenomenon is critical in medical X-ray systems.

Figure 23-1 Electromagnetic spectrum chart (relative wavelengths of radio, light, and X-rays).

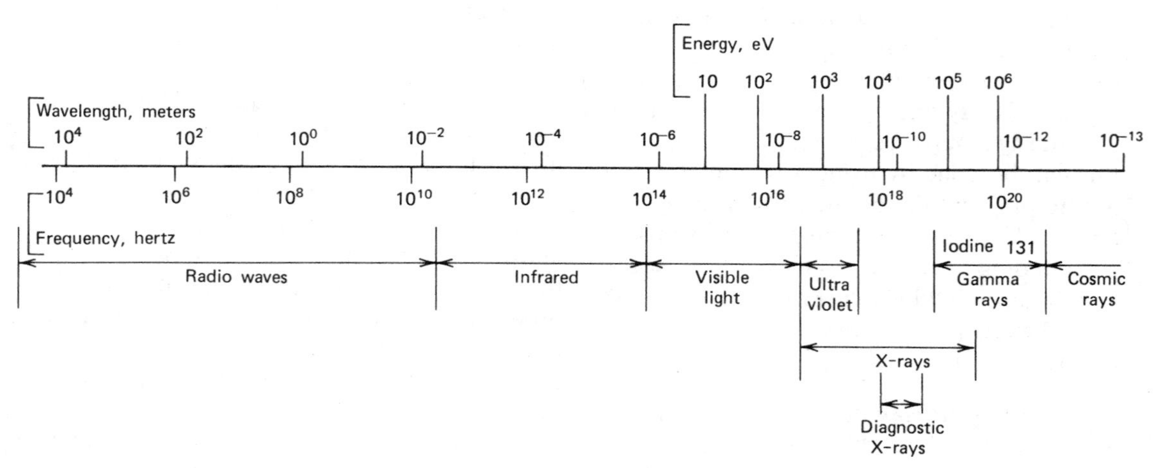

Essentially, three types of radiation exist.

1. *Alpha rays*—positively charged (ionized) particles of helium nuclei whose velocity is moderate (approximately 5 percent the speed of light) and whose penetration depth is small (approximately 5 cm).

2. *Beta rays*—negatively charged electrons of widely varying velocity (may be up to the speed of light) and whose penetration depth is small.

3. *Gamma rays* (10^{14} MHz) and *X rays* (10^{10} MHz)—electromagnetic waves that travel at the speed of light and have high penetration depending upon wavelength (energy). The energy is expressed as:

$$E = hf \tag{23-1}$$

where

E is energy in joules (J)
h is Planck's constant (6.624×10^{-34} J-s)
f is the frequency (Hz)

The electron volt (1 eV $= 1.602 \times 10^{-19}$ J) is the usual expression for gamma ray radiation energy. Furthermore, the radiated wavelength (distance the wave travels in one cycle) can be calculated from the following expression:

$$\lambda = \frac{c}{f} \tag{23-2}$$

where

λ is the wavelength in meters, centimeters, or angstrom units.
f is the X-ray frequency in Hz
c is the speed of light (3×10^8 m/s)

The angstrom unit (Å) is commonly used to measure wavelength.

$$1\ \text{Å} = 10^{-8}\ \text{cm} = 10^{-10}\ \text{m}$$

Higher frequencies (shorter wavelengths) *possess greater energy* than lower frequencies but require higher voltages to produce.

EXAMPLE 23-1

Given the energy level of 6.624×10^{-18} J imparted to an electron stream by an X-ray device, calculate the frequency in MHz and wavelength in m, cm, and Å of the X-ray beam.

SOLUTION

$$f = \frac{E}{h} = \frac{6.624 \times 10^{-18}\ \text{J}}{6.624 \times 10^{-34}\ \text{J-s}} = 10^{10} \times 10^{6}\ \text{Hz}$$

$$f = \mathbf{10^{10}\ MHz}$$

$$\lambda = \frac{c}{f} = \frac{3 \times 10^8\ \text{m/s}}{10^{10}\ \text{MHz}}$$

$\lambda = \mathbf{3 \times 10^{-8}\ m}$ (low-frequency or soft X-ray)

$$\lambda = 3 \times 10^{-8}\ \text{m} \times \frac{10^2\ \text{cm}}{\text{m}} = \mathbf{3 \times 10^{-6}\ cm}$$

$$\lambda = 3 \times 10^{-6}\ \text{cm} \times \frac{\text{Å}}{10^{-8}\ \text{cm}} = \mathbf{3 \times 10^{2}\ Å}$$

Basically, the following quantum effects exist for electromagnetic waves (X-rays included):

1. *Photoelectric Effect*. The photoelectric effect was first noted by Heinrich Hertz in 1887 and won Albert Einstein the Nobel Prize in 1905. The photoelectric effect is the emission of electrons from a clean metallic surface (phototube) when electromagnetic radiation (light waves or X-rays) falls onto that surface. Three other effects are thermionic, field, and secondary emission. Two facts have been noted about the photoelectric effect. First is that the phototube potential is totally *independent of light intensity.* Thus, the energy of the emitted electrons is not a function of light intensity. The light intensity affects only the number of electrons emitted, not their relative energy level. Second, phototube potential is *dependent on the light color.*

2. *Compton Effect*. The Compton effect is a phenomenon that relates to a photon's (packet of energy) ability to transfer only a part of its energy to a charged electrical particle such as an electron in a collision. The

photon still exists after the collision because only part of its energy is transferred to the electron. It does, however, exhibit a lower frequency than it had prior to the collision. This lower frequency or longer wavelength (color change) is caused by lost energy.

3. *Bremsstrahlung.* This phenomenon is especially important to the study of medical X-ray apparatus because it is primarily responsible for the generation of the radiation in the X-ray machine. The word *Bremsstrahlung* in German means *braking radiation.* As shown in Figure 23-2, an electron with an initial kinetic energy E_i approaches and is deflected by the heavy nucleus of a nearby atom. After the deflection, the electron has taken a new (lower) energy level E_d. The law of conservation of energy in this case reveals that the energy of the deflected electron and photon equals that of the incident electron. The energy level of the incident electron minus that of the deflected electron becomes a photon, and if the loss of energy is sufficient the wavelength of the photon will be in the X-ray region of the electromagnetic spectrum. More energy lost means higher X-ray frequency (hard X-rays).

$$E \text{ (incident electron)} - E \text{ (deflected electron)} = E \text{ (photon)} \qquad (23\text{-}3)$$

Ionizing radiation is biologically harmful and is caused by high-energy atomic collision that produces alpha particles (helium nuclei), beta particles (electrons), and gamma rays (photons).

Figure 23-2 Bremsstrahlung collision resulting in X-ray production.

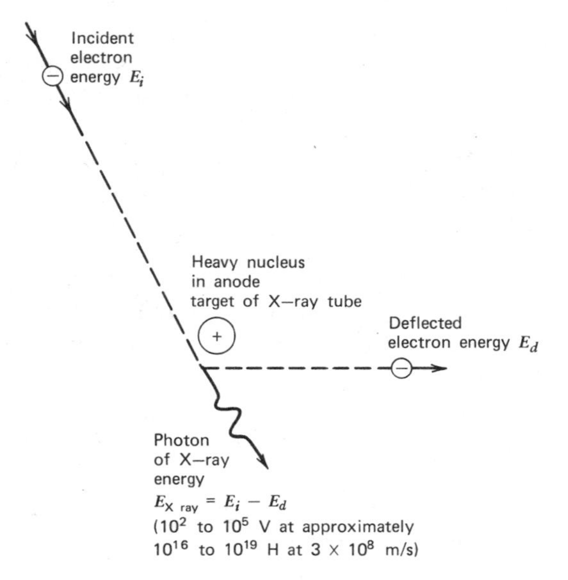

23-5 Nature and Types of Nuclear Radiation

Atomic structure is extensively considered in the field of atomic and nuclear physics. Medical radiation and nuclear physics is a specialty within this field, and many texts contain detailed explanations of the theory. This discussion is directed at equipment technology, and only a brief description of nuclear medicine radioisotopes follows.

In general, radioactivity is grouped into two classes.

1. *Natural radioactivity* results from natural atomic flux (constant motion). Some collisions within an atomic nucleus produce an impact energy that exceeds the nucleus binding energy. Subnuclear particles then escape through *disintegration.* Ths nuclear *decay process* produces nuclear radiation of particles (alpha and beta) and rays (gamma). Natural radiation occurs in materials involving a sequence of steps called a radioactive series. Four series currently exist: thorium (decay to stable lead 208), actinium (decay to stable lead 207), uranium (decay to lead 206), and neptunium (decay to bismuth 209).

2. *Artificial radioactivity* is the same as the natural type, except the radionuclides (radioactive nuclei) are synthetic. These artificial substances (not naturally occurring) are produced by bombarding stable nuclides with neutrons, protons, deuterons, or alpha particles in a linear accelerator or cyclotron. *Ions* (atoms with a net positive or negative charge) and *isotopes* (atoms with more neutrons than protons) are produced. Frequently used medical isotopes (radiopharmaceuticals) are shown in Table 23-1.

23-6 Units for Measuring Radioactivity

The following units are used to indicate the intensity of radioactivity. Essentially, they all indicate how many nuclear disintegrations occur in a given time.

1. *Curie (Ci)*—amount of radioactivity in one gram of radium (3.7×10^{10} disintegrations per second).

2. *Roentgen (R)*—unit of radiation exposure or amount of X-radiation that will produce 2.08×10^9 ion pairs per cubic centimeter of air at standard temperature and pressure (STP). Usable units are the milliroentgen (mR) and microroentgen (μR).

3. *Rad*—radiation absorbed dose. One rad is the radiation dose that will result in an energy absorption of 1.0×10^{-2} joules per kilogram of irradiated material. For practical purposes:

$$1 \text{ R} = 1 \text{ rad}$$

Absorbed dose in rads = exposure in roentgens × 0.834 × factor of absorbing material.

4. *Dose rate*—amount of radiation expressed in roentgens administered or produced per unit time.

5. *X-ray intensity*—dose rate per unit area irradiated.

23-7 Health Dangers from X-Ray and Nuclear Radiation

Radiological safety cannot be understated. The effects of cumulative X-ray dosage of ionizing may result in *mutations*—genetic changes resulting from damage to chromosomes; *physical illness*—vomiting, headache, dizziness, loss of hair, and burns; and *death*—destruction of vital physiological systems such as nervous, cardiovascular, respiratory, renal, and digestive systems and tissues.

23-8 Generation of X-Rays in an X-Ray Tube

X-rays are generated by a *high-vacuum X-ray diode tube* in which electrons are accelerated to high velocities (using a high-voltage power supply up to 100 kV). Radiation intensity is obtained by varying the *high voltage, current,* and *time of exposure*. The radiation is filtered and formed (collimated) to produce optimal contrast relative to the patient dose. The *X-ray tube* (high-vacuum diode) shown in Figures 23-3*a* and *b* operates by emitting electrons from a heated cathode tungsten filament toward a rotating high-voltage anode disc. X-rays arise from the target disc at right angles and are focused by the collimator. Most of the energy is heat. Image *intensifier or photomultiplier tubes* allow greater viewing contrast and, hence, less required patient radiation dose. *Images* are received and viewed on an imaging device (photographic film plate or fluoroscopic screen). Light and dark areas on the film represent high and low tissue penetration, respectively. Dynamic (ongoing) radiographic images when viewed on a *fluoroscopic screen* are termed *fluoroscopy.*

There are two methods used to overcome the heat problem in X-ray tubes (target anode must be able to withstand tremendous heat). One is to construct the target anode of tungsten alloy, which is embedded in a *large mass of copper.* A design trade-off is necessary. Maximum heat capacity results from a large target area, but the best X-ray photographic resolution results from a small

Table 23-1
Some Medical Radioisotopes

Isotope	Radiological Halflife	Predominant Radiation Halflife	Some Target Organs
^{2}H (hydrogen)	Stable	—	—
^{3}H (hydrogen)	12.3 years	Beta	—
^{14}C (carbon)	5770 years	Beta	Pancreas
^{24}Na (sodium)	15 hours	Beta	Blood
^{32}P (phosphorus)	14.3 days	Beta	Liver
^{42}K (potassium)	12.4 hours	Beta	Cardiovascular
^{45}Ca (calcium)	165 days	Beta	Bones
^{85}Sr (strontium)	10.26 years	Gamma	Bones
^{51}Cr (chromium)	27.8 days	Gamma	Urinary
^{58}Co (cobalt)	71 days	Gamma	Urinary
^{82}Br (barium)	36 hours	Gamma	Intestinal
^{131}I (iodine)	8.07 days	Gamma	Thyroid, Urinary, and Gastrointestinal
^{198}Au (gold)	64.8 hours	Gamma	Liver
^{197}Hg (mercury)	65 hours	Beta	Kidney
^{99m}Tc (technetium)	6 hours	Gamma	Spleen, brain, lung, and cardiovascular

target area. Forming a *bevel* on the anode contributes greatly to good heat dissipating and high resolution.

The other solution is to use the *rotating anode* shown in Figure 23-3. The spot target of the previous design is now replaced with a moving (i.e., rotating) target on the rim of a disc. The electron source (cathode) is placed so that it will impinge only a small spot on the rim of the disc. This keeps the focal area down to a few square millimeters, yet the structure is large enough to permit the dissipation of a large amount of heat. Anodes of this type might rotate 10,000 rpm, and the rotating disc shaft spins as a result of magnetic coupling through the sealed glass envelope.

The inside of the X-ray tube consists of a *vacuum* to assure an undisturbed path for electrons and long anode life. Coefficients of thermal expansion of the glass case become important as the tube heats up. Too much heat on these tubes will burn them out quickly. Special microcomputers have been used to warn the operator of excessive anode heat.

Stationary anodes are used mostly in low-energy-level X-ray machines, while the more common rotating anodes are used in higher-energy machines.

The housing used for the X-ray tube must provide three types of protection in addition to the physical or mechanical protection.

1. High voltage—grounded metal housing with high-quality electrical insulation and oil-filled housing to withstand 10 to 150 kV.

2. Heat dissipation—properly constructed anode and oil-filled housing. If excessive heat expands oil too far, a microswitch activates a control circuit to turn high-voltage power off.

3. Radiation shielding—properly constructed metal casing.

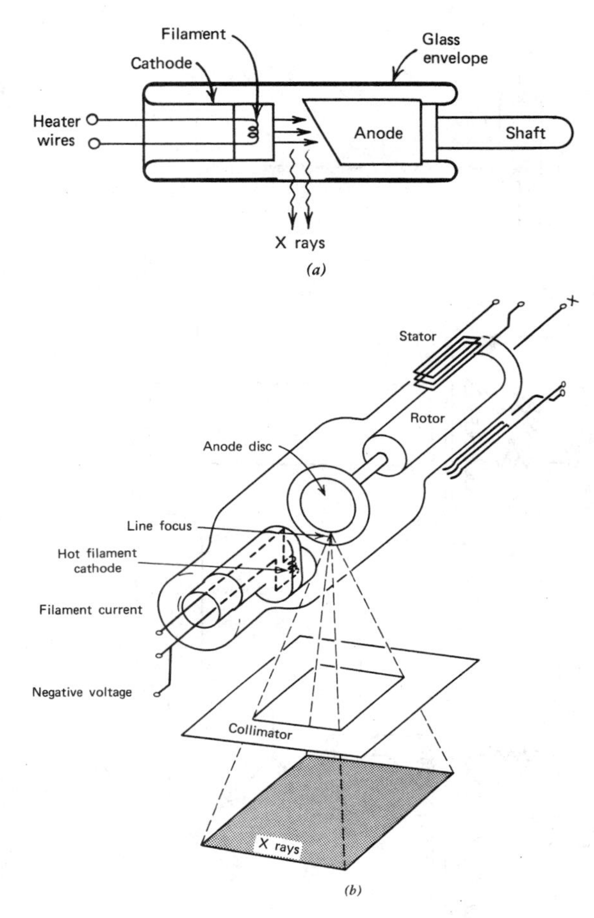

Figure 23-3 X-ray tube with rotating anode. (*a*) Basic tube. (*b*) Basic tube with X-ray output. (From *Medicine and Clinical Engineering,* by Bertil Jacobson and John Webster. Prentice Hall. Used by permission.)

23-9 Block Diagram and Operation of an X-Ray Machine

X-ray machines generate high-energy, high-frequency electromagnetic waves (X-rays) for use in diagnosing and treating disease and physical malfunctions. To accomplish this, X-ray machines have the following major sections, as shown in Figure 23-4:

1. *Multitap ac line autotransformer,* which allows selection of taps to compensate for incoming line variations. These also permit the operator to choose voltages for specific applications.

2. *X-ray tube filament circuit and transformer,* which transforms the ac line to supply power for heating the cathode filament. This power can be selected by taps to change filament heat (filament mA), which changes X-ray tube current (tube mA) and, hence, total X-ray energy delivered to the patient.

3. *X-ray tube high-voltage circuit, transformer, and bridge rectifier,* which transforms the ac line to supply the high dc voltage for accelerating electrons from cathode to anode. The high voltage can be selected by taps to change the kV_p (kilovolts peak) and, hence, total X-ray energy delivered to the patient.

4. *Timing circuit,* which controls turnon, turnoff, and length of X-ray exposure delivered to the patient.

Essentially, *three basic controls* exist on X-ray machines to control patient X-ray dose (penetrating quality, quantity, and timing). These are interrelated and must be properly chosen to suit the slim or obese patient. Good photographic results are sometimes difficult to obtain. These controls are *filament heat* control (mA) for exposure strength, not depth; *kilovolt* control (kV) for penetration depth and contrast; and *timing* devices for time exposure length.

It is extremely important to observe *X-ray tube heat ratings*. Excessive heat will damage a very expensive tube, and the cost and inconvenience of replacement are equally high.

X-ray emission from the tube can be improved by using filters, stationary grids, moving grids (Potter-Buckey diaphragm), cones, cylinders, diaphragms, collimators, and intensifiers (image intensifier tube to

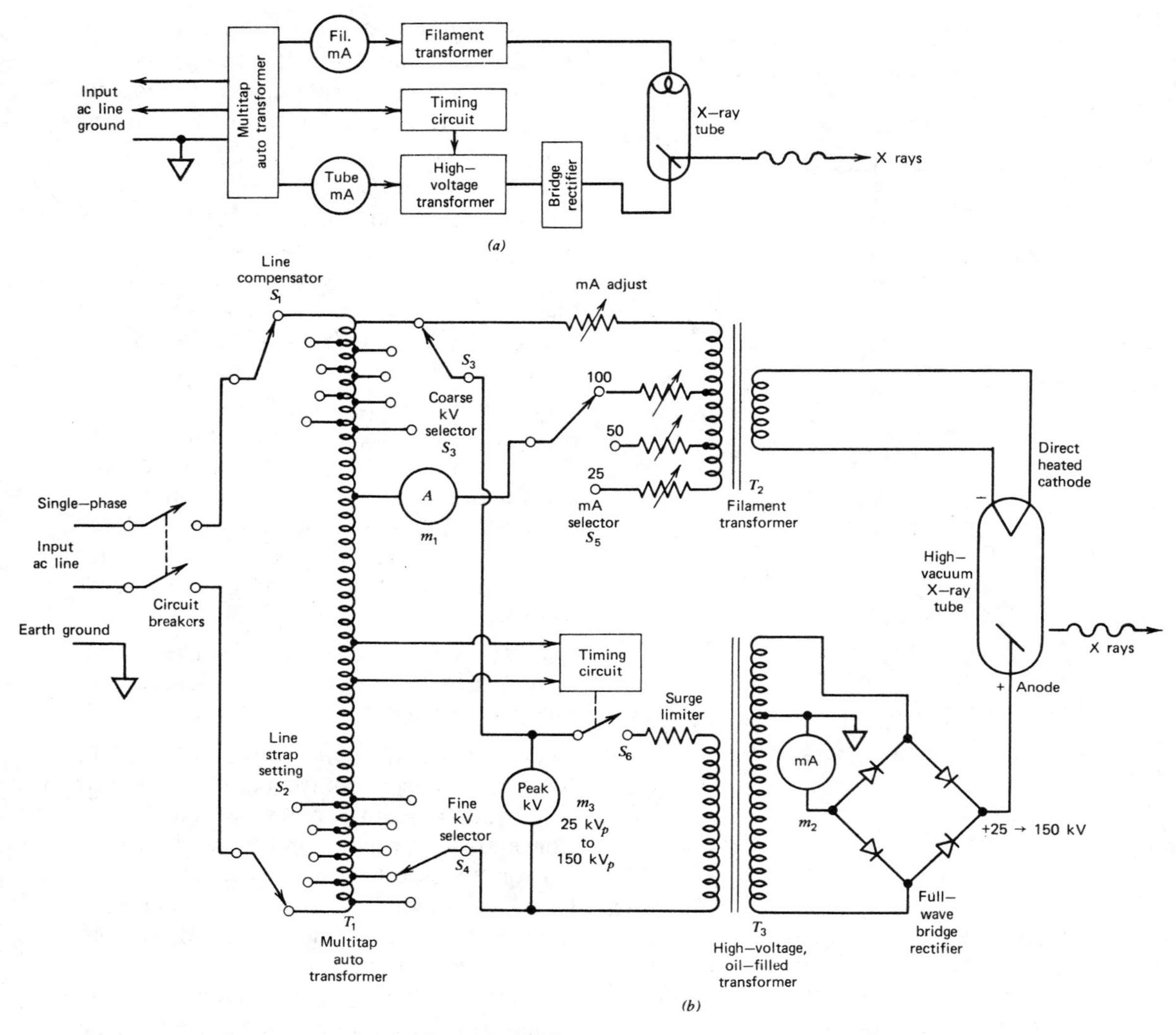

Figure 23-4 (*a*) Simplified block diagram. (*b*) Circuit diagram of an X-ray machine.

increase brightness of the photographic image).

The *multitap ac line autotransformer,* T_1 (shown in Figure 23-4*b*), has several purposes. One is to compensate for normal input line variations by adjusting S_1. When the line is low, S_1 is set near the top until the line-voltage meter indicates normal. The line strap setting, S_2, is set by the installation engineer or technician initially (i.e., 110 V to 450 V ac). The autotransformer also contains switch settings for coarse (S_3, 10 kV) and fine (S_4, 1 kV) high-voltage selection.

The *X-ray tube filament circuit* consists of selector switch S_5, filament transformer, T_2, and the filament of the X-ray tube. T_2 provides isolation from the high-voltage transformer and an added measure of safety. *Elec-*

trocution (see Chapter 19) from as much as 150 kV is a potential danger on X-ray machines. Switch S_5 selects 25, 50, or 100 mA for filament current as adjusted by filament resistors during calibration. As X-ray tubes age, more filament current is required to achieve constant X-ray intensity. A filament current meter, M_1, shows the milliamperes in the X-ray tube cathode.

The *high-voltage circuit* is, essentially, the power supply for the X-ray machine. A separate oil-filled, high-voltage transformer, T_3, receives selectable voltages from coarse and fine kV selector switches(S_3 and S_4) and provides high voltage to the high-voltage diodes. The diode bridge provides full-wave rectified unfiltered dc voltage to the X-ray tube anode. An X-ray tube current meter, M_2, shows the milliamperes passing through the tube, and meter, M_3, indicates the peak kilovoltage

Figure 23-5 Stationary X-ray machine. (Photo courtesy of CGR Medical Corporation.)

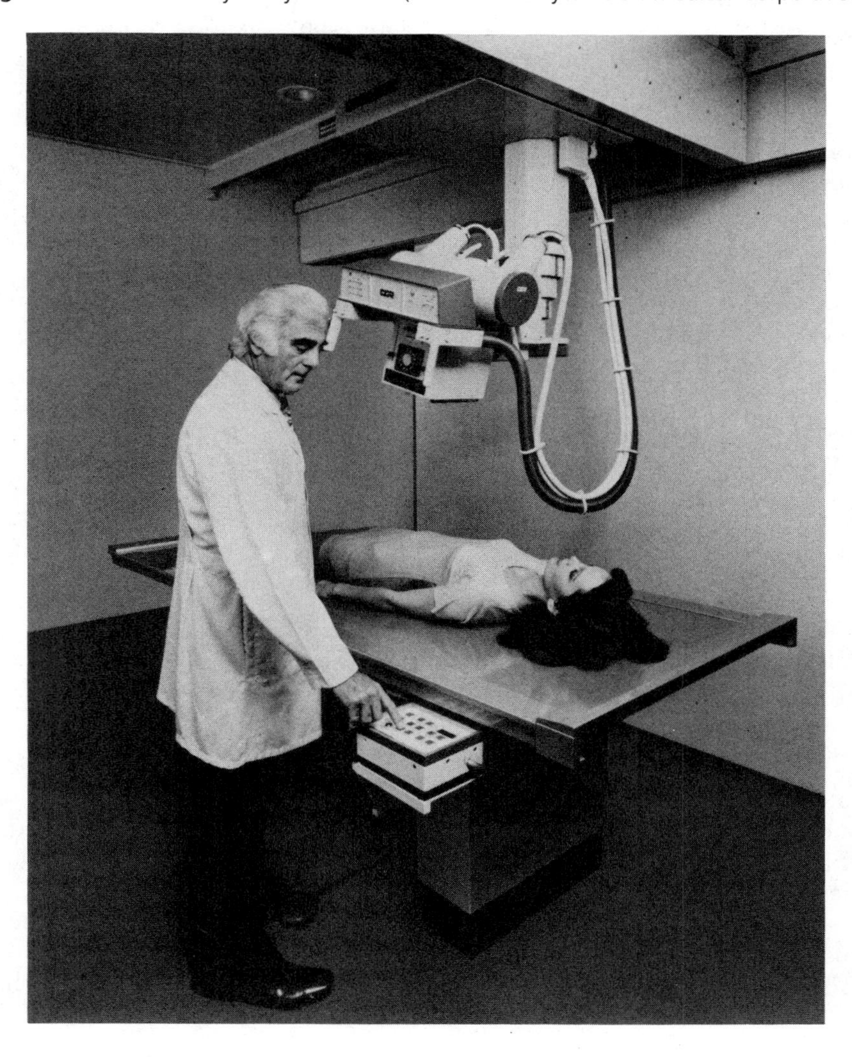

applied to the anode. Higher peak voltages (energy) produce higher-frequency X-rays and greater patient penetration capability (see Example 23-1).

Many larger X-ray machines have *three-phase power* applied. Instead of 120 peaks per second in single phase, 360 peaks per second occur from three-phase. The ripple frequency is higher and, hence, the effective high voltage is greater. Usually the primary windings are connected in the *delta* and the secondary in the *wye* configuration. This gives a more efficient system, and greater sustained energy levels can be obtained. This scheme is, however, more expensive since transformers with three windings and 12 diodes are required.

The *timing circuit* consists of a mechanical motor-driven or electronic counter that closes switch S_6 and applies high voltage to the X-ray tube anode for short periods of time (1/120 s to several seconds). A *mechanical timer* is a hand-wound spring of a clock. A *synchronous timer* is also *mechanical* and consists of a synchronous motor driven from the 60-Hz ac line that closes switch contacts. These contacts are protected by a surge-limiting resistor. An *electronic timer* uses an *electronic switch,* such as a transistor timed by a digital circuit.

Figure 23-5 shows a fixed X-ray machine, and Figure 23-6 shows a portable one.

23-10 Block Diagram and Operation of a Fluoroscopic Machine

Fluoroscopic machines are X-ray machines that generate soft X-rays (reduced frequency and intensity) to produce dynamic visualizations on a fluoroscope. Internal body organs are viewed through the use of a contrast medium that is opaque to X-rays. Patient dosage should not exceed 10 R per minute. Transmitted X-rays fall upon a fluorescent plate or screen as a function of varying tissue density. Fluorescence is the emission of visible light produced when X-rays fall upon crystals in the coating of the screen.

As shown in Figure 23-7, the major sections are the X-ray machine (subsections previously discussed), fluoroscope image pickup, and CRT or closed-circuit video system. The X-ray image falling on a fluorescent grid causes a visible "light" picture to appear. This is optically focused by a lens on the film of a motion picture (automatic) camera. The film can be played back at a later date. The visual image is also focused on a phototube lens and made brighter by an image enhancer. A video camera converts the light image into an electrical video signal, which is delivered to a CRT and displayed through a closed-circuit video system. This gives a "real time" or instantaneous visualization. A fluoroscopic machine is shown in Figure 23-8.

23-11 Block Diagram and Operation of a Nuclear Medicine System

Nuclear medicine systems are used to count radioactive decay from isotopes that have been injected into the body in small amounts and taken up by a target organ to measure its activity. The basic components are a gamma ray camera, rectilinear scanner, fixed detector for in vitro samples, scintillation counter, and survey instrument.

As shown in Figure 23-9*a*, a *Geiger-Mueller tube* radiation detector counts beta particles. Beta particles passing through the gas mixture in the tube cause ionization, and the electrons are collected by the anode and the positive ions by the cathode via a high potential of approximately 1 kV. Therefore, each beta particle causes a brief pulse of current. The total number of current pulses, electronically counted over a given period of

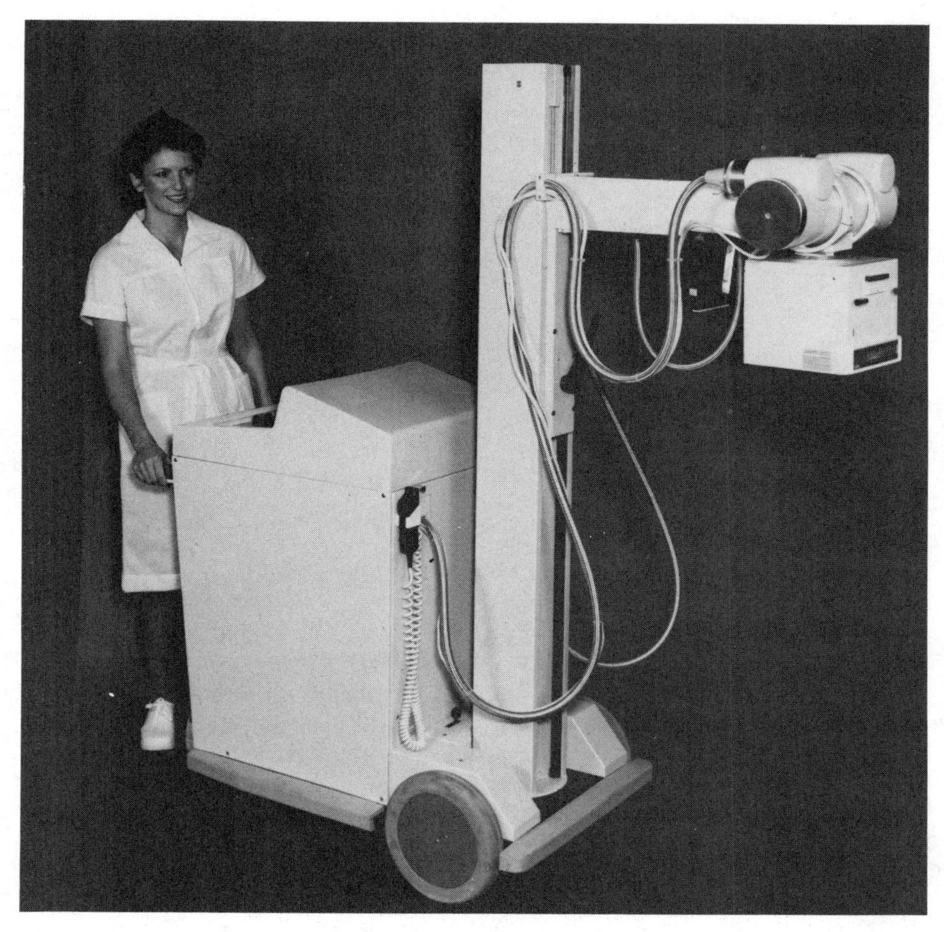

Figure 23-6 Portable X-ray machine. (Photo courtesy of CGR Corporation.)

time, indicate the radiation intensity falling on the mica window.

As shown in Figure 23-9*b*, a scintillation crystal detector with photomultiplier tube measures gamma radiation intensity. The incident gamma rays are detected by the crystal, and flashes of light are produced and reflected onto the cathode of the photoamplifier tube. The dynodes multiply the electrical signal in a secondary emission process by as much as 10 million to produce an appreciable current pulse. Current pulses are produced for rays striking the fixed surface area of the scintillation detector and are then counted to indicate radiation intensity (rays per second).

A *nuclear medicine system* (rectilinear scanner), as shown in Figure 23-10, consists of a detector-collimator, amplifier, analyzer, and recorder. The detector-collimator, driven by the scanner motor assembly, scans back and forth in a linear fashion. A graph or contour map of radioactivity can then be drawn indicating the amount of radioactive

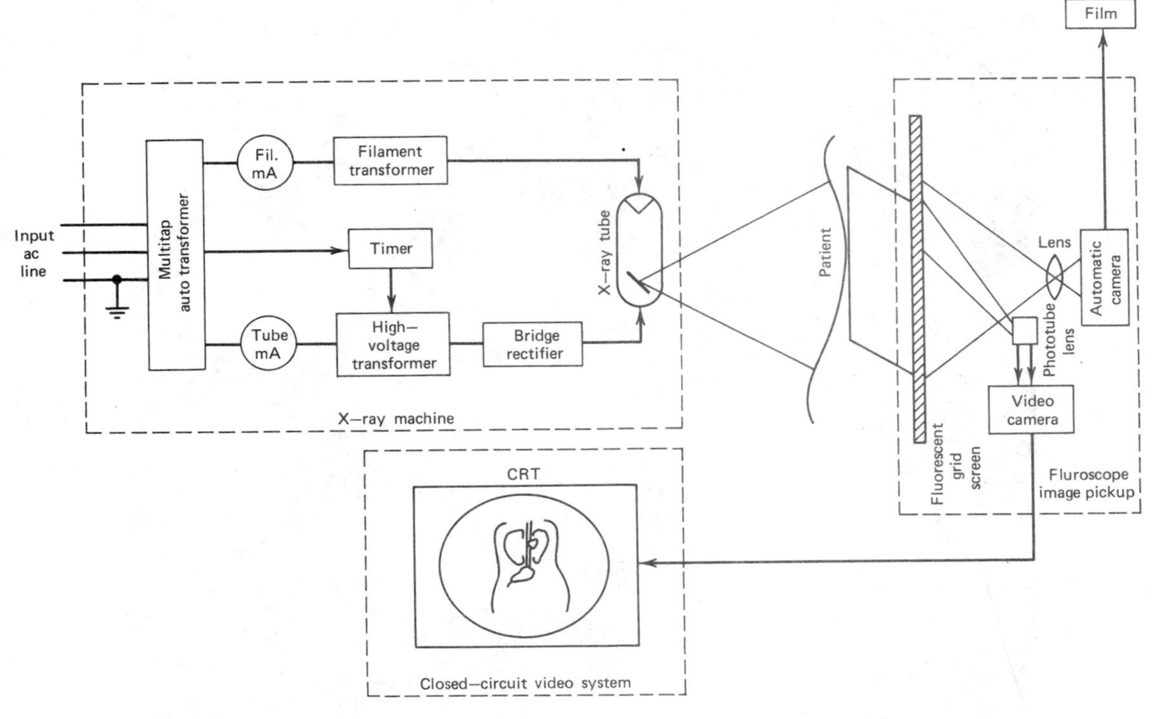

Figure 23-7 Simplified block diagram of a fluoroscope machine.

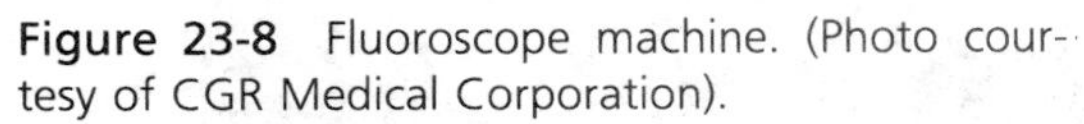
Figure 23-8 Fluoroscope machine. (Photo courtesy of CGR Medical Corporation).

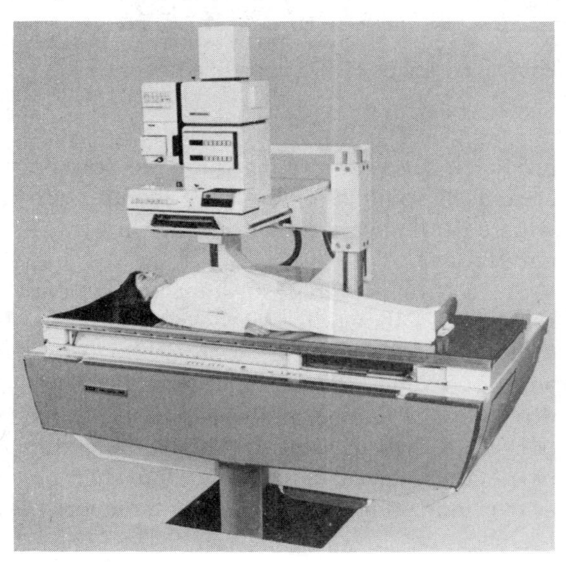

isotope taken up by the target organ. Improper amounts indicate organ malfunction.

The *detector-collimator* typically uses an Na-I crystal to detect the radiation collimated toward its surface. The photomultiplier tube intensifies the signal, after which it is linearly amplified. Its pulse is analyzed for comparisons between successive events. Numbers of pulses per time are important. The dot scan recorder produces a map of dots or dash marks on paper representing the distribution of radioactivity. The photographic recorder produces a photograph of light flashes. Recordings move simultaneously with the scanning device to produce one line scan in unison. Figure 23-11 illustrates a nuclear medicine system.

The *gamma ray camera,* shown in Figure 23-12, produces an image in a different manner than that of a scanner. Gamma rays in-

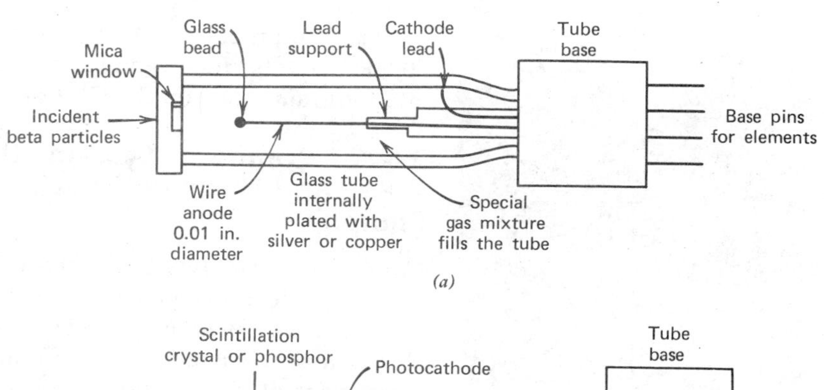

(a)

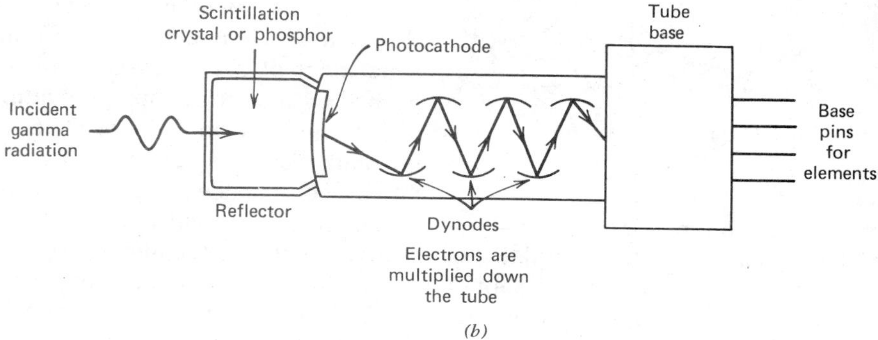

(b)

Figure 23-9 (*a*) Geiger-Mueller tube. (*b*) Scintillation crystal detector with photo multiplier tube.

Figure 23-10 Simplified block diagram of a nuclear medicine system (rectilinear scanner).

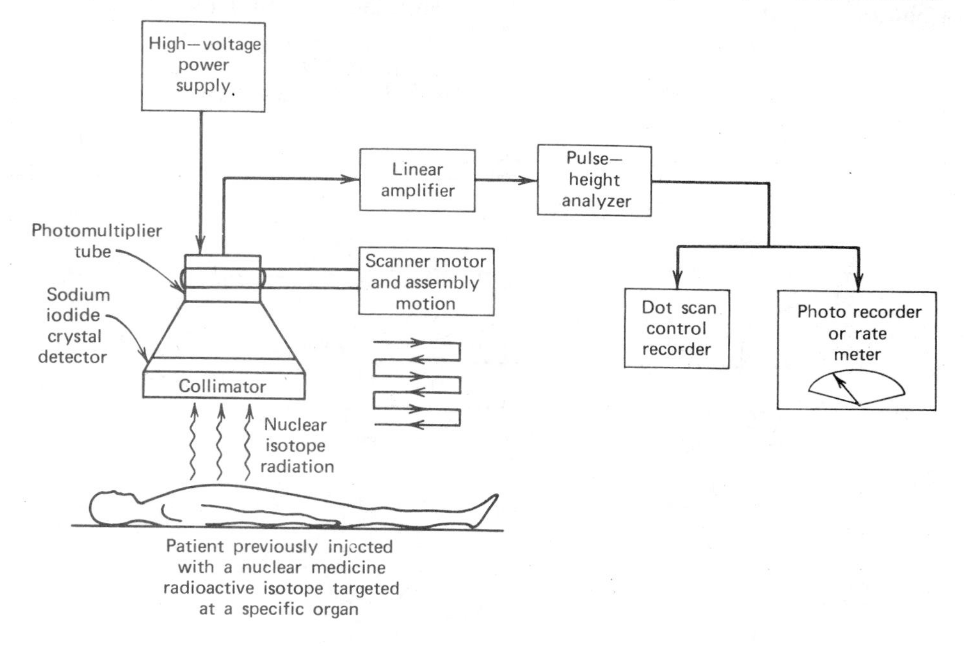

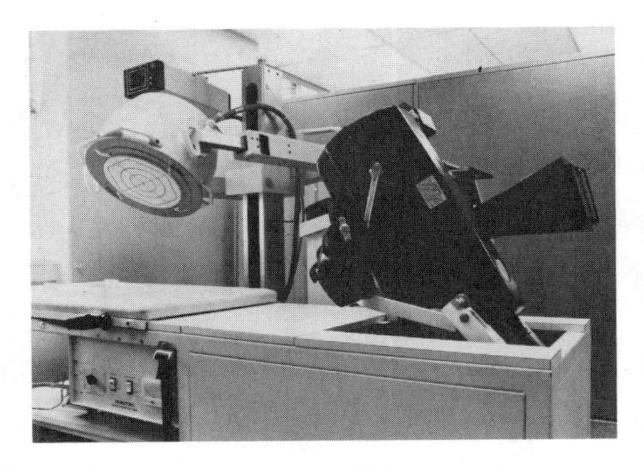

Figure 23-11 Nuclear medicine system.

teract with a large sodium iodide scintillation crystal in the camera, and the scintillations (flashes) are observed by an array of photomultiplier tubes. Typically, 19 tubes are used, and a position analyzer evaluates the flashes from four crystal quadrants. Flashes are produced on an oscilloscope display when the gamma ray meets the pulse-height analyzer requirements. A Polaroid or 35 mm camera photographs the flashes on the oscilloscope to produce a scintiphoto. Up to 500,000 counts, for example, may accumulate for brain scans on the CRT screen.

23-12 Computer Systems Used in X-Ray and Nuclear Medicine Equipment

Computers serve two basic purposes in X-ray and nuclear medicine equipment. First is image *enhancement,* and second is *recording and analysis* of dynamic data. Computer enhancement of X-ray images can be accomplished by measuring fine graduations of exposure intensity on a film or plate. However, the human eye can detect a remarkably small difference in shades of gray, and many physicians prefer to read the X-ray negative directly. Computer techniques are also useful to enhance the response of a gamma-ray camera to correct for the nonuniformity of the crystal face.

Computerized axial tomography (CAT) scanners use computer algorithms for pattern recognition. Each scan must be stored and relative intensities analyzed. For 180° scans, the computer must redraw the three-

Figure 23-12 Simplified block diagram of a gamma-ray camera.

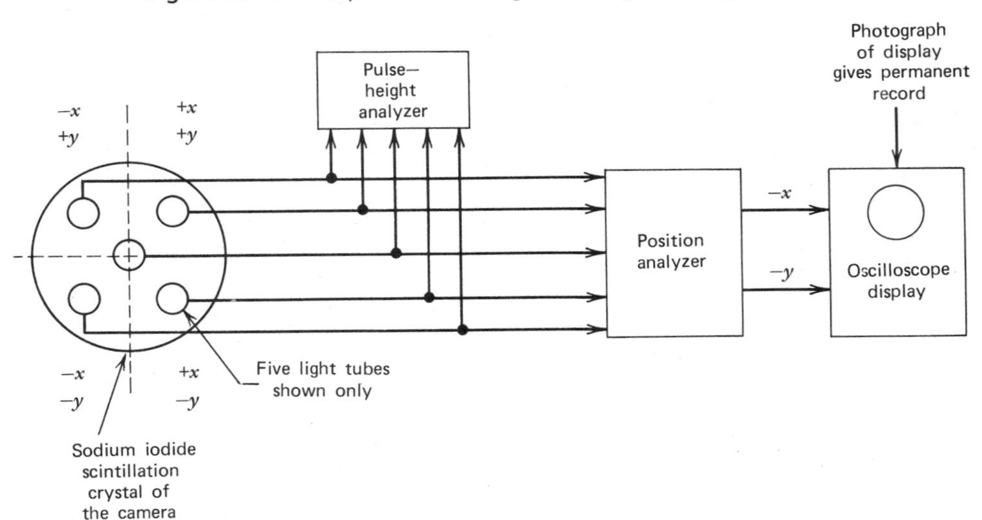

dimensional X-ray information into one two-dimensional picture. The result gives more information than could be obtained from one two-dimensional X-ray picture.

Dynamic imaging systems used to analyze results from gamma cameras involve digitally storing the information and successively analyzing the results. However, the analog analysis involved in viewing the original display may yield a diagnosis as accurate if not more so than by computer. Clearly, the computer is of tremendous value, but, in the final analysis, humans will evaluate results, diagnose, and embark on a treatment procedure. In short, computers analyze, humans think.

23-13 Calibration, Typical Faults, Troubleshooting, and Maintenance Procedures

Calibration of *X-ray machines* involves checking timing against known standards and verifying kilovoltage and milliampere settings on the operating panel. To verify mechanical timing for 1/120 s, observe one pulse on an oscilloscope while synchronizing to the 60-Hz ac power line. Spinning tops are used for longer durations. Special digital devices can also be used. A digital timer is triggered when the X-ray strikes a transducer. When the X-ray pulse has passed, the digital timer is placed on hold. The readout indicates elapsed time or X-ray pulse duration. A *penetrometer* (stack of different lengths of aluminum bars) is used to calibrate the kV and mA. If successive film depths produce proper shaded exposures for changes in the autotransformer taps, the machine is working properly. Prepared plates at fixed distances should receive proper shading for proper calibration.

Calibration of *nuclear medicine equipment* involves verification of proper operation of the lower-level discriminator (LLD) or pulse height analyzer, proper amplification, and display results. A standard nuclear radiation module can be placed under the scanning mechanism and scanned as usual. Observation on the display of a proper map is testimony of proper calibration. Adjustments in the collimator, pulse height analyzer, and synchronous scanning display unit can be made.

Typical faults in *X-ray machines* mostly occur in the mechanical device servomechanisms for positioning the patient tables. Wires break, cables snap, and jams occur. Relays become intermittent and stick open or closed. Most problems, therefore, are electromechanical in nature, not electronic. Occasionally, high-voltage rectifiers open, X-ray tube filaments open, or rotating anodes become pitted. Replacement of the X-ray tube with oil circulation is done by factory-trained personnel. Elimination of radiation leakage paths with special instrumentation and electrical safety are of paramount significance. The electronics in modern machines are highly reliable. Finally, operator error occurs on occasion and can result in injury or death to the patient and operator. Radiation exposure is checked and minimized by periodically measuring exposure tags worn by employees.

Typical faults in *nuclear medicine machines* are also mostly electromechanical. The scanning mechanisms malfunction and recorders occasionally fail. The electronics are reliable, and computers are serviced by manufacturer personnel.

23-14 Summary

1. *X-ray machines* are devices that generate high-frequency, high-energy electromagnetic waves (X-rays). These rays penetrate the body and are detected on a photographic plate. The result on the negative is an exposure or "picture" shown as shades of gray. Dark areas represent high exposure to X-rays (soft structures within the body). Light areas represent bones and dense struc-

tures. Internal body parts can then be observed.

2. *Types of X-ray machines* include "still picture" or chest type, fluoroscope, angiography, tomographic scanner (CAT), and therapeutic radiation therapy units.
3. *Nuclear medicine machines* are devices used to detect radiation from isotopes injected into the bloodstream that accumulate in target organs. Scintillation crystals detect the radiation, and counters measure the amount of disintegration per time. Organ activity can then be measured.
4. *Types of nuclear medicine equipment* include rectilinear scintillation scanner and isotope therapy sources.
5. Nuclear radiation involves *alpha rays* (positive helium nuclei), *beta rays* (negative electrons), *gamma rays* (electromagnetic waves), and *X-rays* (exceedingly high-frequency waves).
6. The frequency of X-rays is proportional to their energy.
7. Three quantum effects are the photoelectric effect, Compton effect, and *Bremsstrahlung*.
8. Units for measuring radioactivity include the curie (Ci), roentgen (R), rad, dose rate, and X-ray intensity.
9. Excessive short-term or long-term (accumulated) X-ray or nuclear radiation dosage may cause mutation, physical illness, or death.
10. X-rays are generated in a high-vacuum diode in which electrons are accelerated to high velocity by a high-voltage power supply.
11. The *major sections* of an *X-ray machine* are a multitap autotransformer, filament circuit and transformer, high-voltage circuit and transformer, and timer.
12. The *major sections* of a *fluoroscopic machine* are X-ray machine, fluoroscopic image pickup, camera, and closed-circuit video system.
13. The *major sections* of a *nuclear medicine system* are detector-collimator, scanner motor assembly, linear amplifier, pulse height analyzer, and recorder.
14. *Computer systems* in radiology include *image enhancement* and *recording* and *analysis* of dynamic data.
15. Calibration on X-ray and nuclear medicine equipment should follow manufacturers' recommendations—typical faults most often include electromechanical failures.

23-15 Recapitulation

Now return to the objectives and self-evaluation questions at the beginning of the chapter and see how well you can answer them. If you cannot answer certain questions, place a check mark next to each and review appropriate parts of the text. Next, try to answer the following questions using the same procedure. When you have answered all of the questions, solve the problems in Section 23-17.

23-16 Questions

23-1 Four classes of X-ray equipment used to measure anatomical features are the ________-________ machine, ________ continuous display, ________ or motion picture, and ________ or tomographic systems.

23-2 CAT scanner refers to ________ ________ ________.

23-3 X-rays are high-frequency ________ waves.

23-4 Alpha rays are positive ________ nuclei; beta rays are negatively charged ________; and gamma rays are ________ waves.

23-5 Higher energy levels imparted to electrons produce *harder/softer* X-rays on the X-ray tube.

23-6 *Bremsstrahlung* refers to ________ radiation.

23-7 The roentgen unit refers to the amount of radiation energy dose per ________, the dose-rate to exposure per ________, and the X-ray intensity to dose-rate per ________.

23-8 List the major hazards from radiation exposure including accumulated doses.

23-9 X-ray tubes use rotating anodes to avoid ________ and prolong tube ________.

23-10 Name and describe the operation of the major sections of an X-ray machine.

23-11 Name and describe the operation of the major sections of a fluoroscopic machine.

23-12 Name and describe the operation of the major sections of a nuclear medicine system.

23-13 In X-ray machines and nuclear medicine systems, why is calibration important? Where do most of the malfunctions occur?

23-17 Problems

23-1 Given an X-ray machine set to 125 kV (on the X-ray tube), which produces X-rays of wavelength 0.01 Å? Calculate the wavelength in meters and energy in joules. Are these hard (diagnostic) or soft (therapeutic) X-rays?

23-2 Given a 136-lb person who receives 0.3 J of radiation energy over his or her entire body, calculate the kilograms of the person and the radiation absorbed dose (rad).

23-3 Given a 136-lb person who accidentally receives a whole body exposure of 200 mR/min dose rate for 30 min, in general, should this person avoid a chest X-ray of 0.2 rad within one year if the physician does not consider it mandatory? *Hint:* for practical purposes 1 R = 1 rad and a maximum of 5 rads per year should be observed (recommended exposure limit by International Committee on Radiation Protection, ICRP).

23-4 Refer to Figure 23-4, simplified block and circuit diagram of an X-ray machine. Given a short in the primary of the filament transformer, T_2, discuss the system fault symptoms from the operator's standpoint. Include meter readings, activation of X-ray machine, circuit breakers, and X-ray tube operation.

23-5 Refer to Figure 23-4, simplified circuit diagram of an X-ray machine. Given system fault symptoms of loud cracking sounds, discuss possible faulty components. Include all transformers, rectifier bridges, and wiring.

23-18 References

1. Hill, D. R. et al., *Principles of Diagnostic X-ray Apparatus,* Phillips Medical Systems (London, 1973).
2. Nave, Carl R. and Brenda C. Nave, *Physics for the Health Sciences,* W. B. Saunders Co. (Philadelphia, 1975).
3. Fernando, Antonio, Conclaves Rocha, and John C. Harbet, *Textbook of Nuclear Medicine,* Lea & Febiger (Philadelphia, 1978).
4. Lange, Robert C., *Nuclear Medicine for Technicians,* Year Book Medical Publishers (Chicago, 1973).
5. Early, Paul J., M. A. Razzak, and Bruce D. Sodee, *Textbook of Nuclear Medicine Technology,* C. V. Mosby (St. Louis, 1975).
6. Sodee, Bruce D. and Paul J. Early, *Technology and Interpretation of Nuclear Medicine Procedures,* C. V. Mosby (St. Louis, 1975).

Chapter 24

Electromagnetic Interference to Medical Electronic Equipment

24-1 Objectives

1. Be able to define *electromagnetic interference* (EMI).
2. Be able to describe the mechanisms of EMI.
3. Be able to describe the effects of EMI on sensors and medical circuits.
4. Be able to describe the basic methods for suppressing EMI.

24-2 Self-Evaluation Questions

These questions test your prior knowledge of the material in this chapter. Look for the answers as you read the text. After you have finished studying the chapter, try answering these questions and those at the end of the chapter (Section 24-11).

1. Define *intermodulation interference*. Use an equation, if needed.
2. Two transmitters operate close to the hospital. One transmitter produces a signal on 188.2 MHz, while the other operates on 146.94 MHz. Is there any danger that either the fundamental frequencies, or their first two harmonics, can combine to produce intermodulation interference?
3. What measures can be taken to prevent radio frequency interference from entering the equipment cabinet via the power lines?
4. What kind of filter can be used to remove a 98.7-MHz FM broadcast signal that is interfering with the operation of an ECG radio telemetry system?

24-3 Introduction

Electromagnetic interference (EMI) is the general term covering a wide variety of interoperability problems in electronic equipment. In one class of EMI problems, a device that produces a large amount of radio frequency (rf) energy (e.g., an electrosurgery machine) interferes with other equipment (such as the ECG monitor). In other cases, unintentional radiations from one piece of equipment affect another. For example, in one instance an FM broadcast radio receiver used by nurses in a post coronary care (telemetry) unit grossly interfered with the operation of the telemetry system. In still other cases, TV or radio reception is affected by nearby radio transmitters that are operating completely legally. In this chapter you will learn to deal with these problems.

EMI problems surface in different forms in medical equipment settings. Part of this problem arises from the fact that there are increasingly large numbers of instruments and other electronic devices used in medicine. Coupled with the proliferation of electronic devices in the medical setting is a tremendous increase in the use of computers and word processors in the hospital. If you doubt that these devices are capable of interference, then try using an AM broadcast band radio receiver in close proximity to a desktop computer. Even further complicating the problem is that there are many different forms of radio transmitters on the premises or nearby. In addition to the walkie-talkies used by the security department, there may also be a hospital page system transmitter on the roof. Nearby buildings and businesses may also have powerful radio transmitters on their own premises. These are capable of interfering with medical equipment. And do not forget the potential interference from nearby broadcast stations. Those stations are among the most powerful emitters of radio frequency energy.

The first problems that we will consider are those that derive from the mixture of a large number of radio signals in the vicinity.

24-4 Intermodulation Problems

There are two interrelated problems that often deteriorate radio reception in communications systems: *intermodulation* and *cross-modulation*. Although there are fine technical differences between these two problems, their cures are about the same, so the authors will take the liberty of calling them all "intermod problems." These problems result in *interference on a given frequency due to heterodyning (mixing) between two other unrelated signals*.

Heterodyning is a term that refers to the nonlinear mixing of two radio frequency signals. Consider the "guitar analogy." Pick one string and make it vibrate to produce a tone. That tone has a frequency of F_1. After that tone dies out, pluck another string and cause a new tone (F_2). Now pluck both strings simultaneously. What happens? How many tones are present? The initial answer might be "F_1 and F_2," but that is wrong. In a nonlinear system (such as hearing), there are additional frequencies created. In addition to F_1 and F_2, there will be the sum frequency ($F_1 + F_2$) and the difference frequency ($F_1 - F_2$). It is the difference frequency, $F_1 - F_2$, by the way, that some guitar players use to tune the instrument. When two frequencies are close, but not exact, the difference frequency is very low, almost subaudible. It is this low frequency that accounts for the slow, wavering tone that one hears as the strings are tuned closer to the same pitch. Exact tune is indicated by the wavering disappearing. This is *heterodyning*.

There are other combination frequencies also produced, and these are determined according to the expression $mF_1 + nF_2$, where m and n are integers. For most purposes, however, these additional frequencies are unimportant (a distinctly important problem with these additional frequencies is discussed here; "intermod problems" need not work on only F_1 and F_2).

There is a hill close to where one author lives that radio technicians and engineers call "Intermod Hill." It happens to be one of the higher locations in the county, so several broadcasters and AT&T have seen fit to build radio towers there. In addition, both of the two main radio towers bristle with land-mobile antennas, whose owners rent space on the tower in order to get better coverage. In total, there are two 50,000-watt FM broadcast stations, a 2000-watt AM broadcast station, a many-frequency microwave relay station, and several dozen 30-mHz to 950-mHz VHF/UHF landmobile stations and radio paging stations. Nearby is a major community hospital that operates its own security system radio station and a hospital radio pager system. The hospital also has a coronary care unit that uses UHF radio telemetry to keep track of ambulatory heart patients. All of those signals can heterodyne together to produce apparently valid signals on other channels.

The frequencies produced in a signal-rich environment are many and roughly follow the aforementioned rule: $F_{unwanted} = [(mF_1) \pm (nF_2)]$; where m and n are integers (1, 2, 3, . . .) and F_1, F_2 are the frequencies present. If the receiver system is linear, then there is little chance for a problem. But nonlinearities do creep in, and when the receiver is nonlinear in a case where extraneous signals are present, then "intermods" show up. Being in the fields of so many radio transmitter signals almost ensures nonlinearity, due to simple overload; hence intermod problems abound on and around "Intermod Hill." Imagine the number of possible combinations when there are literally dozens of frequencies floating around the neighborhood!

The ability (or lack of same) to reject these unwanted signals is a good measure of a receiver's performance. A high-quality, well-designed radio receiver does not respond to either out-of-band or within-band off-channel signals, except in the most extreme cases of overload. The ability to discriminate against these spurious signals is a necessary requirement to specify when requesting quotations and estimates from industry for a new radio communications or telemetry installation.

The linearity or dynamic range of the input RF amplifier of the receiver is the cause of most cases of intermodulation interference. There are other causes, however, and one must consider anything that can cause nonlinearity. For example, a certain type of radio had problems with the automatic gain control (AGC) rectifier diode. It was leaky and would produce a situation where the radio was easily overdriven by external signals. In some sets, the manufacturer used a pair of back-to-back silicon small signal diodes across the antenna terminals in order to shunt possible high-voltage potentials to ground. This method was especially popular a few years ago. A large signal could drive these diodes into conduction and produce severe nonlinearity.

Consider an odd intermod problem that occurred in a hospital that used a VHF radio telemetry unit to monitor patient ECGs in the PCCU, which is the unit that Coronary Care Unit patients go to after they are no longer acute but still bear watching. The portable ECG transmitters generate 1 to 4 milliwatts of VHF RF energy that is frequency modulated with the patient's ECG signal. The signal level is so low level that five or more 17-inch whip antennas, sticking down from the false hanging ceiling, are needed to cover an area that consists of two corridors approximately 150 feet in length. Each antenna is connected directly to a 60-dB master TV antenna amplifier (the VHF ECG radio transmitter channels are located in the "guard bands" between TV video and audio carriers of commercial VHF TV channels); one of the whip/amplifier assemblies is right over the receiver console.

One morning, about 2 A.M., a nurse called the biomedical equipment technician at home complaining that Mr. Jones's ECG was riding in on Mr. Smith's channel. Not quite believing her, the technician nontheless went to the hospital and checked out the situation. Swapping receivers, telemetry transmitters, and amplifiers did no good. Finally, after two hours of trying, he noticed the FM broadcast receiver sitting on top of the telemetry receiver cabinet less than 18 inches from the antenna/amplifier (Figure 24-1), and it was playing. On a sheer hunch, he turned off the receiver and Mr. Jones went back to his own channel! Previously, Jones's signal was showing up on both his own channel and Mr. Smith's channel, but was now only where it belonged. Turning the FM receiver back on caused the situation to return. Also, tuning the radio to another channel made the problem go away.

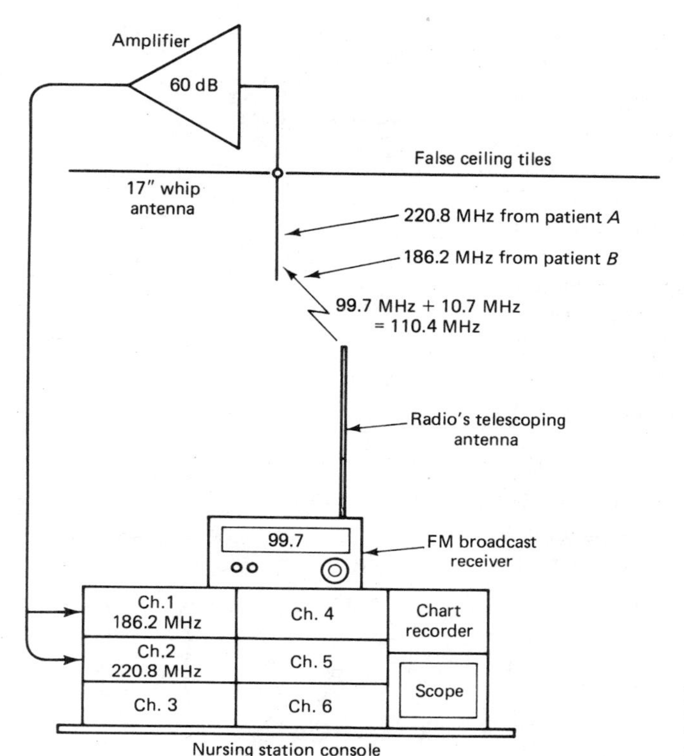

Figure 24-1 Configuration that led to an unusual intermodulation problem.

What happened in that situation? The FM radio is a superheterodyne model (they all are), and its internal local oscillator signal radiated and was heterodyning with Mr. Jones's signal to produce an "intermod" signal on Mr. Smith's channel. The situation is shown graphically in Figure 24-1. The six VHF receivers used in the system are installed in a mainframe rack that forms the nurses' station console (along with an oscilloscope and strip-chart recorder). The FM receiver was placed on top of the receiver rack such that its telescoping whip antenna was only a short distance from the telemetry receiver antenna.

Because the FM receiver is a superheterodyne, it produces a signal from the internal local oscillator that is 10.7 MHz higher than the received frequency. If the radio is tuned to the station at 99.7 MHz on the FM dial, then the local oscillator operates at 99.7 MHz + 10.7 MHz, or 110.4 MHz. This signal is radiated and picked up by the telemetry antenna. Because of its close proximity, it is the strongest signal seen by the input of the 60-dB amplifier. The other signals impinging on the antenna are weak signals at 220.8 MHz from Patient A and 186.2 MHz from Patient B. The mixing action at the input of the 60-dB amplifier, then, consists of 110.4 MHz, 186.2 MHz, and 220.8 MHz.

In the case cited, it was apparently the second harmonic of the FM radio local oscillator signal (i.e., 110.4 MHz × 2, or 220.8 MHz) that caused the problem. Consider the mathematics: 186.2 MHz + 220.8 MHz = 407 MHz. And, also, 407 MHz − 186.2 = 220.8 MHz. In this scenario, the 186.2-MHz transmitter signal will appear on the 220.8-Mhz channel.

The general rule of thumb for which signal will appear is based on the "capture effect." Telemetry transmitters and receivers are frequency modulated (FM). An FM receiver will generally "capture" the strongest of two competing co-channel signals and exclude the other. As a result, the strongest of the two signals (in this example, 186.2 MHz translated to 220.8 MHz by mixing in the 60-dB amplifier) predominates.

Following that night, the hospital banned FM radio receivers, patient-owned TV receivers, and Citizen's Band (CB) and ham radio sets in the CCU/PCCU area for exactly the same reason they are banned on commercial airliners: interference with critical electronic equipment.

24-4.1 Some Solutions

There are a number of ways to overcome most intermod problems. Modification of the receiver is possible, especially since poor design is a basic cause of intermods. But that approach is rarely feasible except for the most technically intrepid. There are, however, a few pointers for the rest of us.

First, make sure that the receiver is well shielded. This problem rarely shows up on costly modern equipment but is a strong possibility on lower-cost receivers. Most radio transceivers have adequate shielding because of the requirements of the transmitter in the same cabinet with the receiver and FCC requirements. But if there are any holes in the shielding, then cover them up with sheet metal or copper foil. For others, the best approach is to use one of the methods shown in Figures 24-2 through 24-4. In all cases, you must identify one of the interfering frequencies or (in some cases) at least the band.

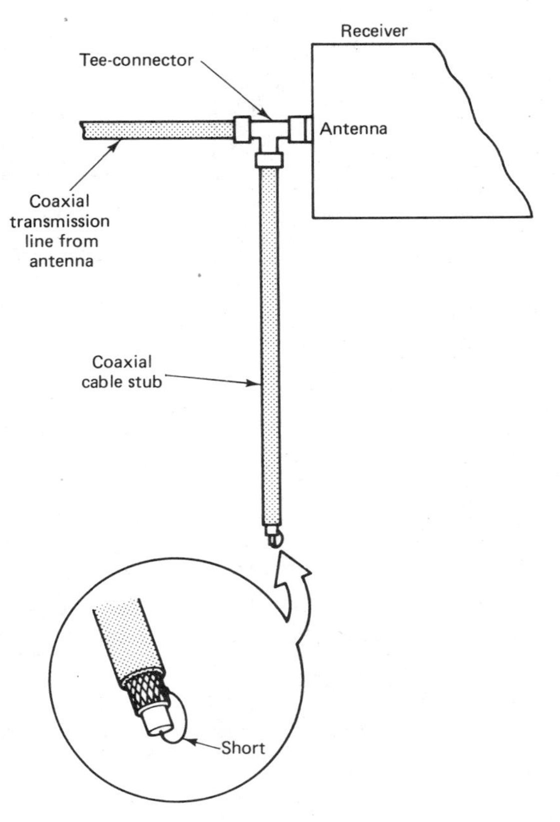

Figure 24-2 Use of coaxial half wavelength stub for EMI suppression.

Halfwave Shorting Stub

A nonmatching load impedance attached to the load end of a transmission line repeats

Figure 24-3 Use of a frequency selective filter to block EMI signals.

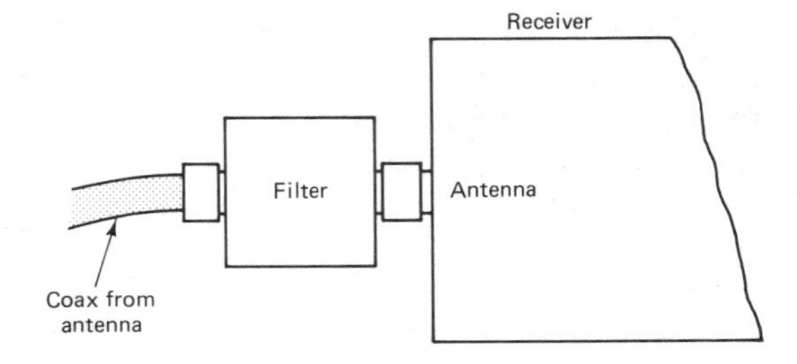

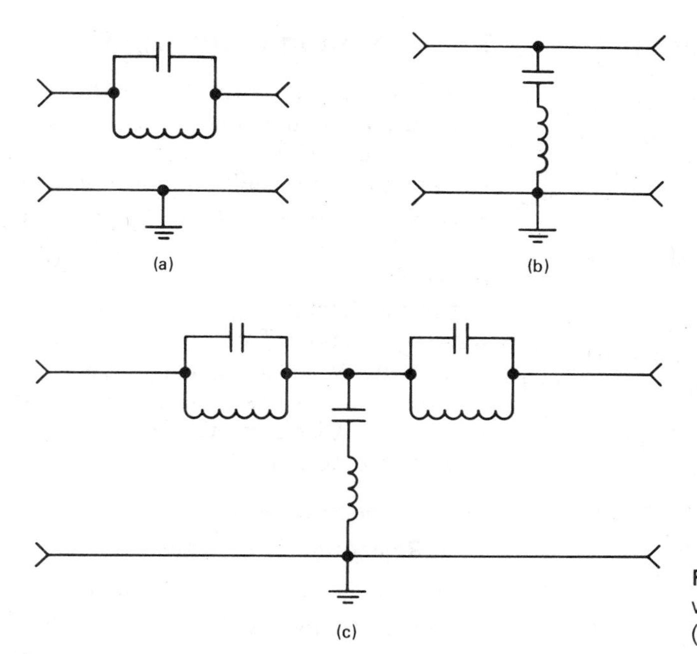

Figure 24-4 Several different wave traps: (*a*) parallel; (*b*) series; (*c*) combination series and parallel.

itself every half wavelength back down the line toward the input end. For example, if a 250-Ω antenna is attached to the load end of a half wavelength piece of 50 Ω coax, an impedance meter at the input end will measure 250 Ω. Lengths other than integer multiples of half wavelength will see different impedances at the input end. This phenomenon is the basis for transmission line transformers used in antenna matching. Therefore, if the end of a piece of coax is shorted (see Figure 24-2), then the input end will see a short circuit at the frequency for which the coax is a half wavelength. The interfering frequency will be shorted to ground, while the desired frequency sees a high impedance, provided that the two are widely separated. The length L is found from $L_{ft} = 492V/F_{MHz}$; where L is in feet, F is in MHz, and V is the velocity factor of the coax (usually 0.66 for regular coax and 0.80 for polyfoam coax).

The method of Figure 24-2 is best suited to cases where the interfering signal is in the VHF region or lower. Because the length of the coax stub is very short relative to the HF wavelength of the desired band, some very untransmission-line-like behavior might take place when transmitting. Therefore, for transceivers some engineers recommend adding a second antenna jack especially for the stub, but connected to the receiver circuitry (RX CKT).

Another method (Figure 24-3) is to use a frequency selective filter. The selection of type of filter, and the cutoff frequency, is determined by the case. In a maze of frequencies like "Intermod Hill," it might be wise to use a bandpass filter on the band of choice. Otherwise, use a low-pass filter if the interfering signal (at least one of them) is higher than the desired band, and a high-pass filter if the interfering signal is lower. For most cases, a low-pass television interference (TVI) filter is desirable on HF, so this solution is automatically taken care of; the 35–45-MHz cutoff point of most such filters will attenuate most VHF signals trying to get back in.

Figure 24-4 shows some of the different types of filter configurations that might be used. The first wavetrap is a parallel resonant circuit (Figure 24-4*a*) in series with the signal line. A parallel resonant trap has a high impedance at the resonant frequency but a low impedance on other frequencies, so it blocks the undesired frequency. Figure 24-4*b* shows a series resonant trap across the signal line. The series resonant circuit is just the opposite of the parallel case: It offers a low impedance at its resonant frequency and a high impedance elsewhere. Thus, it serves much like the coax stub in Figure 24-2. Figure 24-4*c* shows a combination wavetrap that uses both series and parallel resonant elements.

If the interfering station is an FM broadcaster, then most video shops and electronic parts stores carry "FM wavetraps" for 75-Ω antenna systems. These traps cost only a few dollars but provide a tremendous amount of relief from the interference produced by FM broadcast stations.

It might be the case that the interfering signal is entering the equipment on the ac power line. This situation is especially likely if you live very close to a high-power broadcasting station. If there is space inside the receiver, then it might pay to install an ac line filter. Electronic parts distributors sell EMI ac line filters suitable for equipment up to 2000 watts (V-A) or so (look at the ampere rating of the filter). Also, it is possible to install ferrite blocks around the line cord to serve as RF chokes. Some computer stores sell these blocks to reduce EMI from long runs of multiconductor ribbon cable. It is sometimes possible to reduce the EMI pickup by either reducing the cord length or rolling it up into a tight coil. This solution works well where the cord is of a length nearly resonant (quarter wavelength) on the interfering signal's frequency.

24-5 Dealing with TVI/BCI

Television interference, TVI, is the bane of the radio communications and broadcasting communities. True TVI occurs when emissions from a transmitter interfere with the normal operation of a television receiver. Except for solo explorers on the North Slope of Alaska and missionaries among the Indians of the Amazon basin, all radio operators have a potential TVI problem as close as their own TV set or their neighbors' sets.

One way to classify TVI is according to the cause. All electronic devices must perform two functions:

1. Respond to desired signals.
2. Reject undesired signals.

All electronic devices perform more or less in accordance with the first requirement, but many fail with respect to the second. It is often the case that a legally transmitted signal will be received on a neighbor's TV or Hi-Fi set even though the transmitter operation is perfectly normal. On some TV sets, the high-intensity signal from a nearby transmitter drives the RF amplifier into nonlinearity, and that creates harmonics where none existed before. In other cases, audio from the transmission is heard on the audio output of the set, or seen in the video, as a result of signal pickup and rectification inside the set. Improper shielding can cause signals to be picked up on internal leads and fed to the circuits involved.

Remember two rules of thumb: (1) If transmitter emissions are not clean, then getting rid of the TVI/BCI is the radio operator's responsibility; and (2) if the transmitter emissions are clean, and the TVI/BCI is caused by poor medical equipment design, then it is the medical equipment owner's responsibility to fix the problem—not the radio operator's. Unfortunately, in a society that depends too much on lawyers, the solution opted for by some ill-advised administrators is to hire lawyers to intimidate the radio owner into silent submission. Wise managers will shun the lawyers and hire an engineer to solve the problem in a manner that will allow both parties to operate compatibly.

On the other hand, radio transmitter owners must respond responsibly to EMI complaints for several reasons. First, they have a legally imposed responsibility to keep transmitter emissions clean and free of spurious or unnecessary components. They must operate the transmitter legally. Second, it makes for good neighborhood relations if they attempt to help solve the problem. After all, we live in a society where more people seem willing to go running to lawyers than to engineers to solve technical problems. The following steps should be followed.

Step 1. Make sure that the transmitter emissions are clean. Although it requires a spectrum analyzer to be sure that the harmonics are down 40 or more decibels below the carrier, a few simple checks will tell the tale in many cases. An absorption wavemeter is an old-fashioned device that can spot harmonics that are way too high. Also, listening on another receiver from a long distance (1 mile or more) will give some indication of problems: At that distance, if you can hear the second or third harmonic, then it is probably real. Finally, if you own an RF wattmeter, or forward-reading VSWR meter, and a dummy load, then you can make a quick check of the transmitter by measuring the output power with and without a low-pass filter (Figure 24-5) installed in the line. If the power varies between the two readings by appreciably more than the insertion loss of the low-pass filter, then suspect that harmonics are present.

The techniques for making the transmitter clean are simple (Figure 24-6): a low resistance earth ground and adequate filtering. In the station shown in Figure 24-6, there are three frequency selective elements after the transmitter: low-pass filter, antenna tuning unit (in lower-frequency stations), and a resonant antenna. All of these will help in reducing whatever harmonics the transmitter puts out.

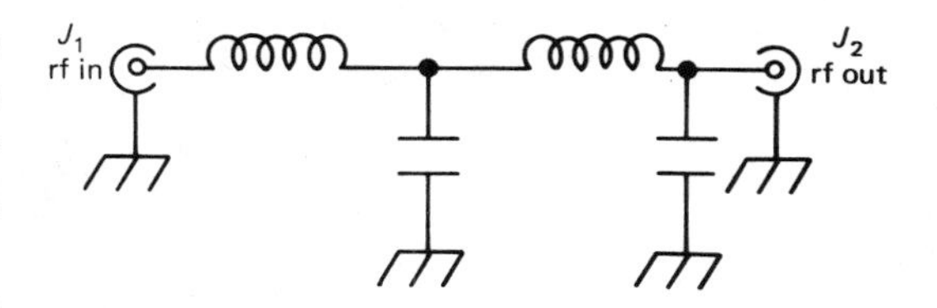

Note: Component values are a function of frequency.

Figure 24-5 Use of a low-pass filter such as this circuit and an RF power meter allows us to check for harmonics.

Step 2. Determine that you really have a TVI problem. Check for TVI both with the transmitter turned on and turned off. Also, make sure the TV is properly adjusted.

Step 3. Try adding a high-pass filter to the TV set. These filters will only pass signals with a frequency greater than about 50 MHz and severely attenuate HF amateur or CB signals. As shown in Figure 24-7, the high-pass filter must be installed as close as possible to the antenna terminals of the TV set. Make the connection wire as short as possible to prevent it from acting as an antenna in its own right.

Counsel the affected TV set owner to install an antenna that uses a coaxial cable transmission line rather than 300-Ω twin-lead. Although theory tells us that there should be no difference, it is nonetheless true that 75-Ω coax systems are less susceptible to all forms of noise, including TVI. Install a high-pass filter and a 75- to 300-Ω TV-type BALUN impedance transformer at the TV's antenna terminals.

Many, perhaps most, TV receivers have very poor internal shielding, so signals can bypass the high-pass filter and get picked up on the leads between the TV tuner (inside

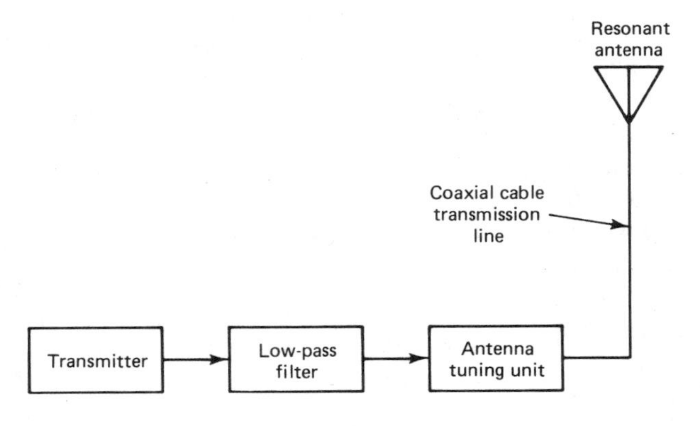

Figure 24-6 Typical transmitter installation with low-pass filter and antenna tuning unit.

the set) and the antenna terminals on the rear of the set. This situation makes the high-pass filter almost useless. A solution for this dilemma is to mount the filter directly on the tuner, inside the set, making the lead length essentially zero.

24-6 Dealing with Signal Overload Problems

Radio receivers are often subject to overload interference. While high-priced, high-quality telemetry receivers will handle overload conditions better than the cheapies, the only real issue is what signal, at what strength, will cause overload problems. In this section we will explore some of the methods that might be used to overcome these problems on shortwave and scanner receivers.

Figure 24-7 TVI filter for interference suppression.

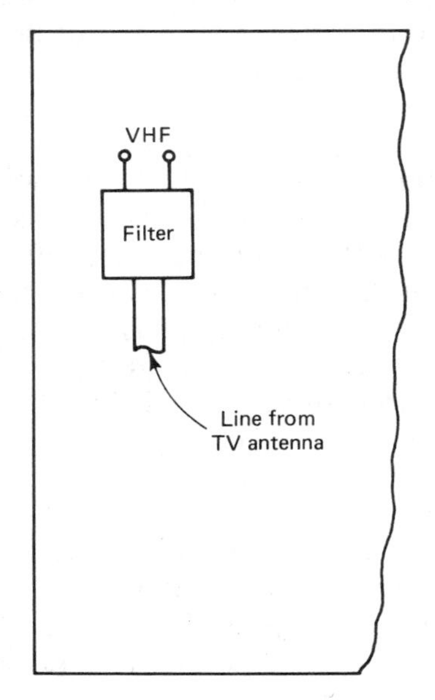

There are four basic types of overload problems seen on radio receivers: *desensitization, generation of unwanted harmonics, distortion of the desired signal,* and *intermodulation frequencies*. These can also be broken into those that involve *co-channel interference, adjacent channel interference,* and *off-channel interference*. Let's take a look at each of these in turn, keeping in mind that actual situations may well involve two or more combinations of both groups—and in fact probably will.

Co-channel interference involves signals on the same channel as the desired signal. Unfortunately, there is next to nothing that can be done for these problems at the receiver. The best solution is to use a directional antenna that favors the desired station and rejects the unfavored signal. Every directional antenna has a direction of least pickup; that is, a null direction. For example, a half wavelength dipole has two nulls off the ends of the radiator element (Figure 24-8), perpendicular to the directions of maximum pickup. A Yagi or quad antenna (Figure

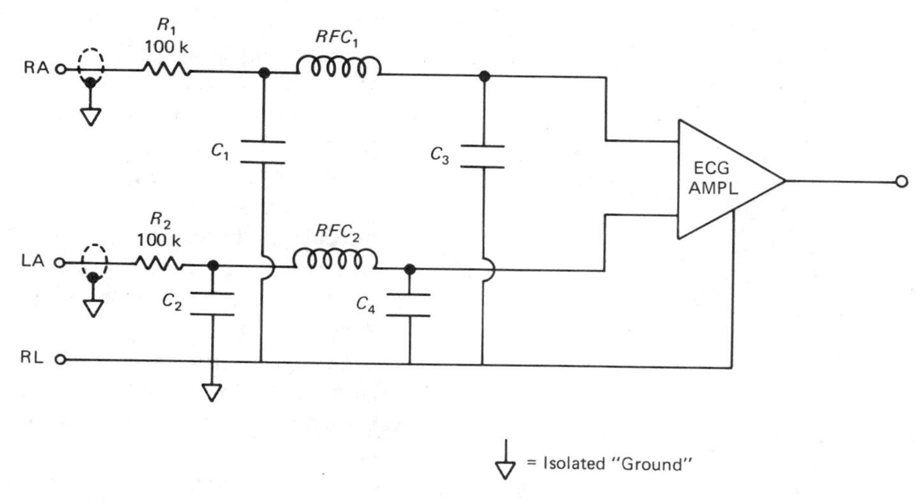

Figure 24-8 Low-pass filtering for EMI protection of ECG amplifier.

24-9) has the null in the direction opposite the direction of maximum pickup (or "off the back"). To overcome co-channel interference, place the null of the antenna in the direction of the interfering signal.

The null directions on beams and dipoles are relatively easy to locate. Unfortunately, many of the antennas used by shortwave listeners are either random length or long-wire types, and the nulls on those antennas are multiple and functions of both frequency and physical length. For those antennas, the prediction of null directions is difficult.

VHF/UHF receivers often use omnidirectional, or at least multidirectional, antennas so the co-channel problem is particularly difficult. For those cases, a change of antenna type might be in order if receiving a particular station is important.

Adjacent channel interference is caused by stations that are on channels close to the frequency being received. In some cases, wavetraps will suffice, but for others the only solution is to have a receiver with excellent dynamic range and selectivity characteristics. We will take a look at wavetraps shortly.

Off-channel or *out-of-band* interference is caused when a strong local signal on a frequency that is unrelated, and not very near the desired frequency, causes interference. These signals are dealt with by using wavetraps, passband filters, or bandstop filters. A wavetrap will attenuate a signal frequency; a passband filter will pass the desired frequency band and attenuate other frequencies; a bandstop filter will attenuate the frequencies in the band of the offending signal, but not other frequencies.

Desensitization occurs when a strong local signal drives the rf amplifier, or other front-end components, very heavily, leaving little dynamic range for desired signals. The effect on the receiver is almost like someone turned down the rf gain control. If the problem comes and goes, it might mean that the local interfering transmitter is being turned on and off. A common cause of desensitization is when the offending rf signal autorectifies in the input rf amplifier device and causes a dc bias that reduces the receiver front-end gain much like automatic gain control.

Generation of unwanted harmonics comes about because a nonlinear electronic

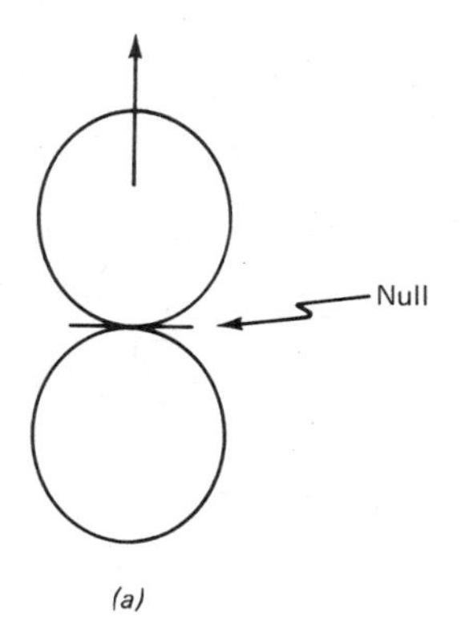

(a)

(b)

Figure 24-9 (*a*) Half wavelength dipole radiation pattern; (*b*) beam antenna radiation pattern.

circuit will cause even a pure, harmonic free sinewave rf signal to become rich in harmonics. This problem is found mostly when an AM broadcast station nearby overloads the rf amplifier of the receiver and drives it into a nonlinear region of operation. Harmonics are integer multiples of the fundamental frequency (2f, 3f, 4f, 5f . . .), so one will find the AM-band station at successively higher frequencies. For example, a station at 1500 KHz AM will be found at 3000 KHz, 4500 KHz, 6000 KHz, and so forth.

Do not call the FCC or the station's chief engineer until you check out the receiver first. The problem is not bad tuning or malfunctions at the radio station, but rather a normal (if undesired) response to nonlinear conditions in *your* receiver.

Rarely are more than the first three or four harmonics generated, but it is quite possible to find cases where the spurious harmonics are generated well into the low-VHF band. In one case, interference existed in the 35-MHz region, but the scanner owner lived close enough to the AM station's tower to worry mightly during hurricane season.

Distortion of desired signals occurs when a receiver lacks the dynamic range to handle a strong local signal. Not only might the audio recovered from the signal be distorted, but there might be some distortion of the rf signal that causes spurious signals to be generated. I have seen a strong FM band signal "re-create itself" up and down the FM band at frequency spacings that are probably explained by modulation theory. The solution is simple: Attenuate the unwanted signal.

24-6.1 Attenuators

An attenuator is a resistor circuit that will provide a loss between input and output. The ratio of the loss is usually expressed in terms of decibels (dB) from the following expressions:

$$\text{dB} = 20 \log_{10} \left(\frac{V_o}{V_{\text{in}}} \right) \qquad (24\text{-}1)$$

and

$$\text{dB} = 10 \log_{10} \left(\frac{P_o}{P_{\text{in}}} \right) \qquad (24\text{-}2)$$

For the sake of reference, a 2:1 ratio of voltage represents -6 dB loss, and a 2:1 ratio of power is a -3 dB loss.

When designing attenuators, it is important to make sure that the devices are truly resistive and contain no reactive components such as inductors and capacitors. Reactive attenuators are sometimes seen but are difficult to make successfully. Second, the attenuator should input and output impedances that are the same as the antenna input impedances of the receiver system; that impedance is usually 50 Ω but in some cases 75 Ω and 300 Ω (the latter are used in TV systems).

Figure 24-10 shows an attenuator that is used in receiver circuits. It is designed for 50-Ω antenna systems and can be used from very low frequency (VLF) up to the mid-VHF region (≈ 200 MHz). The resistors are ¼-watt carbon composition or metal film types; it is important *not* to use wirewound resistors because of their excessive inductance.

Figure 24-10 (*a*) Pi-network attenuator; (*b*) resistor values.

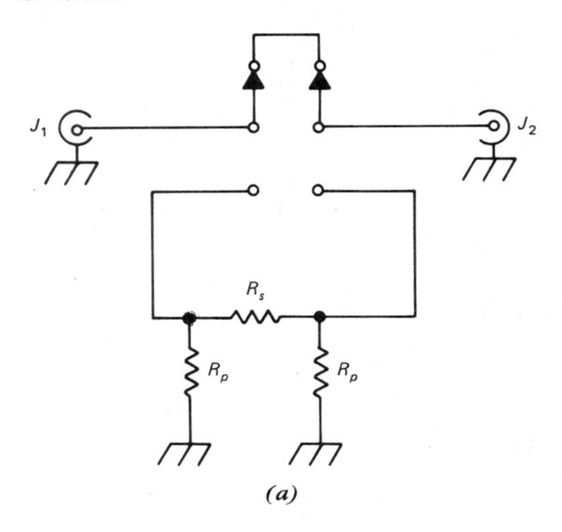

Attenuation (dB)	R_s (Ω)	R_p (Ω)
1	6.2	910
2	12	470
3	18	300
5	33	200
10	75	100
20	270	68

(*b*)

The resistor network in Figure 24-10(*a*) is a pi-network attenuator in which a single series resistance (R_s) is used with two shunt or parallel resistances (R_p). When a receiver has a single-stage attenuator with IN and OUT being the only selections, it is likely that this is the type of circuit used.

The values for the attenuator pad resistors are shown in the table in Figure 24-10(*b*). These values are not exact but rather reflect standard values that are easily obtained. These values cause a slight impedance mismatch, or possibly a slightly different attenuation ratio. Because this attenuator is used to reduce overload rather than to make precision signal strength measurements, we can live with this slight error in order to use commonly available resistance values.

A *double-pole double-throw* (DPDT) switch is used to connect or disconnect the attenuator from the circuit. When the switch is in the up position, the attenuator is disconnected from the circuit. Alternatively, when the switch is in the down position, the attenuator is placed in series with the signal path between jacks J_1 and J_2. The circuit is bilateral, so it does not matter which jack (J_1 or J_2) is connected to the antenna and which is connected to the receiver.

More than one attenuator can be connected in series to form greater attenuation ratios. Each section should be provided with its own shielded enclosure, or at least shielded section of the main enclosure, in order to prevent rf leakage or coupling from providing a ratio other than the design value.

24-6.2 Wavetraps

A *wavetrap* is a circuit that will attenuate either a single frequency or a very narrow band of frequencies. These circuits can be used to attenuate the signal from a single offending station. For example, if you are suffering from interference from an AM radio station, then it might be prudent to wipe out that frequency in the antenna input circuitry. Figure 24-11 shows a pair of wavetraps in a single shielded enclosure. Either the parallel version (C_{1A}, C_{1B}, L_1) or series (C_2, L_2) can be used alone, or both can be used together as shown.

The wavetrap works because of the respective properties of series and parallel resonant inductor-capacitor (*LC*) tank circuits. A parallel resonant tank circuit, such as $C_{1A}/C_{1B}/L_1$, offers a very high impedance to the resonant frequency, but a low impedance to frequencies removed from resonance. Thus, when a parallel resonant tank circuit is placed in series with the signal path, as it is in Figure 24-11*a*, then it provides maximum attenuation at the resonant frequency (solid curve in Figure 24-11*b*). A series resonant circuit, on the other hand, provides a very low impedance to its resonant frequency, so when placed in shunt across the signal line will attenuate the resonant frequency but not the others (dotted-line curve in Figure 24-11*b*).

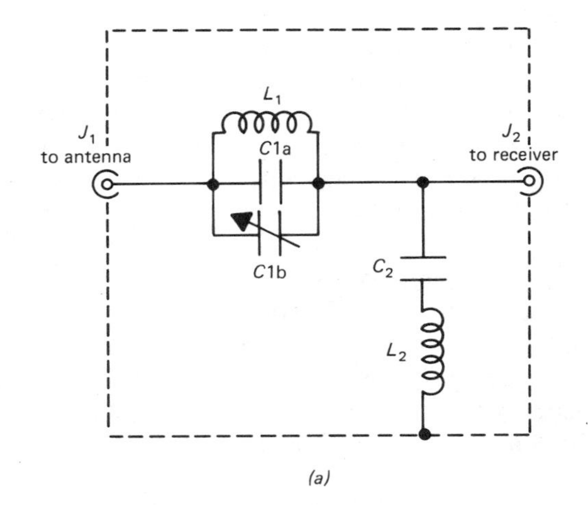

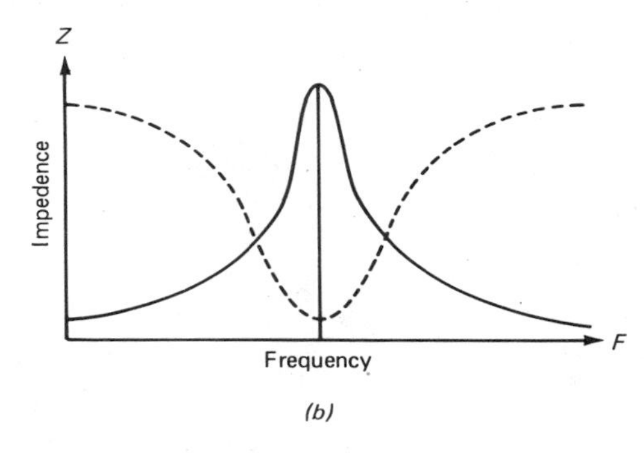

Figure 24-11 (*a*) Parallel and series wave traps; (*b*) frequency response curves.

In practical circuits, using both series and parallel resonant wavetraps will maximize the attenuation of a single frequency if both tank circuits are tuned to the same frequency. In some cases, however, there are two interfering stations, and it might prove necessary to make one tank circuit resonate on one frequency and the other wavetrap resonate on the other frequency. For example, near my home are two strong AM stations: a 2000-watt daytime-only station on 780 KHz, and a 24-hour 5000-watt station on 1390 KHz. Both of these stations are capable of ruining the reception in the low end of the medium-wave and HF shortwave bands. A dual-frequency wavetrap is the solution here.

The values of the components used in the wavetrap depend on the frequency and are found from:

$$F = \frac{1}{2\,\pi\,\sqrt{LC}} \tag{24-3}$$

or, because we are more likely to know the frequency to be attenuated, and want to know the inductance or capacitance:

$$L = \frac{1}{39.5\,F^2C} \tag{24-4}$$

or

$$C = \frac{1}{39.5\, F^2 L} \tag{24-5}$$

where F is in hertz, L is in Henrys, and C is in farads. The actual values will vary from the equation values because of stray capacitances and component tolerances; at low frequencies stray inductances are not usually a factor. Examples of values from Equation 24-5 are provided for the preceding 1390-KHz example:

L*	C
100 μH	131 pF
220 μH	60 pF

* Selected value from catalog

The actual value of the capacitor will be somewhat less than this because of strays, so make either the capacitor or the inductor (or both) adjustable to compensate for the error.

24-6.3 Bandstop Wavetrap

A bandstop wavetrap will attenuate the frequencies within an entire band. These traps are used especially when there are multiple stations in the band that overload the shortwave or scanner receiver. The AM broadcast band is a prime candidate for this treatment because of the large number of high-power signals found on the band in some metropolitan areas. Figure 24-12 shows a bandstop filter designed for the AM broadcast band. It uses components that are relatively easy to obtain and is easy to build. One rule, however: Separate the coils a couple of inches, and place L_3, L_4, and L_5 at right angles to L_1/L_2.

24-6.4 High-Pass Filters

In the cases that we have been dealing with, the offending signals are usually below the desired frequency. Thus, we can use a *high-pass filter* to pass those frequencies that we want, and attenuate those that we do not want.

Scanner operators who worry about AM stations, HF ham stations, and CB stations (not all of which operate at 5-watt legal power) might want to use a low-VHF high-pass filter such as those shown in Figure 24-13. The version shown in Figure 24-13*a* is for coaxial input receivers that require a 75-Ω impedance in the antenna circuit, while Figure 24-13*b* shows one that is for balanced 300-Ω twin-lead systems. In the latter case, sometimes 0.001-μF, 1000-volt disc ceramic capacitors are used at the points marked "X." These capacitors are used for receivers that do not use a power transformer (ac/dc). Such receivers are really too dangerous to use in a hospital environment, so do not (electrical shock hazard is substantial).

These filters are very similar to the TV high-pass filters used to prevent amateur radio and CB interference. You can buy such filters for frequencies > 54 MHz in TV and video stores.

A "universal" high-pass filter is shown in Figure 24-14*a*. The capacitors were 0.001 μF, and the inductor was a 2.7-μH coil. The coil was actually a 10.7-MHz FM IF coil with the capacitor disconnected. Figure 24-14*b* shows the frequency response curve for the filter. The filter was designed to pass frequencies above 3000 KHz and attenuate those frequencies (such as the AM band) below 3000 KHz.

The capacitors and inductors for other frequencies are found from:

$$L = \frac{R}{4\pi F_c} \tag{24-6}$$

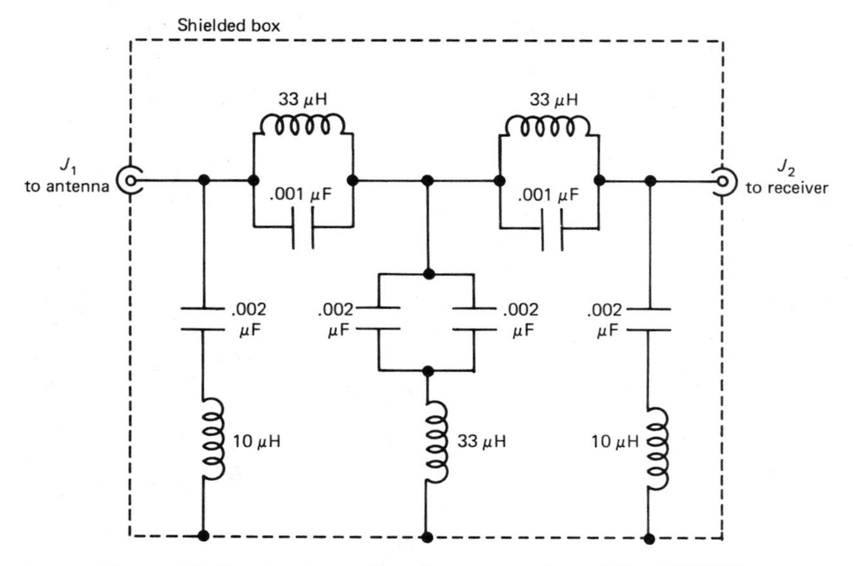

Figure 24-12 Bandstop filter for suppression of the AM BCB.

and

$$C = \frac{2}{4\pi F_c R} \quad (24\text{-}7)$$

where L is in Henrys, C is in farads, F_c is the cutoff frequency in hertz (Hz), and R is the system impedance, either 50 or 75 Ω.

Figure 24-13 High-pass filter circuits: (*a*) unbalanced, (*b*) balanced.

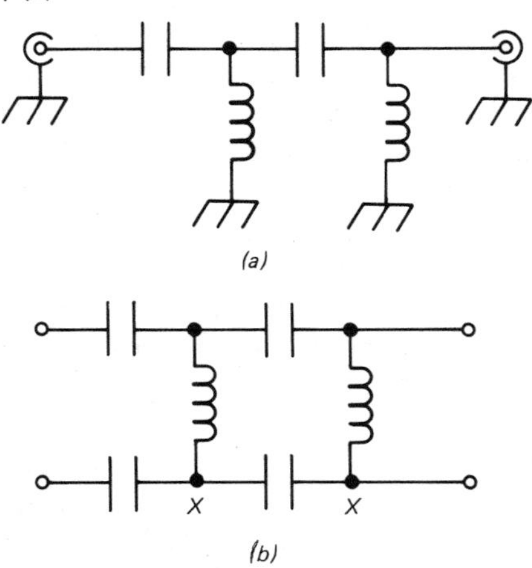

24-7 ECG Equipment and EMI

Medical equipment seems especially sensitive to picking up interfering signals. The ECG electrode wires are exposed at the ends. In addition, the patient's body is basically an electrical conductor, so it makes a relatively decent "antenna" to pick up signals that exist in the air. The solution to that problem is a low-pass filter (see Chapter 5) that blocks radio signals but not the ECG. These filters are not like the 60-Hz filter that is used in the ECG machine to eliminate power line artifact. Those filters take out a segment of the 0.05- to 100-Hz spectrum that is normally part of the ECG waveform and so will always affect the shape of the wave-

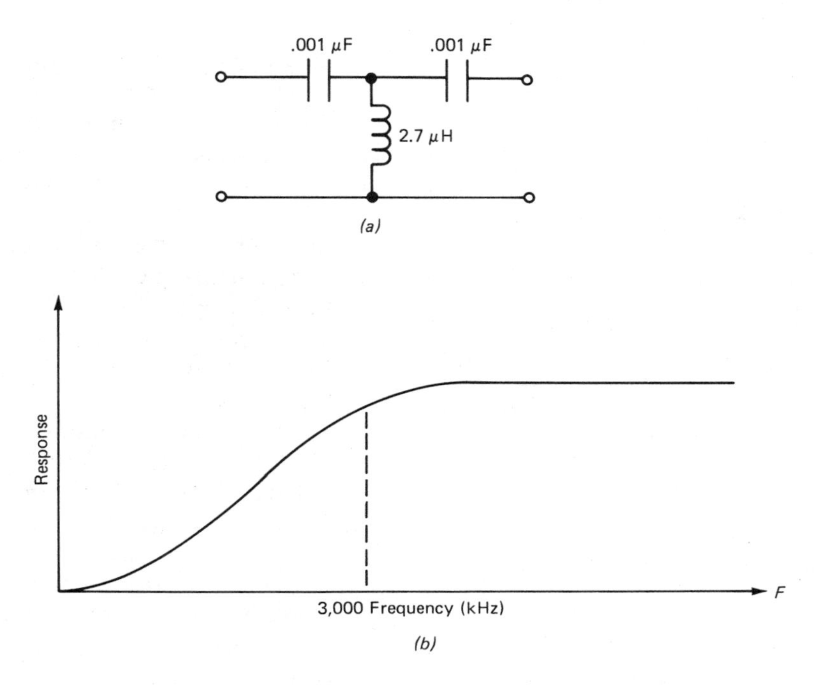

Figure 24-14 Universal high-pass filter.

form presented to the medical person using the machine. The RF low-pass filter has such a high cutoff frequency that the ECG waveform is not affected.

24-8 EMI to Biomedical Sensors

Most biomedical sensors are susceptible to EMI because of the typically very low signal levels produced by those sensors. One authority points to an especially severe EMI problem with Wheatstone bridge strain gages that typically have very low gage factors. It is worthwhile, therefore, to review the causes of, and possible cures for, these types of EMI problems.

Sources of EMI are the same sort of RF generators that also afflicted radio receivers (discussed earlier). One cannot connect a half wavelength shorted stub across a sensor output line, however. At the low (non-RF) frequencies used by sensor signals, all shorted RF stubs are short circuits to all relevant signal frequencies—so the sensor output will be shorted into the stub along with resonant RF frequencies. To identify fixes for this type of EMI case, it is helpful to know the typical routes that EMI signals follow to enter a sensor system.

There are three principal transmission paths for EMI into a sensor system: *penetration, leakage,* and *conduction*. Figure 24-15*a* illustrates penetration. In this case, the impinging electromagnetic (EM) wave cuts across the sensor and its circuit, setting up RF currents in the sensor circuitry. These RF currents become valid signals in the system. The argument is sometimes made that sensor circuits typically have such low fre-

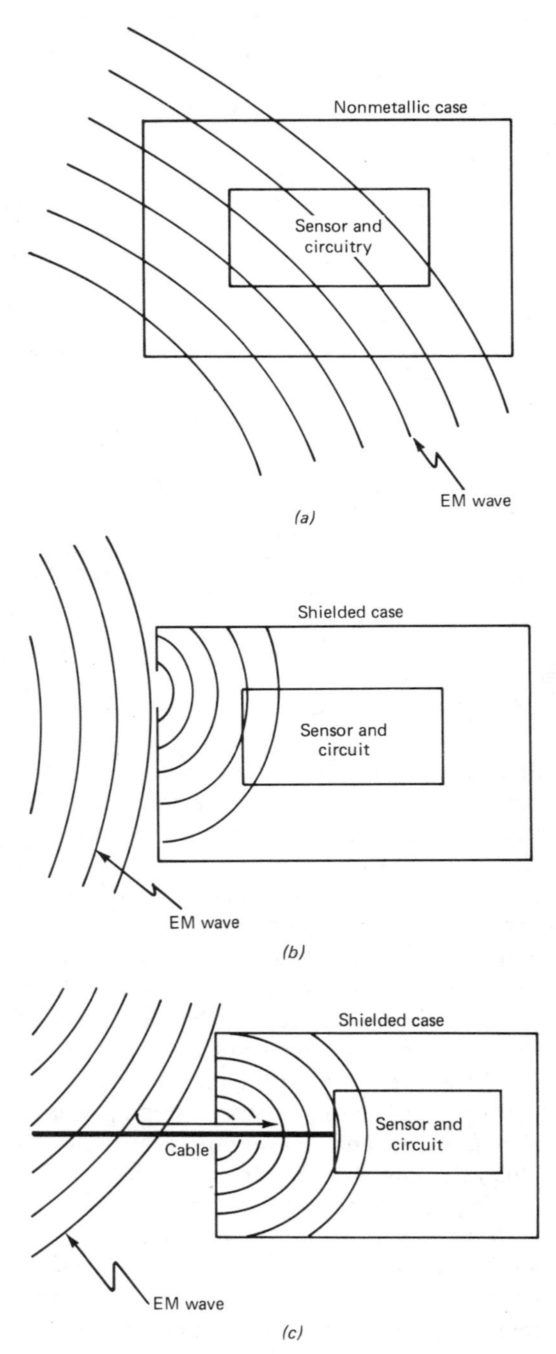

Figure 24-15 Routes for EMI interference to equipment: (*a*) penetration; (*b*) leakage; (*c*) conduction.

quency response that RF signals will not affect them. This argument is, unfortunately, largely specious because of phenomena such as *autorectification.* When this problem occurs, the RF signals rectify (e.g., "detect") in the pn junctions of the semiconductor components used in the sensor electronics. This action produces either a dc bias that distorts the operation of the circuit, or extracts modulation components that are within the frequency response passband of the circuit.

The solution for eliminating penetration EMI is to shield the circuit entirely in a shielded environment. A proper shield is metallic, or another conductive material, and is well sealed. It is often the case, especially where very high frequencies or microwaves are causing the interference, that the shielded box holding the electronics must have additional shielding. In many cases, extra fasteners or screws are needed to seal the lips or edges of the shielded cabinet. The rule of thumb is that the screws should be spaced less than one-half wavelength at the highest frequency expected (or specified for that particular product).

Leakage occurs when the EM wave enters the shielded cabinet through small cracks or spaces in the shielding (Figure 24-15*b*). The standard wisdom is that a half-wavelength slot will admit RF to a shielded compartment. Unfortunately, that is not always true, especially where nonsinusoidal RF waveforms are present. The wavelength of the signal fundamental frequency is:

$$\lambda_{meters} = \frac{300}{F_{\text{MHz}}} \tag{24-8}$$

Thus, a 150-MHz taxicab transmitter creates a 2-meter wavelength signal and so is not a problem in most cases because the necessary gap (1 meter) is wide compared with the size of the instrument cabinet. In the microwave region, however, wavelengths drop to the centimeter and millimeter region, so these become a larger problem.

Even at lower frequencies, however, modulation or other situations can raise the number and strength of the harmonics of the fundamental, and these can cause severe interference under the right circumstances. Pulsed signals, for example, can have significant harmonics of 100 times the fundamental frequency. A 400-MHz signal has a wavelength of 0.75 meters. If it is a sinewave signal, or contains linear voice modulation (or certain other forms), then there will be few harmonics. But if the 400-MHz signal is pulsed, then it might easily have harmonics to 4 GHz or even 40 GHz. At these frequencies, the shielding must be tight.

There are two modes of leakage. The resonant mode occurs when the gap in the sensor shielding is resonant at the interfering frequency. If the gap is a half wavelength, then it will act as a slot antenna and efficiently reradiate the offending signal inside the shielded enclosure.

Conduction EMI (Figure 24-15*c*) occurs when a signal induces current into an exposed power, ground, control, or signal cable entering or leaving the shielded compartment. In this type of interference, the EMI signal reradiates inside the housing and is picked up by the sensor or its electronics. Surprisingly short lengths of wire will act as relatively efficient antennas in the UHF region and above.

24-8.1 Some Cures

Figure 24-16 shows two common fixes for EMI in sensor systems: *filtering* (Figure 24-16*a*) and *shielding* (Figure 24-16*b*). It is likely that both will have to be used in any given system.

The use of filtering to attenuate RF signals before they get to the sensor or electronics circuits is shown in Figure 24-16*a*. Each line in the system (whether power or signal) has a π-section filter consisting of an inductor element called an *rf choke* (RFC) and two capacitors. An example in Figure 24-16*a* is $RFC_1/C_1/C_4$ in the +POWER line. The π-section filter is basically a low-pass RF filter that has a -3 dB cutoff frequency that is well below the offending signal's frequency but well above the highest frequency in the Fourier spectrum of the sensor output signal. A rule of thumb for the filter cutoff frequency is to calculate it from the length of exposed cable in the sensor system:

$$F_c << \frac{300}{2L} \tag{24-9}$$

where

F_c is the -3 dB cutoff frequency of the low-pass filter

L is the length of cable in meters

For example, a 1-foot (0.295 meters) cable is half-wavelength resonant at a frequency of 509 MHz. Signal interference is maximum at that frequency but is still significant at frequencies removed from F_c, especially those that are higher. The "much less than" (<<) symbol in Equation 24-9 can be taken to mean 1/10, so the cutoff frequency, F_c, of the low-pass filter should be $F_c/10$, or in the preceding case 509 MHz/10 = 50.9 MHz.

The filter should be installed as close to the connector in the shielded bulkhead as possible. Otherwise, reradiation can occur from the wiring between the connector and the filter. Commercially available EMI filters usually are chassis or "bulkhead" mounted, so this requirement is easily met most of the time in actual practice.

If the EMI filter is installed in an ac power line, then make absolutely certain that it is of a type that is specifically designed for such service (the voltage rating is not always a sufficient indicator—look for the words *ac line* in the specification). If a filter is not specifically designed for service in 110-V ac

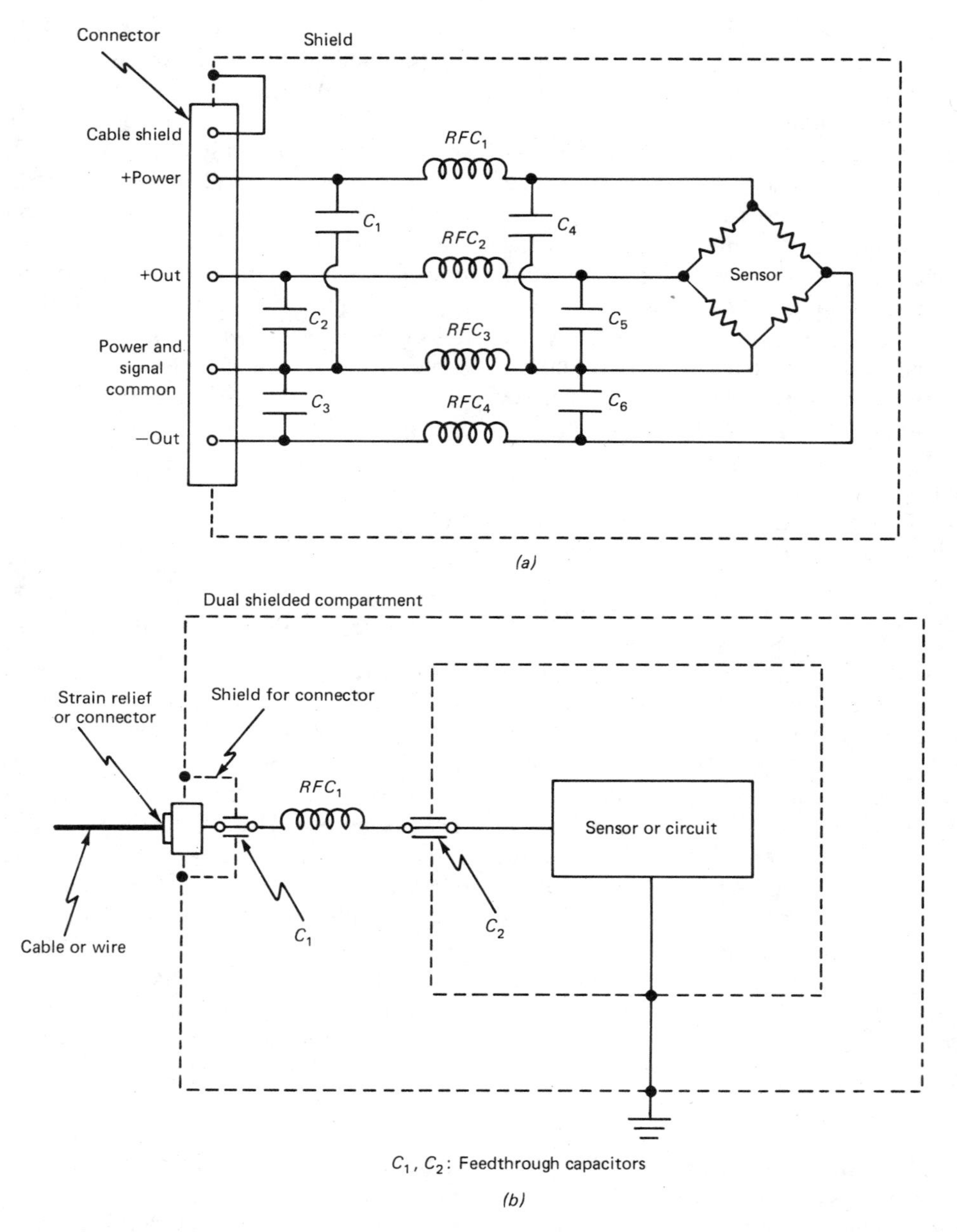

Figure 24-16 Cures for EMI: (*a*) filtering; (*b*) shielding.

or 220-V ac lines, then using them in that service can be dangerous.

Double shielding combined with filtering is shown in Figure 24-16*b*. The sensor is located inside of an inner shielded compartment that is, itself, enclosed within the outer shielded compartment. An EMI filter ($RFC/C_1/C_2$) is placed between the cable strain relief and the inner shield compartment.

Artifacts in sensor data can be largely eliminated by the proper use of the techniques discussed in this chapter. While the topic is worthy of a book in its own right (and indeed, several have been written), these techniques are suitable for many—perhaps most—practical applications.

24-9 Summary

1. Telemetry systems are subject to a number of interference situations, including overload and intermodulation. There are also situations where stations on adjacent and the same channel cause interference.
2. The cure for intermodulation problems is usually filtering to remove one or more offending signals.
3. Medical equipment, such as ECG units, that use exposed leads can suffer interference from electrosurgery and other RF generators. The usual solution is to remove the RF signal by filtering.
4. Filters useful for removing EMI include low-pass, high-pass, and bandpass filters. Sometimes half-wavelength shorted stubs are used.
5. EMI from a variety of sources can enter medical equipment either by direct pickup, by pickup in breaks on the shielding, and via conduction through power lines and signal lines going into the cabinet.

24-10 Recapitulation

Now return to the objectives and self-evaluation questions at the beginning of the chapter and see how well you can answer them. If you cannot answer certain questions, place a check mark next to each and reread appropriate parts of the text. Next, try to answer the following questions using the same procedure. When you have answered all of the questions, solve the problems in Section 24-12.

24-11 Questions

24-1 Define *electromagnetic interference* in your own words.

24-2 ______________ interference is caused by two frequencies, F_1 and F_2, or their harmonics, mixing together in a nonlinear circuit to produce a third frequency.

24-3 The process of mixing together two frequencies (F_1 and F_2) to produce either the sum ($F_1 + F_2$) or difference ($F_1 - F_2$) frequencies is called ______________ing.

24-4 Poor ______________ or ______________ range in the receiver is the cause of most intermodulation interference.

24-5 What circuit inside an FM broadcast receiver has the potential to interfere with radio telemetry units operating in the VHF region?

24-6 The ______________ effect occurs in FM receivers and is characterized by the strongest of two co-channel signals locking the receiver to the exclusion of the weaker signal.

24-7 A ______________ resonant wavetrap in series with the signal line will reject a signal on the trap's resonant frequency.

24-8 Two functions required of all electronic equipment are ______________ and ______________.

24-9 Operators of a transmitter are responsible for ensuring that the output signal is free of ______________ that might interfere with other services.

24-10 A ______________ ______________ filter can be used in series with the leads of an ECG machine to reduce radio frequency RMI.

24-11 List four classes of overload problem seen on radio receivers.

24-12 ______________ occurs when a strong local signal drives the rf amplifier of a receiver very heavily.

24-13 A receiver can generate harmonics of received signals, even when none exist in the actual signal. True or false?

24-14 An rf attenuator produces an output voltage of 50 μV when the input signal level is 750 μV. What is the loss in decibels?

24-15 A ______________ ______________ is a circuit that will attenuate a single frequency.

24-12 Problems

24-1 Find the third order intermodulation products (n = 1, 2, 3 and m = 1, 2, 3) for two frequencies: 54.5 MHz and 27.120 MHz.

24-2 A VHF telemetry transmitter operates on 174.15 MHz. There are also in the same vicinity an FM broadcast station operating at 88.5 MHz and a landmobile station operating at 478.33 MHz. Is there any likelihood of intermodulation interference (up to the third order)?

24-3 It is necessary to remove a 1,390-KHz AM broadcast signal from a VHF telemetry system. Calculate the capacitance that will resonate with a 220-μH inductor used in a parallel resonant wavetrap tuned to this frequency.

24-13 References

1. Carr, Joseph J., *Practical Antenna Handbook*, TAB/McGraw-Hill, Cat. No. 3270 (Blue Ridge Summit, Pa., 17294; 1-800-233-1128).
2. Carr, Joseph J., *Receiver Antenna Handbook*, HighText Publications, Inc., 7128 Miramar Road, #15, San Diego, Calif., 92121.
3. Brad H. Watkins, "Electromagnetic Compatibility of Strain Gage Transducers," *Sensors*, November 1989; pp. 35ff.

Chapter 25

Care and Feeding of Battery-Operated Medical Equipment

25-1 Objectives

1. Be able to describe the different kinds of batteries used in medical equipment.
2. Be able to describe the charging protocols for common batteries.
3. Be able to state the limitations of batteries.
4. Be able to understand the maintenance of battery-operated equipment.

25-2 Self-Evaluation Questions

These questions test your prior knowledge of the material in this chapter. Look for the answers as you read the text. After you have finished studying the chapter, try answering these questions and those at the end of the chapter (Section 25-15).

1. The usual protocol for charging NiCd batteries and cells is to charge at a rate of ___________, to a level of ________ percent of fully charged.
2. NiCd batteries are said to possess ___________ because they will adjust the fully discharged level if shallow cycled too many times.
3. List three different forms of batteries used in medical equipment.
4. The charge available to a battery is rated in units of ___________.

25-3 Introduction

Much medical equipment uses batteries of one sort or another. The "care and feeding" of batteries is a critical element in keeping the equipment ready for use and is thus especially problematical in emergency equipment. Batteries are used for a lot of different reasons. Some equipment is battery powered for reasons of portability. A defibrillator, for example, might be needed almost anyplace; heart attacks do not always occur near an electrical outlet. Although most defibrillators are ac powered (or dual powered), a number are purely battery-powered models. Some patient monitors also run on batteries. These devices are used to keep track of ECG and blood pressure as the patient is transferred between units (e.g., from the emergency room to the intensive care unit). Still other devices use batteries for reasons of patient safety. A cardiac output computer, for example, makes measurements based on a thermistor inserted into the heart. Small amounts of ac "leakage" current could be fatal (Chapter 19), so batteries are used to completely isolate the instrument from the ac power line.

25-4 Cells or Batteries?

A cell is the most basic element in a battery and sets the minimum voltage for that sort of device. Additional voltage is gained by connecting the cells in series, and extra current is available by connecting them in parallel. To be strict, we would refer to single entities as *cells* and multiple-cell entities as *batteries*. But in common usage, it is usually acceptable to be less than rigorous, so all cells and batteries are called *batteries*.

25-5 Nickel Cadmium (NiCd) Cells and Batteries

In this section we will discuss mostly the nickel-cadmium (NiCd) cells and batteries that are commonly used in portable electronics equipment. These batteries have a nominal terminal voltage at full charge of 1.2 volts, except immediately prior to turn-on after a fresh charge (at which time the "open-terminal" voltage is 1.4 volts). Shortly after turn-on, however, the open-terminal voltage drops from 1.4 volts to the nominal value of 1.2 volts for the duration of the operation. As the stored energy is used up, however, the terminal voltage drops lower.

NiCd batteries are rechargeable and will typically sustain a charge-discharge cycle lifetime of 1000 times before becoming unusable. In most cases, manufacturers rate a battery as unusable when the capacity of the battery drops below 80 percent of its original specified value.

25-6 Battery Capacity

The capacity of a battery is measured in *ampere-hours* (i.e., the product of current load, in amperes, and the time required to reach the designated discharge state). The NiCd battery is capable of delivering tremendous currents. For example, the size D (4 A-H) and size F (7 A-H) can deliver short-duration currents of 50 amperes or more. That is why they are used in medical defibrillators and why certain medium-powered portable radio transmitters can use them. As a result of their ability to deliver large currents, NiCd batteries should be fused in order to protect printed wiring tracks, wires, and other conductors. The author has seen copper foil printed wiring board tracks and an on/off switch vaporized by a shorted capacitor across the dc line.

The amount of time that a battery will last is a function of the discharge time, which in turn is determined by the amount of current drawn. Figure 25-1 shows two different discharge scenarios: one for a current of one-tenth the A-H rating, and one for a current equal to the A-H rate. In Figure 25-1*a* the battery will be fully discharged in 10 hours, while in Figure 25-1*b* discharge occurs in one hour. This particular chart is derived from the data published for a size D NiCd cell rated at 4 A-H.

The standard cell ratings for NiCd batteries are as follows:

Battery Size	Ampere-Hour Rating
AA	0.4/0.5/0.7
C	2
D	4
F	7

As you can see in the preceding chart, the AA cells are found in three ratings from 400 to 700 mA-H, depending upon the manufac-

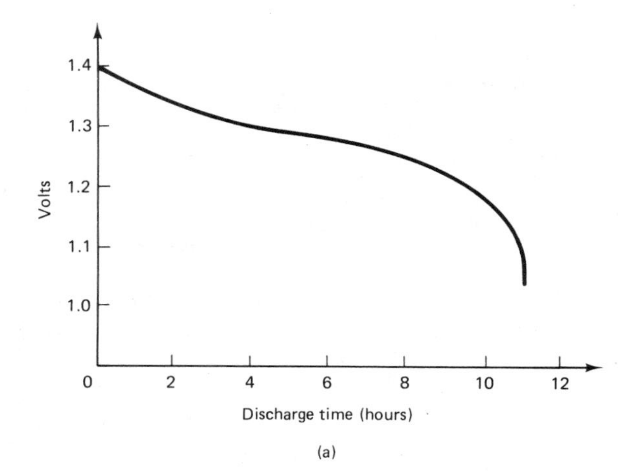

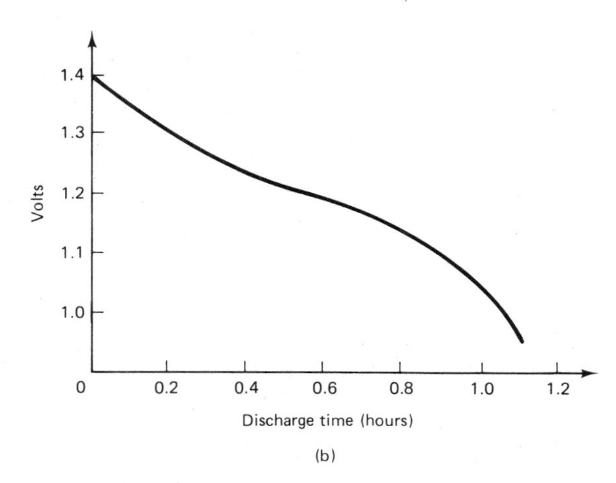

Figure 25-1 Charge/discharge curves for NiCd batteries: (*a*) charging curve, (*b*) discharging curve.

turer and style. You will find a lot of variation from this chart, especially among consumer product NiCd batteries. Some size C cells are rated at both 1 and 1.2 A-H and size D are rated at 2 A-H. Suspect that these are actually lesser cells dressed in size C and size D packages. One manufacturer's representative actually admitted to me that the consumer D cells were actually size C cells inside a D package.

This difference is of little practical consequence to most consumer electronics users and actually results in a lower-cost product. But if you use or service commercial, medical, or communications equipment, make sure that you get the correct ampere-hour rating. A heart patient's chance of survival can be reduced by replacing 4 A-H NiCd cells in the defibrillator with a collection of 2 A-H consumer replacements intended for power toothbrushes and portable radios. *Be sure of the rating of replacement batteries and cells.*

Rating games are also played by some distributors by quoting different discharge rates. One standard method of measuring A-H capacity is the current required to discharge a cell to 1.0 volts in one hour. Some makers, however, define it in terms of the 10-hour discharge rate normalized to ampere-hours. From analyzing Figures 25-1*a* and 25-1*b*, you can see how this might result in a false feeling of full capacity.

25-7 Battery-Charging Protocols

The charging protocol for the NiCd battery depends somewhat on application and manufacturer. In general, though, the charge current must be at least A-H/20, and in many commercial consumer battery chargers it is often A-H/15. For most applications where you can control the charge rate, it is safe to use a charge rate of A-H/10. That is, charge the battery at a current not greater than one-tenth the ampere-hour rating. In addition, the battery must be charged to 140 percent of capacity, so a charge time of 14 hours at A-H/10 is mandated. The general rule is: Charge at 1/10 ampere-hour rating for 14 hours.

Some chargers are designed to fast charge the battery in as little as one hour, with most being three to four hours. Fast charging should not be done unless the battery maker recommends it. Even then, be a little cautious about fast charging (cells can explode from too fast charging). NiCd batteries can be dangerous, so do not ad lib: Follow the maker's recommendations rigorously.

NiCd batteries can also lose energy from merely sitting. Some users find that a battery charged up, and then stored, is unusable at the time when it is eventually turned on. Figure 25-2 shows a storage discharge curve for a typical NiCd battery. The cure for this form of problem is a trickle charge at a rate between A-H/30 and A-H/50. Some commercial battery chargers have a switch that allows either an A-H/10 regular charge rate or an A-H/30 trickle charge.

Another problem with NiCd batteries is the operating temperature and its effects on available capacity. As shown in Figure 25-3, the available current capacity is a function of temperature. As the temperature increases above room temperature (25°C), the available capacity diminishes.

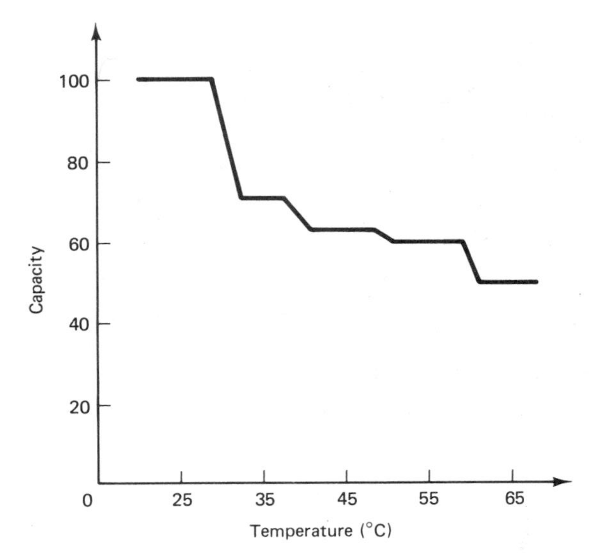

Figure 25-3 Operating capacity vs. temperature curve.

Figure 25-2 Percent full-rated charge vs. storage time curve.

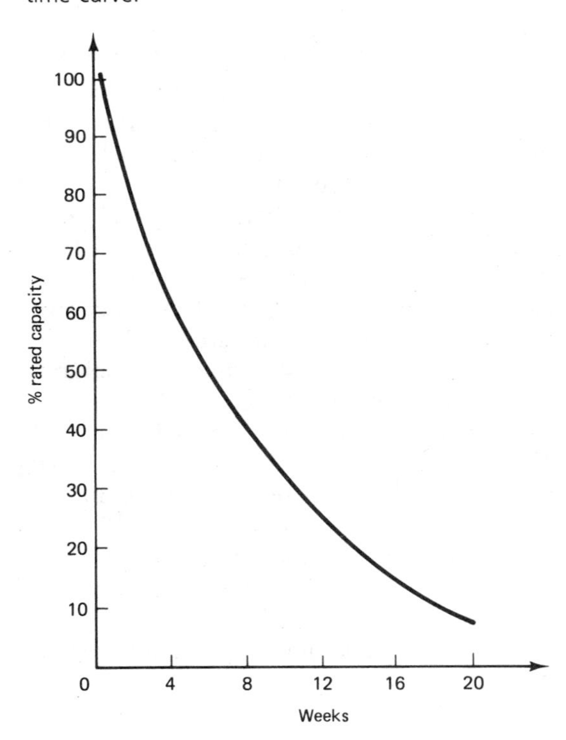

25-8 NiCd Battery "Memory"

A running debate in the industry states that NiCds do, or do not, have a memory problem. *Memory* means that a battery will not allow deep discharge after repeated shallow discharges. For example, if a battery is repeatedly discharged in some particular application to only 80 percent of full capacity, after a while it will "remember" the 80 percent level as the fully discharged point. The battery will then exhibit the fully discharged potential when the charge level is only 80 percent of fully charged. That makes the battery look like a premature failure. A NiCd battery suffering memory problems can sometimes be reformed by repeatedly fully charging it and then immediately deep discharging it. After a while, the memory phenomenon may work itself out.

25-9 Battery Maintenance

When equipment is subject to routine maintenance, it is possible to keep the batteries healthy by following a certain routine. For most equipment the manufacturer recommends that the batteries be periodically discharged and then recharged. The protocol for most is as follows:

1. Fully charge the battery or cell.
2. Discharge it fully with a resistor that draws a current of A-H/10 for about 8 to 9 hours for multicell batteries, and 10 hours for single cells.
3. Recharge the battery at the A-H/10 rate for 14 to 16 hours.

A phenomenon called *polarity reversal* might result if the battery is fully discharged. The cause of this problem is that not all cells have the same terminal voltage at any given time. It might occur that one cell will become charged backwards by the others in the series chain. For this reason, multicell batteries are only discharged to 10 to 20 percent of capacity.

Do not leave the battery in a discharged condition for a lengthy period of time. The battery may then develop interelement shorts. Little metallic or oxide "whiskers" called *dendrites* grow internally from plate to plate and cause a short circuit. The cell potential drops to zero or near-zero, and the cell will refuse to accept a charge. In some cases, we would have to regard the cell as lost and replace it. There are, however, some cells that can be salvaged from short circuits. In medical devices, however, good engineering practice requires replacement with a new cell rather than salvaging the defective cell.

Figure 25-4 shows a revitalization circuit for shorted NiCd cells. It works by vaporizing the internal dendrites that short the plates together. A known-good cell of the same type is placed across the shorted cell through a pushbutton or spring-loaded toggle switch. It is important to use this type of switch instead of a regular switch—you do not want to keep the circuit closed for too long (battery explosion could result). Press the switch several times in succession, and then measure the terminal voltage. If the current from V_1 successfully vaporizes the dendrites inside of V_2, then the terminal voltage will rise.

A word of caution is in order for people using this method: Be sure to wear safety goggles or glasses when performing this operation. NiCd batteries have been known to explode under high current, and explosion could conceivably happen when deshorting a cell. Most equipment maintenance technicians have never seen it happen under this circumstance, but those who are smart would not bet their eyesight on it never happening!

Revitalized batteries and cells should not be regarded as reliable and should only be used for short terms, under emergency con-

Figure 25-4 Flash revitalization circuit.

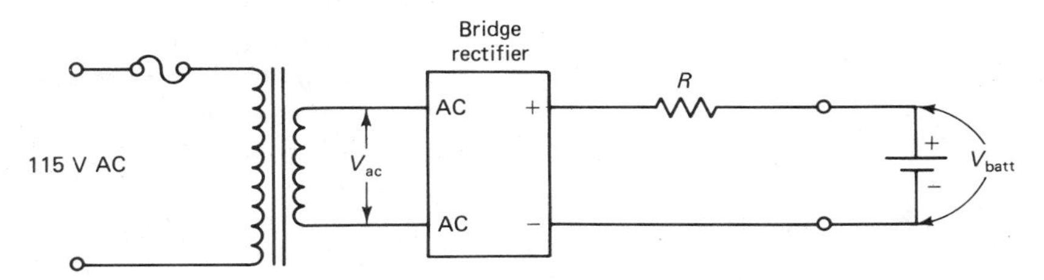

ditions, pending obtaining new batteries or cells for proper replacement.

25-10 Charging NiCd Batteries

There are two basic forms of charger for NiCd batteries: *constant current* (CI) and *constant voltage* (CV). Regardless of the type, it is important not to use a charging current greater than A-H/10 unless specifically instructed to do so by the battery manufacturer (not the equipment maker, by the way, for errors are frequent). The A-H/10 rate is one-tenth of the ampere-hour rating. For a 500-mA-H AA cell, for example, a charging current of 50 mA is used. Similarly, for a 2 A-H size C cell use 200 mA, and for the 4 A-H size D cell use 400 mA. Be cautious not to overcharge batteries using other A-H ratings.

Figure 25-5 shows the basic circuit for a constant current charger of simple design. The transformer secondary voltage should be 2.5 times (or more) the cell or battery voltage. A resistor in series with the rectifier has a value that limits the output current under short-circuit conditions to the official A-H/10 charging rate. This circuit is the basic circuit for most low-cost chargers.

Figure 25-6 shows two electronic constant current chargers based on three-terminal integrated circuit (IC) voltage regulators. A variable circuit is shown in Figure 25-6*a*, and it is based on the LM-317 (up to 1 ampere) or LM-338 (up to 5 amperes). Both circuits require a filtered dc input voltage several volts higher than the battery or cell potential. The actual value is not critical as long as it is sufficiently high enough to turn on the circuit (in general, V_{in} must be equal to or greater than V_{batt} + 3 volts). We can set the charge current by setting the value of resistor R. For example, for a 400-mA charger for 4 A-H size D cells, we would use a resistor value of $1.2/I = 1.2/0.4 = 3$ ohms. Charging currents down to 10 mA can be accommodated by the circuit of Figure 25-6*a*, so both regular and trickle chargers can be designed.

The circuit of Figure 25-6*b* will charge batteries up to 4 A-H with terminal voltages up to 12 V dc. It is similar to the circuit of Figure 25-6*a* but is based on the 5-volt fixed regulator such as the LM-309, LM-340-05, or 7805 devices.

A constant-voltage charger is shown in Figure 25-7. The output voltage of the charger is set by the ratio of R_1 and R_2 and is determined by the equation:

$$V_o = (1.25 \text{ volts})\left(\frac{R_2}{R_1} + 1\right) \quad (25\text{-}1)$$

A series resistor, R_3, prevents the current from exceeding the A-H/10 value and is set to allow the short-circuit current to that value. The required charger output impedance must be the resistance V_o/I_{max}, where V_o is the open-terminal battery voltage and I_{max} is the maximum permissable charging current. For a 12-volt, 4 A-H battery, for example, the required impedance is $12/(4/10) = 12/0.4 = 30$ ohms. We can solve the equation in the figure for R_3 and place that resistor value in series with the output of the regulator. The power rating of the resistor must be $V_o I_{max}$.

Figure 25-5 Basic circuit for a constant current charger.

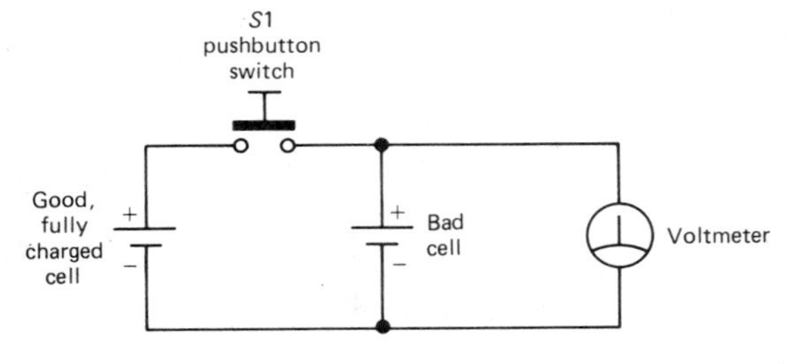

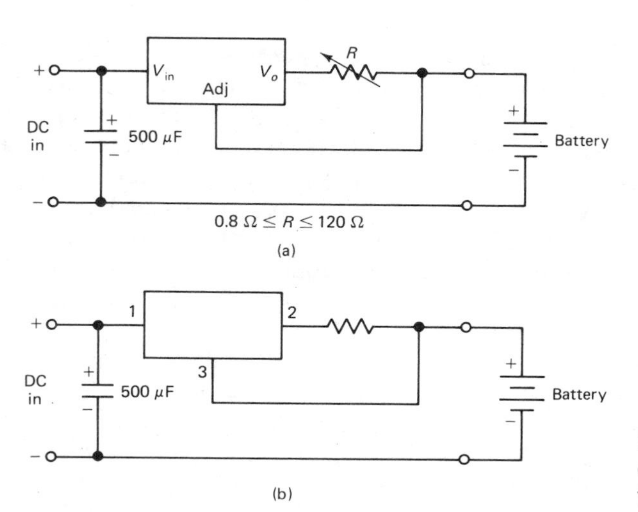

Figure 25-6 Electronic constant current chargers: (*a*) Variable, (*b*) fixed.

25-10.1 Using Bench Power Supplies

A bench power supply should not be used to charge NiCd batteries unless it has both a variable output voltage control and a current-limiting control. Set the output voltage very precisely to the full terminal voltage of the NiCd battery, and adjust the current limiter for a short-circuit current equal to the A-H/10 value. Disconnect the output short from the power supply, and connect it across the battery.

25-11 Multiple-Cell Batteries

A large number of multiple-cell batteries are used in electronic equipment. Most are typically 6-volt, 12-volt, or 24-volt models. In most cases, these batteries are made up of individual AA, C, D, or F cells. It is possible to take apart the original battery packs and replace individual cells to restore the battery to normal operation. Some battery packs are put together with screws or snaps, while others are glued together.

When selecting cells for replacement in a multiple-cell battery, keep several factors in mind. First, of course, is the right size (AA, C, D, F) and the right A-H rating (not all C and D cells are created equal). Also keep in mind whether regular cells or solder-tab cells are needed. Some consumer NiCd cells are in nonstandard packages. One brand of AA

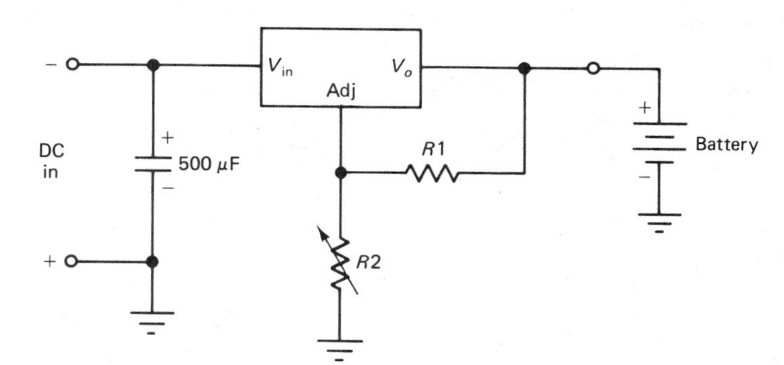

Figure 25-7 Constant voltage charger circuit.

cells is a millimeter or so shorter than standard AA cells. As a result, intermittent operation is sometimes experienced when these replacement cells are used. To avoid the necessity of shimming these cells or retensioning the contact spring, avoid buying them, and use the standards instead. Medical equipment is simply too critical for those batteries.

25-12 Other Batteries

There are several other types of batteries used in medical equipment: *lead-acid, carbon-zinc* and *alkaline dry cells, gel cell, mercury* and *lithium.*

Lead-Acid Batteries

For mobile (e.g., ambulance) and some high-power portable applications, the lead-acid automobile battery is often preferred. These are the familiar batteries used to start automobile engines. Very heavy, and dangerous because of the wet-cell acid content, they are nonetheless popular because they are generally well behaved and easily available. In addition, many radio communications sets are designed to operate from the nominal 13.6 volts dc produced by the typical automobile battery.

In addition to the 13.6-volt (a.k.a., "12-volt") battery, there are also available 6-volt, 24-volt, 28-volt, and 32-volt lead-acid batteries on the market. Some of these are marine (boat) batteries, others are military batteries (28 V dc), and still others are truck batteries. The terminal voltage can be increased by connecting batteries in series, while current availability is increased by connecting batteries in parallel.

Mobile operation is usually carried out with the vehicle battery. But in certain cases, it is wise to have a separate battery for the equipment in order to have battery power in the event of main vehicle power supply failure. Some users might want to consider the type of system shown in Figure 25-8. This system is common in recreational vehicles to power the creature comforts separately from the vehicle battery. The point is to keep your capability, even if you accidentally discharge the vehicle battery by leaving the lights on or suffer some other problem. The back-up battery could then be used to start the vehicle or summon help (depending upon the situation).

The charger will be a generator or alternator installed on the vehicle. Although the ideal system would be to have separate charging and regulating systems, that ideal is not always achievable for certain practical reasons. Thus, we have a single charger and voltage regulator for two (or more) batteries. Isolation between the batteries is provided by a pair of large-current silicon diodes (D_1 and D_2). The rating of these diodes should be at least 1.5 times the maximum charge rate of the charger. In most cases, large stud-mounted diodes are used for D_1 and D_2, and they are mounted on a finned heat sink. Keep in mind on vehicular installations that ambient temperature is high in some locations, and that will affect diode reliability.

For portable operations, some means must be provided to charge the lead-acid battery. In most cases, a small gasoline- or ker-

Figure 25-8 Dual battery charging system.

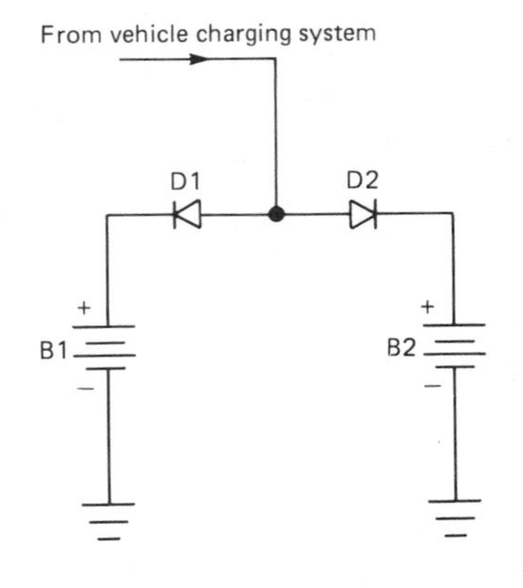

osene-engine-powered generator (called a "light plant" in some catalogs) is used to provide battery power. In some cases, an "auto parts store" type of battery charger is needed to convert the ac output from the generator to dc for the battery. It is increasingly common, however, to find small 500- to 2000-watt generators that include a "12-volt" output that provides from 6 to 35 amperes for purposes of powering radio equipment or charging batteries.

Maintenance of lead-acid batteries is relatively easy but is needed. The water level in each of the cells must be checked periodically. For people with a critical need for the battery, the level should be checked weekly. Although distilled water works best (because of the lack of additional chemicals), ordinary tap water will work in the cells. There are vents in the caps that cover the cells, and these holes must not be blocked. If dirt jams up the opening, then either replace the cap or clean it.

Warning: Lead-acid batteries produce hydrogen as a normal byproduct. If you fail to observe proper procedure, then this hydrogen may blow up the battery and cause serious injury to people and damage to equipment. First, never allow the battery to become overcharged. Second, turn off all circuits connected to the battery (*especially* the charger) before disconnecting the wires to the battery. If current is flowing in those circuits, then a spark will occur, and that spark can create an explosion. This is not a hypothetical possibility, but a real danger.

Carbon-Zinc and Alkaline Dry Cells

These cells and batteries are the ordinary consumer types that you are familiar with in flashlights, radios, and so forth. They are not generally rechargeable and are discarded after use. They are sometimes used in medical devices but for the most part are reserved for flashlights and noncritical devices. One of the problems that limits the usefulness of these batteries and cells is that the terminal voltage drops off over the course of discharging the battery, so device performance may vary as the battery or cell ages.

Mercury (Hg) Dry Cells

Mercury cells are sometimes used in medical devices and laboratory instrumentation because they possess a useful attribute: The terminal voltage remains nearly constant over the life of the battery and drops suddenly at the end of the charge life. This feature means that performance remains relatively constant. The feature also allows Hg batteries (e.g., HG-1 and HG-2) to be used for instrumentation calibration purposes.

As with NiCd cells and batteries, carbon-zinc, alkaline, and mercury cells and batteries should be stored in cool temperatures when not in use. It is common practice to store batteries in a refrigerator, not a freezer, in order to keep them cool. This practice extends the storage life of the cell/battery.

Gel-Cell Batteries

Another form of battery that is popular in portable equipment is the gel-cell. One of the authors has seen these batteries in commercial, medical, and radio communications equipment. Several years ago he worked with a piece of medical equipment used to transport cadaver kidneys to sites where a transplant was performed on a dialysis patient. If you have ever known end-stage renal disease (dialysis) patients, then you know the tragedy of the loss of a donor kidney. The team kept losing kidneys because of battery failure in the transport unit. The manufacturer sent us a new design internal battery charger, but it too was deficient. All of the battery chargers found were high-tech models that depended upon sensing small variations in terminal voltage to determine charge or discharge state. Unfortunately, the analog sensing circuitry drifted enough to give bad results. In desperation, the engineer in our laboratory called the battery maker, instead of the device maker, and asked him. The applications engineer asked if we had ever heard of Kirchoff's Voltage Law. Allowing that we had heard that one before, we let the applications engineering guide us to a solution (see Figure 25-9).

The circuit in Figure 25-9 will allow charging of a gel-cell (and other forms of battery) without resort to a lot of unreliable high-tech circuitry. The charger power supply must have two features: a precisely controlled output voltage, and a current-limit control. With switch S_1 open, set the output voltage to exactly the value of the full-charged voltage of the battery, or perhaps a small amount higher (100 to 200 mV). Make sure S_2 is in the shorting positon, and then close S_1. Adjust the current-limit control for a short-circuit current equal to the maximum permitted charge current of the battery (A-H/10 for many batteries). After the current and voltage are set, place switch S_2 in the "BATT" position and charge the battery. When the battery voltage is less than the power supply output voltage, current flows into the battery. But when the battery voltage equals the power supply voltage, current flow ceases.

Figure 25-9 Electronic power supply for charging gel-cell batteries.

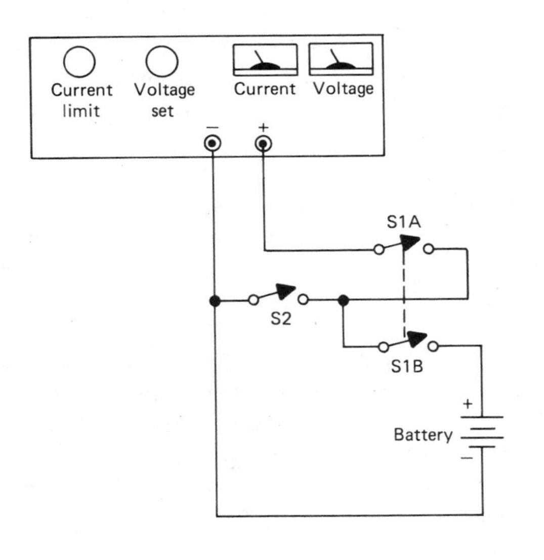

Batteries can provide freedom of operation for medical electronic equipment. But they can also be a nuisance if not maintained correctly. Proper maintenance of the battery will provide long and reliable life.

Lithium Cells

These cells are typically used in computers, watches, and other digital devices as a "keep alive" potential for CMOS memory elements. They typically last quite a long time because only leakage level currents are drawn.

Warning: All sealed batteries and cells are potentially dangerous when overcharged, overheated, or otherwise abused. Under such conditions, the battery or cell may rupture or explode, resulting in considerable hazard. In addition, lithium batteries should not be carried aboard aircraft or otherwise subjected to reduced atmospheric pressures unless specifically designed for the purpose (consult battery/cell manufacturer).

25-13 Summary

1. The types of cells and batteries typically used in medical equipment include nickel-cadmium, carbon-zinc dry cells, alkaline dry cells, mercury and lithium cells, and gel-cels. Lead-acid batteries are used for some mobile or high-powered portable applications.
2. The most common rechargeable cell/battery is the nickel-cadmium (NiCd) type. They are used because they are easy to obtain yet can provide large currents.
3. NiCd cells/batteries are typically charged at a current between A-H/20 and A-H/10 and are kept charged when not in use by a trickle charge of A-H/25 to A-H/30.

25-14 Recapitulation

Now return to the objectives and self-evaluation questions at the beginning of the chapter and see how well you can answer them. If you cannot answer certain questions, place a check mark next to each and reread appropriate parts of the text. Next, try to answer the following questions using the same procedure. When you have answered all of the questions, solve the problems in Section 25-16.

25-15 Questions

25-1 Two reasons to use battery power include portability and the need to provide ______________ from the patient in some instruments (e.g., cardiac output computers).

25-2 List four types of cells used in medical devices.

25-3 The ______________ battery has a typical full charge terminal voltage of 1.2 volts.

25-4 A battery is made of two or more ______________.

25-5 An NiCd battery can typically be cycled ______________ or more times when properly maintained and used.

25-6 A size D cell for industrial and medical use generally is rated at or near ______________ ampere-hours.

25-7 The NiCd battery is typically charged at A-H/______________ but can be charged at A-H/10 if properly controlled.

25-8 When an NiCd cell is trickle charged, the rate will be between A-H/______________ and A-H/______________.

25-9 Proper maintenance of ______________ batteries sometimes calls for the cell to be discharged through a resistor, drawing a current of A-H/10 for 10 hours prior to fully recharging at the same rate for 14 hours.

25-10 Discharged NiCd cells can develop internal ______________ of metal or oxide, and these can cause short circuits between plates.

25-11 The purpose of resistor R_3 in Figure 25-7 is ______________ limiting.

25-12 List two features of a bench-type dc power supply that are absolutely necessary if it is to be used for charging batteries.

25-13 Dry cells and NiCd cells should be stored in a ______________ environment to increase their storage or "shelf" life.

25-16 Problems

25-1 A constant voltage charger is made using the circuit of Figure 25-7 and requires an output voltage of 14.4 volts. If $R_1 = 180\ \Omega$, find the value of R_2.

25-2 If, in Figure 25-7, $R_1 = 240\ \Omega$ and $R_2 = 2200\ \Omega$, the output voltage, V_o, will be ______________ volts.

25-3 A 4 A-H NiCd battery is fully charged when it is connected to a load that draws 110 mA. It has a potential life of ______________ hours before becoming fully discharged.

Chapter 26

Medical Equipment Maintenance: Management, Facilities, and Equipment

26-1 Objectives

1. Be able to define the different types of maintenance repair organization.
2. Be able to define the different skill levels of medical equipment technical personnel.
3. Be able to understand the levels of capability of repair organizations.
4. Be able to describe the management approaches that are appropriate to medical equipment maintenance.

26-2 Self-Evaluation Questions

These questions test your prior knowledge of the material in this chapter. Look for the answers as you read the text. After you have finished studying the chapter, try answering these questions and those at the end of the chapter (see Section 26-11).

1. What are the three levels of maintenance organization?
2. What is the difference between a biomedical equipment technician and a registered professional engineer?
3. What is the difference between a registered professional engineer and a licensed first-class engineer?
4. What is the difference between O-level and D-level tasks?

26-3 Introduction

Several management options exist for maintaining medical equipment. An organization such as a hospital, local government emergency medical service (EMS), or medical practice can look to several forms of maintenance repair organizations (MROs) to take care of its equipment service problems. The general ground rules in this chapter assume either a commercial MRO, an in-house hospital-based MRO, or a shared service MRO. Each of these organizations has several clients (or departments served, in the case of an in-house shop) served by more than two technicians. With either too few clients or such a light work load that only one or two technicians are needed, a somewhat more ad hoc work plan is needed.

Similarly, with regard to city EMS and medical practices, options are somewhat more limited because the total work load is relatively light. For them, service *à la carte*, or a service contract with each manufacturer, may prove both sufficient and cost effective. Alternatively, a local commercial MRO may well serve your needs. While an argument can be made that hospital in-house service can be cheaper, it is rare that a small medical organization needs such staff. However, it is not unprecedented for medical practices to use either the same outside shared services MRO as the local hospital where they have staff privileges, or even the hospital's own in-house department.

26-4 Types of MROs

Although the makeup of MROs varies considerably, there are several important categories: manufacturers' service shops, commercial MROs, in-house hospital-based MROs, shared services MROs, in-house contractors, part-time shops, and single-employee shops. In addition, there are various levels of capability. Borrowing shamelessly from the military logistics world, we will call these levels organizational (O level), intermediate (I level), and depot (D level). The type of service program employed by any given organization should be tailored from these alternatives and may well be a blend of several. It is important for managers and administrators to understand these distinctions in order to make solid decisions on the types of organizations employed.

26-5 Levels of Capability

The concept of levels of capability is not intended to reflect on either the competence or integrity of the people involved, but rather on the design and mission of the repair organization. For example, an organization may find it worthwhile to employ a single highly skilled electronics technician in a partially management and partially technical role. Although the technician could easily perform higher-level tasks (and indeed does on some systems), logistics considerations (e.g., test equipment, spare parts) may preclude higher than O-level repairs despite the personal ability of the technicians(s). The main tasks of that person would be to manage service contracts and be the decision point regarding what is either beyond capability of maintenance (BCM) or beyond economic in-house repair. The person selected for this type of billet must be capable of making such decisions unemotionally. Many highly competent technical people are self-confident (and properly so), capable, and action oriented. If the logistic infrastructure is not in place, however, then such enthusiasm quickly results in an overextended employee.

O Level

This level of capability is the most local, least expensive, requires the smallest amount of training and support but is also the least capable. The O-level shop, however, is not to be disdained, for it offers a "first line of defense" that can:

1. Determine whether a fault actually exists, or whether an operator error was the cause of an anomaly. One of the most common findings in medical equipment shops is "no fault found" (NFF). An O-level worker can reduce the incidence of this finding and therefore the lost time incurred when equipment is sent for repair when none is needed.

2. Perform module or equipment substitution to bring the medical capability back on line rapidly, while the defective equipment is taken back to the shop for repair. The result is rapid return to service of the medical function involved.

3. Perform such minor technical tasks as can be trusted to on-the-job trained nontechnical personnel. For example, on a lot of equipment, nurses and paraprofessionals can perform many "repair" tasks such as stylus replacement, battery replacement, and so forth.

4. Serve as an inventory control point for management to monitor equipment that was referred to higher-level maintenance groups.

5. Keep records pertaining to equipment maintenance histories. Such records are useful in analyzing future procurement options and for defending malpractice actions based on supposedly improperly maintained equipment. In addition, the organization may find the records useful for prosecuting product liability lawsuits or regulatory actions against the manufacture of defective equipment.

The O-level maintenance activity is almost always in-house. Higher capability levels, however, may be either in-house or out-of-house.

I Level

This level of MRO is more highly skilled than the O-level MRO. Electronics and mechanical technicians, biomedical equipment technicians (BMETs), or even graduate professional engineers may be employed, depending upon the scope of the MRO's mission. The I-level MRO can handle any task that can be done at O level, and more. An I-level shop is one that can:

1. Test medical equipment to verify performance and adherence to specifications and safety standards.

2. Diagnose and troubleshoot equipment at least to the subassembly level. These subassemblies are sometimes called shop replaceable assemblies (SRAs) and are those that can be stocked as spare parts. For example, an I-level shop should be competent (and equipped) to replace printed circuit boards, front or rear panel components or controls, dc power supplies, and bolt-on mechanical parts (motors, pumps, etc.).

3. Adjust, align, or "harmonize" internal controls that are not normally available to the operator. This type of operation assumes that the I-level shop has the test equipment required to verify performance and the validity of those settings.

The I-level shop requires sufficient spare parts inventory, or a blanket purchase authority with the vendor, to assure relatively quick repair action turnaround time. It is often the case that the hospital cannot afford to have the technician go through the usual purchasing department in order to obtain parts. At least some limited trust must be placed in the person outside the system, or the hospital must be willing to accept less than rapid repairs. A competent I-level technician who lacks a replacement internal assembly is not able to perform up to the level of expected capabilities. Most locally situated MROs are basically I-level shops (although a few are depot capable).

D Level

The depot is the highest level of repair activity. For many systems, only a major local or regional MRO will qualify. In these cases, the manufacturer's plant may be the only available depot. The D-level shop is capable of troubleshooting to the piece-part level. While the I-level shop works to a subassembly level (e.g., printed circuit boards), the D-level shop can find and replace the faulty component on the subassembly (e.g., the blown transistor of "microchip" that caused the problem).

It is frequently the case that D-level shops work not on equipment as a whole, but rather on the subassemblies that are sent in from I-level shops. In a typical scenario, an I-level technician will replace a printed circuit board (PCB) to bring a piece of equipment back in service. The faulty PCB is then sent to a D-level shop for detailed troubleshooting to the piece-part level.

26-5.1 Which Level to Employ?

A large organization that must grapple with managing a large medical technology base faces the decision regarding level of shop capability required. Although there is a strong temptation to seek easy solutions, that approach is not always the best for any given organization. For example, a commercial MRO may propose taking over all equipment maintenance within the hospital. The facts rarely support such claims.

Similarly, in-house staff, perhaps feeling threatened by the prospect of being replaced by a contractor, may also claim capability that is not supportable by facts. The technical abilities of the employee are rarely an issue, but rather the issue is whether it is economically feasible to support such a facility.

A proper management approach is to inventory the equipment that requires support and then study the staffing and support requirements for each level of support. The decision of whether in-house support is D level, I level, O level, or none at all rests on costs versus the speed with which repairs must be performed.

26-6 Types of Organization

Earlier in this chapter several different broad categories of MRO were identified. Although that list was not intended to be exhaustive, it does serve as a guide to generic types of MROs. Let's discuss each class in turn.

Manufacturer's Service Department

This class of MRO includes those owned and operated by the manufacturer of the equipment. It may be located at the factory (usually D level) or locally (usually I level). As a general rule, the manufacturer can provide among the best and most competent service on its particular equipment. They have better information, deeper experience, and more rapid access to spare parts. However, because of certain problems, the manufacturer MRO may not be the best suited for your situation.

First, there is a potential ''Better is the enemy of good enough'' situation. The question is whether the ''better'' (if it truly exists) is worth the higher cost. Manufacturer's service is usually more expensive than either in-house or commercial MRO service because the manufacturer's overhead burden is typically higher.

Second, the actual proximity of ''local'' service may make a sham out of any claims of rapid service. If the shop serves a region rather than a small locale, then response time may suffer.

Third, the promise of ''local service'' is sometimes met by placing a single technician in either a ''desk space only'' office or the basement of his or her own home. Whether this arrangement works out depends on the size of the service area covered and the number of clients served. One company that used this form of service also used the same people to install equipment when new installations were done. As a result, the single technician in the area may be tied up on an installation and therefore unable to assist you (with your ''down'' coronary care unit) for several days.

Finally, the manufacturer may not have a total commitment to service. To most manufacturers, the money is in sales of new systems and support products (electrodes, gel, supplies); service is a loss center that is viewed as a necessary evil by the company management. In those cases, the service department receives little support or resources in an effort to minimize losses. Make sure the manufacturer views service in a kindly light before committing to its service department. An indicator of the future is your experience during the warranty period, when you have no choice. If the manufacturer seems unwilling to meet warranty obligations, then do not expect it to make a sudden recovery postwarranty.

Commercial MROs

These shops are commercial firms that provide service on medical equipment. Some commercial MROs provide the best solution for cost-effective management of the maintenance problem. As with any type of contractor, the desirability of entrusting work to commercial MROs depends on their integrity, reputation, technical abilities, the depth of their spares inventory, and their ability to respond to customer demands. Each MRO must be evaluated on its own merits. There are, however, a few guidelines to look out for.

First, it is often the case that the marketing claims belie the actual capability of the company. Contract performance is at risk when the technical staff is not actually in place at the time of proposal. Too often, commercial managers shrug off their lack of staff with the promise that staff will be hired as needed. While that flexibility may exist for low-skill-level billets, it is often a pipe dream at the higher-level billets needed for many types of medical equipment service.

Second, the financial condition of the company may not be adequate to support the level of service expected. Even world-class technical staff is unable to perform well if adequate spare parts, service literature, and test equipment are not available. Both the level of technology and the prospect of product liability tend to drive up the cost of medical equipment and the supporting spares. Some electronics components companies are so fearful of product liability suits that they publish a disclaimer against use of their products in medical equipment. In most cases, the price charged for simple components destined for medical equipment, even at the original equipment manufacturer level, is huge compared with the price of the same component for general industrial products. The same problem afflicts the local service company and increases both its own liability and the costs of the spares it sells.

Third, there is frequently an imprudent "can-do" attitude on the part of commercial MROs. Although action-oriented managers often admire such an attitude, it often conceals a lack of real ability to perform. The "can-do" attitude that is not backed up with real or potential capability is a dangerous illusion.

Fourth, the smaller MRO may be insufficiently covered by liability insurance. The author has seen repairs done by inexperienced commercial MRO personnel that could have been called malpractice. In one case, a defibrillator failed during a routine test. The fault was traced to inexcusably poor workmanship practices by the commercial MRO and may have resulted in a patient death if the hospital inspector had not found it first. If that MRO lacked coverage, or the resources to self-insure, then the hospital might have had a severe liability problem.

Hospital and EMS managers are rarely equipped to evaluate the technical capability or the depth of spares inventory of any commercial MRO. However, their armamentarium is not empty if the company has a track record. Demand a list of present and past customers who can be contacted for references. Unless other peer group managers are hopeless liars, a manager should be able to elicit good information about the company's past service record. A string of lost customers, service contract nonrenewals, or dissatisfied clients is not a good indication that the company can do a good job.

In-House MROs

This category of MRO is owned and operated by the organization or hospital that it serves. A properly staffed and equipped department can provide competent service in a very short response time. Experience has shown that equipment emergencies that would otherwise affect patient care can be dealt with promptly by an in-house MRO. One of the authors has repaired equipment in an OR during open-heart surgery procedures. Although repair during a case is bad practice under most situations, there were several cases when the surgeon ordered it done because he believed it critical to have the instruments to bring the patient back off the heart-lung pump, and because the hospital lacked the spare equipment to back up the main system. Neither commercial MROs nor manufacturer service departments can provide that level of service.

The in-house shop can also supply around-the-clock service through the simple expedient of issuing a beeper to the on-call technician. Although other forms of MRO can also offer that service, the reality is often a bit different from the salesperson's claims. For one thing, odd-hours service may be at a higher cost than normal-hours service. Also, the definition of "24-hour service" may be disputed. In one contract, we understood "24-hour service" to mean "around the clock at any hour of the day or night." However, the company responded to our complaints of noncompliance with the suggestion that "24-hour service" meant that they had to provide the service sometime within the next 24 hours after logging in the call from the customer. Ambiguity in a contract can be devastating.

Although cost and convenience advantages are considerable, there are also some disadvantages to in-house service. A big one is the availability of properly trained I-level and D-level technical staff. It is difficult to balance the kind of salaries and benefits required to attract and hold good people with the need to contain costs. In addition, inhouse shops often lack the room for career growth for the technician. As a result, the better person leaves within a short time, often disgruntled. Alternatively, the person settles down into a long term of gray mediocrity that leads to a rut requiring occasional management "attitude adjustment" actions.

Shared Service MROs

Equipment maintenance is an expensive proposition, especially if full I level is required on all systems (plus D level on selected systems). There are many institutions that simply cannot afford their own repair facility beyond the most basic. Even large institutions with adequate resources may find it economically desirable to share the cost of such facilities with other institutions.

A shared service is an MRO that is owned and operated by two or more cooperating institutions that are also the users of the service. In some cases, the shared service is owned by one of the institutions but serves the others on contract. In still other cases, the shared service is a separate entity on its own but is owned and managed by the member institutions. The shared service is sometimes a separate corporation in which the chief officers and/or directors are also officers or directors of the member institutions.

Some shared service MROs are actually more like consortiums. Each institution has its own service facility, but each group is highly specialized in the type of work it handles. For example, one group might be heavy on electronic patient monitoring equipment, while another is expert on anesthesia machines and respirators, while still another is an expert on dialysis machines. The overall capability of the consortium is greater because of nonduplicated efforts.

Shared services can be an effective and low-cost solution to the overall maintenance problem of several institutions. However, as with the other options, there are some problems to solve. Perhaps one of the most difficult problems is the tendency of one institution to dominate the resources of the shared service. Usually, the majority (or most senior) member siphons off service capability that is rightfully due other members.

Another problem with shared services is the tendency to build empires. This problem especially afflicts consortiums because each hospital may want to develop local capability that would ordinarily be allocated to another shop. Service shop managers who see their career potential enhanced by accretion of duties, capability, and power may be most guilty of this practice.

If the problems can be solved and peace maintained, then shared service MROs can be a well-managed, low-cost solution to the equipment maintenance problem. However, the track record of shared services is spotty in this regard because of the aforementioned problems.

In-House Contractor MRO

This category represents a cross between the in-house concept and the commercial MRO concept. In this arrangement, a contractor will place either an entire technical and management team, or a manager who oversees hospital personnel, in the hospital to manage the in-house repair facility. Only a few of these groups exist, so little can be said about them. However, the concept has worked with housekeeping management and so should work well with medical equipment maintenance.

Part-Time Shop

In this case, a small repair organization does medical equipment maintenance for the local hospital on a part-time or ad-hoc basis. In general, it is a bad idea unless the owner of the shop (or key personnel) is familiar with the repair of biomedical equipment. In that case, it might be a good "first line of defense" for a rural or remote hospital. I recall one shop that normally repaired commercial video equipment and two-way radios and also repaired the equipment in the coronary and intensive care units of the local hospital. The nearest commercial MRO or manufacturer service shop was a six-hour drive through the Nevada desert. The manufacturer of the patient monitoring system sold the hospital a stock of spare printed wiring boards and mechanical parts and trained the local shop owner to provide a minimal I-level capability. It worked in that case, but this approach is otherwise fraught with difficulty.

Single-Technician Department

In some smaller hospitals, a single repair technician is employed in plant operations (erroneously called "engineering" in most hospitals, even when there are no college graduate engineers employed there). If adequately supported with test equipment and spares, the properly trained technician can provide O-level, or even I-level, service for a limited number of in-house client departments. However, be careful to ensure the employment of an adequate technician. Many hospitals have been known to employ an electrician who thinks he or she knows something about electronics, or some kid with an amateur radio license as the electronics technician. Insist on credentials for such an employee.

26-7 Technical Personnel

One of the mysteries of medical equipment maintenance for managers and administrators is the types of technical personnel involved and their level of training. Unfortunately, too many titles are less than descriptive to the uninitiated. In addition, there are titles that are shared by different people. For example, consider the title "engineer." If you asked most medical people to define an "engineer," you would elicit images of a person with little education, a blue-collar plant operations uniform, and a leather tool pouch on his or her belt; the use of the term *engineer* by those people is illegal in some states and only tolerated by the law in others because of a long history. The title derives from the use of the word *engineer* to describe highly skilled steam and boiler technicians, whose city licenses often refer to them as first-class, second-class, or third-class "steam engineers." We will discuss true engineers in this section. Various levels and categories of technicians are found in medical equipment maintenance. Some of them are highly specialized, while others are generalists. Several levels of education are common.

1. *On-the-Job or Self-Trained.* This type of person is trained as an apprentice to others and is generally capable of limited tasking. Although there are exceptions, these workers are not generally capable of more than the simplest tasks. However, certain factors are indicators of higher than elementary achievement: status as a certified electronic technician (CET), certification as a biomedical equipment technician (BMET), certain other industrywide certifications, and a diploma from a recognized home-study school. Although often the butt of jokes, the home-study route is an old, established tradition in electronics. There are a number of programs that are well regarded in the electronics industry because they turn out knowledgeable graduates.

2. *Vocational Technical School Graduate.* The vo-tech school is one in which subcollege, post-high-school training is offered. Some vo-tech schools offer the last year or two of high school plus additional work that is normally beyond the high school level. These schools may be local or state government operated, or commercial in nature. Unfortunately, a few commercial schools are little more than diploma mills that rip off guaranteed loan payments offered to immigrants or disadvantaged people to upgrade their economic prospects (check the reputation and accreditation of the school).

The emphasis in vo-tech training is on the practical aspects of electronics theory and differs from the higher levels in the amount and type of mathematics employed in the curriculum. These technicians are usable for O-level maintenance and many I-level tasks depending upon the quality of the school.

3. *Associate Degree Technicians.* The A.Sc. degree requires two years of academic training in a community college or technical institute. The level of training is more theoretical and mathematically oriented than the vo-tech. These graduates can handle O-level, I-level, and most D-level tasks. It is common to find bright, highly talented people in this category who prefer the technical "hands-on" type of work that requires an A.Sc. degree, but who disdain the more paper-oriented tasks of the engineer.

4. *Bachelor's Degree Technicians.* At one time the B.Sc. degree differentiated the technician from the engineer. Today, however, there are many four-year, accredited degree programs that offer a Bachelor of Science in Electronics Technology (BSET) degree. It is perhaps fitting to refer to these people as "technologists" or something similar rather than "technicians." They can handle all tasks and indeed can often manage the shop on a professional level.

5. *Biomedical Equipment Technicians (BMETs).* The title *BMET* is an indicator of certification in medical equipment maintenance. Some applicants may have earned a degree or diploma in medical equipment technology and thereon base a claim to the title. However, in order to avoid confusion, it is perhaps best to reserve the title *BMET* to those who have passed the appropriate certification examinations.

6. *Engineers.* The true engineer is a graduate of a college or university engineering program or a program in a related science that leads one to the same body of knowledge as engineers require. The general rule is simple: no Bachelor of Science degree, no title of "engineer," unless the state government has seen fit to issue the person a professional engineer certificate. (*Note:* Very few ''nondegree'' PE certificates still exist, although a few states still allow the old method of qualification.)

It annoys engineers to be compared with, or thought of in terms of, the plant operations ''engineers,'' because the true engineer holds a degree that is one of the hardest to earn. Indeed, a physician told the author that he switched from engineering to pre-med because he was not smart enough to survive the freshman year in engineering school. Engineers working in hospitals are often looked down on or disdained by nurses with B.S.N. degrees (or even other nurses) as somehow intellectually inferior. There is no nursing degree, even at the graduate school level, that compares in difficulty or intellectual level to even the B.Sc. level in engineering. Anyone who doubts that claim would do well to enroll in the first semester of engineering school to find out the truth (a typical engineering school has a 45 to 55 percent flunk-out or quit rate at the end of the freshman year; it is not uncommon for deans of engineering schools to address incoming freshmen, ''Look at the person next to you . . . in June one of you will not be here—half flunk out or wisely change their major by the end of the first year'').

The Bachelor of Science in Engineering degree is the usual qualification for an engineer. Some applicants will enumerate their degrees as a B.S.E.E. in electrical engineering, a B.S.M.E. in mechanical engineering, a B.S.C.E. in civil engineering, and a B.S.Ch.E. in chemical engineering. In addition to these categories, there are also materials engineers, engineering mechanics degrees (which is a cross-disciplinary degree), and biomedical engineering degrees. In the latter case there is some doubt regarding what curriculum was followed. It is not unusual to find that these graduates are really of the traditional types of engineers with a little biology and a few medical instrumentation courses. For example, one friend of mine has a B.Sc. in biomedical engineering but took only six credit hours' fewer chemical engineering courses than a B.S.Ch.E. graduate. There are other examples of specialty engineering B.Sc. degrees, but in the main specialization beyond the basic traditional engineering disciplines is done in graduate school.

In some cases, graduates with degrees in science (physics particularly) or mathematics are accepted as engineers after a certain amount of relevant training and professional experience:

1. *Certified Clinical Engineer (CCE).* The CCE is a professional board engineering certification. It requires a level of education (B.Sc.) plus four years of relevant experience in a hospital or similar facility. Some older clinical engineers were ''grandfathered'' into the program on the basis of experience, but newer certificates require full compliance with the educational requirements.

2. *Registered Professional Engineer (PE or RPE).* This title indicates that the applicant possesses a state license to practice engineering. It is unfortunate that not all engineers are required to have a state license. In general, only those who are in private practice, who work on public projects, or who must routinely appear as an expert witness in court are required to possess an engineering license. Although some states legally limit those who may use the title *engineer* to describe themselves, there are exceptions. For example, some states exempt engineers employed in manufacturing, as junior engineers to a PE, or (in some cases) those who are employed in institutions such as academia or hospitals. There are also occupational exemptions under which railroad train drivers, crane and heavy equipment operators, and steam mechanics can call themselves ''engineers.'' A ''sanitary engineer,'' by the way, is not a garbage collector with a sense of humor but rather is a speciality subset of civil engineering that designs sewage systems and water treatment plants.

The PE license is based on education, two levels of examination, plus four years of experience. The first level of examination, called the Engineer in Training (EIT) exam, is opened to seniors in engineering programs with 90 hours of acceptable course work completed and to all graduates of engineering programs. After four years of progressively more responsible professional experience, the engineer may take the specialty examination for the type of engineering he or she is qualified to practice. While all EITs take the same exam, the PE exam is tailored to civil, electrical, mechanical, or chemical disciplines. Although not strictly required, some people feel that major hospitals ought to employ a PE who also holds the CCE certificate as the administrator over both biomedical engineering and plant operations.

There are a few states that allow licensing of professional engineers without a degree, or with degrees other than engineering. These states call such qualification *eminence*. If a person demonstrates the body of knowledge required, plus a 12-year record of "progressively more responsible professional experience," then the board may vote to issue the license despite the lack of an engineering school degree. This method of qualification especially benefits foreign graduates whose schools are based on the European model, wherein one "reads engineering" rather than takes specific courses.

Small departments, or the departments in smaller hospitals, can be managed by a technician with a BSET degree or a related degree. Alternatively, with adequate training and experience lesser educated people will also work out well. Larger hospitals, however, should insist on filling the job of Director of Biomedical Engineering Department with a graduate engineer with a certain minimum (two years) relevant experience level. While these people are hard to find, they are also professional people who are able to relate to the medical professionals they serve.

26-8 Management Approaches

Over the years, one of the authors was associated with many maintenance or repair organizations in both technician and engineer-level positions. In addition to practical experience in the workplace, the author's insights have been organized by exposure to certain modern management theories that proved themselves in a wide variety of environments.

Hospitals, and often the associated commercial vendors supplying services, are among the least well-managed enterprises in the country. Indeed, one of the many reasons for the runaway costs of health care is that medical enterprises often suffer from defective management. It is the responsibility of management to find and correct causes of extra or hidden costs in order to make their enterprise more economically viable.

It is not intended here to cover all of the problems in health care management, for indeed the author is not qualified to do so, but rather the problems and issues related to small unit management are discussed. Both biomedical engineering and nursing personnel would be wise to consider some of the issues raised herein.

While one is quick to recognize the inefficiencies of medical enterprise management, one must be just as quick to point out that such inefficiencies are not necessarily the fault of present managers and supervisors. Rather, they are the fault of the system that these people inherited and learned under. Part of the problem these people face is that they are professionals in fields other than management. For example, head nurses and directors of nursing are drawn from the ranks of the registered nurses, just as directors of biomedical engineering are drawn from the ranks of the engineering profession. Their training, the poorly thought-out claims of the proponents of the B.S.N. degree notwithstanding, lacks relevance to management. Yet there is hope, for those people tend to be high-quality individuals who care about the health of the organization—and the skills situation is easily correctible with training.

The principal technique of small unit management in most industries, including health care, is also the most incompetent and least effective. In fact, it is counterproductive and often leads to exactly the opposite effect than intended. Sometimes called "Theory X" management, one might be tempted to call this method the "Drill Sergeant" approach because coercion and intimidation rule the unit. Wherever this method is employed, the supervisor (and usually his or her superiors as well) operate from a set of basic premises that require elements of the following worldview:

1. People are inherently no good; they are lazy and want a free ride. Any quality resulting from their efforts is merely incidental or the result of supervisory browbeatings. This view is exemplified by a cartoon of a tabby cat pulling a piano upstairs under a whip lash: "If you whip the cat hard enough, it will pull a piano upstairs."

2. There is no pride of workmanship. That is, the lazy or incompetent worker (which includes all workers) does not want to do a good job, cannot be convinced to do a good job, and will not do a good job unless coerced.

3. A proper employee is one who keeps quiet, stays in his or her place, never makes waves, never rocks the boat, and never offers any opinion or suggestion whatsoever. "Just do the job, don't comment on it" exemplifies this view.

4. All—or at least the overwhelming majority of—problems in the unit, especially production-related issues, are the fault of workers who will not do their best.

These views are fundamentally flawed. The concept that all employees are somehow lazy is often implicitly accepted by supervisors and managers, even though considerable evidence to the contrary exists. If a set of employees seems to fit the foregoing description, then it is a sure bet that some problem in management is responsible for that morale problem. Perhaps the apparent laziness or incompetence is actually due to severe demoralization brought on by irrational, inconsistent, or insensitive management.

Regarding the belief that the cat will pull the piano upstairs if whipped hard enough, experience suggests exactly the opposite—the "cat" will walk out. This problem is often obscured in the medical, nursing, and medical equipment industries because, in any one locality, it is often the case that other employment opportunities in the field are lacking. Perhaps that is why so many nurses claim to "burn out" so young—they are totally demoralized by the ineffective management of their supervisors.

A key factor to research when a unit has a high turnover rate is the kinds of jobs taken by those who leave. If their new jobs are predominantly career advances, then the problem is mere attrition and no action need be taken. But if the jobs tend to be laterals, dead-ends, or otherwise not career enhancing, this is evidence that the employees are leaving because of poor management.

The only way that any organization can remain economically viable over the long haul is to be in a state of constant improvement. Management must stimulate this process, but it requires the efforts of all employees. Unfortunately, there is a tendency in management (not just in health care) to assume that employees cannot make improvements. In fact, managers often believe that employee suggestions are boat-rocking. There is even record of a head nurse who demanded that the personnel department select interviewees for a position who are "introverted and not the least bit innovative." That prescription is a blueprint for disaster. It is a simple fact that the person who does the job is the one who knows the most about it and is therefore one of the principal people who should be in on solving problems.

If an organization is in such serious trouble that innovative employees are shunned or beaten into demoralization, then theirs is a boat that does not need freedom from boat rockers, but rather needs very much to be rocked—and hard.

One of the most destructive attitudes among managers is the mistaken belief that problems in a unit are due to employees not pulling hard enough. Often accompanied by exhortative slogans, this attitude demonstrates that management does not know its business or is unwilling to think hard about the problem. According to W. Edwards Deming, in *Out of The Crisis* (MIT-CAES, Cambridge, Mass.), the truth is exactly the opposite. In Deming's view, the overwhelming majority of problems are due to the system—which is management's responsibility—not the employee.

Problems should be solved with a view toward making the system better. One of the principal errors involves setting work standards or objectives. One thing that you will reap when setting standards is that you often get exactly what you measure—and no more. Often, what you measure was selected because it is easy, not because it is somehow relevant. Let's consider some examples.

1. In maintenance shops it is common to rate employees on the number of repair jobs completed per week, or the average amount of time spent to repair each unit. The measured parameter that affects the technician's pay next year (and self-esteem this year) is quantity, not quality. Thus, the technician is naturally directed to turning in completed "tickets" rather than ensuring that the hospital's equipment does what it is supposed to do.

2. In many hospitals, nurses are treated severely for medication errors (there is no doubt that medication errors are a serious issue). One hospital fires nurses who make three medication errors in 12 months. As a result, in response to a new nurse who wanted to formally report her medication error, a supervisor said, "On *this unit,* we don't have medication errors!" Which is more dangerous to patients—a system in which medication errors can be freely admitted, or one in which it is in the nurses' best interest to hide the truth and hope that no harm comes to the patient?

In both of the preceding examples, it is the superior management technique to statistically evaluate the work performance and determine if the system is stable. If the technician does not produce enough repair jobs, then perhaps training, equipment, or spare parts supplies are at fault. Similarly, if there are a lot of errors in medication, then perhaps there must be an evaluation of the system to determine how it can overcome the potential for error.

Overcontrol is the last problem which we will address at this time. It is the habit of many supervisors to closely monitor, and frequently tweak, employee performance. An example of this occurs in word processing or data entry departments where computers automatically record the number of keystrokes made per minute. Rather than having the desired effect (i.e., improvement of employee performance), these data actually breed resentment and poorer performance. Keystrokes up, errors up too.

Another insidious form of overcontrol is the habit of some supervisors to make rules or take general disciplinary action on the entire unit for the sins and problems of a few—or even one. Poor managers think that every special problem that arises needs a new regulation. Of course, the new regulation will seem to work for a while, but it is soon ignored. When confronted with special problems, the supervisor needs to evaluate the situation in order to discern whether the problem is systematic or is singular. It never seems to occur to some people that events often are not systematic (the only kind that succumb to general rules) but rather are either random happenings or single-point failures that may never happen again. In the absence of evidence of a wider cause, any generalized action taken will probably be counterproductive and is definitely little more than wheel spinning.

26-9 Summary

1. There are three levels of maintenance repair organizations (MROs): organizational (O), intermediate (I), and depot (D). Each of these levels represents an increasing capability.
2. There are several different types of organizations within the three levels: manufacturer's service department, commercial MRO, in-house MROs, shared service MRO, in-house contractor operated MRO, part-time shop, and single-technician department.
3. Several levels and categories of technical personnel are used in medical equipment maintenance: those trained on the job, vocational-technical school graduates, associate degreed technicians, bachelor's degree technicians, biomedical equipment technicians, graduate engineers, certified clinical engineers (CCEs), and registered professional engineers (RPEs). The BMET and CCE are certifications, while the RPE is a state license.

26-10 Recapitulation

Now return to the objectives and self-test questions at the beginning of the chapter and see how well you can answer them. If you cannot answer certain questions, place a check next to each and review appropriate parts of the text. Next, try to answer the following questions using the same procedure.

26-11 Questions

26-1 A(n) ______________-level maintenance organization troubleshoots to the component level.

26-2 A(n) ______________-level maintenance organization generally repaces subassemblies such as printed wiring boards, pumps, and so forth.

26-3 The most common finding in O-level shops is ______________ ______________.

26-4 A(n) ______________ ______________ maintenance repair organization is owned jointly by several users (e.g., hospitals, medical practices, etc.)

26-5 Three technical training levels include: ______________, ______________ degree, and ______________ degree.

26-6 A graduate engineer is a person who possesses at least a ______________ degree or its equivalent.

26-7 A ______________ ______________ technician is a certified technical worker with experience and training in the repair and maintenance of medical equipment.

26-8 A ______________ ______________ engineer is a professional person who is especially trained, educated, and experienced in managing medical equipment maintenance activities.

26-9 Overcontrol is a symptom of Theory ______________ management style.

26-12 References

1. Aguayo, Rafael, *Dr. Deming: The American Who Taught the Japanese About Quality.*

Simon & Schuster/Fireside (New York, 1990).

2. Amsden, Robert T., Howard E. Butler, and Davida M. Amsden, *SPC Simplified: Practical Steps to Quality,* Quality Resources (White Plains, N.Y., 1986, 1989).
3. Deming, W. Edwards, *Out of the Crisis,* Massachusetts Institute of Technology, Center for Advanced Engineering Study (Cambridge, Mass., 1982, 1986).
4. Dobyns, Lloyd and Clare Crawford-Mason, *Quality or Else: The Revolution in World Business*. Companion to an IBM-funded Public Broadcasting System three-part series of the same name. Houghton Mifflin Co. (Boston, 1991).
5. Gabor, Andrea, *The Man Who Discovered Quality,* Random House/Times (New York, 1990).
6. Walton, Mary, *The Deming Management Method,* Putnam Publishing Company/Perigree (New York, 1986). Paperback edition.
7. Walton, Mary, *Deming Management At Work* (six case studies), Putnam Publishing Company/Perigree (New York, 1990, 1991). Paperback edition.

Appendix A

Some Math Notes

This textbook was originally designed for use by students who may not have studied *calculus*. But as work progressed, it became apparent that we could not adequately explain matters such as *cardiac output* and *mean arterial pressure* without using the *notation* used in calculus mathematics. Keep in mind that you need not know how to *work* calculus problems (after all, the electronic instruments *do* the calculus), but you should understand what is meant when the *symbols* used in calculus are encountered in the text.

There are two fundamental operations in calculus: *integration* and *differentiation*. In electronic instruments, circuits are used to perform these functions.

In brief, integration is the mathematical process of finding the total *area* under a curve on a graph. Differentiation, on the other hand, is the process of finding the *instantaneous rate of change* of a curve.

The "curves" in electronic circuits would be the graph of *voltage-versus-time,* in most cases. We would say, then, a graph of the voltage as a function of time.

A-1 Differentiation

Differentiation is the process of finding the instantaneous rate of change of the curve. Rate of change is given by the *slope* of the curve. If the curve is a straight line, like the "curve" shown in Figure A-1*a*, then finding the rate of change (i.e., slope) is very easy. In that simple case, it is the *change* in E, divided by the *change* in T:

$$\text{Slope} = \frac{E_2 - E_1}{T_2 - T_1} \tag{A-1}$$

We use a special notation to indicate the quantities $(E_2 - E_1)$ and $(T_2 - T_1)$. These quantities are *changes,* so we can use the Greek letter "delta" (Δ) to replace the expressions.

$$\Delta E = E_2 - E_1 \tag{A-2}$$

$$\Delta T = T_2 - T_1 \tag{A-3}$$

It is standard mathematical practice to use Δ whenever we wish to denote a *small change* in some quantity. If we use the delta notation in Equation A-1, then, we would write:

$$\text{Slope} = \frac{\Delta E}{\Delta T} \tag{A-4}$$

The preceding equations work very well if the "curve" is a straight line. But what if the curve is not a straight line? Figure A-1*b* show such a situation. What do we do if it is necessary to know the instantaneous rate of change at some specific point on that curve? Although the answer to that question goes right to the heart of calculus mathe-

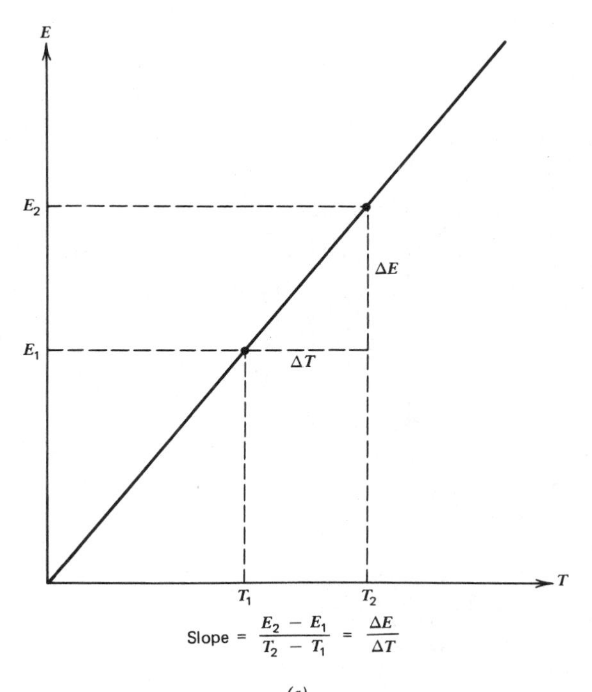

(a)

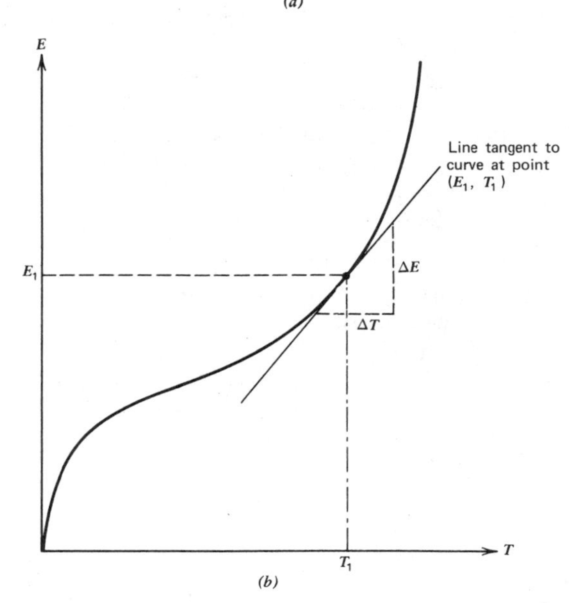

(b)

Figure A-1

matics, we can sum it up adequately for our purposes very simply. The instantaneous rate of change, also known as the *derivative*, can be found by taking the slope of a line *tangent to the curve at the point of interest*. Any calculus book will be able to prove to you that this is true.

When speaking of derivatives, we still use the concept of Equation A-1, but change the notation from "$\Delta E/\Delta R$" to

$$\text{Derivative} = \frac{dE}{dT} \tag{A-5}$$

When you see dE/dT, then, you should recognize that we are asking for the instantaneous rate of change of E with respect to T. This may sometimes be written in the form E, but this is merely shorthand for dE/dT:

$$E = \frac{dE}{dT} \tag{A-6}$$

An operational amplifier circuit that will differentiate signals is discussed in Section 4-8.3.

A-2 Integration

Integration is the process of finding the *area* under a curve. This is shown in Figure A-2, where we see two *voltage-versus-time* curves (E_1 and E_2). Assume that we want to measure the areas under these curves over the time interval T_1 to T_2. In the case of E_1, the problem is very easy because the region of interest forms a rectangle. The area is

$$A = (E_1 - 0)(T_2 - T_1) \tag{A-7}$$
$$A = (E_1)(T_2 - T_1) \tag{A-8}$$
$$\text{A} = E_1 \times \Delta T \tag{A-9}$$

But what do we do when confronted with a curve such as the signal labeled E_2 in Fig-

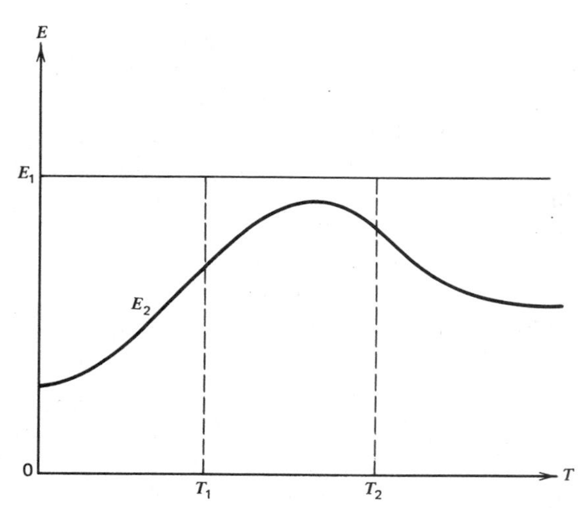

Figure A-2

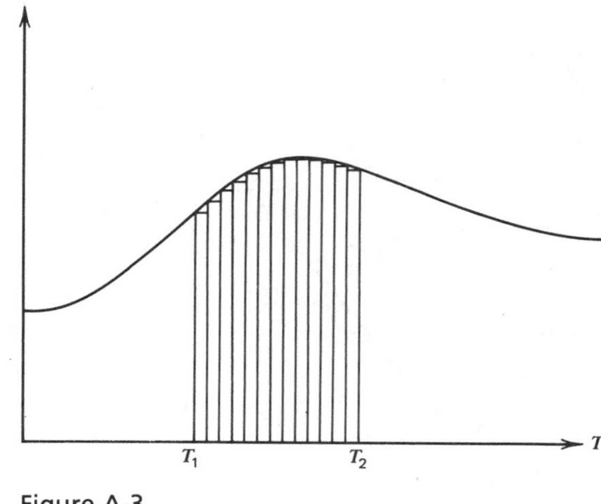

Figure A-3

ure A-2? This curve is not so well behaved, so we must revert to calculus.

The mathematical techniques of integration tell us that the solution is to break up the area of interest into a lot of rectangles (see Figure A-3). We can easily calculate the area of each rectangle as ($E_1 \times \Delta T$). By summing up the areas of all of the rectangles, we gain an approximation of the area under the curve.

If the rectangles are large, then the approximation is poor. But as the width of the rectangles becomes smaller, the approximation becomes better. Finally, when ΔT becomes very, very small (i.e., as $\Delta T \rightarrow 0$), the approximation is exact.

The symbol used for integration is shown in Equation A-10.

$$\text{Area} = \int_{T_1}^{T_2} E_2 dT \qquad \text{(A-10)}$$

The S-shaped symbol is the integral symbol. T_1 and T_2 are used to denote the boundaries, or limits, over which integration is desired. The dT tells us that we are integrating with respect to time (T). Equation A-10 is read: "integral of E_2 with respect to T, over the interval T_1 to T_2."

When dealing with electronic signals, integration is also called *time averaging;* hence the use of integrators to calculate quantities such as mean arterial blood pressure from the arterial waveform.

Electronic integration is performed by accumulating a charge in a capacitor. An example of an electronic integrator, using an operational amplifier, is discussed in Section 4-8.2.

Appendix B

Medical Terminology

Medical terminology sounds like a lot of jargon to the layperson but is a reasonable, concise method of communication. Medical terms are made of combinations of roots, suffixes, and prefixes with highly specific meaning. For example, *hepatitis,* meaning an inflammation of the liver, is made of a root *hepar* (meaning *liver*), and a suffix *itis* (inflammation). Many of the long, polysyllabic words used in medical environments become quite simple in their meaning when broken down into the roots, prefixes, and suffixes. The following lists give some of the more commonly encountered terms.

Roots	Meaning
aden	gland
arteria	artery
arthros	joint
auris	ear
brachion	arm
cardium	heart
cephalo	brain
cholecyst	gall bladder
colon	intestine
costa	rib
cranium	skull, head
derma	skin
enteron	intestine
epithelium	skin
esophagus	gullet
gaster	stomach
hema, hemo	blood
hepar	liver

Roots	Meaning
hydro	water
hystera	womb
kypsis	bladder
larynx	throat
myelos	marrow
nasus	nose
nephros	kidney
neuron	nerve cell
odons	tooth
odynia	pain
opsikas	eye
os	bone
osteon	bone
ostrium	mouth
otis	ear
pes	foot
pharynx	throat
phlebos	vein
pleura	chest
pneumones	lungs
psyche	Mind
pulmones	lungs
pyelos	kidney
pyretos	fever
ren	kidney
rhin	nose
rhythmos	rhythm
spondylos	vertebra
stoma	mouth
thorax	chest
trachea	windpipe
trophe	nutrition
vene	vein
vesica	bladder

Prefixes	Meaning
a-	absence, not
ab-	away from, off
ad-	toward
amphi-	on both sides, bilateral
an-	absence of
ante-	before, in front of
antero-	in front
anti-	against
ap-	separation
apo-	separation
bi-	two
bio-	pertaining to life
brady-	slow
cardio-	pertaining to the heart
cephalo-	head
chiro-	hand
chole-	bile
co-	together
con-	with
costo-	rib
cysto-	sac, bladder
dactylo-	digit (finger, toe)
derma-	skin
dermato-	skin
di-	twice
dia-	apart, through
dys-	difficult, painful
ec-	outside of
ecto-	outside
ex-	outside
en-	within
endo-	within
ento-	within
entero-	intestines
epi-	upon
ex-	away from
exo-	outside of
extro-	outside
eu-	well, good
gastro-	stomach
hema-, hemo-	blood
hemato-	blood
hetero-	different
homo-	the same, or the same sort
hydro-	water
hypno-	sleep
hypo-	beneath, deficient, lower than
hystero-	uterus
ileo-	ileum
in-, il-, ir-	within, not, inside
infra-	beneath
inter-	between
intra-	within
iso-	equal, same
kilo-	1000
leuko-	white, clear
litho-	stone
macro-	abnormally large, large
mal-	bad
media-	middle
mega-	great size; 1,000,000
melano-	black
meso-	middle
meta-	more than, change, after, next
micro-	small; 0.000001 (i.e., 1/1,000,000)
mono-	one
morpho-	form
multi-	many
myelo-	bone marrow; pertaining to the spinal cord
myo-	muscle
neo-	new
nephro-	kidney
neuro-	nerves
ob-	in front of
odonto-	tooth
ophthalmo-	eye
ortho-	straight, normal
osteo-	bone
oto-	ear
pan-	all
para-	beside

Prefixes	Meaning
patho-	pertaining to disease
peri-	around
pneumo-	lungs, respiration
pod-	foot
poly-	many
pre-	before
pro-	before
procot-	rectum
pseudo-	false
pyo-	pus
pyr-	fire, heat
quadra-	four
retro-	located behind, backwards
rhino-	nose
semi-	half
sphygmo-	pulse
sub-	near, moderately, under
super-	excessive, above
supra-	above
sym-	union
syn-	union
tachy-	fast, extremely fast
trans-	across
tri-	three
ultra-	beyond
uni-	single

Suffixes	Meaning
-agogue	inducing agent
-agra	sudden acute pain
-algia	painful

Suffixes	Meaning
-cele	tumor, swelling
-centesis	puncture into
-clasia	remedy
-ectomy	surgical excision of
-ecstasis	dilatation
-edema	swelling
-emia	blood
-graph	graphical record
-ia	diseased condition
-iasis	process or procedure
-itis	inflammation
-logy	study, or science, of
-mania	abnormally excessive preoccupation
-meter	resembling
-oid	resembling
-oma	tumor
-opia	vision
-osis	fullness, excess
-pathy	morbid disease
-phobia	dread, fear
-plasty	plastic surgery repair
-rrhea	discharge or flow
-sclerosis	hardening
-scope	instrument for examining
-scopy	visual examination of
-stomy	artificial opening
-tomy	incision

Appendix C

Glossary

accretion growth or enlargement
alveolus air sac or cell in lungs
amnion thin membrane around fetus
amniotic pertaining to the amnion
angstrom unit of length (i.e., 1 Angstrom = 10^{-10} meters)
anterior situated in front of
aorta great artery carrying blood from the left side of heart
aortic pertaining to the aorta
arborizations form resembling a tree
arrhythmia alteration in rhythm
arteriole one of the smallest arteries that becomes a capillary
artifacts error in a test result, graph, or written record
atria (pl.) see atrium
atrioventricular located between the upper and lower chambers of the heart
atrium upper chamber of the heart
auricle see atrium
autonomic action independent of free volition
axon long, thin portion of a nerve cell that carries the impulse away from the neuron
bioelectricity electrical activity pertaining to a living cell
biophysics branch of science that applied the concepts of physical science to biology and medicine
brachial related to, or pertaining to, the arm
bradycardia slow heart rate
bronchi (pl.) see bronchus
bronchus tube leading from trachea to either left or right lung
capillaries smallest vessels in the body
cardiac pertaining to the heart
cardiology study of the heart and its diseases
cardiovascular relating to the circulatory system
catheter small tube that is inserted into the body to permit injections of fluids, introduction of medication, withdraw fluids, or keep a vessel open
cell smallest body capable of life
cephalic pertaining to the skull or head
cerebellum large dorsal brain structure
cerebrum anterior portion of the brain
cornea transparent covering of the center portion of the eye
cortex outer layer of tissue on an organ
cortical pertaining to the cortex
cranium portion of the skull covering the brain
curare drug that produces muscle relaxation
cytoplasm the matter inside of a cell, except to the nucleus
defibrillator electrical machine used to stop fibrillation of the heart by application of an electrical shock
dendrite portion of the nerve cell that conducts impulses toward the cell
depolarized state of being partially, or totally, depolarized
diastole expansion of the chambers of the heart so that they may fill with blood
diastolic pertaining to diastole
dicrotic double humped waveform
dicrotic notch feature of arterial pressure waveform
dorsal situated near, or toward, the back
ECG (abbr.) electrocardiograph
ectopic located in other than normal position
EEG (abbr.) electroencephalograph
EKG (German abbr.) used in place of ECG
electrocardiogram tracing of the electrical signals produced by the heart
electrode conductor used to make electrical contact between a wire and a conductive surface

electrodermograph recorder for measuring the galvanic skin resistance
electroencephalogram recording of brain biopotentials
electroencephalograph machine for making electroencephalograms
electrogastrogram recording of the simultaneous electrical and physical activity of the stomach
electrolyte solution in which electrical current is due to ion mobility
electromyogram recording of the biopotentials produced by skeletal muscles
electromyograph machine for making electromyograms
embolus abnormal solid or gaseous particle in bloodstream
embryo undeveloped stage of fetus
EMG (abbr.) electromyograph, electromyogram
extracellular outside of the cells
extracorporeal outside of the body
fibula smaller of the two bones in the leg
galvanic that which produces a direct current
hemisphere half of a spherical object
homogeneity all of the same sort, state of
homogeneous of the same sort
infarct area of necrotic tissue due to loss of blood perfusion
inhomogeneity not homogeneous
intracellular inside of the cell
ion atom or molecule that carries an electrical charge
iris colored portion of the eye behind the cornea
isoelectric having the same electrical charge (i.e., a state of zero potential difference)
isothermal having the same temperature in all portions
isotropic having the same properties in all directions
latency apparent inactivity
lobe rounded portion of an organ
lumen hollow portion of a tubular organ
manometer gas pressure meter
membrane thin layer of tissue
metabolism total of all life processes
micron unit of length 1/1,000,000 meters
mitochondria small granules or rods
mitral stenosis narrowing of the orifice between left atrium and left ventricle
myocardium a muscle layer of the heart
myograph instrument for the measurement of muscle contraction
necrosis death of cells or tissue
neuron nerve cell
nucleus central structure (i.e., in cells, atoms, etc.)
occipital related, or pertaining to, the rear portion of the head
organ group of specialized cells that perform a special function
orthogonal at right angles to, normal to
parietal pertaining to the upper, rear, portion of head
permeable ability to pass through pores
peroneal pertaining to the outer side of lower leg
piezoelectric electrical activity due to deformation of a crystalline structure
plethysmograhy recording volume changes due to blood flow
pneumatic pertaining to, or operated by, gases
pneumograph measuring instrument for recording respiration
pneumotachygraph instrument that measures respiration rate
posterior pertaining to the rear
protoplasm material making up portions of the cell
psychogalvanic electrical activity produced by mental or emotional stress
pulmonary pertaining to the lungs
pupil variable diameter aperture of the eye
radical group of atoms that can be replaced by a single atom
radioisotope artificially produced radioactive element
retina light-sensitive membrane in the eye
rheobase smallest electrical current that will produce stimulation
sagittal pertaining to, or parallel to, the midline of the body
scalp skin of the head covered by hair
semipermeable permeable only to certain substances
sinoatrial (SA) node collection of heart cells that automatically discharge to function as a natural pacemaker
sinus irregular cavity
sphygmomanometer apparatus used to measure blood pressure

spirometer instrument that measures respiratory volumes

stereotaxic precision positioning

synapse junction where impulse transmits from one nerve cell to another

systemic pertaining to the entire body

systole period during which heart contracts

tachycardia excessively fast heart rate

thermistor electrical component that exhibits changes in electrical resistance with changes in temperature

thermocouple device that uses a bimetallic strip to produce voltage proportional to temperature

thoracic pertaining to the thorax

thorax section of the body between the abdomen and neck

thrombus clot of blood remaining at its site of origin

tibia large bone in the leg

tissue collection of similar cells that perform a specific function or have a similar form

torso trunk of the human body

trachea main tube passing air from the atmosphere into the lungs

transducer device that converts energy from one form to another for purposes of measurement or control

ulnar pertaining to the larger of the two bones in the forearm

utero Latin dative for uterus

uterus organ in the female body for the protection and nourishment of the fetus

vasoconstrictors agents that narrow blood vessels

vasodilators agents that widen blood vessels

vasomotor agent affecting the size of a blood vessel

ventricle lower chambers of the heart

venule small vein connected to capillaries

viable capable of living

Index

1-Hz isolation mode rejection, 97
1-mV calibration pulse, 124, 131, 451-452
1-mV reference, 139
10-20 system, 302
4-20 mA current loop, 104
60-Hz, 66, 100, 134, 137, 143, 445
60-Hz interference, 138, 154
60-Hz common mode noise, 141
60-Hz noise, 137, 139

AAMI, 96, 135, 136, 139, 410–411
A/D converter, 105, 119, 139, 143, 144, 337, 340, 460
A/D, high resolution, 143
A-scan, 388
Absolute pressure, 161
Absorption, ultrasound, 378
Absorption losses, 475
Ac carrier amplifier, 170, 177
Ac-coupled amplifier, 107, 108, 130
Ac coupling, 87
Ac mains (power), 91
Ac motor, 149
Acceleration signal, 449
Acceptance angle, 470
Acid-base balance, 243, 245, 364
Acoustical impedance, 378
Acoustical wave, 375
Acquisition time, 114
ACTA scanner, 299
Action potential, 10, 17, 19, 123, 301
Adrenalin, 292
Aerosol therapy, 264
Afferent neuron, 283
After hyperpolarization, 301
Ag-AgCl, 28, 31, 35, 302, 366
AGC, 532
Air flow rate, 254
Air tube, 274
Airway resistance, 242
Alarm, 325, 329
AlGaAs, 485
Alpha activity, 307
Alpha rays, 515
Alternator, 149
Alveoli, 3, 238, 244
AM detector, 251
Amateur radio, 397
Ambient air, 271
Amino acid, 356
Amplifier, 54, 55
  ac carrier, 170
  antilog, 81, 83
  bandpass, 107
  basic configuration, 58
  bioelectric, 30, 31, 54, 55, 67, 87, 100, 108, 110, 123
  biopotential, 30, 31
  buffer, 129
  carrier, 91
  chopper stabilized, 107, 108
  dc, 30, 130, 170, 176–177
  difference, 144
  differential, 70, 71, 87, 108, 130, 312
  driven right-leg, 426
  ECG, 54
  EEG, 108
  electronic, 103, 163
  front-end, 91, 135
  gain, 177
  gating, 456
  ground reinforced differential, 426
  high gain, 55
  input, 30, 88, 459
  instrumentation, 71–77, 80, 104
  isolation, 54, 88, 91, 105, 111, 127
  logarithmic, 81, 83, 111
  low-gain, 55
  low-level, 55
  medium gain, 55
  optically coupled, 93
  PMMC, 449
  pressure, 170, 173
  programmable gain, 135
  receiver, 477
  summation, 185
  transimpedance, 60
Amplitude modulator, 488
Anabolism, 355
Analog multiplexer (MUX), 135
Analog waveform, 340
Analog-to-digital converter (A/D), 79, 80, 100, 105, 119, 139, 337
Anemometer, 78
Anesthesia, 348, 349
  machine, 352
  technician, 350
Anesthesiologist, 349
Anesthetist, 349
Aneurysms, 393
Angiography, 512
Angle of diffraction, 361
Anode, 518
ANSI, 479–480
Antecubital space, 164
Antenna, 375
Anterior-anterior set, 223
Anterior paddle, 223
Anterior-posterior set, 223
Antialiasing filter, 144
Antibody, 354
Antigen, 354
Aorta, 12
Aortic valve, 14
Aperture jitter, 115
Aperture uncertainty, 114
Apex phonocardiography, 214
Apnea, 244
Apnea alarm, 249, 250, 252
Application Specific Integrated Circuit (ASIC), 74
Aqueous heparin, 169
Arrhythmias, 22, 151, 220
  detection software, 127
Arterial blood pressure, 159
Arterial cutdown, 169
Arterial monitor, 176
Arterial pulse, 207
Arterial system, 199
Arteries, 6, 10, 13, 22, 290
Arterioles, 13
Articulations, 3

Artifact, movement, 33
Artificial mechanical ventilation, 270
Artificial radioactivity, 516
Artificial respiratory therapy, 261
Artificial ventilatory control, 260–261
Aseptic procedures, 348
ASIC, 74
Association areas (brain), 289
Astrup method, 366
Atmospheric pressure, 45, 162, 163
Atria, 14, 199, 219
Atria tachycardia, 229
Atrial contraction, 20, 22
Atrial fibrillation, 22, 220
Atrial flutter, 229
Atrial-ventricular synchronized pacemaker, 229
Atrioventricular block, 22
Atrioventricular (AV) node, 10, 17, 219
Atrium, 6, 11, 14, 207
Attenuation, 89, 540
Attenuation, X-ray, 299
Audio output, 381
Auditory (tone burst) evoked potential, 315
Augmented limb leads, 125
Auscultation, 159, 166–167
Autoanalyzer, 358, 369
Autoclave, 274, 349–350
Automatic control system, 6
Automatic gain control (AGC), 532
Automatic zero circuit, 185
Automatic zeroing, 87
Autonomic nervous system (ANS), 6, 278, 280, 290
Autotransfusion, 231
AV node, 10, 19, 219
AV pacer, 229
Axon, 281, 282

Backscatter reflection, 381
Bacterial filter, 268
Bacteriological cultures, 274
Balance, 285
Balloon, 192
Band-aid incision, 467
Bandpass filter, 139
Bandstop wavetrap, 543
Bandwidth, 67, 77, 144, 472
Baroreceptors, 8, 243
Barrier, capacitive, 115
Baseline, flat, 133
Batteries, 201, 226, 228, 551ff
Battery maintenance, 91
Beam intensity, 379
Bel, 473
Bell jar, 255
Bennett valve, 266
Bernoulli principle, 264
Beta activity, 307
Beta rays, 515
Bicuspid valve, 14
BiFET, 31
BiMOS, 31
Biochemical activity, 293
Bioelectric sensing, 28
Bioelectric signal, 26, 30, 32, 115
Bioelectricity, 16
Bioelectrode, 26
Biomedical:
  electrode, 30
  equipment, 66, 67
  equipment technician, 186, 326
  filter, 66
  instrumentation, 111
  strain gage transducers, 43
  transducer, 37
Biophysical sensing, 25
Biophysical signal, 68
Biopotential pickup, 300
Biopotentials, 26
Biphasic pulse, 229
Bipolar arrangement, 303
Bipolar electrode, 397
Bipolar limb leads, 125
Bleeding, 348
Blood, 12, 17, 353
  bacteriological test, 357
  blockage, 263
  cell, 236
  cell analyzer, 357
  cell counter, 362
  chemical test, 356
  chemistry, 243, 353
  circulation, 290, 410
  circulation system, 1, 6
  flow, 209, 241, 379
  flow measurements, 198, 199, 209
  flow rate, 12, 200
  gases, 250
  gas analyzers, 258, 357, 366, 368
  gas measurement, 245
  pressure, 8, 135, 159, 290
  pressure cuff, 385
  pressure device, ultrasonic, 387
  pressure kits, 164
  pressure measurement, 163, 385
  pressure transducer, 350
  pressure waveform, 144
  serological tests, 356
  sugar, 290
  temperature, 202
  test, 356
  velocity, 381
BMET, 270, 432
Body systems, 1
Bonded strain gage, 25, 42
Bouncing-ball display, 459
Bovie, 351
Boyle's law, 233, 234
Brachial artery, 164
Bradycardia, 22, 340
Brain scan, 298
Brain stem, 243, 283
Brain ventricles, 299
Breathing, mechanics of, 239
Bremsstrahlung, 516
Brewster's angle, 484
Bridge amplifier, 76, 79
Bridge circuit, capacitive, 47
Bridge signal, 81
Broadcast, 530
Bronchi, 3, 238
Bronchioles, 238–239
Bronchitis, 266
BSM controller module, 337
Bundle branches, 17
Bundle of His, 17, 34
Bunn tent, 264
Burns, 404

Cable:
  ECG, 151
  monomodal, 473
  multiconductor, 335
  patient, 149
  ribbon, 536
  shielded, 109
Calibration:
  circuit, 81, 173, 174
Calomel reference cell, 365
Cannula, 169, 230, 231, 264
Capacitance:
  interelectrode, 400
  parasitic, 77, 109
  stray, 77, 135, 138
Capacitive barrier, 115
Capacitive leakage current, 417
Capacitive reactance, 70
Capacitor, 47, 48, 222
  parallel plate type, 48
Capacitor microphone, 48
Capillaries, 3, 6, 13, 238
Carbaminohemoglobin, 354
Carbon dioxide electrode, 366
Cardiac activity, 317, 416
Cardiac cycle, 385
Cardiac output, 14, 198, 200
Cardiac output computer, 201, 202, 204
Cardiopulmonary resuscitation (CPR), 266, 416
Cardiotachometer, 327, 329
Cardiovascular measurement, 198, 256
Cardiovascular technician, 350
Cardioversion, 225
Cardiovert mode, 225, 226
Carotid artery, 290
Carrier signal, 93, 99
Carrier oscillator, 91

Carrier noise, 99
Cartilage, 3
CAT scan, 299, 513, 526
Catabolism, 355
Catheter, 164, 169, 187, 203, 207, 409
Catheterization, 217, 409, 426
Catheterization laboratory, 179, 216–217
Cathode ray tube (CRT), 299, 454
Cauterization, 398
Cavitation, 379–380
CCD, 467
CCU, 166
Cell 1–2, 12, 353
Cell body (soma), 281
Cell, red blood, 236
Central monitoring console, 333
Central nervous system, 6, 278, 280
Central venous pressure (CVP), 159, 164, 168, 205, 206, 207
Centrifuge, 355
Cerebellum, 283, 285
Cerebral angiography, 298
Cerebral cortex, 288
Cerebrospinal fluid, 290
Cerebrum, 283, 288, 303
Certified Registered Nurse Anesthetist, 349
Charge button, 224
Charge-coupled diode (CCD), 467
Charge gradient, 26
Charles' law, 233, 234
Chart recorder, 255
Chemoreceptors, 243
Chicken Heart (signal generator), 344
Chopped sinewave, 401
Chopper vibrator, 107
Chordae tendinae, 14
Chromatograph, 358, 368
Chronic hypersomnia, 307
Chyme, 3
Circulatory system, 6, 10, 161
Citizen's Band, 534
Clad fiber optic, 469
Cladding layer, 475
Clark electrode, 367
Clark-Severinghaus electrode, 367
CM, 100
CM noise, 77, 128
CM voltage, 90, 128
CMOS, 91, 119
CMOS analog switch, 456
CMOS switch, 177
CMR, 70, 77, 90, 134, 135
CMRR (*See* Common mode rejection ratio)
CMV, 127
CNS, 6, 8
Coagulation, 398, 401
Cochannel interference, 538
Coefficient of reflection, 379
Coefficient of thermal expansion, 518
Coherence, 480
Cold junction compensation, 103
Collimated light, 444
Colorimeter, 353, 357, 358–359, 368, 369
Common electrode, 124
Common indifference point, 303
Common mode (CM), 100
Common mode input voltage, 81
Common-mode interference, 109, 137–138
Common mode noise, 77, 137
Common mode rejection, 70, 77, 90, 127
Common mode rejection ratio (CMRR), 57, 87
Common-mode signal, 57, 109, 154
Common mode voltage, 71, 137
Communication link, 465
Communications system, fiber optic, 476
Compound-B scan, 388
Compressed air, 262
Compressed Gas Association, 262, 263
Compton effect, 515
Computer, 87, 451, 495ff
  analog, 56
  programmable digital, 496
Computerized axial tomography, 299, 513, 526
Computerized monitoring system, 338
Conduction:
  electronic, 25, 26
  ionic, 25, 26
  thermal, losses, 379
Constant flush infusion system (CFIS), 169
Constantan, 41
Contact button, 31
Control logic, 203
Core, 401
Coronary Care Unit (CCU), 323
Corrective maintenance, 411
Corundum, 483
Coulter counter, 364
Counterpulsation, 219
Coupling losses, 475
Cranial nerves, 290
Critical angle, 468
Critical damping, 194, 448
Critical mode, 472
CRNA, 349
CRO (cathode ray oscilloscope), 144, 388, 391, 405, 440, 454, 455, 459
Cross-modulation, 531
CRT, 299, 454, 456, 513
CRT camera recorder, 444
CRT phosphor, 455
CRT screen, 325, 444, 459, 461
Crystal, 380, 384
Crystalline structure, 380
CSF, 290
Cuff, 164, 385
Curie, 517
Current, 79, 93, 96
  resistive leakage, 417
Current density, 397
Current feedback, 426
Current loop (4-20 mA), 104
Current-to-voltage converter, 60
Cut-off frequencies, 141
CVP (*See* Central venous pressure), 159, 164, 168, 205, 206, 207
Cyclopropane, 409, 466

D/A converter, 143, 144, 460
D'Arsonval movement, 146, 441, 445
D-ring paddle, 223
Dalton's law, 233, 234, 236
Damage to equipment, 409
Damping, 189, 194
Data acquisition system, 115, 256
Data driver, 489
Data error rate, 417
dB, 473
Dc level, 458
Dc offset bias, 478
Dc restoration, 87, 130, 139
Dc-to-Dc converter, 99
Dead space, 241, 261
Deadband, 447
Decay, 516
Decibel, 473
Decoder, 476
Defect losses, 475
Defibrillation, 96, 100, 102, 416
Defibrillator, 127, 131, 170, 217, 219, 220, 223, 226
  circuit, 224
  damage to, 133
  dc, 222
  Lown, 221
  paddle, 223
  portable, 222
  protection, 151
  testing, 227
Defibrillator/cardioverter modes, 225
Deflection yoke, 454
DeForest, 466
Delay line, 227
Delay line analogy, 18
Delta activity, 307
Delta-sigma converter, 144
Demodulator, 88, 89, 91, 97, 476

Dendrites, 281
Deoxyribonucleic acid, 293
Depolarized cell, 17
Derivative, 184
Desensitization, 538
Detector:
  am, 251
  diastolic, 180
  fault, 429
  flow, 380
  pressure, 180
Detector-collimeter, 523ff
Dialyzer, 369
Diaphragm, 45, 48, 49, 239, 241
Diastole, 14, 20
Diastolic display, 387
Diastolic pressure, 20, 165, 166
Dicrotic notch, 22
Diencephalon, 283
Difference frequency, 384
Differentiation, 81
Differentiator, 81, 82, 226, 327
Digestion, 290
Digital computer, 466
Digital counter circuit, 329
Digital ground, 119
Digital panel meter, 203
Digital recorder, 345
Digital telemetry, 342
Digital voltmeter, 177
Digital-to-analog converter, 143, 149, 459
Digitization, 337
Digitization, high resolution, 139, 141
Dilution curve, 200
Dilution techniques, 199
Diode, 51, 81, 132
Diode, laser, 486
Diode, PIN, 478
DIP IC package, 93, 97
Discharge button, 222
Discretes, 335
Discriminator, 384
Disintegration, 516
Dispersion, 471, 480, 483
Distortion, 540
Distortion, plumbing system, 189
DNA, 293
Domain, frequency, 481
Doppler, 380
Doppler components, 388
Doppler effect, 380
Doppler flow detector, 380ff
Doppler flow meter, 387
Doppler flow transducer, 381
Doppler shift, 167, 256, 257, 381, 385, 388
Dose rate, 517
Dot matrix analog recorder, 452
Dot matrix printer, 452
Dot scan converter, 524
Double-beam spectrophotometer, 362
Double-humped waveform, 223
dP/dT circuit, 184
Drift, 55, 107
Drift cancellation, 177
Drive, motor, 445
Drive pulley, 445
Driver, transmitter, 477
Droop, 186
Drug, overdose, 266
Dummy load, 396, 404
Duplex, 477
Duty cycle, 379
Dye dilution, 200
Dynamic equilibrium, 290
Dynamic LIM monitor, 429
Dynamic range, 80
Dyspnea, 245

ECG (electrocardiogram):
  abnormalities, 138
  amplifier, 30, 54, 55, 100, 131, 136, 226, 325, 427
  applications, 105
  bandwidth, 133
  cable, 151
  chest surface signal, 303
  circuit, 66
  data, 149
  digitized, 345
  display, 144
  electrode, 31, 32, 345, 404
  electrode signal, 143
  equipment and EMI, 544
  esophageal, 125
  faults, 152
  features, 20
  fetal, 387
  frequency, 66
  front-end, 135
  interdigital, 125
  isolated-input, 427
  laboratory, 149
  leads, 138, 433
  lead selector switch, 124
  machine, 74, 96, 123, 125, 146, 149, 451
  machine maintenance, 152
  monitor, 93, 217, 250, 337, 443
  monitor, EMI to, 530
  monitoring, 119, 326
  paper, 144
  potentials, 215
  preamplifier, 127, 129, 131, 154, 224
  readout device, 144
  recorder, 147
  recording, 124, 127, 131, 132
  signal, 55, 66, 68, 113, 125, 127, 129, 133, 137, 143, 338
ECG (electrocardiogram) *(Contd.)*
  stress testing, 33
  system, 87, 113, 135, 138, 139, 224
  technician, 152
  telemetry, 342
  toilet seat, 125
  trace, 422
  troubleshooting, 152
  waveform, 23, 33, 55, 94, 108, 123, 129, 138, 143, 217, 325
Echocardiography, 390
Echoencephalography, 299, 391
Ectopic beats, 22
Ectopic foci, 416
EEG, (*See also* Electroencephalograph) 26, 28, 34, 55, 144, 283, 298, 302
  amplifier, 108
  changes, 293
  diagnostic uses, 307
  electrode, 34
  electrode 10–20 system, 302
  frequency bands, 305
  frequency spectrum, 307
  machine, 293
  multichannel, 308
  output signals, 311
  pacemaker, 302
  patterns, 307
  preamplifier, 311
  record, 300
  scalp potentials, 30
  signal, 297, 302, 303
  signal analysis, 300
  system, 310
  system maintenance, 316
  technician, 308
  telemetry, 316
  waveform, 293
Effector, 283
Efferent neuron, 283
Efferent nerves, 282
Einthoven triangle, 123, 125, 138
EKG (German init. for ECG), 20, 123
Electric fields, 98
Electric shock, 220, 404, 409, 431
Electrical axis, of the heart, 124
Electrical defects, 203
Electrical hazards, 411, 436
Electrical rhythms, 302
Electrical stimulus, 14
Electro-optics, 465
Electrocardiogram (*See also* ECG), 20, 23, 26, 123, 416
Electrochemical activity, 293
Electroconduction system, of the heart, 17, 19, 123, 219, 227
Electroconductive gel, 403

Electrocorticographic electrode, 303
Electrode, 20, 25, 26, 132, 227, 250, 333
  column, 33
  cup, 303
  defects, 154
  defibrillator, 404
  double layer, 27
  ECG, 32
  EEG, 34
  effect, 143
  glass, 365, 367
  hydrogen-hydrogen, 27
  impedance, 33
  indifferent, 125
  indwelling, 33
  metallic, 26, 32
  model circuit, 28
  needle, 30, 31, 33
  offset potential, 25, 27–28, 101, 143
  passive, 396
  perfectly nonpolarized, 28
  perfectly nonreversible, 28
  perfectly polarized, 28
  perfectly reversible, 28
  pill, 125
  potential, 28, 31
  problems, 32
  right-arm, 130
  sample, 75
  scalp, 34, 303, 388
  scalpdisc, 303
  sites, 32
  slippage, 33
  surface, 30, 31, 32
Electrode/electrolyte interface, 27
Electroencephalography (*See also* EEG), 293, 297, 300
Electroencephalogram (EEG), 26, 34
Electrolyte, 27
Electrolytic paste, 252
Electrolytic solution, 26
Electromagnetic interference (EMI), 65, 66, 99, 466, 530ff
Electromagnetic spectrum, 516
Electromagnetic wave, 375, 512, 514
Electrometer, 49, 74, 75
Electromyogram, 26
Electronic devices, 25
Electronic switch, 131
Electrostatic field, 454
Electrosurgery, 151, 401
Electrosurgery equipment, 351
Electrosurgery machine, 348, 396, 397, 399
Electrosurgery unit (ESU), 97, 133, 404
Electrosurgery waveform, 399
EMG, 26, 65, 79, 99, 100, 466, 530ff
EMI/RFI rejection, 100
Emotional reactions, 290
Encephalon, 283
Encoder, 476
End tidal volume, 258
End-of-conversion (EOC) pulse, 460
Endocardial lead, 228
Endocrine system 6, 285
Endoscope, 467
Energy:
  delivered, 222
  kinetic, 484
  stored, 222
  ultrasonic, 380, 391
Engineer, 75
Envelope recording, 214
Environmental impact statement, 245
Epicardium, 14
Epilepsy, 300, 316
Epinephrine, 292
Equilibrium, 285
Equipotential ground system, 409, 429
Erase bar display, 459
Error, data, 471
Erythrocyte, 353, 367
Esophagus, 3, 467
ESU 97, 133, 134
Ether, 409
Ethylene oxide gas, 274, 349, 351
Evoked potentials, 314–315
Excitation voltage, 177
Excitatory postsynaptic potential, 282, 301
Expiration, 241, 243, 245
Expiratory centers, 243
Expiratory reserve volume, 242, 250
Explosion, 262, 409
Exponential decay curve, 204
Extinction voltage, 98
Extracorporeal blood pump, 219

Fast flush lever, 169
Fatty acids, 356
Fault current, 136
Fault detector, 429
FDA, 262
Federal Communications Commission (FCC), 316, 338, 534
Feedback, 85
Feedback circuit, 130
Feedback control system, 86
Feedback current, 479
Feedback diode, 68
Feedback error integrator, 87
Feedback integrator 130
Feedback loop, 66, 79, 100, 130, 138, 243, 279
Feedback system, 243
Feedthrough noise, 99
Femoral artery, 217
Ferrite core, 401
FET transistor diode, 68
FET IA, 76, 102
Fetal monitor, 387
Fetal scan, 388
Fetus, 388
Fiber link, 466
Fiber optics, 465, 469, 473
Fiber optic communications system, 476
Fiber optic element, 471
Fiber optic isolation, 99, 466
Fiber optic receiver (FOR), 100
Fiber optic transmitter, 99
Fibrillation, 22, 220
Fibrinogen, 354
Fick's method, 199
Field strength meter (FSM), 342, 344
Fight-or-flight state, 292
Filter, 6, 66, 99, 154
  air, 274
  active, 67
  biomedical, 66
  capacitor, 96
  high pass, 83, 130, 327, 543
  multipole, 66
  photometer, 357
Filtering, 77, 133, 155, 388
Flame photometer, 353, 357, 360
Floating current source, 113
Flow detector, 380
Flow meter, 260, 264, 381, 387
Flow volume, 254
Fluid pressure, 159
Fluid transport system, 1
Fluoroscopy, 517, 522
Flush method, 166
Flyback loading, 88
FM, 530
FM signal generator, 344
FM wavetrap, 536
Food & Drug Administration, 262
FOR, 100
Force, 160
Force transducer, 44
Formalin gas, 351
Fourier series, 192
Frank electrode system, 216
Frank lead system, 198
FRC (*See* Functional residual capacity)
Frequency, 379, 380, 514
Frequency domain, 481
Frequency response, 55, 88, 93, 192, 441
Frequency spectrum (EEG), 307, 398
Fresnel reflection losses, 476
Frontal lobes, 288

FSM, 342, 344
Functional mean pressure, 182
Functional residual capacity (FRC), 242, 250

GaAs, 485
Gage factor (GF), 42
Gage manometer, 162
Gage, pressure, 161–163
Gage, vacuum, 163
Gain, 88
Gain buffer, 143
Gain error, 60, 70, 80
Galvanometer, 146, 440, 443, 445
Gamma activity, 307
Gamma radiation intensity, 523
Gamma rays, 515
Gamma ray camera, 524
Gas chamber, 484
Gas cylinders, 262
Gas laws, 234
Gas regulator, 260, 263
Gas sterilization, 194
Gases, medical, 262
Gases, therapeutic, 263
Gastrointestinal system (GI), 3, 6
Gate, SCR, 226
Gaussian distribution, 471
Geiger-Mueller tube, 522
Gel, 403
Gel, conductive, 31
German standard VDE-0884, 98
GFI (*See* Ground fault interrupter)
Glass bulb, 365
Gloves, sterile, 349
Glow lamp, 131
Graded fibers, 472
Graded Index Fibers, 472
Ground fault interrupter (GFI), 409, 431
Ground loop, 104
Ground system, 431
Ground wire loop resistance tester, 432
Grounding 109
Guard, 65
Guard bands, 338, 532
Guard shield, 65, 109
Guarding, input, 108
Gynecology, 349

Halfcell potential, 25–28, 365
Half-duplex, 477
Halfpower point, 471
Halfwave stub, 534
Harmonics, 189, 530
He-Ne laser, 484ff
Heart rate ruler, 146
Heart, 10, 13, 123, 161, 219
  block, 22, 219, 227
  function, 290
Heart *(Contd.)*
  murmurs, 213
  muscle, 220
  pressures, 205
  problems, 22
  rate, 22, 144, 198, 327
  rate, fetal, 388
  rate, monitoring, 96
  sounds, 213ff
Heart-lung machine, 229
Heat, dissipation, 518
Heating bath, 369
Hematocrit, 356
Hemispheres, 288
Hemoglobin, 353ff, 364, 367
Hertz, Heinrich, 515
Heterodyning, 531
Hi-Fi, 536
High frequency transient, 97
High voltage, 518
High voltage spike, 132
Histological tests, 357
Hollerith, Herman, 496
Homeostasis, 1, 243, 279
Hooke's law, 45
Hormones, 6, 293
Humidifier, 268
Hybrid function module, 73
Hydrodynamic system, 160
Hydrostatic pressure, 159, 186, 290
Hyperapnea, 245
Hypercapnia, 245, 261
Hyperkinetic syndrome, 289
Hyperventilation, 243
Hypopharynx, 238
Hypotensive, 166
Hypothalamus, 283
Hypoventilation, 244
Hypoxia, 244, 261

IA CMR, 129
Idler roller, 147
Idler pulley, 445
Impedance, 70, 379, 396
  acoustical, 378
  electrode, 33
  high, 398
  infinite, 89
  input, 55, 85, 87
  load, 534
  optical, 476
  single-ended input, 314
  skin, 31
  source, 55
  thoracic, 250
Impedance matching, 476
Impedance pneumograph, 250, 252
Inclusions, 475
Index of refraction, 377, 467, 476
Indocyanine green, 199
Induction coil, 228
Infection, 370
Inferior vena cava, 11
Infrared detector, 67
Infrasound, 166
InGaSaP, 485
Inhibitory post-synaptic potential, 282, 301
Injectate bolus, 200, 202
Injection laser, 485
Ink jet recorder, 443
Ink manifold, 443
Ink pen recorder, 449
Ink pen writer, 443
Input, 57
  differential, 111, 445
  differential amplifier, 111, 124
  noninverting, 61
  single-ended, 87
Input bias currents, 85, 87
Input guarding, 108
Input impedance, 55, 85
Input resistance, 70
Inspection, 467
Inspiration, 236, 239, 241
Inspiratory capacity (IC), 242, 250
Inspiratory centers, 243
Inspiratory reserve volume, 242, 249
Instrument designer, 30
Instrumentation, biomedical, 25, 111, 353
Integrated circuit (IC), 55
Integration, 81, 182, 203–204
Integrator, 81, 130, 131, 200, 205, 327
Intensity, beam, 379
Intensity modulation, 388
Intensive Care Unit (ICU), 166, 323
Intercostal muscle, 241, 243
Interelectrode capacitance, 400
Interference, 67, 77, 109, 538
Interference, ESU, 134
Intermittent Positive Pressure Breathing (IPPB), 260, 266, 268
Intermod Hill, 531
Intermodal dispersion, 470, 471
Intermodulation, 530, 531
Interpleural space, 229
Intraalveolar pressure, 243, 245
Intracellular fluid, 3
Intracellular motion, 379–380
Intracerebral electrode, 303
Intrathoracic pressure, 243
Intravenous (IV), 169, 187
Intraventricular (brain) pressure, 159
Inverse square law, 475, 514
Inverting follower, 54, 58
Inverting input, 57
Involuntary neuronal activity, 243
Ionization potential, 16, 132

Ions, 3
Irradiation time, 379
Ischemia, 32
Iso-amp, 102
Iso-amp barrier, 96
Isolated common, 119
Isolated signal, 100
Isolation barrier, 88
Isolation mode, 89
Isolation-mode rejection (IMR), 89, 90, 97, 101, 133
Isolation-mode rejection ratio, 89
Isolation-mode voltage (IMV), 89
Isolation techniques, 91

JCAHO, 410
Jelly, 379
JFET, 55, 85, 93, 131
JFET amplifier, 55
Jitter, aperture, 115
Jitter, muscle, 155
Johnson noise, 64
Joule, 222

Kidneys, 6, 11
Kirchoff's current law, 56, 58, 61
Knife edge, 441
Korotkoff sounds, 164, 166, 167, 385
Korotkoff, N.S., 166
Kymograph, 255

L-C pi filter, 99
L-C section, 223
Laprascope, 467
Laryngopharynx, 238
Larynx, 238
Laser, 149, 480, 483
- am-modulated, 490
- argon gas, 485
- carbon dioxide, 485
- excimer, 484
- gas, 484
- injection, 485
- insulating crystal, 465, 483
- insulating solid, 482
- Nd:YAG, 483
- optically pumped, 482
- PN junction, 485
- ruby 482
- solid crystal, 482
- types of, 482

Laser action, 481
Laser diode, 486
Laser diode driver, 486
Laser diode receiver circuit, 490
Laser Safety Officer 480
Laser trimming, 70, 96
Lateral ventricles, 391
Lead AVF, 125
Lead AVL, 125
Lead AVR, 125
Lead fault indicator, 331
Lead selector switch, 124, 131
Lead-I, 125, 450
Lead-II, 125
Lead-III, 125
Leads, patient, 128
Leakage current, 409, 417
Leakage current tester, 433
LED (light-emitting diode), 93, 94, 96, 99, 207, 477, 486
Ledley, Robert S., 299
Left ventricle, 20, 207
Leucocytes, 354
Level detector, 327
Light, 377
Light densitometer, 200
Light-emitting diode (*See* LED)
Light source, 352
Light-to-voltage circuit, 60
Limb leads
- augmented, 125
- bipolar, 125
- unipolar, 125

Limits-of-travel stops, 448
Line frequency, 422
Line isolation monitor (LIM), 429
Line isolation system, 428
Line isolation transformer, 409
Linear-B scan, 388
Linear ramp, function, 447
Linearity, 51
Lipid, 356
Liquid crystal readout, 205
Lissajous pattern, 447
Lithium iodine cell, 228
Lobes, 288
Longitudinal coherence, 480
Longitudinal fissure, 288
Losses:
- absorption, 475
- coupling, 475
- defect, 475
- fiber optic, 473
- fresnel reflection, 476
- numerical aperture coupling, 475
- reflection, 476
- transmission, 475

Low bias current, 74
Low-pass filter, 82, 101
Lown defibrillator, 221
Lown waveform, 220, 222
Luer lock, 173
Lungs, 238
- capillaries, 238
- compliance, 243
- elasticity, 243
- volume, 262

LVDT, 45, 47

Macroshock, 409, 415
Magnetic deflection, 454
Magnetic field, 146, 441, 481
Magnetic flow measurement, 212
Main memory, 459
Manometer, 161, 164, 169, 176, 386
Manometer, water, 168
Manometry, electronic, 169
Marconi, 466
MASER, 479
Master antenna television (MATV), 338
Master tissues, 293
Maximum expiratory flow rate, 245
Maximum voluntary ventilation (MVV), 245
Mean arterial pressure (MAP), 166, 167, 182
Mean cell hemoglobin concentration, 356
Mean cell volume, 356, 364
Mean detector, 180
Mean flow velocity, 384
Mean pressure, functional, 182
Mechanical wave, 375
Medical monitoring system, 170
Medulla, 243, 283, 287
Membrane, 281, 365
Membrane, semipermeable, 2, 16
Memory, 459
Meninges, 290
Mercury cell, 228
Mercury column, 162
Mercury strain gage, 209, 252
Meridional rays, 469
Metabolism, 2, 355
Metastable state, 482
Meter movement, DC, 406
Meter pointer, 146
Microcomputer, 495, 502
Microelectrode, 26, 31, 34, 35, 36
Microelectrode capacitance, neutralizing, 37
Microorganism, 349, 350
Microphone, 48, 75, 214
Microprocessor, 326, 337, 495, 502
Microshock, 88, 409, 415, 416, 417, 420
Microtome, 357
Minimal cerebral dysfunction, 289
Minimal volume, 250
Minute volume, 250, 254
Mitosis, 3
Mitral valve, 20
Mixing tube, 370
Mode, critical, 472
MODEM, 100, 149, 337, 345
Modes, transmission, 470
Modular design, 326
Modulation, 88, 99
Modulator, 89, 91, 96, 115, 476
Modulator, amplitude, 488
Monitor, 325, 335, 443

Monitoring, 131
Monitoring technician, 350
Monochromiticity, 480, 481
Monochronometer, 361
Monolithic circuits, 71
Monolithic ICIA, 73
Monopulse waveform, 220, 222
Monostable multivibrator, 17, 220, 327
Montages, 303, 308
MOSFET, 85, 113, 366
MOSFET amplifier, 55
Motor drive, 445
Motor homunculus, 288
Motor nerves, 290
Mouth, 3
Movement artifact, 32, 33
Multimode fiber, 471
Multiple-input circuits, 68
Multipole filter, 66
Muscle, 281
  activity, 317
  jitter, 155
  movements, 285
Muscular system, 5
Musculoskeletal system, 3
Myelin sheath, 281
Myocardial infarction, 22, 228
Myocardium, 14, 416

Nasal cannula, 264
Nasion, 303
Nasogastric tube, 125
Nasopharyngeal electrode, 303
Nasopharynx, 238
National Bureau of Standards, 162
National Electric Code (NEC), 419
National Fire Protection Agency (NFPA), 262, 410
Natural gas, 466
Nebulizer, 260, 264, 270
NEC, 410
Needle, 164
Needle electrode, 30–33
Negative feedback loop, 7, 55, 130, 279
Negative feedback network, 83
Negative temperature coefficient, 51
Neonatal oxygen monitor, 368
Nerve, 282
Nerve conduction, 282
Nerve impulse, 281, 293
Nerve impulse conduction, 278
Nervous system, 6, 278, 280, 293
  parasympathetic, 280, 293
  peripheral, 278, 280, 290
  sympathetic, 292
Neurilemma, 281
Neuron, 281, 288, 292
Neurosurgery, 349
Neutral ground wire, 418
Neutralization capacitance, 37
Nichrome, 70
Nickel-Chromium, 70
Nodes of Ranvier, 281
Noise, 55, 64, 107
Noise-cancelling voltage, 127
Noise current, 79
Noise pickup,109
Noise rejection, 97
Noise, rms, 139
Noise voltage, 79
Nomograph, 366
Noninvasive monitoring, 387
Noninvasive pressure measurement, 167
Noninverting follower, with gain, 58, 61
Noninverting input, 54, 57, 58, 61, 68, 70
Noninverting terminal, 87
Normal distribution, 471
Nose, 3
Nuclear medicine, 298, 513
Nuclear radiation, 511, 516, 517
Nucleus, 2
Null condition, 25, 38
Numerical aperture, 470, 475
Nurses' station, 338, 340
Nutrients, 3

O.R., equipment, 351
O.R., personnel, 349
Obstetrics, 349
Occipital lobes, 289
Occlusion, 164
Off-channel interference, 538
Offset control, 252
Offset error, 80
Offset null control, 94
Offset null terminals, 86
Offset voltage, 86
Offset voltage error, 86
Ohm's law, 12, 56, 61
Ohmmeter, 51
Olfactory bulb, 289
On-resistance, 113
Open-loop voltage gain, 57, 85
Operational amplifier, 54, 55, 71
  ideal properties, 57, 58
  specifications, 85
Ophthalmology, 349
Optic chiasma, 289
Optical fiber, 472
Optical impedance, 476
Optical medium, 377
Optical recorder, 443–444
Optical transmission, 99
Optically coupled circuits, 93
Optoisolator, 93
Oral surgery, 349
Orderly, 350
Organism, 1
Orthopedics, 349
Oscillation, 129, 138, 380
  amplifier, 77
  damped, 398
  free, 192
  pressure 167
Oscillator, 49, 91, 399
Oscillograph, 441
Oscillometric method, 167
Oscilloscope, 68, 91, 144, 146, 227, 273, 329, 335, 338, 340, 454, 462
  cathode ray (CRO), 388
  digital storage, 459
  high frequency, 301
  multibeam, 456
  monitor, 224
  nonfade, 459
OSHA, 262, 410
Otolaryngology, 349
Output amplifier, 459
Output impedance, 85
Output offset voltage, 86
Overcritically damped, 448
Overload, signal, 538
Overshoot, 447
Overvoltage, 81
Overvoltage protection diode, 74
Oxidizing reaction, 27
Oximeter, pulse, 61
Oxygen adder, 268
Oxygen analyzer, portable, 273
Oxygen disassociation curve, 236
Oxygen electrode, 366
Oxygen hoods, 264
Oxygen mask, 264
Oxygen saturation, 367
Oxygen tent, 264
Oxygen therapy, 260, 263, 264
Oxygenator, 230
Oxyhemoglobin, 354

P-Q delay circuit, 229
P-wave, 20
Pacemaker, 17, 23, 125, 219, 226, 227, 229
  batteries, 228
  demand, 229
  implantable, 228
  lead wire, 228
Pacer pulse, 138–139
Paddle, 223
Pain, 348
Palpation method, 166
Paper drive motor, 445
Paper tray, 147
Parade display, 459
Parasitic capacitance, 77, 109
Parasympathetic, 290
Parietal lobes, 289
Partial discharge testing, 98

Partial pressure, 236
Pascal, Blaise, 160
Pascal's principle, 160, 169
Paste, conductive, 31
Patient cable, 149
Patient-cycled artificial ventilation, 261
Patient electrode, 139
Patient leads, 128
Patient plate, 396
PCB (*See* Printed circuit board)
PCCU, (*See* Post-Coronary Care Unit)
Peak pressure, 261
Peak-reading circuit, 406
Pen motor assembly, 441
Pen position, 445
Pen position transducer, 448
Peptide, 356
Perception, 281
Percutaneous puncture, 169
Perfusion, 22
Perfusion technician, 350
Perfusionist, 350
Pericardium, 14
Period, 184, 376
Peripheral nervous system (PNS), 278, 280, 290
Peripheral resistance units, 12
Peristaltic pump, 230, 369
Peristaltic waves, 3
Permanent magnet moving coil (PMMC), 146, 440, 441, 443, 445, 449
Permeability, 16
Pemeable core, 45
Persistence, 455
pH, 243
pH/blood gas analyzer, 357, 364
pH/blood gas monitor, 367
pH electrode, 364
pH meter, 245, 364, 368
pH probe, 65, 75, 243
Pharynx, 238
Phase detector, 93
Phase-locked loop (PLL), 93
Phonocardiography, 213
Photoamplifier tube, 523
Photocell, 207
Photodetector, 60, 358, 360
Photodetector circuit, 61
Photodiode, 94, 96, 491
Photodiode amplifier, 61
Photoelectric effect, 515
Photometric analyzer, 367
Photon, 481
Photoplethysmograph (PPG), 207–208, 325, 345
Photoresistor, 93, 207
Photosensitive paper, 444
Photosensor, 491
Phototransistor, 93
PHS, 410
Physiology, 1
  control system, 8
  monitoring, 139
  saline, 34
  signal, 455
Piezoelectric crystal, 381
Piezoelectric resonator, 380
Piezoelectric ultrasound sensor, 167
Piezoresistivity, 41
  element, 37
  semiconductor material, 45
  strain gage, solid-state, 44, 252
Piping system, 262
Pituitary gland, 285
Planck's constant, 481, 515
Planck's theory of energy quanta, 481
Plasma, 12, 353
Plasma nutrient, 354
Plasma protein, 354
Platelets, 354
Platelet count, 356
Platinum-platinum black, 28
Platinum plug, 35
Platinum wire, 253, 367
Plethysmography, 207
Pleural cavities, 238
PLL (*See* Phase-locked loop)
Plumbing analogy, 12
PMMC (*See* Permanent magnet moving coil)
PN diode, 478
PN junction laser, 485
Pneumograph, 249, 253
Pneumograph, impedance, 250, 252
Pneumonia, 263
Pneumotachometer, 245, 249
Pneumotaxic centers, 243
Pollutants, 245
Polygraph machine, 443
Polypnea, 245
Pons, 243, 283
Population inversion, 482, 485
Positive ions, 26
Positive temperature coefficient, 51
Post-Coronary Care Unit, 338, 532
Postdeflection acceleration electrode, 455
Postfogging, 444
Poststerilization period, 351
Potentiometer:
  recording, 440, 445
  self-nulling, 445
Power supply, 56, 99, 130
Power supply oscillator, 99
Power thyratron, 398
Power wiring, 431
PPG (*See* Photoplethysmograph)
Preamplifier, 202
Preauricular points, 303
Precipitating reagents, 360
Precision colorimeter, 368
Precordial signal, 139
Prefrontal lobes, 289, 293
Presets, 270
Pressure, 159, 160
  atmospheric, 45, 163
  cuff, 385
Pressure amplifier, 170, 173, 176, 327
Pressure capacitor, 47
Pressure-compensated flowmeter, 264
Pressure-cycled ventilation, 261
Pressure differentiation, 161, 184
Pressure dome, 169, 170
Pressure gage, 163, 450
Pressure measurements, 159, 167
Pressure measurement, electronic, 387
Pressure monitor, 173, 179, 443
Pressure oscillations, 167
Pressure preset, 270
Pressure sensor, quartz, 47
Pressure transducer, 37, 163, 169
Presynaptic spike, 301
Preventive maintenance, 274, 411
Priestley, Joseph, 262
Primary effects, 415
Print hammer, 451
Printed circuit board (PCB), 135
Propagation, 378
Proportional mass, 356
Proportioning pump, 369
Proprioception, 287
Protection circuit, diode clamp, 81
Pseudorectilinear writing, 441
Pulmonary abnormalities, 261
Pulmonary artery, 12, 14, 20, 159, 207, 238
  wedge pressure, 207
Pulmonary capacity, 250
Pulmonary edema, 263
Pulmonary function, 245
Pulmonary instrumentation, 256
Pulmonary measurement, 256
Pulmonary semilunar valve, 12
Pulmonary valve, 20
Pulsatile flow, 160
Pulse circuit, 131
Pulse-counting detector, 93
Pulse oximeter, 61
Pulse rate meter, 340
Pulse velocity, 207
Pulse width, 88
Pulsed ultrasound, 388
Pulser, laser diode, 489
Pump head, 229
Purkinje fibers, 17, 19
Push-pull, 93

Pyroelectric infrared heat detector, 67

QRS complex, 20, 23, 138
Quartz pressure sensor, 47
Quartz transducer, 47
Quasistatic system, 160

R-C network, 222
R-R interval, 229, 329
R-wave, 20, 147, 225, 226, 229
R-wave discriminator, 327
Rad, 517
braking, 516
Radiation
exposure to, 512
health dangers of, 517
ionizing, 516
low-level, 513
shielding, 518
Radio, 534
Radio frequencies (RF), 396
ammeter, 396, 404, 405
amplifier, 532
choke, 398
coupling, 404
energy, 532
generator, 397
power, 344, 397
power amplifier, 401
signal, 401
tank circuit, 398
Radio receiver, 534
Radio telemetry, 338
Radio transmission, 316, 338
Radio wave, 375
Radio-opaque dyes, 298
Radioactivity, measuring, 517
Radionuclides, 229
Radiosensitivity, 513
Ramp function, 447
Ramp generator, 185
Range, 88
Range compression, 83
Ranvier, 281
Rapid eye movement (REM), 307
Rate meter, 91
RBC, 290, 353, 355, 368
Reactance, 70
Receiver circuit, laser diode, 490
Receiver, troubleshooting, 344
Receptacle polarity tester, 433
Receptor, 283
Recirculating memory, 462
Recirculation artifact, 200–204
Recorder, 369
ink pen, 449
problems, 447
Rectifier bridge, 96
Rectilinear motion, 441
Red blood cell, 12, 236, 290, 353, 356
Reducing reaction, 27
Reference capacitor, 47
Reference electrode, 34, 75
Reference photodetector, 360
Reference point, 130
Reflected wave, 379
Reflection, 377
coefficient, 379
losses, 476
Reflex action, 283
Reflex mechanism, 243
Refraction, 377, 467
Refractory period, 17, 20, 220
Regurgitation (blood), 14
Rejection, EMI/RFI, 100
Relay, high voltage, 222
Repeater system, 344
Reproduction, 3
Residual volume, 250
Resistance, 35, 433
RTD, 104
tip spreading, 36
Resistance wire, 146, 446
Resistor:
input, 85
metal oxide variable, 132
ratio-matched, 80, 112
series protection, 136
Resistor temperature device (RTD), 104
Resonance, 189, 380
Resonant tank circuit, 542
Respiration, 233ff
cellular, 235
diseases, 260
failure, 261
instrumentation, 249
lung, 236
monitor, 250, 273
organs of, 236
rate of, 261
regulation of, 243
system, 233ff, 264
therapist, 266
therapy, 261
transducer, 250
Respirator, 253
Respiratory bronchioles, 239
Respiratory centers, 243
Respiratory minute volume, 262
Respiratory organs, 234
Respiratory therapy equipment, 260ff
Resting potential, 16, 17
Reticular activation system (RAS), 285
Retrograde catherization, 217
Ribbon cable, 536
Right-leg drive, 128, 138, 144
Ringing, 189
Rms noise, 139
RN, 350
Rod, ruby, 483
Roentgen, 517
Rotor, 49

SA (sinoatrial) node, 10, 17, 19, 23, 219, 229
Safety, 265, 409
electrical, 408ff
electrosurgery, 401
Samples, 369
Sample-and-hold, 114, 215
Scanner, 524
Schwann cells, 281
Scintillation counter, 200, 298, 299
SCR gate, 226
Scratch pad memory, 459
Scrub nurse, 350
Self-excitation, 17
Self-heating, 177
Semiconductor diode, 81
Semilunar valve, 14
Semipermeable membrane, 2, 16
Sensitivity, 44
strain gage, 43
transducer, 44
Sensitivity control, 173
Sensitivity factor, 44
Sensitivity (gain), 88
Sensor, 37
Sensory homunculus, 289
Sensory nerves, 282, 290
Sensory nervous system, 6
Septum pellucidum, 391
Serum, 354
Servoamplifier, 448
Servomechanism, 447
Servomotor, 445
Servorecorder, 445
Set energy, 221
Set level control, 224, 225
Settling time, 114
Severinghaus electrode, 366
Shear forces, 380
Shearing action, 379
Shield driver, 109
Shield, guard, 109
Shielded cable, 109
Shielding, 518
Shock, 220, 404, 409
Shock hazard, 88
Shorting stub, 534
Signal:
bioelectric, 115
differential, 71
Signal acquisition, 25, 30
Signal generator, 344
Signal overload, 538
Signal processing circuits, 81
Signal source, 108
Silver, 28
Silver-silver chloride electrode, 28, 302, 366

Simplex, 477
Single-breath diffusion system, 256
Single-point fault, 335
Skeletal muscle, 32, 281
Skeleton, 4
Skew rays, 470
Small-Outline Integrated Circuit (SOIC), 75
Smooth muscle, 281
Snell's law, 468
Sodium-potassium pump, 16
Software, 87, 495
SOIC (*See* Small-Outline Integrated Circuit)
Solid-state circuits, 400
Solid-state PN junction, 50
Somatic artifact, 131
Somatic system, 280
Somatic tremor, 155
Sonacide, 274
Sound waves, physics of, 376
Source impedance, 55
Span, 88
Spark gap, 397ff
Spark gap generator, 398
Spectrophotometer, 353, 357, 361, 362
Spectrum, 475
Spectrum analyzer, 344
Sphenoidal electrode, 303
Sphygmomanometer, 159, 163, 164
Sphygmomanometry, 166, 385
Spike potential, 301
Spinal column, 290
Spinal cord, 290
Spinal fluid measurement, 168
Spinal fluid pressure, 159, 164
Spinal manometer, 169
Spinal nerves, 290
Spinal tap, 164, 290
Spirometer, 245, 255, 256, 268
Spot welding (laser), 485
Standard Lead System, 124
Standing waves, 476
Stator, 49
Step-function frequency response test, 192
Step-function input signal, 448
Step index, 471
Stepper motor, 445
Sterilization, 194, 273–274, 350, 351, 370
Steroids, 356
Stethoscope, 164, 387
Stimulated emission region, 482
Stopcock, 169, 173, 185, 192
Strain gage, 25, 40, 44, 71, 170, 177, 209, 252
  types of, 42
Stray capacitance, 77, 135, 138
Stress testing, 33
Strip-chart recorder, 91, 131, 144, 209, 227, 273, 338, 340
Stroke volume, 198
Stub, half wave shorting, 534
Stylus, 147, 449
Stylus pressure gage, 450, 451
Subcritical angle, 468
Subcritical damping, 194
Suction apparatus, 352
Suction pump, 367
Summation junction, 109
Summation node, 58
Summed response potentials, 315
Summing junction, 58, 60
Supercritical angle, 469
Superior vena cava, 11
Surface macroelectrodes, 26
Surgery, 348, 404
  oral, 349
  types of, 349
Swan-Ganz, 206–207
Sylvian fissure, 300
Sympathetic, 290
Sympathetic discharge, 292
Synapse, 281, 283
Synaptic ultrastructure, 281
Sync/Defib, 225
Synchronized/Instantaneous, 225
Systole, 10, 14
Systole beeper, 328
Systole lamp, 328
Systolic detector, 180
Systolic manometer, 386
Systolic pressure, 20, 165, 166, 386

T-M scan, 390
T-wave, 20, 139, 225, 226
Tachometer, 149
Tachycardia, 22, 340
Tachypnea, 245
Tapered delay, 222
Tapered delay machine, 223
Taut-band meter, 445
TCG, (*See* Time-compensated gain)
Tee-piece, 253
Telemetry, 316, 338, 342
  landline, 338, 345
  portable, 344
  radio, 338
  troubleshooting, 342
Telemetry transmitter, 344
Temperature, 8
  coefficient, 51
  controller, 230
  measurement, 103
  transducer, 50, 51
Temporal lobes, 289, 293
Tension roller, 147
Tension tester, 432
Testing defibrillators, 227
  high-voltage, 97
Thalamus, 283
Thermal contact writing, 441
Thermal effect, 107, 379
Thermal expansion, 518
Thermal noise, 64–65
Thermal recorder, 147, 443, 444
Thermal resistor, 51
Thermal writing, 146
Thermistor, 37, 50, 78, 202, 253
Thermistor, bead, 253
Thermocouple, 50, 103, 229
Thermocouple RF ammeter, 404, 405
Thermodilution method, 200
Thermoplastic dielectric, 417
Theta activity, 307
Thorax, 13
  cavity, 229, 238
  impedance, 250
  surgery, 349
Thorpe flowmeter, 264
Thrombin, 354
Tidal volume, 242, 249, 261, 262
Time average, 82, 182
Time-compensated gain (TCG), 391
Time-motion (T-M) plot, 390
Time rate of change circuit, 82
Time-vs-ampitude display, 388
Timing mechanism, 271
Tip spreading resistance, 36
Tomogram, 299, 388, 512
Toricelli, Evangelista, 161
Toroid core, 401
Total internal reflection, 469
Total lung capacity (TLC), 242, 250
Tracer material, 199
Trachea, 3, 238, 270
Tracheostomy, 270
Transducer, 25 (def.), 37ff, 42, 83, 111, 169, 177, 187, 391, 392
  biomedical, 37
  capacitive, 48
  care of, 194
  force, 44
  inductive, 45
  mesh, 254
  pen position, 448
  pressure, 37, 163, 169
  quartz, 47
  receiver, 380
  respiratory, 250
  transmit, 380
  ultrasonic, 379, 380
  ultrasound, 375
Transfer function, 58, 83
Transformer, 405
  -coupled device, 96
  isolation, 427
  linear voltage differential, 45
  magnetic, 115
  power, 102
  power isolation, 428

Transistor, 51
  unijunction (UJT), 398
Transmission:
  losses, 475
  modes, 470
  optical, 99
  system, 476
Transmit crystal, 167
Transmitter, 340
  radio, 338
  two-wire, 104
Transmitter driver, 477
Transmitting fiber, 476
Transparency, 475
Transverse coherence, 480
Tremor, somatic, 155
Tricuspid valve, 10, 12, 14, 20
Tumors, 298, 393
Tungsten-iodine lamp, 483
TV camera, 467
TV couplers, 342
TVI/BCI, 536
Twisted pair, 104

UL (*See* Underwriter's Laboratories)
Ultrasonic blood pressure device, 387
Ultrasonic energy, 380
Ultrasonic equipment, 298, 299
Ultrasonic fetal scan, 388
Ultrasonic flow meter, 255
Ultrasonic method, 167
Ultrasonic nebulizer, 264
Ultrasonic waves, 257
Ultrasound, 299, 375, 378
  absorption, 378
  biological effects, 379
  energy, 391
  pulsed, 388
Unbonded strain gage, 25, 42, 43
Undercritical damping, 194, 448
Undershoot, 447
Underwiter's Laboratories (UL), 136, 416
Unipolar arrangement, 303
Unipolar chest leads (V1-V6), 125
Unipolar electrode, 397
Unipolar limb leads, 123, 125
Unity gain noninverting follower, 58, 63
Universal test box, 404
Upscale burn-out, 104
Urine, 354
Urology, 349

Vacuum gage, 163
Vacuum period, 350
Vacuum tube, 397
Vacuum tube sinewave circuit, 399
Valve, 14, 163
Varistor, 132
VC (*See* Vital capacity)
VCG (*See* Vectorcardiography)
VCO (*See* Voltage-controlled oscillator)
VDE-0884 (German standard), 98
Vectorcardiography (VCG), 215
Veins, 6, 10
Velocity, 376
Vena cava, 11, 207
Venous return, 262
Ventilation, 243
  mechanically assisted, 261
  wasted, 241
Ventilation failure device, 273
Ventilator, 78, 253
  failure, 262
  typical faults, 274
Ventricle, 6, 11, 12, 14, 20, 354
  of the brain, 299
Ventricular contraction, 20, 22
Ventricular fibrillation, 22, 220, 224, 228, 409, 410, 416
Ventricular pressure, 14, 20
Vertebral canal, 290
Vertical amplifier bandwidth, 144
VFC (*See* Voltage-to-frequency converter)
Video terminals, 327
Virtual ground, 113
Visceral pleura, 238
Vital capacity (VC), 242, 250
Voltage comparator, 224
Voltage-controlled current source, 111
Voltage-controlled oscillator (VCO), 93, 345
Voltage divider, half bridge, 398
Voltage gain, 57
Voltage-to-frequency converter (VFC), 99, 100
Volume-cycled ventilation, 261
Volume preset, 270
Volumetric flow, 385
Volumetric ventilation control, 268
VSWR meter, 537

Wall outlets, 264
Walter, Carl, 410
Water, sterile, 264
Water manometer, 168
Wattmeter, 537
Watt-seconds, 222
Wave, 375
Wave velocity, 376
Waveform, 189, 398
  anomalies, 124
  tapered dc delay, 220
  trapezoidal, 220, 223
Wavelength, 376, 514
Wavetrap, 399, 536, 542, 543
Wedge pressure, 207
Wheatstone bridge, 25, 37, 40, 45, 50, 71, 75, 111, 170, 198, 201, 203, 252, 253, 545
White blood cell (WBC), 12, 354, 356
White noise, 66
Whole body scanner, 299
Wideband noise, 66
Wilson Central Resistor network, 125, 135, 137, 138, 141
Work function, 50
Writing edge, 147
Writing knife edge, 147
Writing pen, 441
Writing systems, 441ff

X-ray, 511ff, 519
  attenuation, 299
  beam, 298
  equipment, 298
  generation of, 517
  images, 298
  intensity, 517
  still-picture, 512
  techniques, 513
  therapeutic, 513
  tomographic, 299
  tube, 518
X-Y recorder, 446, 447
X-Y servo, 447
Xenon flash tube, 482

Y-time recording, 441
YAG laser, 483

Z-fold paper, 147
Zener diode, 132, 133
Zero baseline, 40
Zero-crossing detector, 384
Zero reference, 163
Zero suppression, 88
Zero stimulus offset error, 174